Thermodynamics: A Smart Approach

Thermodynamics: A Smart Approach

Ibrahim Dincer
Ontario Tech. University
Canada

This edition first published 2021

Registered Offices
John Wiley & Sons, Inc., 111 River Street, Hoboken, NJ 07030, USA
John Wiley & Sons Ltd, The Atrium, Southern Gate, Chichester, West Sussex, PO19 8SQ, UK

Editorial Office
The Atrium, Southern Gate, Chichester, West Sussex, PO19 8SQ, UK

For details of our global editorial offices, customer services, and more information about Wiley products visit us at www.wiley.com.

Wiley also publishes its books in a variety of electronic formats and by print-on-demand. Some content that appears in standard print versions of this book may not be available in other formats.

Library of Congress Cataloging-in-Publication Data

Names: Dincer, Ibrahim, 1964– author.
Title: Thermodynamics : A smart approach / Ibrahim Dincer.
Description: Hoboken, NJ : Wiley, [2021] | Includes bibliographical references and index.
Identifiers: LCCN 2020015376 (print) | LCCN 2020015377 (ebook) | ISBN 9781119387862 (cloth) | ISBN 9781119387855 (adobe pdf) | ISBN 9781119387879 (epub)
Subjects: LCSH: Thermodynamics.
Classification: LCC TJ265 .D56 2021 (print) | LCC TJ265 (ebook) | DDC 621.402/1–dc23
LC record available at https://lccn.loc.gov/2020015376
LC ebook record available at https://lccn.loc.gov/2020015377

Cover Design: Wiley
Cover Image: Designed by Ibrahim Dincer and Drawn by Muhammed Iberia Aydin

Set in 9.5/12.5pt STIXTwoText by SPi Global, Pondicherry, India
Printed and bound by CPI Group (UK) Ltd, Croydon, CR0 4YY

C9781119387862_050124

Contents

Preface *xi*
Acknowledgements *xiii*

1 Fundamentals *1*

1.1 Introduction *1*
1.2 The Spectrum of Energy *3*
1.3 Two Pillars of Thermodynamics *4*
1.4 Units and Dimensions *6*
1.5 The Zeroth Law of Thermodynamics *8*
1.6 The First Law of Thermodynamics *9*
1.7 The Second Law of Thermodynamics *11*
1.8 Thermodynamic System *13*
1.9 Seven-step Approach *14*
1.9.1 Property *15*
1.9.2 State *31*
1.9.3 Process *32*
1.9.4 Cycle *34*
1.9.5 First Law of Thermodynamics *35*
1.9.6 Second Law of Thermodynamics *38*
1.9.7 Performance Assessment *40*
1.10 Engineering Equation Solver as a Potential Software *46*
1.11 Closing Remarks *49*
Study Questions/Problems *49*
a) Concept *49*
b) True or False Type Questions *50*
c) Multiple Choice Type Questions *51*
d) Problems *53*

2 **Energy Aspects** *61*
2.1 Introduction *61*
2.2 Macroscopic Thermodynamics versus Microscopic Thermodynamics *62*
2.3 Energy and the Environment *64*
2.4 Forms of Energy *65*
2.5 The First Law of Thermodynamics *75*
2.5.1 Energy Balance Equations *78*
2.5.2 Energy Losses *80*
2.6 Pure Substances *83*
2.6.1 Phases *84*
2.6.2 Phase Changes of Water *85*
2.6.3 Property Diagrams *90*
2.6.4 Property Tables *96*
2.7 Ideal Gas Equation *108*
2.7.1 When is Water Vapor an Ideal Gas? *109*
2.7.2 Compressibility Factor *110*
2.8 Closing Remarks *113*
Study Questions/Problems *113*
Reference *118*

3 **Energy Analysis** *119*
3.1 Introduction *119*
3.2 Thermodynamic Systems *120*
3.3 Closed Systems *121*
3.4 Modes of Energy Transfer *121*
3.5 Types of Works *122*
3.5.1 Boundary Movement Work *123*
3.6 Energy Balance Equation for Closed Systems *131*
3.7 Specific Heat Capacities *133*
3.8 Open Systems *139*
3.9 Closing Remarks *157*
Study Questions/Problems *157*
Multiple Choice Questions *157*
Problems *159*

4 **Entropy and Exergy** *169*
4.1 Introduction *169*
4.2 The Second Law of Thermodynamics *170*
4.3 Reversible and Irreversible Processes *173*
4.4 The Carnot Concept *174*
4.4.1 The Carnot Principle *176*
4.4.2 Temperature Ratio *176*

4.5 Entropy *178*
4.6 Entropy Balance Equations *181*
4.7 Isentropic Processes *188*
4.8 Isentropic Processes for Ideal Gases *191*
4.9 Isentropic Efficiencies for Ideal Gases *193*
4.10 Exergy *199*
4.11 Energy vs Exergy *200*
4.12 The Different Forms of Exergy *204*
4.12.1 Flow Exergy *204*
4.12.2 Thermal Exergy *204*
4.12.3 Exergy of Work *204*
4.12.4 Exergy of Electricity *204*
4.13 Exergy Destruction *205*
4.14 Reference Environment *205*
4.14.1 Natural-Environment-Subsystem Models *205*
4.15 Exergy Balance Equation for Closed Systems *205*
4.16 Exergy Balance Equation for Open Systems *211*
4.17 Exergy Efficiency *216*
4.18 Concluding Remarks *219*
Study Questions/Problems *219*
Multiple Choice Questions *219*
Problems *221*
References *232*

5 System Analysis *233*
5.1 Introduction *233*
5.2 Thermodynamic Laws *234*
5.3 Closed Systems *236*
5.3.1 Nonflow Exergy with Specific Heat Capacity *241*
5.3.2 Moving Boundary Closed Systems *245*
5.4 Open Systems *254*
5.4.1 Steady-state Steady-flow Systems *254*
5.4.2 Unsteady-state Uniform-flow Processes *277*
5.5 Exergy Efficiency *285*
5.6 Closing Remarks *289*
Study Questions/Problems *290*
Concept Questions *290*
Problems *290*

6 Power Cycles *301*
6.1 Introduction *301*
6.2 Carnot Concept for Power Generation *302*
6.3 Heat Engines *303*
6.3.1 Performance Assessment *305*

6.4 Otto Cycle *309*
6.5 Diesel Cycle *320*
6.6 Dual Cycle *332*
6.7 Brayton Cycle *342*
6.7.1 Regenerative Brayton Cycle *358*
6.8 Rankine Cycle *369*
6.8.1 Ideal Reheat Rankine Cycle *380*
6.8.2 Cogeneration Rankine Cycle *393*
6.8.3 Combined Brayton–Rankine Cycles *409*
6.9 Concluding Remarks *424*
Study Questions and Problems *424*
Questions *424*
Problems *426*

7 Refrigeration and Heat Pump Cycles *441*
7.1 Introduction *441*
7.2 Refrigerants *443*
7.3 Basic Refrigeration Cycle *445*
7.4 Air-Standard Refrigeration Systems *459*
7.5 Cascade Refrigeration Systems *464*
7.6 Heat Pumps *471*
7.7 Absorption Refrigeration Cycles *488*
7.8 Closing Remarks *497*
Study Questions and Problems *497*
Concept Questions *497*
Problems *499*

8 Fuel Combustion *515*
8.1 Introduction *515*
8.2 Fuels *516*
8.3 Fossil Fuels *518*
8.3.1 Coal *520*
8.3.2 Crude Oil *522*
8.3.3 Natural Gas *522*
8.4 Forms of Chemical Energy *524*
8.5 First Law of Thermodynamics Analysis *524*
8.6 Combustion Reactions *527*
8.7 Combustion in a Closed System *532*
8.8 Combustion in Open Systems *535*
8.8.1 Incomplete Combustion *539*
8.8.2 Adiabatic Flame Temperature *543*

8.9 SLT Analysis of Fuel Combustion Processes *547*
8.10 Combustion Efficiency *549*
8.11 Concluding Remarks *569*
Study Questions and Problems *569*
Questions *569*
Problems *571*

Nomenclature *575*

Appendix 1: Thermodynamic Tables *579*

Appendix 2: *T-s* Diagrams *655*

Index *661*

Preface

Writing a preface for a textbook, such as this one, is not easy. Should it just highlight the importance of the subject and summarize each chapter accordingly? I have done this in many of my research-focused books and have decided to do something different here. I will give four pieces of advice and share a quick anecdote that I have been telling my thermodynamics students for the past 20+ years.

- My first piece of advice: Do not be scared of thermodynamics; it is not a monster! The more you like it the better friend it becomes.
- My second piece of advice: Understand thermodynamics better! The better you understand thermodynamics the better you will manage your relationships with anything and everything around you.
- My third piece of advice: Minimize entropy generation and exergy destruction in yourself and in your interactions with the people and things around you! The better you minimize such entropy generation and exergy destruction, the better life you will have.
- My fourth piece of advice: Do not compete against others! Compete always against yourself! If you always compete against yourself, you will become better and better. If you compete against others, you may become envier.

I always tell my students in my first lecture that you should not allow any thermodynamic crimes around you and act immediately to call crime stoppers to request a thermodynamic expert to deal with the matter. This kind of connection or example is given to make students understand the importance of thermodynamic laws and concepts and the role of these in keeping things in a better order.

In closing, I hope this book will help instructors, students, engineers, and readers in general better understand the laws and concepts, better design, analyze, evaluate, and improve the systems, and better manage the relationships with resources, environment, and sustainability.

Ibrahim Dincer
July 2020

Acknowledgements

The first person that I should mention is Dr. Maan Al-Zareer, who has been my course teaching assistant in thermodynamics courses during his masters and doctoral studies under my supervision. Maan helped with several things for almost every chapter. The second person is Dr. Muhammed Iberia Aydin, who has been my visiting scholar from Turkey, for helping out with many drawings and system designs, as well as cover image drawing. The next two people are Osamah Siddiqui and Haris Ishaq, who have been my masters and doctoral students, for their distinct help in preparing chapter-end questions and problems and their solutions in addition to the power point presentations. Another graduate student of mine, Mert Temiz, has prepared the appendices in the Engineering Equation Solver under my supervision. Furthermore, my PhD student Merve Ozturk has helped a lot in checking the example problems, repeating the solutions and correcting them accordingly. I sincerely appreciate what Maan, Muhammed, Osamah, Haris, Mert and Merve have done for me in this book.

In addition, there have been numerous other masters and doctoral students of mine, namely:

- Khaled Al-Hamed
- Aida Farsi
- Yarkin Gevez
- Ahmed Hasan
- Ali Ismael
- Sherif S. Rashwan
- Faran Razi
- Farid Safari

who have prepared some examples and chapter end problems. I greatly appreciate their assistance.

Also, I would to take this opportunity to acknowledge the financial contribution provided by the Turkish Academy of Sciences to support some students who have helped me in research and preparing materials for this book.

Last but not least, I warmly thank my biological family members and academic family members who have been a great source of support and motivation.

Ibrahim Dincer
July 2020

1

Fundamentals

1.1 Introduction

Energy has always been, historically, one of the most critical issues for humankind, which first started using wood (e.g. $C/H = 9.2$ for oak bark) as the source of energy; this was followed by coal (e.g. $C/H = 2.7$ for anthracite), oil (e.g. $C/H = 0.9$ for Alberta oil), and natural gas (e.g. $C/H = 0.26$ for Canadian natural gas). We are now moving into a low- and/or no-carbon era where hydrogen and other carbon free fuels (such as ammonia) become very critical solutions for implementation in our daily life. This is nicely illustrated in Figure 1.1 by providing a graph of carbon/hydrogen ratio versus types of fuels. It is also important to mention that humankind needs cleaner solutions, with carbon-free fuels (such as hydrogen and ammonia. These will result in significantly reduced environmental impact, particularly much lesser air, water and soil pollution which will apparently help improve human health and human welfare. Recently we have found ourselves in the Covid-19 coronavirus pandemic that has impacted every human being directly or indirectly. Of course, it has most harshly affected the elderly and people with weak immune system and those inflicted with various respiratory and cardiovascular illnesses. It has been evident that pollution, particularly air pollution, is recognized as a major risk to such illnesses and health problems. Here, the bottom line is that improving environmental quality will help improve human health and that people can better cope with such virus pandemics.

Another example of the importance of energy in humankind history is that energy sources have always been, are being, and will be the source of the main issues, ranging from conflicts to wars and peace. The competition around energy matters has been even more critical since the industrial revolution, when industry and other aspects of human life shifted its main driving fuel from human and animal power to fuel-based power and industrial activities.

Since we have begun facing many challenges, in particular ranging from the economy to the environment and technology to sustainability, it has even become more apparent that humankind needs more efficient, more cost effective, more environmentally benign, and more sustainable energy options and solutions. Figure 1.2 shows the key targets of sustainable development with respect to better design, analysis, and assessment, better management, better efficiency, better resources use, better environment, and better energy security, which are critical for any place/country to achieve better sustainability. The requirements for attaining such tasks come down to the thermodynamic fundamentals,

Thermodynamics: A Smart Approach, First Edition. Ibrahim Dincer.

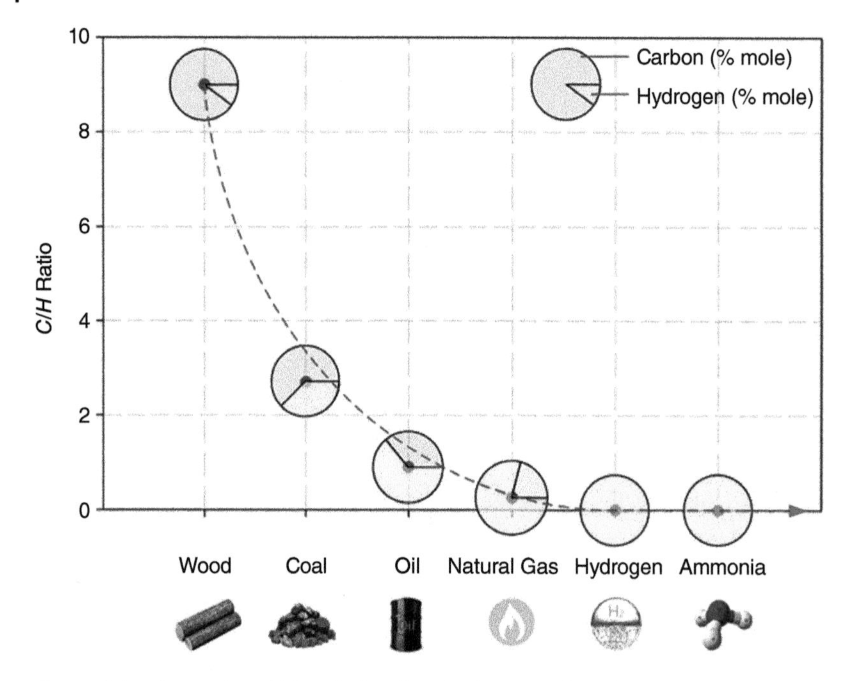

Figure 1.1 Illustration of historical carbon-hydrogen ratios of various fuels used by humankind.

concepts, and laws. That is why we need to understand, learn, and teach thermodynamics in a better way – to better tackle the issues and provide better solutions.

If we look at the environmental dimensions of the challenges through for example global warming, the phenomena can only be better understood with thermodynamics and analyzed and assessed by thermodynamic tools. This is another clear example of the power of thermodynamics.

Since, in this book, thermodynamics is defined as the science of energy, which comes from the first law of thermodynamics (FLT), and exergy, which comes from the second law of thermodynamics (SLT), it clearly shows that the subject sits on these two laws, namely two pillars, just like the way in which a person has two legs.

Going back to the earlier discussion, the requirements of more adequate energy, better environment, and better sustainability have been the main motivation behind going beyond traditional analysis methods and techniques. Traditionally, the FLT, which is recognized as the conservation law, has been the only tool comprehensively used in design, analysis, and assessment of thermodynamic systems. However, it became more apparent in the 1970s and 1980s that the FLT does not achieve much and has limited ability to help achieve things due to the fact that it is insufficient and incapable of addressing practical systems with irreversibilities (or losses, inefficiencies, etc.) and unable to quantify these for assessment and improvement. This is the key reason to have the SLT brought into the picture to account for irreversibilities or destruction through entropy and exergy.

Exergy is distinguished to be a primary tool under the SLT. Thermodynamics is defined as the science of energy and exergy. There are various definitions for energy; however, the definition chosen here is that energy is what causes changes or has the ability to cause changes.

Figure 1.2 Key targets for sustainable development.

Comparing the definition of thermodynamics in this book with the literature, this book follows a more correct approach and more consistent approach, as it considers both energy (coming from the first law of thermodynamics) and exergy (from the second law of thermodynamics) concepts with the same units consistently, and highlights two key efficiencies for performance assessment as the energy concept brings energy efficiency and exergy brings exergy efficiency. This way the concept of efficiencies dwell on two correct pillars of the FLT and the SLT for practical applications and complement each other. The exergy efficiency becomes more important for practical systems and applications since it is a true measure of system performance and indicates how much the actual performance deviates from the ideal performance.

In order to understand thermodynamics, it is essential to understand the four laws of thermodynamics: the zeroth law, first law, second law, and third law of thermodynamics. Each of these four laws is described later with details.

In closing, this chapter aims to provide the introductory aspects of thermodynamics, the basic principles, the main concepts, and the key points to better illustrate thermodynamics along with numerous examples.

1.2 The Spectrum of Energy

In the introduction it is mentioned that energy is critical for humanity. There have been many individuals and organizations ranking the world's key challenges where energy has always been among the top three issues – the first in many –followed by environment,

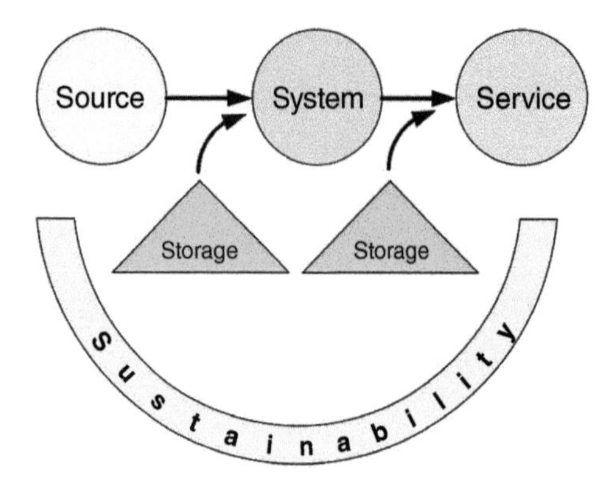

Figure 1.3 Illustration of the 3S + 2S ≅ **S** approach.

economy, water, food, poverty, etc. Of course, energy challenges require energy solutions in a smartly diversified portfolio, although many just propose the renewable energy sources, such as renewable energy technologies, clean fuels (e.g. hydrogen, ammonia), cleaner technologies for fossil fuels, efficient energy use and energy conservation, nuclear energy, and waste to energy technologies. Such smart energy solutions require a holistic approach to see the complete spectrum and understand the key dimensions. As presented in many events, there is a 3S + 2S ≅ **S** approach as illustrated in Figure 1.3 which clearly shows that everything related to energy comes down to 3S, source, system, and service. For any system we need a source, which could be a fossil fuel source or a renewable source or a nuclear source. Based on the services needed in terms of useful outputs (commodities), such as electricity, heat, hot water, cooling, air conditioning, fresh water, drying air, fuel, etc., we need to design the system that will be fed by a source. This system may be a single-, co- or tri-generation system or a multigeneration system with more than three useful outputs. The next 2S is illustrated in the form of storage as needed in this energy spectrum. For example, one may have solar energy not available all the time; what is needed is storage to offset the mismatch. For the second part after the system, one may have more useful output produced that needed. What is required is storage. Therefore, the energy sustainability **S** requires 3S and 2S provided accordingly.

1.3 Two Pillars of Thermodynamics

Thermodynamics, as described in the previous section, can be defined as the subject of both energy and exergy which illustrated in Figure 1.4, based on the previously mentioned two pillars. Of course, it shows that the weight of exergy is more due to its role. The first column of scale is energy, which is brought in by the FLT and concerned with energy as a quantity that is conserved throughout in any system. However, energy alone cannot support thermodynamics, and energy should not be treated as a quantity as it has a quality, which is defined by exergy. Exergy is derived or is based on the SLT as it defines the quality of energy and provides a more meaningful rational for the flow of energy from one reservoir to another. There is another significant thermodynamic property coming from the SLT, namely

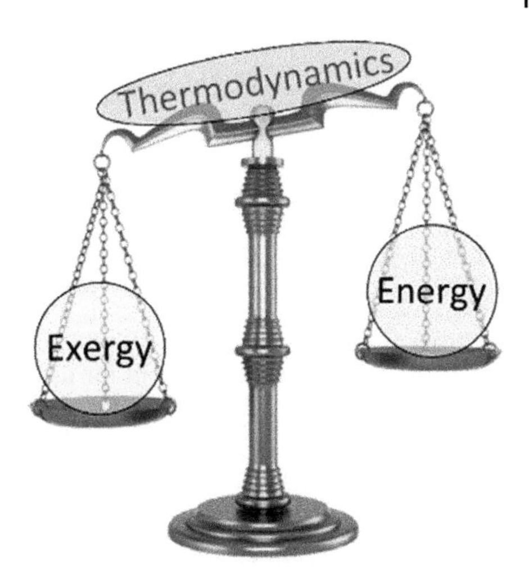

Figure 1.4 Thermodynamic pillars of energy and exergy.

entropy, which is defined as the degree of disorder. This is also be discussed later. We can easily connect entropy as a literal approach to our daily life and call some situations where things are messy as entropic. Figure 1.5 illustrates two cases where we have low-entropy and high-entropy cases. Of course, the high-entropy case is the more messy.

As defined earlier, the FLT essentially brings the concept of energy, with the fact that energy is neither created nor destroyed but is always conserved; this is recognized as the conservation of energy principle. However, the SLT provides something more meaningful and more important with the concept of exergy: the fact that exergy is always destroyed/consumed; this is recognized as the principle of the nonconservation of exergy.

Figure 1.5 Graphical illustration of entropy for a daily situation as low-entropy (less-entropic) and high-entropy (more entropic) cases.

One should keep in mind that thermodynamics governs many aspects of our daily life, and to better explain the importance of thermodynamics and energy engineering various aspects of energy and thermodynamics in the world are presented next.

1.4 Units and Dimensions

In thermodynamics, the basics are important. In this regard, we need to look at units and dimensions and understand them for use. **Dimensions** are defined as physical quantities and the dimensions and the magnitudes assigned to these dimensions are referred to as **units**. There are many quantities used in our daily life where some are measurable, such as temperature and temperature, and some that are not measurable, such as enthalpy and entropy. Figure 1.6 shows a daily example where we have a glass of soda at 20 °C and can directly measure this by using a simple thermometer.

The unmeasurable quantities are normally calculated from measurable quantities. The quantities are often presented in terms of one or more of the basic dimensions, which are presented in Table 1.1. The dimensions shown in Table 1.1 are often referred to as the primary or fundamental dimensions; other quantities where the dimensions are a combination of two or more of primary dimensions are referred to as secondary or derived dimensions.

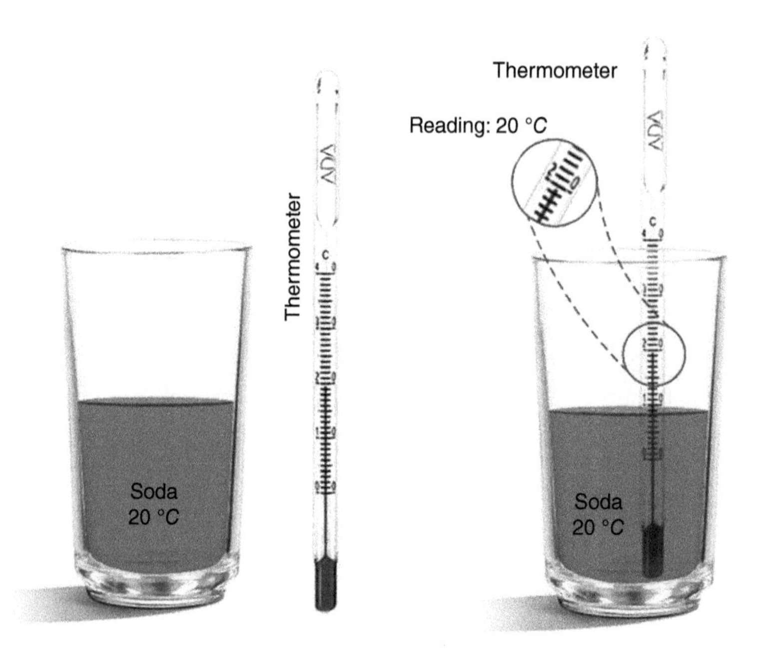

Figure 1.6 A thermometer can be used to measure the temperature of a drink by having both in direct contact.

Table 1.1 Some basic dimensions in SI units.

Dimension	Unit
Length	meter (m)
Time	second (s)
Mass	kilogram (kg)
Temperature	Celcius ($°C$), Kelvin (K)
Electric current	ampere (A)
Amount of matter	mole (mol)
Amount of light	candela (cd)

⇒ Reminder

It is a common practice that one needs to write thermodynamic equations as well as balance equations in a dimensionally correct and consistent manner to have the final result correct; only in this way can we achieve **dimensional homogeneity**. For example, only an elephant can make a baby elephant or, in other words, having an elephant on one side of the equation means that there should be another elephant on the other side of the equation to be able to consistently balance. What we can make of this as a principle for engineers is the "3C rule": make things **C**orrect, **C**omplete and **C**onsistent.

An example of a secondary dimension is velocity, which has the dimension of length over time. Everyone knows the fact that we cannot add apples to potatoes to find the total amount of what? This is illustrated in Figure 1.7 which must be kept in mind. Note that the dimensional homogeneity can be used to solve engineering problems just based on the dimensions and the units given in Example 1.1.

Figure 1.7 Illustration to highlight the importance of dimensional homogeneity.

Example 1.1 Calculate the mass of an object with a density of 1 kg/m^3 and a volume of 1 m^3

Solution

$$Mass = (Density) \times (Volume)$$

$$m = \rho V$$

$$m = 1\frac{kg}{m^3} \times 1\,m^3 = \mathbf{1\,kg}$$

1.5 The Zeroth Law of Thermodynamics

In thermodynamics, there are four laws:

- Zeroth law of thermodynamics
- First Law of Thermodynamics (FLT)
- Second Law of Thermodynamics (SLT)
- Third law of thermodynamics

The first and second laws of thermodynamics are recognized as **governing laws** like the constitutional laws for a state or country or institution which are known as the primary rules for regulating the functioning of a state or country or institution. When we look at the zeroth and third laws of thermodynamics, these are seen more as **guiding policies** for any state or country or institution. After this linkage, one may clearly understand that the first and second laws of thermodynamics are governing laws and the zeroth and third laws of thermodynamics are guiding laws depending on special/specific situations.

After these introductory points, we introduce the zeroth law of thermodynamics in this section with a supporting example. The zeroth law of thermodynamics states that when there are two bodies/objects (such as A and B) in contact in thermal equilibrium and another body/object (such as C) that is in thermal equilibrium with the body/object A, it will be in thermal equilibrium with the body/object B if brought into a contact. In brief, it states that if two bodies are each in thermal equilibrium with a third one, they will then be in thermal equilibrium with each other. This is illustrated in Figure 1.8. One may think that such a definition seems trivial and raises question of why such an obvious conclusion is considered one of the main laws of thermodynamics. The answer to this question is that the conclusion resulting from the zeroth law cannot be concluded from the other thermodynamics laws; another reason is that this zeroth law can be used to validate the concept of measuring the temperature.

The zeroth law of thermodynamics was originally formulated and stated by R.H. Fowler in 1931. Although it was formulated around more than half a century after the first and the second laws of thermodynamics were introduced, scientists felt that the law was a fundamental one and should come before the first and second laws of thermodynamics, so naming it the zeroth law.

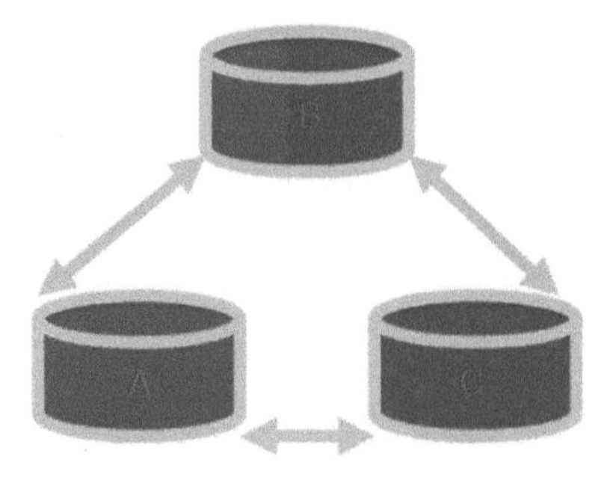

Figure 1.8 An illustration of the zeroth law of thermodynamics.

Example 1.2 Three fluid tanks in different sizes with different fluids are shown in Figure 1.9. The fluids in tanks A and B have the same temperature. If tank C, which is in thermal equilibrium with tank B is connected to tank A, find the temperature of tank C?

Solution

Since both of tanks A and B are in thermal equilibrium, both are expected to have the same temperature of 20 °*C* based on the zeroth law of thermodynamics. Once tank C, which is in thermal equilibrium with tank B, is connected to tank A, all three tanks will be in thermal equilibrium based on the zeroth law. So, the final temperature of tank C will be 20 °*C* accordingly.

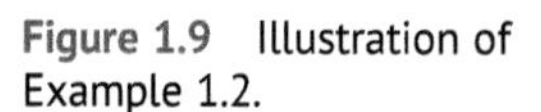

Figure 1.9 Illustration of Example 1.2.

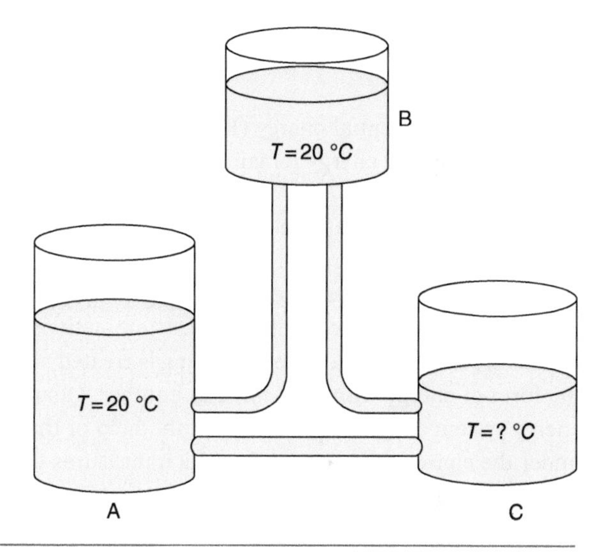

1.6 The First Law of Thermodynamics

The FLT defines one of the fundamental governing laws of life, which is recognized as the principle of conservation of energy. This principle states that no matter what happens energy cannot be created nor destroyed, but it can change from one form to another. For example, energy is converted from the thermal form of energy (so called: heat) to the mechanical form of energy (so-called: work) in a thermal power plant. We have another example where we have a refrigeration system. In this case, the electricity in the form of work is given to the system to operate it in order to produce a cooling effect in the evaporator which is in the form of heat.

It should be kept in mind energy is always conserved throughout any process, system, or application, and hence the total amount of energy will always remain constant. Figure 1.10 shows an example that graphically illustrates the conservation of energy of a soccer ball rolling down the hill where values are given for both kinetic energy (KE) as the function

Figure 1.10 Illustration of the first law of thermodynamics concept through a soccer ball rolling down a hill with a constant total energy of 100 energy units.

of velocity and potential energy (PE) as the function of height (elevation). One can easily notice that the total energy remains constant as 100 units from the beginning to the end at every position.

The FLT is a presentation of the conservation of energy principle, where it defines energy as one of the properties of thermodynamics. From the definition of the FLT it enables us to analyze the energy exchange and interactions of power plants and energy devices since the FLT accounts for these interactions and makes sure that all these interactions are within the balance as nothing is created nor destroyed. It can, however, change the form of energy from one form to another. Quantifying the performance of such an energy system can also be done with the help of the FLT through the energy efficiency under the conservation principle, which measures the performance of an energy system through evaluating the amount of energy that is converted to the desired form and comparing it to the amount of energy that is supplied to the energy system or energy device. For example, consider a diesel power generator. The amount of energy fed into the system (power generator) is in the form of chemical energy after burning the diesel fuel. The amount of energy that is converted to a useful form is the amount of electricity produced by this generator. However, with all the advantages and the benefits we gain with the FLT, there is a clear disadvantage and insufficiency of not considering the irreversibilities, losses, inefficiencies, and quality destructions.

Example 1.3 The rock shown in Figure 1.11 freely falls down to the ground. The *PE* and *KE* of the rock are shown on the rock at different positions in time. Find what the total energy of the rock will be when it eventually hits the ground.

Solution

It is clearly seen that at any point the total energy ($PE + KE$) is **10 units**. The rock will hit the ground with a total energy of **10 units**, which fully comes from the *KE*. The rock will eventually stop on the ground and one may question what will happen to the 10 units of energy. It will be transferred to the ground, causing some damage.

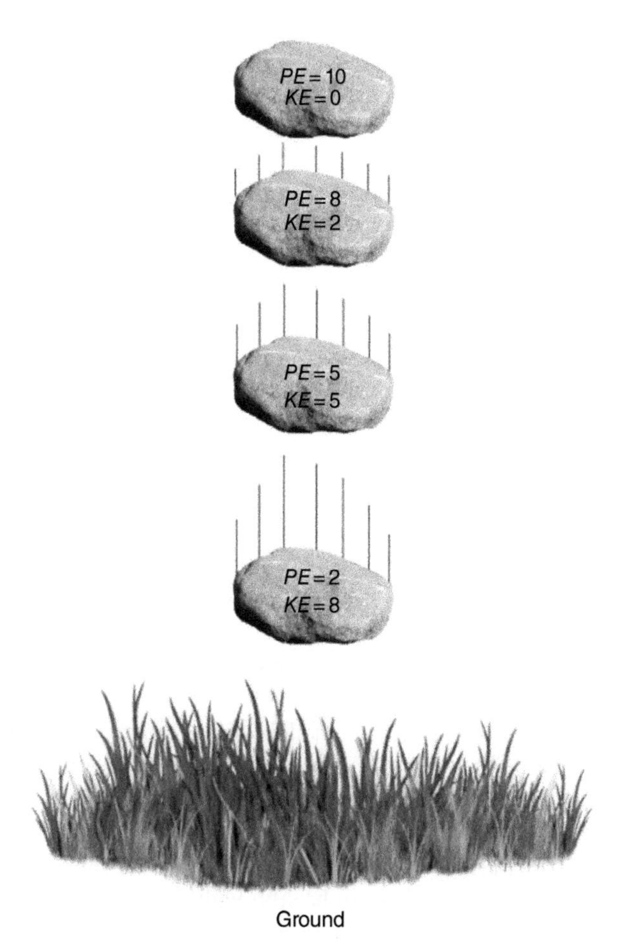

Figure 1.11 A rock falling with the same total energy content.

1.7 The Second Law of Thermodynamics

The deficiencies of the FLT in not taking into account the irreversibilities, losses, inefficiencies, and destructions are compensated by the SLT. This can easily be done through entropy and exergy approaches under the SLT. In this book, the primary focus becomes on exergy rather than entropy due to the fact that exergy units are consistent with energy units and that exergy takes us to exergy efficiency while energy leads to energy efficiency. This way we have the opportunity to compare both FLT and SLT results. The SLT is about both quantity and quality while the FLT is about quantity only.

For example: consider 1 *kJ* of superheated vapor, 1 *kJ* of liquid water, 1 *kJ* of heat, 1 *kJ* of electrical work, and 1 *kJ* of mechanical work and comment on if these will have the energy qualities and hence exergy contents. They are energetically the same under the FLT. However, they will not be the same under the ST as each type of energy has different energy quality which refers to the exergy content. Thus, their energy qualities become different

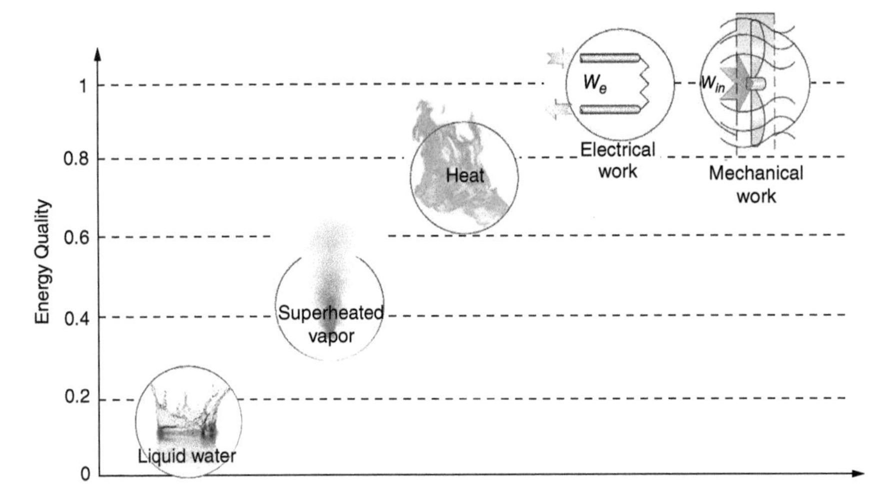

Figure 1.12 Energy quality diagram of various forms of heat and work.

as shown in Figure 1.12. Liquid water has the lowest and both electrical and mechanical works have the highest due to the fact that their actual energy contents become the same as the exergy contents.

Another example is the upper part of the water in the ocean since it is at ambient temperature and pressure, its quality is quantified as zero since it is already at ambient conditions, which are considered dead state (or reference state). Since the heat transfer naturally takes place in the direction of decreasing quality, an example is that thermal energy will naturally transfer from a high quality source (higher temperature source) to low quality thermal energy source (lower temperature source). This can be seen with the illustration in Figure 1.13 where a snowman is made out of snow at $-10\,°C$ and the next day the

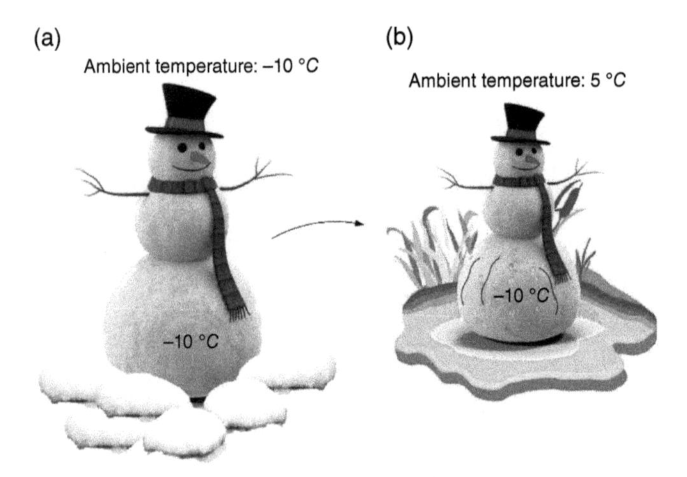

Figure 1.13 Illustration of the SLT with a snowman example where that heat transfer takes place in the direction of decreasing quality: (a) where there is no temperature difference between the snowman and the ambient to cause any heat transfer and hence melting; (b) where there is a temperature difference to cause heat transfer and help melt the snowman.

ambient temperature becomes 5 °*C*. Guess what will happen! The snowman will start melting, and the melted snow will become liquid water and eventually reach an ambient temperature of 5 °*C*. There are two key points to make here: (i) heat is transferred from a higher temperature to a lower temperature and (ii) the thermal equilibrium is achieved eventually where both have the same temperature and no more heat transfer takes place.

Example 1.4 From each of the following two pairs select which energy reservoir or source has higher quality.

a) Ice cube and warm water
b) Hot tea and cold water

Solution

a) The energy in the **warm water** is at higher quality (at a higher temperature) than that of the ice cube (at a lower temperature), since if both are brought into contact, thermal energy (so-called: heat) will transfer from the warm water to the ice cube, which eventually will melt it. Based on the SLT, heat will transfer from a high quality source to a lower quality source, the warm water will then have a higher quality energy.
b) The energy in the **hot tea** has higher quality (at a higher temperature) than that of the cold water (at a lower temperature) for the same reason mentioned in the above point. If both are brought into contact in a similar way heat will transfer from the hot tea to the cold water, which eventually will melt it. Based on the SLT, heat will transfer from a high quality source to a lower quality source, the hot tea will then have a higher quality energy.

1.8 Thermodynamic System

A thermodynamic *system* is defined as a mass of matter or a region in space or a device, which is separated from its surroundings, selected specifically for analysis and assessment. The system selected may be fixed in a stationary form or moving in a nonstationary form. The interaction of system with its surroundings is important in studies where one needs to define an imaginary boundary to start showing all inputs and outputs. *Surroundings* is anything surrounding the system and becoming external to it. It may be a region or mass depending on the application selected. A *boundary* is treated as a real or imaginary closed surface separating the system from its surroundings where mass and energy may enter in and go out of the system. Figure 1.14 shows a soda can as closed system and a pump as an open system with the boundaries and surroundings. Of course, the following chapters will dwell on such open and closed systems in detail with many examples.

For the thermodynamic system that involves mass entering and leaving its boundary then such systems are referred to as *open systems*, while the systems that have a constant, fixed amount of mass, where mass cannot enter or leave its boundary, are referred to as *closed systems*.

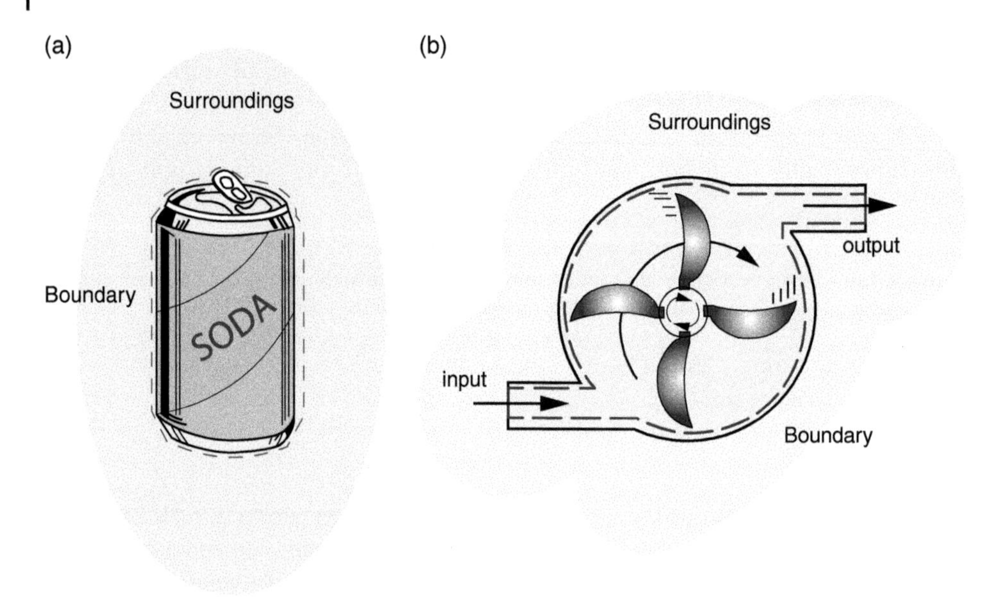

Figure 1.14 Two examples of (a) a soda can to illustrate a closed system and (b) a pump to illustrate an open system.

1.9 Seven-step Approach

This section introduces a seven-step approach to better understand the scope of thermodynamics in the book and clearly shows how thermodynamic concepts are orchestrated in a logical order. In thermodynamics everything starts with property, which is defined as any characteristic of a system (such as temperature, pressure). The next step is state, which is defined as any condition of a system (such as inlet and exit conditions) where there is a need to define at least two properties for each state. The third step is process, which is defined as a change from one state to another (such as isothermal [constant temperature], isobaric [constant pressure] processes) where there is a need to define at least two state points. It is really a beauty of "2," which means there is a need for a pair (couple) in almost everything. The fourth step is cycle, which is defined as the series of processes where the final state is expected to finally reach the initial state, which will make it complete a cycle. There is a need for at least two cycles to achieve this. One should keep in mind that the common power generating and refrigerating cycles have four processes to achieve the final desired outputs (for example, power and cooling effect). The next step is FLT, which was defined earlier as one of the main laws of thermodynamics (conservation of energy law) and which is followed by the SLT (nonconservation of exergy law). These are two critical laws that are the main pillars of thermodynamics. The seventh step is performance assessment, which is essential for all thermodynamic systems and their assessments. This can only be done through a pair of energy and exergy efficiencies or a pair of energetic and exergetic coefficients of performance. The seven step approach is summarized graphically in Figure 1.15; each step in the approach is defined in more details in the coming subsections.

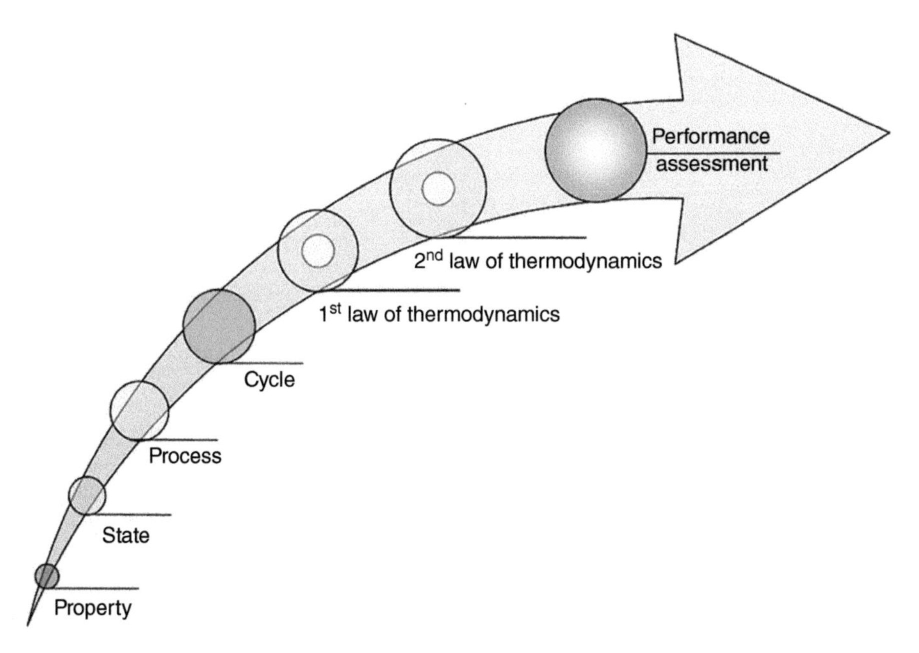

Figure 1.15 A schematic illustration of the seven-step approach.

1.9.1 Property

Property, as defined earlier, is recognized as any characteristic of a system and can be distinguished based on the relation to the mass and volume of the system. The properties that are independent of the mass and volume of a system are recognized as intensive properties, for example: temperature and pressure as well as specific properties. The properties that are dependent on the mass (m) and volume of the system are referred to extensive properties, for example: mass and volume. This is illustrated in Figure 1.16. Figure 1.16a shows a single apple, which is then cut into two halves as shown in Figure 1.16b. The questions are: Which properties will change? Which ones will not change? The replies: the properties, such as m and V, will change and are to be called "extensive properties." The properties, such as P and T, will not change from full apple to half apple which are to be called "intensive properties."

Furthermore, one should note that all lower case letters, such as specific volume (v), specific internal energy (u), specific energy (e), specific enthalpy (h), specific work (w), specific entropy (s) and specific exergy (ex) become intensive properties, except for temperature (T) and pressure (P).

The upper case letters, such as volume (V), internal energy (U), energy (U), enthalpy (H), work (W), entropy (S) and exergy (Ex), are extensive properties, except for m.

a) Specific Volume

Let us look at an example where we obtain an intensive property, specific volume (v), from an extensive property, Volume (V), by dividing it by mass (m) as follows:

$$v = V/m \tag{1.1}$$

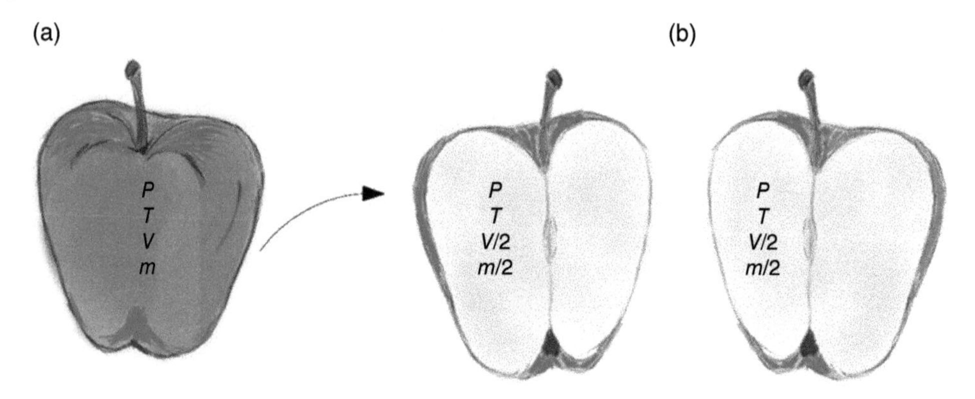

Figure 1.16 A comparison of intensive vs. extensive properties: (a) a full apple and (b) two half apples with intensive and extensive properties.

Note that the specific volume as well as other properties depends normally on the temperature and pressure of the system. For most gases (which are commonly treated as *compressible substances*), the specific volume is proportional to temperature and inversely proportional to pressure due to the ideal gas equation: $Pv = RT$. On the other hand, the specific volumes of liquids and solids tend to remain constant with a variation in pressure; these are referred to *incompressible substances*. Table 1.2 presents a comparison between the specific volumes of water and air with a variation of pressure at a temperature of 25 °*C*. Furthermore, one can see in this table that increasing the pressure from 100 to 200 *kPa* decreases the specific volume by more than 50% while the specific volume of water

Table 1.2 Variation of the specific volumes of incompressible liquid water and ambient air with the variation of pressure at a constant temperature of 25 °*C*.[a]

Pressure (*kPa*)	Specific volume of water (m^3/kg)	Specific volume of air (m^3/kg)
100	0.001 003	0.8558
110	0.001 003	0.778
120	0.001 003	0.7132
130	0.001 003	0.6583
140	0.001 003	0.6113
150	0.001 003	0.5706
160	0.001 003	0.5349
170	0.001 003	0.5034
180	0.001 003	0.4755
190	0.001 003	0.4504
200	0.001 003	0.4279

[a] Data are generated using the Engineering Equation solver (EES).

Table 1.3 Variation of the specific volumes of incompressible liquid water and ambient air with the variation of temperature at a constant pressure of 100 *kPa*.[a]

Temperature (°*C*)	Specific volume of water (m^3/kg)	Specific volume of air (m^3/kg)
10	0.001	0.8128
20	0.001 002	0.8415
30	0.001 004	0.8702
40	0.001 008	0.8989
50	0.001 012	0.9276
60	0.001 017	0.9563
70	0.001 023	0.985

[a] Data are generated using the Engineering Equation solver (EES).

remains constant. This is absolutely a clear indication that water is incompressible substance whereas air is compressible substance.

In addition, Table 1.3 tabulates the specific volume values of both water and air this time with varying temperatures from 10 to 70 °*C* at a constant pressure of 100 *kPa*. It can be seen from the table that the specific volume of water increases about 2.3% while the specific volume of air it increases about 21% with increasing temperature from 10 to 70 °*C*. This clearly shows that compressible substances experience larger differences in both pressure and temperature changes.

b) Density

Density is defined as the reciprocal (inverse form) of specific volume, which is considered another useful thermodynamic property as it is more commonly used in daily applications. It can also be used to determine whether an object will sink or float in addition to many other benefits that arise from knowing the density. As shown in Figure 1.17, if the density of an object is greater than the density of water then the object or body will sink; however, if the density of a body is less than the density of water, then the object will then float.

When the density of an object is measured relative to the density of water, it is referred to the *specific gravity*. The water that is usually used as the reference point to the specific gravity is the at a temperature of 4 °*C* and atmospheric pressure, which will result in a specific volume of 0.001 m^3/kg and a density of 1000 kg/m^3. The specific gravity is denoted in this book as SG and is calculated as follows:

$$SG = \rho_i/\rho_{ref} \tag{1.2}$$

Here, ρ_i denotes the density of any fluid and ρ_{ref} denotes the density of a reference substance or material, where, for example, water at a temperature of 4 °*C* with a density of 1000 kg/m^3 is often used.

Note that the temperature and the pressure are intensive properties as introduced earlier, and they are further discussed in upcoming sections.

Figure 1.17 If the object density is greater than the sea's density the object will sink and if the object density is less then the object will float, depending on the contact surface area.

Example 1.5 Consider the following objects and find out which one of these will float or sink in a water bath:

a) a metallic bolt with a density of 7800 kg/m^3
b) mercury with a density of 13 600 kg/m^3
c) wood with a density of 750 kg/m^3

Solution

There is a need to check the specific gravity with the following equation to see what will happen.

$$SG = \frac{\rho_i}{\rho_{ref}}$$

a) Calculate the specific gravity of metallic bolt:

$$SG = \frac{\rho_i}{1000\ kg/m^3} = \frac{7800\ kg/m^3}{1000\ kg/m^3} = 7.8 > 1 \text{ which will then sink.}$$

b) Calculate the specific gravity of mercury:

$$SG = \frac{\rho_i}{1000\ kg/m^3} = \frac{13600\ kg/m^3}{1000\ kg/m^3} = 13.6 > 1 \text{ which will then sink.}$$

c) Calculate the specific gravity of wood:

$$SG = \frac{\rho_i}{1000\ kg/m^3} = \frac{750\ kg/m^3}{1000\ kg/m^3} = 0.75 < 1 \text{ which will then float.}$$

Example 1.6
Consider this time that the specific gravity value is given as 13.6. Find the density of this substance.

Solution

We need to use specific gravity equation as given below to extract the density of the unknown substance.

$$SG = \frac{\rho_i}{\rho_{ref}}$$

$$13.6 = \frac{\rho_i}{1000\ kg/m^3} \rightarrow \rho_i = 13\,600\ kg/m^3 \text{ which appears to be mercury.}$$

---

c) Temperature
In our daily life, the term "property" is mostly used for temperature to determine out how cold or hot a thing is. Although there is no a commonly accepted exact definition of temperature, it is generally defined as a measure of hotness and coldness. Some examples are then: the weather is colder today; the water is warmer this time; I feel colder; and many more. In addition, we usually understand or comprehend temperature based on our own experience and sensation. An example of the sensation part of determining the temperature of an object at room temperature – a metallic chair makes us feel colder than a wooden one, although both are in the same room at the same temperature. The scientific reasoning behind the metallic chair making someone feeling colder is due to the thermal conductivity, which is a property of materials that measures the effectiveness of the material to conduct heat. Sensation is not the only factor affecting our qualitative judgment on the degree of hotness or coldness is our own experience. For example, let us consider two people living in two distinctly different locations, such as one experiencing cold weather in Canada and the other one experiencing hot weather in the Middle East. Of course, they will adapt themselves to their living environments and their senses will be quite different for cold and hot. This clearly indicates that living species easily adapt themselves to the conditions in their living environments.

The key question here is that how accurately one can measure temperatures. This will definitely depend on various criteria, namely the materials being use to quantify it and their sensitivities to the temperature variations as well as their sizes, shapes, and phases. It is also important to consider repeatability and predictability in measuring the temperatures which will lead to accuracy. This will then bring us to the zeroth law of thermodynamics, which plays a critical role in temperature measurement devices. For example, a basic thermometer is recognized as a temperature measuring device, as illustrated in Figure 1.18. Note that the thermometer measures the temperature through the zeroth law of thermodynamics, as when it reaches thermal equilibrium with the object that it is in contact it will have the same temperature. One of the interesting facts of temperature is shown in Figure 1.19, where it shows the variation of the atmospheric temperature for the condition when the starting temperature from the earth surface is 25 °*C*.

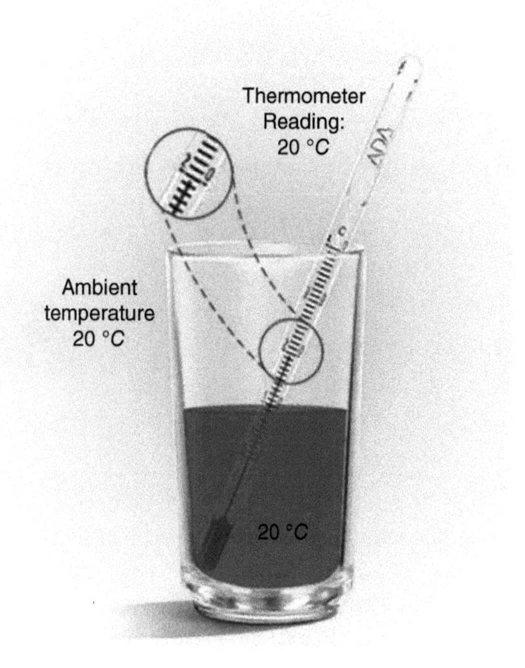

Figure 1.18 Illustration of how thermal equilibrium exists between a glass of liquid and its surroundings at the same temperature.

Note that the temperature reduces through the troposphere which extends from the Earth's surface to nearly 11 *km*. Following the froposphere, the temperature is well below zero, nearly around $-56\,°C$. The temperature remains constant as the elevation increases from 11 to 20 *km* above Earth's surface (Figure 1.19. Moving further away from the Earth's surface also causes variations in the temperature, where it increases in some sphere bands while it decreases in others. Note the maximum temperature the atmosphere has in the Earth's surface to an elevation of 105 *km* is illustrated in Figure 1.19.

- ***Temperature Scales***

The most common temperature scales may be divided into two categories. In the first category are those that are relative to two points, namely boiling point and triple point, while the second category is based on the scientific fact that the lowest possible temperature point (so-called: absolute zero) is when the molecules stops moving (zero *KE*). The two temperatures that the first category of temperature scales are based on (the triple point of water and boiling of water) are the Celsius and Fahrenheit scales. The triple point temperature is the temperature where all of the three phases of a substance exists at the same time. The triple point of water can be found at a temperature of $0.01\,°C$, $273.16\,K$, $32.02\,°F$, and $491.69\,R$. The second temperature that the Celsius and the Fahrenheit scales is based on is the boiling temperature of water, which was experimentally determined to be $100\,°C$ and $212\,°F$.

The second category is based on the scientific fact that the lowest possible temperature point is when the molecules stops moving, since the temperature is a quantification of

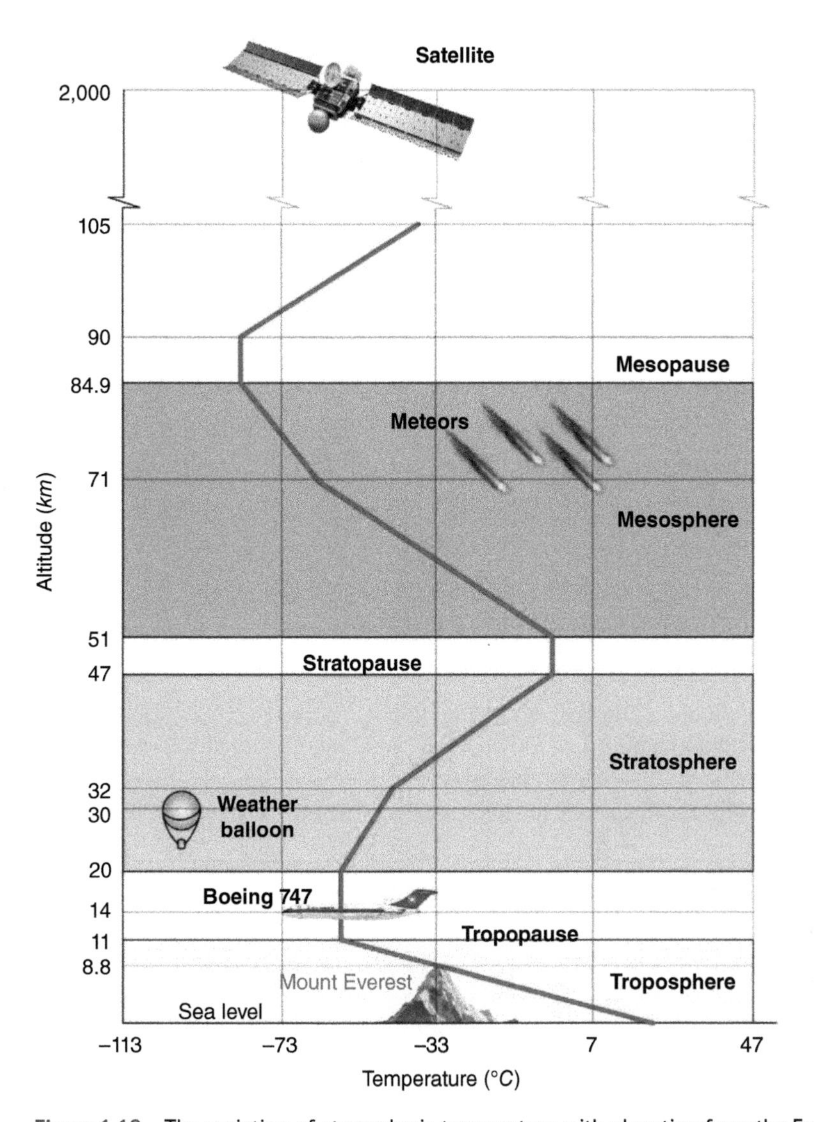

Figure 1.19 The variation of atmospheric temperature with elevation from the Earth's surface.

the molecules *KE*. It means that the lowest temperature would be at the lowest molecules *KE* and the lowest *KE* is zero. Kelvin and Rankine are absolute temperature scales, indicating that they depend on the zero temperature corresponding to a zero molecule *KE*. In this regard, Figure 1.20 shows the groups of temperature scales with the Celsius and Kelvin scales under the "*le Système International d'unités* (SI), referring to the International System of Units. The SI system is a simple and logical unit system that is based on the decimal relationship between the various units of the same dimensions. The second group of temperature scales with the Fahrenheit and Rankine scales is known as the English System of Units, which is also known as the *United State Customary System* (USCS). Although one

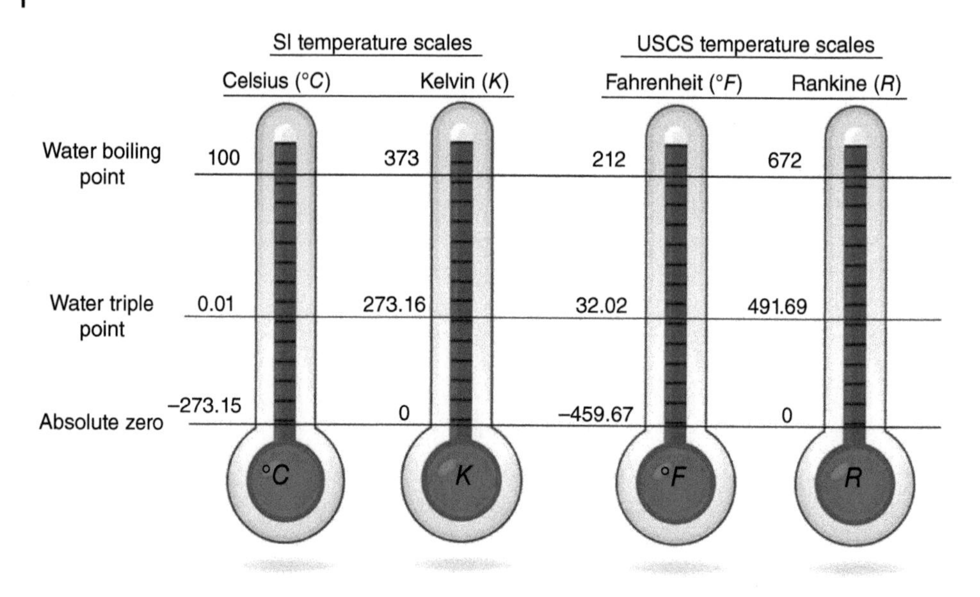

Figure 1.20 A comparative illustration of two temperature scales.

may refer to Figure 1.18 for a graphical illustration of the four common temperature scales, we essentially consider both Celsius and Kelvin as the most common scales to use in the international unit system (SI) throughout this book.

The different temperature scales can be related to one another by using one or more of the following:

$$T(K) = T(^oC) + 273.15 \tag{1.3}$$

$$T(^oF) = 1.8T(^oC) + 32 \tag{1.4}$$

$$T(R) = T(^oF) + 459.67 \tag{1.5}$$

$$T(R) = 1.8T(K) \tag{1.6}$$

Example 1.7 Let us determine the temperature of a cup of coffee at 80 °*C* on other temperature scales, in a) Kelvin (*K*), b) Fahrenheit (°*F*) and c) Rankine (*R*).

Solution

a) $T(K) = T(^oC) + 273.15 = 80 + 273.15 = \mathbf{353.15\ K}$
b) $T(^oF) = 1.8T(^oC) + 32 = 1.8 \times 80 + 32 = \mathbf{176\ ^oF}$
c) $T(R) = T(^oF) + 459.67 = 176 + 459.67 = \mathbf{635.67\ R}$

---

d) Pressure

Pressure is defined as the force that a fluid exerts per unit area and is written as:

$$P = F/A \tag{1.7a}$$

Accordingly, the pressure at a point that is located on an area that has an area of A and has a body with mass m covering it is calculated as follows:

$$P = mg/A \tag{1.7b}$$

Here, m denotes the mass and g denotes the gravity acceleration. Obviously, mg is the force.

Note that the counterpart of pressure in solids is referred to as stress.

Over the years different pressure units have been developed and introduced; some of them are more commonly used than others. Here, some the pressure units are listed in terms of each other to have a way to convert between them:

$$1\ bar = 10^5\ Pa = 0.1\ MPa = 100\ kPa$$

$$1\ atm = 101{,}325\ Pa = 101.335\ kPa = 1.01325\ bar$$

$$1\ kgf/cm^2 = 9.807\ N/cm^2 = 9.807 \times 10^4\ N/m^2 = 9.807 \times 10^4\ Pa = 0.9807\ bar = 0.96788\ atm$$

$$1\ atm = 14.696\ psi$$

$$1\ kgf/cm^2 = 14.223\ psi$$

$$1\ Pa = 2.953 \times 10^{-4}\ \text{in}\ Hg = 4.015 \times 10^{-3}\ \text{in water}$$

As shown in Figure 1.21, the gage pressure is measured relative to the atmospheric pressure, and the gage pressure is used to present pressures that are higher than the atmospheric pressure. The gage pressure can be calculated from the absolute pressure and the atmospheric pressure as follows:

$$P_{gage} = P_{abs} - P_{atm} \tag{1.8}$$

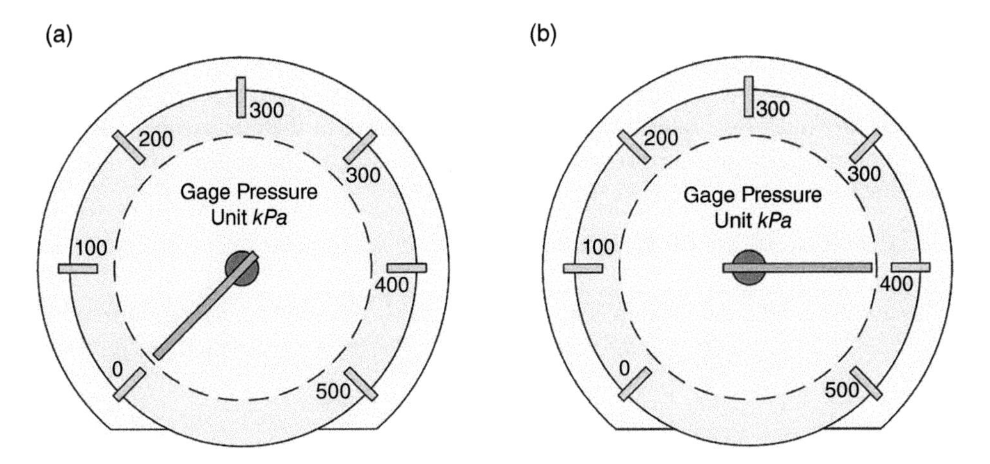

Figure 1.21 A pressure gage example (a) the gage pressure reads zero for a reading for an atmospheric pressure (b) the gage pressure reads 400 *kPa*, which is 400 *kPa* above the atmospheric pressure.

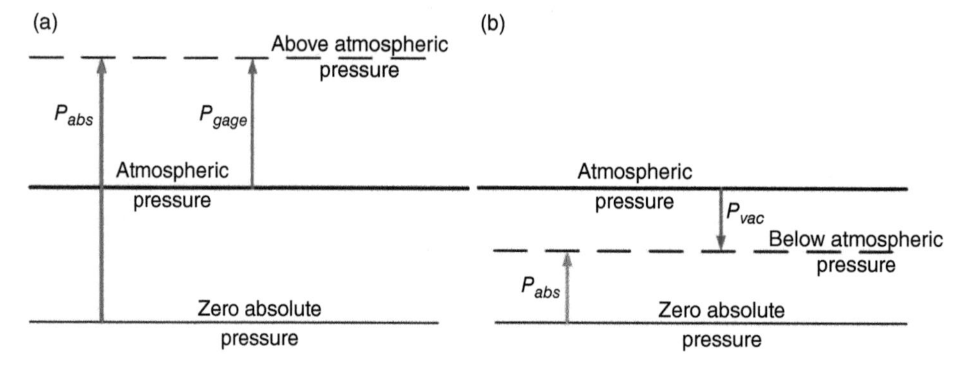

Figure 1.22 Graphical illustration of (a) the gage pressure used when the pressure is above the atmospheric pressure and (b) the vacuum pressure used when the pressure is below the atmospheric pressure.

Here, the absolute pressure is referred to as P_{abs} and it presents the actual pressure at a given point. However, pressure values that are less than the atmospheric pressure are referred to as vacuum pressure and can be calculated as follows:

$$P_{vac} = P_{atm} - P_{abs} \tag{1.9}$$

Equations (1.7) and (1.8) are graphically presented in Figure 1.22, where in Figure 1.22a the gage pressure is the absolute pressure subtracted from it the atmospheric pressure. The gage pressure is used when the absolute pressure is higher than the atmospheric pressure, while the vacuum pressure is used when the absolute pressure is lower than the atmospheric pressure as shown in Figure 1.22b.

Example 1.8 The absolute pressure in a gas container is found to equal to 50 kPa. Express the pressure of the container relative to the atmospheric pressure.

Solution

Since the absolute pressure of the gas container is lower than the atmospheric pressure, then when the gas container pressure is presented relative to the atmospheric pressure we will use Eq. (1.9), which is the vacuum pressure.

$$P_{vac} = P_{atm} - P_{abs} = 101.325 - 50 = \mathbf{51.325\ kPa}$$

However, in a fluid the pressure varies with the depth and does not change in the horizontal direction as gravity is in the vertical direction. The pressure variation in a fluid as a function of the depth is calculated as follows:

$$P_{abs}(h) = P_{atm} + \rho g h \tag{1.10}$$

or

$$P_{gage}(h) = \rho g h \tag{1.11}$$

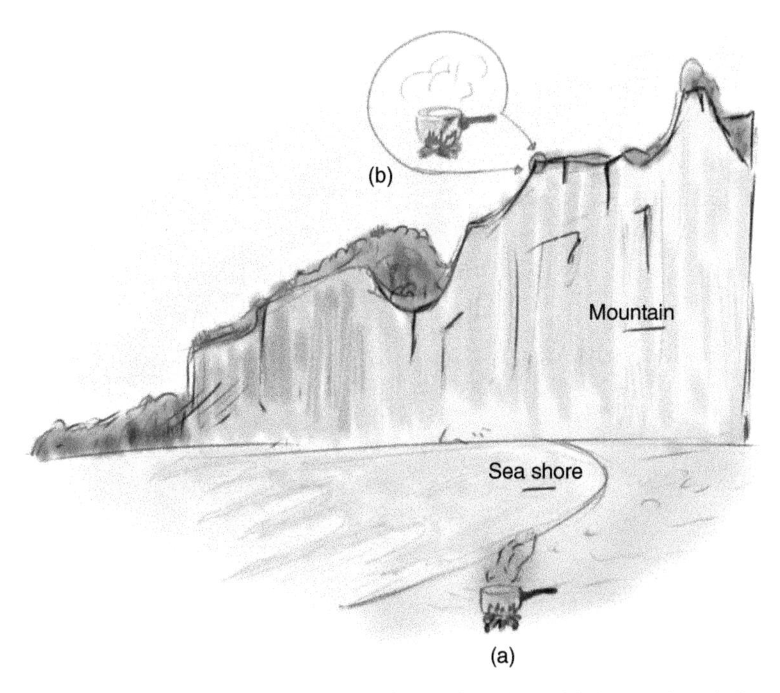

Figure 1.23 Cooking a chicken soup at lower elevation at (a) the sea shore is faster than on (b) the top of the mountain.

Based on Eqs. (1.10) and (1.11) the pressure in fluids varies with depth, and since the atmospheric pressure on the Earth's surface is the result of the weight of the atmosphere above a specified area on the Earth's surface, then the atmospheric pressure varies with the elevation from the Earth's surface. One should note that, as will be discussed in the next chapter, the thermophysical properties of water in liquid state vary with pressure; an example of such a property is the water boiling temperature. Cooking a chicken soup depends on the water boiling temperature, since the boiling temperature is the highest temperature one can cook in water. Since the boiling temperature depends on the pressure, then cooking at lower elevations results in faster cooking compared to cooking at higher elevation, as shown in Figure 1.23, since the boiling temperature of water increases as the pressure of the water increases.

Figure 1.24 shows that points that are at the same depth from the surface have the same pressure, for example points C and B have the same pressure, which is greater than the pressure at the points A, E, and D. Since points A and E are at the same depth then both have equal pressure, which is higher than the pressure at point D. The conclusion that was drawn from Eqs. (1.10) and (1.11) and was highlighted in Figure 1.24 has led to Pascal's principle.

Pascal's law (or Pascal's principle) highlights that the pressure in a fluid remains constant in the horizontal direction and, as a conclusion, the pressure applied to a fluid in a confined space is transmitted throughout the fluid at the same amount.

$$P_1 = P_2 \rightarrow \text{where } P = \frac{F}{A} \text{ then } \frac{F_1}{A_1} = \frac{F_2}{A_2} \tag{1.12}$$

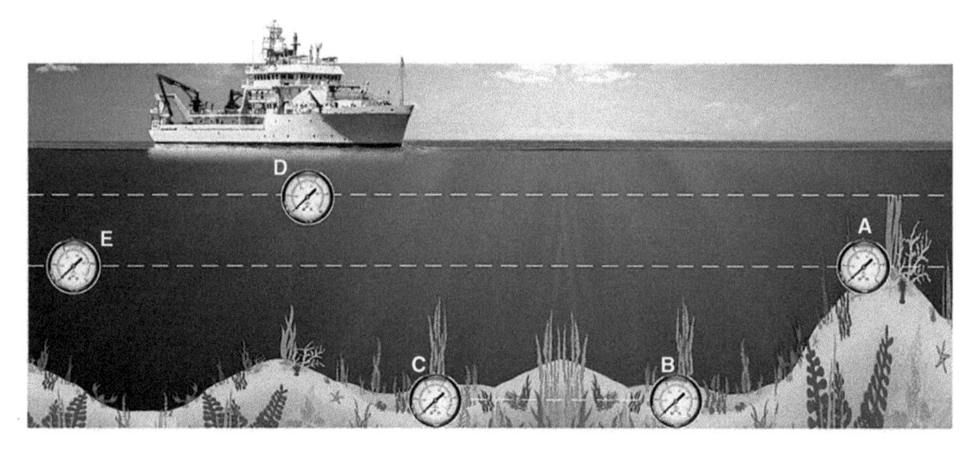

Figure 1.24 The pressure at points that are at the same depth from the free surface is equal.

Here, the subscripts 1 and 2 refers to the points in Figure 1.25, F refers to the force exerted on the piston, and A is the area of the piston.

Figure 1.25 shows an example of Pascal's principle, which uses the fact that if there are two points at the same height that are connected within the same fluid domain, they have the same pressure. In this case, a man standing on the side (at point 2), where the piston cross-sectional area is A_2 with a pressure of P_2, will be able to lift a much heavier group of containers sitting on the piston (at point 1) with a cross-sectional area of A_1 at pressure P_1.

Example 1.9 The lift shown in Figure 1.25 is designed to lift a number of containers that weigh 1590 kg and sit on a piston with an area of 15 m^2. Calculate the mass that needs to be used on piston 2, which has an area of 0.1 m^2, to lift these containers.

Solution

Based on the Pascal's principle:

$$P_1 = P_2 \rightarrow \frac{F_1}{A_1} = \frac{F_2}{A_2}$$

$$\frac{1590\ kg \times 9.81\ m/s^2}{15\ m^2} = \frac{F_2}{0.1\ m^2}$$

$$F_2 = 103.99\ N = m \times 9.81\ m/s^2$$

$$\mathbf{m_2 = 10.6\ kg}$$

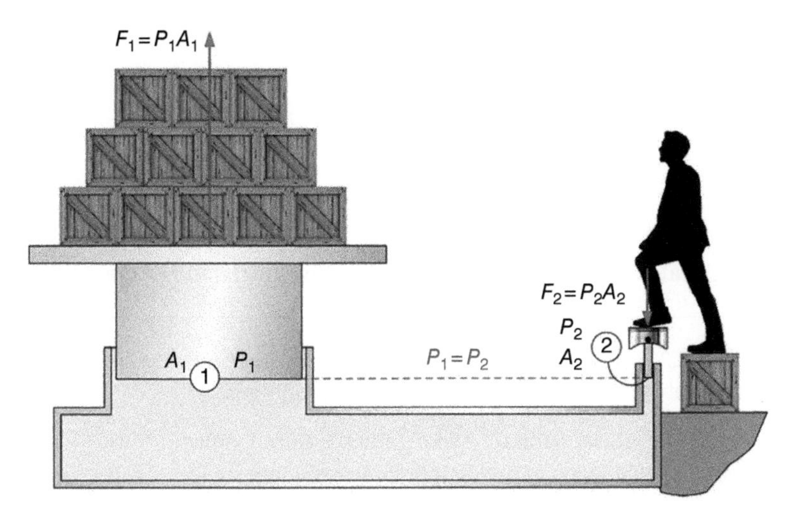

Figure 1.25 By using Pascal's principle many containers are lifted with a small force being applied at point 2, which has a small area.

- **Manometer**

One of the most commonly used pressure measuring devices is the manometer, which is employed to measure the pressure difference through the height of a fluid column. Manometers are often made of a glass or plastic U-tube containing one or more type of fluids that vary in density. Materials that will be encountered in this book are mercury, water, oil, and alcohol; however, mercury is often the fluid of choice in such devices aiming to maintain a small and a compact manometer because of its density, which is 13.6 times higher than that of water. Figure 1.26 shows a basic one link manometer is used to measure the pressure of a gas tank.

As shown in Figure 1.26, the pressure in the gas spherical tank can be expressed in terms of the height of the fluid column filling up the manometer tube as follows:

$$P_1 = P_2 = \rho gh + P_{atm} \tag{1.13}$$

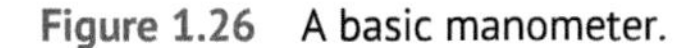
Figure 1.26 A basic manometer.

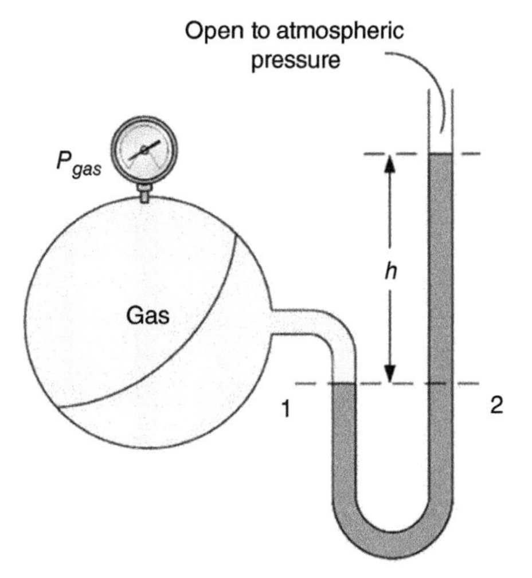

where the absolute pressure P_1 is calculated by summing the pressure produced by the column of fluid filling the manometer tube and the pressure of the atmosphere acting on the free surface of the tube fluid open to the atmosphere. However, the gage shown in Figure 1.26 measures the gage pressure, which is calculated as follows:

$$P_1 - \rho gh = P_{atm} \rightarrow P_{gas} = P_{1,gage} = P_1 - P_{atm} = (\rho gh + P_{atm}) - P_{atm} = \rho gh \tag{1.14}$$

Although the manometer shown in Figure 1.26 is a simple/basic one, there are manometers in a more complicated manner with multiple tubes, as illustrated in Figure 1.27.

⇒ Remember

There is a basic rule in solving these manometer pressure problems:

a) if we go above the sea level, we subtract hydrostatic pressure (ρgh), but
b) if we go below the sea level, we add hydrostatic pressure (ρgh).

This is illustrated in Figure 1.28.

Based on the manometer system given in Figure 1.27, one may need to find the pressure after going through the solution methodology to write the pressure balance equation:

$$P_1 - \rho_A g h_1 - \rho_B g h_2 + \rho_C g(h_1 + h_2) - \rho_D g(h_3 + h_4 + h_5) = P_{atm}$$

which comes out of the following steps:

Step 1: Start with the point assigned in the system (usually preferable to start with the gas container pressure), which is point 1 in Figure 1.27. Note that the pressure variation with depth is negligible in the case where the fluid is the gaseous state, which means the pressure in the gas container is equivalent to the pressure at point 1.

Step 2: Write the pressure for each fluid or segment. While moving through the tubes as a maze, for each segment that has a height h and constant density (constant fluid) a ρgh is added to the pressure selected in step 1 if the you are going down the tube segment while

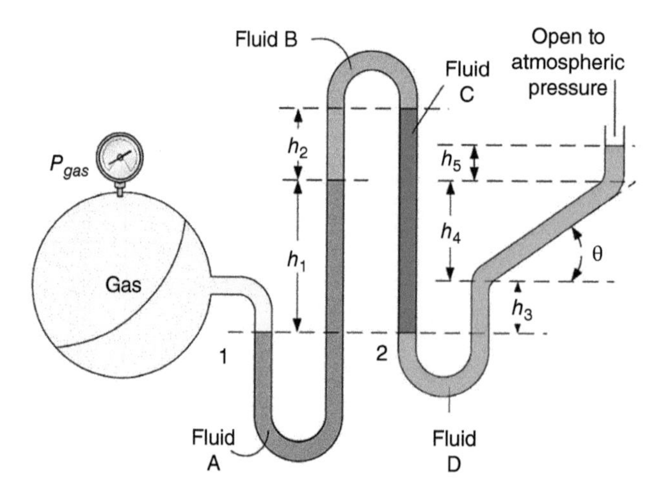

Figure 1.27 An advanced manometer with four different types of fluid, where each has a different density.

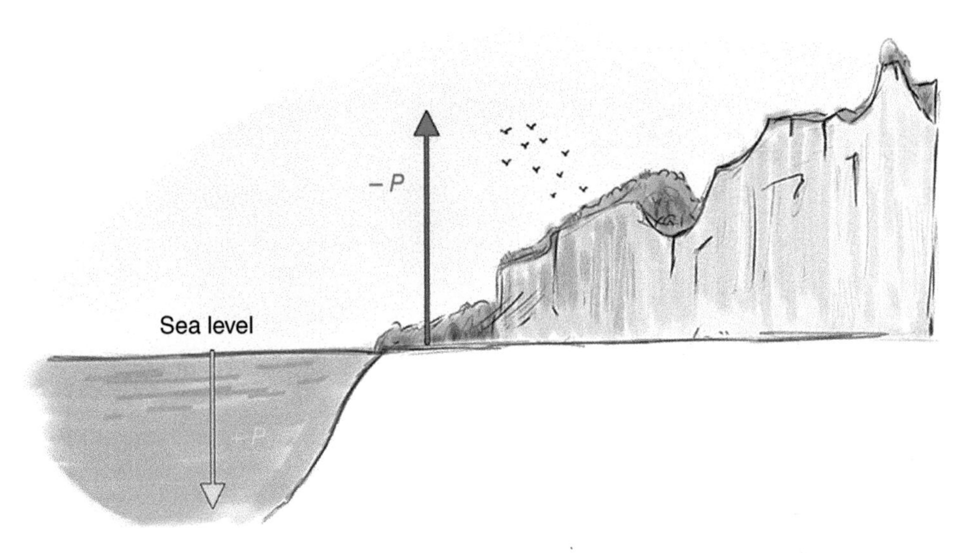

Figure 1.28 Illustration of how pressure will be taken into account based on the reference level (i.e., sea level): going down is positive, as we sum, which is illustrated with +P and going up is negative, as we subtract, which is illustrated with −*P*.

solving the maze; a ρgh is subtracted if you are moving up the tube segment while solving the maze. In this regard, we go down to the bottom and make a U turn up to point 2 where the pressure will remain the same since there is no elevation difference. After going beyond point 2, the pressure becomes $P_1 - \rho_A g h_1$ for fluid A.

Step 3: In a similar fashion, the pressure becomes $P_1 - \rho_A g h_1 - \rho_B g h_2$ for fluid B.

Step 4: The pressure becomes $P_1 - \rho_A g h_1 - \rho_B g h_2 + \rho_C g(h_1 + h_2)$ for fluid C.

Step 5: In a similar fashion, the pressure becomes

$$P_1 - \rho_A g h_1 - \rho_B g h_2 + \rho_C g(h_1 + h_2) - \rho_D g(h_3 + h_4 + h_5)$$

since the vertical heights are only consider in writing the balance equation.

Step 6: It finally reaches the last point where it is open to the atmosphere where the total pressure is equivalent to the atmospheric pressure as follows:

$$P_1 - \rho_A g h_1 - \rho_B g h_2 + \rho_C g(h_1 + h_2) - \rho_D g(h_3 + h_4 + h_5) = P_{atm}$$

Example 1.10 There are two pipes in a system connected by a double U-tube manometer as shown in Figure 1.29 where the brine pipe is connected to a tank filled with different fluids. Oil and brine are flowing in parallel horizontal pipelines. The pressure at the center of the oil pipe is 200 *kPa*. Take the density of water is 1000 kg/m^3.

a) Write the pressure balance equation between points 1 and 2.
b) Find the pressure at point 2.
c) Write the pressure balance equation between points 2 and 3.
d) Find the pressure of point 3.

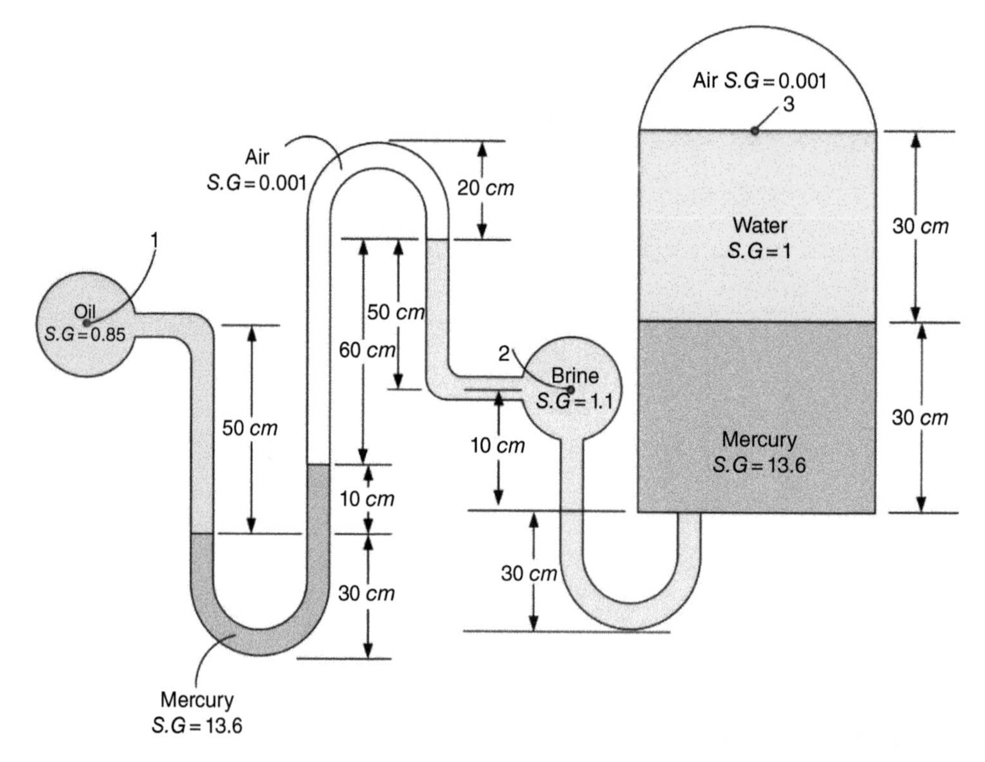

Figure 1.29 Graphical illustration of the system discussed in Example 1.10.

Solution

a) Write the pressure balance equation between points 1 and 2

$$P_1 + \rho_o g(0.5) - \rho_m g(0.1) - \rho_a g(0.6) + \rho_B g(0.5) = P_2$$

b) Find the pressure at point 2

$$200\ kPa \times \frac{1000\ \text{Pa}}{1\ kPa} + 850\frac{kg}{m^3} \times 9.81\frac{m}{s^2} \times (0.5\ m) - 13600\frac{kg}{m^3} \times 9.81\frac{m}{s^2} \times (0.1\ m)$$
$$-1\frac{kg}{m^3} \times 9.81\frac{m}{s^2} \times (0.6\ m) + 1100\frac{kg}{m^3} \times 9.81\frac{m}{s^2} \times (0.5\ m) = P_2$$

$\boldsymbol{P_2 = 196{,}217.3\ Pa}$

c) Write the pressure balance equation between points 2 and 3

$$P_2 + \rho_B g(0.1) - \rho_m g(0.3) - \rho_w g(0.3) = P_3$$

d) Find the pressure of point 3

$$196217.3\ Pa + 1100\frac{kg}{m^3} \times 9.81\frac{m}{s^2} \times (0.1\,m) - 13600\frac{kg}{m^3} \times 9.81\frac{m}{s^2} \times (0.3\ m)$$
$$-1000\frac{kg}{m^3} \times 9.81\frac{m}{s^2} \times (0.3\ m) = P_3$$

$\boldsymbol{P_3 = 154{,}328.6\ Pa}$

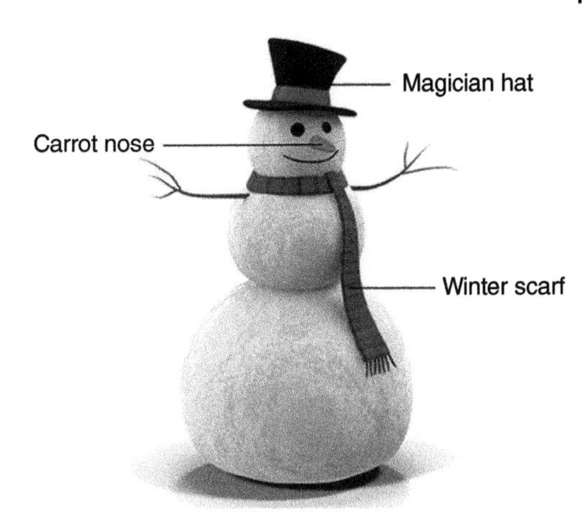

Figure 1.30 A snowman with some features/ details to describe it.

1.9.2 State

State is defined as the condition that can be described by at least two thermodynamic properties (or sometimes one property and some information such as saturated liquid or saturated vapor type would be sufficient). The state properties may either be measured or calculated at the respective state points for thermodynamic analysis and calculations. For anything and hence everything around us in our daily life we always define their states by describing two to three specific details about the person to distinguish from others, such as wearing eye glasses and a hat and having mustache. These specific details are like the properties provided for a system in thermodynamics. Figure 1.30 shows a drawing of a snowman with three specific details, such as magician hat, carrot nose, and winter scarf to be able to distinguish and/or describe it.

a) Equilibrium

In thermodynamics, we also deal with equilibrium states, which refer to a state of balance since there are no driving forces or unbalanced potentials within the system that might drive the system to change state by changing properties. There are different modes or types of thermodynamic equilibriums; for example:

- Thermal equilibrium is a state where the same temperature is reached throughout the entire system. Figure 1.31 shows two cases. The first one (Figure 1.31a shows no thermal equilibrium while the second (Figure 1.31b shows thermal equilibrium within the system at the same, uniform temperature. Note that both system and its surroundings may also reach thermal equilibrium if they have the equivalent temperatures.

There are also numerous other types of equilibrium, ranging from mechanical to chemical types as briefly listed below:

- Mechanical equilibrium is defined as a state where the system has the same pressure throughout and does not change during the process.
- Phase equilibrium is defined as a condition where the system has reached the same phase, such as liquid or vapor, that does not change during the process.

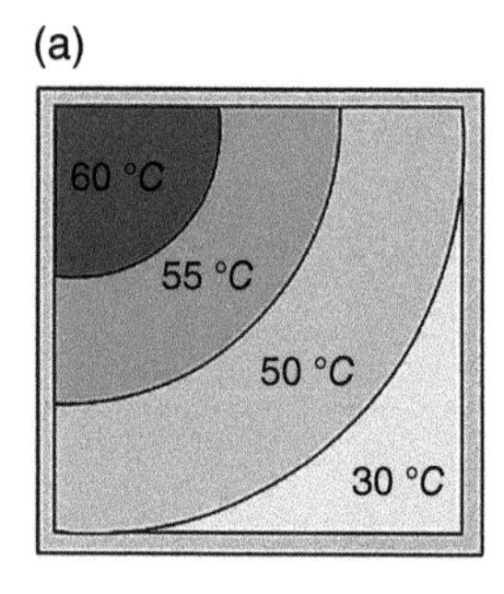

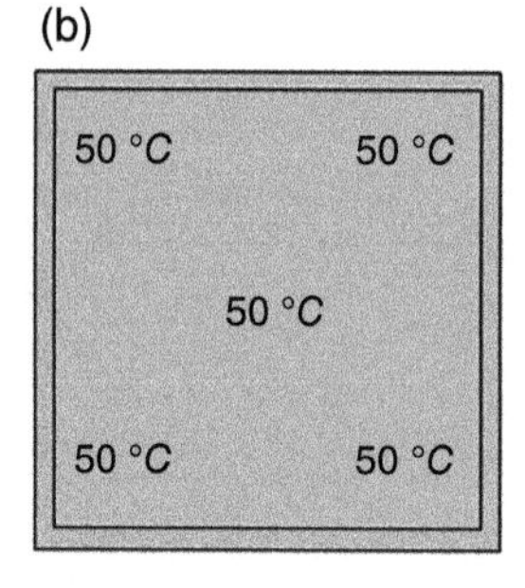

Figure 1.31 Graphical illustrations of (a) a system which is not in thermal equilibrium and (b) a system which has reached thermal equilibrium.

- Chemical equilibrium is defined as a state where the system has uniform chemical composition that does not change during the process.

Note that a quasi-static or quasi-equilibrium process is a process processing in a sufficiently slow manner that the system remains infinitesimally close to an equilibrium state at all times.

For a system where there are no electrical, gravitational, magnetic, motional, and surface tension effects, it is referred to as simple compressible system and the state of such system can be completely described by specifying two independent intensive properties. A final note on the equilibrium is that a system will not be in equilibrium unless all the relevant equilibrium criteria are satisfied.

1.9.3 Process

Process is defined as a change from one state point to another state point for a chosen system to study. There are many processes in thermodynamics, for example:

- isothermal (constant temperature) process
- isobaric (constant pressure) process
- isochoric (constant volume) process
- isenthalpic (constant enthalpy) process
- isentropic (constant entropy) process
- adiabatic (no heat transfer) process

Shown in Figure 1.32 is an example to illustrate a process from state 1 to state 2 on a thermodynamic diagram, which is a useful method to present a process especially for a common two-coordinate system in x and y (here x refers to property A and y refers to property B), such as pressure, temperature, volume, enthalpy, and entropy. Since the process is treated as a path function, some paths can take the system from one equilibrium state to another while having the temperature of the system remaining constant, such process is called an isothermal process. The system can also go from one equilibrium state to another while having a constant volume, such process is called an isochoric process, whereas if the system changed its state while having a constant pressure then the process is referred to as an isobaric process.

In addition, Figure 1.33 shows the four types of configurations as illustrated by two types of closed and two types of open systems. Figure 1.33a shows a closed system with a fixed

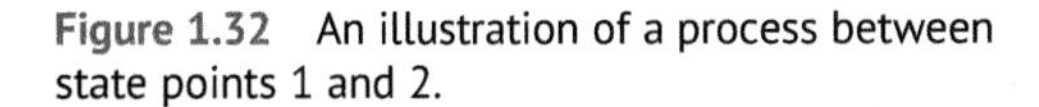

Figure 1.32 An illustration of a process between state points 1 and 2.

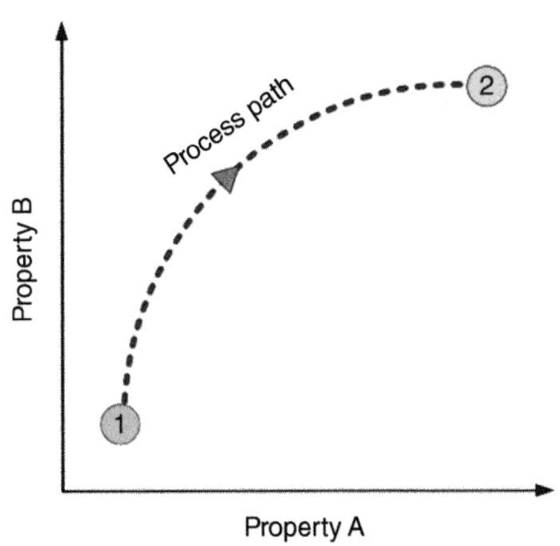

amount of mass within the system where there is no mass flow crossing the boundary while Figure 1.33b exhibits a closed system with moving boundary. In both these systems there are energy (heat and work) transfers crossing the boundary. The other two illustrations (Figure 1.33c and d) show open systems with mass flows and energy transfers crossing the boundary.

A process can be defined as a system going from one state to another, where a system is a quantity of matter or a region in space chosen for analysis and study, while this quantity of matter or region in space can be fixed or it can move and accelerate. Figure 1.30 shows the definition of a system as described earlier. Note that as shown Figure 1.30 a system is often bounded and separated from what is outside the system by a real or an imaginary line referred to as the boundary. Everything else outside the boundary is referred to as the surroundings.

a) Steady-state Steady-flow and Uniform-state Uniform-flow Processes

When we deal with systems, we need to look at the processes and see how the properties behave during these processes. This is especially important if open systems are considered for investigation and when there will be two types of processes, such as steady-state steady-flow (SSSF), where the properties within the system do not change with time and become

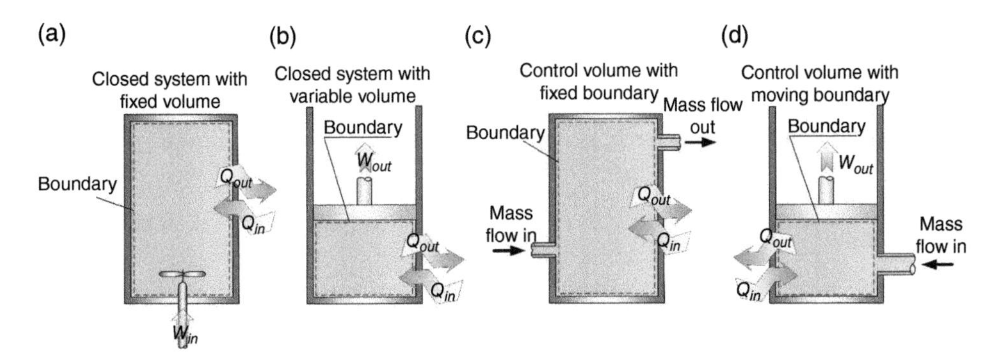

Figure 1.33 Four types of systems in closed and open forms.

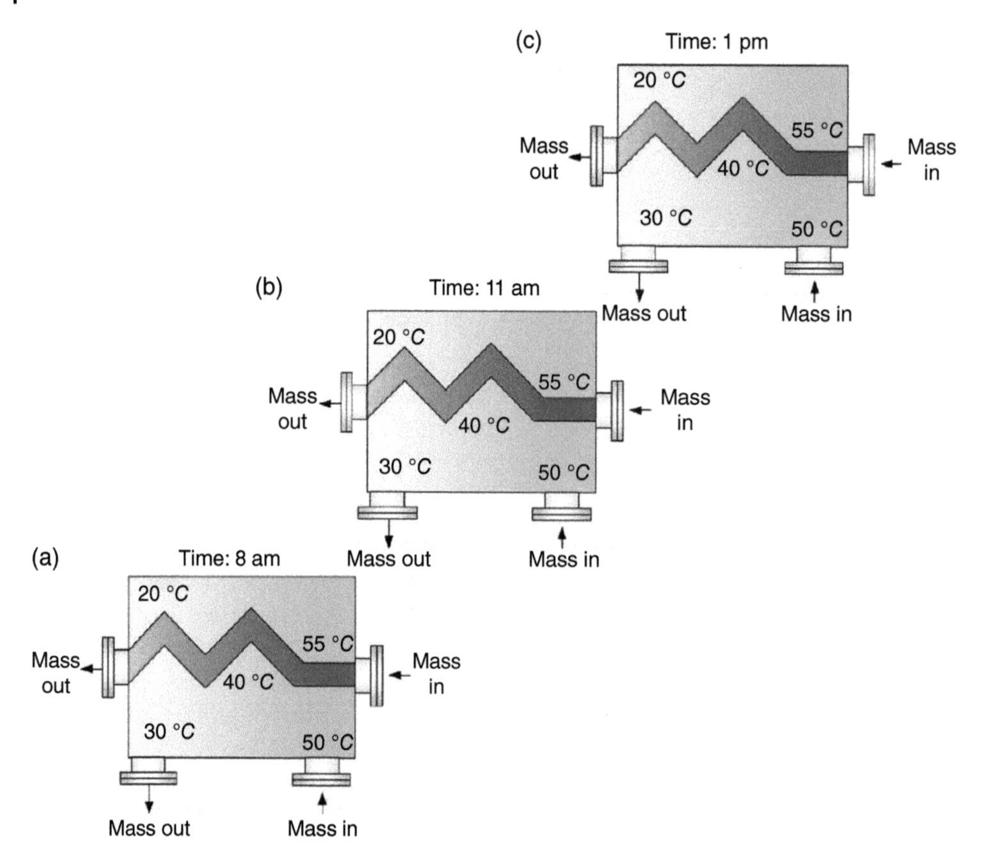

Figure 1.34 Illustration of a steady-state steady-flow (SSSF) process for an open system at various times (a) 8:00 a.m., (b) 11:00 a.m., and (c) 1:00 p.m. where there are mass flows cross the boundary.

steady state, and uniform-state uniform-flow (USUF), where the properties and mass within the system change with time and become time dependent (i.e. transient). Figure 1.34 shows an SSSF open system where we have the incoming and outgoing mass flow rates remaining the same and the temperatures at various points within the system remaining the same as time progresses from 8 : 00 a.m. until 1 : 00 p.m.

Figure 1.35 shows a typical USUF process where one has a container filled with water over time, which becomes time dependent (unsteady) since the mass within the system increases from 8:00 a.m. to 1:00 p.m. Such examples are very common in daily life where we see filling and discharging tanks, containers, reservoirs, etc. We will further elaborate on these systems along with examples in the forthcoming chapters.

1.9.4 Cycle

A thermodynamic cycle is defined as the operation of coming back to the original starting state point, after going through a number of processes, to achieve generating the desired output. Thermodynamically, we need a minimum of two processes to be able to complete a cycle (Figure 1.36). In thermodynamic there is nothing fishy! Everything is logical. Here is

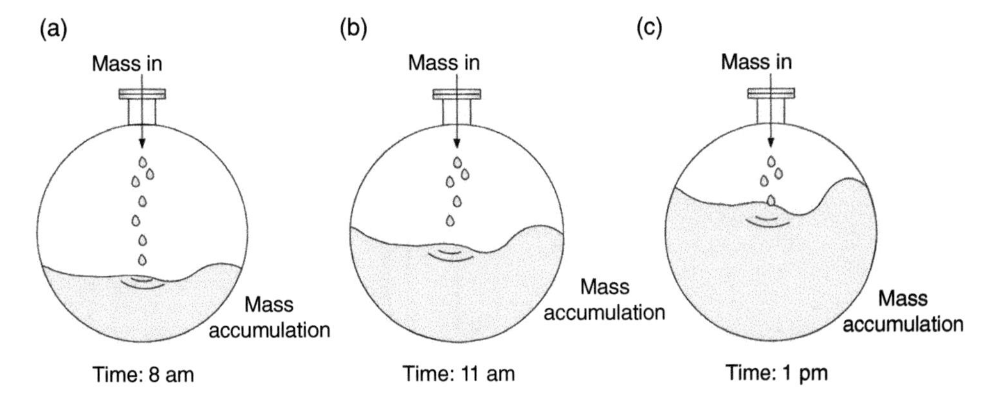

Figure 1.35 Illustration of a uniform-state uniform-flow (USUF) process for an open system at various times (a) 8:00 a.m., (b) 11:00 a.m., and (c) 1:00 p.m. where there is mass accumulation over time within the system.

an observation: there is a need for two properties to define a state point, two state points to define a process, and two processes to define a cycle, respectively.

Note that most of the basic power and refrigeration cycles, in addition to Carnot cycles, consist of four processes to make a cycle. Figure 1.37a shows a cycle where there are four processes on a property diagram, while Figure 1.37b is a T-s diagram of a Carnot heat engine cycle (which is treated as an ideal, reversible cycle) that consists of two isothermal processes (1–2: isothermal expansion and 3–4: isothermal compression) and two isentropic processes (2–3: isentropic expansion and 4–1: isentropic compression).

1.9.5 First Law of Thermodynamics

The FLT is recognized as one of the fundamental guiding laws of life, which is the principle of conservation of energy. The conservation of energy principle states that energy cannot be created nor destroyed, but it can just change from one form to another. The FLT has been

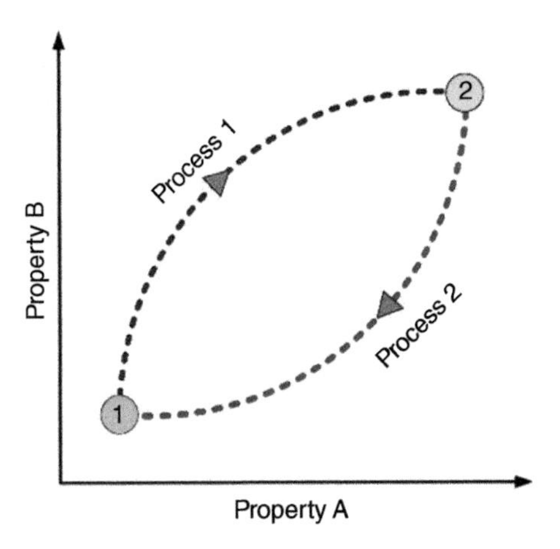

Figure 1.36 A property diagram of a cycle consisting of two processes.

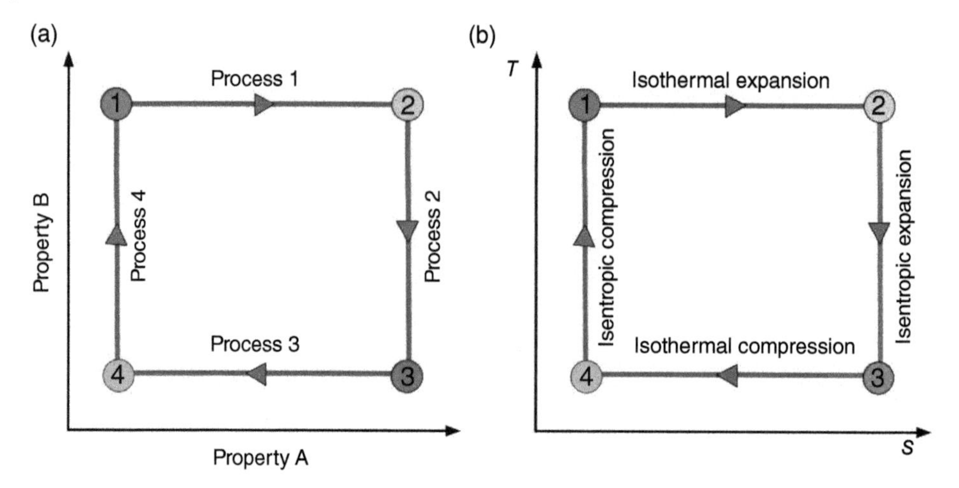

Figure 1.37 Illustration of (a) a cycle with four processes on a property diagram and (b) a Carnot heat engine cycle with four processes.

extensively used for design and analysis of energy systems as it maintains the energy balance and keeps track of all energy interactions. The FLT can then be expressed as follows:

$$E_{in} - E_{out} = \Delta E_{system} \tag{1.15}$$

Here, the subscripts in and out refer to the energy interactions crossing the system boundary into and out of the system domain. If the system operates as a steady-state process, then Eq. (1.15) can reduce to:

$$\dot{E}_{in} = \dot{E}_{out} \tag{1.16}$$

The dot above the energy symbol (E) refers to a rate property. Note that the system does not gain or lose energy, meaning that the rate change of the energy of the system equal to zero.

Since energy is introduced as the fifth step in the seven-step approach, we will have two forms of energy, namely heat and work, to talk about in thermodynamics. These are thoroughly discussed in subsequent chapters.

One of the forms of energy is heat that transfer from one body to another based on the temperature of each of the two bodies, which means the temperature difference is the driving force of the heat transfer. For example, when a body gains a total of 10 kJ of heat then we can present this as:

$$Q_{in} = 10\,kJ = mc_p(T_2 - T_1) \tag{1.17}$$

Here, Q is the heat, c_p is the specific heat capacity of the body being heated, and T is the temperature of the body, with subscript 1 for initial and subscript 2 for final state. Note that Eq. (1.17) is known as a simplified form with assumptions for a specific case.

Another common form of energy is work, and there are generally two types of work considered in thermodynamics, namely (i) boundary movement work (such as compression and expansion processes in piston–cylinder mechanisms) and (ii) shaft work (such as rotating machineries, like a pump). Some simple examples are given here and details are discussed later within specific sections.

Example 1.11 A tea glass with approximately 175 g of tea initially at a temperature of 80 °C losses heat to the surrounding air in a room temperature of 25 °C as shown in Figure 1.38. Calculate the amount of heat lost to surrounding air to be able to reach the surrounding air temperature. Consider tea with sugar specific heat capacity $c_p = 2000\ J/kg\,°C$.

Solution

The amount of heat lost to the surrounding air to be able to reach the surrounding air temperature:

$$\dot{E}_{in} = \dot{E}_{out}$$

$$m_{tea} c_p T_{initial} = m_{tea} c_p T_{final} + Q_{loss}$$

$$0.175\ kg \times 2 \frac{kJ}{kg\,^oC} \times 80\,^oC = 0.175\ kg \times 2.0 \frac{kJ}{kg\,^oC} \times 25\,^oC + Q_{loss}$$

$$Q_{loss} = 0.175\ kg \times 2 \frac{kJ}{kg\,^oC} \times (80\,^oC - 25\,^oC)$$

$$\mathbf{Q_{loss} = 19.25\ kJ}$$

Figure 1.38 Tea glass at a temperature 80 °C loses heat to the surrounding air at a temperature of 25 °C.

Example 1.12

One kilogram of warm milk, as shown in Figure 1.39, at a temperature of 60 °C is cooled by stirring resulting in a heat loss of 100 kJ from the milk to its surroundings to reach a temperature of 25 °C. Calculate the work input to the stirrer to achieve the required cooling. Consider milk specific heat capacity $c_p = 4180 \frac{J}{kg\,^oC}$

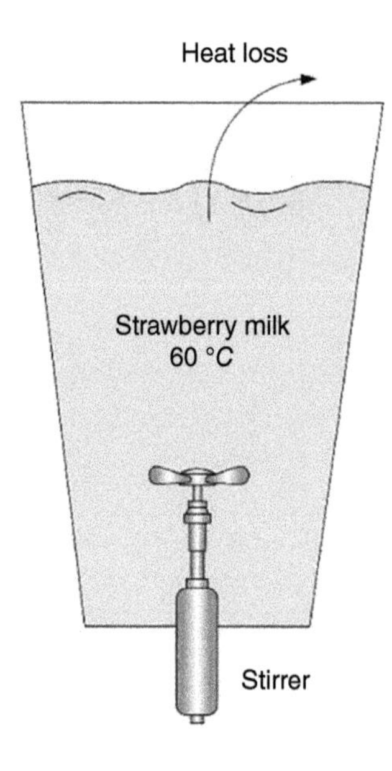

Figure 1.39 Warm milk being cooled by stirring.

Solution

The amount of work added to the milk to enhance the cooling and achieve a heat loss of 100 kJ and final temperature 25 °C is calculated as follows:

$$m_{milk}c_pT_{initial} + W_{in} = m_{milk}c_pT_{final} + Q_{loss}$$

$$1\ kg \times 4.18\frac{kJ}{kg\ ^oC} \times 60\ ^oC + W_{in} = 1\,kg \times 4.18\frac{kJ}{kg\ ^oC} \times 25\ ^oC + 100\ kJ$$

$$W_{in} = kg \times 4.18\frac{kJ}{kg\ ^oC} \times (25\ ^oC - 60\ ^oC) + 100\ kJ$$

$$\boldsymbol{W_{in} = -46.3\ kJ}$$

1.9.6 Second Law of Thermodynamics

The SLT is considered the most significant law of thermodynamics for practical systems and applications. Without this law, nothing can easily and truly be explained in the universe. The SLT is about several critical things, such nonconservation of available energy, irreversible behavior, entropy generation, exergy destruction, etc.

While the FLT brings energy analysis as a solution methodology for analysis and energy efficiency for system performance assessment, the SLT offers both entropy and exergy analyses for all irreversible, practical systems and applications in order to better design and analyze and exergy efficiency for a true system performance assessment. In order to analyze any system one must write balance equations for mass, energy, entropy, and exergy after placing the system boundary and identifying all inputs and outputs. These balance equations will be introduced later for different types of systems, including closed and open systems.

The SLT becomes instrumental in providing insights into environmental impact and sustainable development. The most appropriate link between the second law and environmental impact has been suggested to be exergy, in part because it is a measure of the departure of the state of a system from that of the environment. The magnitude of the exergy of a system depends on the states of both the system and the environment. This departure is zero only when the system is in equilibrium with its environment.

Energy has always been the most critical issue for humanity and is recognized as the first priority for all activities in the world. Humans started using wood as the first source of energy; this was then followed by coal, oil, natural gas and nuclear, as shown in the energy source historical ladder in Figure 1.40. We are now moving into an era where clean fuels are badly required to minimize environmental impact and increase sustainability.

Note that historically energy has been the source and cause of conflicts, wars, and peace. Since the industrial revolution, energy competition has been even stiffer. It has become more apparent that humanity needs more efficient, more cost effective, more environmentally benign, and more sustainable energy options and solutions. Such a requirement has been the main motivation behind going beyond traditional analysis methods and techniques. Traditionally, the FLT, which is recognized as the conservation law, has been the only tool comprehensively used in design, analysis, and assessment of thermodynamic systems. It is now crystal clear to everyone that the FLT is insufficient and incapable in addressing practical systems with irreversibilities (or losses, inefficiencies, etc.). That is why the SLT has been brought into the picture to account for irreversibilities or destruction through entropy and exergy. Exergy is distinguished itself to be a primary tool under the SLT.

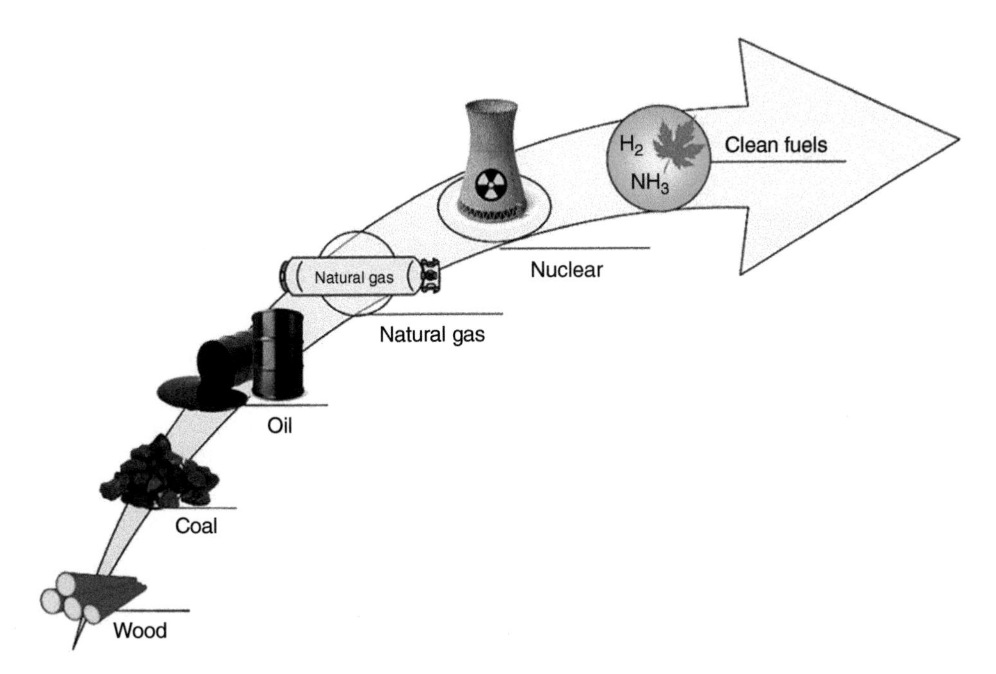

Figure 1.40 The development of energy sources throughout humanity past and present, and prediction of future sources.

Generally, thermodynamics is defined by many as the science of energy (referring to the FLT) and entropy (referring to the SLT). Here, thermodynamics is redefined as the science of energy and exergy, which makes it more correct to cover both laws of thermodynamics, make both energy and exergy quantities of the same unit consistently, and reach out two key efficiencies through energy and exergy efficiencies. This way the concepts dwell on the right pillars of practical applications. The exergy efficiency becomes important for practical systems and applications as it is a true measure of system performance and indicates how much the actual performance deviates the ideal performance.

Exergy analysis is now a recognized thermodynamic analysis technique based primarily on the SLT and appears to be the only tool for assessing and comparing processes and systems rationally and meaningfully. Consequently, exergy analysis can assist in improving and optimizing designs and analyses. Two key features of exergy analysis become more attractive, such that (i) it determines the true locations, types, and magnitudes of irreversibilities, losses, destruction and inefficiencies, and (ii) it identifies the potential for a system to be more efficient.

1.9.7 Performance Assessment

In thermodynamics, the description of an energy conversion system is usually followed by an appropriate efficiency definition of the system. A concentrated study of thermodynamics may be accomplished by the study of various efficiencies and other performance criteria and possible ways to improve these.

For an engineering system, efficiency, in general, can be defined as the ratio of desired output to required input.

In general, the efficiency of any system can be defined as follows:

$$Efficiency = \frac{Useful\ output}{Required\ input} \tag{1.18}$$

The above general efficiency definition can be written in more specific form for an energy system as follows:

$$Energy\ efficiency = \frac{Useful\ energy\ output}{Required\ energy\ input} \tag{1.19}$$

In thermodynamics, the description of an energy conversion system is usually followed by an appropriate efficiency definition of the system that evaluates the system conversion performance in terms of a ratio which presents how much of the inputted energy was successfully converted to the desired form in the outputs. A concentrated study of thermodynamics may be accomplished by the study of various efficiencies and ways to increase them.

For an engineering system, efficiency, in general, can be defined as the ratio of desired output to required input.

$$\eta = \frac{Desired\ output}{Required\ input} \tag{1.20}$$

Although this definition provides a simple general understanding of efficiency, a variety of specific efficiency relations for different engineering systems and operations have been developed. Different efficiency definitions based on the first and second laws of thermodynamics have been the subject of a large number of publications. Various efficiency definitions used for common energy conversion systems is the topic of this book. Many approaches that can be used to define efficiencies are provided and their implications are discussed. This book uses a logical and intuitive approach in defining efficiencies, and it is intended to provide a clear understanding of various efficiencies used in many common energy systems. In an exception to the energy efficiency definition, the performances of heat pumps and refrigeration units are assessed using the coefficient of performance (*COP*) principle. The coefficient of performance of a heat pump and a refrigeration unit can be written respectively as follows:

$$COP_{HP} = \frac{Heat\ output}{Work\ input} \tag{1.21}$$

$$COP_{R} = \frac{Cooling\ output}{Work\ input} \tag{1.22}$$

where *COP* denotes the coefficient of performance and the subscripts *HP* and *R* denotes the heat pump and the refrigeration unit respectively. Here, heat output is obtained from the condenser for heating purposes while the cooling output is the useful output in the evaporator where both systems need work input to drive them.

a) Energy Efficiency of a Compressor

The compressor is a mechanical device that consumes power to increase the pressure of a gas. From the simple definition of the compressor that was introduced in the previous sentence and the general energy efficiency definition, the energy efficiency of a compressor can be written as follows:

$$\eta_{comp} = \frac{\dot{m}(h_{out} - h_{in})}{\dot{W}_{in}} \tag{1.23}$$

where $\dot{m}$ is the mass flow rate of the gas being compressed, $(h_{out} - h_{in})$ is the amount of specific energy that the compressed air received when it was being compressed, and $\dot{W}_{in}$ the amount of power supplied to the compressor. Note that h is a thermophysical property known as enthalpy and it will be introduced in detail in the forthcoming chapters. Similar to the above compressor energy efficiency, the energy efficiency definitions of the most common thermodynamic devices are tabulated in Table 1.4.

Table 1.4 Energy efficiency definitions of some common thermodynamic devices.

Device	Energy efficiency
Pump	$\eta_{pump} = \frac{\dot{m}(h_{out} - h_{in})}{\dot{W}_{in}}$
Turbine	$\eta_T = \frac{\dot{W}_{out}}{\dot{m}(h_{in} - h_{out})}$
Electric motor	$\eta_m = \frac{\dot{W}_{elec}}{\dot{W}_{shaft}}$

Table 1.5 Energy efficiencies of some devices and cycles.

Device	Energy efficiency (%)
Pump	85
Gas turbine	33
Screw compressor	73
Throttling valve	100
Wind turbine	38
Photovoltaic (monocrystalline)	22
Reheat Rankine Cycle	38
Combined cycle (Rankine and Brayton)	56
Automobile (gasoline)	18
Automobile (electric)	60
PEM fuel cell	55
Electric kettle	85
Thermoelectric generator	5

Table 1.5 summarizes the average energy efficiencies of devices used in day to day activities, as well as power production modules and cycles that are used to produce electricity daily. These are not of course solid numbers, but are intended to give an idea about these for practical applications.

Example 1.13 A propane fueled water heater, as shown in Figure 1.41 receives energy from the supplied propane fuel, which heats the water by burning the propane and producing heat. Then the hot water is used to heat up the house. Define the energy conversion efficiency of the propane fueled water heater and make the necessary calculations to find it.

Solution

The desired output of the propane fueled water heater is the heat the hot water provides for the house.

The required input is the amount of fuel supplied to the water heater.

Then the conversion efficiency is written as:

$$\eta = \frac{Desired\ output}{Required\ input} = \frac{(Heat\ supplied\ to\ the\ \mathrm{hot}\ water)}{(Fuel\ supplied\ to\ the\ water\ heater)}$$

The energy delivered with the fuel can be defined through the heating value of the fuel, which is the amount of heat released when a specific amount of fuel at room temperature is completely burned and the combustion products cool down to the room temperature if it is the higher heating value (*HHV*). However, for the case where the water leaves the combustion in the form of water vapor then the amount of heat released is referred to as

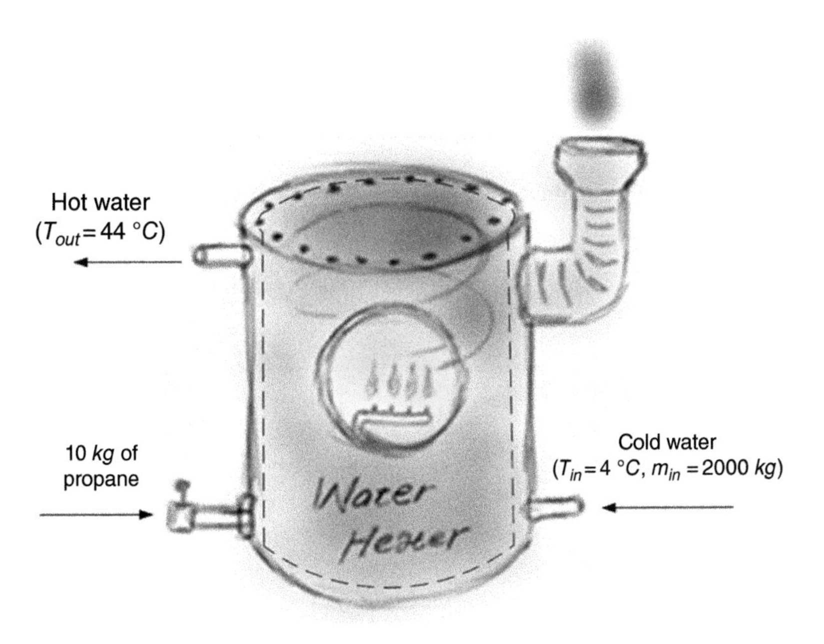

Figure 1.41 Propane fueled water heater.

the lower heating value (*LHV*). The *HHV* and the *LHV* have the unit of energy units over mass units (*kJ/kg* or *Btu/lb*). Then the above-mentioned efficiency can be presented as follows:

$$\eta = \frac{Q}{m_{propane} \times LHV_{propane}}$$

Here, *LHV* is used in the efficiency equation as the water from the combustion leaves the combustion chamber in the form of water vapor. The *LHV* of propane can be taken as 46,400 *kJ/kg* and the mass can be taken as 1 *kg*.

$$Q_{in} = m_{propane} \times LHV_{propane}$$

$$Q_{in} = 10\ kg \times 46{,}400 \frac{kJ}{kg}$$

$$Q_{in} = 464{,}000\ kJ$$

One can then find the output of the system to determine the energy efficiency of the system. *Q* is defined as the output of the system, which is the heat gained by the water to reach 44 °*C*. *Q* can be defined by the following equation:

$$Q_{in} = m \times Cp \times \Delta T$$

where m is the mass of the water, *Cp* is the specific heat capacity (*kJ/kg* °*C*), and ΔT is the change in temperature of the water.

$$Q_{out} = 2000\ kg \times 4.2 \frac{kJ}{kg\,°C} \times (44-4)$$

$$Q_{out} = 336{,}000\ kJ$$

Finally, the efficiency can be calculated as follows:

$$\eta = \frac{336{,}000\ kJ}{464{,}000\ kJ}$$

$$\eta = 0.724 = 72.4\%$$

The energy efficiency of the boiler is rated at 72.4%, meaning that not all the heat was transferred to the water and/or some of the heat was lost to the surroundings.

Example 1.14 A coal fueled power plant converts 156.25 MW of coal energy into 50 MW of electrical power through three steps. Define the conversion efficiency for each step and the overall efficiency of this coal fueled thermal power plant. Note that the main energy conversion steps of the power plant, as shown in Figure 1.42, are:

Step 1: Burning the coal to generate the amount of heat needed.
Step 2: Converting thermal energy into mechanical energy through a power generating cycle.
Step 3: Converting mechanical energy into electrical power through a generator.

These three conversion steps are illustrated in Figure 1.42 with the efficiency definitions. We now follow the steps given above to finalize the solution.

Solution

The first energy conversion step converts the chemical energy to thermal energy by combustion of coal. Similar to the propane fed water heater efficiency, the first step energy conversion efficiency is written as follows:

$$\eta_{S1} = \frac{\dot{Q}_{thermal}}{\dot{E}n_{in,coal}}$$

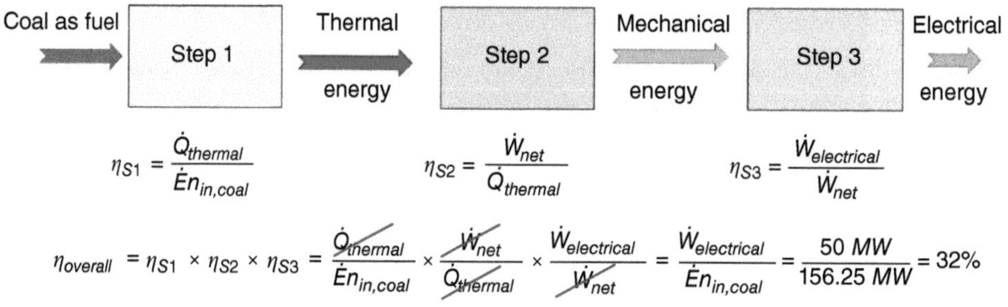

Coal as fuel → Step 1 → Thermal energy → Step 2 → Mechanical energy → Step 3 → Electrical energy

$$\eta_{S1} = \frac{\dot{Q}_{thermal}}{\dot{E}n_{in,coal}} \qquad \eta_{S2} = \frac{\dot{W}_{net}}{\dot{Q}_{thermal}} \qquad \eta_{S3} = \frac{\dot{W}_{electrical}}{\dot{W}_{net}}$$

$$\eta_{overall} = \eta_{S1} \times \eta_{S2} \times \eta_{S3} = \frac{\dot{Q}_{thermal}}{\dot{E}n_{in,coal}} \times \frac{\dot{W}_{net}}{\dot{Q}_{thermal}} \times \frac{\dot{W}_{electrical}}{\dot{W}_{net}} = \frac{\dot{W}_{electrical}}{\dot{E}n_{in,coal}} = \frac{50\ MW}{156.25\ MW} = 32\%$$

Figure 1.42 Illustration of power generation from coal with efficiencies.

The second energy conversion step converts the thermal energy to rotational mechanical energy. The second energy conversion step efficiency is written as:

$$\eta_{S2} = \frac{\dot{W}_{net}}{\dot{Q}_{thermal}}$$

Here, net refers to the rotational mechanical energy. The third and final step in the coal power plant is the conversion of rotational mechanical energy into electrical energy; its efficiency can be expressed as follows:

$$\eta_{S3} = \frac{\dot{W}_{electrical}}{\dot{W}_{net}}$$

However, by looking at the overall system, where it has a single input, coal, and a single output, which is electrical energy, then the overall system conversion efficiency is expressed as follows:

$$\eta_{ov} = \frac{\dot{W}_{electrical}}{\dot{E}n_{in,coal}}$$

Note that the overall power plant efficiency can also be presented in terms of the energy conversion efficiency of its main components or main energy conversion steps as follows:

$$\eta_{ov} = \eta_{S1} \times \eta_{S2} \times \eta_{S3} = \frac{\dot{W}_{electrical}}{\dot{E}n_{in,coal}} = \frac{50\ MW}{156.25\ MW} = \mathbf{32\%}$$

Example 1.15 This example is about how to make profit from different electricity cost in day and night as illustrated in Figure 1.43. The electricity cost during the night is \$0.06/kWh and the electricity cost during the day is \$0.12/kWh. The water volume flow rate in the system in both directions is 2 m^3/s. Assume the energy efficiency of the motor and pump combination is 100% (working during the night), whereas the energy efficiency of the turbine and generator is 90% (working during the day). Neglecting the frictional losses in the pipes and assuming that the pump and motor combination is used from 10 : 00 p.m. to 8 : 00 a.m., and the turbine and generator are used from 10 : 00 a.m. to 8 : 00 p.m, calculate the potential revenue generated by the proposed system per year.

Solution

At night:

The system consumes electrical energy, which means that the system owner is paying money to the electrical distribution company.

The total mass pumped during the 10 hours' night (10 p.m. to 8 a.m.) is calculated as follows:

$$m_{tot} = \rho_w \dot{V}(operation\ time)$$

$$m_{tot} = 1000\,\frac{kg}{m^3} \times 2\,\frac{m^3}{s} \times 10\ h \times 3600\,\frac{s}{h} = 72{,}000{,}000\ kg/day$$

Writing the energy balance equation:

$$E_{in} = E_{out}$$

$$m_{tot} gh = W_{pump}$$

$$W_{pump} = 72{,}000{,}000 \frac{kg}{day} \times 9.81 \frac{m}{s^2} \times 50\ m = 3.53 \times 10^{10} \frac{J}{day}$$

The total amount of money paid during the night:

$$C_{night} 2wsx = 3.53 \times 10^{10} \frac{J}{day} \times \frac{1\ kJ}{1000\ J} \times \frac{\$0.06}{kWh} \times \frac{1\ h}{3600\ s} \times = \frac{\$588.6}{day}$$

The amount of energy recovered during the day is the same as the energy stored during the night. However, we have to consider the efficiency of the motor–turbine, which is 90% compared to 100% efficiency in motor–pump system:

$$\eta_{motor-turbine} = \frac{electrical\ power\ produced}{potential\ energy\ stored\ in\ elevated\ water}$$

$$0.9 = \frac{Electrical\ power\ produced}{3.53 \times 10^{10}\ J/day}$$

$$Electrical\ power\ produced = 3.18 \times 10^{10} \frac{J}{day}$$

$$The\ revenues = 3.18 \times 10^{10} \frac{J}{day} \times \frac{1\ kJ}{1000\ J} \times \frac{\$0.12}{kWh} \times \frac{1\ h}{3600\ s} = \frac{\$1060}{day}$$

Then the profit per year is calculated as follows:

$$\frac{Profit}{year} = \frac{revenue}{year} - \frac{cost}{year}$$

$$\frac{Profit}{year} = \left(\frac{\$1060}{day} \times \frac{365\ days}{year}\right) - \left(\frac{\$588.6}{day} \times \frac{365\ days}{year}\right) = \frac{\mathbf{\$172{,}061}}{\mathbf{year}}$$

1.10 Engineering Equation Solver as a Potential Software

EES (pronounced "ease") is a general equation-solving program that can numerically solve thousands of coupled nonlinear algebraic and differential equations. The program can also be used to solve differential and integral equations, do optimization, provide uncertainty analyses, perform linear and nonlinear regression, convert units, check unit consistency, and generate publication-quality plots. A major feature of EES is the high accuracy thermodynamic and transport property database that is provided for hundreds of substances in a manner that allows it to be used with the equation solving capability.

Next in this section, two examples of the use of EES in solving thermodynamics problems are presented. However, a good practice is to make sure that once a new EES window is open, as shown in Figure 1.43, the units selected to run the simulation are also

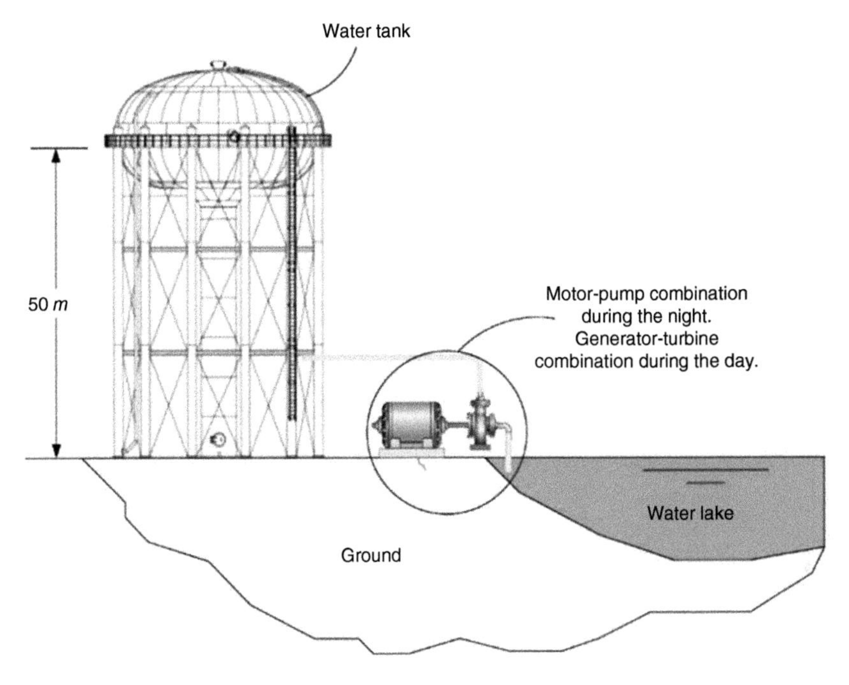

Figure 1.43 Night pumping and day turbine system.

checked as well. In order to open a file, go to file and then select from the dropdown menu new.

In order to open the window, go to options and then select unit system. The window shown in Figure 1.44 will open. After setting up the desired unit system and units, click on the OK button. Table 1.6 presents the procedure of writing Greek letters in EES accordingly.

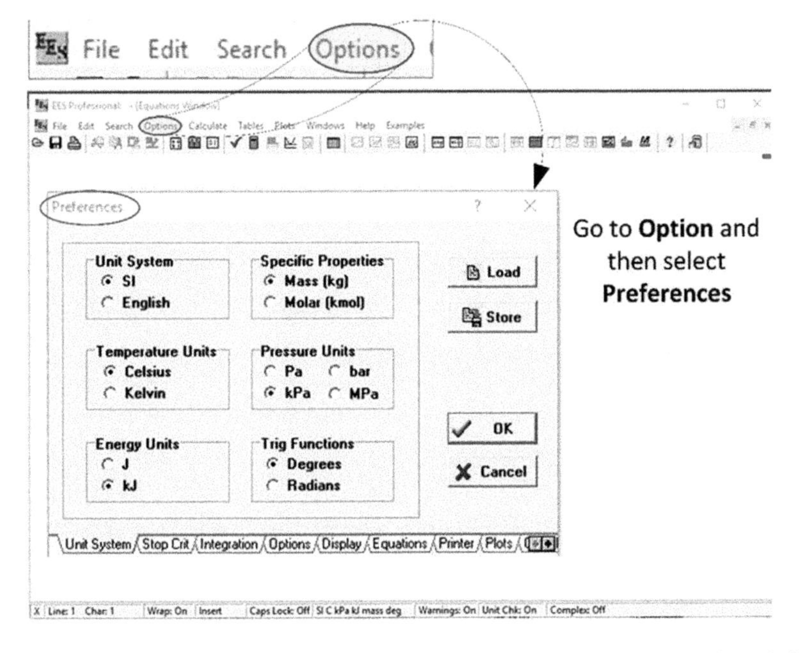

Figure 1.44 An EES window with unit system and parameter unit selection window in it.

Table 1.6 Some of the Greek letters and the corresponding EES expression.

EES expression	Corresponding Greek letter
rho	ρ
mu	μ
gamma	γ
delta (case sensitive)	δ
DELTA (case sensitive)	Δ
zeta	ζ
beta	β
alpha	α

Example 1.16 Find the following properties at the specified temperature and pressure:

a) Density of water at a temperature of 25 °*C* and a pressure of 1 *bar*,
b) Density of water at a temperature of 90 °*C* and a pressure of 101 325 *Pa*,
c) Viscosity of water at a temperature of 90 °*C* and a pressure of 101 325 *Pa*.

Solution

The following are the steps to follow to find material properties where the first two steps are shown in Figure 1.45:

a) Density of water at a temperature of 25 °*C* and a pressure of 1 *bar*:

   Line 1: *T_1* = 25 [*C*]
   Line 2: *P_1* = 1∗*convert(bar,Pa)*
   Line 3: *Rho_1* = *density(water,T* = *T_1,P=P_1*)

   The density at the previous mentioned temperature and pressure is equal to 997.1 kg/m^3.
b) Density of water at a temperature of 90 °*C* and a pressure of 101 325 *Pa*:

   Line 1: *T_1* = 90 [*C*]
   Line 2: *P_1* = 101325 [*Pa*]
   Line 3: *Rho_1* = *density(water,T* = *T_1,P=P_1*)

   The density at the previous mentioned temperature and pressure is equal to 1038 kg/m^3.

c) Viscosity of water at a temperature of 90 °*C* and a pressure of 101 325 Pa:

   Line 1: *T_1* = 90 [*C*]
   Line 2: *P_1* = 101325 [*Pa*]
   Line 3: *mu_1* = *viscosity(water,T* = *T_1,P=P_1*)

   The viscosity at the previous mentioned temperature and pressure is equal to 0.0008903 kg/ms.

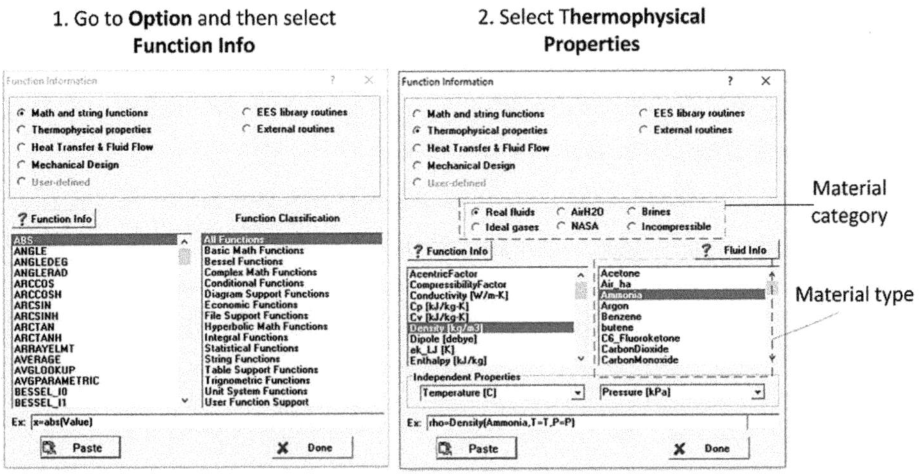

1. Go to **Option** and then select **Function Info**

2. Select **Thermophysical Properties**

Material category

Material type

Figure 1.45 The first two steps in the solution in Example 1.16.

1.11 Closing Remarks

This introductory chapter has discussed energy and its importance, thermodynamic fundamentals and basic definitions, a seven-step approach in teaching thermodynamics, thermodynamic laws and performance assessment, and it has provided numerous examples to better understand the concept and recognize the systems and applications from the thermodynamic perspectives.

Study Questions/Problems

This section contains practice questions and problems, including short answer questions, proof questions, true, and false, multiple choice and calculation problems.

a) Concept

1 Although the zeroth law of thermodynamics was introduced after the first and second laws of thermodynamics, it was referred to as the zeroth law, explain why?

2 Which thermodynamics law considers the quality as well as the quantity of energy?

3 Provide an example where the quality of energy is important not just the quantity.

4 Prove that dividing the volume of a system (an extensive property) over the mass of the same system will provide us with an intensive property called the specific volume. (Hint: you can use the dividing space example.)

5 Which feels colder for a bare human touch, a metallic chair or a wooden chair in an ambient temperature space, or will they both feel the same, and why?

6 Provide the definition of energy efficiency and discuss its basic foundation for thermal power plant.

7 Write energy efficiencies for both pump and turbine and discuss the key difference between them.

8 Describe both extensive and intensive properties and compare them at least with three examples.

9 Explain why 3C rule is important.

10 Explain why seven step approach is important in thermodynamics.

b) True or False Type Questions

1 Thermodynamics is the science of energy and entropy.

2 Thermodynamics is the science of energy and enthalpy.

3 FLT considers both the quantity as well as the quality of energy.

4 Based on the FLT, the amount of energy available is constant and will never change.

5 You need a minimum of three independent properties in order to fully define a state.

6 You need a minimum of four processes to make a cycle.

7 One property is enough to define the state of a thermodynamic system.

8 Temperature is an intensive property.

9 Properties that are independent of the size of the system are called extensive properties.

10 Energy cannot be created nor destroyed, however exergy can be destroyed.

c) Multiple Choice Type Questions

1 If objects A and B are in thermal equilibrium with object C, and object C is at 100 °C, then the temperature of objects A and B are:
  - **a** 50 °C and 50 °C.
  - **b** 100 °C and 100 °C.
  - **c** It is impossible to determine since it depends on the sizes of objects A, B, and C.
  - **d** It is impossible to determine since it depends on the sizes of objects A and B.
  - **e** It is impossible to determine since it depends on the size and material.

2 Which of the following is the hottest?
  - **a** 80 °C cup of coffee.
  - **b** 100 °F cup of milk.
  - **c** 273 K water.
  - **d** 400 R metallic structure.

3 Why does a bicyclist pick up speed on a downhill road even when he is not pedaling?
  - **a** The *PE* is converted to *KE*.
  - **b** The *KE* is converted to *PE*.
  - **c** This should not happen since it is a violation of the conservation of energy principle (FLT).
  - **d** The mass of the bicyclist is converted into a *KE* since the bicyclist is loosing weight.
  - **e** The bicyclist is picking up speed since $KE = \frac{1}{2}mv^2$.

4 People living in the mountains develop lungs with a volume:
  - **a** Larger than those living at low elevation.
  - **b** Smaller than those living at low elevation.
  - **c** The same as those living at low elevation.

5 Why when vehicle tires have a hole in them, does air comes out of the tires?
  - **a** Because the pressure inside the tires is higher than the atmospheric pressure.
  - **b** Because the pressure inside the tires is lower than the atmospheric pressure.
  - **c** Because the tires are made of rubber and the rubber is squeezing the air out.
  - **d** Because the vehicle is heavy and it squeezes the tires to release the air out.
  - **e** Air will not come out, the air someone feels is just an external air circulation.

6 Energy can not be created nor destroyed it can be converted from one from to another, is the definition of the:
  - **a** FLT
  - **b** SLT
  - **c** Zeroth law of thermodynamics
  - **d** Third law of thermodynamics
  - **e** Fourth law of thermodynamics.

7 A rock falling from a certain elevation above the ground is an example of the:
  a FLT
  b SLT
  c Zeroth law of thermodynamics
  d Third law of thermodynamics
  e Fourth law of thermodynamics.

8 Why would a cyclist pick up speed as they are going downhill even when he/she is not pedaling?
  a Due to the conversion of *PE* to *KE*.
  b Due to the conversion of *KE* to *PE*.
  c It will never happen because it is violating the FLT.
  d It will never happen because it is violating the SLT.

9 A student claims that a cup of coffee on his/her table warmed up to 80 °*C* by picking up energy form the surrounding air, which is at 25 °*C*.
  a It will never happen because it is violating the SLT.
  b It will never happen because it is violating the FLT.
  c Due to the conversion of *PE* to *KE*.
  d Due to the conversion of *KE* to *PE*.

10 Thermodynamics is the science of
  a Energy and exergy
  b Energy and entropy
  c Exergy and entropy
  d Enthalpy and energy
  e None of the above.

11 Which of the following is not listed on the seven-step approach?
  a Zeroth law of thermodynamics
  b FLT
  c SLT
  d State
  e Property.

12 Which of the following is systems is not a thermodynamics system?
  a Ordinary
  b Open
  c Closed
  d Controlled mass
  e Controlled volume.

13 A system boundary is:
  a A real surface
  b An imaginary surface
  c A surface that separate the system from its surroundings

d A surface with zero thickness
e All of the above.

14 The Carnot cycle includes
a Two isentropic and two isothermal processes
b Two isenthalpic and two isentropic processes
c Two isobaric and two isenthalpic processes
d None of the above

15 The temperature change in Celsius is equivalent to the temperature change in:
a Rankine
b Fahrenheit
c Kelvin
d None of the above

d) Problems

1 Calculate the temperature of a $50\,^{\circ}C$ cup of coffee in:
a Rankine (R)
b Kelvin (K)
c Fahrenheit ($^{\circ}F$).

2 An oil fed power plant converts $160\,MW$ of oil into $55\,MW$ of electrical power through various steps. The main energy conversion steps of the power plant are (starting from the component receiving the oil and ending up with the production of electricity):
- Chemical energy to thermal energy.
- Thermal energy to mechanical energy.
- Mechanical energy to electrical energy.

3 A tea glass with $175\,g$ of tea initially at a temperature of $90\,^{\circ}C$ losses heat to the surrounding air in a room at a temperature of $25\,^{\circ}C$. Calculate the amount of heat lost to the surrounding air to be able to reach the surrounding air temperature. Consider tea with sugar specific heat capacity $c_p = 2000\,\dfrac{J}{kg\,^{o}C}$

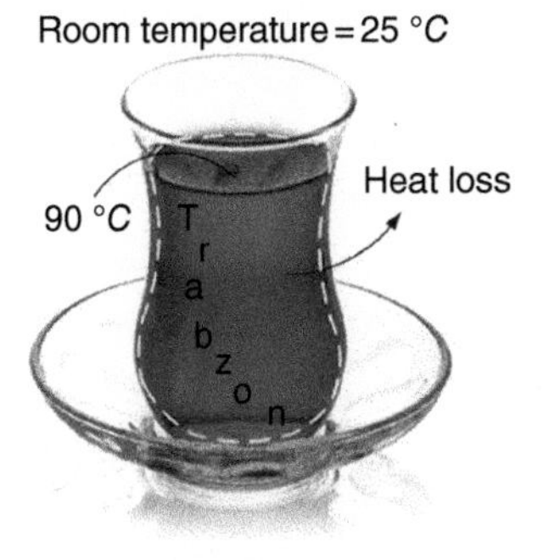

4 1 kg of cold milk at a temperature of 10 °C is mixed with strawberries and bananas. Calculate the work input without increasing the temperature of the milkshake. Consider milk specific heat capacity $c_p = 4180\ \frac{J}{kg\ ^oC}$ and the masses of strawberry and banana as 0.5 kg each.

5 Calculate the gage pressure of the spherical tank when fluid A is mercury ($\rho_{Hg} = 13600\ kg/m^3$), fluid B density is $\rho_B = 900\ kg/m^3$, and fluid C has a density of $\rho_C = 760\ kg/m^3$, and if $h_1 = 10\ cm$, $h_2 = 5\ cm$, and $h_3 = 6\ cm$. Take the difference in higher and lower levels of fluid B as 1 cm.

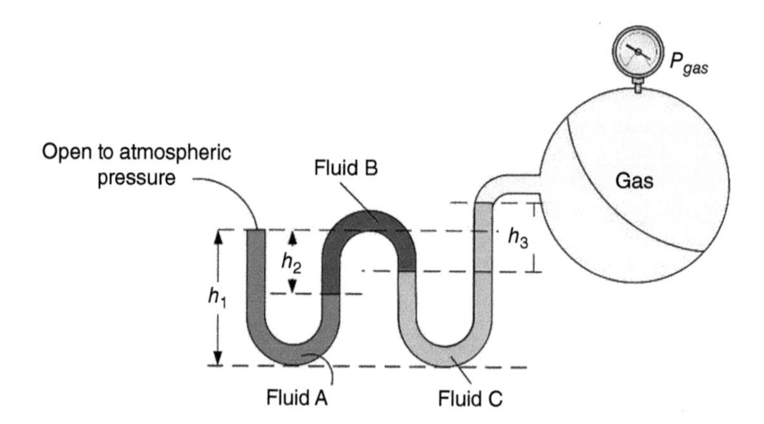

6 What is the gage pressure of the gas tank that is connected the manometer as shown below if the heights values are $h_1 = 10\ cm$, $h_2 = 5\ cm$, and $h_3 = 2\ cm$, where fluid A is mercury with a density of 13 600 kg/m^3, fluid B is water with a density of 1000 kg/m^3, and fluid C is oil with a density of 760 kg/m^3:

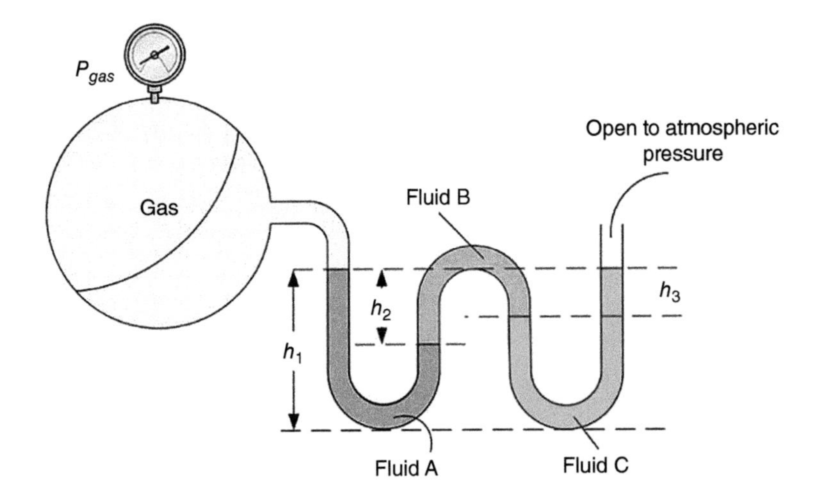

7 What is the gage pressure of the gas tank that is connected the manometer as shown below if the heights values are $h_1 = 15\ cm$ and $h_2 = 5\ cm$, where fluid A is mercury with a density of 13 600 kg/m^3 and fluid B is water with a density of 1000 kg/m^3:

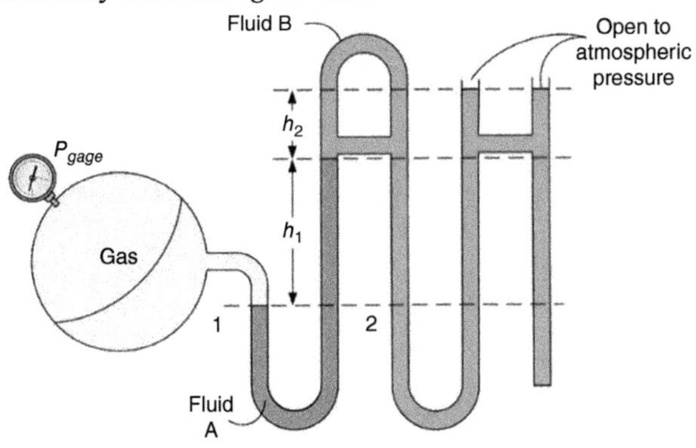

8 What is the gage pressure of the cylindrical gas tank that is connected the manometer as shown below if the heights values are $h_1 = 15\,cm$, $h_2 = 10\,cm$, and $h_3 = 10\,cm$ when fluid A is water with a density of 1000 kg/m^3 and fluid B is oil with a density of 760 kg/m^3:

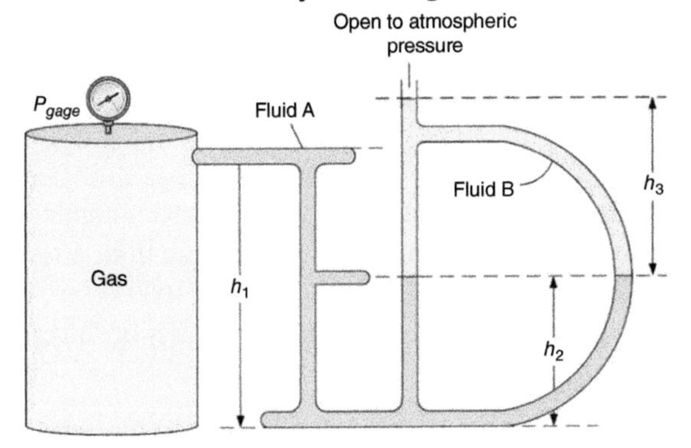

9 What is the gage pressure of the gas tank that is connected the manometer as shown below if the height values are $h_1 = 15\ cm$, $h_2 = 5\ cm$, $h_3 = 2\ cm$, $h_4 = 10\ cm$, and $h_5 = 7\ cm$ when fluid A is mercury with a density of $13\,600\ kg/m^3$, fluid B is water with a density of $1000\ kg/m^3$, fluid C is gasoline with a density of $720\ kg/m^3$, and fluid D is oil with a density of $760\ kg/m^3$ (note that the angle $\theta = 30^o$). Consider the height of fluid C as $1.5\ cm$.

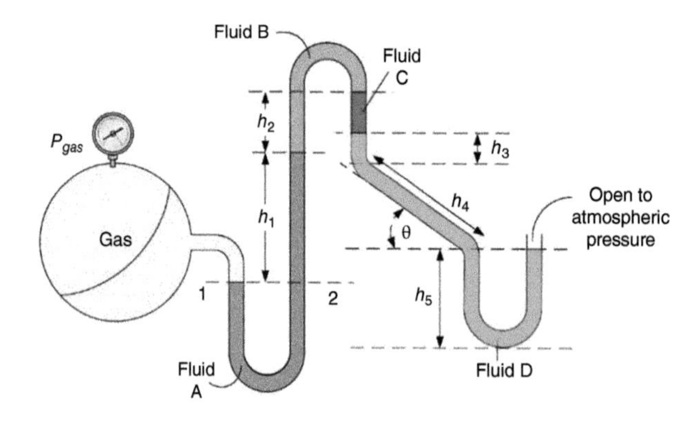

10 What is the gage pressure of the gas tank that is connected to the manometer as shown below if the heights values are $h_1 = 15\ cm$, $h_2 = 5\ cm$, and $h_3 = 12\ cm$. Fluid A is mercury with a density of $13\,600\ kg/m^3$ and fluid B is water with a density of $1000\ kg/m^3$:

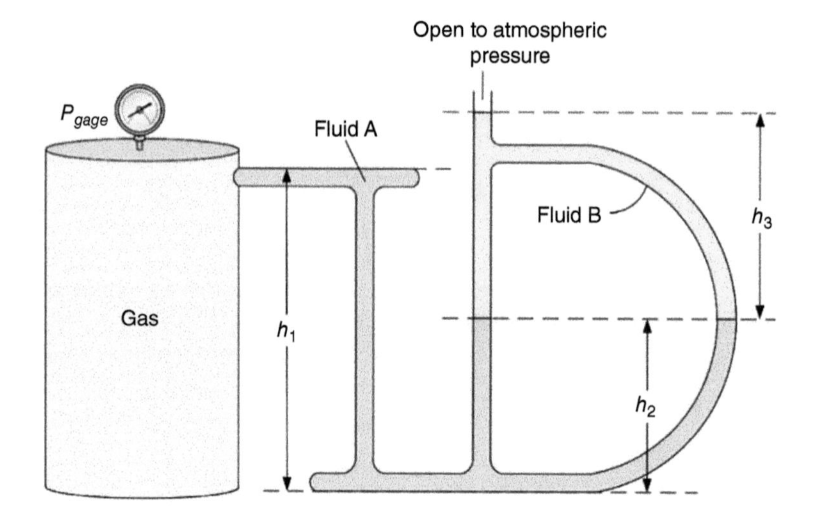

11 What is the gage pressure of the gas tank that is connected the manometer as shown below if the heights values are $h_1 = 15\ cm$ and $h_2 = 5\ cm$, where fluid A is mercury with a density of $13\,600\ kg/m^3$, fluid B is water with a density of $1000\ kg/m^3$, fluid C is gasoline with a density of $720\ kg/m^3$, and fluid D is oil with a density of $760\ kg/m^3$:

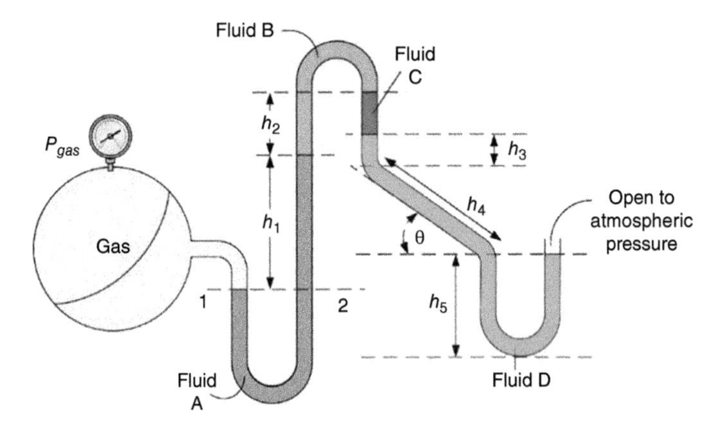

12 What is the gage pressure of the gas tank that is connected the manometer as shown below if the heights values are $h_1 = 15\ cm$ and $h_2 = 5\ cm$, where fluid A is mercury with a density of 13 600 kg/m^3 and fluid B is water with a density of 1000 kg/m^3:

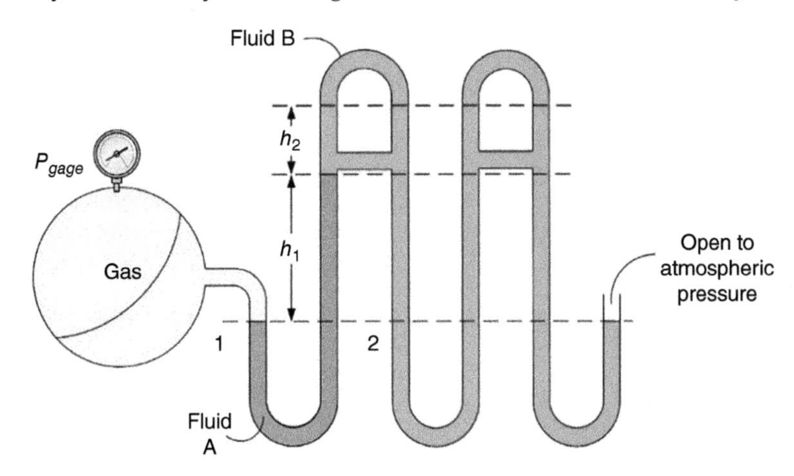

13 What is the gage pressure of the gas tank that is connected the manometer as shown below if the heights values are $h_1 = 15\ cm$ and $h_2 = 4\ cm$, where fluid A is mercury with a density of 13 600 kg/m^3, fluid B is water with a density of 1000 kg/m^3, fluid C is gasoline with a density of 720 kg/m^3, and fluid D is oil with a density of 760 kg/m^3:

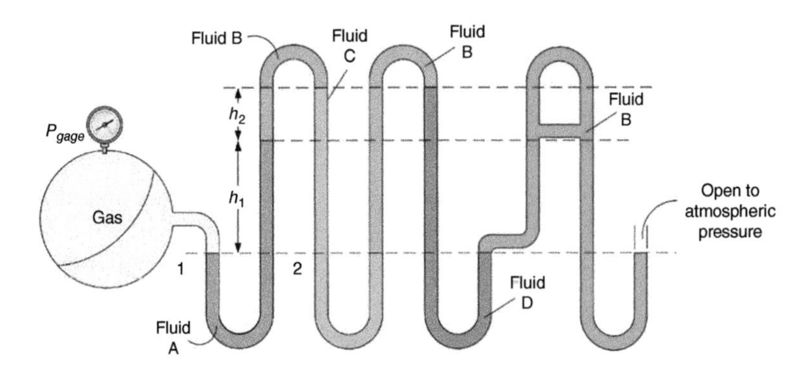

14 What is the gage pressure of the cylindrical gas tank that is connected the manometer as shown below if the heights values are $h_1 = 15\,cm$, $h_2 = 10\,cm$, and $h_3 = 10\,cm$ when fluid A is water with a density of $1000\,kg/m^3$ and fluid B is oil with a density of $760\,kg/m^3$:

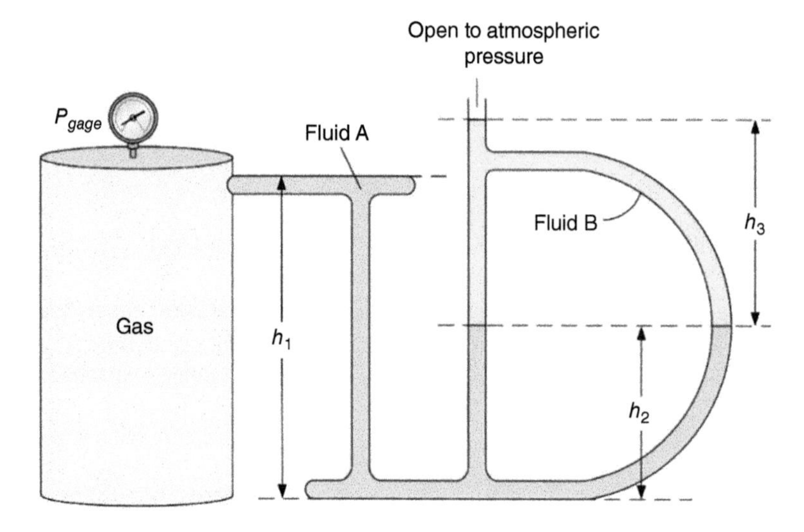

15 Find the potential and kinetic energies of the rolling snowball in the intermediate position as shown in the figure below (note that you can neglect the increase of the mass of the snowball):

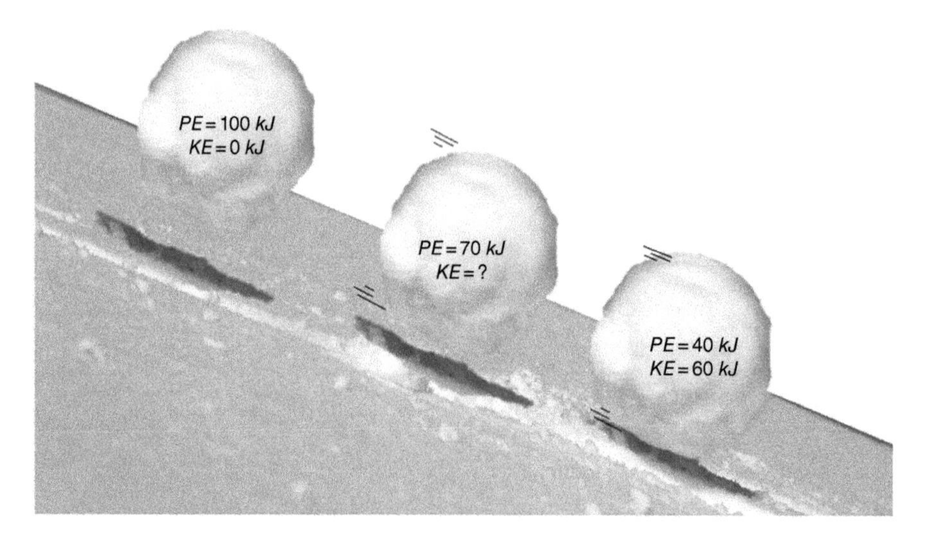

16 A natural gas fueled water heater tank works by burning natural gas and utilizing the thermal energy released to heat water. Consider a water heater tank that contains $50\,kg$ of water and heats it from a temperature of $25\,°C$ to $70\,°C$. If the water heater is known to have an efficiency of 70%, determine the amount of natural gas required. (Take the lower heating value of natural gas as 47.1 MJ/kg and the specific heat capacity of water as 4.2 $kJ/kg\,°C$.)

17 According to Figure 1.25, a lift is designed to lift containers with the help of a piston. How much weight can be lifted on piston 1 if area of 10 m^2 is available by applying 200 N force on the piston 2 area of 0.2 m^2.

18 Calculate the density of a liquid if the mass and volume are 1 kg and 1 m^3 and convert the density units to g/cm^3.

19 The specific density of two substances are 0.65 and 5.8. Find the density of each substance and describe if the objects will sink or float.

20 The pressure of a gas container relative to the atmospheric pressure is 60 kPa. Find the absolute pressure in the gas container.

21 According to Figure 1.39, 1.5 kg of milk is cooled by stirring which results in the heat loss of 75 kJ from milk to the atmospheric temperature 25 $°C$ surroundings. If 30 kJ of work input in provided for stirring, calculate the temperature of the hot milk before stirring. Consider milk specific heat capacity of 4180 $J/kg\,°C$.

22 According to Figure 1.41, a propane fueled water heater receives energy from supplied propane fuel which heats the water using fuel and hot water is used to heat up the house. If energy conversion efficiency is 76%, make the necessary calculations and calculate heat output.

23 An electric water heater is used to heat water from a temperature of 25 $°C$ to a temperature of 80 $°C$. If the water tank contains 50 L of water, determine the time required by the electric heater if the power rating of the electric heater is 20 W. Consider the density of water as 1000 kg/m^3 and the specific heat as 4.2 $kJ/kg\,°C$.

24 Reconsider Problem 23 this time with a final temperature of 110 $°C$ using an electric heater of 50 W and find the time required for this heating process.

25 Reconsider Problem 23 this time for thermal oil with a final temperature of 250 $°C$ using an electric heater. Calculate the capacity of heater required to heat this in 15 minutes. Take the density of thermal oil as 810 kg/m^3 and the specific heat as 2.14 $kJ/kg\,°C$.

26 An air compressor with an energy efficiency of 80% consumes 50 kW of power input. If the mass flow rate of air is known to be 1.2 kg/s, determine the change in air enthalpy during the compression process.

27 Reconsider Problem 26 this time with an energy efficiency of 91%. Find the change of enthalpy and compare with the result obtained in Problem 26.

28 A steam turbine with an energy efficiency of 85% produces 120 kW work rate. If the mass flow rate of air is known to be 1 kg/s, calculate the change in the enthalpy during this expansion process.

29 A coal-fueled power plant converts 100 MW of coal energy into 25 MW of electrical power through three steps. Define the conversion efficiency for each step and find the overall energy efficiency of the power plant.

30 Reconsider Problem 29 this time for multiple fuels, oil, natural gas, wood and hydrogen and discuss their impacts on the environment.

2

Energy Aspects

2.1 Introduction

Energy has always been vital for humanity for almost everything, ranging from cooking meals to heating houses and running cars to operating plants. The driver of today's economic sectors, such as industrial, residential, commercial, public, utility, agricultural, and transportation, is really energy (Figure 2.1). Outside of the residential sector, in the industrial sector energy is required for power to manufacture goods that we use on daily basis. The transportation sector is also dependent on energy to power cars, trucks, aircrafts, and ships extensively. These are of course only a few abstract examples to reflect the importance.

There was a question in the previous chapter why should we study thermodynamics? If you need energy for anything and everything, this brings a huge necessity to design, analyze, and evaluate all these energy systems and applications. That's why thermodynamics is linked to energy, environment, and sustainable development, as illustrated in Figure 2.2. Anything that you do with thermodynamics will have an impact on these three domains. By learning thermodynamics right and practicing right one can help create achieve:

- better design, analysis, assessment, and improvement
- better management
- better resources use
- better efficiency
- better cost effectiveness
- better energy security
- better environment
- better sustainability

which makes this particular discipline really unique and necessary. Furthermore, thermodynamically conscious people will help achieve a better world for everyone. So, it is really important to learn the energy concepts and aspects related to all systems and applications and use thermodynamics as a key vehicle in this regard. This is exactly what this chapter aims to do.

Thermodynamics: A Smart Approach, First Edition. Ibrahim Dincer.

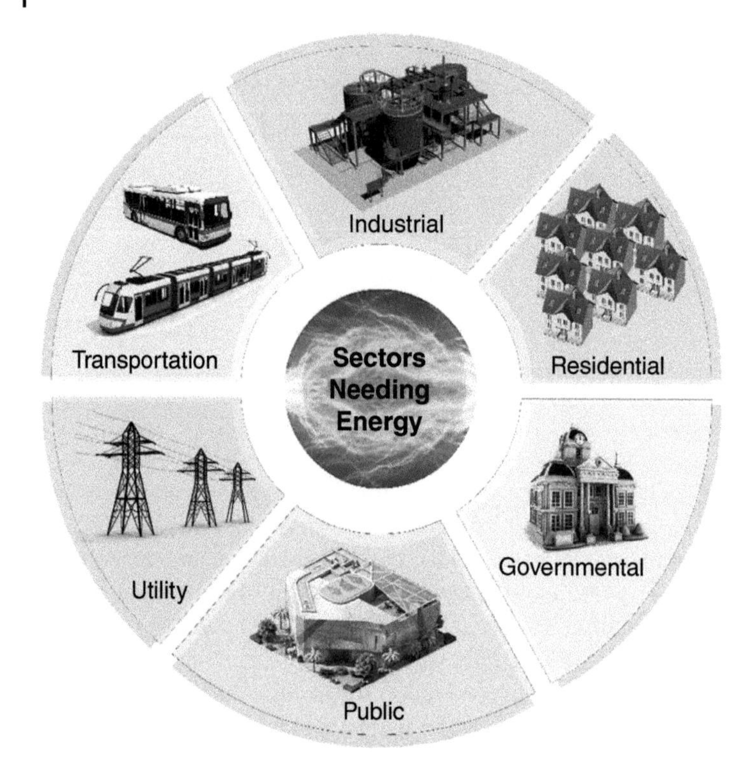

Figure 2.1 Various sectors driven by energy.

2.2 Macroscopic Thermodynamics versus Microscopic Thermodynamics

Energy can be in various forms: thermal, potential, kinetic, mechanical, electric, magnetic, nuclear, chemical, etc. In order to quantify the energies, one needs to thermodynamically study each separately and carefully. However, this is not easy and one can classify these forms into two main categories:

- *Microscopic thermodynamics*: This is something that can be done by at the microscopic level, as to be observed by microscopes. It is really related to atomic/molecular level thermodynamic activities.
- *Macroscopic thermodynamics*: This is something that can be done by at the macroscopic level, as to be observed by eyes of people. It is really related to practical level thermodynamic activities. This is technically what engineering thermodynamics considers and studies.

These two concepts are clearly observed in Figure 2.3 where Figure 2.3a shows water in a glass bowl with no activity observed by eyes. However, there is activity microscopically at the atomic/molecular level kinetically and potentially. Figure 2.3b shows a bowl of water

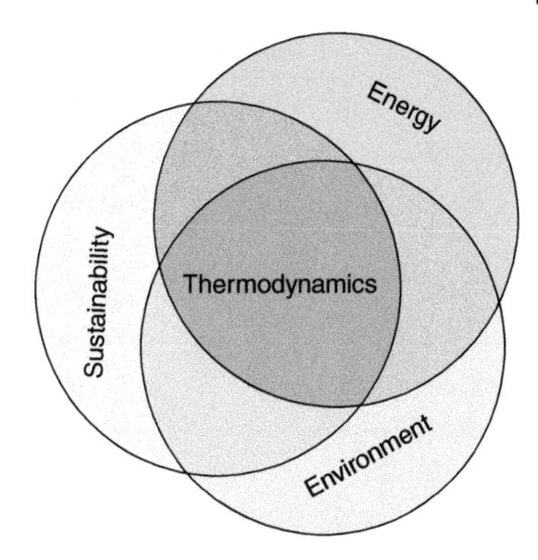

Figure 2.2 Thermodynamics as the heart of the three key domains of energy, environment, and sustainability.

being heated and boiled; this is clearly observed by eyes and makes it a nice example for macroscopic level of thermodynamics. In conjunction with the above given descriptions, one may categorize them into microscopic and macroscopic forms of energies as well. Macroscopic forms of energy are those where, in fact, we look deep into the internal energies and behavior of the material, such as those that are either measured or calculated based on the changes caused by kinetic and potential energies as shown in Figure 2.3b.

In the macroscopic approach, one needs to consider three types of energy – internal energy, kinetic energy, and potential energy – under total energy (E) of a specific mass (m), which is the summation of these three energies as follows:

$$E = U + KE + PE = mu + \frac{mV^2}{2} + mgz\ (kJ) \tag{2.1}$$

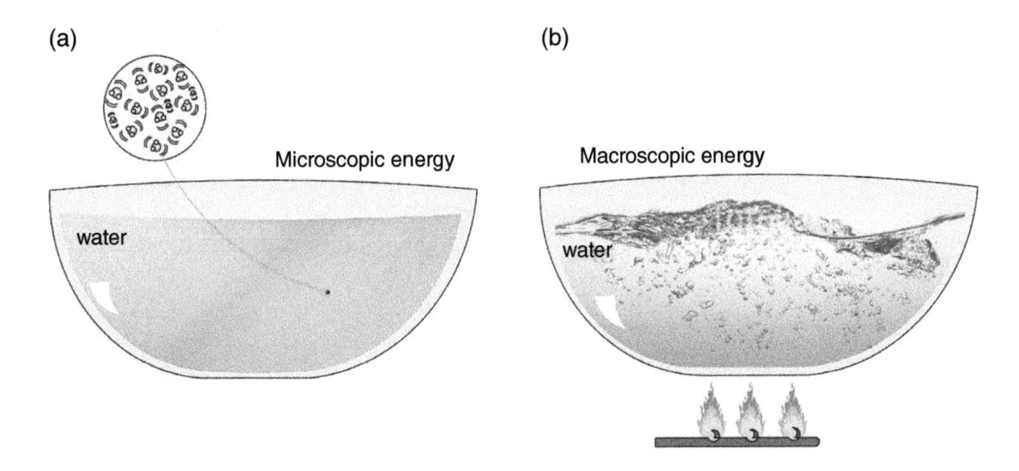

Figure 2.3 Illustration of (a) microscopic and (b) macroscopic energy behaviors.

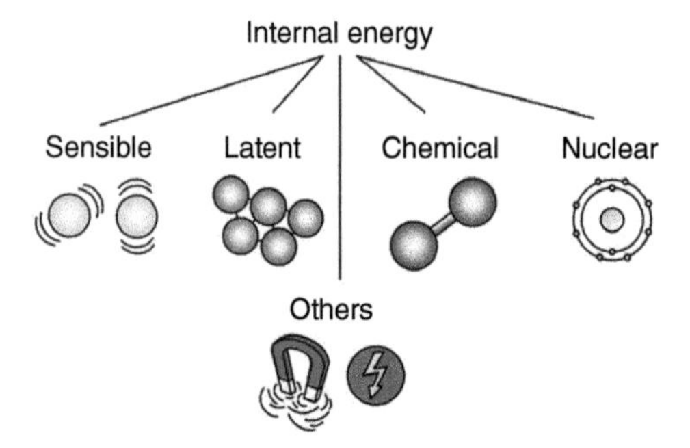

Figure 2.4 Breakdown of the internal energy.

where E denotes the total energy, U denotes the internal energy, KE denotes kinetic energy, PE is the potential energy, u is the specific internal energy, V is the velocity, g is the gravitational acceleration (9.81 m/s^2 and 32.17 ft/s^2), and z is the elevation of the mass above the earth surface or the reference point.

The total energy can also be presented in the specific form as follows:

$$e = u + ke + pe = u + \frac{V^2}{2} + gz \ (kJ/kg) \tag{2.2}$$

Here, e denotes the specific total energy, u is the specific internal energy, ke is the specific kinetic energy and pe denotes the specific potential energy.

The internal energy presented in the total energy equations is the sum of all microscopic forms of energy of a system or a mass, which is related to the molecular structure and the degree of molecular activity. The internal energy can be defined as the sum of the potential and kinetic energies of the molecules or, in a more general description, the summation of all the microscopic forms of energy of a system. Internal energy may generally be classified into five different forms of microscopic energies, as shown in Figure 2.4. As shown in the figure, these are the sensible, nuclear, latent, chemical, and possibly other types of energies (such as magnetic). Chemical energy represents the part of the internal energy that is associated with the chemical bonds holding the atoms together to form a molecule. Latent energy is the part of the internal energy that it is stored in the bonds holding the structure of the phase, in other words in order to phase change the system the latent energy must be supplied or removed from the system. Nuclear energy is associated with the energy responsible for holding the material atom nuclear together. Finally, sensible energy, represents the part of the internal energy associated with the kinetic energy of the molecules, where the system temperature is usually a measure of this energy.

2.3 Energy and the Environment

One of the most critical challenges affecting the relationship between ourselves and the environment is energy. Anything we do with energy, in terms of production, distribution, conversion, transportation, and consumption determines how much and how

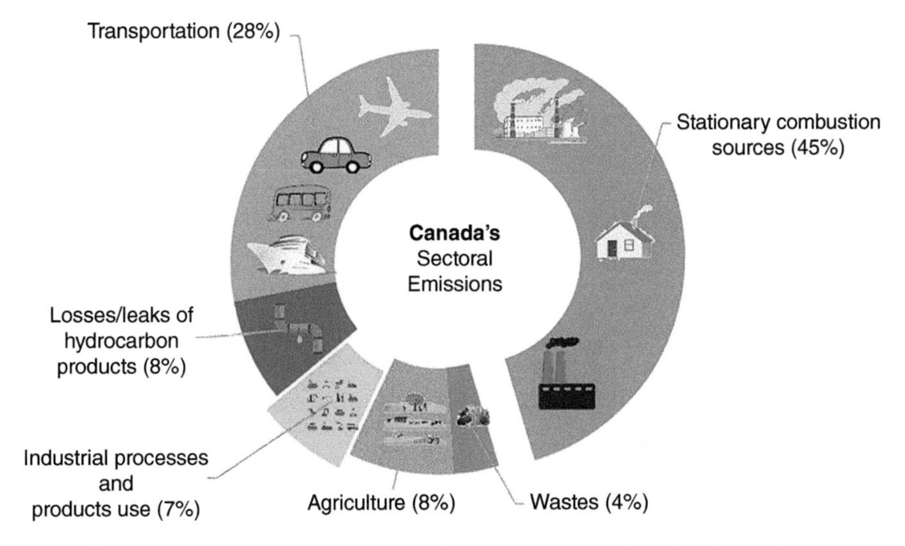

Figure 2.5 A breakdown of Canada's emissions by source in units of unit of million-ton (*Mt*) CO_2 equivalent. *Source:* data from Ref [1].

severely we can affect our environment. The levels of interactions play a critical role, depending on the sources and systems deployed, and can represent how damaging or healing such interactions with it might be. An example of our damaging interactions with the environment through energy production and utilization is the use of fossil fuels. They are hydrocarbon based fuels and contain both carbon and hydrogen, and greenhouse gases (GHGs) become unavoidable emissions in every process we burn them. Of course, the main contributors of GHGs in the world are carbon dioxide (CO_2) (accounting for about 75%), methane (CH_4) (accounting for 15%), and nitrous oxide (N_2O) (accounting for about 9%); the remaining 1% covers other GHGs. Of course, carbon dioxide has the biggest share among the GHGs and is essentially creating the biggest global concern.

Here, we provide an example from Canada to see what sources cause GHG emissions in the unit of million-ton (*Mt*) CO_2 equivalent, as illustrated in Figure 2.5. It shows that stationary combustion sources, covering residential buildings, power plants, factories, etc., emit the largest amount of GHGs, about 45%, while transportation sector contributes about 28%, including on-road and off-road applications, and the rest contribute about 27% through industrial processes, products use, agricultural applications, wastes and losses and leaks of hydrocarbon products (including chemical and fuels).

2.4 Forms of Energy

As mentioned in Chapter 1, there are three common types of energy, namely heat transfer, work transfer, and flow energy to consider in system analysis and assessment.

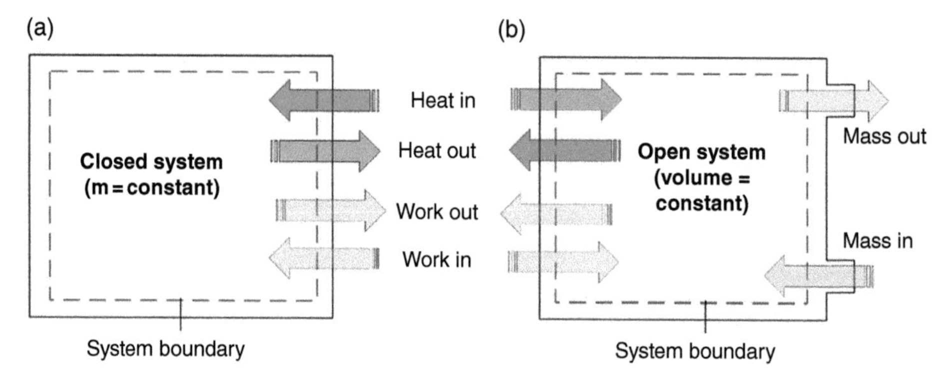

Figure 2.6 (a) Energy can cross a closed system (controlled mass or constant mass process) boundary in the form of work and heat; (b) energy can cross an open system (controlled volume or constant volume process) boundary in the form of work, heat, and flow energy (with mass).

- In dealing with a closed system we may have heat and work crossing the system boundary as the mass remains constant within the system due to no mass flow crossing the boundary.
- In dealing with an open system we may have all three – flow energy, heat, and work – crossing the system boundary. The flow energy may have three key parts: flow enthalpy, flow kinetic energy, and flow potential energy, which will be illustrated later.

The above listed definitions are illustrated in Figure 2.6 for both closed and open systems. Note that we will use a system boundary for both closed and open systems throughout the book although some authors have used control surface, rather than system boundary, for open systems.

A quick summary of the above is provided in Table 2.1. Note that kinetic and potential energies become zero if the closed system is a stationary system. For open systems, the changes in kinetic and potential energy equal zero if the changes between inlet and exit kinetic and potential energies are negligible.

Table 2.1 A summary of possible energies for closed and open systems.

Type of Energy	Closed System	Open System
Non-flow (internal) energy	✓	×
Flow energy	×	✓
Heat transfer	✓	✓
Work transfer	✓	✓
Kinetic energy	✓	✓
Potential energy	✓	✓

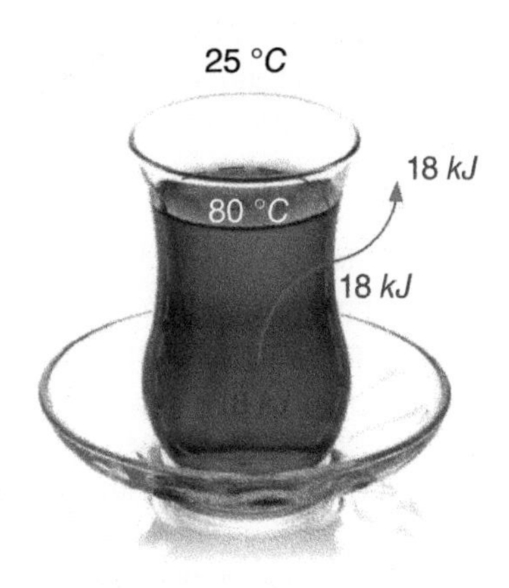

Figure 2.7 An illustration of heat transfer from hot tea to its surroundings due to temperature difference, which is the driving force.

(a) Energy transfer in the form of heat

Heat is a form of energy (so-called: thermal energy) that is transferred from one system to another or from the system to its surrounding based on the temperature difference between the two systems or between the system and its surroundings, respectively. Note that energy is in the form of heat only when the energy crosses the system boundary due to the temperature difference (where the temperature of tea within the boundary is about $80\,°C$ and surrounding temperature is $25\,°C$, respectively), as shown in Figure 2.7. The temperature difference drives the heat transfer from the hot tea crossing the boundary to the surroundings.

Figure 2.8 shows four modes of heat transfer to nicely illustrate the possible combinations. In Figure 2.8a the hot tea is loosing heat at $7\,kJ$ with a temperature difference of

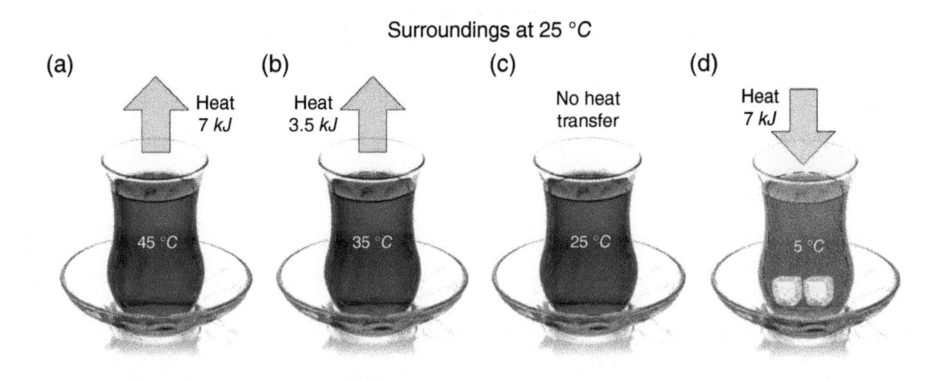

Figure 2.8 The heat transfer driving force is the temperature difference, which also determines the direction of the heat transfer.

20 °*C* between tea and its surroundings. In Figure 2.8b the hot tea is loosing heat at 3.5 *kJ* with a temperature difference of 10 °*C* between the tea and its surroundings. Figure 2.8c shows no heat transfer since both the system and its surrounding have the same temperature, which leads to thermal equilibrium. This particular process is also called an adiabatic process, a process in which no heat transfer takes place. In Figure 2.8d this time the tea is colder than the ambient temperature, which results in heat penetration into the system. Since the temperature difference is 20 °*C*, the heat transfer will take place in an opposite direction from the surroundings to the system at the same amount (i.e. 7 *kJ*). Of course, in three cases the heat transfer, which is the transfer of the thermal energy due to the temperature difference, takes place where the temperature difference is the driving force. The other conclusion is that the larger the temperature difference the larger the heat transfer rate.

Heat transfer is a form of energy transfer; thus, the most common unit of measure is kJ in the SI system or Btu in the English unit system. The amount of heat transferred during a process that takes the system from one state (e.g. 1) to another (e.g. 2) is denoted by Q_{12} where the order of the subscripts comes from the starting state to the final state. The heat transfer Q_{12} can be presented per unit mass of the system as follows:

$$q_{12} = \frac{Q_{12}}{m} \left(\frac{kJ}{kg}\right) \tag{2.3}$$

Sometimes it is more convenient to present the heat transfer as a heat transfer rate, which is the amount of heat transferred into or out of a system per unit time. An example where it is recommended to present the heat transfer as its rate is for a process where the properties and the operating conditions do not depend on time. The heat transfer rate is denoted by placing a dot on top of the heat transfer rate symbol $(\dot{Q})$, which has the units of *kJ/s* (1 *kJ*/*s* = 1 *kW*). Note that the dot placed on top of the heat transfer rate symbol (*Q*) presents the per unit time or in mathematical terms presents the time derivative. The heat transfer rate can then be calculated as follows:

$$\dot{Q} = \frac{d}{dt}(Q) \ (kW) \tag{2.4}$$

In specific cases where the heat transfer rate is constant throughout the process, Eq. (2.4) can be reduced to the following:

$$\dot{Q} = Q/\Delta t \ (kW) \tag{2.5}$$

Here, Δt denotes the time interval of the process with a constant heat transfer rate.

Note that there are three different modes of the thermal energy transfer in the form of heat; these are classified based on how the contributing systems are in contact and are conduction, convection, and radiation. Figure 2.9 shows a schematic diagram of these different modes of heat transfer. Conduction is the mode of heat transfer where the contributing systems are in physical contact, as shown in the figure where someone is touching the fire with a stainless steel tong. Convection is the second mode of heat transfer, where the heat transfers through a fluid that is the connecting link between the contributing systems, as shown the figure where the person is heating his hands with the hot air where the fluid assisting in transferring the heat is referred to as the medium. The third mode of heat transfer is

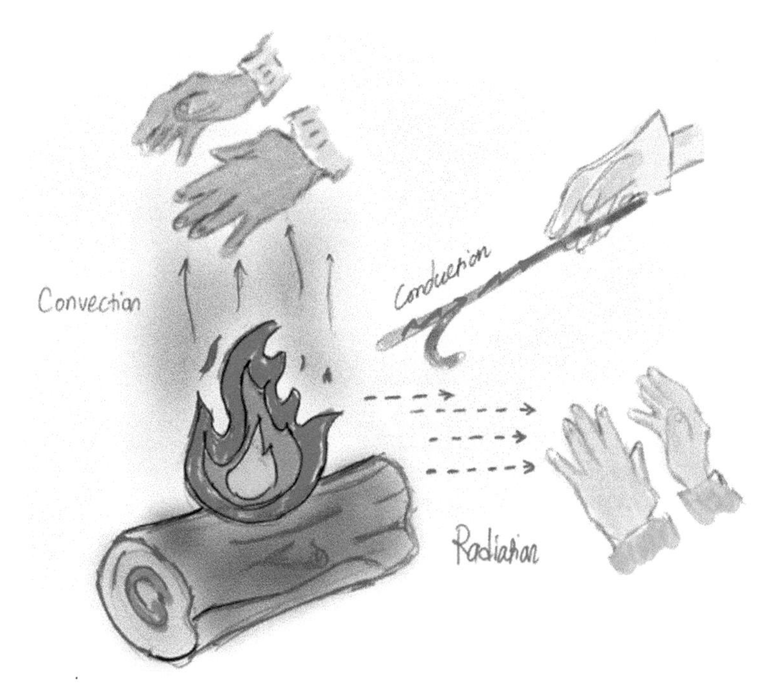

Figure 2.9 Illustration of three modes of heat transfer in the form of conduction, convention, and radiation.

radiation, where the heat transfers through the emissions of electromagnetic waves, as shown in the figure which helps heating things radiatively. Note that from the three different modes of heat transfer, the only mode of heat transfer that does not require a material medium to transfer heat is the radiation mode of heat transfer. These modes of heat transfer are the key subjects of heat transfer courses, rather than thermodynamics.

Considerations for the System Boundary

Defining the system boundary is a very critical step in determining the energy interactions of the system with its surroundings. Consider an insulated oven (Figure 2.10), where the *shish* kebab (ground meat around a barbeque skewer) is roasted in an oven by the heat

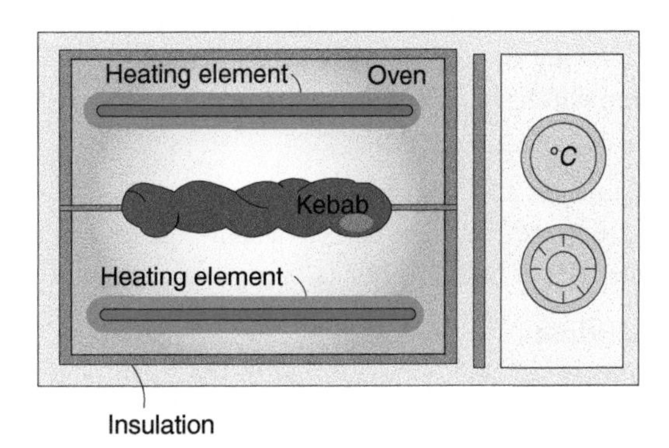

Figure 2.10 An insulated (adiabatic type) oven where the *shish* kebab is roasted in the oven.

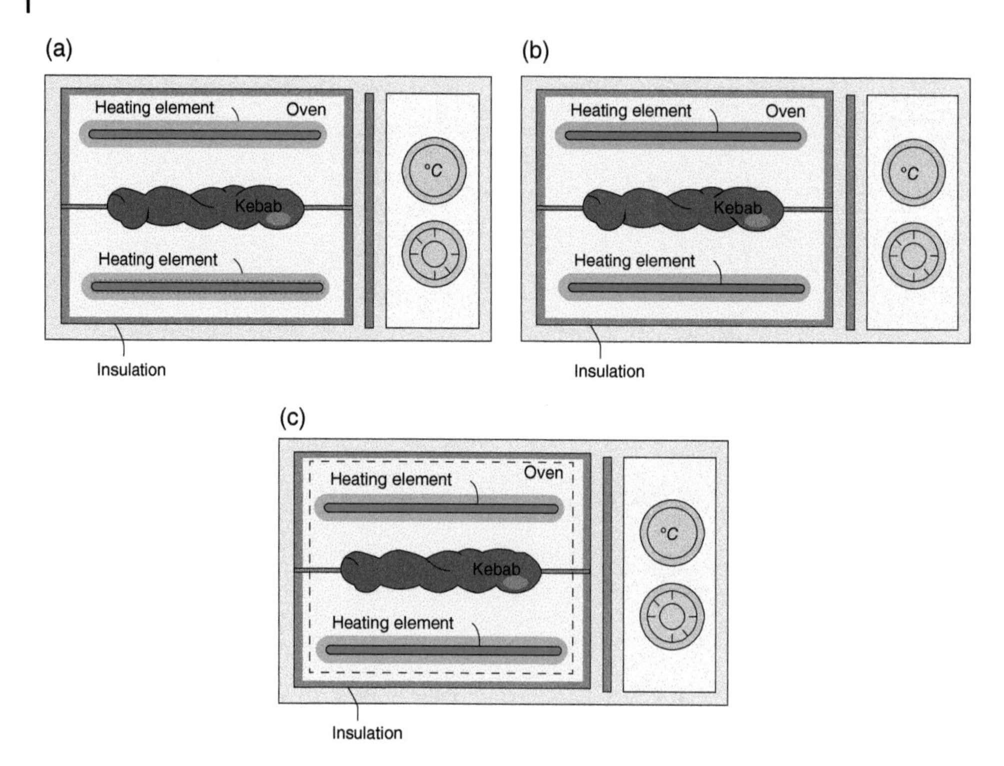

Figure 2.11 Three options of defining system boundary for analysis (a) where the boundary is around the heating element, (b) where the boundary is around the kebab skewer, and (c) where the boundary is inside the oven having all heating elements and *shish* kebabs covered inside.

released from the heating element. It is not important to approach this problem for analysis. The first thing needed is to define the system boundary, where there are energy interactions that have to be determined.

One may come with three approaches, as illustrated in Figure 2.11. Figure 2.11a shows the case where one may have the boundary placed just around the heating element to show that the heat is leaving the heating element to essentially roast the kebab. Also, Figure 2.11b shows a case where this time one may have boundary just around the *shish* kebab to determine how much heat will cross the boundary to cook the kebab. Furthermore, one last option here, as shown in Figure 2.11c, is that one may place the boundary inside the oven by taking all heating elements and *shish* kebabs inside. This last one with an adiabatic system will have no energy crossing the boundary as all will remain inside of it.

Example 2.1 Figure 2.12 is provide to illustrate two cases, namely: (a) the sun is arriving at the earth's surface and (b) a split-type air-conditioning unit working in a room. Identify the type of heat transfer taking place in each of these two cases.

Solution

In case (a) heat is transferred to the earth by traveling through the space in the form of radiation.

In case (b) there is cooling taking place, by the cold air sent into the room, and the heat transfer occurs in the form of convection through the indoor air movement.

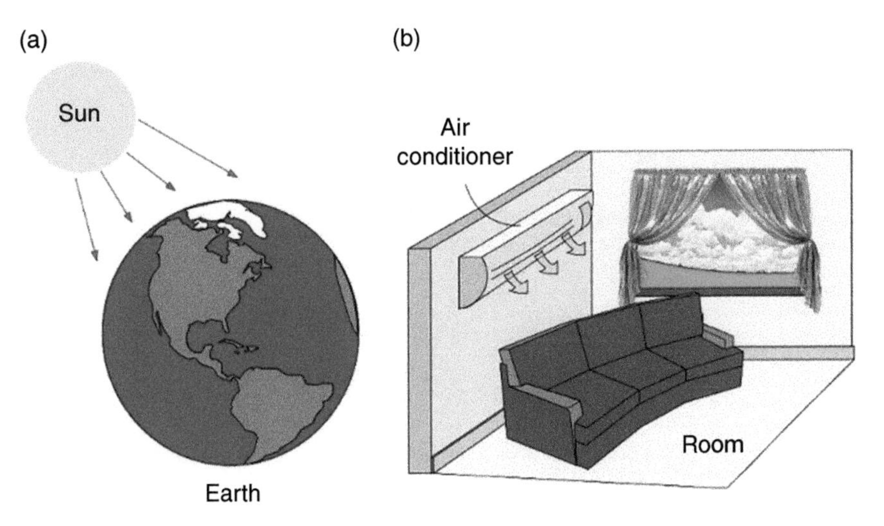

Figure 2.12 (a) Heat from the sun to the earth and (b) air-conditioning unit working in a room.

(b) Energy transfer in the form of work

Similar to heat, energy can cross the boundary of the system in the form of work. Work is also an energy interaction between a system with its surroundings. For a closed system and as shown in Figure 2.6a, the only two forms of energy that can cross its boundary are work and heat. Since heat transfers' driving force is the temperature difference, any transfer of energy from and to a closed system that is not caused by the temperature difference is work. Work can be defined as the energy transfer that is associated with a force acting on an object to move it for a distance to create work, which is defined as the force times distance that the object traveled. What if someone applies a force to an object but is not able to move it? Does it creates work? If one goes by the definition, it is simply "no" (see an example in Figure 2.13 where a baby tries to move a wooden object, but cannot achieve any result). Examples of energy transferred in the form of work include a rising piston with boundary movement work and a rotating shaft with shaft work. This comes as the force moves the piston through a vertical distance and the shaft rotates to move the fluid a circumferential distance.

Work transfer is a form of energy transfer; thus, the most common unit of measure is kJ in the SI system or Btu in the English unit system. The amount of work transferred during a process that takes the system from one state (1) to another (2) is denoted by W_{12}, where the order of the subscripts comes from the starting state to the final state. The work *transfer* W_{12} can be presented per unit mass of the system as follows:

$$w_{12} = \frac{W_{12}}{m} \left(\frac{J}{kg}\right) \tag{2.6}$$

The work transfer rate (power) is denoted by placing a dot on top of the work transfer symbol $(\dot{W})$ which has the units of J/s $(1\,J/s = 1\,W)$. Note that the dot placed on top of

Figure 2.13 The baby pushing a box is not able to move it, resulting in no work done.

the work transfer rate symbol ($\dot{W}$) presents the per unit time or in mathematical terms presents the time derivative. The work transfer rate can be calculated as follows:

$$\dot{W} = \frac{d}{dt}(W) \tag{2.7}$$

Since the work transfer is defined as the force done over a distance, then the work can be calculated as follows for the case where the force applied through that distance is constant:

$$W = Fs \ (J) \tag{2.8}$$

However, for the case where the force applied through the distance varies through the distance then the work transfer calculated as follows:

$$W = \int_1^2 F\,ds \ (J) \tag{2.9}$$

One can conclude that the two requirements needed to have between the system and its surroundings for a work interaction are (i) to have a force acting on the system boundary and (ii) the system boundary must move. There are different forms of mechanical work, including moving boundary work and shaft work as commonly used in thermodynamics. There may be other types of works, such spring work, studied as special cases.

Example 2.2 A snowball with a mass of 1 kg (Figure 2.14a) rolls down a mountain. Calculate how much work is spent to roll the snowball down the hill by 1 m.

Solution

Before starting to solve the problem, the following assumptions are made: (i) friction force from the ground is negligible and the nature of the rolling of the snowball is not considered; (ii) air resistance is neglected to further simplify the problem and compensate for the lack of required data.

The forces acting on the rolling snowball are shown in Figure 2.14b. Since the snowball roles in the downhill direction then the only force that leads to work done on the rolling

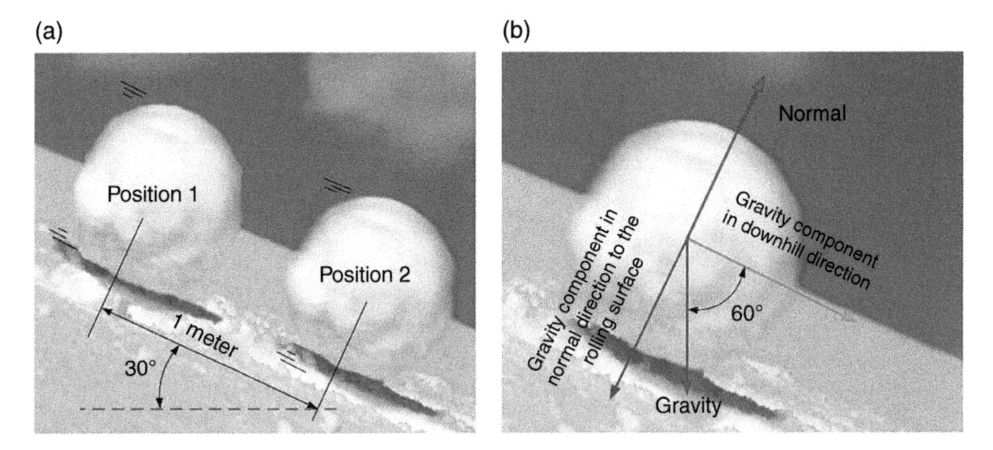

Figure 2.14 (a) A schematic diagram of Example 2.2, and (b) the forces acting on the rolling snowball.

snowball is the gravity component in the downhill directions (rolling direction). The work done by the downhill gravity component is calculated as follows:

$$W_{sb} = m_{sb}g\cos(60^o)z$$

Here, the subscript sb denotes snowball, g is the gravity acceleration (9.81 m/s^2) and z is the distance where the downhill direction gravity component is acting on.

The amount of work in terms of potential energy becomes

$$W_{sb} = 1\times 9.81\times 0.5\times 1 = \mathbf{4.905\ J}$$

(c) Energy transfer in the form of flow energy

As stated earlier, there are three forms of energies – heat, work, and flow energy – to consider in system analysis and assessment. This is important for the cases where mass flows and changes its position, and hence the boundary changes. This is represented by enthalpy (h), which is another thermodynamic property.

Richard Mollier (1863–1935) was a German professor of Applied Physics and Mechanics in Göttingen and Dresden and spent most of his life in investigating thermodynamic properties and their property diagrams for practical utilization. He was a pioneer of experimental research in thermodynamics, and his major contribution was the introduction of enthalpy as flow energy:

$$h = u + Pv \tag{2.10}$$

where u is fluid internal energy (J/kg or kJ/kg), P is fluid pressure (Pa or kPa) and v is fluid specific volume (m^3/kg).

Enthalpy is a unique property that plays a role in closed systems where there is boundary movement work, for example, in piston–cylinder mechanisms. In addition, it is more critical in open systems where there is mass flow crossing the boundary. One may wonder

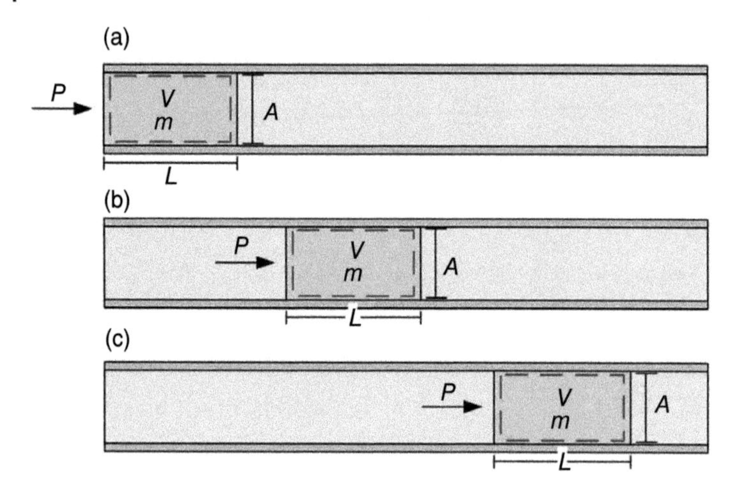

Figure 2.15 Illustration of flow energy through.

about flow kinetic and potential energies! There may be cases where both energies are involved, which may require total flow energy, introduced as follows:

$$h = u + Pv + ke + pe = u + Pv + \frac{1}{2}V^2 + gz \tag{2.11}$$

where V is flow velocity (m/s), g is gravitational acceleration (m/s^2), and z is the elevation (m).

Figure 2.15 shows a fluid element identified in a pipe flow (which is treated as an open system) crossing the boundary and proceeding accordingly as illustrated with (a), (b), and (c). If one wants to define the total flow energy, it will result in flow energy only: $h = u + Pv$ due to the fact that both exit kinetic and potential energies will be same as both inlet kinetic and potential energies.

Example 2.3 Let us have a nozzle, as an open system, where air enters a nozzle steadily at a velocity of 20 m/s. The exit area of the nozzle is 0.4 m^2. The flow rate of air through the nozzle is given as 78.8 kg/s, where the enthalpy values are specified as given in Figure 2.16. Calculate the velocity of the air at the nozzle exit using the energy balance for this system.

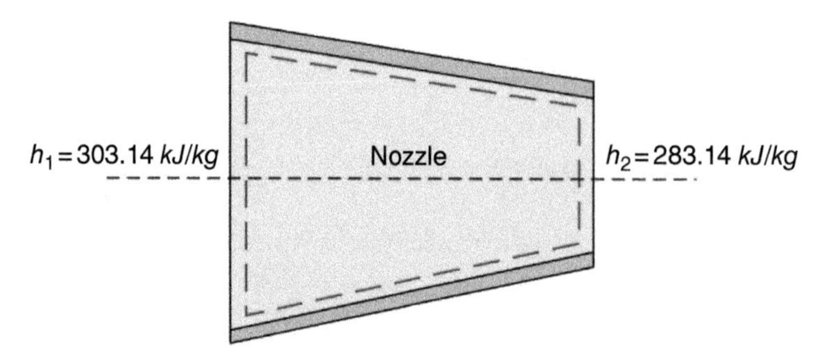

Figure 2.16 Schematic diagram of an air nozzle for Example 2.3.

Solution

First, we look at flow enthalpies, velocities, and elevations. The enthalpy values are already given and the potential energy remains unchanged since there is no difference in elevation between inlet and exit, while the velocities at the inlet and exit become different due to different cross-sectional areas based on:

$$\dot{m} = \rho VA$$

where $\dot{m}$ is mass flow rate (kg/s), ρ is fluid density (kg/m^3), V is fluid velocity (m/s), and A is cross-sectional area (m^2).

By applying the energy balance on the boundary of the nozzle, the energy balance equation can be written as:

$$\dot{E}_{in} = \dot{E}_{out} \rightarrow \dot{E}_1 = \dot{E}_2$$

$$H_1 + KE_1 + PE_1 = H_2 + KE_2 + PE_2$$

$$\dot{m}_1 h_1 + \frac{\dot{m}\left(V_1^2\right)}{2} = \dot{m}_2 h_2 + \frac{\dot{m}_2\left(V_2^2\right)}{2}$$

Since the mass flow rate at the inlet equals that at the outlet, then the above energy balance can be reduced to:

$$h_1 + \frac{V_1^2}{2} = h_2 + \frac{V_2^2}{2}$$

$$303.14\frac{kJ}{kg} + \frac{(20\,m/s)^2}{2} \times \frac{1\ kJ/kg}{1000\ m^2/s^2} = 283.14\frac{kJ}{kg} + \frac{V_2^2}{2} \times \frac{1\ kJ/kg}{1000\ m^2/s^2}$$

$$\mathbf{V_2 = 199\ m/s}$$

2.5 The First Law of Thermodynamics

In the previous chapter and at the beginning of this chapter, various forms of energy, such as heat Q, work W, and total energy E, are considered individually and discussed for various systems. Some introduction to the first law of thermodynamics (FLT) is made. It is now important to look at what crosses the system boundary, how it crosses the boundary, and what the interaction with its surroundings is. Of course, one may add some more questions to these. FLT, which is known as the conservation of energy principle, provides a sound basis for studying the relationships among the various forms of energy and energy interactions. Based on experimental observations, the FLT states that energy can be neither created nor destroyed during a process; it can only change forms. Therefore, every bit of energy should be accounted for during a process. Everyone knows that a rock at some elevation possesses some potential energy, and part of this potential energy is converted to kinetic energy as the rock falls. Experimental investigations show that the decrease in potential energy (mgz) exactly equals the increase in kinetic energy when the air resistance is negligible, thus confirming the conservation of energy principle for mechanical energy.

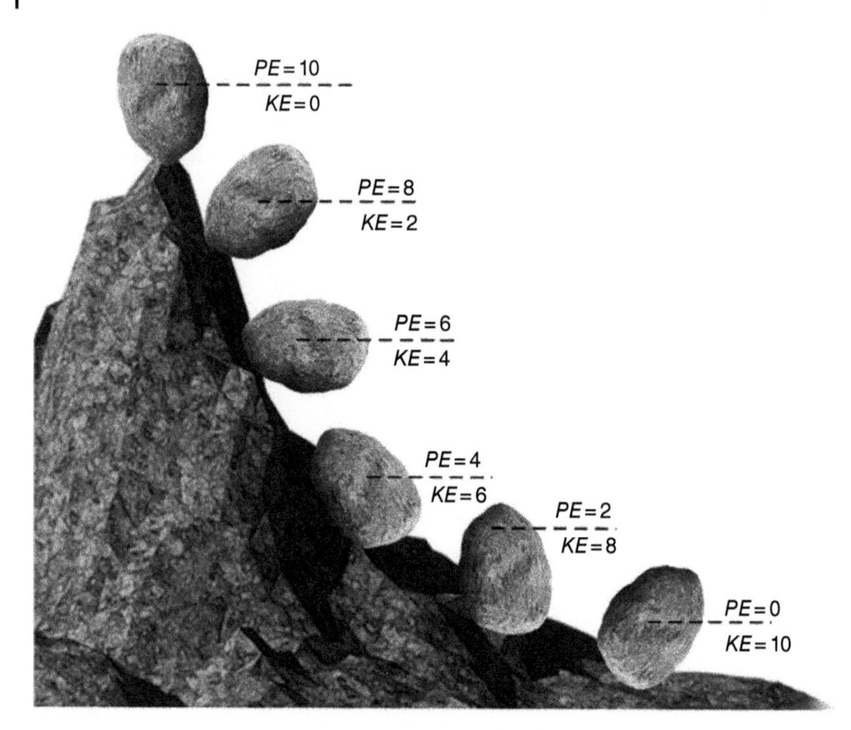

Figure 2.17 Illustration of energy conservation concept through a rolling rock with a total of 10 units of energy.

Figure 2.17 illustrates an eye catching type example of a rolling rock to show the variation of total energy content in every step. Therefore, this example of the conservation of energy principle presents the change of energy kinetically and potentially throughout the process while having the total energy remain constant. As shown in Figure 2.17 when the rock was on top of the hill and not moving, the energy that rock possessed was entirely potential energy at 10 units, making the total energy content 10 units only. Then when the rock started to roll down the hill the potential energy started to convert to kinetic energy with the assumptions that there are no friction losses to either air and ground friction. Through the rolling process the total energy, which is, in this case, the summation of the potential energy plus the kinetic energy remained constant throughout the process. At the bottom of the hill, one can say that all the gravitational potential energy has now been converted to kinetic energy. This comes as the reference point for the initial height of the rock is the bottom of hill.

The FLT is a presentation of the conservation of energy principle, where it defines energy as one of the properties of thermodynamics. From the definition of the FLT it enables us to analyze the energy exchange and interactions of power plants and energy devices. This come as the FLT accounts for these interactions and makes sure that all these interactions are within the balance. Furthermore, it asserts that no energy was created nor destroyed; however, it can be converted from one form of energy to another. Measuring the energetic performance of such an energy system can also be done with the help of the FLT through the energy efficiency, which measures the performance of an energy system by evaluating the amount of energy that was converted to the desired form to the amount of energy that was supplied to the energy system or energy device. For example, considering a diesel power generator, the amount of energy fed to the system (generator) can be defined as the amount of diesel. Diesel before combustion in this case is chemical energy. The amount of energy that was converted to a useful form is the amount of electricity generated by the generator. However, with all the advantages and the benefits we gain with the FLT, there is the disadvantage of not considering the irreversibilities, losses, inefficiencies, and quality destructions.

Example 2.4 A 10 kg metallic weight (with initial kinetic energy of zero and potential energy of 981 J) is freely falling from a high of 10 m, as shown in Figure 2.18. The velocity of the object at state point 2 is 9 m/s. Calculate both the kinetic and potential energies at state points 2 and 3 with the heights and velocities given in the figure. Also, calculate z_2 and V_3.

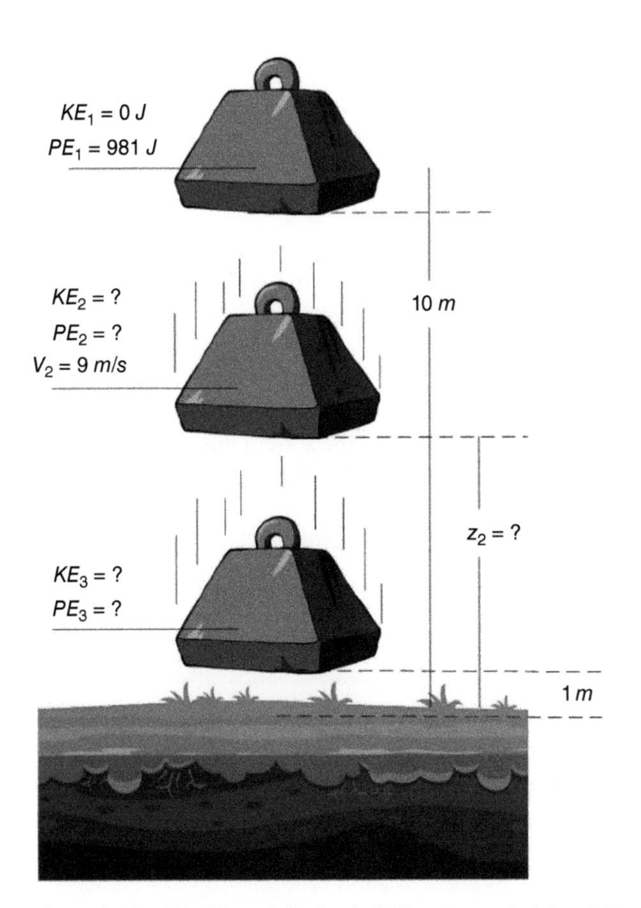

Figure 2.18 Metallic weight freely falling from a height of 10 m.

Solution

From the FLT we can write that the total energy is conserved at every state and results in:

$$E_{t1} = E_{t2} = E_{t3}$$

Between states 1 and 2:

$$E_{t1} = E_{t2}$$

$$KE_1 + PE_1 = KE_2 + PE_2$$

$$KE_1 + mgh_1 = \frac{1}{2}mV^2 + mgh_2$$

$$0\ kJ + 981\ J = \frac{1}{2} \times 10\ kg \times 9^2 + 10 \times 9.81\ \frac{m}{s^2} \times h_2$$

$$\boldsymbol{h_2 = 5.87\ m}$$

Between states 2 and 3:

$$E_{t2} = E_{t3}$$

$$KE_2 + PE_2 = KE_3 + PE_3$$

$$\frac{1}{2} \times 10\ kg \times 9^2 + 10 \times 9.81\ \frac{m}{s^2} \times 5.87\ m = KE_3 + 10 \times 9.81\ \frac{m}{s^2} \times 1\ m$$

$$\boldsymbol{KE_3 = 882.9\ J}$$

$$KE_3 = 882.9\ J = \frac{1}{2} \times 10\ kg \times V_3^2$$

$$\boldsymbol{v_3 = 13.29\ m/s}$$

2.5.1 Energy Balance Equations

The conservation of energy principle is essentially derived from the FLT, which highlights the net change (increase or decrease) in the total energy of the system during a process is equal to the difference between incoming and outgoing total energies.

$$(\textit{Total energy entering the system}) = (\textit{Total energy leaving the system})$$

which can be presented in terms of symbols as:

$$E_{in} = E_{out} \tag{2.12}$$

This equation presents the FLT and represents the overall systems energy interaction processes. E presents the energy of the system, which can be broken into internal energy, kinetic energy, and potential energy. Note that the initial amount of energy the system has is considered as an input while the final amount of energy the system has is considered as energy out.

$$\Delta E = \Delta U + \Delta KE + \Delta PE \tag{2.13}$$

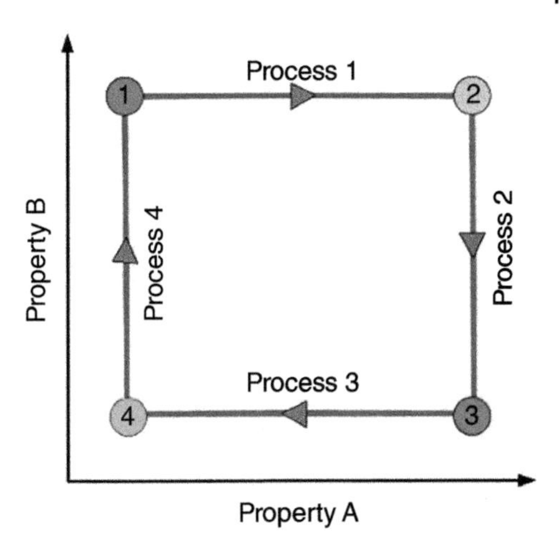

Figure 2.19 For a cycle $\Delta E_{system} = 0$ since $E_1 = E_2$, thus $E_{in} = E_{out}$.

and

$$\Delta U = m(u_2 - u_1) + \frac{m\left(V_2^2 - V_1^2\right)}{2} + mg(z_2 - z_1) \tag{2.14}$$

From the definition of the energy of the system and its breakdown shown in Eqs. (2.11)–(2.14), a stationary system will have the kinetic energy and the potential energy equal to zero, thus the change of the energy of the system is equal to the change in the internal energy as follows:

$$\Delta E = \Delta U \text{ since } \Delta KE = 0 \text{ and } \Delta PE = 0$$

The rate form of the energy balance can be written as:

$$\sum \dot{E}_{in} = \sum \dot{E}_{out} \tag{2.15}$$

Further note that the energy balance can also be written for a system undergoing a cycle as shown on a property diagram given in Figure 2.19. It provides a properties plot of a cycle, where the starting point is the same as the end point. Such plots are the diagrams that show the variations of usually two properties of the system throughout the process that the system undergoes. Since the starting point is the same as the last point of the system then the total energy change of the system is equal to zero. Having the total change in the energy of the system equal to zero, then the amount of energy entering the system equals that leaving the system.

Example 2.5 Water is heated in a water tank with an electrical heating element at 20 *kJ* as shown in Figure 2.20. There is also additional heat of 40 *kJ* supplied thermally from another source. The water tank losses some of the heat at 5 *kJ*. Calculate how much heat the water gained in the tank.

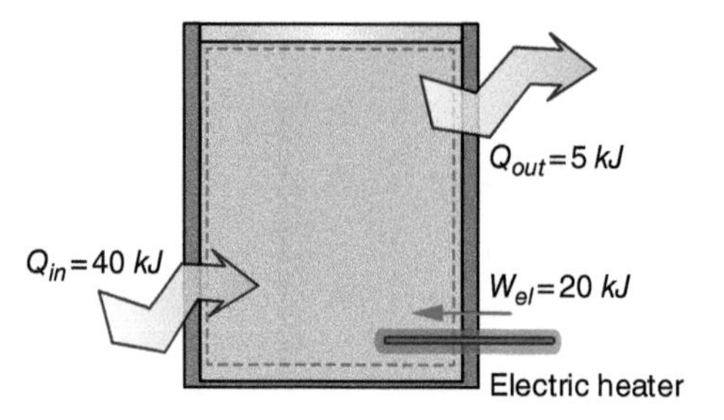

Figure 2.20 Schematic diagram of Example 2.5.

Solution

Here, there is a need to find the change in the internal energy of water. One can, in this regard, write the energy balance equation as follows:

$$\sum E_{in} = \sum E_{out}$$

$$U_1 + Q_{in} + W_e = U_2 + Q_{out}$$

$$U_1 + 40\ kJ + 20\ kJ = U_2 + 5\ kJ$$

$$U_2 - U_1 = 40\ kJ + 20\ kJ - 5\ kJ$$

$$\Delta U_{12} = 40 + 20 - 5 = \mathbf{55\ kJ}$$

2.5.2 Energy Losses

There is another important aspect to mention here: that there are losses that take place in various applications in terms of heat losses and/or work losses. An important task is to identify and quantify these for system design and analysis. In daily life, there are so many applications where we end up with energy losses, primarily heat losses. Depending on the applications, energy losses occur through energy leaving the subject matter system into surrounding environment. Such losses will be included as parts of energy outputs, so the prime energy balance equation is kept same as before:

$$\sum E_{in} = \sum E_{out}$$

Here, the term $\sum E_{out}$ covers $\sum E_{losses}$ if they leave the system. The question might be what if they penetrate into the systems, such as fridges or refrigerators or cold stores where we do provide cooling and keep the temperature less than the surrounding environment temperature. How are they accounted for? In such cases, the heat losses will be in the form of energy inputs. In this case, $\sum E_{in}$ will cover $\sum E_{losses}$. One may question: why do we call them losses if they come in? The point is that the system is kept at a lower temperature, such heat inputs will cause losses. That is why calling them losses would be more meaningful.

Figure 2.21 shows three examples where there are heat losses taking place. Figure 2.21a shows a paper cup filled with hot coffee at 80 °C in a room at 25 °C. The cup with hot coffee will lose heat until its temperature reaches room temperature (so-called: thermal equilibrium as mentioned in the previous chapter). The second example, Figure 2.21b is about a hot water tank where the water is heated. The water is heated up to 80–90 °C. Consider a

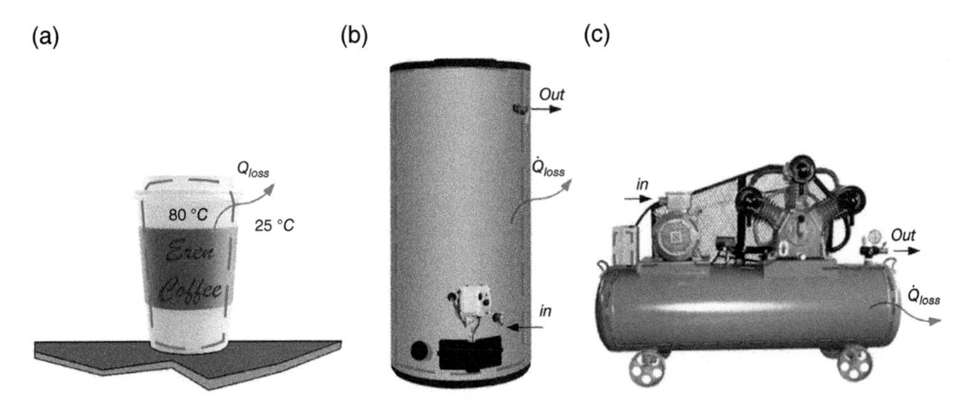

Figure 2.21 Three examples for heat losses: (a) a paper cup filled with hot coffee, (b) a hot water tank, and (c) an air compressor.

basement at about 20 °*C*, the hot water tank will lose heat. The next example, Figure 2.21c is about an air compressor where it losses some heat since the system gets warmer than its surroundings and, hence, it rejects some heat. The last two examples are examples of open systems while first one was a closed system.

One should keep in mind that heat is not always rejected or lost. There are practical applications, such as cooling and freezing applications where the systems are colder than their

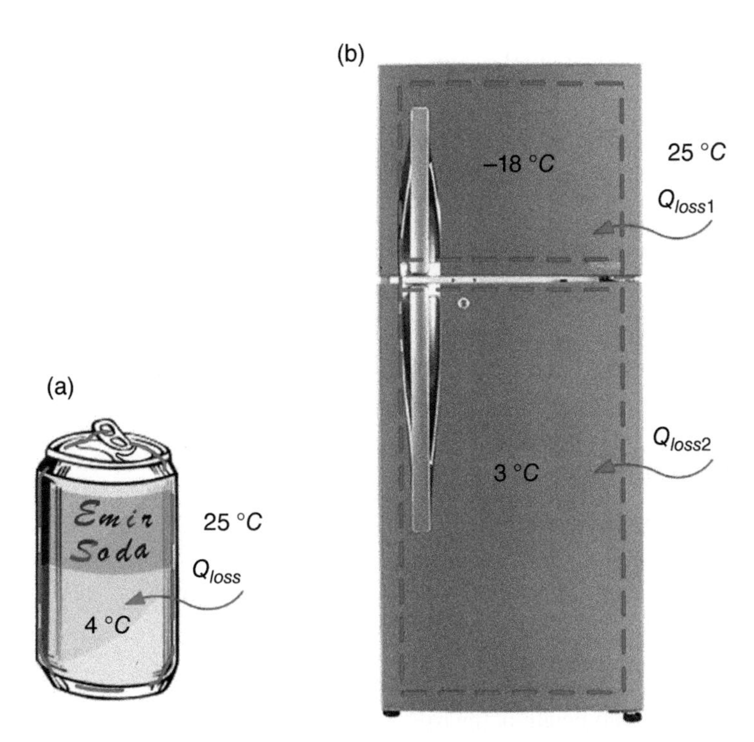

Figure 2.22 Two examples of heat losses in the form of heat penetration where there are colder systems: (a) a paper cup filled with hot coffee and (b) an air compressor.

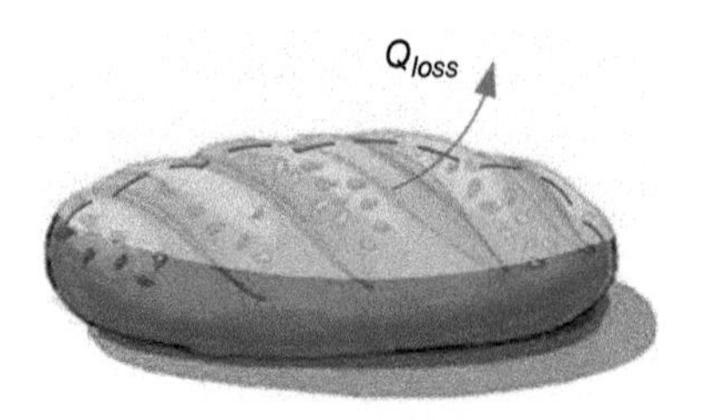

Figure 2.23 A freshly baked bread is losing heat, along with an energy balance equation written underneath.

surrounding environments. Figure 2.22 shows two examples where heat is lost (penetrated) into the system itself. Figure 2.22a shows a cold soda which might be taken out of a fridge at $4\,°C$ in a room at $25\,°C$. Obviously, there will be heat penetration into the system that causes losses. Figure 2.22b shows a household fridge that consists of two sections: one cooler section at $3\,°C$ and one freezer section at $-18\,°C$, respectively. If these are assumed to be in a room at $25\,°C$, there will be heat penetrations (losses) into these systems. There are of course many more examples to show from our daily applications. For example, Figure 2.23 shows a freshly baked loaf of bread (at about $60\,°C$) which was taken out of an oven and kept in a room at about $25\,°C$. The bread will lose heat accordingly and the energy balance equation will take place as written in the figure.

Example 2.6 A basketball freely falls from rest, as shown in Figure 2.24, at a height of 2 m with a mass of 0.5 kg. The basketball loses 10% of its total energy (in the form of mechanical energy which work) at its first collision with the court floor, and then 15% of its total energy at the second collision with the court floor. Calculate the maximum height of the basketball from the court floor after the first and second collisions.

Solution

As before we need to write energy balance equation as:

$$\sum E_{in} = \sum E_{out}$$

Since the basketball is freely falling from rest, then we can find the total energy at the first state as follows:

$$E_1 = KE_1 + PE_1$$

$$E_1 = 0 + mgh_1 = 0 + 0.5\ kg \times 9.81\ \frac{m}{s^2} \times 2\ m = 9.81\ J$$

From the energy balance:

$$E_1 = E_2 + E_{loss,10\%}$$

$$E_1 = E_2 + 0.1 \times E_1$$

$$E_1 = KE_2 + PE_2 + 0.1 \times E_1$$

Since we are finding the maximum height of the ball, then the velocity of the ball at that height is zero and the above equation can be reduced to:

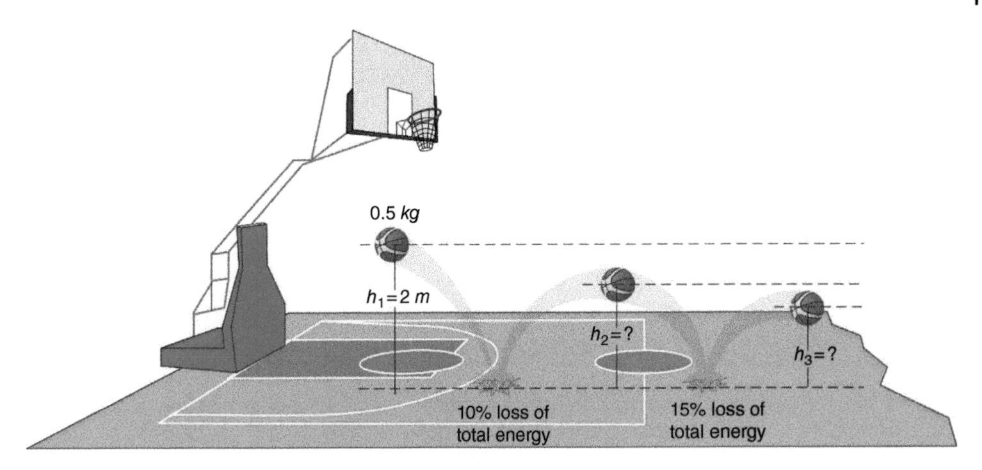

Figure 2.24 A basketball freely falling from rest at a height of 2 *m* with a mass of 0.5 *kg*, while losing part of its total energy in every collision with the court floor.

$$E_1 = PE_2 + 0.1 \times E_1$$

$$E_1 = mgh_2 + 0.1 \times E_1$$

$$0.9 \times 9.81\ J = 0.5\ kg \times 9.81\ \frac{m}{s^2} \times h_2$$

$$\boldsymbol{h_2 = 1.8\ m}$$

From the energy balance:

$$E_2 = E_3 + E_{loss,15\%}$$

$$E_2 = E_3 + 0.15 \times E_2$$

$$E_2 = KE_3 + PE_3 + 0.15 \times E_2$$

Since we are finding the maximum height of the ball, then the velocity of the ball at that height is zero and the above equation can be reduced to:

$$E_2 = PE_3 + 0.15 \times E_2$$

$$E_2 = mgh_3 + 0.15 \times E_2$$

$$0.85 \times 8.829\ J = 0.5\ kg \times 9.81\ \frac{m}{s^2} \times h_3$$

$$\boldsymbol{h_3 = 1.53\ m}$$

2.6 Pure Substances

A pure substance is defined as a substance that has a fixed chemical composition. There are additional definitions available, one of which is a substance made of only one type of atom or only a group of atoms bonded together (so-called one type of molecule). A measurement of a pure substance can confirm purity. Some common examples of pure substances are

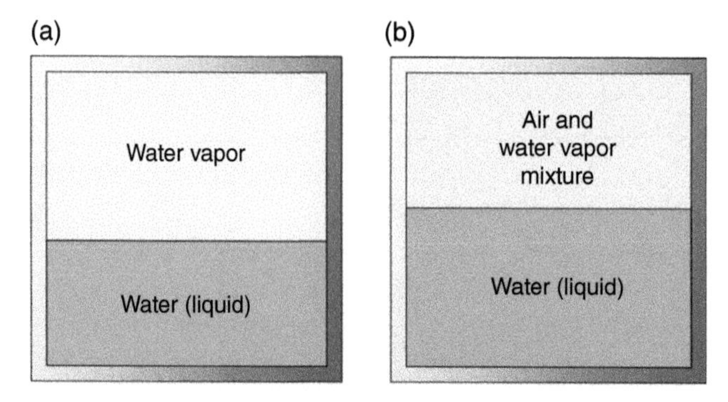

Figure 2.25 (a) Vapor and liquid mixture having the same chemical composition and (b) vapor and liquid mixture each have a different chemical composition.

water, hydrogen, helium, oxygen, nitrogen, copper (with only copper atoms), etc. A mixture of two or more phases in a substance makes it still a pure substance as long as they remain homogeneous, such as ice slush, due to the same chemical composition. However, a mixture of olive oil and water will not be a pure substance, due to different chemical compositions.

This section starts by introducing the concepts of the pure substance, and the physics of phase changing processes. Then it will dwell on the properties details by illustrating how the property diagrams can be used to present thermodynamic processes, which also shows how the system moves or goes from one state point to another. This section also includes how to use traditional thermodynamic tables to extract the properties of pure substances. Finally, the section concludes with a discussion on ideal gases and on what basis we can assume a gas to behave ideally.

Any substance with a fixed chemical composition throughout such as water, nitrogen, helium, and carbon dioxide is a pure substance. Note that having a substance with a fixed chemical composition but in two physical phases, the substance is still referred to as pure substance, as shown in Figure 2.25a. However, in a vapor and liquid mixture where the vapor and the liquid have different chemical compositions, then the mixture is not a pure substance (Figure 2.25b). However, if various chemical elements or compounds make a homogeneous mixture then the mixture can be considered as a pure substance.

2.6.1 Phases

In general, phase is defined as a quantity of matter that may consist of a single substance or a mixture of substances. Such matters are expected to be chemically and physically uniform or homogeneous within the defined system.

Substances can exist in different phases such as solid, liquid, gas, and plasma. For example, at room temperature and pressure water exists as a liquid, while at the same conditions oxygen is in the gaseous phase. The first three mentioned phases are the main phases and those that will be discussed in detail in this book. The molecular structure of the material that is held by intermolecular forces is what defines the phase of the material. The three main phases mentioned earlier are usually referred to as the three principal phases; however, a substance

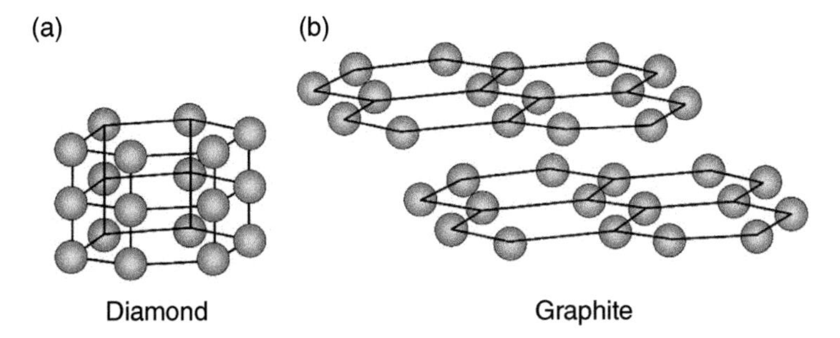

Figure 2.26 Illustration of the arrangement of carbon atoms in (a) diamond and (b) graphite.

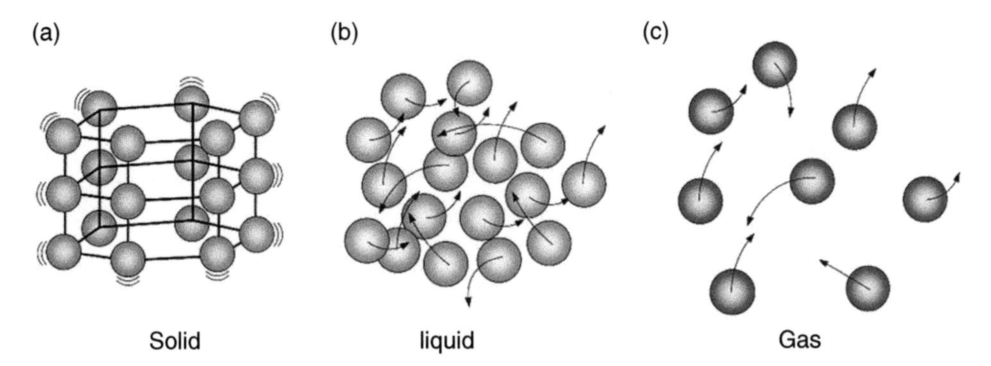

Figure 2.27 (a) The atoms of the substance in the solid phase are fixed in location; (b) in the liquid phase the molecules move around each other but maintain a close fixed distance between them; and (c) molecules of a substance in the gaseous phase moves randomly and freely with little attraction force between them.

can have more than one structure at its single principle phase. For example, carbon has more than one solid phase, such as diamond and graphite (Figure 2.26).

It is important in thermodynamics to study the science of phase change at the macroscale. However, understanding the changing of the intermolecular forces between the molecules during the phase changes is not necessary in the field of thermodynamics. The intermolecular forces are not discussed in this book, but a brief schematic presentation of the effect of these forces on the molecular structure in the three principal phases is shown in Figure 2.27.

2.6.2 Phase Changes of Water

Further to the phase changes, the prime focus is made on water as a most common pure substance. This section shows how water, as an example of a pure substance, changes phase between liquid and vapor. It is given as an example of a pure substance to explain the phase changes since water is the working fluid of a most common power generating cycle (the so-called: Rankine cycle), which is discussed later. Keep in mind that all pure substances exhibit the same general behavior. Next, we are going to explain the phases and the steps water goes through when a fixed amount of water is heated under the constant pressure of 1

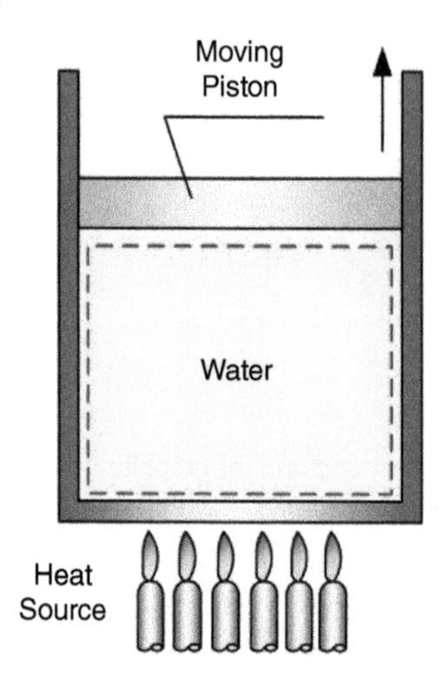

Figure 2.28 A piston–cylinder mechanism to maintain constant pressure.

atm (i.e. 100 *kPa*) and the starting phase is the compressed liquid phase at a temperature of 5 °*C*. Note that at 1 atm pressure water boils at a temperature of 100 °*C*. Maintaining a constant pressure for the water being heated can be done by having the water in a piston–cylinder device, where the weight of the piston is constant and maintains a constant pressure of 1 atm (Figure 2.28).

(a) Compressed liquid

Compressed liquid (or sub-cooled liquid) is defined as a liquid pressurized (i.e. compressed) by force and subcooled. In this regard, their pressures are higher than the saturation pressures and their temperatures less than the saturation temperatures. This is clearly seen in Figure 2.29. Also, either the water is kept in this state due to having a higher pressure than the saturation pressure at a temperature of 5 °*C*, or due to having a lower temperature than the saturation temperature at a pressure of 1 atm. Furthermore, a compressed liquid or subcooled liquid can be characterized by the following; when the substance is at higher pressure than the saturation pressure at a given temperature, or lower temperature than the saturation temperature at a given pressure. For example, if water is at a temperature of 100 °*C*, and a pressure of 2 *atm*, which is higher than the saturation pressure (1 *atm*) of water at the given temperature (100 °*C*), then the water is compressed liquid or subcooled liquid. Also, for water that is at a given pressure of 1 atm, and a temperature of 50 °*C*, which is lower than the saturation temperature of water (100 °*C*) at the given pressure, then water is also considered as compressed liquid.

(b) Saturated liquid

Continuous heating of the sub-cooled liquid water at a constant pressure will raise the temperature of the water while maintaining it in the liquid phase. When the temperature of the liquid water reaches 100 °*C*, it starts to experience a phase change. This temperature where a liquid begin vaporizing is defined as the saturation temperature. Therefore, at the moment where the water temperature has just reached a temperature of 100 °*C* (the saturation temperature of water at a pressure of 1 *atm*) and it is completely liquid, it is referred to as

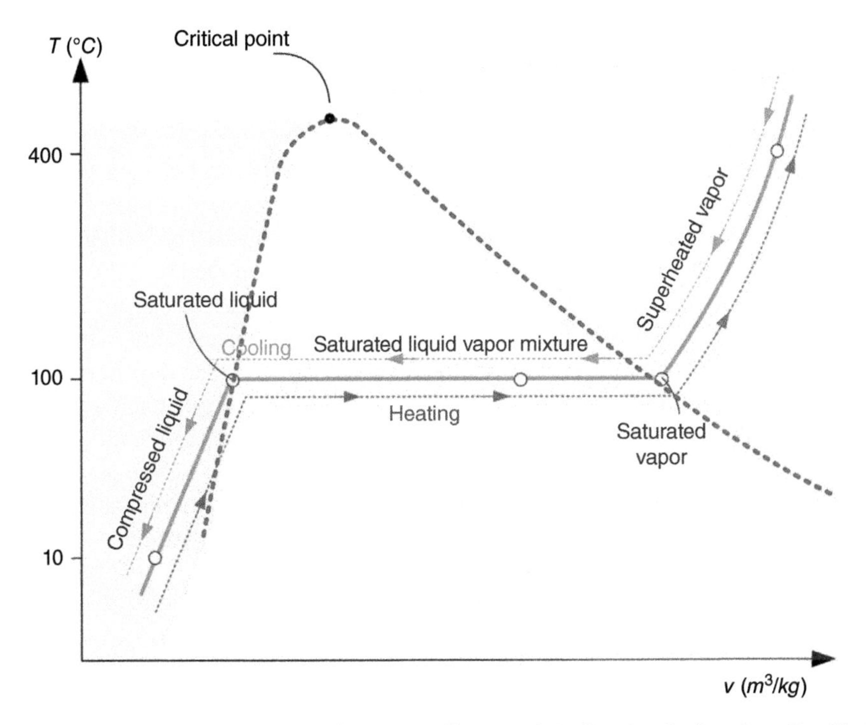

Figure 2.29 Temperature-specific volume diagram of the heating (red) and cooling (dark blue) process of water at a constant pressure of 1 *atm*.

saturated liquid. The saturated liquid is the state where any small addition of heat will result in generation of water vapor while maintaining a constant saturation temperature of 100 °C for this case where the pressure is maintained constant at 1 *atm*. The saturated liquid line is what connect between all the statured liquid points as shown in Figure 2.29.

(c) Saturated liquid and vapor mixture

When further heat is added to the water in the saturated liquid state, it starts vaporizing as mentioned above. Where there is both liquid and vapor then the state of the water is referred to as saturated water liquid vapor mixture. Continuous heating of the saturated water liquid vapor mixture will reduce the liquid fraction in the mixture and increases the vapor fraction in the mixture, while the temperature of the mixture remains constant. The area under the dome shape shown in Figure 2.29 present the saturated liquid and vapor mixture region. In this region, both temperature and pressure remain constant while the phase changes.

(d) Saturated vapor

Further heating of the saturated water liquid vapor mixture will result in converting all the liquid water into vapor. When all the liquid in the mixture becomes vapor it is called saturated vapor. Note that the temperature of saturated vapor is the same as the temperature of the saturated water liquid vapor mixture, which is 100 °C for the considered case of a constant pressure of 1 *atm*. Any further heating of the saturated vapor leads to increasing the temperature of the vapor, and that vapor which has a temperature above the saturation temperature is referred to as superheated vapor. The line connecting all the statured vapor points is shown in Figure 2.29.

(e) Superheated vapor

As mentioned earlier further heating of the saturated vapor will lead to having a superheated fluid. Superheated vapor can be defined as vapor that has a temperature higher than the saturation temperature. The different phases and states pure substances go through are summarized in Figure 2.29. Note that in Figure 2.29 the result and the direction of the heating process and the cooling process are indicated on the plot. Furthermore, a superheated vapor can be characterized by the following: the substance is at lower pressure than the saturation pressure at a given temperature, or it is at higher temperature than the saturation temperature at a given pressure. For example, if water is at a temperature of 100 °*C*, and a pressure of 0.5 *atm*, which is lower than the saturation pressure (1 *atm*) of water at the given temperature, then the water is superheated vapor. Also, for water that is at a given pressure of 1 atm and a temperature of 150 °*C*, which is lower than the saturation temperature of water (100 °*C*) at the given pressure, then water is also considered as superheated vapor.

Saturation pressure and saturation temperature

The saturation temperature is the temperature where the pure substance changes phase from liquid to vapor or from vapor to liquid at a specified pressure (saturation pressure), where the pure substance maintains a constant temperature during the phase change process. For example, the saturation temperature of water at a saturation pressure of 101.42 *kPa* is 100 °*C*. The saturation temperature varies with the pressure of the liquid. Going back to water as an example of a pure substance, the saturation temperature of water increases to 151.8 °*C* when the pressure increases to 500 *kPa*. However, at a saturation pressure of 1.23 *kPa* the saturation temperature is 10 °*C*. Further examples of the variation of the saturation temperature and its dependence on the pressure for pure substances are presented in Table 2.2, which shows data for the pure substances water and propane. It presents the variation of the saturation temperature, which can also be referred to as the boiling temperature of water and propane, at different pressures. Note that at a water

Table 2.2 Saturation temperatures of water and propane at different pressures.

Pressure (*kPa*)	Water saturation temperature (°*C*)	Propane saturation temperature (°*C*)
0.40	−5	−120.1
0.61	0	−116.3
0.87	5	−112.9
1.23	10	−109.4
101.3 (1 *atm*)	100	−42.1
1554	200	45.6

pressure of 1.23 kPa the water can boil or condense at a temperature of 10 °C, where at the same pressure propane will boil or condense at a temperature of −109.4 °C. Going back to Figure 2.29 we can see that heating the water from the compressed liquid state to saturated liquid state results in an increase in its temperature. The type of heat that increases (gaining heat) or reduces (rejecting heat) the temperature of a material is called the sensible heat. However, during the phase change process between liquid and vapor or between solid and liquid, the heat rejected or gained is called latent heat of vaporization and latent heat of fusion, respectively. The latent of heat of vaporization is the amount of heat absorbed during evaporation or released during condensation, while the latent heat of fusion is the amount of heat absorbed during melting or released during solidification.

One of the consequences of the dependence of the saturation temperature of a pure substance on the pressure is the variation of the boiling temperature of water with elevation above the sea surface. Figure 2.30 shows that as the elevation above the sea level increase the atmospheric pressure decreases, which result in a lower water boiling temperature. As shown in Figure 2.30 at an elevation of 1000 m above sea level the atmospheric pressure is 89.55 kPa, where at this pressure the boiling temperature of water is 96.3 °C. The highest mountain top in the world is Mount Everest, which stands up to a height of 8848 m above sea level, at such height the boiling temperature of water is around 70 °C.

Cooling with pressure drop

Cooling with pressure drop for fruits and vegetables is a common cooling method in the food refrigeration industry, especially for soft fruits, such as strawberries and blueberries, and leafy vegetables, such as lettuce and leek. Vacuum cooling is one of the applications that is based on the dependence of the saturation temperature and pressure. Vacuum cooling takes advantage of the relationship between the saturation temperature and pressure. The process takes place by reducing the pressure of a sealed cooling chamber to a pressure

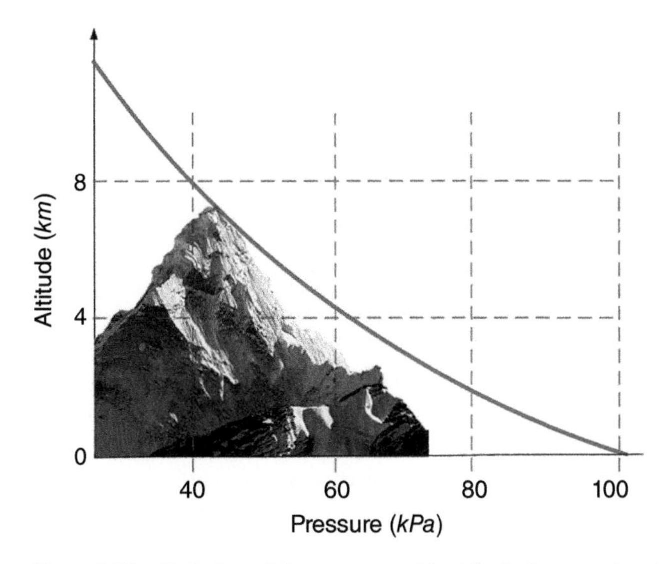

Figure 2.30 Variation of the pressure with altitude from sea level.

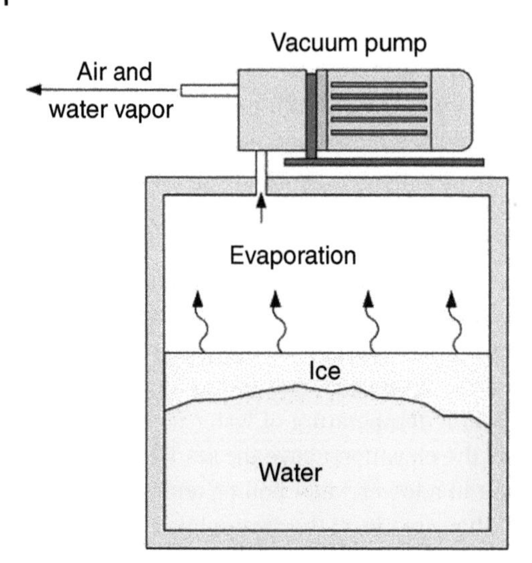

Figure 2.31 Schematic of a vacuum cooler.

that has a saturation temperature corresponding to the desired cooling temperature. Lowering the pressure as well as saturation temperature will result in evaporating some water from the product being cooled, resulting in cooling the product to the saturation temperature. In the specific case where the pressure of the sealed cooling chamber is dropped to 0.6 kPa, which has a corresponding saturation temperature of $0\,°C$, the vacuum cooling is referred to as vacuum freezing. A schematic illustration of a vacuum cooler is shown in Figure 2.31.

2.6.3 Property Diagrams

Property diagrams have been used as useful tools to illustrate the phases and phase changes specific to each substance and to better understand the behavior of such phase changes for systems. It is really necessary to comprehend the variations of properties during processes, including phase change processes. In this section, the property diagrams, such as temperature and pressure and specific volume (P-v and T-v), are further discussed for various phases and phase changes. Note that these diagrams are presented for water as an example on pure substances, and after presenting them they are compared to other pure substances.

(a) The temperature-specific volume (T-v) diagram

We are going to start with the temperature-specific volume (T-v) diagram as it was introduced earlier in Figure 2.29, which displays a T-v diagram of water for a pressure of 1 atm including the phase change. In this section, we now go one step ahead to study multiple constant pressure lines under various phases and phase changes and relation to the critical point along the saturation line.

Figure 2.32 shows the T-v diagram for water for seven constant pressure processes at various pressure values ranging from 10 to 25 000 kPa. It is shown in this figure that as the pressure increases the corresponding saturation temperature increases as well. As the saturation pressure increases the saturated liquid specific volume increases while the saturated vapor specific volume decreases and both values become closer to each other (Figure 2.32).

The straight horizontal lines in the plot represent constant temperature lines which present the phase change from saturated liquid to vapor. The lines of the saturated liquid and vapor become shorter as the pressure increases until a pressure value of 22 060 *kPa* is reached. The horizontal phase change line becomes shorter to a point where the saturated liquid and vapor points become one point. Having a smaller horizontal phase change line results in less energy per unit mass to phase change the material. For water, at a pressure of 22 060 *kPa* the horizontal phase change line becomes a point, which means that no gradual phase change occurs, it suddenly changes from liquid to vapor or from vapor to liquid, this is known as the **critical point**. This critical point is the point where both saturated liquid and saturated vapor overlap and is an unique point for every substance and different for every substance. At this critical point the saturated liquid turns into saturated vapor, literally from liquid to vapor.

For example, the critical point for water has the following properties: a temperature of 373.95 $°C$ and a specific volume of 0.003 106 m^3/kg (Figure 2.32b). The left half starting from the critical point of red dotted line passing through the saturated liquid and vapor lines is referred to as the saturated liquid line, while the right half is referred to as the saturated vapor line as shown in Figure 2.32a. Figure 2.33 shows a *T-v* property diagram of a pure substance, in this case water was used for the corresponding pressure values. This figure is important for one additional thing as it also shows the pseudocritical line, which is an extension of the saturation line in a temperature pressure plot as shown in Figure 2.34.

Figure 2.35a compares the critical pressure of various pure substances including refrigerants such as ammonia and Freon and fuels such as benzene. As shown in

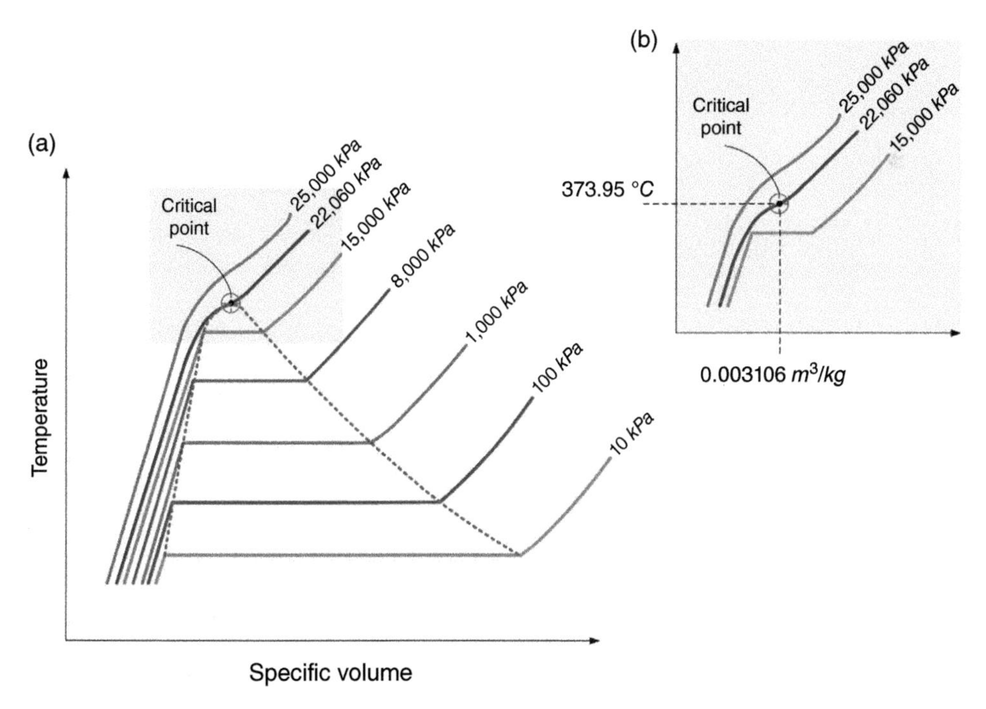

Figure 2.32 (a) *T-v* property diagram of various constant pressure phase changes of water as a pure substance and (b) its critical point with values.

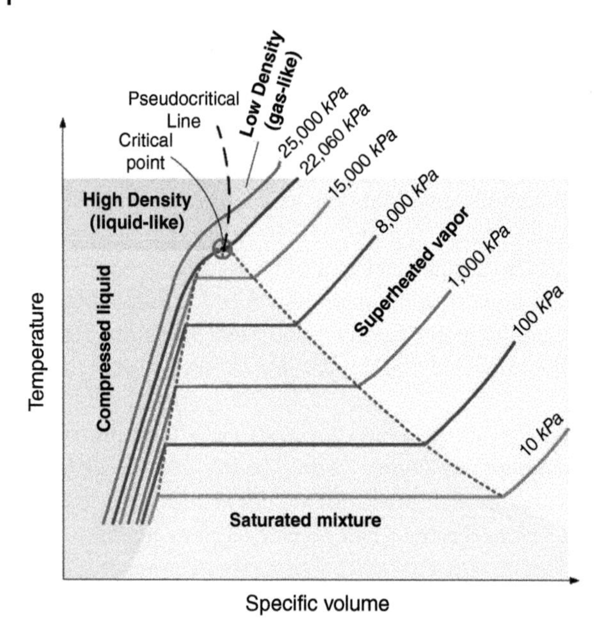

Figure 2.33 *T-v* property diagram of water highlighting the different water phases.

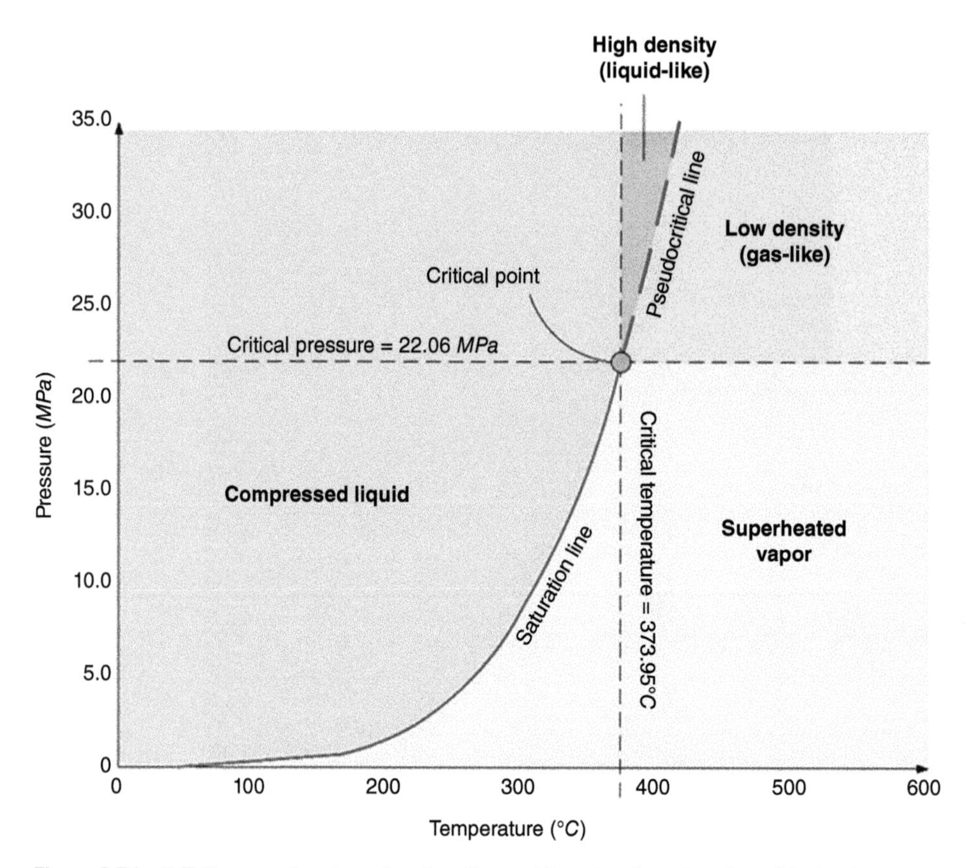

Figure 2.34 *P-T* diagram of water saturation line and its extension after the critical point.

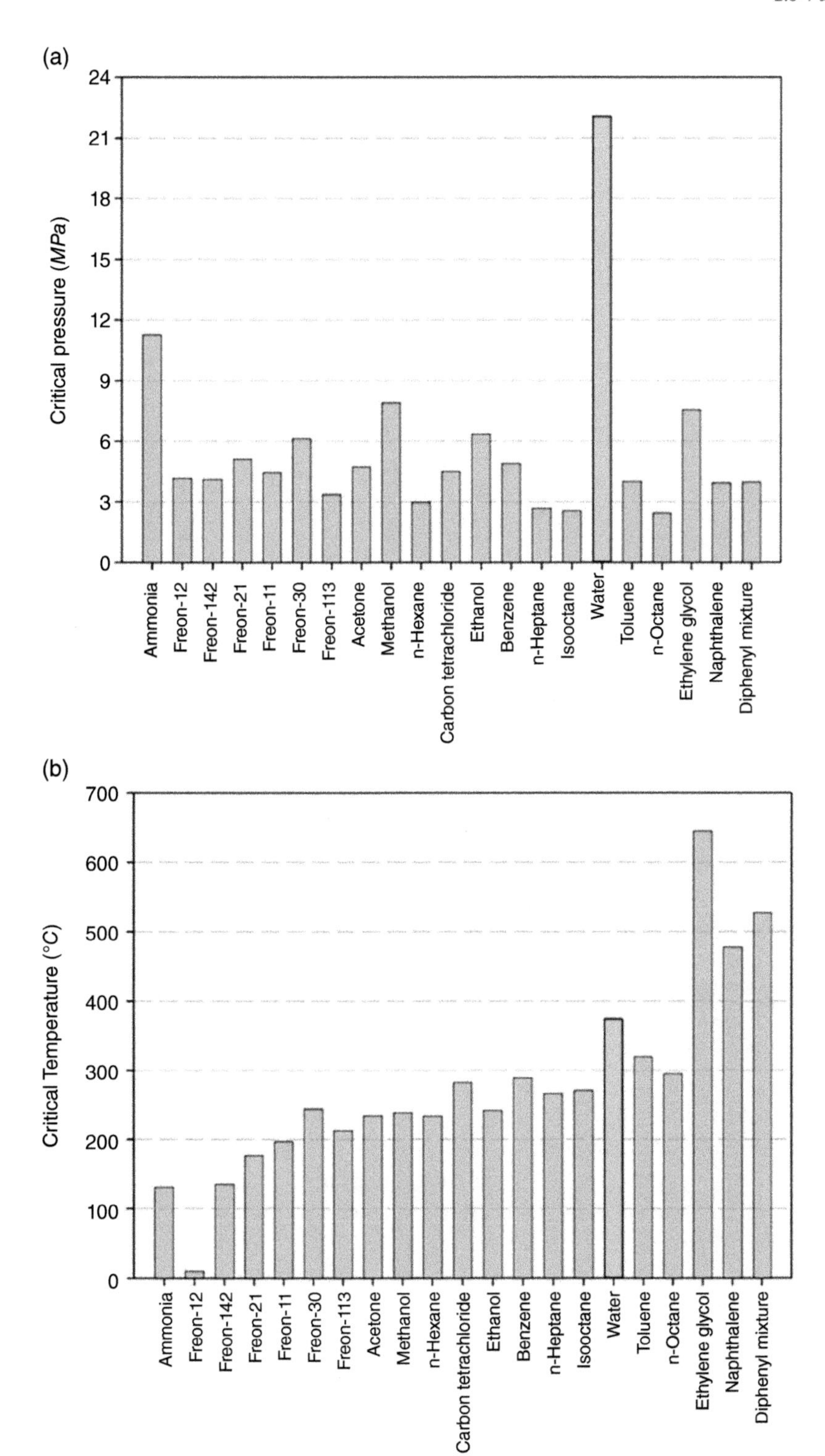

Figure 2.35 (a) Critical pressures of various fluids including water and (b) critical temperatures of various fluids including water.

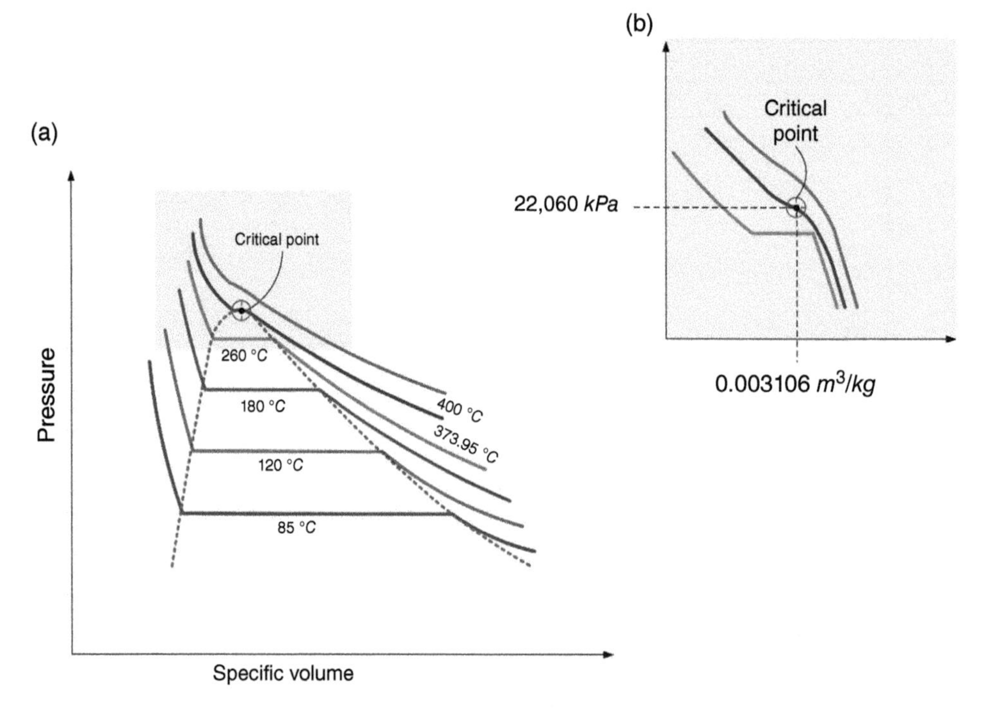

Figure 2.36 (a) *P-v* property diagram of various constant temperature phase changes of water as a pure substance and (b) its critical point with values.

Figure 2.35a, of the considered pure substances, ammonia has a critical pressure almost half that of water. In conjunction with this, Figure 2.35b compares the critical temperature of various pure substances.

(b) The pressure-specific volume (*P-v*) diagram

The overall shape of the pressure-specific volume diagram (*P-v* diagram) is similar to the *T-v* diagram with one exception – there are constant temperature lines instead of constant pressure lines. In addition, the constant temperature lines have a trend that is opposite to the constant pressure lines in the *T-v* property diagram shown in Figure 2.32a. In a similar fashion, we plot a *P-v* diagram, as shown in Figure 2.36, for various constant temperature lines, ranging from 85 to 400 °*C* with the critical point indicated. The temperature at the critical point is 373.95 °*C* with critical pressure of 22 060 *kPa* and specific volume of 0.003 106 m^3/kg.

Triple point **versus** ***critical point***

As described earlier in the chapter, a case where the two phases of liquid and vapor coexists, described as *critical point*. However, there may be specific conditions where all three phases of solid, liquid, and vapor coexist in equilibrium in a system, the so-called: **triple point** (Figure 2.37). On a *P-v* diagram a horizontal line where the three phases coexist is referred to as the triple line, which have a single pressure and temperature value. The

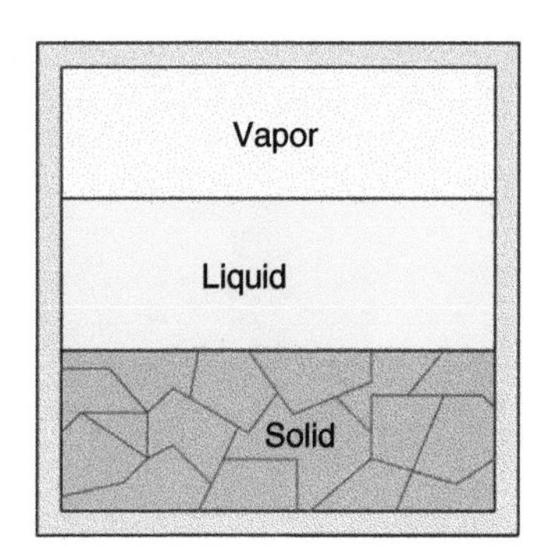

Figure 2.37 The triple point where all three phases coexist.

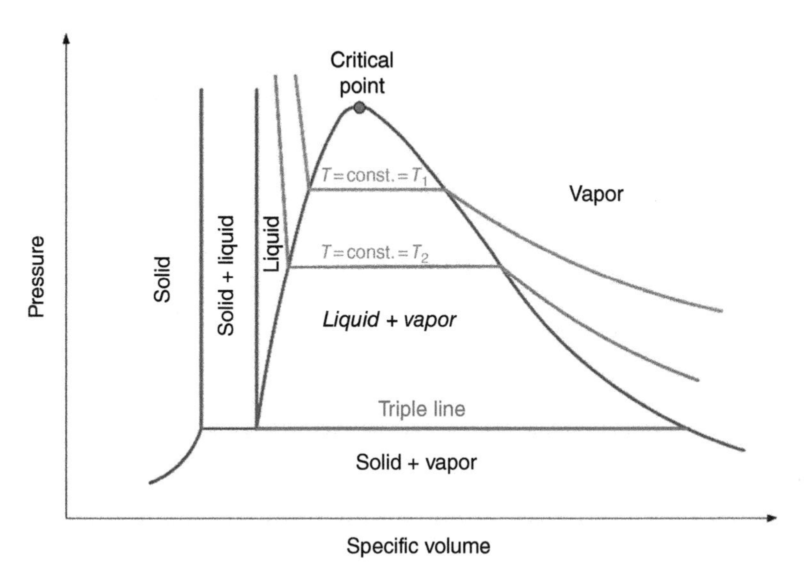

Figure 2.38 *P-v* property diagram for water showing the three phases of solid, liquid, and vapor, and where they all coexist on the triple line.

corresponding pressure and temperature are 0.6117 *kPa* and 0.01 °*C* for water, respectively. Figure 2.38 shows the triple line on a *P-v* diagram. As shown in Figure 2.38, the triple line connects the solid and vapor region with the liquid and vapor mixture region. In a *P-T* diagram the triple line will appear as a point, which is called the triple point. The triple point varies from one pure substance to another; for example the triple point temperature and pressure of ammonia are −80.75 °*C* and 6.08 *kPa*. Another example of the triple point of pure substance is 3626.85 °*C* and 10 100 *kPa* for temperature and pressure, respectively, for carbon (graphite).

(c) Pressure-specific volume-temperature (*P-v-T*) surface diagrams

For a simple compressible substance, its state can be fixed by any two independent intensive properties. This means for a simple compressible substance if two independent properties

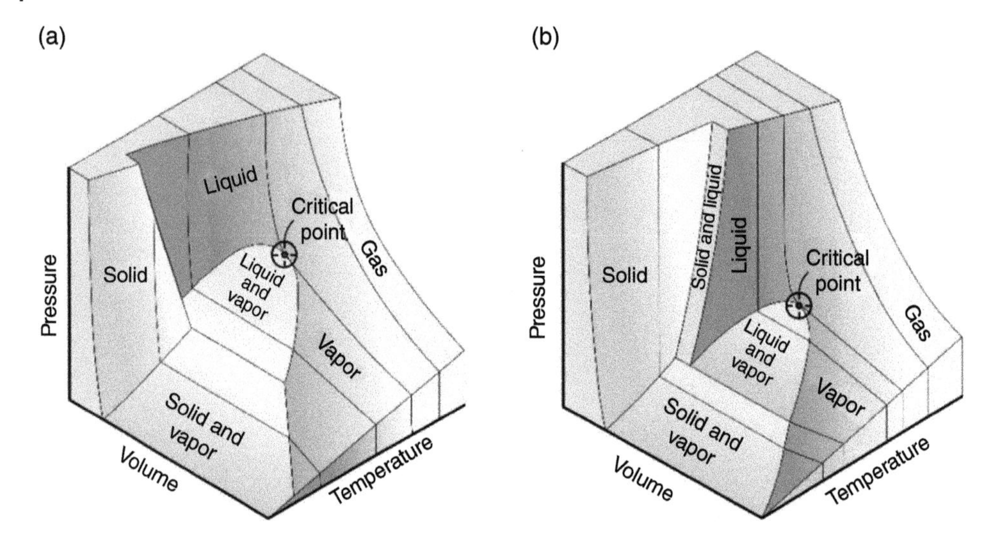

Figure 2.39 Water pressure as a function of two independent properties, specific volume and temperature (a) for water (expands on freezing) (b) for a pure substance that contracts on freezing.

are known and defined, all other properties become dependent properties. The properties of a pure substance are a function of only two independent properties, considering the two independent properties as x and y, then any third property can be called z, then $z = f(x, y)$. If we selected the specific volume and the temperature as the two independent properties and from these two independent values, we can get the pressure, which is now is a dependent property.

The pressure as a function of the temperature and specific volume is presented in Figure 2.39. Figure 2.39a shows that of water, which is an example of pure substances that expands on freezing, while Figure 2.39b shows that of a pure substance that contracts on freezing. Although these property surface plots are not commonly used in engineering thermodynamics, they are useful in advanced studies to better understand the phases and phase changes for practical applications.

2.6.4 Property Tables

As shown in Figure 2.40, which is a T-v property diagram for water (Figure 2.40a) and for ammonia (Figure 2.40b), the relationship between the thermodynamic properties is complicated and a number of complicated correlations are used to calculate these properties. In order to make it easier for thermodynamics engineers to analyze systems and processes, the properties are often presented in the form of tables, as shown in Table 2.3. The thermodynamic properties of water are usually presented in three main tables. The first is for the compressed liquid, where the pressure and the temperature are independent. The second table is for the saturated water tables, which contains the properties of the saturated liquid, saturated vapor and saturated liquid vapor mixture, where the temperature and the pressure are dependent. The third table is for the superheated vapor properties which are available in various sources for direct use.

(a)

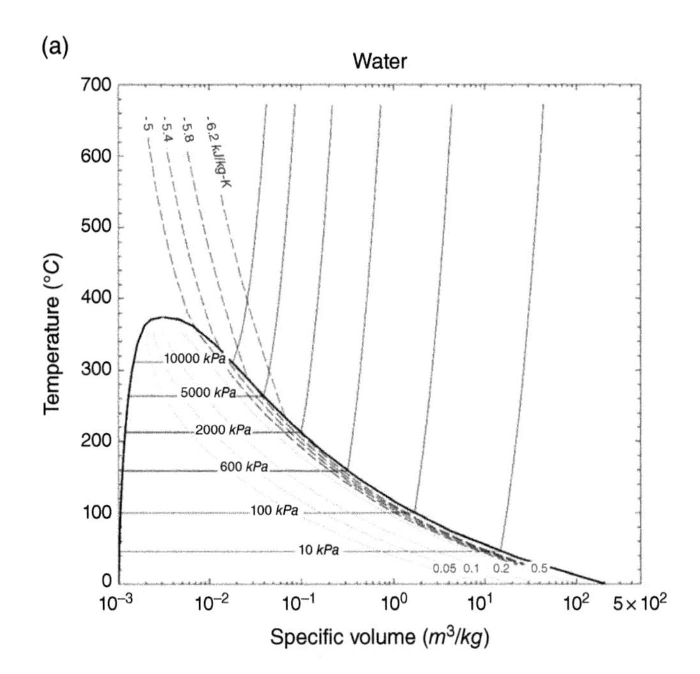

(b)

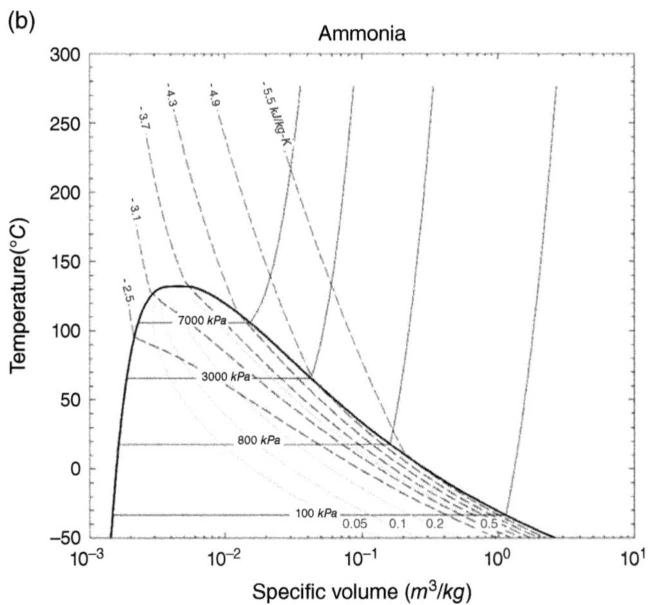

Figure 2.40 *T-v* diagrams of water and ammonia as generated by EES.

(a) Compressed liquid

Compressed liquid tables are used to determine the thermodynamics properties of water in the compressed liquid state (Table 2.4), where the pressure and the temperature are independent. Note that the compressed liquid tables often list properties for high pressure

Table 2.3 Example of a property table for saturated water.

Saturated water (function of pressure)				
		Specific volume (m^3/kg)		
Pressure (*kPa*)	**Saturation temperature (°*C*)**	**Saturated liquid, v_f**	**Difference, v_{fg}**	**Saturated vapor, v_g**
100	99.61	0.001 043	1.6931	1.6941
125	105.97	0.001 048	1.3740	1.3750
150	111.35	0.001 053	1.1583	1.1594
200	120.21	0.001 061	0.884 72	0.885 78

Table 2.4 Example of compressed liquid water property table.

Compressed liquid water				
	Temperature (°*C*)	**Specific volume (m^3/kg)**	**Specific internal energy (*kJ/kg*)**	**Specific enthalpy (*kJ/kg*)**
Pressure = 5 *MPa*	263.94 (saturation)	0.001 286 2	1148.1	1154.5
	0	0.000 997 7	0.04	5.03
	20	0.000 999 6	83.61	88.61
	40	0.001 005 7	166.92	171.95
	60	0.001 014 9	250/29	255.36
	80	**0.001 026 7**	**333.82**	338.96

values, as the thermodynamic properties of water in the compressed liquid state are slightly affected by the variation of pressure. Since the compressed liquid tables are often limited, an approximate solution is to take the compressed liquid property from the saturated water table as saturated liquid water based on the temperature of the compressed liquid. However, we can reduce the error of the saturated liquid water assumption by considering the pressure difference as follows:

$$h = h_f + v_f \times (P - P_{sat}) \tag{2.16}$$

where v_f, h_f, and P_{sat} are taken at the temperature of the compressed liquid. An example of a compressed liquid water table is presented in Table 2.4. Note that h is the specific enthalpy, which is one of the intensive properties that is based on the combination of internal energy of the pure substance and its flow energy; it can be expressed as:

$$h = u + Pv \quad (kJ/kg) \tag{2.17}$$

where h is the specific enthalpy, u is the specific internal energy, and the combination of Pv (pressure multiplied with the specific volume) is called the flow energy. The concept of enthalpy is often used to present the energy of flowing pure substance.

The saturation tables are used to determine the state of the system, whether it is compressed liquid or saturated liquid, vapor, and liquid vapor mixture or superheated vapor, and the following are the different cases where the state of the system is compressed liquid.

- For a given temperature, when the pressure is higher than the saturation pressure.
- For a given pressure, when the temperature is lower than the saturation temperature.
- For a given pressure or a temperature, when the specific volume is lower than the saturated liquid specific volume.
- For a given pressure or a temperature, when the specific internal energy is lower than the saturated liquid internal energy.
- For a given pressure or a temperature, when the specific enthalpy is lower than the saturated liquid specific enthalpy.

Example 2.7 A 10 L pressurized container at a pressure of 5 MPa has water at a temperature of 80 °C. Determine the following for the water in the tank:

a) State of water in the tank.
b) Specific volume.
c) Specific internal energy.

Solution

a) The state of water is found by comparing the temperature of water in the tank with the saturation temperature of water at the given pressure. From Table 2.4, the temperature of water in the tank is less than the saturation temperature of the water at the given pressure:

$$80\ {}^oC < 263.94\ {}^oC \text{ saturation temperature of water at given pressure 5 } MPa$$

Then the state of water in the pressurized tank is **compressed liquid.**

b) The specific volume of water is found from the compressed liquid table at the given pressure and temperature

$$\boldsymbol{v = 0.0010267\ \frac{m^3}{kg}}$$

c) The internal energy of water is found from the compressed liquid table at the given pressure and temperature

$$\boldsymbol{u = 333.82\ \frac{kJ}{kg}}$$

(b) Saturated liquid and saturated vapor mixtures

The vaporization process starts with saturated liquid and ends up with the saturated vapor as described in the earlier sections and as shown in Figure 2.41. To determine the properties of saturated liquid, vapor, and liquid vapor mixture we use the saturation tables. The details of the saturated tables are shown in Figure 2.42. In order to determine the properties of the

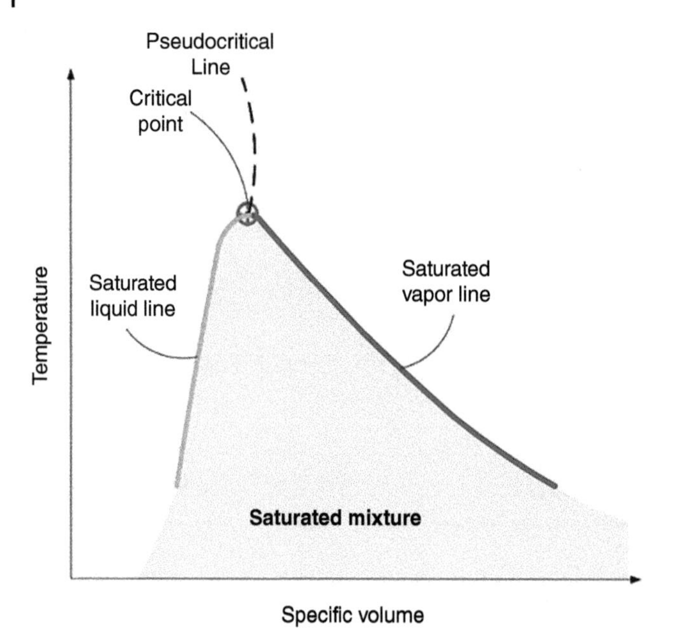

Figure 2.41 Between the saturated liquid line and the saturated vapor line is the saturated mixture region.

saturated liquid water (the left part of the saturated line shown in Figure 2.41 [light blue]) we select the properties from the saturated table that has the subscript f as shown in Figure 2.42. Selecting the properties of saturated liquid is explained in Example 2.8.

Example 2.8 A 10 L pressurized container at a pressure of 200 kPa has water in the saturated liquid state. Determine the following properties of the water in the tank: (a) specific volume, (b) temperature and (c) mass of the water in the container.

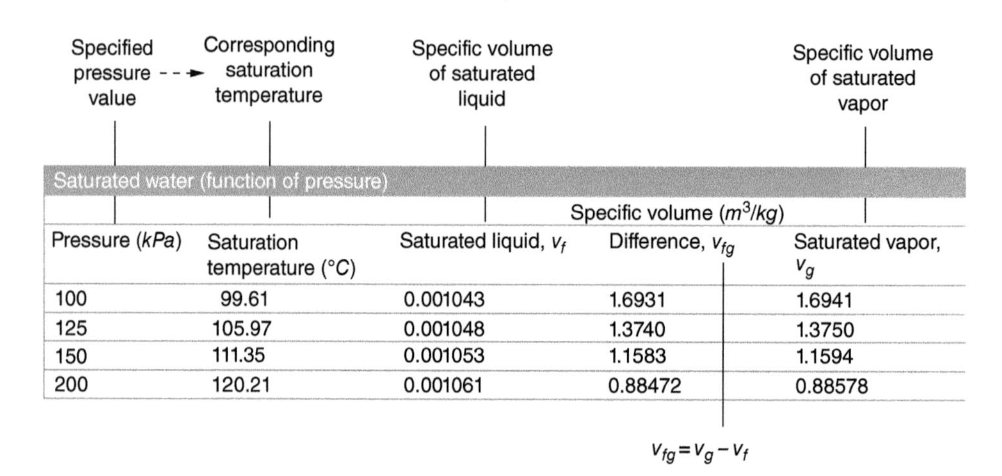

Saturated water (function of pressure)				
		Specific volume (m^3/kg)		
Pressure (kPa)	Saturation temperature (°C)	Saturated liquid, v_f	Difference, v_{fg}	Saturated vapor, v_g
100	99.61	0.001043	1.6931	1.6941
125	105.97	0.001048	1.3740	1.3750
150	111.35	0.001053	1.1583	1.1594
200	120.21	0.001061	0.88472	0.88578

Figure 2.42 Details of the saturated table (saturated water).

Solution

Since the water is in the saturated liquid state then we go to the saturated water table (such as Appendix B-1b) to obtain the properties of that water in the container.

a) $v_f = 0.001061\ m^3/kg$ from Table 2.5 since the water is in the saturated liquid phase.
b) For pure substances that are in the saturated phase, including saturated liquid, liquid vapor mixture and vapor, the temperature and pressure are dependent. Since we know the pressure value, which is 200 kPa then the corresponding saturation temperature is the temperature of the container, which is based on Table 2.5 120.21 °C.
c) From the definition of the specific volume

$$v = V/m$$

$$0.001061\ \frac{m^3}{kg} = \left(10\ L \times \frac{1\ m^3}{1000\ L}\right)/(m)$$

$$\boldsymbol{m = 0.01/0.001061 = 9.425\ kg}$$

Vapor quality

Vapor quality was defined earlier as the ratio of the mass of water vapor to the total mass of the mixture:

$$x = \frac{m_{vapor}}{m_{mixture}}$$

which is the most significant parameter for saturated liquid and saturated vapor mixtures. It ranges between 0 and 1 (or 0 and 100% in percentage illustration).

If one wants to obtain the specific volume of the mixture, the following approach will work out:

$$V_{mixture} = V_{vapor} + V_{liquid} \rightarrow (mv)_{mixture} = (mv)_{vapor} + (mv)_{liquid}$$

Table 2.5 Part of the saturated water table.

Saturated water (function of pressure)				
		Specific volume (m^3/kg)		
Pressure (kPa)	Saturation temperature (°C)	Saturated liquid, v_f	Difference, v_{fg}	Saturated vapor, v_g
100	99.61	0.001 043	1.6931	1.6941
125	105.97	0.001 048	1.3740	1.3750
150	111.35	0.001 053	1.1583	1.1594
200	120.21	0.001 061	0.884 72	0.885 78

Dividing each term by $m_{mixture}$ and substituting ($m_{liquid} = m_{mixture} - m_{vapor}$) one finds:

$$v_{mixture} = xv_{vapor} + (1-x)v_{liquid} \rightarrow v_{mixture} = v_{liquid} + x(v_{vapor} - v_{liquid})$$

With the notation of f for liquid and g for vapor (as well as $v_{fg} = v_g - v_f$), it results in:

$$v_{mixture} = v_f + xv_{fg} \tag{2.18}$$

For mixtures where both liquid and vapor exist at the same time, the saturated tables can be used. Furthermore, with the help of above given approach, the following equations the properties of mixtures such as the specific volume, internal energy and the enthalpy values can be calculated as follows:

$$u_{mixture} = u_f + xu_{fg} \tag{2.19}$$

$$h_{mixture} = h_f + xh_{fg} \tag{2.20}$$

$$s_{mixture} = s_f + xs_{fg} \tag{2.21}$$

One may further obtain the quality out of the above written equations as:

$$x = \frac{v_{mixture} - v_f}{v_{fg}} \tag{2.22}$$

$$x = \frac{u_{mixture} - u_f}{u_{fg}} \tag{2.23}$$

$$x = \frac{h_{mixture} - h_f}{h_{fg}} \tag{2.24}$$

$$x = \frac{s_{mixture} - s_f}{s_{fg}} \tag{2.25}$$

Example 2.9 In a pressurized container at a pressure of 150 *kPa* as shown in Figure 2.43, two thirds of the mass is water vapor and the remaining portion is liquid water. Find both (a) specific volume and (b) temperature for the liquid water in the container.

Solution

Since both states of vapor and liquid coexist then the state of the system of the water in the pressurized container is the saturated liquid vapor mixture and the saturated tables (such as Appendix B-1b) are used with the above given quality equations.

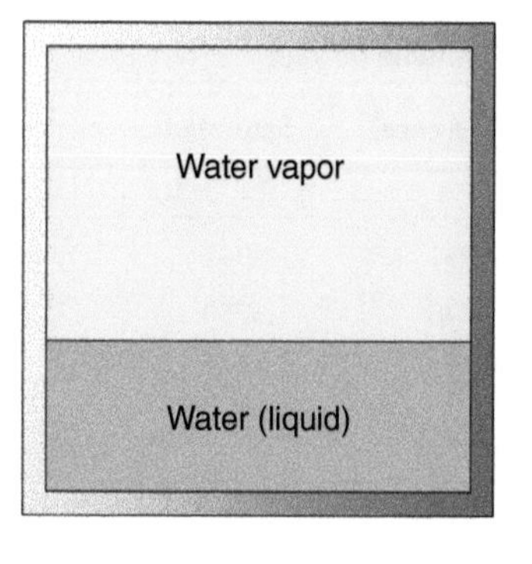

Figure 2.43 A schematic diagram for Example 2.9.

Table 2.6 Part of the saturated water table.

Saturated water (function of pressure)				
		Specific volume (m^3/kg)		
Pressure (kPa)	Saturation temperature (°C)	Saturated liquid, v_f	Difference, v_{fg}	Saturated vapor, v_g
100	99.61	0.001 043	1.6931	1.6941
125	105.97	0.001 048	1.3740	1.3750
150	111.35	0.001 053	1.1583	1.1594
200	120.21	0.001 061	0.884 72	0.885 78

a) From Table 2.6 and based on the pressure of the container the values are plugged into Eq. 2.22 as follows, where x is called the quality and it is calculated by $x = m_{vapor}/m_{mix}$:

$$v_{mix} = v_f + xv_{fg} = 0.001053 + \left(\frac{2}{3} \times 1.1583\right) = \mathbf{0.773253\ m^3/kg}$$

b) Since the system is in the saturated liquid vapor mixture then the temperature is dependent on the pressure and it is equal to the saturation temperature that corresponds to the container pressure of 150 kPa, which is **111.35 °C** from Table 2.6.

(c) Superheated vapor

To the right of the saturated vapor line shown in Figure 2.33 is the superheated region. The saturation tables are used to determine the state of the system in the following cases: compressed liquid or saturated liquid, vapor, and liquid vapor mixture or superheated vapor. Furthermore, the following are the different cases where the state of the system is determining to be superheated vapor:

- For a given temperature, when the pressure is lower than the saturation pressure.
- For a given pressure, when the temperature is higher than the saturation temperature.
- For a given pressure or temperature, when the specific volume is higher than the saturated liquid specific volume.
- For a given pressure or temperature, when the specific internal energy is higher than the saturated liquid internal energy.
- For a given pressure or temperature, when the specific enthalpy is higher than the saturated liquid specific enthalpy.

Example 2.10 A cylinder shown in Figure 2.44 has water at a pressure of 10 kPa and a temperature of 150 °C. For the water in the cylinder determine the following: (a) specific volume, (b) specific internal energy, and (c) specific enthalpy.

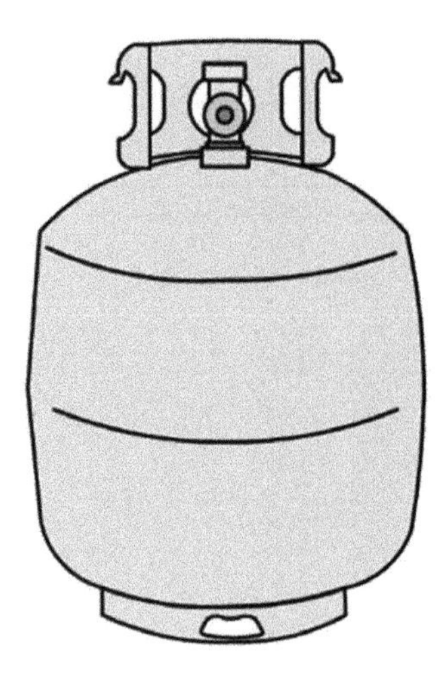

Figure 2.44 Schematics for Example 2.10.

Solution

Since the example did not mention what the state of the water is inside the cylinder, first we must determine the state by referring back to the saturation table (such as Appendix B-1b).

Comparing the cylinder temperature with the saturation temperature corresponding to the cylinder pressure from Table 2.7, we can see that the cylinder temperature is higher than the saturation temperature at the given pressure. Since the system temperature is higher than the saturation temperature at the given pressure then the state of the water in the system is superheated vapor. For water in the superheated region, one needs to use the superheated vapor table (Table 2.8).

a) From Table 2.8, at a pressure of 10 kPa we select the properties that corresponds to the system temperature (150 °C). The specific volume of the system is equal to **19.513 m^3/kg**.
b) Similar to part (a) the specific internal energy is equal to **2587.9 kJ/kg.**
c) Similar to part (a) the specific enthalpy is equal to **2783.0 kJ/kg.**

Table 2.7 A partial table for saturated water.

Saturated water (function of pressure)				
		Specific volume (m^3/kg)		
Pressure (kPa)	Saturation temperature (°C)	Saturated liquid, v_f	Difference, v_{fg}	Saturated vapor, v_g
10	45.81	0.001 010	14.669	14.670
20	53.97	0.001 017	7.6379	7.6481
30	69.09	0.001 022	5.2277	5.2287

Table 2.8 A partial table for superheated water vapor.

Compressed liquid water				
	Temperature (°C)	**Specific volume (m^3/kg)**	**Specific internal energy (kJ/kg)**	**Specific enthalpy (kJ/kg)**
Pressure = 0.01 MPa	45.81 (saturation)	14.670	2437.2	2583.9
	50	14.867	2443.3	2592.0
	100	17.196	2515.5	2687.5
	150	19.513	2587.9	2783.0
	200	21.826	2661.4	2879.6
	250	24.136	2736.1	2977.5

Example 2.11 A 4 L piston–cylinder device, as shown in Figure 2.45, contains 2 kg of a saturated liquid-vapor mixture of water at 70 °C. The water is slowly heated until it exists in a single phase of vapor. Find (a) final pressure, (b) final temperature, and (c) vapor quality for the initial state.

Solution

a) Since the device is a piston–cylinder device then the pressure remains constant throughout the process. And since the heat was used only to change phase then from Table 2.9:

$$P_2 = P_1 = P_{sat@70\ °C} = 31.202\ kPa$$

b) Since the heat was used only to change phase then from Table 2.9:

$$T_1 = T_2 = 70\ °C$$

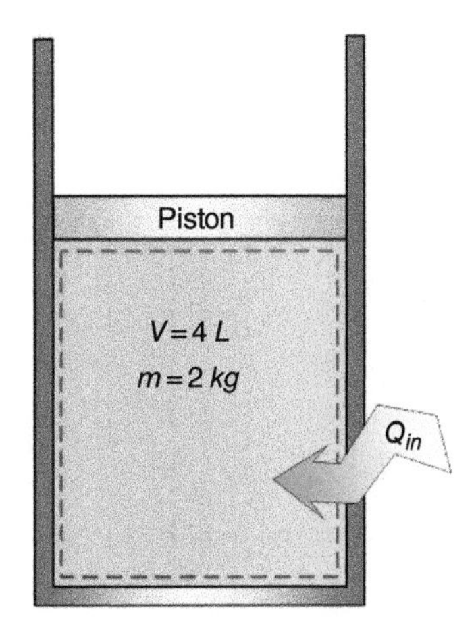

Figure 2.45 A piston–cylinder device for Example 2.11.

Table 2.9 Saturated water table.

Saturated water (function of temperature)				
		Specific volume (m^3/kg)		
Temperature (°C)	**Saturation Pressure (kPa)**	**Saturated liquid, v_f**	**Difference, v_{fg}**	**Saturated vapor, v_g**
50	12.34	0.001 012	12.025	12.026
70	31.202	0.001 023	5.0386	5.0396
100	101.42	0.001 043	1.6710	1.6720

c) The subject matter water is a mixture of saturated liquid and vapor in the beginning. So, one can use Eq. (2.26) to find the quality as follows:

$$x = \frac{v_{mixture} - v_f}{v_{fg}} = \frac{\frac{4\ L}{2\ kg} \times \frac{1\ m^3}{1000\ L} - 0.001023\ \frac{m^3}{kg}}{5.0386\ \frac{m^3}{kg}} = \mathbf{0.0002}$$

which clearly indicates that it is almost saturated liquid with some negligible amount of vapor.

Example 2.12 Water inside a piston–cylinder device, as shown in Figure 2.46, operates initially at a pressure of 200 kPa and a temperature of 350 °C. The piston–cylinder stops are made to prevent the piston going further down. The water in the piston–cylinder device is cooled at a constant pressure until it reaches the saturated vapor state and the piston is resting on the stops. Then the water is cooled further until the pressure in the tank reaches 100 kPa. (a) Sketch the process on a T-v diagram, (b) find T_2, and (c) find the overall change in internal energy per unit mass of water.

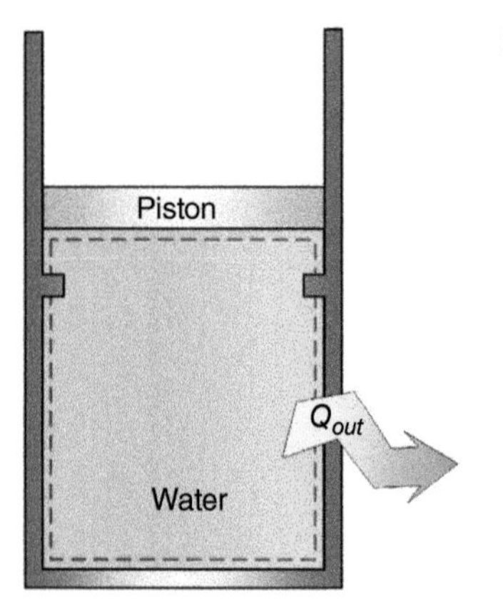

Figure 2.46 A piston–cylinder device for Example 2.12.

Solution

a) In order to draw the *T-v* we need to find the temperature and the specific volume of each of three points.

State 1: $T_1 = 350\ ^oC,\ P_1 = 200\ kPa$

From the saturated water table (such as Appendix B-1b) we can identify, since the temperature of the water is higher than the saturation temperature at P_1, that the state of water is superheated steam and the properties are obtained from the superheated water vapor tables (such as Appendix B-1c).

$$h_1 = 3172\ \frac{kJ}{kg},\ u_1 = 2887\ \frac{kJ}{kg}\ and\ v_1 = 1.433\ \frac{m^3}{kg}$$

State 2 is defined at the instant the piston just reached the stops.

$P_2 = P_1 = 200\ kPa$ since the process occurs in a piston–cylinder device where the piston is freely moving from state 1 to 2.

The quality $x_2 = 1$, since it is saturated vapor.

Then the properties are taken at $200\ kPa$ as follows:

$$T_2 = T_{sat@P_2 = 200\ kPa} = 120.2\ ^oC$$

$$h_2 = 2707\ \frac{kJ}{kg},\ u_2 = 2529\ \frac{kJ}{kg}\ and\ v_2 = 0.8859\ \frac{m^3}{kg}$$

State 3 is defined as the final state of the system.

$$P_3 = 100\ kPa$$

And since after state 2 the system is now a rigid tank, then the specific volume remains constant throughout the process from state 2 to 3.

$$v_3 = 0.8859\ \frac{m^3}{kg}$$

Going to the saturated water table (such as Appendix B-1b) to determine the state of the water at state 3 by comparing the value of the specific volume at state 3 with the specific volumes of saturated liquid and saturated vapor at the pressure P_3.

$$v_f < v_3 < v_g$$

Then the water at state 3 is saturated liquid vapor mixture, so:

$$T_3 = T_{sat@P_3} = 99.61\ ^oC$$

Since the state of the water in state 3 is a saturated liquid vapor mixture, then in order to find any quality we must first find the quality of the water in that mixture, which can be found as follows:

$$v_3 = v_{f@P_3} + x_3\left(v_{g@P_3} - v_{f@P_3}\right)$$

$$x_3 = \frac{v_3 - v_{f@P_3}}{v_{g@P_3} - v_{f@P_3}} = \frac{0.8859 - 0.001043}{1.6941 - 0.001043} = 0.5226$$

$$u_3 = u_{f@P_3} + x_3\left(u_{fg@P_3}\right) = 417.4 + (0.5226 \times 2088.2) = 1509\ \frac{kJ}{kg}$$

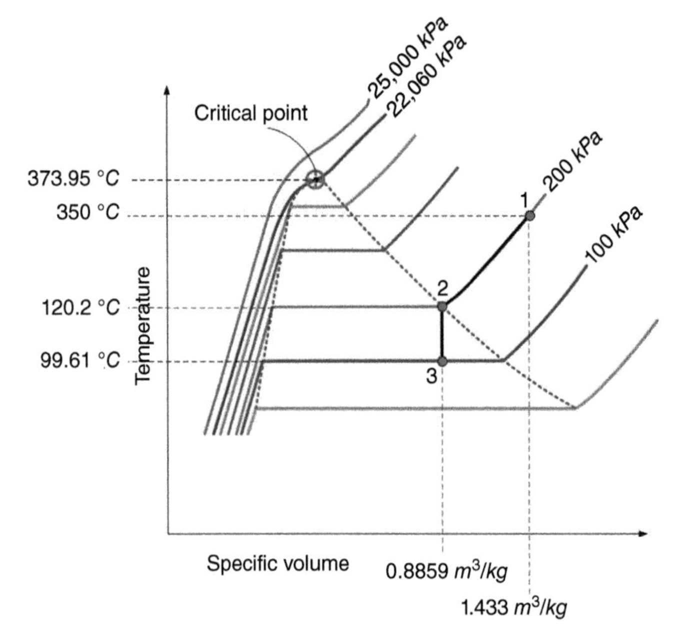

Figure 2.47 *T-v* plot of the processes occurring in Example 2.12.

Then by using the temperatures and the specific volumes we can draw the diagram with the support of information about going from one state to another (Figure 2.47).

b) Find the overall change in the internal energy of the water per unit mass of water:

$$\Delta u = u_3 - u_1 = 1509 - 2887 = \mathbf{-1378}\ \mathbf{\frac{kJ}{kg}}$$

2.7 Ideal Gas Equation

The property tables introduced earlier for pure substances contains very accurate information; however, one of the disadvantages they have is that they are bulky and consume time when extracting the properties from them. A smarter approach is to use simple relations that relate the properties together and provide acceptable and accurate results. Equations that relate temperature, pressure, and specific volume are known as equations of state. The first to observe the inverse proportional relationship between pressure and temperature of gases was Robert Boyle in 1662. In 1802 Charles and Gay-Lussac experimentally determined that the relationship between the pressure, temperature, and the specific volume can be defined as:

$$Pv = RT \tag{2.26}$$

where, R is called the ideal gas constant, and Eq. 2.26 is referred to as idea gas equation of state. The ideal gas constant is specific for each gas and is calculated as:

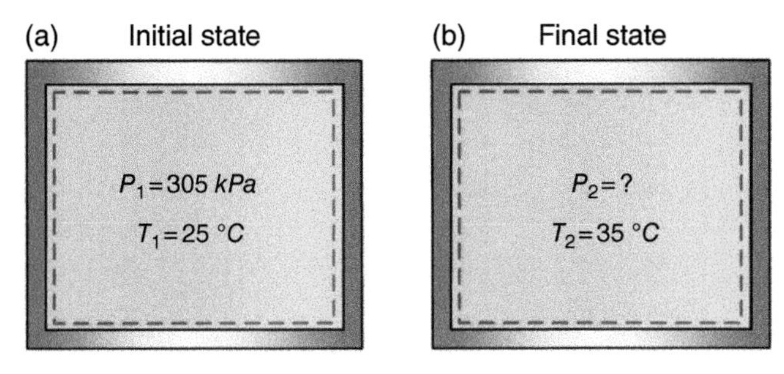

Figure 2.48 The pressurized rigid tank from Example 2.13 (a) initial state of the tank and (b) final state of the tank.

$$R = \frac{R_u}{M} \tag{2.27}$$

where, R_u is the universal gas constant (8.31 447 *kJ/kmol K*) and M is the molar mass of the gas.

Example 2.13 A pressurized rigid container is initially kept at a pressure of 305 *kPa* and a temperature of 25 °*C* as shown in Figure 2.48. If the temperature of gas in the container, which is air in this example, is increased to 35 °*C* find the final pressure of the gas container.

Solution

Since the container is rigid then the volume of the container remains constant through the heating process, and since there was no mention of any gas entering or escaping the container then the specific volume of the gas in the container is constant as the temperature increases. Then the initial state of the gas (referred to as state 1) and the final state of the gas in the container (referred to as state 2) can be related by equating the specific volume in the equations for each state as follows:

State 1: $P_1 v_1 = RT_1 \rightarrow v_1 = RT_1/P_1$
State 2: $P_2 v_2 = RT_2 \rightarrow v_2 = RT_2/P_2$

As mentioned earlier $v_1 = v_2$ then

$$\frac{RT_1}{P_1} = \frac{RT_2}{P_2} \rightarrow P_2 = \left(\frac{T_2}{T_1}\right) P_1 = \left(\frac{35+273}{25+273}\right) \times (305) = \mathbf{315\ kPa}$$

2.7.1 When is Water Vapor an Ideal Gas?

The accurate source of properties of water vapor can be obtained from the property tables. However, the convenience of the ideal gas equations introduced the question of whether the ideal gas equation of state can be used to calculate the properties of water vapor. If we use the ideal gas equations to calculate the specific volume of water vapor (superheated water vapor and including the saturated water) we will find that at the following condition water vapor can be fairly treated as an ideal gas:

- Pressure lower than 10 *kPa* and at any temperature.

Table 2.10 Superheated water vapor.

Compressed liquid water				
	Temperature (°C)	Specific volume (m^3/kg)	Specific internal energy (kJ/kg)	Specific enthalpy (kJ/kg)
Pressure = 0.01 MPa	45.81 (saturation)	14.670	2437.2	2583.9
	50	14.867	2443.3	2592.0
	100	17.196	2515.5	2687.5
	150	19.513	2587.9	2783.0
	200	21.826	2661.4	2879.6
	250	24.136	2736.1	2977.5

Note that a percentage error of 0.1% is considered acceptable. Another important conclusion from carrying out comparison calculations throughout different points in the property plot is that the highest deviation from the ideal gas behavior is in the region around the critical point.

Example 2.14 Show that at a pressure of 10 kPa and a temperature of 100 °C water vapor can be treated as an ideal gas.

Solution

The basis of the comparison will be performed based on specific volume. First the specific volume of the water vapor at the given pressure and temperature is determined using the superheated water vapor table (Table 2.10); the value is 17.196 m^3/kg.

The next step is to calculate the specific volume of the water vapor by using the ideal gas equation of state:

$$v = \frac{RT}{P} = \frac{0.4615\ kJ/kgK \times (100 + 273)K}{10\ kPa} = 17.21395\ m^3/kg$$

The percentage error is calculated as follows:

$$\%\text{error} = \frac{|\text{v}_{\text{tables}} - \text{v}_{\text{equ}}|}{\text{v}_{\text{equ}}} \times 100 = \frac{|17.196 - 17.21395|}{17.21395} \times 100 = \mathbf{0.1044\%}$$

2.7.2 Compressibility Factor

As was introduced earlier and shown in Example 2.13, the ideal gas equation of state is a very convenient way of calculating the properties of gases. It was shown in Example 2.13 that at a pressure of 10 kPa the error reaches 0.1% with the tables, which is acceptable and can be considered accurate enough. However, at pressures higher than 10 kPa the percentage error increases to reach 152.7% close to the critical point and on the saturation vapor line. The increase in the percentage error shows the deviation of real gas behavior from ideal gas behavior. This can be accounted for through the compressibility factor (Z), which is used as a correction factor to the ideal gas equation as follows:

$$Pv = ZRT \tag{2.28}$$

where Z is can be calculated as:

$$Z = v_{actual}/v_{deal} \tag{2.29}$$

Since the gases behave differently at different temperatures, plotting the compressibility factor with the pressure and temperature will result in a different plot for each gas. However, if the temperature and pressure of the gases are normalized against their critical temperature and pressure respectively, then the result is a single compressibility chart for all gases, as shown in Figure 2.49.

For a given temperature and a pressure of any gas the reduced pressure and reduced temperature (the normalized temperature and pressure respectively) are calculated as follows:

$$T_R = \frac{T}{T_{cr}} \tag{2.30}$$

$$P_R = \frac{P}{P_{cr}} \tag{2.31}$$

where P_R is the reduced pressure, T_R is the reduced temperature, P_{cr} is the critical pressure, and T_{cr} is the critical temperature. From the generalized compressibility chart, the previous conclusion that was presented earlier can be extended and presented as follows:

- For any given temperature and a reduced pressure that is much less than 1 the behavior of the gas can be considered as an ideal gas
- For any given pressure and high temperatures with a reduced temperature more than 2, the behavior of the gas can be considered as an ideal gas as well

To reduce the tedious calculations and iterations in certain cases where the two known properties about a gas are not the temperature and pressure, a third reduced parameter is introduced:

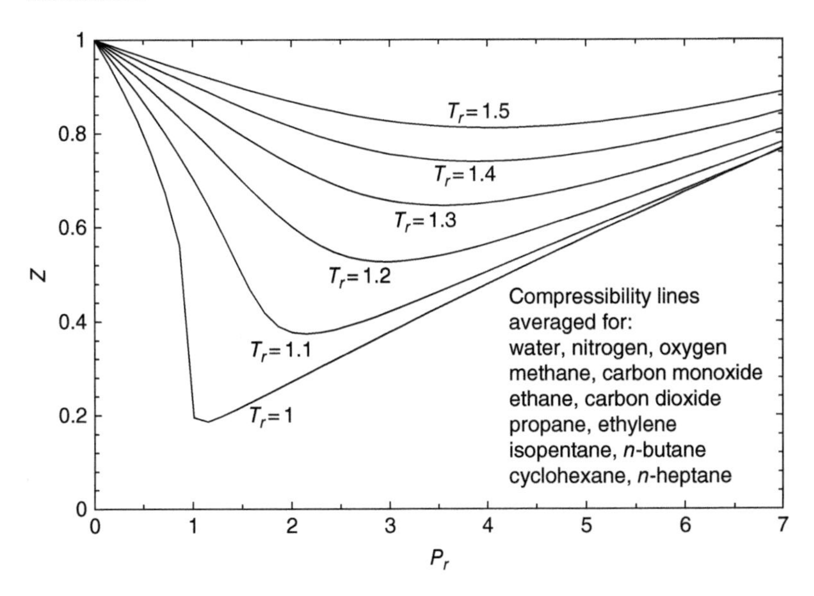

Figure 2.49 A generalized compressibility chart obtained for 13 fluids (generated through Engineering Equation Solver [EES] software which is used for Z prediction with real gas equations of state).

$$v_R = \frac{v_{actual}}{RT_{cr}/P_{cr}} \tag{2.32}$$

where, v_R is the reduced specific volume.

Example 2.15 A refrigerant cylinder contains *R*-134*a* at a pressure of 1 *MPa* and 50 °*C* (Figure 2.50). Calculate:

a) The specific volume using the ideal gas equation of state.
b) The corrected specific volume using the compressibility chart to find.
c) The percentage error between the specific volume in (a) and (b) with an experimental determined value of 0.021796 m^3/kg.

Solution

a) In this part of the problem we will be using the ideal gas equation of state to find the specific volume of the refrigerant in the tank:

$$v_{ideal} = \frac{RT}{P} = \frac{0.0815\ kPam^3/kgK \times (50+273)K}{1000\ kPa} = \mathbf{0.026325\ m^3/kg}$$

b) The second part of the problem recommends the use of the compressibility factor from the compressibility chart, and in order to use the compressibility chart we need the reduced pressure and reduced temperature, which are calculated as follows:

$$T_R = \frac{T}{T_{cr}} = \frac{(50+273)\,K}{(374.2\ K)} = 0.863$$

$$P_R = \frac{P}{P_{cr}} = \frac{1000\ kPa}{4059\ kPa} = 0.246$$

From the compressibility chart one can find that Z = 0.84, then the specific volume of the refrigerant tank while considering the compressibility factor is calculated as follows:

$$v_{corrected} = Z\frac{RT}{P} = 0.84 \times \frac{0.0815\ kPam^3/kgK \times (50+273)K}{1000\ kPa} = \mathbf{0.022113\ m^3/kg}$$

c) The percentage difference for parts (a) and (b)

$$\%error_a = \frac{|0.026325 - 0.021796|}{0.021796} \times 100 = \mathbf{20.8\%}$$

$$\%error_b = \frac{|0.022113 - 0.021796|}{0.021796} \times 100 = \mathbf{1.45\%}$$

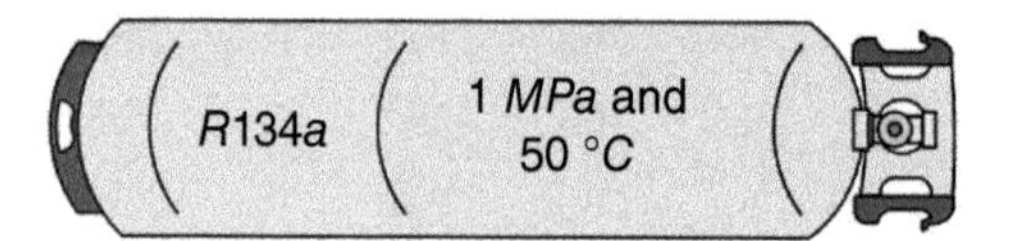

Figure 2.50 Refrigerant cylinder for Example 2.15.

2.8 Closing Remarks

This chapter comes after the fundamentals and makes a prime focus on energy and environmental aspects of thermodynamic systems, the FLT and its applications to closed and open systems, phases and phase changes of pure substances, property diagrams and tables, ideal gas relation, and real gases with compressibility factor. There are illustrative examples presented in every section to illustrate the concepts and definitions and make them clearer to the readers.

Study Questions/Problems

1 Classify which is the dominant mode of heat transfer as shown in Figure 2.12(a) in the following.
  a A spoon getting hot while stirring the food in the pot.
  b Cooling a cup of tea when it is left in a cold windy space.
  c Air conditioning the space.
  d Wind chill effect.
  e Heat from the sun to earth.

2 Classify the following into intensive and extensive properties.
  a Mass
  b Volume
  c Specific volume
  d Specific internal energy.

3 Is it possible to boil water at a temperature of $50\,°C$?

4 Is it possible to boil water at a temperature of $300\,°C$?

5 If a water pot that is open to the atmosphere was first used to boil water in a house on the sea level and then used to boil water on a mountain top, at which location would the water boil at a lower temperature? Explain why.

6 For each of the following determine the phase of water.
  a Temperature of $140\,°C$ and specific volume $0.05\ m^3/kg$.
  b Temperature of $125\,°C$ and pressure of $750\ kPa$.
  c Temperature of $500\,°C$ and specific volume $0.140\ m^3/kg$.
  d Temperature of $190\,°C$ and pressure of $2500\ kPa$.

7 For each of the following determine the pressure of water.
  a Temperature of $140\,°C$ and specific volume $0.05\ m^3/kg$.
  b Temperature of $500\,°C$ and specific volume $0.140\ m^3/kg$.
  c Temperature of $50\,°C$ and specific volume of $7.72\ m^3/kg$.

8 For each of the following determine the temperature of water.
- **a** Pressure of 400 kPa and an internal energy of 1450 kJ/kg.
- **b** Pressure of 4000 kPa and an internal energy 3040 kJ/kg.

9 For each of the following determine the specific internal energy of water.
- **a** Temperature of 140 °C and specific volume 0.05 m^3/kg.
- **b** Temperature of 125 °C and pressure of 750 kPa.
- **c** Temperature of 500 °C and specific volume 0.140 m^3/kg.
- **d** Temperature of 190 °C and pressure of 2500 kPa.
- **e** Pressure of 400 kPa and a specific volume of 0.2 m^3/kg.
- **f** Pressure of 4000 kPa and a specific volume of 0.5 m^3/kg.

10 For each of the systems shown in the figure, using the FLT:
- **a** Write energy balance equation.
- **b** Find the change in the energy of the system.

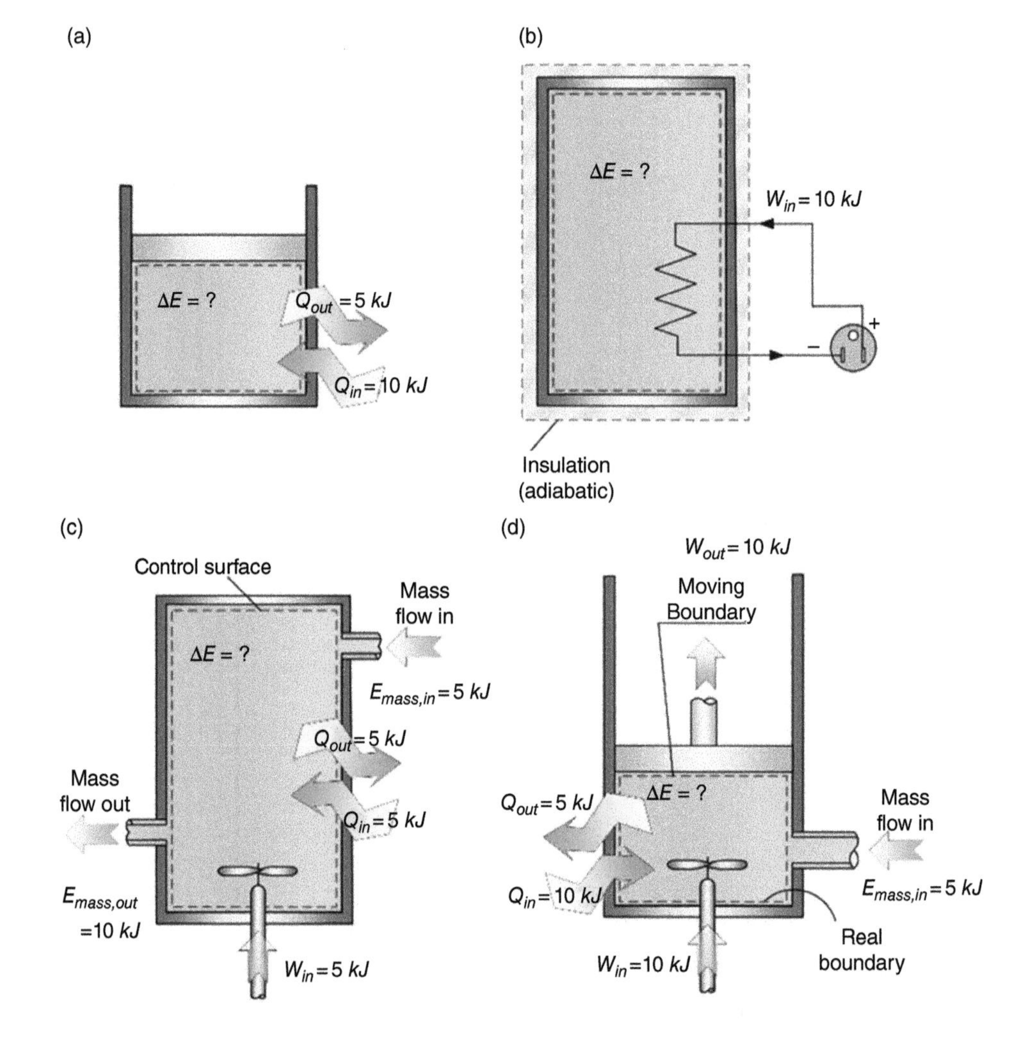

11 Water is heated by an electrical heater by 10 kJ in a piston–cylinder device. The water in the heater undergoes a heating process, where 40 kJ of thermal energy is added from an external heat source to the water and 5 kJ escapes the heater. Calculate the boundary work.

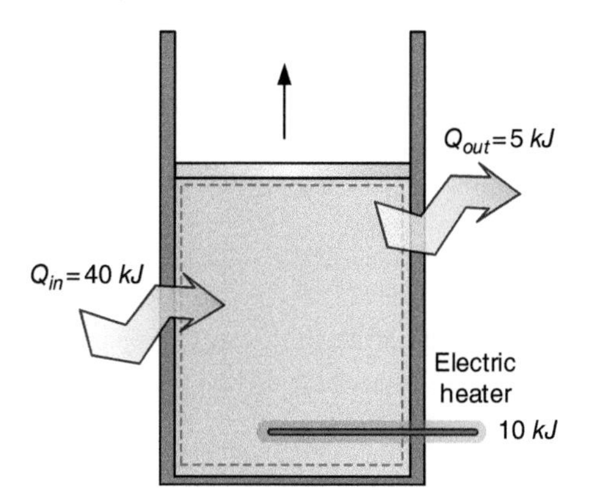

12 Water is heated in a water heater with a stirring paddle wheel. The water heater pan is closed as shown in example schematic. The water in the heater undergoes a heating process, where 40 kJ of thermal energy is added to the water and 5 kJ escapes the heater. If the change in the system energy is 35.5 kJ, then calculate the energy input from the rotating paddle.

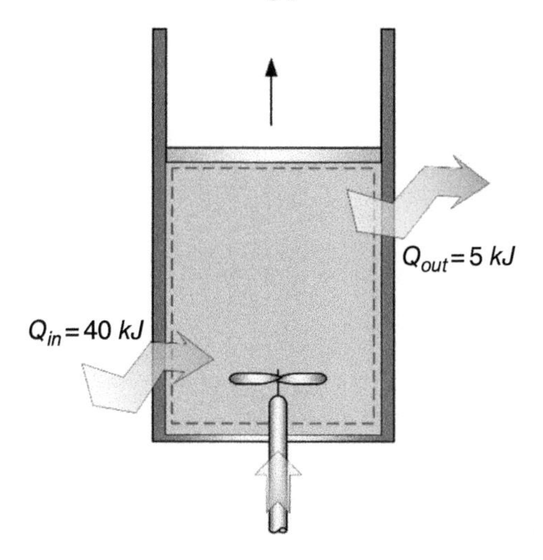

13 A pressurized rigid container is filled with gas and heated by two sources: thermally with 10 kJ and electrically with another 10 kJ. Calculate the amount of increase in internal energy (in kJ).

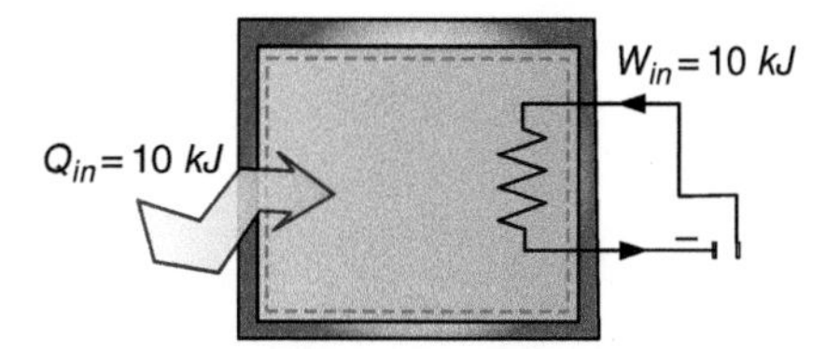

14 A 1 L pressurized container at a pressure of 200 kPa has water in the saturated liquid state. Determine the following properties of the water in the tank.

a Specific volume
b Temperature
c Mass of the water in the container.

15 A 10 L pressurized container at a pressure of 100 kPa has water in the saturated vapor state. Determine the following properties of the water in the tank.

a Specific volume
b Temperature
c Mass of the water in the container.

16 In a container at a pressure of 100 kPa, two thirds of the mass in it is water vapor and the remaining portion is liquid water. Calculate the following for the water in the container.

a Specific volume
b Temperature.

17 The cylinder shown in the figure has water at a pressure of 100 kPa and a temperature of 150 °C. For the water in the cylinder determine the following.

a Specific volume
b Specific internal energy
c Specific enthalpy.

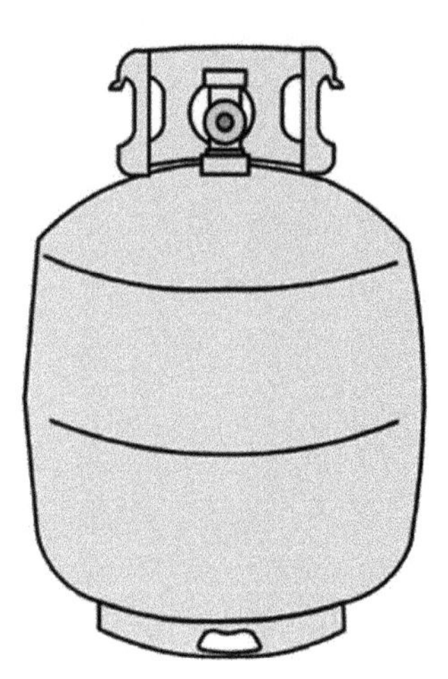

18 A pressurized container at a pressure of 150 kPa has water in the saturated liquid state. Determine the following properties of the water in the tank.

a Specific volume
b Temperature
c Mass of the water in the container if the container volume is 15 L.

19 A 10 *L* piston–cylinder device contains 2 *kg* of saturated liquid-vapor mixture of water at 70 °*C*. The water is slowly heated until it exists in a single phase.

a Determine the final pressure.

b Determine the final temperature.

c Calculate both the final pressure and temperature this time with the same initial properties.

d Plot the process on a *T*-*v* property diagram.

20 A 1 m^3 piston–cylinder device contains 2 *kg* of water at 70 °*C* and a pressure of 1 *atm*. The water is slowly heated/cooled until it exists in a single phase.

a Determine the final pressure.

b Determine the final temperature.

c Calculate both the final pressure and temperature this time with the same initial properties.

d Plot the process on a *T*-*v* property diagram.

21 A pressurized container at a pressure of 100 *kPa* has water in the saturated vapor state. Determine the following properties of the water in the tank.

a Specific volume

b Temperature.

22 A pressurized rigid container initially at a pressure of 400 *kPa* and a temperature of 25 °*C* is shown in the figure. If the temperature of the gas in the container, which is air in this example, is increased to 100 °*C*, calculate the final pressure of the gas container.

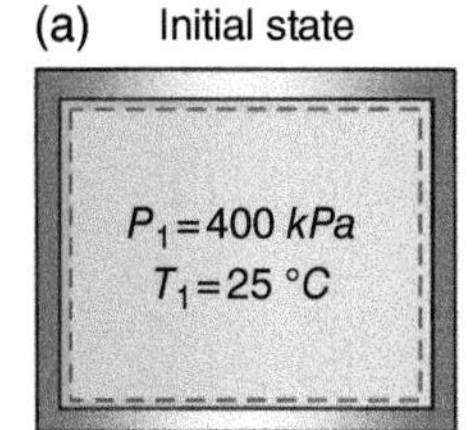

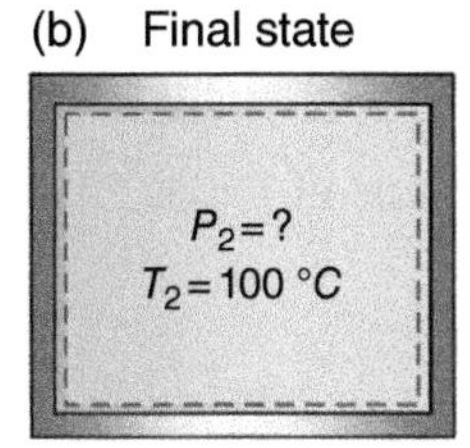

23 In the pressurized container shown in the figure, at a pressure of 160 *kPa* two thirds of the mass is water vapor and the remaining portion is liquid water. Find the saturation temperature and the specific volume for the mixture in the container.

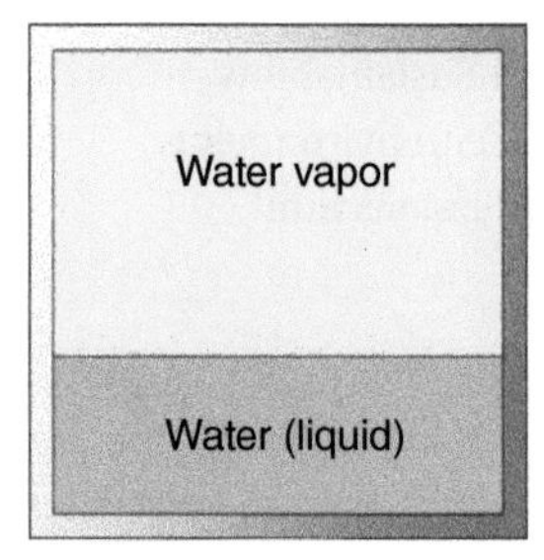

24 A 4 L piston–cylinder device contains 2 kg of saturated liquid-*vapor* mixture of water at 100 °C. The water is slowly heated until it exists in a single phase.
  a Determine the final pressure.
  b Determine the final temperature.
  c Calculate both the final pressure and final temperature this time with the same initial properties.

25 A pressurized rigid container is at a pressure of 305 kPa and a temperature of 30 °C. If the temperature of an ideal gas in the container (air in this example) is increased to 35 °C, determine the final pressure of the gas container.

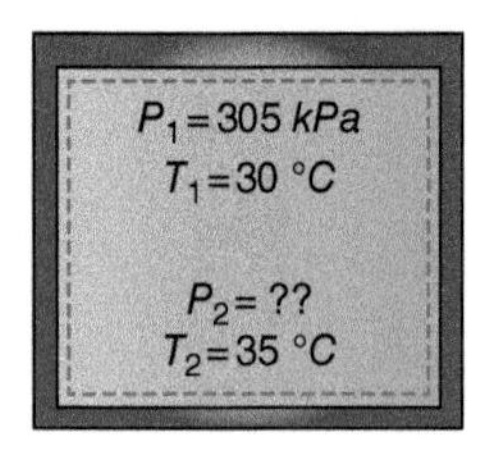

26 Determine whether or not water at a pressure of 1 kPa and a temperature of 100 °C water vapor can be treated as an ideal gas.

27 Determine whether or not water at a pressure of 100 kPa and a temperature of 300 °C water vapor can be treated as an ideal gas.

28 Determine whether or not water at a pressure of 200 kPa and a temperature of 500 °C water vapor can be treated as an ideal gas.

29 Determine whether or not water at a pressure of 10 kPa and a temperature of 100 °C water vapor can be treated as an ideal gas.

30 A refrigerant cylinder contains R-134a at a pressure of 3.5 MPa and a temperature of 75 °C. Calculate: (a) the specific volume using the ideal gas equation of state and (b) the corrected specific volume using the compressibility chart.

Reference

1 Environment and Climate Change Canada, Canadian Environmental Sustainability Indicators: Greenhouse Gas Emissions, 2019. https://www.canada.ca/en/environment-climate-change/services/environmental-indicators/greenhouse-gas-emissions.html (last accessed 13 March 2020).

3

Energy Analysis

3.1 Introduction

Both of the previous chapters have made some introduction to energy and its importance for various systems and applications and discussed both the first and second laws of thermodynamics and their requirements. We really need to go a bit deeper into the systems and start classifying and analyzing them thermodynamically through the first law of thermodynamics (FLT) approach (using the energy principle). This is the main focus in this chapter. There is a need to introduce a step by step approach to reach the goal, which is illustrated in Figure 3.1. The approach starts with a definition of system boundary, then with identification of all inputs and outputs (for mass and energy), and is followed by writing mass and energy balance equations. These are primarily covered in this chapter, and writing entropy and exergy balance equations will come into the picture in forthcoming chapters.

As mentioned in Chapter 1, there are different forms of the energy including heat, work, potential energy, kinetic energy, etc. However, for closed systems energy can cross its boundary in two forms of energy only: heat and work. Open systems can also experience work and heat transfer, in addition to also having energy crossing their boundaries in the form of flow energy with mass leaving and entering the system. So far, we have considered various forms of energy such as heat Q, work W, and total energy E individually, and no attempt has been made to relate them to each other during a process. The FLT, also known as the conservation of energy principle, provides a sound basis for studying the relationships among the various forms of energy and energy interactions.

Based on experimental observations, the FLT states that energy cannot be created or destroyed during a process; it can only change from one form to another. Therefore, every bit of energy should be accounted for during a process. We all know that a rock at some elevation possesses some potential energy, and part of this potential energy is converted to kinetic energy as the rock falls. Experimental data show that the decrease in potential energy (mgz) exactly equals the increase in kinetic energy when air resistance is negligible, thus confirming the conservation of energy principle.

Thermodynamics: A Smart Approach, First Edition. Ibrahim Dincer.

Figure 3.1 A step by step approach to better engage and learn thermodynamics.

3.2 Thermodynamic Systems

As mentioned earlier both in Chapters 1 and 2, thermodynamic systems can generally be classified into two main categories, depending on whether or not mass crosses the boundary of the system, as shown in Figure 3.2. However, there may be special systems, such as isolated systems, where there is no mass transfer and no energy transfer across the boundary. For thermodynamic systems that involves mass entering and leaving its boundary, then such systems are referred to as open systems. *Closed systems* have a constant, fixed amount of mass within the system where mass cannot enter or leave its boundary. *Open systems* are systems that have both mass and energy transfers across the system boundary, so differentiating them from the closed systems.

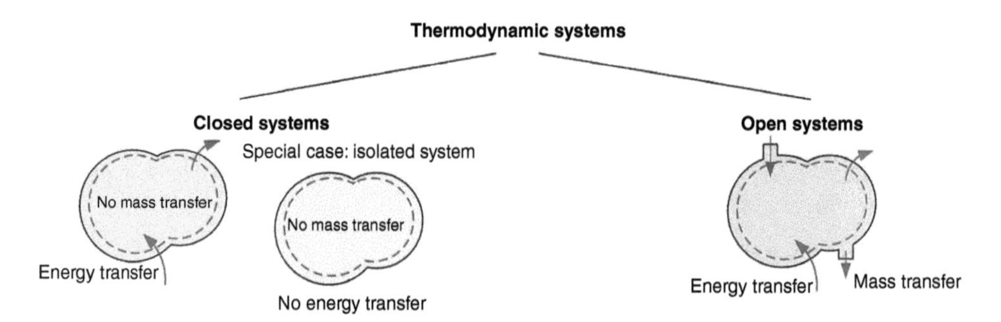

Figure 3.2 Thermodynamics systems can be classified into two main categories.

3.3 Closed Systems

Closed systems can be categorized into two main categories dependent on the nature of the boundary of the system. The two categories are named fixed and moving boundary systems as shown in Figure 3.3.

As shown in Figure 3.3 closed systems can be classified based on the nature of the system boundary. The systems illustrated under fixed boundary mainly refer to rigid tanks where various processes take place, ranging from heating to cooling. Moving boundary systems, from what the name suggests, involve a boundary of the system that can move, or in other words the size of the system boundary can change. Some examples of such devices are a balloon and a piston–cylinder device. The balloon shown in Figure 3.3 is a moving boundary system, since the air or the gas inside the balloon when heated will expand and will push the boundary of the system to the outside. The category of closed systems is the fixed boundary systems, such as a rigid tank.

3.4 Modes of Energy Transfer

In the previous chapter, the types of energy transfers were briefly introduced and discussed from the energy point of view. In conjunction with this, we further dwell on the subject with more systems and examples. Having said that energy transfers in to and out of the system boundary in mainly three different forms, which are shown and summarized in Figure 3.4. The first mode of energy transfer, which occur only in open systems, is associated with mass flow rate. Here, total mass flow energy covers flow enthalpy, flow kinetic energy, and flow potential energy. The second and third images are used to illustrate work transfer for a pump through shaft work and the next for piston–cylinder mechanism through boundary movement work. The last image is related to an open system where the heat transfers are indicated as shown in Figure 3.4.

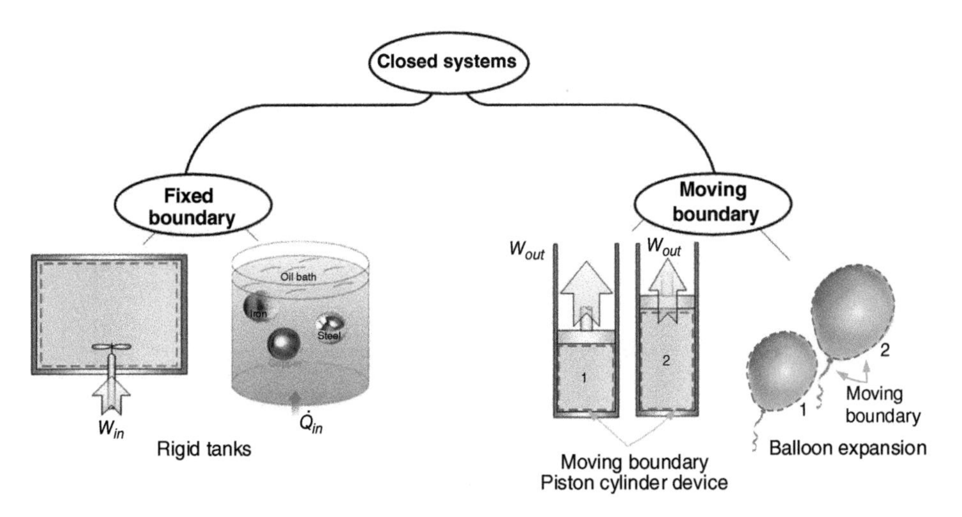

Figure 3.3 Classification of the closed systems based on the nature of their boundary.

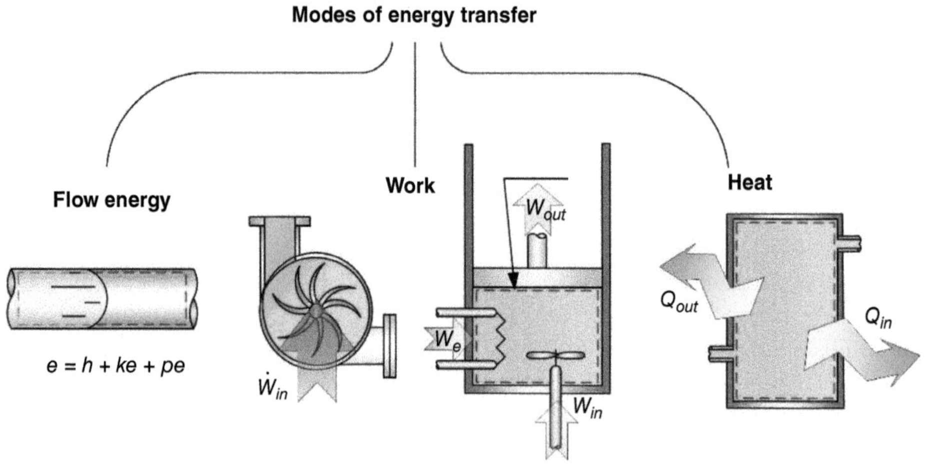

Modes of energy transfer

Flow energy

Work

Heat

$e = h + ke + pe$

$\dot{W}_{in}$ W_{out} W_e W_{in} Q_{out} Q_{in}

Figure 3.4 Common modes of energy transfer where there is a flow energy illustrated for a pipe flow; there are pump and piston-cylinder mechanism with shaft, electrical and boundary movement works involved; and there is a tank with incoming and outgoing heat given..

3.5 Types of Works

There are different forms of how the energy in the form of work can enter or leave the boundary of a closed thermodynamic system; these different forms are presented in Figure 3.5. Shaft work, electrical work and boundary work are considered three different forms of work that can enter or leave the system boundary. In regards to shaft and electrical works, one can define them respectively as follows:

$$W_{sh} = V\Delta P \tag{3.1}$$

$$W_e = VI\Delta t \tag{3.2}$$

where the pressure difference is defined in two forms: $\Delta P = (P_2 - P_1)$ for pumping and compression, and $\Delta P = (P_1 - P_2)$ for expansion. In addition, the time difference is defined as $\Delta t = (t_2 - t_1)$.

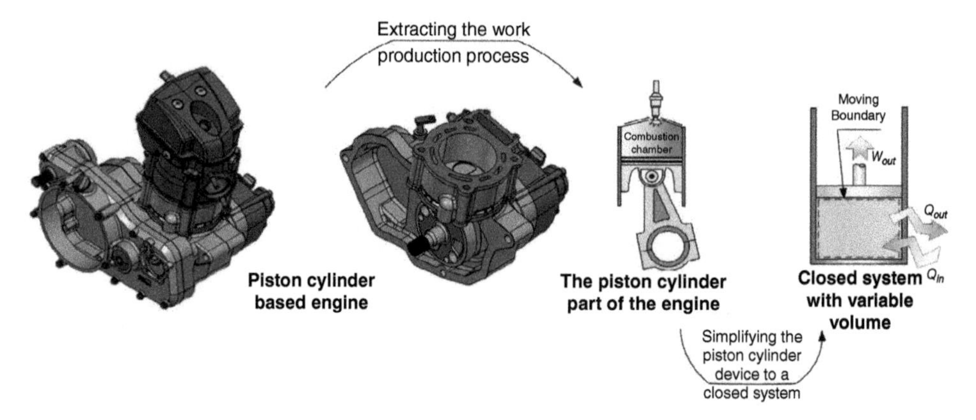

Figure 3.5 Simplification of complex energy systems to closed systems, where the example here is an internal combustion engine based on piston–cylinder devices to a closed system with a moving a boundary.

Regarding the boundary work the next subsection is dedicated toward the boundary work and how to calculate it as well as its different cases.

3.5.1 Boundary Movement Work

Internal combustion vehicles and power production processes are based on mechanical work produced by the movement of the boundary of what can be approximated as closed systems. We can see in Figure 3.5 how a piston–cylinder device such as an internal combustion engine can be simplified to a closed system with a moving boundary. The closed systems with moving boundaries can either produce or consume boundary work.

The boundary work or work consumed by a piston–cylinder device that is approximated as a closed system is derived in this section. First the work produced or consumed by the movement of the boundary of a closed system is referred to as boundary work. For a piston–cylinder device shown in Figure 3.6, the mechanical work can be presented in term of small displacement of the piston as follows:

$$\delta W_b = Fds \tag{3.3}$$

Here, W denotes work and the subscript b denotes boundary, F denotes force, and ds is the small finite differential increment of distance.

As shown in Figure 3.6, usually the area of the piston is constant in most piston–cylinder devices. For a constant piston cross-sectional area (A), Eq. (3.3) can be reduced by realizing that the pushing force on the piston comes from the pressure inside the system contained between the piston and the cylinder (Figure 3.6) as follows:

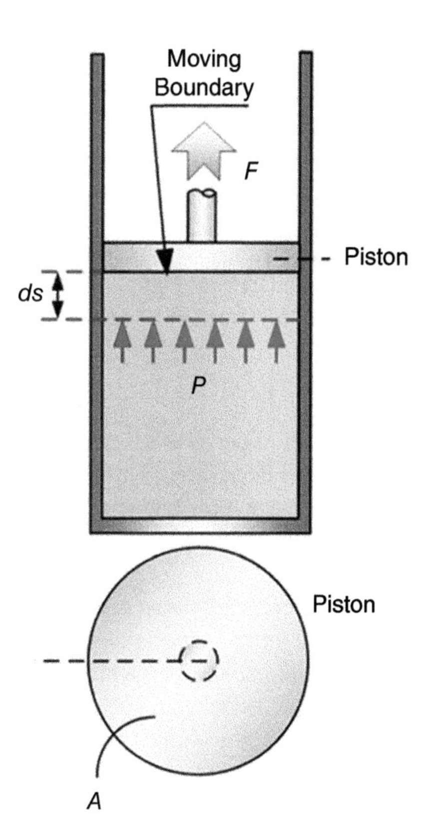

Figure 3.6 Piston–cylinder device expansion or contraction is associated with δW_b mechanical boundary work.

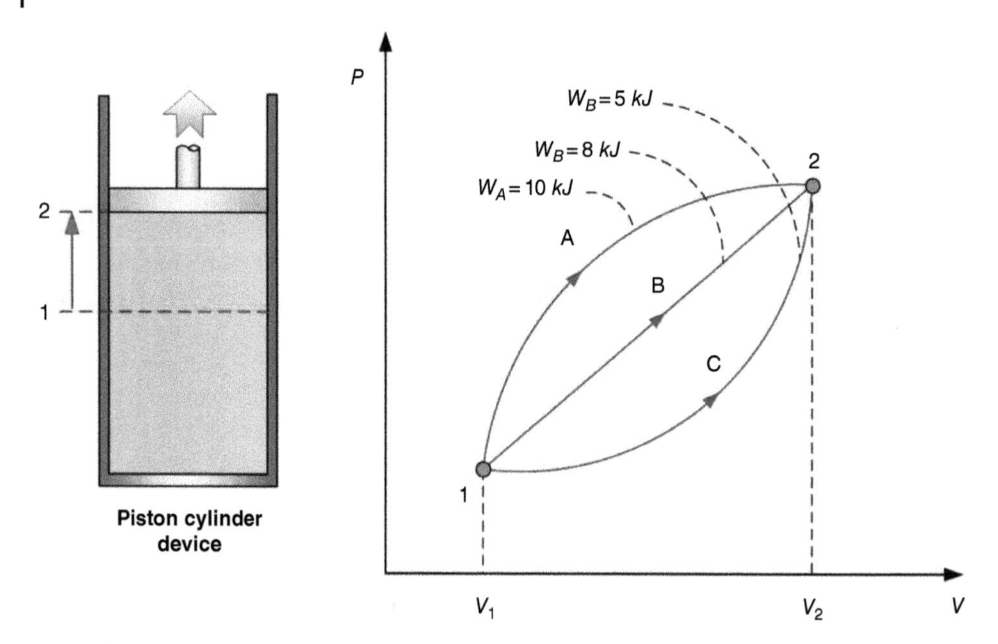

Figure 3.7 The work done by the expanding piston–cylinder device is a path function.

$$\delta W_b = PAds \tag{3.4}$$

where P is pressure applied to the cross-sectional area of the piston (kPa), A is cross-sectional area of the piston (m^2) and ds is small distance traveled by the piston (m).

Here, W denotes work and the subscript b denotes boundary, F denotes force, ds is the small finite differential increment of distance, P is for pressure, and A is for the area of the piston.

Equation (3.4) can be further manipulated to combine the piston area with the finite piston displacement to present the boundary work in terms of pressure and volume:

$$\delta W_b = PdV \tag{3.5}$$

The boundary work done by or done on a piston–cylinder device can be calculated through integrating Eq. (3.5) from one state point to another as follows:

$$W_b = \int_1^2 P_i dV \tag{3.6}$$

where the boundary work is calculated between states 1 and 2. Note that since the boundary work is calculated through integration it means that the boundary work is a path function, which means that the boundary work produced changes between the same two states with different paths followed between the two states as shown in Figure 3.7.

3.5.1.1 Processes Related to Boundary Movement Work

In this section, four key processes – isobaric (constant-pressure), isothermal (constant-temperature), polytropic (which is defined as a process obeying $PV^n = constant$), and adiabatic (which is defined a process with no heat transfer as obeying $PV^k = constant$) – are considered for the boundary movement work, particularly related to the systems where ideal gases are employed.

(a) Isobaric compression and expansion

During a compression or expansion process of gases, the pressure and volume are often related by the ideal gas law if we assume the gases behave as an ideal gas as follows:

$$PV = mRT \tag{3.7}$$

However, there are special cases where the compression or the expansion process occurs at a constant pressure or what can be referred to as **isobaric** compression or expansion, which leads to having Eq. (3.7) rearranged to:

$$\frac{V}{T} = \frac{mR}{P} = C \tag{3.8}$$

Here, C is constant. Since it is a piston–cylinder device then the boundary work can be calculated through integrating the pressure inside the piston–cylinder device with respect to the variation of the piston–cylinder volume:

$$W_b = \int_1^2 P_i dV \tag{3.9}$$

Since the piston–cylinder device in the example (Figure 3.4) expands at a constant pressure then the above equation can be represented as:

$$W_b = P\int_1^2 dV = P(V_2 - V_1) \tag{3.10}$$

Since the mass of the water inside the system does not change throughout the process:

$$W_b = mP(v_2 - v_1) \tag{3.11}$$

(b) Isothermal compression and expansion

In special cases the compression or expansion is carried out at a constant temperature, either by having the system rejecting or receiving heat, in such cases these systems are referred to undergo an **isothermal** process. In the case where fluid in the piston–cylinder device is an ideal gas, since it is an ideal gas and the temperature throughout the process then the following is true:

$$PV = mRT_c = C \tag{3.12}$$

We can see that in an isothermal expansion or compression process in a piston–cylinder device, the multiplication of the pressure and the volume at any point through the expansion or the compression process remains constant for the case where the fluid inside the device is an ideal gas, then:

$$W_b = \int_1^2 P_i dV = \int_1^2 \frac{C}{V} dV = C\int_1^2 \frac{1}{V} dV = C \times \ln\left(\frac{V_2}{V_1}\right) = P_1 V_1 \times ln\left(\frac{V_2}{V_1}\right)$$

$$W_b = P_1 V_1 \times \ln\left(\frac{V_2}{V_1}\right) \tag{3.13}$$

(c) Polytropic compression and expansion

For actual compression and expansion processes in a gas-based piston–cylinder device, the pressure and the volume inside the device are related by the following equation:

$$PV^n = C \tag{3.14}$$

Here, n is the ploytropic constant. Eq. (3.14) can be rearranged to have the pressure as a function of volume as:

$$P = CV^{-n} \tag{3.15}$$

By substituting the above equation in the boundary work we get:

$$W_b = \int_1^2 PdV = \int_1^2 CV^{-n}dV = C\frac{V_2^{-n+1} - V_1^{-n+1}}{-n+1} = \frac{P_2V_2 - P_1V_1}{1-n}$$

The above boundary work equation of polytropic processes can be rewritten using the ideal gas flow to present the work in terms of temperature and mass as:

$$W_b = \frac{mR(T_2 - T_1)}{1-n} \tag{3.16}$$

(d) Adiabatic compression and expansion

For actual compression and expansion processes in a gas-based piston–cylinder device, the pressure and the volume inside the device are related by the following equation:

$$PV^k = C \tag{3.17}$$

Here, k is the adiabatic constant. Eq. (3.17) can be rearranged to have the pressure as a function of volume as follows:

$$P = CV^{-k} \tag{3.18}$$

By substituting the above equation in the boundary work we get:

$$W_b = \int_1^2 PdV = \int_1^2 CV^{-k}dV = C\frac{V_2^{-k+1} - V_1^{-k+1}}{-k+1} = \frac{P_2V_2 - P_1V_1}{1-k}$$

The above boundary work equation of polytropic processes can be rewritten by using the ideal gas flow to present the work in terms of temperature and mass as:

$$W_b = \frac{mR(T_2 - T_1)}{1-k} \tag{3.19}$$

Finally, it is important to compare these four different processes to show the trend lines of their behaviors, as illustrated in Figure 3.8.

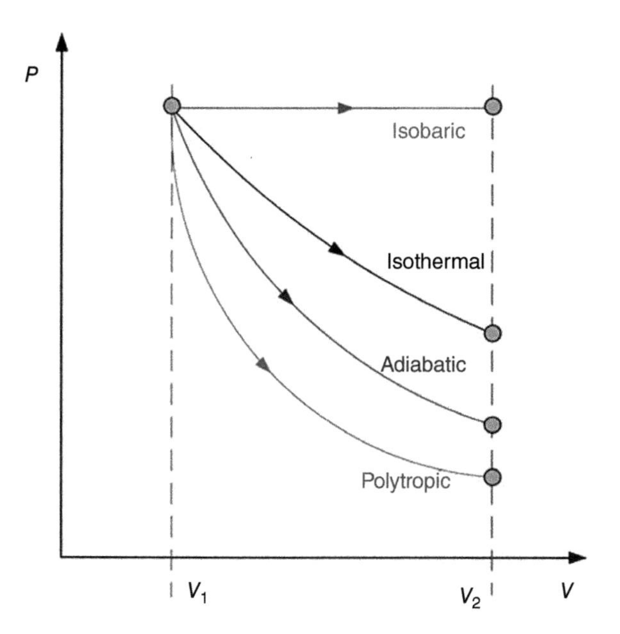

Figure 3.8 Comparisons of polytropic, adiabatic, isothermal, and isobaric expansion processes on a P–V diagram.

Example 3.1 Consider a piston–cylinder device expanding at a constant pressure of 100 *kPa* as shown in Figure 3.9. The piston–cylinder device contains 1 *kg* of water. The water inside the piston is initially superheated water at a temperature of 200 °*C*. The water inside the piston is heated by an external heat source until the water in the system reaches a temperature of 300 °*C*. Calculate the boundary work produced by the system based on an isobaric process.

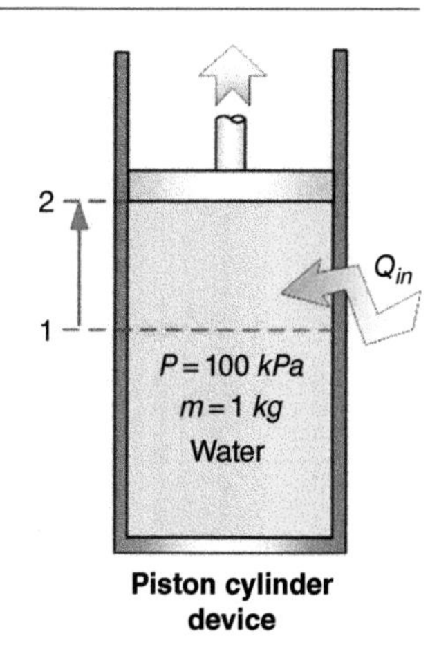

Figure 3.9 Schematic diagram of the piston–cylinder device discussed in Example 3.1.

Solution

Since it is a piston–cylinder device then the boundary work can be calculated through integrating the pressure inside the piston–cylinder device with respect to the variation of the piston–cylinder volume as follows:

$$W_b = \int_1^2 P_i dV$$

Since the piston–cylinder device in the example, shown in Figure 3.9, expands at a constant pressure then the above equation can be represented as:

$$W_b = P\int_1^2 dV = P(V_2 - V_1)$$

Since the mass of the water inside the system does not change throughout the process:

$$W_b = mP(v_2 - v_1)$$

State 1:

$$\left.\begin{aligned} P_1 &= 100\ kPa \\ T_1 &= 200\ ^\circ C \end{aligned}\right\} v_1 = 2.172\ \frac{m^3}{kg}$$

State 2:

$$\left.\begin{aligned} P_2 &= 100\ kPa \\ T_2 &= 300\ ^\circ C \end{aligned}\right\} v_2 = 2.639\ \frac{m^3}{kg}$$

$$W_b = mP(v_2 - v_1) = 1\ kg \times 100\ kPa \times (2.639 - 2.172)\frac{m^3}{kg} = \mathbf{46.7\ kJ}$$

Example 3.2 Consider this time a piston–cylinder device filled with air which is compressed at a constant temperature to change its volume from 0.4 m^3, where the pressure of the piston is 200 kPa, to a volume of 0.1 m^3. Figure 3.10 shows both the piston–cylinder mechanism along with its P–V diagram for the process. Assume that the air in the piston–cylinder device can be treated as an ideal gas through the compression process. Find the boundary work consumed by this compression process of the piston–cylinder device discussed.

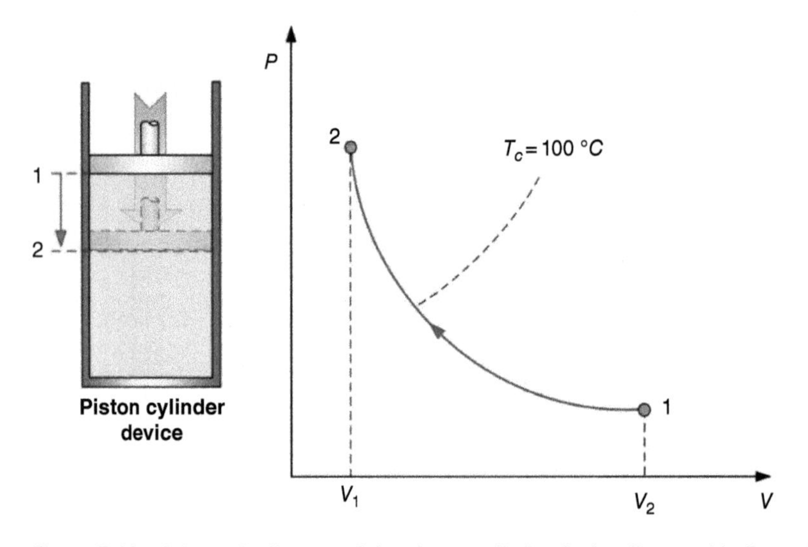

Figure 3.10 Schematic diagram of the piston–cylinder device discussed in Example 3.2.

Solution

Since we can treat air as an ideal gas and the temperature is constant throughout the process then the following is true:

$$PV = mRT_c = constant = C$$

Then we can see that in an isothermal expansion or compression process in a piston–cylinder device the multiplication of the pressure and the volume at any point through the expansion or the compression process remains constant. The relation is only true for the case where the fluid inside the device is an ideal gas, therefore the relation can be written as:

$$W_b = \int_1^2 P_i dV = \int_1^2 \frac{C}{V} dV = C\int_1^2 \frac{1}{V} dV = C \times ln\left(\frac{V_2}{V_1}\right) = P_1 V_1 \times ln\left(\frac{V_2}{V_1}\right)$$

$$W_b = P_1 V_1 \times ln\left(\frac{V_2}{V_1}\right) = 200\ kPa \times 0.4\ m^3 \times \left(ln\left(\frac{0.1}{0.4}\right)\right)$$

$$\mathbf{W_b = -111\ kJ}$$

Example 3.3 Consider this time an adiabatic piston–cylinder device filled with air, as shown in Figure 3.11, which is compressed to change its volume from 0.4 m^3, where the pressure of the piston is 200 kPa, to a volume of 0.1 m^3. The temperature is initially kept at 100 °C. Assume that the air in the piston–cylinder device can be treated as an ideal gas through the compression process. Find the boundary work consumed by the compressing piston–cylinder device based on an adiabatic process.

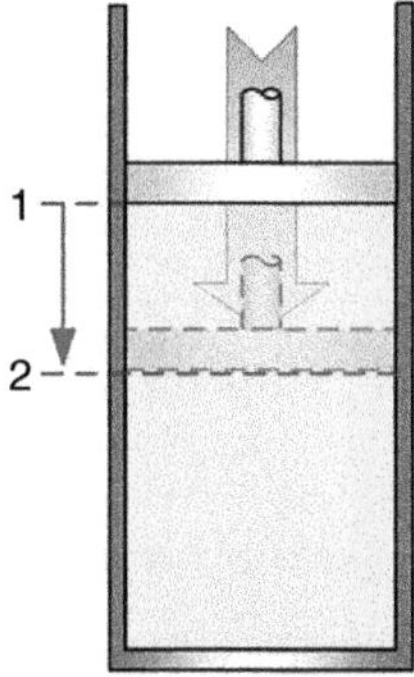

Figure 3.11 Schematic diagram of the piston–cylinder device discussed in Example 3.3.

Solution

Since the process is an adiabatic process:

$$P_2 V_2^k = P_1 V_1^k$$

By rearranging it:

$$\left(\frac{P_2}{P_1}\right) = \left(\frac{V_1}{V_2}\right)^k \rightarrow \frac{P_2}{200} = \left(\frac{0.4}{0.1}\right)^{1.4}$$

$$P_2 = 1392.9\ kPa$$

For an adiabatic process:

$$W_{in} = W_b = -\frac{P_2 V_2 - P_1 V_1}{1-k} = \mathbf{148.23\ kJ}$$

Example 3.4 A piston–cylinder device with an initial volume of 0.2 m^3 contains helium gas initially at 10 °C and 100 kPa (Figure 3.12). The helium is compressed through a polytropic process ($PV^n = constant$) to a pressure of 700 kPa and a temperature of 290 °C. Calculate the boundary work done on this system during the polytropic process.

Solution

The properties of the system at the initial state of the system are found at 10 °C and 100 kPa using the ideal gas equation:

$$v_1 = \frac{RT_1}{P_1} = \frac{2.0769\ \frac{kJ}{kgK} \times (10 + 273.15)K}{100\ kPa} = 5.881\ \frac{m^3}{kg}$$

$$v_1 = \frac{V}{m}$$

$$5.881\ \frac{m^3}{kg} = \frac{0.2\ m^3}{m}$$

$$m = 0.034\ kg$$

$$\frac{P_1V_1}{T_1} = \frac{P_2V_2}{T_2} \rightarrow V_2 = \frac{T_2P_1}{T_1P_2}V_1 = \frac{563\ K}{283\ K} \times \frac{100\ kPa}{700\ kPa} \times 0.2\ m^3 = 0.05684\ m^3$$

Since the process is a polytropic process:

$$P_2V_2^n = P_1V_1^n$$

By rearranging it:

$$\left(\frac{P_2}{P_1}\right) = \left(\frac{V_1}{V_2}\right)^n \rightarrow \frac{700}{100} = \left(\frac{0.2}{0.05684}\right)^n$$

Solving for n:

$$n = 1.547$$

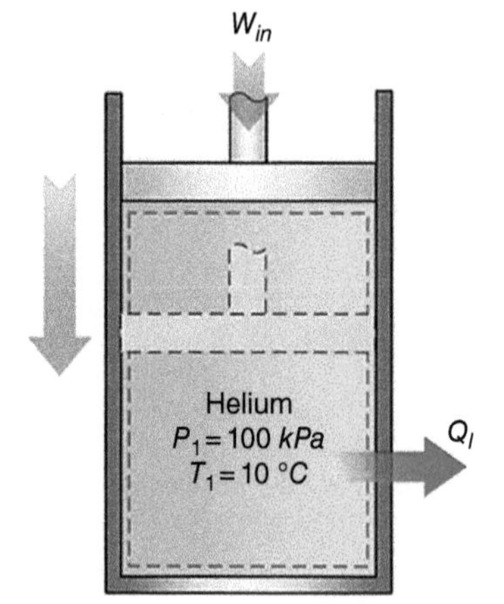

Figure 3.12 Schematic diagram of the piston–cylinder device discussed in Example 3.4.

For a polytropic process:

$$W_{in} = W_b = -\frac{mR(T_2 - T_1)}{1-n} = \mathbf{36.19\ kJ}$$

3.6 Energy Balance Equation for Closed Systems

In this section, the main focus will be on the FLT and its applications to closed systems through energy balance equations (EBEs). As stated earlier, the mass remains constant within the system without crossing the boundary while either work or heat or both can cross the boundary, as illustrated in Figure 3.13 for a moving boundary system. Since there is conservation of energy due to the FLT, all energy inputs will be equivalent to all energy outputs.

As was shown in the earlier sections, a closed system can either be a fixed boundary system or an open boundary system. In this section the energy balance will be applied to various closed systems with fixed boundaries as well.

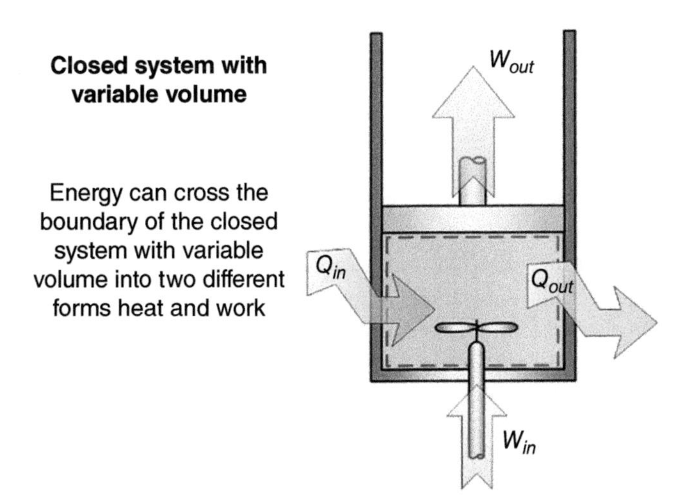

Figure 3.13 Different forms of energy can cross the boundary of the closed system given here while mass cannot.

Example 3.5 An electrically heated and mechanically stirred rigid tank is shown in Figure 3.14. The water in this rigid tank is initially at a pressure of 1 *atm* and a temperature of 25 °*C* with a volume of 10 *L*. A total of 50 *kJ* of heat is transferred to the water while losing 10% of it during the heating process. An electrical heater is used to heat the water for 10 seconds and draws 1 *A* of current at a voltage of 120 *V*. If the rotating shaft adds a total of 100 *kJ* of shaft work to the water in the tank while stirring it. (a) Write the mass and energy balance equations and (b) calculate the final temperature of the tank.

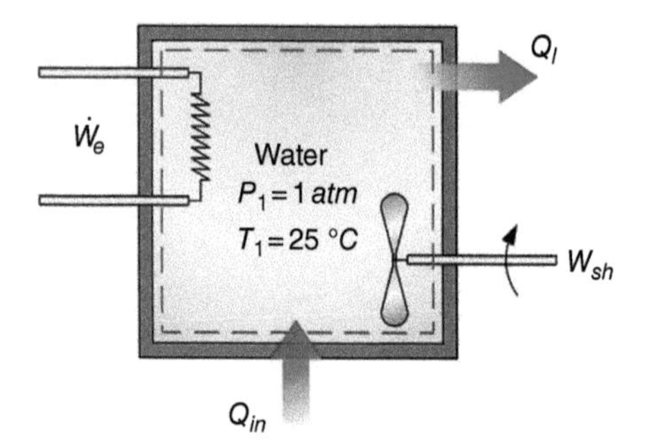

Figure 3.14 A schematic diagram of the rigid tank discussed in Example 3.5.

Solution

a) Write the mass and energy balance equations.

The change of the kinetic energy of the system can be neglected since the system velocity does not change (not accelerating nor decelerating) throughout the process. The change in the potential energy of the system can be neglected as well since the system elevation does not change throughout the process. Based on the aforementioned assumptions the mass and energy balance can be written as follows:

$$\text{Mass Balance Equation (MBE)} : m_1 = m_2 = constant$$

$$\text{Energy Balance Equation (EBE)} : Q_{in} + W_e + W_{sh} + m_1 u_1 = m_2 u_2 + Q_l$$

where $W_e = \dot{W}_e \times \Delta t$

b) Calculate the final temperature of the tank.

From the property tables (such as Appendix B-1b) or by using a software that provides properties such as Engineering Equation Solver (EES), the saturation temperature of water at a pressure of 1 atm = 101.325 kPa is found to be 99.61 °C. Since the temperature of water is lower than the saturation temperature, then it is in the compressed liquid state. Then we go to the compressed liquid tables or another properties source to extract the properties for the subcooled water. In most published subcooled property tables (such as Appendix B-1d), the tables commonly begins with high pressures of 5 MPa. In such cases either a property database software such as EES or NIST refprop can be used. Another option is to approximate the properties as the saturated liquid properties at the given temperature (25 °C)

$$u_1 = u_f @25\ ^oC = 104.7\ kJ/kg$$

$$v_1 = v_f @25\ ^oC = 0.001003\ m^3/kg$$

$$V = 10\ L \times 1\ \frac{m^3}{1000\ L} = 0.01\ m^3$$

$$v_1 = \frac{V_1}{m_1} \rightarrow 0.001003 = \frac{0.01}{m_1}$$

$m_1 = 9.970\ kg = m_2 = m$ since the system is a closed system where no mass enters or leaves the system.

There is a need to substitute these values into the EBE to further calculate what is needed to be found in terms of energy.

Since the process occurred in a rigid tank than it is a constant volume process, which means we can use the following: $u = c_v T$

$$Q_{in} + \dot{W}_e \times \Delta t + W_{sh} + m_1 u_1 = m_2 u_2 + Q_l$$

$$Q_{in} + \dot{W}_e \times \Delta t + W_{sh} - Q_l = \Delta U = m(u_2 - u_1)$$

$$Q_{in} + \dot{W}_e \times \Delta t + W_{sh} - Q_l = mc_v(T_2 - T_1)$$

Using the property table (such as Appendix A1-a) values, we can find the specific heat at a constant volume, which is taken at the initial temperature (25 °*C*)

$$c_v = 4.18 \frac{kJ}{kg\ °C}$$

$$50\ kJ + \left(1A \times 120\ V \times \frac{1\frac{kJ}{s}}{1000\ AV} \times 10\ s\right) + 100\ kJ - 0.1 \times 50\ kJ$$

$$= 9.970\ kg \times 4.18 \frac{kJ}{kg\ °C}(T_2 - 25\ °C)$$

$$\mathbf{T_2 = 28.51\ °C}$$

3.7 Specific Heat Capacities

The amount of energy that is required to raise the temperature of unit mass of a substance by one degree is different for each substance. For example, 1 *kg* of water needs to be supplied with 4.18 *kJ* of energy to raise its temperature from 25 to 26 °*C*, which is an increase in temperature by one degree Celsius. However, if we take 1 *kg* of octane only half of the energy provided to water is needed for it to experience the same increase in temperature. For both examples of specific heat capacity mentioned, they are considered to have maintained a constant pressure. Thus, the amount of energy required to raise the temperature of 1 *kg* of a substance by one degree Celsius or Kelvin is called *specific heat capacity*. There are two kinds of specific heat capacities, *constant volume specific heat* (at constant volume with no boundary change) c_v and *constant pressure specific heat* (at a constant pressure with boundary change) c_p.

However, if the heating or cooling process occurs at a constant volume rather than a constant pressure then the specific heat capacity at a constant volume will be lower than that at a constant pressure. The specific heat capacity at a constant volume is:

$$du = c_v dT \tag{3.20}$$

Here, c_v is the specific heat capacity at a constant volume. However, the enthalpy is presented in terms of the specific heat capacity at a constant pressure as:

$$dh = c_p dT \tag{3.21}$$

For specific cases where the specific heat capacity does not change with the variation of temperature or there is a negligible change in the specific heat capacity with the temperature.

a) Specific Heat Capacity Relations for Ideal Gases

By using the definition of enthalpy, it is seen that the enthalpy is dependent on internal energy; and in order to define the internal energy and enthalpy in terms of specific heat capacities, we can form a relationship between the two specific heat capacities as:

$$h = u + Pv \tag{3.22}$$

and for ideal gases we can relate the temperature, pressure, and specific volume by using the ideal gas law as:

$$Pv = RT \tag{3.23}$$

Then we can replace the energy flow term from the enthalpy definition with RT as follows:

$$h = u + RT \tag{3.24}$$

Then by substituting in Eq. (3.24) the enthalpy and the internal energy by their specific heat capacity-based function the resulting expression will be as:

$$c_p T = c_v T + RT \tag{3.25}$$

By dividing Eq. (3.25) with the temperature we will obtain a relationship between the two specific heat capacities for ideal gases as follows:

$$c_p = c_v + R \tag{3.26}$$

Example 3.6 Air that behaves as an ideal gas is heated at a constant pressure of 200 *kPa* from a temperature of 300 to 600 *K*. Calculate the change in the internal energy of the air.

Solution

Finding the change of the internal energy by using the specific heat capacities can be done as follows:

$$du = c_v dT$$

$$\int_{u_1}^{u_2} du = \int_{T_1}^{T_2} c_v dT$$

If we assume that the specific heat capacity does not vary with temperature or the variation of the specific heat capacity from temperature T_1 to T_2 is negligible, then the above integration will be solved as follows:

$$\Delta u = \int_{T_1}^{T_2} c_v dT = c_v (T_2 - T_1)$$

where dT can be taken as the average temperature between T_1 and T_2. However, dT can also be taken as the given temperature, if one of the state temperatures is missing. Further iterations can also be performed to obtain a more accurate result. In this example both the initial and final temperatures are given then the specific heat is taken from tables (such as Appendix A3) or from EES based on the average temperature:

$$T_{avg} = \frac{T_2 + T_1}{2} = \frac{300 + 600}{2} = 450\ K$$

$$\Delta u = c_{v,avg}(T_2 - T_1) = 0.733\ \frac{kJ}{kgK} \times (600 - 300)K = \mathbf{220}\ \frac{\mathbf{kJ}}{\mathbf{kg}}$$

Example 3.7 A 10 *kg* oil bath, as shown in Figure 3.15, contains three balls: a 1 *kg* steel ball, a 1 *kg* copper ball, and a 1 *kg* iron ball. The oil and three balls are initially at 25 °*C*. If the bath is heated to a temperature of 75 °*C* at a constant pressure of 101.325 *kPa*, (a) write the mass and energy balance equations and (b) find the amount of heat transferred to the oil bath.

Solution

a) Write the mass and energy balance equations as follows:

$$\text{MBE}: m_{i,1} = m_{i,2} = m_i, m_{s,1} = m_{s,2} = m_s, m_{c,1} = m_{c,2} = m_c, and\, m_{o,1} = m_{o,2} = m_o$$

$$\text{EBE}: m_i u_{i,1} + m_s u_{s,1} + m_c u_{c,1} + m_o u_{o,1} + Q_{in} = m_i u_{i,2} + m_s u_{s,2} + m_c u_{c,2} + m_o u_{o,2}$$

where *i* for iron, *s* for steel, *c* for copper, and *o* for oil.

b) Find the amount of heat transferred to the bath from the following equation:

$$m_i u_{i,1} + m_s u_{s,1} + m_c u_{c,1} + m_o u_{o,1} + Q_{in} = m_i u_{i,2} + m_s u_{s,2} + m_c u_{c,2} + m_o u_{o,2}$$

$$m_i c_{v,i} T_1 + m_s c_{v,s} T_1 + m_c c_{v,c} T_1 + m_o c_{v,o} T_1 + Q_{in} = m_i c_{v,i} T_2 + m_s c_{v,s} T_2 + m_c c_{v,c} T_2 + m_o c_{v,o} T_2$$

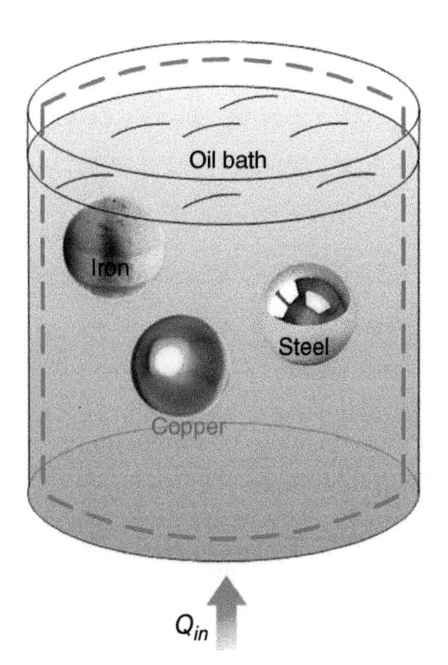

Figure 3.15 A schematic diagram of the rigid tank discussed in Example 3.7.

Since solids and liquids are usually considered incompressible substances, then $c_v = c_P = c$

$$(1 \times 0.45 \times 298.15) + (1 \times 0.5 \times 298.15) + (1 \times 0.386 \times 298.15) + (10 \times 1.8 \times 298.15) + Q_{in} = (1 \times 0.45 \times 348.2) + (1 \times 0.5 \times 348.2) + (1 \times 0.386 \times 348.2) + (10 \times 1.8 \times 348.2)$$

$$\mathbf{Q_{in} = 967.8\ kJ}$$

Example 3.8 Here, reconsider Example 3.4 with a piston–cylinder device with an initial volume of 0.2 m^3 containing helium gas initially at 10 °C and 100 kPa. The helium is now compressed through a polytropic process ($PV^n = constant$) to a pressure of 700 kPa and a temperature of 290 °C. (a) Write the mass and energy balance equations and (b) calculate the heat loss.

Solution

a) Write the mass and energy balance equations.

$$\text{MBE}: m_1 = m_2 = constant$$
$$\text{EBE}: m_1u_1 + W_{in} = m_2u_2 + Q_l$$

b) Calculate the heat loss.

From the solution of Example 3.4 we found that the boundary work was equal to 36.19 kJ, which is an input to the system. Then by using the EBE and some further rearrangement of the terms we can find the heat loss as follows:

$$m_1u_1 + W_{in} = m_2u_2 + Q_l$$

Taking the initial internal energy to the right side of the equation:

$$W_{in} = m_2u_2 - m_1u_1 + Q_l$$
$$W_{in} = mc_v(T_2 - T_1) + Q_l$$
$$Q_l = W_{in} - mc_v(T_2 - T_1)$$
$$\mathbf{Q_l = 6.51\ kJ}$$

Example 3.9 A piston–cylinder device, as shown in Figure 3.16, contains a total of 12 kg of refrigerant $R134a$ in saturated vapor state at a pressure of 240 kPa. An electrical heater is turned on for 360 *seconds* while connected to 110 V electrical source having a 12.8 A current running through the heater; in addition to the electrical heater 300 kJ of heat is transferred to the refrigerant during the same duration. (a) Write the mass and energy balance equations, (b) calculate the final state temperature and pressure, and (c) calculate the boundary work done by the system during the expansion process.

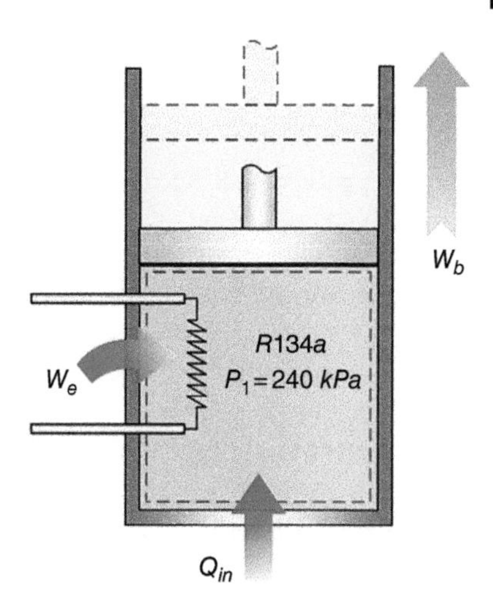

Figure 3.16 Schematic of the rigid tank discussed in Example 3.9.

Solution

a) Write the mass and energy balance equations.

The change of the kinetic energy of the system can be neglected since the system velocity does not change (not accelerating nor decelerating) throughout the process. The change in the potential energy of the system can also be neglected as well since the system elevation does not change throughout the process. Based on the above-mentioned assumptions the mass and energy balance can be written as:

$$\text{MBE}: m_1 = m_2 = const.$$

$$\text{EBE}: Q_{in} + W_e + m_1 u_1 = m_2 u_2 + W_b$$

where $W_e = \dot{W}_e \times \Delta t$

The boundary work is achieved in a constant pressure process, then:

$$W_b = P \times (V_2 - V_1) = PV_2 - PV_1$$

and by using the definition of the enthalpy:

$$H = U + PV$$

Then, using the EBE:

$$Q_{in} + \dot{W}_e \times \Delta t + U_1 = U_2 + (PV_2 - PV_1)$$

$$Q_{in} + \dot{W}_e \times \Delta t = U_2 - U_1 + (PV_2 - PV_1)$$

$$= (U_2 + PV_2) - (U_1 + PV_1)$$

$$Q_{in} + \dot{W}_e \times \Delta t = H_2 - H_1$$

$$\boldsymbol{Q_{in} + \dot{W}_e \times \Delta t = m \times (h_2 - h_1)}$$

b) Calculate the final state pressure and temperature.

$$Q_{in} = 300\ kJ$$

$$\dot{W}_e = IV = 12.8\ A \times 110\ V \times \frac{1\ kJ/s}{1000\ AV} = 1.408\ kJ/s$$

The $R134a$ properties are at the initial state, with a saturated vapor ($x = 1$) and a pressure of 240 kPa, then:

$$h_1 = h_g@240\ kPa = 247.32\ \frac{kJ}{kg}$$

By substituting the values of terms in the EBE:

$$300\ kJ + \left(1.408\ \frac{kJ}{s} \times 360\ s\right) = 12\ kg \times \left(h_2 - 247.32\ \frac{kJ}{kg}\right)$$

$$h_2 = 314.56\ kJ/kg$$

Since the device is a piston–cylinder device, then the process occurs at a constant pressure

$$\boldsymbol{P_2 = 240\ kPa}$$

In order to find the temperature and pressure of the final state, we need to define the state.

There are various ways of finding the final states temperature and pressure in the example; as it depends on what data are available to obtain the corresponding properties. If the property tables are available as discussed and presented in Chapter 2 then the following procedure can be followed to determining the state:

1) Compare h_2 with the $h_f\ and\ h_g$ at the pressure of 240 kPa.
   a) If $h_2 > h_g$ then the fluid is in a superheated state
   b) If $h_2 = h_g$ then the fluid is in a saturated vapor state
   c) If $h_f < h_2 < h_g$ then the fluid is in a saturated liquid vapor mixture state
   d) If $h_2 = h_f$ then the fluid is in a saturated liquid state
   e) If $h_2 < h_f$ then the fluid is in a subcooled liquid state
2) If the fluid is superheated, then the superheated tables are used to determine the temperature.
3) If the fluid is subcooled liquid, then the subcooled liquid tables are used to determine the temperature.
4) If the liquid is saturated, then the temperature is equal to the saturation temperature and the saturated tables are used to find the remaining properties.

   Note that for further explanation and information on how to find properties and states from tables please refer back to Chapter 2.

   However, in other cases where the properties database software is available, we enter the two independent properties and the software will provide us with the state and all the other properties that we need.

   By following one of the above two methods, the final temperature of the piston–cylinder device is obtained to be **70 °C**.

c) Calculate the boundary work done by the system.

In order to find the boundary work, we go back again to the EBE:

$$Q_{in} + \dot{W}_e \times \Delta t + m_1 u_1 = m_2 u_2 + W_b$$

We need u_1 *and* u_2 to find the boundary work (W_b).

The property data are obtained either from the tables (such as Appendix B-3b) or from the EES to find:

$$u_1 = u_g @240\ kPa = 227.17\ \frac{kJ}{kg}$$

$$u_2 = 287.38\ \frac{kJ}{kg}$$

$$Q_{in} + \dot{W}_e \times \Delta t + m_1 u_1 = m_2 u_2 + W_b$$

$$300\ kJ + \left(1.408\ \frac{kJ}{s} \times 360\ s\right) + 12\ kg \times \frac{227.17\ kJ}{kg} = 12\ kg \times \frac{287.38\ kJ}{kg} + W_b$$

$\mathbf{W_b = 84.36\ kJ}$

3.8 Open Systems

In this section, the focus will be given to open systems where both mass and energy cross the system boundary. We will avoid using control volume for the open systems since the system boundary will be sufficient to use like closed systems. We basically bring everything down to a two-dimensional sketch for system design, analysis, and assessment. Figure 3.17

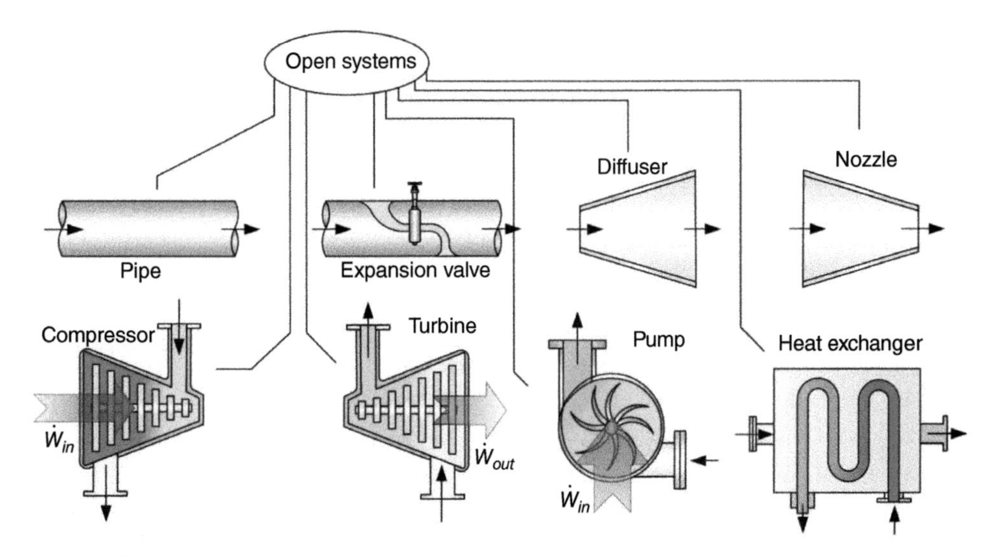

Figure 3.17 Some common types of open systems.

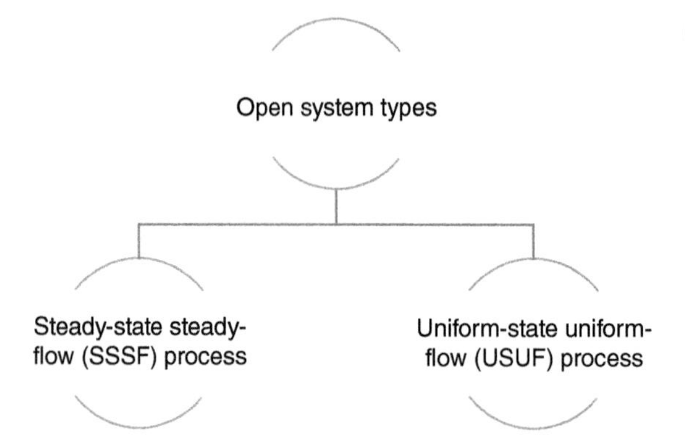

Figure 3.18 Types of open systems

illustrates some commonly used open-type devices (for example, pipe, expansion valve, diffuser, nozzle, compressor, turbine, pump and heat exchanger) in various systems and applications, ranging from power generating systems to refrigerators and heat pumps. We will specifically dwell on each of these systems with examples.

When one deals with open systems, there are two common types of the open systems to discuss as illustrated in Figure 3.18, as follows:

(a) Steady-state steady-flow (SSSF) process

This is a process which is essentially recognized as a steady-flow process. It is defined as a process where the flow crosses the boundary steadily with no change with time. This is the most commonly used type. Almost all systems and devices are treated as SSSF systems and devices. For example, a thermal power plant is treated this way since the start-up and shut-down periods are not considered. When one looks at it, it operates steadily throughout the year. There are, of course, many more examples to add to this, ranging from gas turbines to refrigeration and heat pump systems. Such SSSF processes are the ones where the properties inside the boundary of the system do not change with time. This section will provide examples on the application of the EBE on SSSF.

(b) Uniform-state uniform-flow (USUF) process

This is a process which is essentially recognized as an unsteady-flow process. It is defined as a process where the flow crosses the boundary unsteadily and time-dependently. Such systems are also common in our daily applications wherever one filling and discharging tanks, they end up with such processes. One can add more to this, by providing further examples: filling fuel tanks of the car, filling and discharging the reservoirs and fuel tankers. Unsteady state uniform flow processes are described in this section through various examples. The section will also introduce how the FLT is also applied to such processes. Of course such systems will be further discussed through the forthcoming examples.

Example 3.10 Water is flowing through a 3 m long pipe at a rate of 1 kg/s, the water enters at a temperature of 95 °C and a pressure of 100 kPa (Figure 3.19). The pipe ends up with the heat loss occurring at a rate of 1 kW per meter of the pipe length. (a) Write both mass and energy balance equations and (b) find the temperature of the water leaving the pipe.

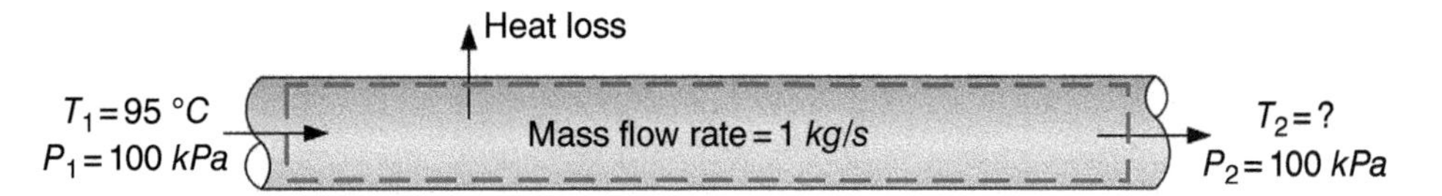

Figure 3.19 Schematic for pipe in Example 3.10.

Solution

a) Write the mass and energy balance equations as follows:

$$\text{MBE}: \dot{m}_1 = \dot{m}_2 = \dot{m}$$

$$\text{EBE}: \dot{m}_1 h_1 = \dot{m}_2 h_2 + \dot{Q}_{out}$$

b) Calculate the temperature of the water leaving the pipe after 3 m of the pipe length as follows:

$$\dot{m}_1 h_1 = \dot{m}_2 h_2 + \dot{Q}_{out}$$

$$1\,\frac{kg}{s} \times h_1 = 1\,\frac{kg}{s} \times h_2 + 1\,\frac{kW}{m} \times 3\ m$$

$$at\ P_1 = 100\ kPa \text{ and } T_1 = 95\ ^oC,\ h_1 = 398.1$$

By solving the EBE, one finds $h_2 = 395.1\,\frac{kJ}{kg}$

with $h_2 = 395.1\,\frac{kJ}{kg}$ and $P_2 = 100\ kPa$ (*by assuming no pressure losses occurs in the pipe*), one obtains the exit temperature of water as

$$\boldsymbol{T_2 = 94.29\ ^oC}$$

Example 3.11 Consider a diffuser and a nozzle separately as shown in Figure 3.20. Air enters a diffuser steadily at a velocity of 199 m/s. The mass flow rate of air through the nozzle is found to be 10 kg/s, where the air inlet and outlet temperatures are 290 and 305 K, respectively. (a) Find the velocity of the air at the diffuser exit using the energy balance on the diffuser. (b) Assume that the air leaving the diffuser enters a nozzle that has a heat gain of 100 kW, find the temperature of air leaving the nozzle if it has a velocity of 199 m/s.

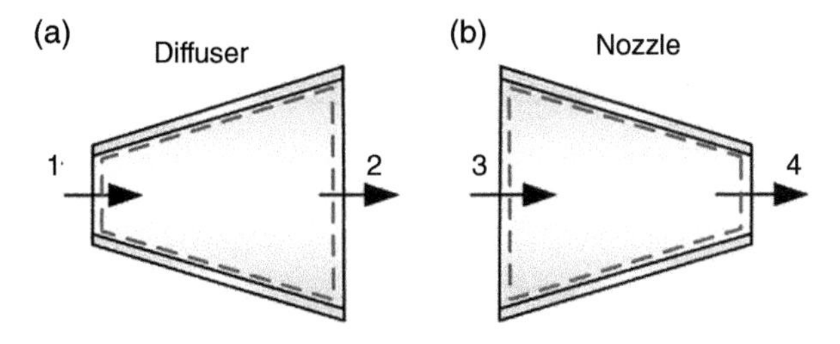

Figure 3.20 Illustration of (a) a diffuser and (b) a nozzle for Example 3.11.

Solution

a) Find the velocity of the air at the diffuser exit using the energy balance on the diffuser.
By applying the energy balance on the boundary of the diffuser, the EBE can be written as:

$$\text{MBE}: \dot{m}_1 = \dot{m}_2 = const$$

$$\text{EBE}: \dot{m}_1 h_1 + \frac{\dot{m}_1\left(V_1^2\right)}{2} = \dot{m}_2 h_2 + \frac{\dot{m}_2\left(V_2^2\right)}{2}$$

Since the mass flow rate at the inlet equals that at the outlet then the above energy balance can be reduced to:

$$h_1 + \frac{V_1^2}{2} = h_2 + \frac{V_2^2}{2}$$

Assuming air is an ideal gas, the enthalpies of air at the inlet and outlet of the diffuser are found as follows:

$$T_1 = 290\ K\} h_1 = 290.16\ \frac{kJ}{kg}$$

$$T_2 = 305\ K\} h_2 = 305.22\ \frac{kJ}{kg}$$

$$290.16\ \frac{kJ}{kg} + \frac{(199\ m/s)^2}{2} \times \frac{1\ kJ/kg}{1000\ m^2/s^2} = 305.22\ \frac{kJ}{kg} + \frac{V_2^2}{2} \times \frac{1\ kJ/kg}{1000\ m^2/s^2}$$

$$\mathbf{V_2 = 97.4\ m/s}$$

b) If the air leaving the diffuser enters a nozzle with a heat gain of 100 kW, find the temperature of air leaving the nozzle if it has a velocity of 199 m/s.

$$\text{MBE}: \dot{m}_3 = \dot{m}_4 = const.$$

$$\text{EBE}: \dot{m}_3 h_3 + \frac{\dot{m}_3\ V_3^2}{2} + \dot{Q}_{in} = \dot{m}_4 h_4 + \frac{\dot{m}_4\ V_4^2}{2}$$

$$10\ \frac{kg}{s} \times 305.22\ \frac{kJ}{kg} + 10\ \frac{kg}{s} \times \left(\frac{\left(97.4\ \frac{m}{s}\right)^2}{2} \times \frac{1\ \frac{kJ}{kg}}{1000\ \frac{m^2}{s^2}} \right) + 100\ kW$$

$$= 10\ \frac{kg}{s} \times h_4 + 10\ \frac{kg}{s} \times \left(\frac{(199\ m/s)^2}{2} \times \frac{1\ \frac{kJ}{kg}}{1000\ \frac{m^2}{s^2}} \right)$$

$$h_4 = 300.2\ \frac{kJ}{kg} \rightarrow \text{Since } h_4 = C_p T_4 \rightarrow T_4 = \frac{h_4}{C_p}$$

with C_p is 1 kJ/kgK, one obtains it as

$$\mathbf{T_4 = 300\ K}$$

Example 3.12 Consider that superheated steam enters an adiabatic steam turbine, as shown in Figure 3.21, at a mass flow rate of 10 *kg/s*. The superheated steam enters at a temperature and pressure of 500 °*C* and 10 000 *kPa*. The steam turbine converts part of the energy from the steam to shaft work. The steam exits the turbine as a saturated liquid and vapor at a pressure of 10 *kPa* with a mixture quality of 0.95. Calculate the turbine work output.

Solution

In order to find the work rate produced by the steam turbine, one writes both mass and energy balance equations as follows:

$$\text{MBE}: \dot{m}_1 = \dot{m}_2$$

$$\text{EBE}: \dot{m}_1 h_1 = \dot{m}_2 h_2 + \dot{W}_{out}$$

Using the EBE, we calculate the specific power produced by the steam turbine as follows:

$$\dot{m}_1 h_1 = \dot{m}_2 h_2 + \dot{W}_{out}$$

From the MBE: $\dot{m}_1 = \dot{m}_2 = \dot{m}$ and by dividing by the mass flow rate, the EBE becomes:

$$h_1 = h_2 + w_{out}$$

$$\left.\begin{array}{l} P_1 = 10\ MPa \\ T_1 = 500\ °C \end{array}\right\} h_1 = 3374\ kJ/kg$$

$$P_2 = 10\ kPa\} \begin{array}{l} h_f = 191.81\ kJ/kg \\ h_{fg} = 2392.1\ kJ/kg \end{array}$$

$$h_2 = h_f + x_2 h_{fg}$$

$$h_2 = 191.81 + 0.95 \times 2392.1 = 2464\ kJ/kg$$

$$3374 = 2464 + w_{out}$$

$$w_{out} = 909.8\ kJ/kg$$

$$\dot{W}_{out} = \dot{m} \times w_{out} = 10\ \frac{kg}{s} \times 909.8\ \frac{kJ}{kg} = \mathbf{9.098\ MW}$$

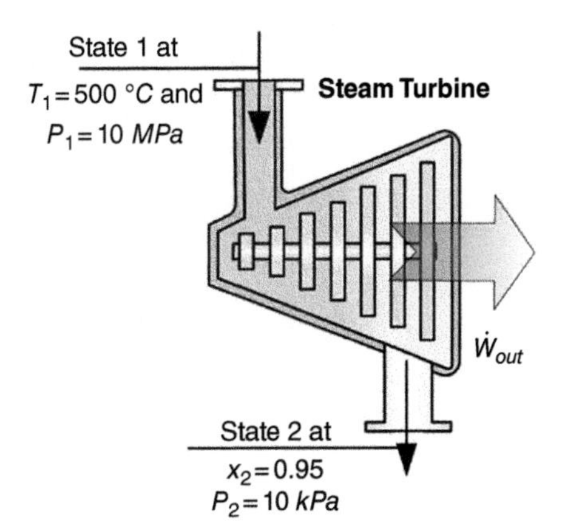

Figure 3.21 Schematic diagram of the steam turbine for Example 3.12.

Example 3.13 Consider this time that steam enters a two-stage steam turbine, as shown in Figure 3.22, at a rate of 13 *kg/s*. The steam enters at a temperature and pressure of 873.15 *K* and 8000 *kPa*. The steam expands in the turbine to saturated vapor state at a pressure of 300 *kPa*, where 10% of the flow is taken out of the turbine at state point 2. The remaining flow is further expanded to saturated liquid vapor mixture with a quality of 85% and pressure of 10 *kPa*. Neglect the changes in kinetic and potential energies and assume a heat loss of 1000 *kW*. (a) Write the mass balance equation (MBE); (b) write the energy balance equation (EBE); and (c) calculate the turbine work rate.

Solution

a) Write the MBE as follows:

$$\dot{m}_1 = \dot{m}_2 + \dot{m}_3$$

where $\dot{m}_2 = 0.1 \times \dot{m}_1$ since 10% of the incoming flow rate is taken out.

b) Write the EBE as follows:

With the negligible changes in kinetic and potential energies, one can write the EBE as:

$$\dot{m}_1 h_1 = \dot{m}_2 h_2 + \dot{m}_3 h_3 + \dot{Q}_{loss} + \dot{W}_{out}$$

c) Calculate the turbine work rate as follows:

$$\left.\begin{array}{r} P_1 = 8000\ kPa \\ T_1 = 600\ ^oC \end{array}\right\} h_1 = 3642.4\ kJ/kg$$

$$\left.\begin{array}{r} P_2 = 300\ kPa \\ x_2 = 1 \end{array}\right\} h_2 = 2724.9\ kJ/kg$$

$$h_3 = h_f + x\left(h_{fg}\right)$$

$$h_3 = 191.81 + 0.85 \times (2392.1) = 2225.095\ kJ/kg$$

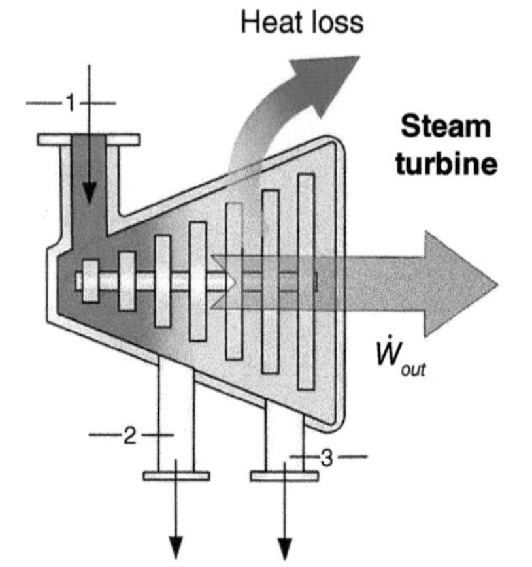

Figure 3.22 A schematic sketch of the steam turbine in Example 3.13.

$$\left(13\,\frac{kg}{s}\times 3642.4\,\frac{kJ}{kg}\right)$$

$$=\left(0.1\times 13\,\frac{kg}{s}\times 2724.9\,\frac{kJ}{kg}\right)+\left(0.9\times 13\,\frac{kg}{s}\times 2225.095\,\frac{kJ}{kg}\right)$$

$$+\left(1000\,\frac{kJ}{s}\right)+\dot{W}_{out}$$

$$\boldsymbol{\dot{W}_{out}=16{,}775\ kW}$$

Example 3.14 Consider an unmixed type heat exchanger, as shown in Figure 3.23a, where a water flow rate of 115,200 kg/h is heated at a constant pressure by $4.8\times 10^6\ kg/h$ of exhaust gases as shown in the figure. Assume the exhausts consist only of air where you can use ideal gas equations. $\left(c_{p,air}=1.005\,\frac{kJ}{kgK}\right)$. (a) Write the MBEs; (b) write the EBE; (c) calculate the exit temperature of water; (d) calculate the exit temperature of exhaust; and (e) calculate the amount of heat transferred to the water.

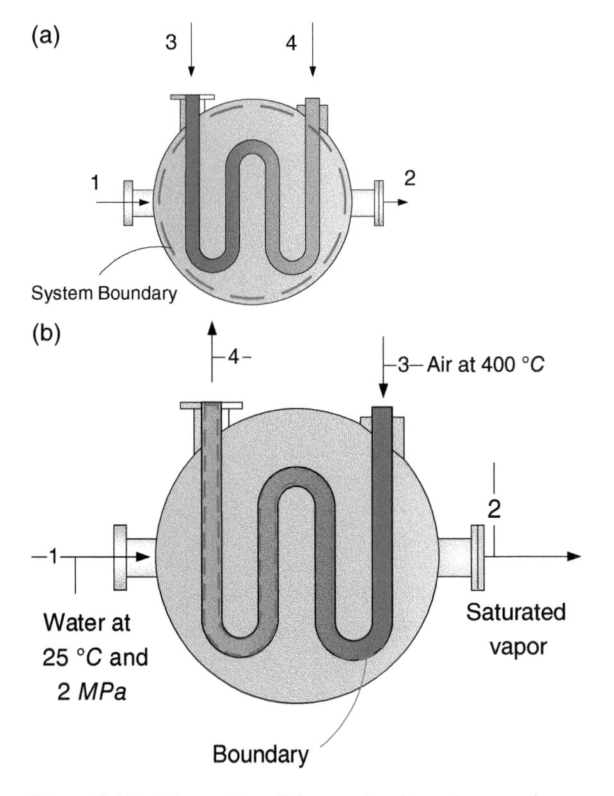

Figure 3.23 Illustration of the unmixed type heat exchanger considered in Example 3.14: (a) with the system boundary placed in the inner surface of the shell and (b) with the system boundary selected on the inner pipe only.

Solution

a) Write the MBEs as follows:

$$\dot{m}_1 = \dot{m}_2 = \dot{m}_w \text{ and } \dot{m}_3 = \dot{m}_4 = \dot{m}_a$$

b) Write the EBE as follows:

With the negligible changes in kinetic and potential energies, one can write the EBE as:

$$\dot{m}_1 h_1 + \dot{m}_3 h_3 = \dot{m}_2 h_2 + \dot{m}_4 h_4$$

c) Calculate the exit temperature of water as follows:

$$T_2 = T_{sat}@P = 2\ MPa = \mathbf{212.38\ °C}$$

d) Calculate the exit temperature of the exhaust as follows:

$$\dot{m}_2 h_2 - \dot{m}_1 h_1 = \dot{m}_3 h_3 - \dot{m}_4 h_4$$

$$\dot{m}_2 h_2 - \dot{m}_1 h_1 = \dot{m}_a c_p (T_3 - T_4)$$

$$h_1 = h_{f@T=25\ °C} = 104.83\ kJ/kg$$

$$h_2 = h_{g@2MPa} = 2798.3\ kJ/kg$$

$$\dot{m}_w (h_2 - h_1) = \dot{m}_a c_p (T_3 - T_4)$$

$$115{,}200\ \frac{kg}{h} \times \frac{1\ h}{60\ min} \times \frac{1\ min}{60\ s} (2798.3 - 104.83)\ \frac{kJ}{kg}$$

$$= 4.8 \times 10^6\ \frac{kg}{h} \times \frac{1\ h}{3600\ s} \times 1.005\ \frac{kJ}{kgK} (400 + 273 - \mathrm{T}_4)\ K$$

$$\mathbf{T_4 = 609\ K = 335\ °C}$$

e) Calculate the amount of heat transferred to the water as follows:

$$\mathrm{EBE}: \dot{m}_3 h_3 = \dot{m}_4 h_4 + \dot{Q}_{out}$$

$$\dot{m}_3 h_3 - \dot{m}_4 h_4 = \dot{Q}_{out}$$

$$\dot{m}_a c_p (T_3 - T_4) = \dot{Q}_{out}$$

$$4.8 \times 10^6\ \frac{kg}{h} \times \frac{1\ h}{60\ min} \times \frac{1\ min}{60\ s} \times 1.005\ \frac{kJ}{kgK} (400 - 335)\ K = \dot{Q}_{out}$$

$$\mathbf{\dot{Q}_{out} = 87100\ kW}$$

Example 3.15 Consider another heat exchanger case with a recuperator, as shown in Figure 3.24, where the exhaust gases leave an internal combustion engine at a pressure and temperature of 150 *kPa* and 400 °*C* with a flow rate of 0.8 *kg/s*. The exhaust gases are used to produce saturated steam (with vapor quality of 1) at a temperature of 200 °*C* in an insulated heat exchanger. The exhaust gases are mainly air and can be treated as ideal gas in this problem. The inlet conditions of the water and the exit conditions of the exhaust gases at the boundaries of the heat exchanger are: (i) the water enters the heat exchanger at

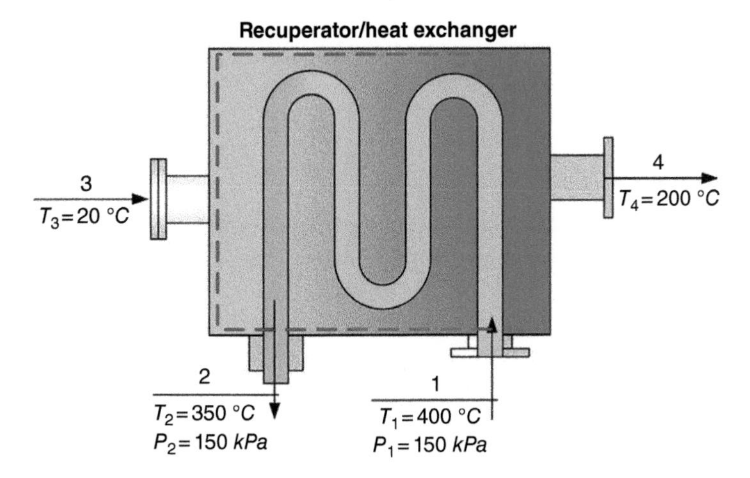

Figure 3.24 Schematic diagram of the cross-flow heat exchanger used to produce steam in Example 3.15.

the ambient temperature of 20 °*C* and (ii) the exhaust gases leave the heat exchanger at 350 °*C*. (a) Write both the mass and energy balance equations and (b) calculate the heat exchange rate of the generated steam.

Solution

a) The inlet and exit states of the exhaust gases are denoted (1) and (2) and for water they are denoted as (3) and (4). In order to determine the heat rate produced by the heat exchanger, one needs to write the balance equations as follows:

$$\text{MBE}: \dot{m}_3 = \dot{m}_4 \,\text{and}\, \dot{m}_1 = \dot{m}_2$$

$$\text{EBE}: \dot{m}_1 h_1 + \dot{m}_3 h_3 = \dot{m}_2 h_2 + \dot{m}_4 h_4$$

b) The properties of water passing through the heat exchanger are obtained from steam tables (such as Appendix B-1a):

$$T_3 = 20\ ^{\circ}C, liquid \rightarrow h_3 = 83.91\,\frac{kJ}{kg},\ s_3 = 0.29649\,\frac{kJ}{kgK}$$

$$T_4 = 200\ ^{\circ}C, saturated\ vapor \rightarrow h_4 = 2792.0\,\frac{kJ}{kg},\ s_4 = 6.4302\,\frac{kJ}{kgK}$$

In order to calculate the mass flow rate of the produced steam, the EBE is applied to the heat exchanger as follows:

$$\dot{m}_{exh} h_1 + \dot{m}_w h_3 = \dot{m}_{exh} h_2 + \dot{m}_w h_4$$

$$\dot{m}_w = \frac{\dot{m}_{exh}(h_1 - h_2)}{h_4 - h_3}$$

Note that air is assumed to behave as an ideal gas. The change of the enthalpy of the flow can be expressed in terms of the temperatures and the heat capacity at a constant pressure (c_p):

$$\dot{m}_w = \frac{\dot{m}_{\text{exh}} c_p (T_1 - T_2)}{h_4 - h_3} = \frac{0.8\ \frac{kg}{s} \times 1.063\ \frac{kJ}{kg\ ^\circ C} \times (400 - 350)\ ^\circ C}{(2792.0 - 83.91)\ \frac{kJ}{kg}}$$

$$\dot{m}_w = \mathbf{0.01570\ kg/s}$$

The heat exchange rate is calculated by applying the energy balance on the exhaust side of the heat exchanger. Note that the choice of applying the balance to the gas side of the heat exchanger is done to make use the given data from the question, to avoid using calculated numbers.

$$\dot{Q} = \dot{m}_{exh} h_1 - \dot{m}_{exh} h_2 = \dot{m}_{exh} c_p (T_1 - T_2)$$

$$\dot{Q} = \dot{m}_{exh} c_p (T_1 - T_2) = 0.8\ \frac{kg}{s} \times 1.063\ \frac{kJ}{kg\ ^\circ C} \times (400 - 350)\ ^\circ C = \mathbf{42.52\ kW}$$

Example 3.16 An air compressor, as shown in Figure 3.25, compresses air from atmospheric pressure and temperature of 280 *K* to a discharge condition at a pressure and temperature of 600 *kPa* and 400 *K*, respectively. The mass flow rate of the air entering the compressor is 1 *kg/s*. The compressor losses heat through its walls at a rate of 40% of the total work rate running the compressor. Assume that the changes in both kinetic and potential energies are negligible and that air is treated as ideal gas. Take C_p for air as 1 *kJ/kgK*. (a) Write both mass and energy balance equations and (b) calculate the work rate consumed by the compressor.

Solution

a) Write the mass and energy balance equations.

$$\text{MBE}: \dot{m}_1 = \dot{m}_2 = \dot{m}$$

$$\text{EBE}: \dot{m}_1 h_1 + \dot{W}_{in} = \dot{m}_2 h_2 + \dot{Q}_{out}$$

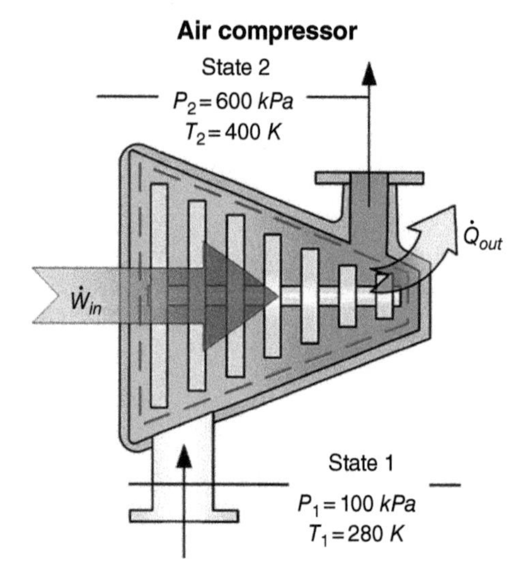

Figure 3.25 Schematic diagram of the air compressor for Example 3.16.

b) The work rate consumed by the compressor.
One can express the work rate consumed by the compressor using the EBE:

$$\dot{W}_{in} = \dot{Q}_{out} + \dot{m}_2 h_2 - \dot{m}_1 h_1$$

Here, the properties of the stream entering and leaving the compressor are found from the air property tables (such as Appendix E-1) or by using the EES software.

In order to find the work rate consumed by the compressor while compressing the air from atmospheric pressure to a pressure of 600 kPa, the EBE is rearranged as follows:

$$\dot{W}_{in} - (0.4 \times \dot{W}_{in}) = \dot{m}_2 h_2 - \dot{m}_1 h_1 = \dot{m} C_p (T_2 - T_1)$$

$$0.6 \times \dot{W}_{in} = 1 \times 1 \times (400 - 280)$$

$$\dot{W}_{in} = \mathbf{200\ kW}$$

Example 3.17 The pressure of water at ambient conditions is increased in a pump to 900 kPa and a temperature of 30 °C, as shown in Figure 3.26. The water entering the pump is at a temperature of 15 °C and atmospheric pressure, and the mass flow rate is 0.5 kg/s. Determine the amount of power the pump consumes during the pressurization process. If the efficiency of the electrical motor is 95%, find the electrical power delivered to the electric motor.

Solution

It is first necessary to write two balance equations for the pump as follows:

$$\text{MBE}: \dot{m}_{in} = \dot{m}_{out} = \dot{m}$$

$$\text{EBE}: \dot{m}_{in} h_{in} + \dot{W}_{in} = \dot{m}_{out} h_{out}$$

Note that one may also use shaft work equation for this as $\dot{W}_{in} = \dot{W}_p = \dot{m}_v (P_2 - P_1)$. Here, v is taken v_1 since the change in specific volume for saturated liquid and compressed water is negligibly small.

One needs to go one step ahead and use the above given EBE to calculate the pump work rate, which is also known as the power required for water pumping (pressurization).

Here, the properties of the stream entering and leaving the pump are found from water properties tables (such as Appendix B-1a) or by using the EES software, which contains the

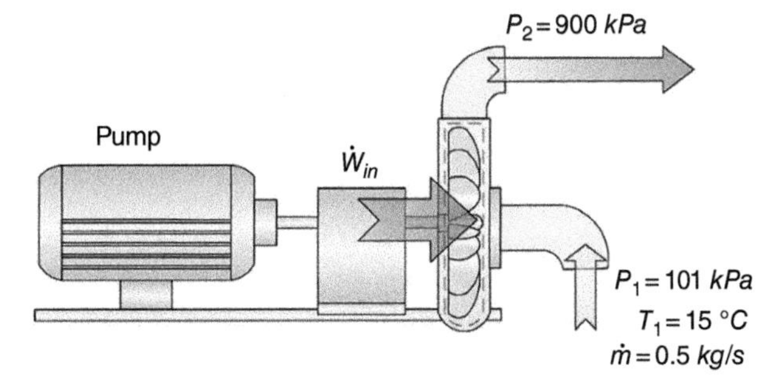

Figure 3.26 A pump–motor system used for water pumping in Example 3.17.

database for the properties of most of the fluids through a large range of temperatures and pressures; the properties are:

$$\left.\begin{array}{r} P_1 = 101\ kPa \\ T_1 = 15\,^{o}C \end{array}\right\} h_1 = 62.98\ \frac{kJ}{kg}$$

$$\left.\begin{array}{r} P_1 = 900\ kPa \\ T_1 = 30\,^{o}C \end{array}\right\} h_2 = 126.5\ \frac{kJ}{kg}$$

The work consumption rate of the pump can be calculated by rearranging the EBE as:

$$\dot{W}_{in} = \dot{m}_2 h_2 - \dot{m}_1 h_1$$

Substituting the given mass flow rate and the specific enthalpy values from the thermodynamic tables into the rearranged EBE yields the following work consumption rate:

$$\dot{W}_{in} = (0.5 \times 126.5) - (0.5 \times 62.98)$$

$$\boldsymbol{\dot{W}_{in} = 31.76\ kW}$$

The electrical power consumed by the electric motor to produce the required pumping power is calculated through the efficiency of the motor as follows:

$$\eta_{motor} = \frac{\dot{W}_{in,pump}}{\dot{W}_{elec}} \rightarrow 0.95 = \frac{31.76}{\dot{W}_{elec}} \rightarrow \boldsymbol{\dot{W}_{elec} = 33.43\ kW}$$

Example 3.18 As shown in Figure 3.27, refrigerant *R*-134*a* flowing through a pipe at a pressure of 0.8 *MPa* is in a saturated liquid state. The refrigerant is throttled down by a throttling valve to a pressure of 0.12 *MPa*. (a) Write the mass and energy balance equations and (b) find the exit temperature at state 2.

Solution

a) Write the mass and energy balance equations.

$$\text{MBE}: \dot{m}_1 = \dot{m}_2 = const.$$

$$\text{EBE}: \dot{m}_1 h_1 = \dot{m}_2 h_2$$

b) Calculate the exit temperature at state 2.

By using a properties-based software or the properties tables (such as Appendix B-3b), the enthalpy at the inlet of the throttling valve is found as follows:

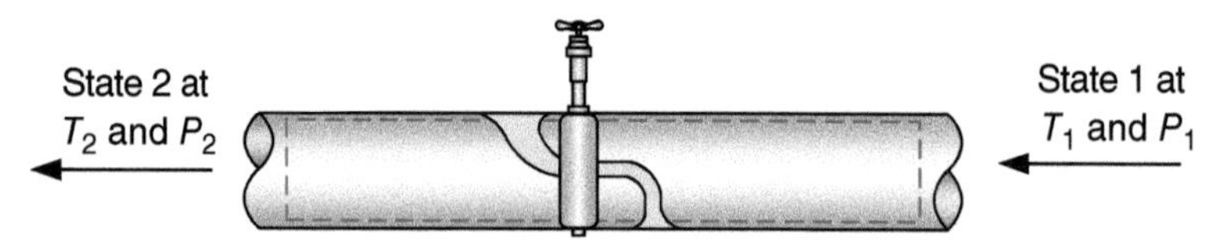

Figure 3.27 Throttling valve reducing the pressure of the refrigerant in Example 3.18.

$$\left.\begin{aligned} P_1 &= 0.8\ MPa \\ x_1 &= 0\,(saturated\ liquid) \end{aligned}\right\} h_1 = 95.5\,\frac{kJ}{kg}$$

Then, by using the EBE h_2 is calculated as follows:

$$\dot{m}_1 h_1 = \dot{m}_2 h_2 \rightarrow by\ using\ MBE \rightarrow h_1 = h_2 = 95.5\,\frac{kJ}{kg}$$

$$\left.\begin{aligned} P_2 &= 0.12\ MPa \\ h_2 &= 95.5\,\frac{kJ}{kg} \end{aligned}\right\} \mathbf{T_2 = -22.32\,^{o}C}$$

Example 3.19 Four different streams of water enter a mixing chamber at a flow rate of 1 *kg/s*, as shown in the Figure 3.28. The mixing chamber rejects heat at a rate of 10 *kW*. (a) Write the MBE, (b) write the EBE, (c) identify the state of the water exiting the mixing chamber, and (d) find the temperature and specific volume of the exiting water.

Solution

a) Write the MBE.

$$\dot{m}_1 + \dot{m}_2 + \dot{m}_3 + \dot{m}_4 = \dot{m}_5$$

b) Write the EBE.

The changes in kinetic and potential energies are negligible. The EBE is written as

$$\dot{m}_1 h_1 + \dot{m}_2 h_2 + \dot{m}_3 h_3 + \dot{m}_4 h_4 = \dot{m}_5 h_5 + \dot{Q}_{out}$$

c) Identify the state of the water exiting the mixing chamber.

$$\left.\begin{aligned} P_1 &= 100\ kPa \\ T_1 &= 10\,^{o}C \end{aligned}\right\} h_1 = 42.08\ kJ/kg$$

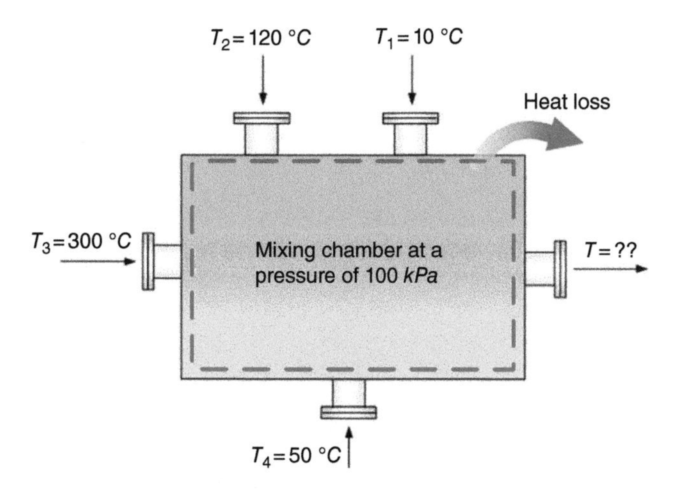

Figure 3.28 A schematic diagram of the mixing chamber in Example 3.19.

$$\left.\begin{array}{r} P_2 = 100\ kPa \\ T_2 = 120\,^oC \end{array}\right\} h_2 = 2716\ kJ/kg$$

$$\left.\begin{array}{r} P_3 = 100\ kPa \\ T_3 = 300\,^oC \end{array}\right\} h_3 = 3074\ kJ/kg$$

$$\left.\begin{array}{r} P_4 = 100\ kPa \\ T_4 = 50\,^oC \end{array}\right\} h_4 = 209.4\ kJ/kg$$

$$\left(1\frac{kg}{s} \times 42.08\frac{kJ}{kg}\right) + \left(1\frac{kg}{s} \times 2716\frac{kJ}{kg}\right) + \left(1\frac{kg}{s} \times 3074\frac{kJ}{kg}\right) + \left(1\frac{kg}{s} \times 209.4\frac{kJ}{kg}\right)$$

$$= \left(4\frac{kg}{s}\right) \times h_5 + 10\ kW$$

$$h_5 = 1507\ kJ/kg$$

$$\left.\begin{array}{r} P_5 = 100\ kPa \\ h_5 = 1507\ kJ/kg \end{array}\right\} \textbf{\textit{saturated liquid vapor mixture}}$$

d) Find the temperature and specific volume of the exiting water.

$$\left.\begin{array}{r} P_5 = 100\ kPa \\ h_5 = 1507\ kJ/kg \end{array}\right\} T_5 = 99.6\,^oC$$

$$h_5 = h_f + x_5\left(h_{fg}\right)$$

$$x_5 = \frac{h_5 - h_f}{h_{fg}} = \frac{1507 - 417.5}{2675 - 417.5} = 0.48$$

$$v_5 = v_f + x_5\left(v_{fg}\right)$$

$$v_5 = 0.001043 + 0.48 \times (1.694 - 0.001043) = \mathbf{0.814\ m^3/kg}$$

Example 3.20 A tank filled with helium is heated at a constant rate of heat. Helium then transfers heat to the water flowing in through the pipe. The water flow rate is fixed at 1 *kg/s*. The process is illustrated in Figure 3.29. (a) Write the MBEs, (b) write the EBE, and (c) calculate the amount of heat transfer rate to water.

Solution

a) Write the MBE for both fluids.

$$\dot{m}_1 = \dot{m}_2 = \dot{m}$$

$$m_{o,1} = m_{o,2} = m_o$$

b) Write the EBE.

The changes in kinetic and potential energies are negligible. The EBE is therefore written as:

$$\dot{m}_1 h_1 + \dot{Q}_{in} = \dot{m}_2 h_2$$

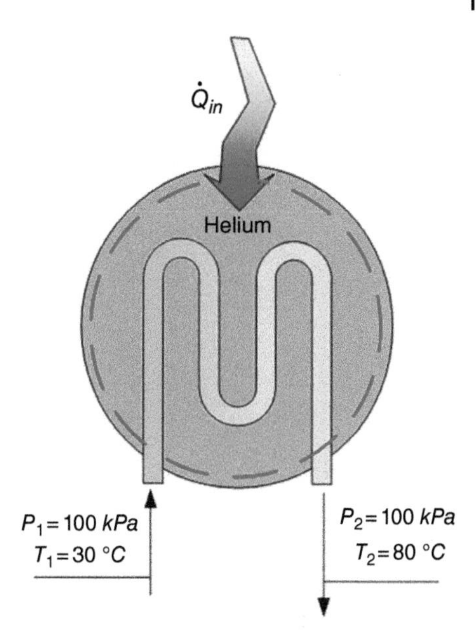

Figure 3.29 Schematic of the mixing chamber in Example 3.20.

c) Find the amount of heat rate transferred to the water.

$$\left.\begin{aligned} P_1 &= 100\ kPa \\ T_1 &= 30\,^{o}C \end{aligned}\right\} h_1 = 125.8\ kJ/kg$$

$$\left.\begin{aligned} P_2 &= 100\ kPa \\ T_2 &= 80\,^{o}C \end{aligned}\right\} h_2 = 335\ kJ/kg$$

$$\dot{m}_1 h_1 + \dot{Q}_{in} = \dot{m}_2 h_2$$

$$\dot{Q}_{in} = 1\frac{kg}{s} \times (335 - 125.8)\frac{kJ}{kg} = \mathbf{209.2\ kW}$$

Example 3.21 Consider one high-pressure turbine and one low-pressure turbine connected as shown in Figure 3.30. Each turbine rejects some heat as indicated. The state point data are provided in the figure. Assume that the changes in kinetic and potential energies are negligible. (a) Write mass and energy balance equations for each turbine, (b) calculate the turbine work outputs, and (c) find the temperature of water leaving the second turbine.

Solution

a) Write the mass and energy balance equations for each turbine separately.

$$\text{MBE} : \dot{m}_1 = \dot{m}_2 = \dot{m}_3 = \dot{m}_w$$

$$\text{EBE (turbine 1)} : \dot{m}_w h_1 = \dot{m}_w h_2 + \dot{W}_{out,1} + \dot{Q}_{out,1}$$

$$\text{EBE (turbine 2)} : \dot{m}_w h_2 = \dot{m}_w h_3 + \dot{W}_{out,2} + \dot{Q}_{out,2}$$

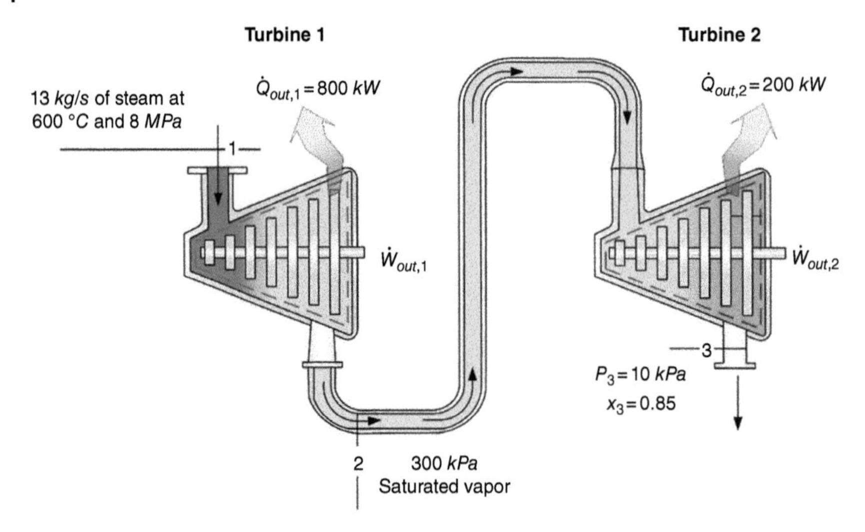

Figure 3.30 A schematic of two connected turbines in Example 3.21.

b) Calculate the work rates produced by both turbines.

From properties tables (such as Appendix B-1b and B-1c):

$$\left.\begin{matrix} P_1 = 8000\ kPa \\ T_1 = 600\ ^oC \end{matrix}\right\} \begin{matrix} h_1 = 3642.4\ kJ/kg \\ s_1 = 7.0221\ kJ/kg.K \end{matrix}$$

$$\left.\begin{matrix} P_2 = 300\ kPa \\ x_2 = 1 \end{matrix}\right\} h_2 = 2724.9\ kJ/kg$$

$$\dot{m}_w h_1 = \dot{m}_w h_2 + \dot{W}_{out,1} + \dot{Q}_{out,1}$$

$$\left(13\frac{kg}{s} \times 3642.4\frac{kJ}{kg}\right) = \left(13\frac{kg}{s} \times 2724.9\frac{kJ}{kg}\right) + \left(800\frac{kJ}{s}\right) + \dot{W}_{out,1}$$

$$\mathbf{\dot{W}_{out,1} = 11128\ kW}$$

$$h_3 = h_f + x(h_{fg})$$

$$h_3 = 191.81 + 0.85 \times (2392.1) = 2225.095\ kJ/kg$$

$$\dot{m}_w h_2 = \dot{m}_w h_3 + \dot{W}_{out,2} + \dot{Q}_{out,2}$$

$$\left(13\frac{kg}{s} \times 2724.9\frac{kJ}{kg}\right) = \left(13\frac{kg}{s} \times 2225.095\frac{kJ}{kg}\right) + \left(200\frac{kJ}{s}\right) + \dot{W}_{out,2}$$

$$\mathbf{\dot{W}_{out,2} = 6297\ kW}$$

$$\mathbf{\dot{W}_{out} = \dot{W}_{out,1} + \dot{W}_{out,2} = 17425\ kW}$$

c) Find the temperature of the water leaving turbine 2.

From the property tables (such as Appendix B-1b) or the database available in the EES software:

$$\left.\begin{matrix} P_3 = 10\ kPa \\ x_3 = 0.85 \end{matrix}\right\} \mathbf{T_3 = 45.82\ ^oC}$$

Example 3.22 Consider an insulated rigid tank, as shown in Figure 3.31, that is initially empty. It is now filled with steam from a supply line at a temperature of $300\,^{\circ}C$ and a pressure of $1\,MPa$. The tank is filled with steam until the pressure in the tank reaches $1\,MPa$. Determine the final tank temperature.

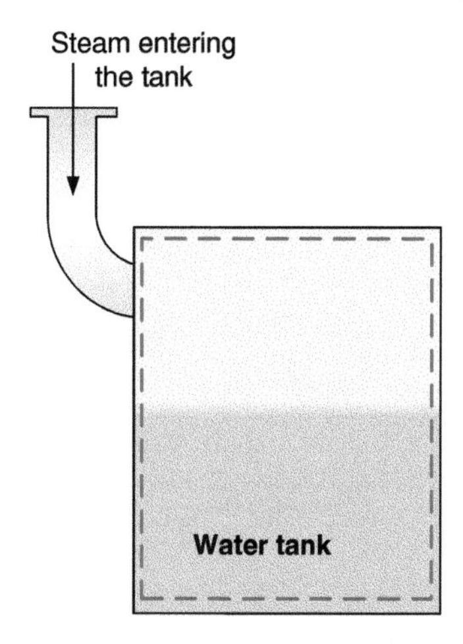

Figure 3.31 Schematic diagram of the tank being filled with steam in Example 3.22.

Solution

By using the information given, the initial mass is zero as it was mentioned that the tank is empty. Noting that the tank is initially empty, the MBE and EBE can be written as:

$$MBE : m_1 + \dot{m}_{in} \times t = m_2 \rightarrow \dot{m}_{in} \times t = m_2 \text{ since } m_I \text{ is zero.}$$

$$EBE : m_1 u_1 + \dot{m}_{in} h_{in} \times t = m_2 u_2$$

By using the note that the tank initially was evacuated then the initial mass is zero, with the help of the MBE the EBE as follows:

$$\dot{m}_{in} h_{in} \times t = m_2 u_2$$

And from the MBE $\dot{m}_{in} \times t = m_2$ then:

$$u_2 = h_{in}$$

$$\left.\begin{array}{l} P_{in} = 1\ MPa \\ T_{in} = 300\,^{o}C \end{array}\right\} h_{in} = 3051.6\,\frac{kJ}{kg}$$

$$u_2 = 3051.6\,\frac{kJ}{kg}$$

Then based on the temperature and pressure of the final state, the final temperature is found to be:

$$\mathbf{T_2 = 456.1\,^{o}C}$$

A special application of unsteady-flow process: thermal energy storage

A USUF process analysis is usually applied to thermal energy storage systems given their operational nature. Figure 3.32 shows a simple thermal energy storage system, where the thermal energy supplied by a heater is stored by the storage medium, which in this case is water. The hot water retains most of its thermal energy, as the tank is well insulated resulting in low heat losses. When the heat needs to be recovered, a heat transfer fluid flows through a pipe in the tank and absorbs the heat available from the system.

The thermal energy storage process considered in Figure 3.32 is a simple case of a thermal energy storage system undergoing a complete storage cycle, where the final state of the system is the same as the starting state. The system undergoes a total of three processes in a single cycle. Usually in the analysis of such systems the heat exchangers are assumed to undergo no heat losses during the charging and discharging processes. The assumption

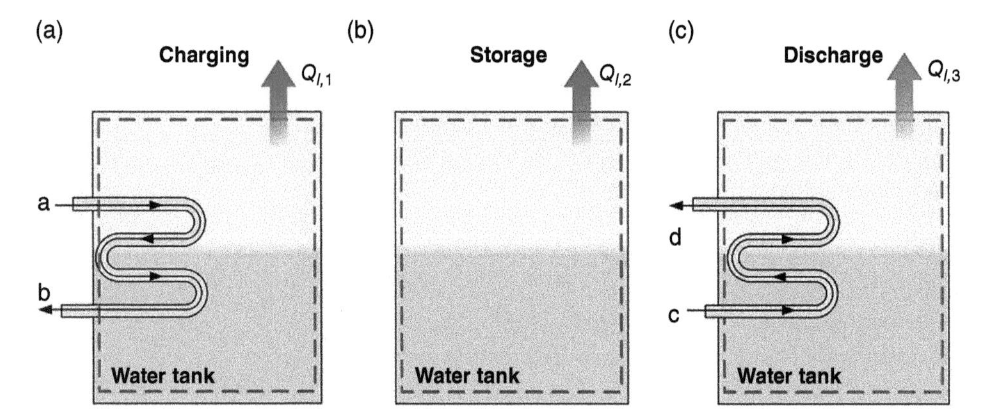

Figure 3.32 Three stages of the thermal storage system (simple heat storage process): (a) charging phase, (b) storing phase, and (c) discharging phase.

of no heat loss is valid due to the scenario where the heat losses from the storage medium is much larger and more significant than the amount of energy lost from the charging fluid.

Example 3.23 Consider a hot water storage tank, as shown in Figure 3.33, where three storage processes take place: charging as shown in Figure 3.33a, storing in Figure 3.33b, and discharging as shown in Figure 3.33c. Write mass and energy balance equations for each of the process.

Solution

For the charging stage, from state 1 to state 2:

$$MBE : m_1 = m_2 = m = const., \text{ and } \dot{m}_a = \dot{m}_b = const.$$

$$EBE : m_1 u_1 + \dot{m}_a h_a t_{ch} = m_2 u_2 + \dot{m}_b h_b t_{ch} + Q_{l,1}$$

For the storage stage, from state 2 to state 3:

$$MBE : m_2 = m_3 = m = const.$$

$$EBE : m_2 u_2 = m_3 u_3 + Q_{l,2} \text{ (assuming } \textit{no} \text{ heat losses occurs)}$$

For the discharging stage, from state 3 to state 4:

$$MBE : m_3 = m_4 = m = const., \text{ and } \dot{m}_d = \dot{m}_c = const.$$

$$EBE : m_3 u_3 + \dot{m}_d h_d t_{disch} = m_4 u_4 + \dot{m}_c h_c t_{disch} + Q_{l,3}$$

The overall efficiency of the complete storage cycle is calculated by dividing the amount of energy recovered during the discharging phase over the amount of energy that was supplied to the system during the charging phase as follows:

$$\eta = \frac{\dot{m}_c h_c t_{disch} - \dot{m}_d h_d t_{disch}}{\dot{m}_a h_a t_{ch} - \dot{m}_b h_b t_{ch}}$$

where, t_{ch} is the charging time and t_{disch} is the discharging time.

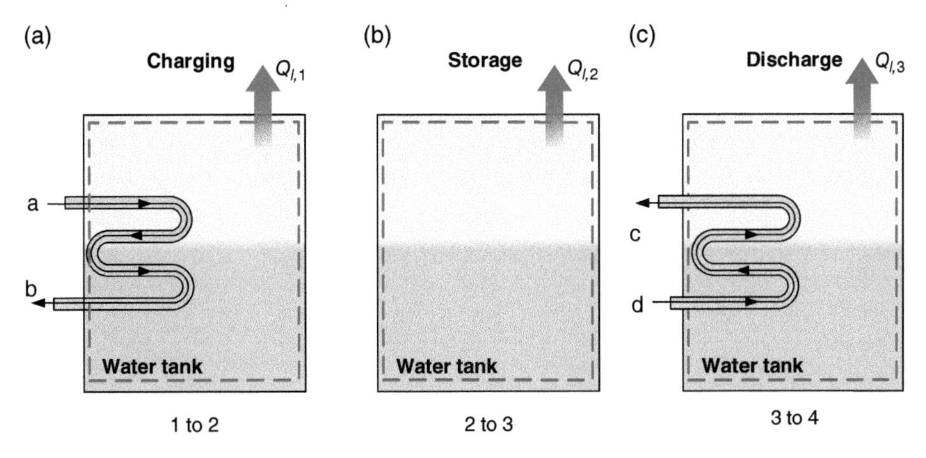

Figure 3.33 Three stages of the thermal storage system for Example 3.23.

3.9 Closing Remarks

This chapter has focused on energy analysis in general and more specifically on closed and open thermodynamic systems. Under closed systems both fixed boundary and moving boundary systems are considered, and the energy balance equations are introduced and discussed. In regards to the open systems, both steady-flow and unsteady-flow processes are considered for energy analysis through the balance equations. In this regard, the FLT is essentially applied to all types of the closed and open systems. Numerous examples are presented for almost every type of system to provide better understanding of concepts and balance equations. Finally, a thermal energy storage system with its three subprocesses (charging, storing, and discharging) is presented to illustrate how to analyze time-dependent uniform-state uniform-flow processes.

Study Questions/Problems

Multiple Choice Questions

1 A piston–cylinder device expands in a polytropic process, which of the following remains constant:
   a PV^n
   b RT
   c V
   d T
   e PV.

2 A rigid tank filled with liquid water is being heated from an external power source until all the liquid water evaporates to superheated vapor, which of the following remains constant during the previous mentioned process:
   a Volume
   b Density

- **c** Pressure
- **d** Internal energy of the water
- **e** All of the above.

3 A piston–cylinder device filled with water being heated from an external power source until all the liquid water evaporates to superheated vapor. Which of the following remains constant during the previous mentioned process:
- **a** Volume
- **b** Density
- **c** Pressure
- **d** Internal energy of the water
- **e** All of the above.

4 When the volume inside a piston–cylinder device expands:
- **a** It generates boundary work
- **b** It generates shaft work
- **c** It consumes boundary work
- **d** All of the above
- **e** None of the above

5 Which one of the following processes is associated with the throttling process
- **a** Isothermal
- **b** Isenthalpic
- **c** Isochoric
- **d** All of the above
- **e** None of the above

6 A system which has a mass flow in and out is referred to as
- **a** Closed System
- **b** Isolated System
- **c** Open System
- **d** None of the above

7 A process in which the pressure remains constant is known as
- **a** Isochoric
- **b** Isothermal
- **c** Isotropic
- **d** Isobaric

8 The USUF process stands for
- **a** United system, United Flow
- **b** Uniform system, Uniform Flow
- **c** United Steady, Uniform Flow
- **d** Unsteady system, Unsteady Flow

9 Which one of the following is not an example of open system devices
   a Piston-cylinder
   b Turbine
   c Pump
   d Nozzle

10 Heat transfer cannot not occur in which one of the following processes:
   a Isobaric
   b Isothermal
   c Isochoric
   d Adiabatic

Problems

1 Air that behaves in ideal gas behavior is heated at a constant pressure of 200 *kPa* from a temperature of 300 to 600 *K*. Calculate the change in the internal energy of the air using (a) average specific heat capacity, (b) specific heat capacity that is a function of temperature, (c) using data from the tables. (d) Compare (a) and (b) with the answer in part (c) in terms of error difference.

2 Consider the above given problem and change air to carbon dioxide (CO_2) to find out how the results will change.

3 Nitrogen that behaves in ideal gas behavior is heated at a constant pressure of 200 *kPa* from a temperature of 300 to 600 *K*. Calculate the change in the internal energy of the air using (a) average specific heat capacity, (b) specific heat capacity that is a function of temperature, (c) using data from the tables. (d) Compare (a) and (b) with the answer in part (c) in terms of error difference.

4 Consider the above given problem and change nitrogen to hydrogen (H_2) to find out how the results will change.

5 A 5 *L* rigid tank, as shown in the figure, contains air at 100 *kPa* and 20 °*C*. 66 *kJ* of heat is transferred to the air; 10% of the heat supplied is lost. 100 *kJ* of mechanical work through shaft 1 and 50 *kJ* of mechanical work through shaft 2 are used to run the fans inside the tank. The properties of air will be taken from the tables. Write mass and energy balance equations and find the final pressure and temperature of the tank.

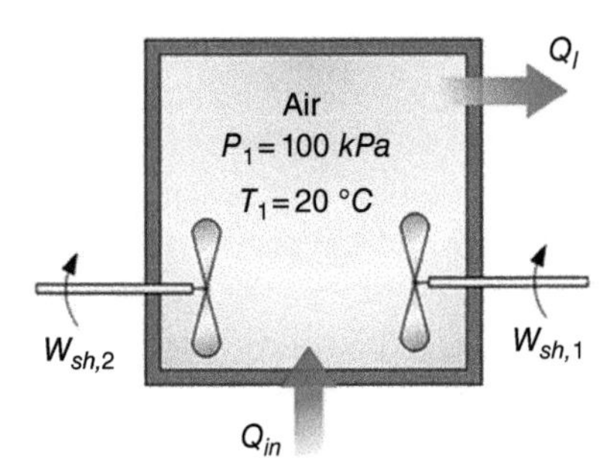

6 Consider the above given problem and change air to carbon dioxide (CO_2) to find out how much the results will change.

7 The water in a rigid tank, as shown in the figure, initially at a pressure of 1 *atm* and a temperature of 25 °*C* has a volume of 10 *L*. 50 *kJ* of heat is transferred to the water; 50% of the supplied heat is lost. An electrical heater runs on 0.5 *A* for 20 seconds while being connected to 120 *V* electric source. A rotating shaft adds 120 *kJ* of shaft work to the water in the tank while stirring it.

a Write the mass and energy balance equations.

b Calculate the final temperature of the tank.

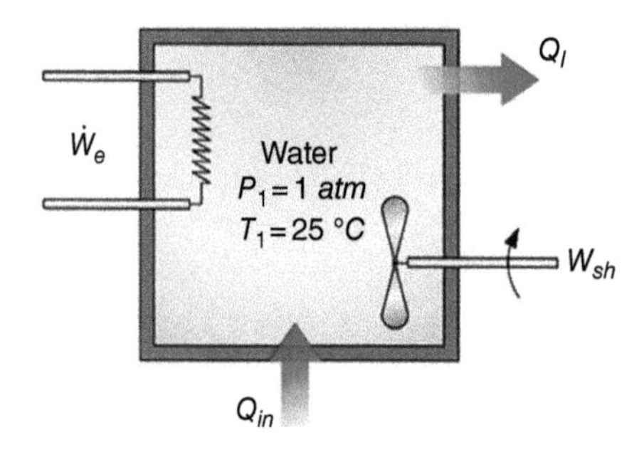

8 Consider the above given problem this time as an adiabatic rigid tank and find out how the results will change.

9 Consider a rigid tank filled with helium which is heated by a heat source at a temperature of 100 °*C*. The rigid tank has a volume of 1 m^3 and contains a 1 *kg* of real gas helium. Calculate both the pressure and temperature of the helium in the rigid tank after 10 *kJ* of heat is transferred to the system while knowing that the system initially was at an ambient temperature of 25 °*C*.

10 As shown in Figure 3.11, an adiabatic piston–cylinder filled with air is compressed from 150 *kPa* to 1000 *kPa*. If the initial volume is 0.5 m^3, and initial temperature is 80 °*C*. Assume that the air in the piston–cylinder device can be treated as an ideal gas through the compression process. Find the final volume in the piston–cylinder after compression.

11 A piston–cylinder device contains air that is initially at a pressure of 400 *kPa*, as shown in the figure. A paddle wheel is used to add work energy to the air inside the piston–cylinder device with an amount of 50 *kJ* per each *kg* of air inside the piston–cylinder. Thermal energy is added to the system from a heat source with a temperature of 100 °*C*, transferring the system from state 1 to state 2 and the piston moves so that the temperature of the system remains constant at 17 °*C*, through the process the volume contained in the system increases to three times its original volume. Calculate:

a The boundary work
b The thermal energy added to the system.

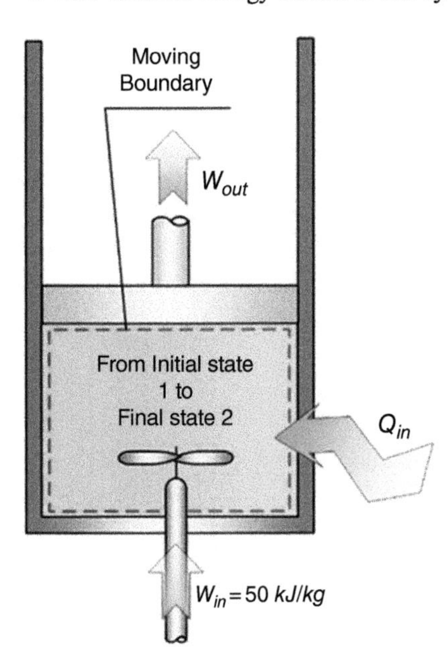

12 A piston-cylinder device with an initial volume of 0.1 m^3, as shown in the figure, contains saturated water vapor at a temperature of 100 °C. The water is compressed so that the final volume is 80% of the initial volume. If 40 kJ of heat is lost to the environment through the compression process:
a Write mass and energy balance equations.
b Calculate the electrical power supplied.

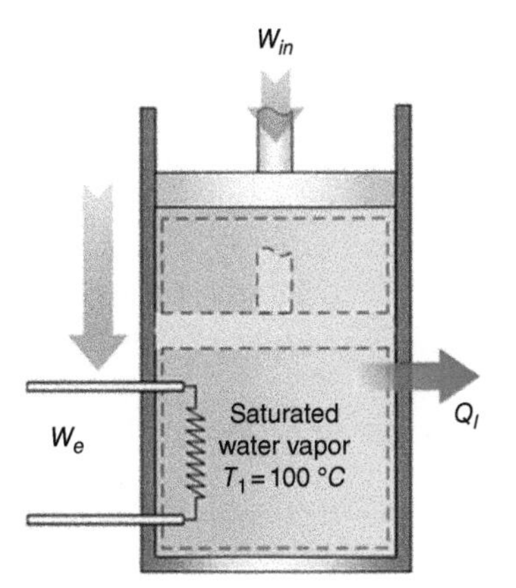

13 As shown in Figure 3.16, a piston–cylinder device, as shown in Figure 3.16, contains a total of 20 *kg* of *R*134*a* refrigerant in saturated vapor state at a initial pressure of 180 *kPa* and final pressure of 140 *kPa*. An electrical heater supplies 50 *kJ* of electric work.

a Write the mass and energy balance equations

b Calculate the electrical heat supplied through electric heater if the boundary work done is 165 *kJ*.

14 Water flows at a flow rate of 1 *kg/s* and enters a water pipe at a temperature of 100 °*C* and a pressure of 100 *kPa*, as shown in the figure. The pipe is insulated such that the pipe loses heat at a rate of 3 *kW* per each meter of pipe length.

a Write the balance equations.

b Find the temperature at the outlet of the pipe.

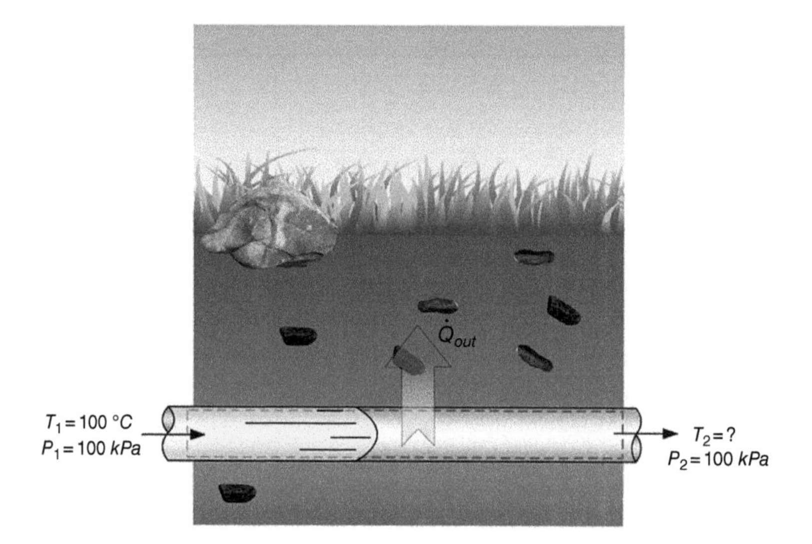

15 Air enters a diffuser steadily at a velocity of 250 *m/s*. The exit area of the diffuser is 0.4 m^2. The mass flow rate through the diffuser is found to be 78.8 *kg/s*, where the air thermophysical properties are shown in a diffuser diagram as illustrated in Figure 3.20a.

a Write the balance equations.

b Find the velocity of the air at the diffuser exit using the energy balance on the diffuser.

16 Consider that a gas flow of oxygen (O_2) with a velocity of 270 *m*/s enters an adiabatic diffuser at a pressure of 60 *kPa* and a temperature of 7 °*C* and exits the diffuser at a pressure of 85 *kPa* and a temperature of 27 °*C*. Calculate:

a The exit velocity of the fluid.

b The ratio of the inlet area to the exit area of the nozzle.

17 An adiabatic steam turbine receives 10 kg/s of superheated steam at a temperature of 550 °C and a pressure of 10 000 kPa, as shown in the figure. The steam turbine converts part of the energy in the steam to shaft work. The steam exits the steam turbine as a saturated vapor at a pressure of 10 kPa and a mixture quality of 1.

a Write the balance equations.

b Find the power output.

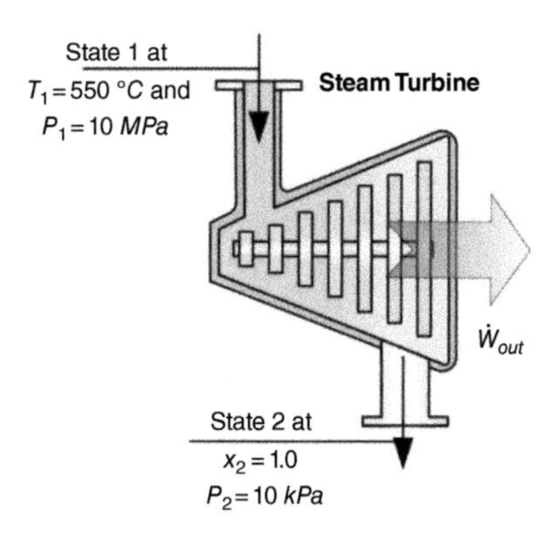

18 Consider an adiabatic steam turbine, as shown in the figure, with the following inlet and exit states: $P_1 = 12\,000\,kPa$, $T_1 = 625\,°C$, $P_2 = 10\,kPa$, $x_2 = 0.95$. Taking the dead-state temperature of steam as saturated liquid at 25 °C:

a Write the balance equations.

b Determine work rate output by the steam turbine if the mass flow rate is known to be 2 kg/s.

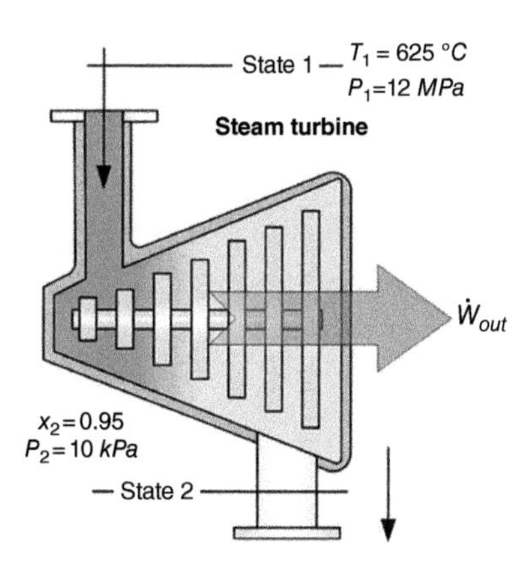

19 For the heat exchanger shown in the figure, write the mass and energy balance equations and find the heat rate required by the heat exchanger if the fluid is water flowing at a flow rate of 0.5 kg/s.

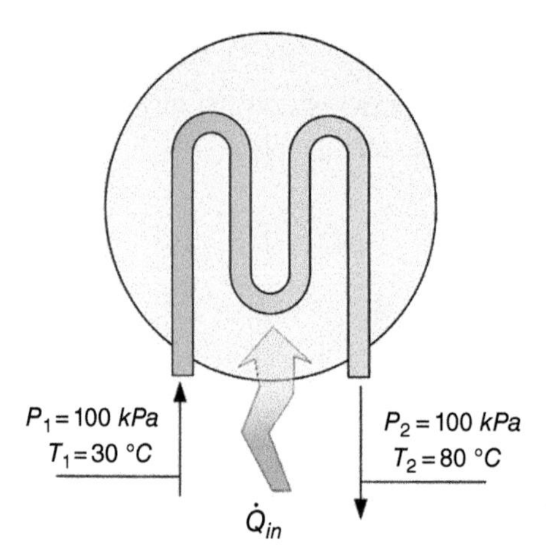

20 A water pump raises the pressure of water from 100 *kPa* to 2.0 *MPa*, as shown in the figure. The initial temperature of the water is 25 °*C*.

a Write the balance equations.

b Calculate the work rate consumed by the pump if the mass flow rate is 2.5 *kg/s*.

c Find the temperature of the water leaving the pump.

d Calculate the rate of entropy change of water during the process.

e Calculate the entropy generation rate.

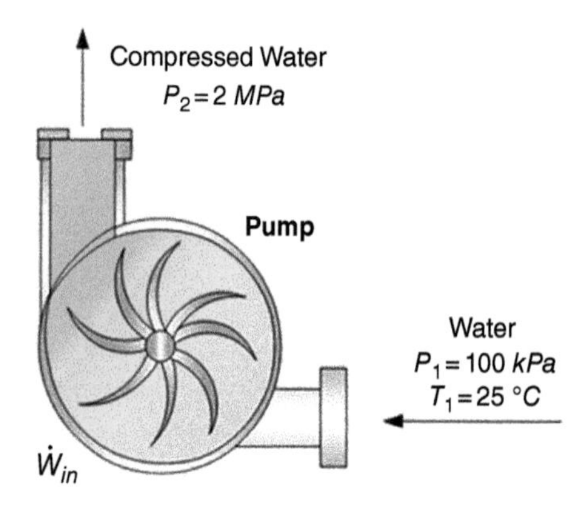

21 A 32 *kg/s* of water is heated at a constant pressure by 4.8×10^6 *kg/h* of exhausts gases, as shown in the figure. Assume the exhaust gases consist only of air where you can use ideal gas equations. $\left(c_{p,air} = 1.005 \frac{kJ}{kgK}\right)$.

a Write the balance equations.

b Calculate the exit temperature of water.

c Calculate the exit temperature of exhaust.

d Calculate the amount of heat rate transferred to the water.

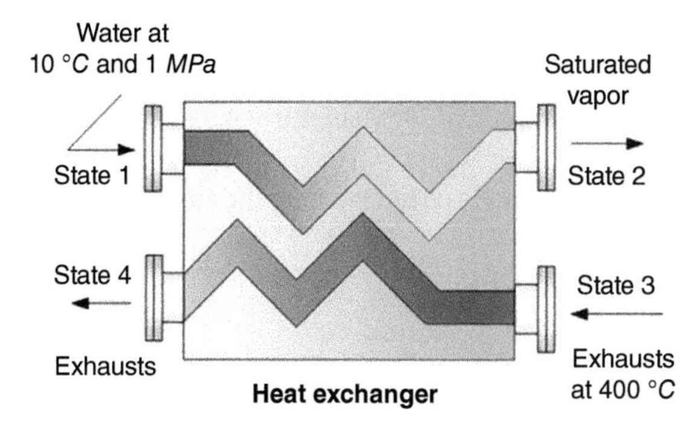

22 The pressure of water at ambient conditions is increased in a pump to 800 *kPa* and a temperature of 30 °*C*, as shown in the figure. The water entering the pump is at a temperature of 10 °*C* and atmospheric pressure, and the mass flow rate is 1.2 *kg*/*s*.

a Write the balance equations.

b Determine the amount of power the pump consumes during the pressurization process.

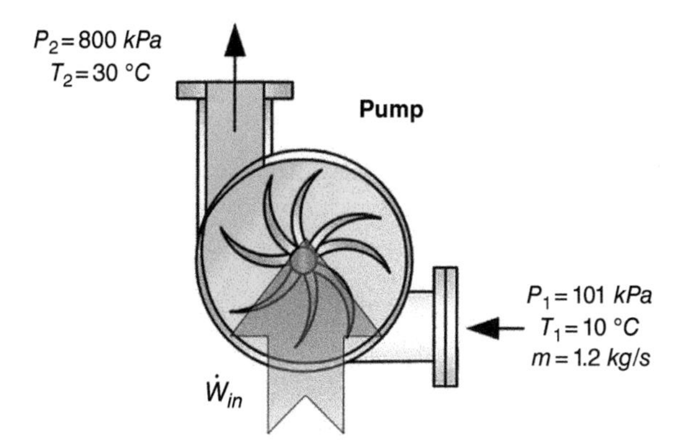

23 An air compressor (as shown in Figure 3.25) compresses air from atmospheric pressure and a temperature of 280 *K* to the discharge conditions at a pressure and a temperature of 600 *kPa* and 400 *K*, respectively. The mass flow rate of the air entering the compressor is 1 *kg*/*s*.

a Write the balance equations.

b Find the work rate consumed by the compressor.

24 Consider that an adiabatic nozzle (as shown in Figure 3.20) has accelerating steam entering with a pressure of 600 *kPa* at a temperature of 500 *K* and velocity of 120 *m*/*s*. The adiabatic nozzle has a cross-sectional area ratio of 2 : 1, which results in an exit velocity of 380 *m*/*s*. Assume a constant specific heat of 1.95 *kJ/kgK* and negligible changes in density. Calculate the following:

a The exit temperature.

b The exit pressure.

25 Consider an unmixed heat exchanger as shown in the figure. Water at an ambient pressure and a temperature of 99 °*C* is used to maintain the temperature of a building at 30 °*C*; the water exits the building heating system at a temperature of 35 °*C*.

a Write the balance equations.

b Find the amount of thermal energy released from the hot water to the ambient of the building if the mass flow rate of water is 2.8 *kg/s*.

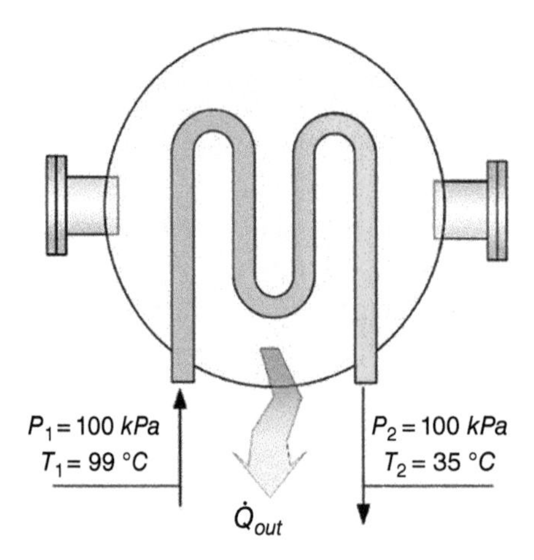

26 In an expansion valve (as shown in Figure 3.27) propane enters the pipe at a pressure of 11 *bar* and a temperature of 10 °*C*. The expansion valve reduces the pressure of the propane to 3 *bar*.

a Write the balance equations.

b Determine the exit temperature of the propane.

27 In an expansion valve (as shown in Figure 3.27), refrigerant *R*134*a* enters at 7 *bar* and temperature of 20 °*C*. The expansion valve drops the pressure to 1 *bar*.

a Write the balance equations.

b Determine the temperature at which the refrigerant exits the valve.

28 Consider an adiabatic steam turbine (as shown in Figure 3.22) that operates between the pressures of 1 *MPa* and 10 *kPa*. If steam enters at a temperature of 350 °*C* with a mass flow rate of 5 *kg/s* and the turbine has an isentropic efficiency of 85%.

a Write all balance equations for the turbine.

b Determine the work output rate from the turbine.

29 Consider the mixing chamber shown in Figure 3.28 in Example 3.19. The mass flow rate of input stream 1 is 1 *kg/s*, input stream 2 is 2.5 *kg/s*, input steam 3 is 0.5 *kg/s*, input stream 4 is 0.75 *kg/s*. If the input temperatures and pressures of each stream are as described in the example and the rate of heat loss is 8 *kW*.

a Write the mass and energy balance equations for the mixing chamber.

b Identify the state of water exiting the mixing chamber.

c Find the temperature of exiting water.

30 Water is flowing at the flow rate of 2 kg/s through a 5 m long pipe (as shown in 3.19). Water enters the pipe at temperature of 55 $°C$ and ambient pressure. The heat losses of 10 kW occurred through the pipe length.

a Write both mass and energy balance equations.

b Find the enthalpy of the water leaving the pipe.

4

Entropy and Exergy

4.1 Introduction

Energy and its sources have always been on the top of the list of the most critical issues for humanity, which started using wood and then coal as the first source of energy, subsequently followed by oil then natural gas. Throughout history it was one of the main reasons for conflicts, wars, and peace as well. A corner stone in energy importance and use was the industrial revolution, which changed factors from the source of power, to the numbers of workers, to the use of machines that run on power and energy. Since the industrial revolution and the changes that it caused to the infrastructure of cities, the competition over energy has been even stiffer. With the rise of competition for energy and its resources it is has become more apparent that humanity needs more efficient, more cost effective, more environmentally benign, and more sustainable energy options and solutions. The rise of these requirements have become the main motivation behind going beyond traditional analysis methods and techniques. Traditionally, systems design, analysis, and performance assessment was carried out through a single tool, which is the first law of thermodynamics (FLT). It is now crystal clear to everyone that the FLT is insufficient and incapable of addressing practical systems with irreversibilities (or losses, inefficiencies, etc.). That is why the future of thermodynamics analysis should not stop at the FLT, and the second law of thermodynamics (SLT) has been brought into the picture to account for irreversibilities or destructions through entropy and exergy. Exergy has distinguished itself to be a primary tool under the SLT.

Generally, thermodynamics is usually defined in various books and scientific articles as the science of energy (referring to the FLT) and entropy (referring to the SLT). Since one can consider thermodynamics as the key subject in studying the concepts of the FLT and the SLT and their applications to various types of closed and open systems, the subject of thermodynamics is redefined here as the science of energy and exergy, which is more correct and consistent with the subject matter of the laws. Here, the new definition covers both laws of thermodynamics, gives both energy and exergy quantities in the same unit consistently, and embraces two key efficiencies, as the energy concept relates to energy efficiency and the exergy efficiency relates to performance assessment. This way the concepts dwell on the right pillars for practical applications. The exergy efficiency becomes important for practical systems and applications since it is a true measure of system performance and indicates how much actual performance deviates from ideal performance.

Thermodynamics: A Smart Approach, First Edition. Ibrahim Dincer.

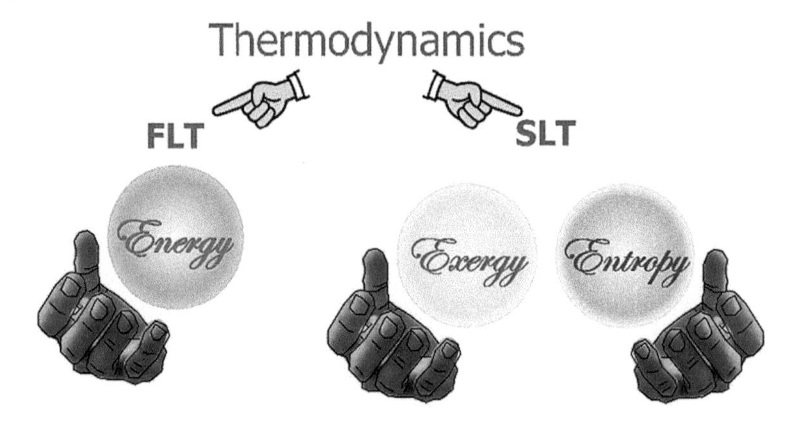

Figure 4.1 An illustration of thermodynamic pillars.

Figure 4.1 illustrates the stated points, indicating that the FLT brings energy only and that the SLT brings both entropy and exergy, which makes this chapter the heart of this particular book. Since energy is covered in the previous chapters, this chapter focuses on entropy and exergy and related concepts.

Through the brief introduction about energy and energy systems, and how thermodynamics can in aid in assessing their performance and how efficiently we are utilizing our energy resources, as well as how the thermodynamics performance assessment tools are based on the first and second laws of thermodynamics, we can now understand the importance of the second law. Although the author believes that exergy is the key property from the SLT, the chapter will discuss entropy with brief examples, followed by exergy. More importantly, this chapter introduces the balance equations for entropy and exergy in addition what were introduced previously in Chapter 3 for mass and energy balance equations for closed and open systems. Therefore, all four balance equations for mass, energy, entropy and exergy which are applied to various types of closed and open types of systems.

4.2 The Second Law of Thermodynamics

Referring back to the previous examples that have explained what the FLT lacks in terms of presenting the maximum obtainable work or work potential or energy quality or maximum amount of energy or available energy of a system or a medium, the key weakness of the FLT is that is does not account for irreversibilities, losses, inefficiencies, and destructions. Since all practical systems are irreversible and end up with irreversibilities, losses, inefficiencies, and destructions, there is a strong need to go beyond this law. This can only be done by considering the SLT as the prime tool. If one further elaborates on the issues, the following example is helpful. The FLT presents that the energy in the upper part of the water in the ocean has more energy than the combustion products that result from a complete combustion of 1 *kg* of natural gas. When it presented in this way the disadvantages of the FLT clearly appear. In order to tackle with the shortcomings of the FLT, one has to go to the SLT for system design, analysis, and assessment. The SLT considers both quantity and quality of energy, and it hence states that actual processes occur in the direction of decreasing quality of energy.

Following the example that has been stated regarding the upper part of the water in the ocean, since it is at ambient temperature and pressure then its quality is quantified and it is zero as it is already at ambient conditions which are considered dead state (or reference state).

In the SLT, there are historically two key statements made earlier to emphasize what is possible. Or what is impossible. These are known as Kelvin–Plank statement and the Clausius statement.

(a) Kelvin–Plank statement

Before starting with the definitions of the two crucial properties derived from the SLT, which are entropy and exergy, it is important to introduce the Kelvin–Plank statement of the SLT, which can be stated as follows: "*no system can produce a net amount of work while operating in a cycle and exchanging heat with a single thermal energy reservoir.*" The subject matter of the Kelvin–Plank statement simply means that there is no heat engine possible of converting the entire heat input to work output with a conversion efficiency of 100%, with no waste or loss of energy. The illustration of this statement is given in Figure 4.2a for a heat engine that is practically impossible. It shows that conversion of heat input to the heat engine completely to work output is impossible, there always should be heat rejection to the surrounding environment.

(b) Clausius statement

The second statement linked to the SLT is the Clausius statement, which is also related to the Kelvin–Plank statement, with the difference that it is specifically for a refrigerator. It states that "*it is impossible to construct a refrigerator that transfers heat from a cold reservoir to a hot reservoir without the aid of work input.*" or "*no refrigeration system can operate without work input to provide the desired cooling load.*" It simply states that there is not possible for any refrigerator to produce a cooling effect with no external energy input. The illustration of this statement is given in Figure 4.2b for a refrigerator.

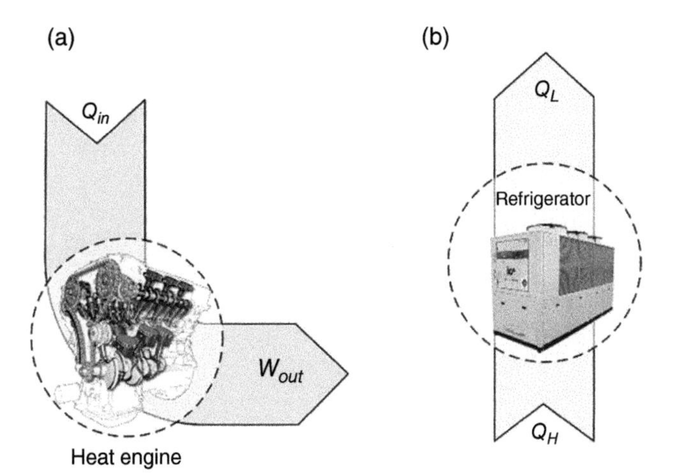

Figure 4.2 Impossibilities stated by (a) the Kelvin–Plank statement for a heat engine and (b) the Clausius for a refrigerator.

Clausius inequality

The Clausius inequality is known as the Clausius theorem with a mathematical explanation of the SLT through entropy. It was essentially introduced by Rudolf Clausius in order to link the entropy of the system to the heat transfer within the system and/or crossing the system boundary through the interaction with the surroundings. Of course, his ultimate target was to develop an entropy relationship, understand its role, and determine its magnitude quantitatively. The relationship gives a quick determination if a cyclical process is reversible or irreversible.

Furthermore, the Clausius statement is about the entropy and its inequality, which is valid for all thermodynamic cycles in specific, and systems in general. If entropy changes, it is isentropic or reversible. Moreover, it always becomes greater than zero, which represents irreversibility due to irreversibilities. The concept of inequality (out of the Clausius statement) is applicable to all cycles, including refrigeration cycles. A cycle with no irreversibilities is ideally reversed and considered a reversible cycle. Of course, this is something practically impossible. One may consider a system or cycle reversible for the sake of analysis where the results of it will not reflect the reality, as all systems (including cycles) are irreversible. In closing, one can conclude that the equality in the Clausius statement (the so-called Clausius inequality) holds for totally reversible or just internally reversible cycles, and the inequality for all the irreversible ones. From the Clausius inequality one can write the following cyclic integration:

$$\oint \frac{\delta Q}{T} \leq 0$$

One can also break the cyclic integration to:

$$\int_1^2 \frac{\delta Q}{T} + \int_2^1 \left(\frac{\delta Q}{T}\right)_{int} \leq 0$$

A mathematical manipulation may be done by opening up the second integration as follows:

$$\int_1^2 \frac{\delta Q}{T} + S_1 - S_2 \leq 0$$

Rearranging the terms to have the system change in entropy at one side of the equation gives:

$$\int_1^2 \frac{\delta Q}{T} \leq S_2 - S_1$$

which results in:

$$dS \geq \frac{\delta Q}{T}$$

Example 4.1 From each of the pairs select which energy reservoir or source has higher quality: (a) ice cube and warm water; and (b) hot tea and cold water:

Solution

a) The energy in the **warm water** is a higher quality than that of the ice cube, since if both are brought into contact energy will transfer from the warm water to the ice cube, which eventually will melt. Based on the SLT energy will transfer from a high quality source to a lower quality source, the warm water will then have a higher quality energy.
b) The energy in the **hot tea** has higher quality than that of the cold water for the same reason mentioned in point (a).

4.3 Reversible and Irreversible Processes

Based on the definitions of the Kelvin–Plank statement about the SLT, no heat engine will have an efficiency of 100%. A question may be arise in this regard: if a 100% efficient heat engine is not possible, what efficiency would then be possible? or, what is the maximum efficiency that a heat engine can have? In order to answer these questions, there is a need to describe ideal (maximum) efficiency and practical (actual) efficiency, which will require one to dwell on reversible and irreversible processes to be able to address these points.

A **reversible process** is a process that can be reversed without leaving any trace on the surroundings, in particular with no losses. On the contrary, an **irreversible process** is a process where reversibility cannot be achieved, meaning that the process ends up with irreversibilites. Figure 4.3 show both reversible and irreversible processes illustrated with two examples. Figure 4.3a shows a pendulum with no air friction, where the system and its surroundings are brought back to the system initial state at the end of the reversing the process. Another definition is when the net work and the net heat transfer exchanged between the surroundings and the system are equal to zero for the full cycle. Figure 4.3b shows a canned

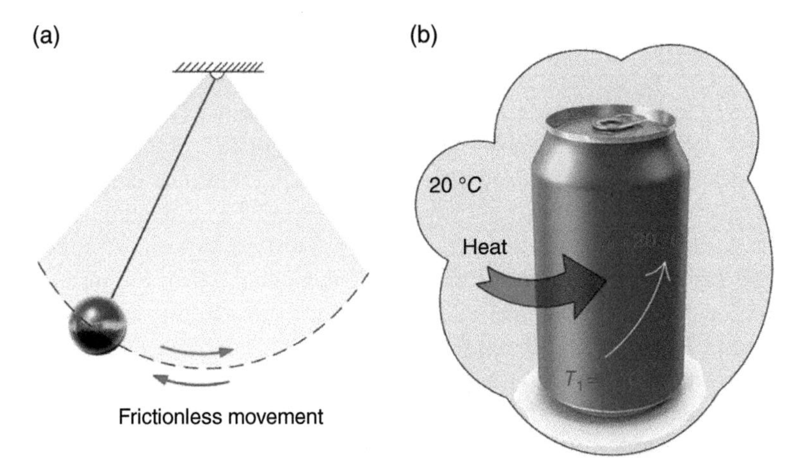

Figure 4.3 Graphical illustrations of (a) a reversible process and (b) an irreversible process.

drink initially at 0 °*C* is placed in an environment at 20 °*C*. Heat is transferred from the surroundings to the cold drink until it reaches the same temperature at 20 °*C*, which leads to thermal equilibrium. Then we can define real processes that do not follow the principles of reversible processes as irreversible processes.

Referring back to the start of the start of the section, where the definition of a reversible processes was introduced as an answer to what maximum efficiency a heat engine can have since 100% is not possible. This leads to the definition of **the second law efficiency** for actual processes, which is a measure of how far the actual system performance resembles the performance of a reversible process. Of course, it will indicate how much an actual process will deviate from the ideal one. For the sake of consistency, energy and exergy efficiencies will be used throughout for heat engine type systems and energetic and exergetic coefficients of performance for refrigeration and heat pump systems.

4.4 The Carnot Concept

Although the Carnot cycle is usually used in many thermodynamics textbooks, it is about a **Concept of Ideality**. In fact, with this concept Sadi Carnot aimed to define the limit for power generating and refrigeration cycles. Anyway, we should not be confused about this since there is no cycle introduced by Carnot. In conjunction with this, the Carnot concept is the concept of ideality, where it shows the maximum potential performance a power engine or a thermodynamics cycle can achieve. To show how the Carnot concept is applied to power generation cycles, especially the power generation cycle that can be approximated as a piston–cylinder device, will be introduced later the chapter. The result of the application of the Carnot concept on the previously mentioned power cycle is a Carnot cycle that results in the highest efficiency of any cycle of heat engines under power generating systems. This comes as the Carnot cycle neglects all irreversibilities, such as friction, heat losses, etc. This makes the Carnot cycle a completely reversible cycle that consists of the following processes, as shown in Figure 4.4, where the first stage is a **reversible isothermal expansion** in which the gas absorbs thermal energy q_{in} and expands at a constant temperature T_h. As the gas expands, work is done on its surrounding, as the cylinder head is moving. Theoretically, as the air expands the gas should experience a temperature drop. However, a heat source is provided to maintain the gas's isothermal state. The reversible isothermal expansion is followed by a **reversible adiabatic expansion**, in which the gas expands and experiences a temperature drop from T_h to T_l. The gas continues to expand however; the thermal heat source is not provided anymore but replaced by insulation. The piston in this process is assumed frictionless. Furthermore, the process is assumed to be quasi-equilibrium. The **reversible isothermal compression** follows, where the piston is going back to its original position as the external forces from the surroundings are doing work on the gas. The insulation in this process is removed and a heat sink at a temperature of T_l. is provided to the cylinder's head. The heat sink ensures the gas's isothermal state as it is compressed. Finally, during **reversible adiabatic compression** in this process insulation replaces the heat sink, and as the piston continues to compress the gas the temperature of the gas rises to its original temperature of T_h.

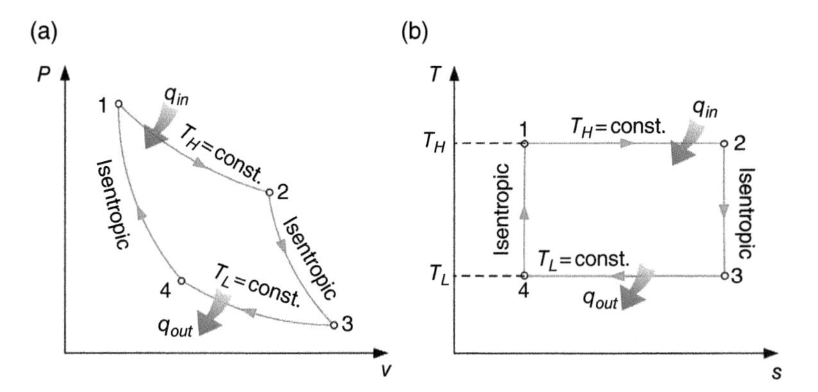

Figure 4.4 Process illustrations of (a) *P-v* diagram and (b) *T-s* diagram of the Carnot cycle applied to heat engines that are based on a closed type processes.

A demonstrative *P-v* diagram is shown for the Carnot cycle in Figure 4.4. The temperatures at state 1 and 2 are assumed to be equal to T_h and the temperatures at state 3 and 4 are assumed to be equal to T_l. The area inside the *P–v* diagram represents the network output of the system along with the quasi-static equilibrium boundary. During the processes 1-2-3 work is being done by the gas on its surroundings as the gas is expanding, as seen in Figure 4.4. As for process 3-4-1 work is being done on the gas as the gas is being compressed and the correlating decreasing volume can be seen in Figure 4.5.

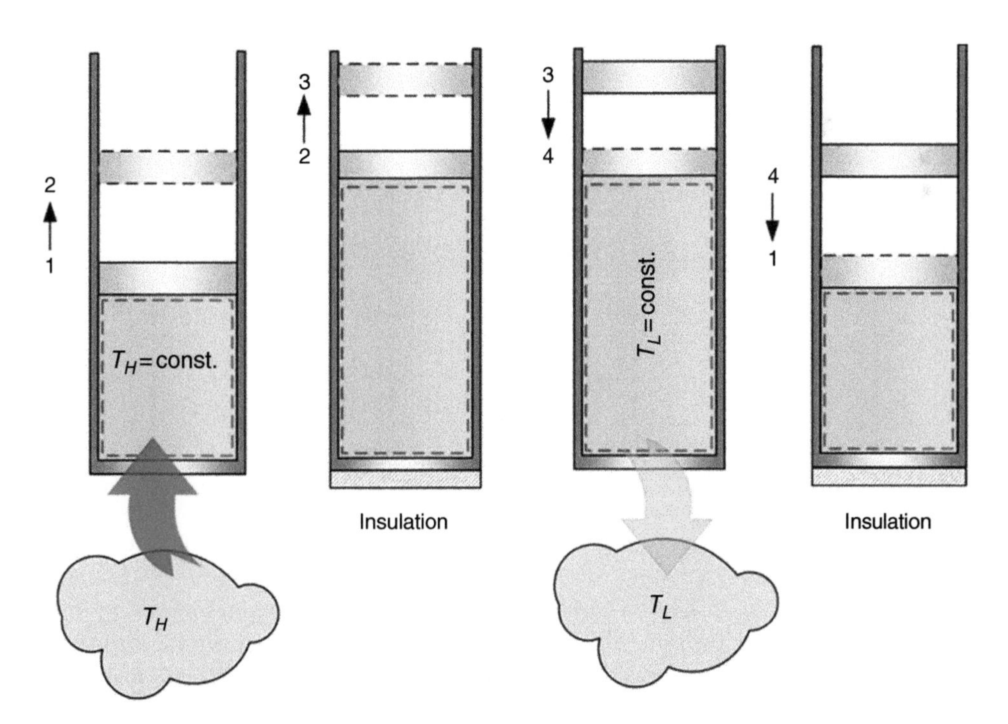

Figure 4.5 Carnot cycle demonstration using an ideal piston.

4.4.1 The Carnot Principle

The previously introduced Kelvin–Plank and Clausius statements impose the second law limits on the operation of the cyclic devices. As shown previously in Figure 4.2, a heat engine cannot produce work output only exchanging heat with a single thermal reservoir as well as a refrigeration unit cannot operate without receiving work input. From the earlier statements one may extract the Carnot principles into two main outcomes:

- **First outcome**: this states that the efficiency of a real heat engine will always be less than a reversible heat engine where both operate between the same two thermal reservoirs (such as high-temperature reservoir at T_H and low-temperature reservoir at T_L).
- **Second outcome**: this states that the efficiency of any number of reversible heat engines that operate between the same two thermal reservoirs is always equal.

The above stated outcomes of the Carnot principle are graphically shown in Figure 4.6, where the irreversible heat engine will always have a lower efficiency than the reversible heat engine when both are operating between the two same thermal reservoirs.

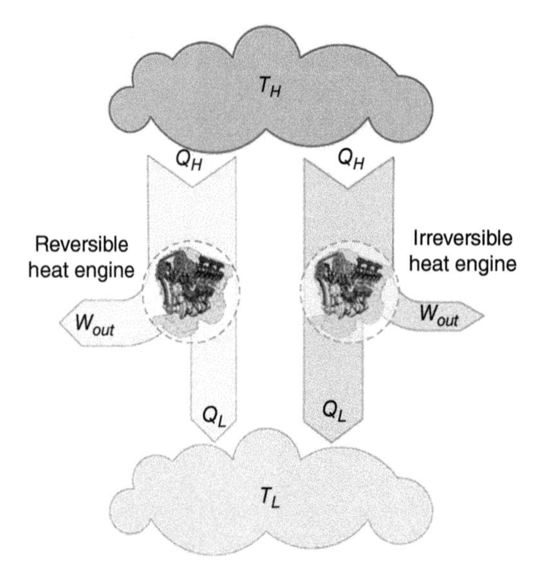

Figure 4.6 Demonstration of the Carnot principle.

4.4.2 Temperature Ratio

A thermodynamics temperature scale is a temperature scale that does not depend on the properties of substances that are used to measure the temperature. The second statement of the Carnot principle discussed earlier mentions that all heat engines operating in a reversible manner have the same efficiency if they were operating between the same two thermal reservoirs. We can derive from the previous statement that the performance of reversible engines does not depend on the working fluid and its properties, as well as being independent of the type, components, and execution of the cycle. Then the efficiency of reversible heat engines can be presented as:

$$\eta_{rev} = f(T_H, T_L)$$

And, as will be presented in the coming chapters, the efficiency of heat engines can be presented as:

$$\eta_{HE} = f(Q_H, Q_L)$$

The energy efficiency of heat engines can be calculated as:

$$\eta_{HE} = 1 - \frac{Q_L}{Q_H}$$

If we now integrate the area bounded by the *T-s* diagram of the Carnot cycle shown in Figure 4.3b it will give us the net work output of the Carnot cycle as:

$$W = \int_{S_1}^{S_2} TdS = (T_H - T_L) \times (S_H - S_L)$$

$$W = T_H(S_H - S_L) - T_L(S_H - S_L)$$

Since the difference between the heat input to the heat engine and the heat rejected from the heat engine is equal to the net work out of the heat engine:

$$W = Q_H - Q_L = T_H(S_H - S_L) - T_L(S_H - S_L)$$

From the above equation:

$$Q_H = T_H(S_H - S_L)$$

$$Q_L = T_L(S_H - S_L)$$

Then the energy of efficiency of a Carnot heat engine can be presented as:

$$\eta_{rev} = 1 - \frac{Q_L}{Q_H} = 1 - \frac{T_L(S_H - S_L)}{T_H(S_H - S_L)} = 1 - \frac{T_L}{T_H}$$

Then, for a reversible heat engine, the following is true:

$$\left(\frac{Q_H}{Q_L}\right)_{rev} = \frac{T_H}{T_L}$$

Based on the earlier equation, then the maximum possible efficiency of a heat engine can be calculated based on the temperatures of the two reservoirs it is operating between:

$$\eta_{rev} = 1 - \frac{Q_L}{Q_H} = 1 - \frac{T_L}{T_H}$$

Example 4.2 A Carnot heat engine receives heat from a source at a temperature of 1000 °*C*, and then releases the remaining heat to a cold thermal reservoir at a temperature of 25 °*C*. Find the maximum thermal efficiency the heat engine can achieve

Solution

The maximum possible (ideal) efficiency which may be achieved by a heat engine is the Carnot efficiency, which depends only on the temperatures of the reservoirs it is operating between and it is calculated as follows:

$$\eta_{rev} = 1 - \frac{T_H}{T_L} = 1 - \frac{25 + 273\ K}{1000 + 273\ K} = \mathbf{0.77}$$

4.5 Entropy

The SLT introduces two concepts of entropy and exergy along with two properties of entropy and exergy. The first to be introduced is entropy, which is defined as a degree of disorder or a measure of randomness, and is not a measurable quantity; however, it is calculated through other measurable quantities. Figure 4.7 is a comparative illustration of high entropy and low entropy room in terms of order. The left room is a messy one with high entropy which is recognized as an entropic place where the degree of disorderliness is high.

The entropy as a specific property is an intensive property. However, the entropy as a quantity is an extensive property, where the total entropy is the summation of the change of the system and the change of entropy due to the interactions with the surrounding environment as shown in Figure 4.8 and mathematically defined as follows:

$$\Delta S_{total} = \Delta S_{sys} + \Delta S_{surr} \tag{4.1}$$

Here, S denotes entropy the extensive property, and the subscripts *sys* and *surr* denote system and surroundings. Note that ΔS_{surr} refers to the changes in entropy of the system due to its interactions with the surroundings, while ΔS_{sys} refers to the changes in entropy

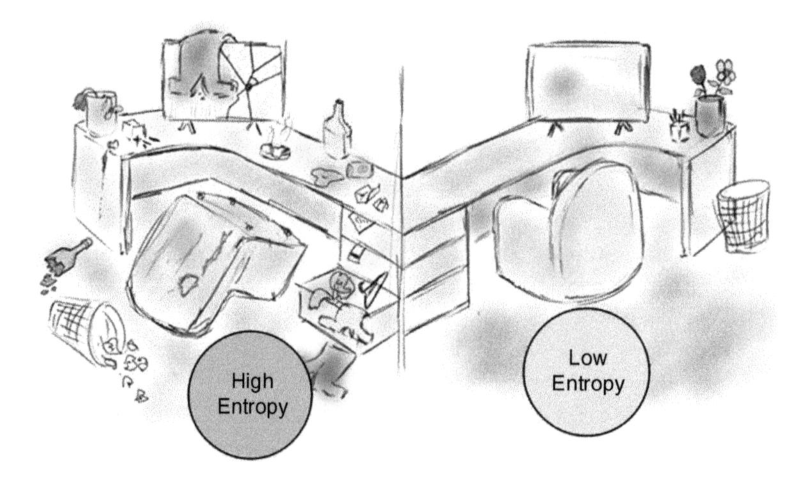

Figure 4.7 Comparative illustrations of low and high entropy cases in a daily life.

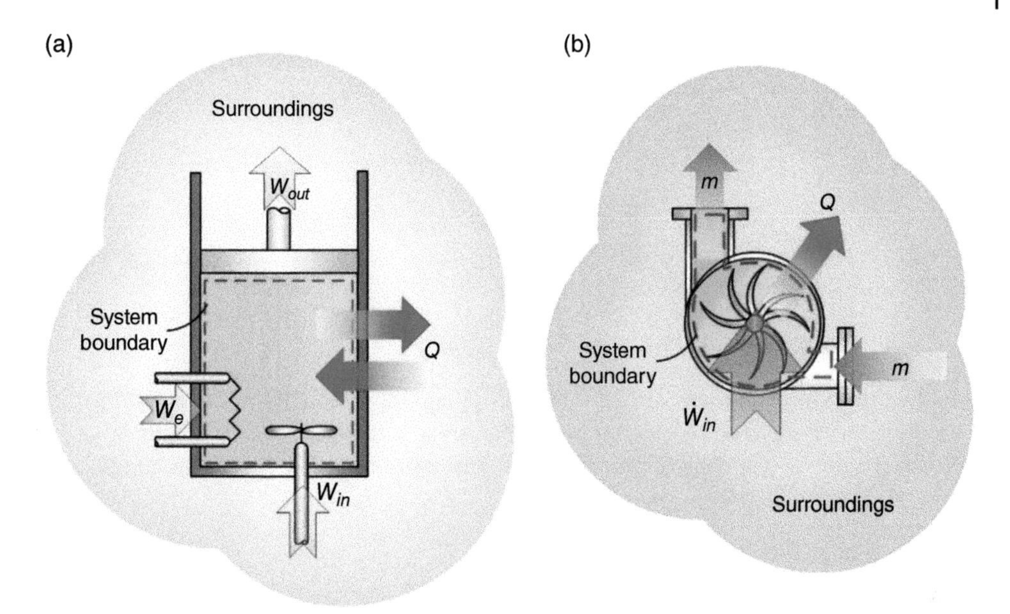

Figure 4.8 Illustration of a system and its interaction with its surroundings (a) for a closed system and (b) for an open system.

of the system due to the changes in the system properties. Note the total change in entropy can also be referred to as entropy generation, then the following equation can be written:

$$S_{gen} = \Delta S_{sys} + \Delta S_{surr} \tag{4.2}$$

Note that since the entropy can be created but not destroyed then entropy generation is always positive for a process that has irreversibilities and zero for completely reversible process. The size of entropy generation indicates the size of the irreversiblities of the system. From the SLT and the concepts of entropy generation, we can write the general entropy balance equation for closed energy systems as:

Initial amount of entropy the system has + the amount of entropy entering the system + the amount of entropy generated = Final amount of entropy the system has + the amount of entropy leaving the system

which can be simplified to:

$$\Sigma S_{in} + S_{gen} = \Sigma S_{out}$$

Entropy may enter and leave the system in one form, which is the entropy associated with heat transfer only. One should, in this regard, note the following:

"Entropy is not associated with any types of work and is not therefore a function of work."

This is also illustrated in Figure 4.9 to indicate that entropy is only associated with heat transfer, but not work.

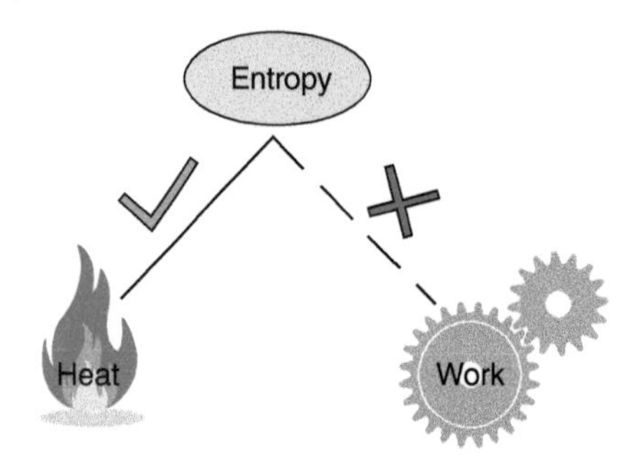

Figure 4.9 Entropy's association with heat, but not with work.

In the case of open systems, there is flow entropy coming in and flow entropy leaving the boundary; these are to be considered in addition to heat transfer across the boundary.

One can note the following common forms of entropy: s is specific entropy in *kJ/kgK*, *S* is entropy in *kJ/K*, and $\dot{S}$ is entropy rate in *kW/K*.

These relationships are then defined as follows:

$$S = m \times s; \ \ \dot{S} = \dot{m} \times s \text{ and } \dot{S} = S/t.$$

The entropy generation rate can be defined as follows regarding the changes in the system entropy rate due to the changes in the properties of the system and the changes in the system entropy rate due to the interactions of the system with its surroundings as shown in Figure 4.8b. It is therefore written as

$$\dot{S}_{gen} = \Delta\dot{S}_{sys} + \Delta\dot{S}_{surr} \tag{4.3}$$

Here, the dot refers to the rate change of the entropy. Then as we have defined a general entropy balance equation for the closed systems, the general entropy balance equation for open systems:

The amount of entropy rate entering the system + the entropy generation rate

= the amount of entropy rate leaving the system

However, for open systems entropy can leave or enter the open system in two forms, entropy associated with heat transfer and that associated with mass flow rate. Finally before ending the section and going to entropy balance equations we will introduce the entropy associated with a heat transfer as follows:

$$\text{One with high-temperature heat source} : \Delta S_H = \frac{Q_H}{T_H}$$

$$\text{One with low-temperature heat source} : \Delta S_L = \frac{Q_L}{T_L}$$

Here, the temperature is the temperature of the source for the case when heat is transferred from a thermal reservoir to the system, and in the case where heat is leaving the system then we chose the temperature of the thermal reservoir receiving the heat from the system.

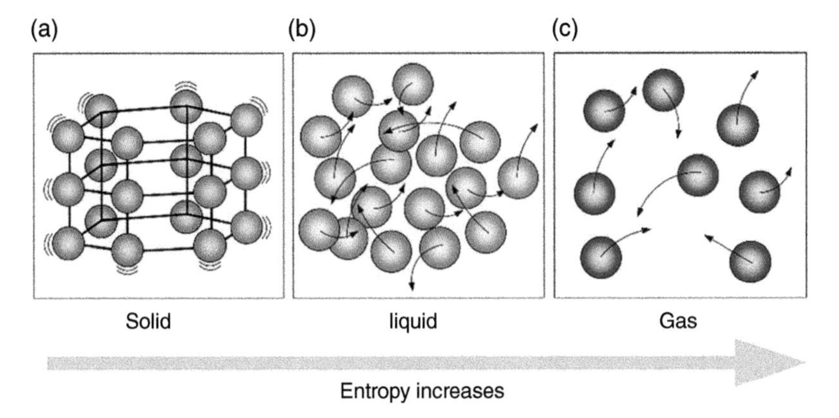

Figure 4.10 Magnitude of entropy (a) for solid, (b) for liquid and (c) for gas.

There is an important thing to highlight here, that is how the entropy will change from solid to liquid, liquid to vapor, etc. As illustrated in Figure 4.10, entropy increases from solid to liquid and from liquid to vapor. The lowest entropy will be in the solid while the highest is in the gaseous form.

If we continue on the illustration given in Figure 4.10 and provide some values for specific conditions:

- Specific entropy for ice at $-2.5\,°C$: $s_s = -1.24\,kJ/kgK$ (this is negative due to the fact that heat is removed and makes it negative).
- Specific entropy for saturated liquid at $25\,°C$: $s_l = 0.367\,kJ/kgK$.
- Specific entropy for superheated vapor at $250\,°C$: $s_v = 8.03\,kJ/kgK$.

This clearly shows that the highest entropy takes places in superheated vapor.

The next section introduces different examples that deals with entropy balance equations of closed and open systems.

4.6 Entropy Balance Equations

The mass energy balance equations are written earlier in Chapter 3 for various closed and open types of thermodynamic systems. We now proceed one more step ahead and write entropy balance equation for closed systems:

$$\Sigma S_{in} + \frac{\Sigma Q_{in}}{T_s} + S_{gen} = \Sigma S_{out} + \frac{\Sigma Q_{out}}{T_{surr}}$$

where subscripts in denotes input, s source, gen generation, out output, and surr immediate surrounding.

The above equation is now written for open systems in a rate form as:

$$\Sigma \dot{S}_{in} + \frac{\Sigma \dot{Q}_{in}}{T_s} + \dot{S}_{gen} = \Sigma \dot{S}_{out} + \frac{\Sigma \dot{Q}_{out}}{T_{surr}}$$

In the following section, the main focus will be given to how to write entropy balance equations for various types of closed and open types of systems under various conditions. Note that there will be mass balance equations, energy balance equations, and entropy balance equations written for them.

In regards to the ideal gases, the work consumption rate of the pump was not included in the entropy balance equation of the pump since there is no entropy associated with the work rate. The next example analysis is of an heat exchanger in terms of entropy in addition to mass and energy, where the example also features the presence of an ideal gas, air. Note that for ideal gases equations can be used to find the entropy change of the substance rather than relying on property tables or thermophysical property database software such as EES. For ideal gases the change in entropy can be calculated using the following equations:

$$\Delta s = s_2 - s_1 = \left(c_p \, ln \left(\frac{T_2}{T_1}\right) - R ln \left(\frac{P_2}{P_1}\right)\right) \tag{4.4}$$

$$\Delta s = s_2 - s_1 = \left(c_v \, ln \left(\frac{T_2}{T_1}\right) + R ln \left(\frac{v_2}{v_1}\right)\right) \tag{4.5}$$

Some examples are now presented using the above given equations for ideal gases to exemplify the solution methodology.

Example 4.3 A 1 m^3 rigid tank being heated and stirred, as shown in Figure 4.11, contains air at 100 kPa and 20 °C. 66 kJ of heat is transferred to the air where 10% of the heat supplied is lost. 100 kJ mechanical work through shaft 1 and 50 kJ mechanical work through shaft 2 are provided to run the fans inside the tank. Take the source temperature for incoming heat as 600 °C and the immediate surrounding (boundary) temperature as 60 °C. The properties of air can be taken from tables or materials properties software. (a) Write the mass balance equation (MBE), energy balance equation (EBE), and entropy balance equation (EnBE); (b) find the tank final pressure and temperature; and (c) calculate the entropy generation.

Solution

a) Write the mass, energy, and entropy balance equations.

$$\text{MBE}: m_1 = m_2 = m$$

$$\text{EBE}: m_1 u_1 + W_{sh,1} + W_{sh,2} + Q_{in} = m_2 u_2 + Q_l$$

$$\text{EnBE}: m_1 s_1 + Q_{in}/T_s + S_{gen} = m_2 s_2 + Q_l/T_{surr}$$

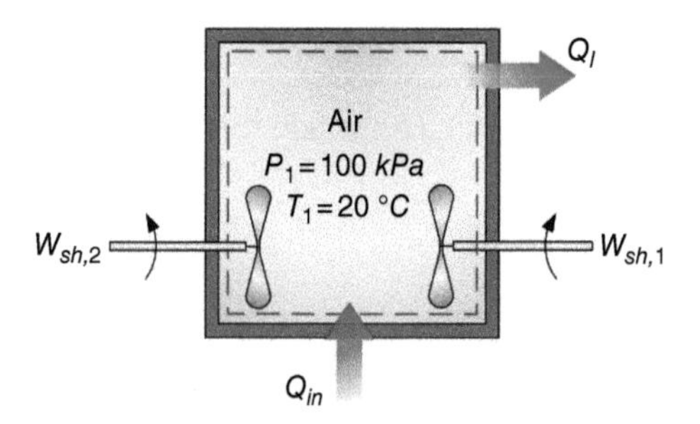

Figure 4.11 Schematic of the rigid tank discussed in Example 4.3.

b) Find the tank final pressure and temperature.

$$P_1V = mRT_1$$

$$m = \frac{P_1V}{RT_1} = 100\ \frac{kPa \times 1\ m^3}{0.287\ kJ/kgK \times 293\ K} = 1.189\ kg$$

From property tables or materials and thermophysical property software:

$$u_1 = 209.3\ kJ/kg$$

$$m_1u_1 + W_{sh,1} + W_{sh,2} + Q_{in} = m_2u_2 + Q_l$$

$$1.189\ kg \times 209.3\frac{kJ}{kg} + 100\ kJ + 50\ kJ + 66\ kJ = 1.189\ kg \times u_2 + 0.1 \times 66\ kJ$$

$$u_2 = 385.4\frac{kJ}{kg}$$

From the property tables or thermophysical property software, one obtains T_2 using u_2 as follows:

$$\mathbf{T_2 = 534.3\ K}$$

$$P_2 = \frac{mRT_2}{V} = \frac{1.189\ kg \times 0.287\ kJ/kgK \times 534.3\ K}{1\ m^3} = \mathbf{182.3\ kPa}$$

c) Calculate the entropy generation from the above given equation.

$$S_{gen} = m(s_2 - s_1) + \frac{Q_l}{T_{surr}} - \frac{Q_{in}}{T_s}$$

with c_v = 0.718 kJ/kgK

For ideal gases, one can write the following for entropy change:

$$s_2 - s_1 = c_v ln\left(\frac{T_2}{T_1}\right) + Rln\left(\frac{v_2}{v_1}\right) = 0.718\frac{kJ}{kgK} \times ln\left(\frac{534.3}{293}\right) = 0.4313\frac{kJ}{kgK}$$

$$S_{gen} = 1.189 \times (0.4313) + \frac{0.1 \times 66}{60 + 273} - \frac{66}{600 + 273}$$

$$S_{gen} = \mathbf{0.457\ kJ/K}$$

Example 4.4 A piston–cylinder device with an initial volume of 0.1 m^3, as shown in Figure 4.12, contains saturated water vapor at a temperature of 100 °C. The water is compressed so that the final volume is 80% of the initial volume. A total of 40 kJ of heat is lost to the surroundings during the compression process. Take the immediate boundary temperature (T_b) as 25 °C. (a) Write the mass, energy, and entropy balance equations, (b) find the tank final pressure and temperature, (c) calculate the compression work, and (d) calculate the entropy generation.

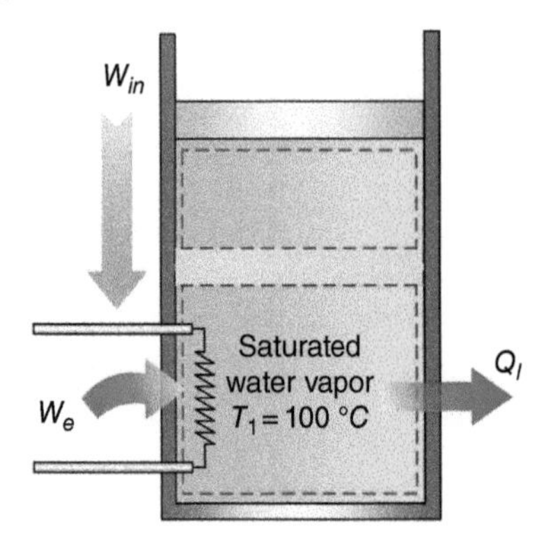

Figure 4.12 Schematic of the piston–cylinder mechanism considered in Example 4.4.

Solution

a) Write the mass, energy, and entropy balance equations.

$$\text{MBE}: m_1 = m_2 = m$$

$$\text{EBE}: m_1u_1 + W_e + W_{in} = m_2u_2 + Q_l \text{ or } mh_1 + W_e = mh_2 + Q_l$$

where $W_{in} = m(P_1v_1 - P_2v_2)$ and $h = u + Pv$

$$\text{EnBE}: m_1s_1 + S_{gen} = m_2s_2 + (Q_l/T_b)$$

where T_b is considered. One may consider the immediate surrounding temperature or immediate boundary temperature or surface temperature or average temperature or environmental (reference) temperature. Any is fine, but needs to made clear in the problem or in the assumptions made.

b) Find the tank final pressure and temperature.

From the property tables (such as Appendix B1a) or thermophysical property software, one obtains:

$$\left.\begin{array}{r} T_1 = 100\,^\circ C \\ x_1 = 1 \text{ (sat.vap.)} \end{array}\right\} \begin{array}{l} h_1 = 2676\ kJ/kg \\ u_1 = 2506\ kJ/kg \\ v_1 = 1.674\ m^3/kg \\ P_1 = P_{sat} = 101.3\ kPa \end{array}$$

$$v_1 = \frac{V_1}{m} \rightarrow 1.674\,\frac{m^3}{kg} = \frac{0.1\ m^3}{m} \rightarrow m = 0.05975\ kg$$

$$v_2 = \frac{V_2}{m} \rightarrow v_2 = \frac{0.8 \times 0.1\ m^3}{0.05975\ kg} = 1.339\,\frac{m^3}{kg}$$

From the property tables (such as Appendix B1a) or thermophysical property software, one obtains:

$$\left.\begin{array}{r} P_2 = P_1 = 101.3\ kPa \\ v_2 = 1.339\ m^3/kg \end{array}\right\} \begin{array}{l} u_2 = 2088\ kJ/kg \\ h_2 = 2224\ kJ/kg \end{array}$$

with

$$v_2 = v_f + xv_{fg} = v_f + x(v_g - v_f)$$

$$x_2 = \frac{v_2 - v_f}{v_g - v_f} = (1.339 - 0.001043)/(1.6734 - 0.001043) = 0.8$$

$$mh_1 + W_e = mh_2 + Q_l$$

$$0.05975\ kg \times 2676\ \frac{kJ}{kg} + W_e = 0.05975\ kg \times 2224\ \frac{kJ}{kg} + 40\ kJ$$

$$\boldsymbol{W_e = 12.99\ kJ}$$

c) Calculate the compression work.

$$m_1u_1 + W_e + W_{\text{in}} = m_2u_2 + Q_l$$

$$0.05975\ kg \times 2506\ \frac{kJ}{kg} + 12.99\ kJ + W_{in} = 0.05975\ kg \times 2088\ \frac{kJ}{kg} + 40\ kJ$$

$$\boldsymbol{W_{\text{in}} = 2.004\ kJ}$$

d) Calculate the entropy generation.

$$\left.\begin{array}{r} T_1 = 100\,^\circ C \\ x_1 = 1\ (\text{sat.vap.}) \end{array}\right\} s_1 = 7.354\ \frac{kJ}{kgK}$$

with the data obtained:

$$\left.\begin{array}{r} P_2 = 101.3\ kPa \\ v_2 = 1.339\ m^3/kg \end{array}\right\} s_2 = 6.144\ \frac{kJ}{kgK}$$

$$m_1s_1 + S_{gen} = m_2s_2 + (Q_l/T_b)$$

$$0.05975\ kg \times 7.354\ \frac{kJ}{kgK} + S_{gen} = 0.05975\ kg \times 6.144\ \frac{kJ}{kgK} + \frac{40\ kJ}{25 + 273\ K}$$

$$\boldsymbol{S_{gen} = 0.062\ kJ/K}$$

This section introduces different examples that deal with entropy balance equations of open energy systems, starting with a pump problem showing how the work is never associated with entropy then followed by a heat exchanger problem for heat transfer analysis and entropy analysis of processes involves heat exchange. Note that in the coming examples, mass and energy balance equations are also solved.

Example 4.5 A water pump consumes 500 kW, as shown in Figure 4.13, and raises the pressure of 10 kg/s of saturated liquid water from 100 kPa to 5.0 MPa. (a) Write the mass, energy, and entropy balance equations, (b) find the pump exit temperature, and (c) calculate the entropy generation rate.

Solution

a) Write the mass, energy, and entropy balance equations.

$$\text{MBE}: \dot{m}_1 = \dot{m}_2 = \dot{m}$$

$$\text{EBE}: \dot{m}h_1 + \dot{W}_{in} = \dot{m}h_2$$

$\text{EnBE}: \dot{m}s_1 + \dot{S}_{gen} = \dot{m}s_2$ (one has to remember that entropy is not associated with any of type of work).

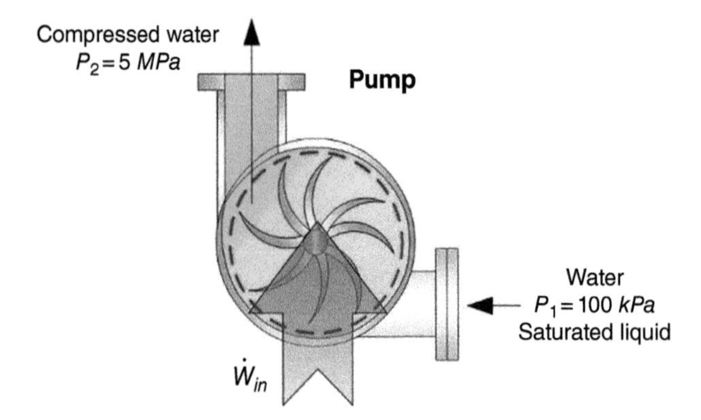

Figure 4.13 Schematic diagram of the water pump discussed in Example 4.5.

b) Find the pump exit temperature.

From the property tables (such as Appendix B1a) or thermophysical property software, one gets

$$\left.\begin{array}{l} P_1 = 100\ kPa \\ x_1 = 0\ (sat.liq.) \end{array}\right\} \begin{array}{l} s_1 = 1.303\ kJ/kgK \\ h_1 = 417.5\ kJ/kg \end{array}$$

$$10\frac{kg}{s} \times 417.5\frac{kJ}{kg} + 500\ kW = 10\frac{kg}{s} \times h_2$$

$$h_2 = 467.5\frac{kJ}{kg}$$

Furthermore, with additional data, it is obtained as:

$$\left.\begin{array}{l} P_2 = 5000\ kPa \\ h_2 = 467.5\ kJ/kg \end{array}\right\} \begin{array}{l} \mathbf{T_2 = 383.8\ K} \\ s_2 = 1.421\ kJ/kgK \end{array}$$

c) Calculate the entropy generation.

$$\dot{m}s_1 + \dot{S}_{gen} = \dot{m}s_2$$

$$10\frac{kg}{s} \times 1.303\frac{kJ}{kgK} + \dot{S}_{gen} = 10\frac{kg}{s} \times 1.421\frac{kJ}{kgK}$$

$$\mathbf{\dot{S}_{gen} = 1.186\ kW/K}$$

Example 4.6 A 32 *kg/s* of water is heated at a constant pressure by 4.8×10^6 *kg/h* of exhausts gases as shown in Figure 4.14. Assume that the exhausts consist only of air where you can use ideal gas equations. (a) Write the mass, energy, and entropy balance equations, (b) calculate the exit temperature of the water and the exhaust air, (c) calculate the heat transfer rate to the water, and (d) calculate the entropy generation.

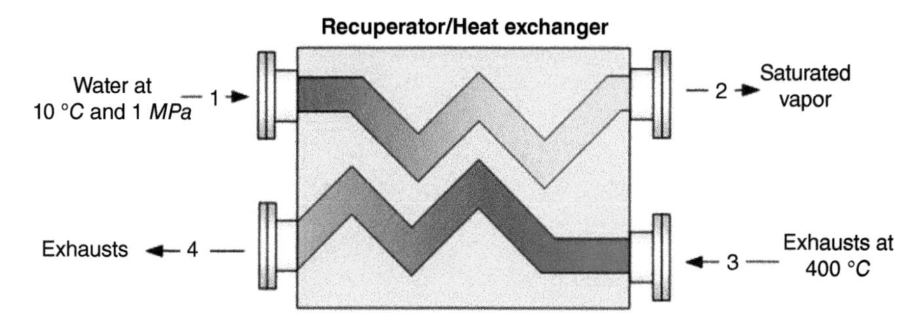

Figure 4.14 Schematic diagram of the heat exchanger discussed in Example 4.6.

Solution

a) Write the mass, energy, and entropy balance equations.

$$\text{MBE}: \dot{m}_1 = \dot{m}_2 = \dot{m}_w, \dot{m}_3 = \dot{m}_4 = \dot{m}_a$$

$$\text{EBE}: \dot{m}_1 h_1 + \dot{m}_3 h_3 = \dot{m}_2 h_2 + \dot{m}_4 h_4$$

$$\text{EnBE}: \dot{m}_1 s_1 + \dot{m}_3 s_3 + \dot{S}_{gen} = \dot{m}_2 s_2 + \dot{m}_4 s_4$$

b) Calculate the exit temperature of the water and the exhaust air.

From the property tables (such as Appendix B1a) or thermophysical property software, one gets:

$$\boldsymbol{T_2 = T_{sat@P=1\ MPa} = 179.9\,^oC}$$

$$\dot{m}_2 h_2 - \dot{m}_1 h_1 = \dot{m}_3 h_3 - \dot{m}_4 h_4$$

$$\dot{m}_2 h_2 - \dot{m}_1 h_1 = \dot{m}_a c_p (T_3 - T_4)$$

Furthermore, one gets the following to use in the EBE:

$$h_1 = h_{f@T=10\,^oC} = 41.99\ kJ/kg$$

$$h_2 = h_{g@1MPa} = 2778\ kJ/kg$$

$$\dot{m}_w (h_2 - h_1) = \dot{m}_a c_p (T_3 - T_4)$$

$$32\,\frac{kg}{s} \times (2778 - 41.99)\frac{kJ}{kg} = 4.8 \times 10^6\,\frac{kg}{h} \times \frac{1\ h}{3600\ s} \times 1.005\,\frac{kJ}{kgK}(400 + 273 - T_4)\ K$$

$$\boldsymbol{T_4 = 334.5\,^oC}$$

c) Calculate the heat transfer rate to the water.

$$\dot{m}_3 h_3 = \dot{m}_4 h_4 + \dot{Q}_{\text{out}}$$

$$\dot{m}_3 h_3 - \dot{m}_4 h_4 = \dot{Q}_{\text{out}}$$

$$\dot{m}_a c_p (T_3 - T_4) = \dot{Q}_{\text{out}}$$

$$4.8 \times 10^6\,\frac{kg}{h} \times \frac{1\ h}{60\ \text{min}} \times \frac{1\ \text{min}}{60\ s} \times 1.005\,\frac{kJ}{kgK}(400 - 334.5)\ K = \dot{Q}_{\text{out}}$$

$$\boldsymbol{\dot{Q}_{\text{out}} = 87770\ kW}$$

d) Calculate the entropy generation.

$$\dot{m}_1 s_1 + \dot{m}_3 s_3 + \dot{S}_{gen} = \dot{m}_2 s_2 + \dot{m}_4 s_4$$

Similarly, one gets:

$$s_1 = s_{f@T=10\,^\circ C} = 0.151\ kJ/kgK$$

$$s_2 = s_{g@1MPa} = 6.586\ kJ/kgK$$

$$\dot{S}_{gen} = \dot{m}_a \times (s_4 - s_3) + \dot{m}_w (s_2 - s_1)$$

$$\dot{S}_{gen} = \dot{m}_a \times \left(c_p\, ln\left(\frac{T_4}{T_3}\right) - Rln\left(\frac{P_4}{P_3}\right)\right) + \dot{m}_w (s_2 - s_1)$$

$$\dot{S}_{gen} = 4.8 \times 10^6\ \frac{kg}{h} \times \frac{1\ h}{3600\ s} \times \left(1.005\ \frac{kJ}{kgK} \times ln\left(\frac{334.5 + 273}{400 + 273}\right)\right)$$

$$+ 32\ \frac{kg}{s} \times (6.586 - 0.151) \frac{kJ}{kgK}$$

$$\mathbf{\dot{S}_{gen} = 68.71\ kW/K}$$

4.7 Isentropic Processes

A special process occurs when the entropy of what is entering a steady state open system is equal to the entropy of what is leaving the open system; it is known as an isentropic process. When processes are not isentropic then we can measure how close the process is to the isentropic case by defining the isentropic efficiency. The isentropic efficiency of a turbine, for example, as shown in Figure 4.15a, can be defined as the specific actual work output (w_a) divided by the specific isentropic work output (w_s):

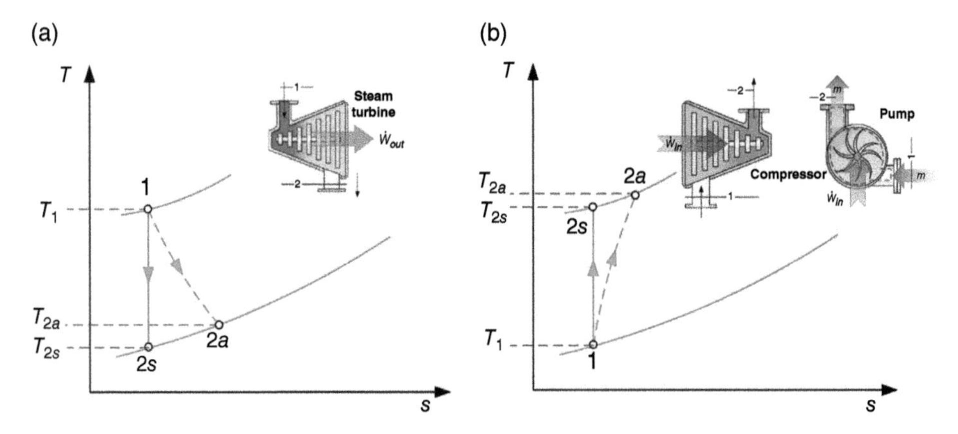

Figure 4.15 Illustration of actual and isentropic expansion and compression processes for (a) a turbine and (b) a compressor and pump.

$$\eta_{is,t} = \frac{w_a}{w_s} = \frac{h_{2a} - h_1}{h_{2s} - h_1} \tag{4.6}$$

where the subscript refers to the isentropic and the subscripts a and s refers to the actual and isentropic (constant entropy) cases of the turbine. However, the definition of the isentropic efficiency for work consuming devices such as compressors and pumps as shown in Figure 4.15b can be made as the ratio of the specific isentropic work input divided by the specific actual work input:

$$\eta_{is,c} = \frac{w_s}{w_a} = \frac{h_{2s} - h_1}{h_{2a} - h_1} \tag{4.7}$$

$$\eta_{is,p} = \frac{w_s}{w_a} = \frac{h_{2s} - h_1}{h_{2a} - h_1} \tag{4.8}$$

Here, the subscripts c and p refer to the compressor and pump.

We can see that in the case of power producing devices such as the turbine, the power produced by the actual turbine is less than the power that a reversible (isentropic) turbine would produce, which has resulted in the isentropic efficiency definition as that shown in Eq. (4.6). However, for power consuming devices, they would have consumed less power if they were isentropic.

Example 4.7 Two steam turbines are connected as shown in Figure 4.16. If the changes in kinetic and potential energies are neglected then (a) write mass, energy, and entropy balance equations for each turbine separately, (b) calculate the work rates produced by both turbines, (c) find the temperature of the water leaving turbine 2, and (d) calculate the total entropy generation rate. Take the boundary temperature (T_b) as 45 °C.

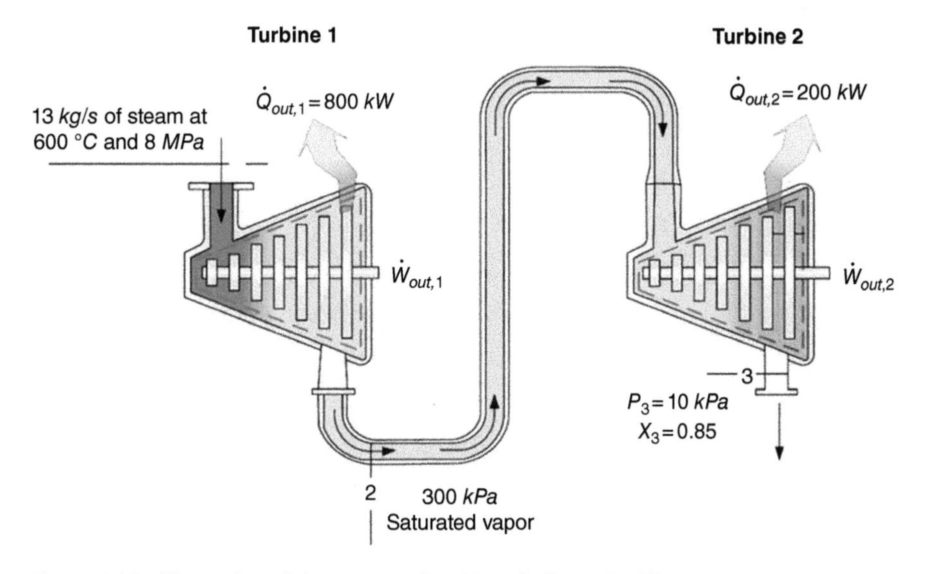

Figure 4.16 Illustration of the connected turbines in Example 4.7.

Solution

a) Write mass, energy, and entropy balance equations for each turbine separately.

$$\text{MBE}: \dot{m}_1 = \dot{m}_2 \text{ and } \dot{m}_3 = \dot{m}_w$$

$$\text{EBE (turbine 1)}: \dot{m}_w h_1 = \dot{m}_w h_2 + \dot{W}_{out,1} + \dot{Q}_{out,1}$$

$$\text{EBE (turbine 2)}: \dot{m}_w h_2 = \dot{m}_w h_3 + \dot{W}_{out,2} + \dot{Q}_{out,2}$$

$$\text{EnBE (turbine 1)}: \dot{m}_w s_1 + \dot{S}_{gen},1 = \dot{m}_w s_2 + \dot{Q}_{out,1}/T_b$$

$$\text{EnBE (turbine 2)}: \dot{m}_w s_2 + \dot{S}_{gen},2 = \dot{m}_w s_3 + \dot{Q}_{out,2}/T_b$$

b) Calculate the work rates produced by both turbines.

From the property tables (such as Appendix B1a) or thermophysical property software, one gets:

$$\left.\begin{aligned} P_1 &= 8000\ kPa \\ T_1 &= 600\,^{o}C \end{aligned}\right\} \begin{aligned} h_1 &= 3642.4\ kJ/kg \\ s_1 &= 7.0221\ kJ/kgK \end{aligned}$$

$$\left.\begin{aligned} P_2 &= 300\ kPa \\ x_2 &= 1 \end{aligned}\right\} h_2 = 2724.9\ kJ/kg$$

$$\dot{m}_w h_1 = \dot{m}_w h_2 + \dot{W}_{out,1} + \dot{Q}_{out,1}$$

$$\left(13\frac{kg}{s} \times 3642.4\frac{kJ}{kg}\right) = \left(13\frac{kg}{s} \times 2724.9\frac{kJ}{kg}\right) + \left(800\frac{kJ}{s}\right) + \dot{W}_{out,1}$$

$$\mathbf{\dot{W}_{out,1} = 11128\ kW}$$

$$h_3 = h_f + x(h_{fg})$$

$$h_3 = 191.81 + 0.85 \times (2392.1) = 2225.095\ kJ/kg$$

$$\dot{m}_w h_2 = \dot{m}_w h_3 + \dot{W}_{out,2} + \dot{Q}_{out,2}$$

$$\left(13\frac{kg}{s} \times 2724.9\frac{kJ}{kg}\right) = \left(13\frac{kg}{s} \times 2225.095\frac{kJ}{kg}\right) + \left(200\frac{kJ}{s}\right) + \dot{W}_{out,2}$$

$$\mathbf{\dot{W}_{out,2} = 6297\ kW}$$

$$\mathbf{\dot{W}_{out} = \dot{W}_{out,1} + \dot{W}_{out,2} = 17425\ kW}$$

c) Find the temperature of the water leaving turbine 2.

From the property tables (such as Appendix B1a) or thermophysical property software (such as EES), one gets

$$\left.\begin{aligned} P_3 &= 10\ kPa \\ x_3 &= 0.85 \end{aligned}\right\} \mathbf{T_3 = 45.82\,^{o}C}$$

d) Calculate the total entropy generation rate.

$$s_3 = s_f + x(s_{fg})$$

$$s_3 = 0.6492 + 0.85 \times 7.4996 = 7.024\ kJ/kg$$

By adding the following two equations:

$$\dot{m}_w s_1 + \dot{S}_{gen,1} = \dot{m}_w s_2 + \frac{\dot{Q}_{out,1}}{T_b}$$

and

$$\dot{m}_w s_2 + \dot{S}_{gen,2} = \dot{m}_w s_3 + \frac{\dot{Q}_{out,2}}{T_b}$$

This results in:

$$\dot{m}_w s_1 + \dot{S}_{gen} = \dot{m}_w s_3 + \dot{Q}_{out,1}/T_b + \dot{Q}_{out,2}/T_b$$

$$\dot{S}_{gen} = 13\,\frac{kg}{s} \times (7.024 - 7.0221)\frac{kJ}{kgK} + \frac{800\ kW}{(45+273)} + \frac{200\ kW}{(45+273)}$$

$$\mathbf{\dot{S}_{gen} = 3.17\ kW/K}$$

4.8 Isentropic Processes for Ideal Gases

The relations that were introduced earlier to calculate the entropy change of ideal gases can be set to equal zero for isentropic processes of ideal gases and thus new isentropic relations for ideal gases can be developed. Assuming that the specific heat capacity is constant or the changes in it are negligible, then the relation for the change in entropy for an isentropic process can be written as:

$$0 = s_2 - s_1 = \left(c_p\, ln\left(\frac{T_2}{T_1}\right) - R ln\left(\frac{P_2}{P_1}\right)\right)$$

Then by rearranging the above expression:

$$c_p\, ln\left(\frac{T_2}{T_1}\right) = R ln\left(\frac{P_2}{P_1}\right)$$

$$ln\left(\frac{T_2}{T_1}\right) = \frac{R}{c_p}\, ln\left(\frac{P_2}{P_1}\right)$$

$$\left(\frac{T_2}{T_1}\right) = \left(\frac{P_2}{P_1}\right)^{\frac{R}{c_p}} \tag{4.9}$$

Since R, which is the gas constant, is equal to the difference between the specific heat capacity at a constant pressure and the specific heat capacity at a constant volume then we can replace the exponent of the pressure ratio in Eq. (4.9):

$$\frac{R}{c_p} = \frac{c_p - c_v}{c_p} = \frac{c_p}{c_p} - \frac{c_v}{c_p} = 1 - \frac{1}{k} = \frac{k-1}{k}$$

Then by substituting the above expression back into Eq. (4.9), the following expression in terms of the specific heat capacity ratio is obtained:

$$\left(\frac{T_2}{T_1}\right)=\left(\frac{P_2}{P_1}\right)^{(k-1)/k}$$

Similarly, by equalizing Eq. (4.5) to zero as follows, isentropic relations can be derived between

$$0=s_2-s_1=\left(c_v\, ln\left(\frac{T_2}{T_1}\right)+Rln\left(\frac{v_2}{v_1}\right)\right)$$

$$c_v\, ln\left(\frac{T_2}{T_1}\right)=-Rln\left(\frac{v_2}{v_1}\right)$$

$$ln\left(\frac{T_2}{T_1}\right)=-R/c_v\, ln\left(\frac{v_2}{v_1}\right)$$

$$\left(\frac{T_2}{T_1}\right)=\left(\frac{v_2}{v_1}\right)^{-\frac{R}{c_v}}$$

$$\left(\frac{T_2}{T_1}\right)=\left(\frac{v_1}{v_2}\right)^{\frac{R}{c_v}} \tag{4.10}$$

Then similarly, by replacing the gas constant with the difference between the specific heat capacities we obtain the following expression for the exponent of the volume ratio:

$$\frac{R}{c_v}=\frac{c_p-c_v}{c_v}=k-1$$

Then, by substituting the above expression back to Eq. (4.10), we get the following expression in terms of the specific heat capacity ratio:

$$\left(\frac{T_2}{T_1}\right)=\left(\frac{v_1}{v_2}\right)^{k-1}$$

Finally, from the above two relations (Eqs. 4.9 and 4.10), we can obtain a third isentropic, ideal gas relationship between the pressure and the specific volume:

$$\left(\frac{P_2}{P_1}\right)=\left(\frac{v_1}{v_2}\right)^{k} \tag{4.11}$$

The importance of the previously derived equations from the entropy relations of ideal gases for isentropic processes is highlighted below.

For isentropic processes:

$$\left(\frac{P_2}{P_1}\right)=\left(\frac{v_1}{v_2}\right)^{k}$$

$$\left(\frac{T_2}{T_1}\right)=\left(\frac{v_1}{v_2}\right)^{k-1}$$

$$\left(\frac{T_2}{T_1}\right) = \left(\frac{P_2}{P_1}\right)^{(k-1)/k}$$

For isothermal processes:

$$P_1 v_1 = P_2 v_2$$

For isobaric processes:

$$v_1 T_2 = v_2 T_1$$

For isochoric processes:

$$P_2 T_1 = P_1 T_2$$

For polytropic processes:

$$\frac{P_1}{P_2} = \left(\frac{V_2}{V_1}\right)^n$$

4.9 Isentropic Efficiencies for Ideal Gases

In this section, we will measure how close a device is behaving to the isentropic case, where the device will have no internal irreversibilites, in other words both internally and externally reversible. However, the story differs in actual cases, and the performance is quantified by so-called: isentropic efficiency, which needs to be specifically written for each device. The following subsections cover these in detail.

(a) Turbine

A fluid or gas is expanded in a turbine to produce power. There are of course both steam (i.e. using water) and gas (i.e. using air) turbines; these are considered for analysis. Turbines are normally assumed to be adiabatic for analysis purposes. The performance of an adiabatic turbine is usually expressed by isentropic efficiency.

Consider a turbine with inlet state 1 with temperature T_1 and pressure P_1 and an exit state 2 with temperature T_2 and pressure P_2 as shown in Figure 4.17. The power output from this

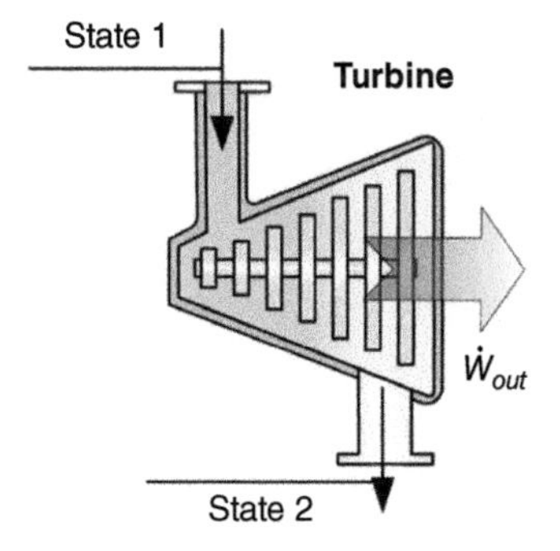

Figure 4.17 Schematic diagram of a turbine.

compressor would be maximum if the fluid is expanded reversibly and adiabatically (i.e. isentropically) between the given initial state and given exit pressure. The isentropic efficiency is then the ratio of actual work rate to the isentropic work rate:

$$\eta_{is,t} = \frac{\dot{W}_a}{\dot{W}_s} = \frac{\dot{m}(h_1 - h_{2a})}{\dot{m}(h_1 - h_{2s})} = \frac{\dot{m}C_p(T_1 - T_{2a})}{\dot{m}C_p(T_1 - T_{2s})} = \frac{(T_1 - T_{2a})}{(T_1 - T_{2s})} \tag{4.12}$$

where $\dot{m}$ is the mass flow rate of gas (*kg/s*), C_p is the constant pressure specific heat of gas (*kJ/kgK*), h_1 is inlet enthalpy of the gas (*kJ/kg*), h_{2s} is the enthalpy of the gas at the turbine outlet (*kJ/kg*) if the process was isentropic while h_{2a} is the actual enthalpy of the gas (*kJ/kg*), and the subscript t refers to turbine. T_1, T_{2a}, and T_{2s} are the inlet, actual exit and isentropic exit temperatures of the gas.

A few reminders are listed for the above outlined methodology for turbine as follows:

- The relationship $h = C_pT$ is used for ideal gases to further simplify the equation.
- The isentropic enthalpy at the exit may directly be obtained from exit pressure and exit entropy (since s_1 equals s_{2s}).
- The changes in kinetic and potential energies are generally neglected unless the information and/or data are provided in the problem.

(b) Compressor

A compressor is used to increase the pressure of a gas, as shown in Figure 4.18. A power (essentially work) input is needed for the compression process, and the compressor is treated as a work consuming device. The performance of an adiabatic compressor is usually expressed by isentropic (adiabatic) efficiency.

Consider an adiabatic compressor with inlet state 1 and an exit state 2. The power input to this compressor would be minimum if the gas is compressed reversibly and adiabatically (i.e. isentropically) between the given initial state and given exit pressure. The isentropic efficiency is then defined as the ratio of the isentropic work rate to the actual work rate of the compressor:

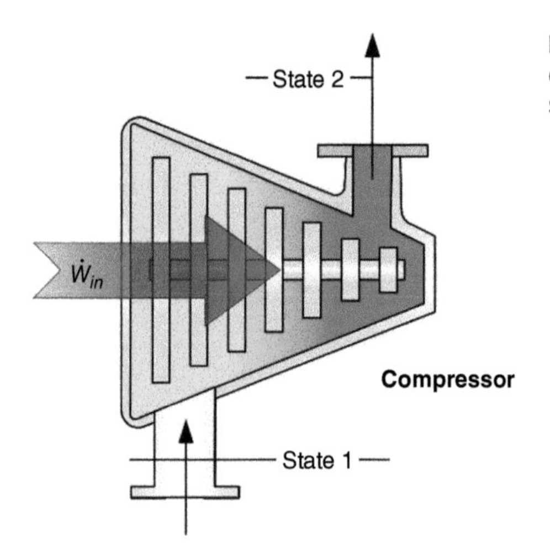

Figure 4.18 A schematic diagram of a compressor showing the inlet streams, exit streams, and the energy interactions.

$$\eta_{is,c} = \frac{\dot{W}_s}{\dot{W}_a} = \frac{\dot{m}(h_{2s} - h_1)}{\dot{m}(h_{2a} - h_1)} = \frac{\dot{m}C_p(T_{2s} - T_1)}{\dot{m}C_p(T_{2a} - T_1)} = \frac{(T_{2s} - T_1)}{(T_{2a} - T_1)} \tag{4.13}$$

where $\dot{m}$ is the mass flow rate of gas (*kg/s*), C_p is the constant pressure specific heat of the gas (*kJ/kgK*), h_1 is inlet enthalpy of the gas (*kJ/kg*), h_{2s} is the enthalpy of the gas at the compressor outlet (*kJ/kg*) if the process was isentropic while h_{2a} is the actual enthalpy of the gas (*kJ/kg*), and the subscript *c* refers to compressor. T_1, T_{2a}, and T_{2s} are the inlet, actual exit, and isentropic exit temperatures of the gas.

A few reminders are listed for the above outlined methodology for turbine as follows:

- The relationship $h = C_pT$ is used for ideal gases to further simplify the equation.
- The isentropic enthalpy at the exit may be directly obtained from exit pressure and exit entropy (since s_1 equals s_{2s}).
- The changes in kinetic and potential energies are generally neglected unless the information and/or data are provided in the problem.

If the gas is modeled as an ideal gas with constant specific heats, the isentropic work input rates is determined from:

$$\dot{W}_s = \dot{m}\frac{kR(T_2 - T_1)}{k-1} = \dot{m}\frac{kRT_1}{k-1}\left[\left(\frac{P_2}{P_1}\right)^{\frac{k-1}{k}} - 1\right] \tag{4.14}$$

where k is the specific heat ratio ($k = c_p/c_v$). Its value is 1.4 for air at room temperature.

The gas is sometimes cooled as being compressed in a nonadiabatic compressor to reduce power input. This is because the work input is proportional to specific volume of the gas and cooling the gas decreases its specific volume. The isentropic efficiency cannot be used in such nonadiabatic compressors. Instead, an isothermal efficiency may be defined as:

$$\eta_{ist,c} = \frac{\dot{W}_{ist}}{\dot{W}_a} \tag{4.15}$$

where the subscript *ist* refers to an isothermal process.

The work input rate for the reversible, isothermal case is given for an ideal gas with constant specific heats as:

$$\dot{W}_{ist} = \dot{m}RT_1 \ln\left(\frac{P_2}{P_1}\right) \tag{4.16}$$

where R is the gas constant, T is the inlet temperature of the gas, and P_1 and P_2 are the pressures at the inlet and exit of the compressor, respectively. In some cases, a reversible, polytropic process may be used as the ideal process for compression applications. Then, a polytropic efficiency may be defined as:

$$\eta_{pol,c} = \frac{\dot{W}_{pol}}{\dot{W}_a} \tag{4.17}$$

where the subscript *pol* refers to polytropic process and $\dot{W}_a = \dot{m}(h_{2a} - h_1)$.

Furthermore, the work input rate for a reversible, polytropic process is given for an ideal gas with constant specific heats as:

$$\dot{W}_{pol} = \dot{m}\frac{nR(T_2 - T_1)}{n-1} = \dot{m}\frac{nRT_1}{n-1}\left[\left(\frac{P_2}{P_1}\right)^{\frac{n-1}{n}} - 1\right] \tag{4.18}$$

Here, n is the polytropic exponent, which is similar to adiabatic exponent (k), but its value is slightly higher than k.

(c) Pump

While discussing ideal gases, we insert the subsection on a pump here, particularly for liquids. A pump is used to increase the pressure of a liquid. A mechanical work input is needed for this process. The liquid may be considered to be an incompressible fluid and the power input for the isentropic case may be determined from specific volume and pressure data. A schematic diagram of a pump is shown in Figure 4.19.

When the changes in potential and kinetic energies of a liquid are negligible, the isentropic efficiency of a pump is defined as:

$$\eta_{is,p} = \frac{\dot{W}_s}{\dot{W}_a} = \frac{\dot{m}v(P_2 - P_1)}{\dot{m}(h_{2a} - h_1)} \tag{4.19}$$

where v is the specific volume of the liquid; it is usually taken at the pump inlet and the subscript p refers to pump.

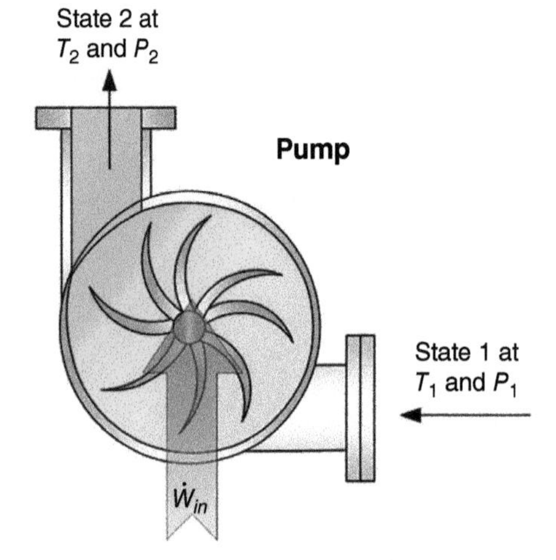

Figure 4.19 Schematic diagram of a pump.

(d) Nozzle

A nozzle is essentially an adiabatic device, as shown in Figure 4.20, because of the negligible heat transfer and is used to accelerate a fluid or gas. Therefore, the isentropic (i.e. reversible

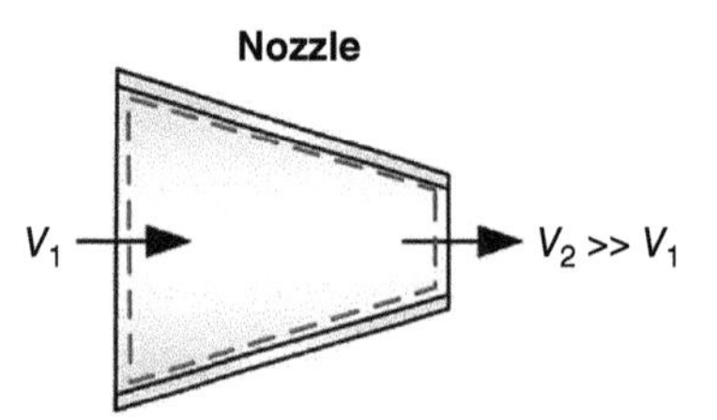

Figure 4.20 Nozzles are shaped so that they can convert pressure energy into kinetic energy.

and adiabatic) process serves as a suitable model for nozzles. The **isentropic efficiency of a nozzle** is defined as *the ratio of the actual kinetic energy of the fluid at the nozzle exit to the kinetic energy value at the exit of an isentropic nozzle for the same inlet state and exit pressure* as follows:

$$\eta_{is,nz} = \frac{KE_a}{KE_s} = \frac{V_{2a}^2}{V_{2s}^2} \tag{4.20}$$

where the subscript *nz* refers to nozzle.

When the inlet velocity is negligible, the isentropic efficiency of the nozzle can be expressed in terms of enthalpies:

$$\eta_{is,nz} = \frac{h_1 - h_2}{h_1 - h_{2s}} \tag{4.21}$$

(e) Diffuser

Diffusers, as shown in Figure 4.21, are steady flow devices that increase the pressure of fluids by reducing their kinetic energy or, in other words, reducing the fluid moving velocity. These devices usually do not introduce any extra energy such as heat or work and they also do not extract work or heat out of the fluids that pass through them. What they do is just change the form of energy that the fluid possesses and usually they are designed to be adiabatic.

The **isentropic efficiency of a diffuser** is the ratio of the final kinetic energy of fluid exiting the diffuser if the process was isentropic to the actual kinetic energy of the fluid exiting the diffuser and is:

$$\eta_{is,df} = \frac{KE_s}{KE_a} = V_{2s}^2 / V_{2a}^2 \tag{4.22}$$

Here, the subscript df refers to the diffuser. Diffusers are built to convert the kinetic energy of the fluid into enthalpy increase of that fluid or gas.

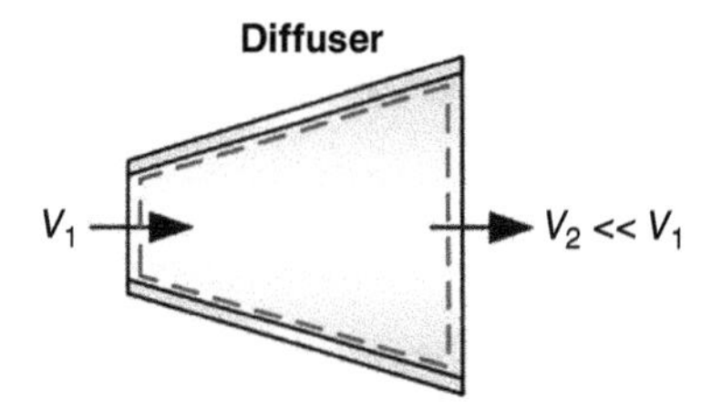

Figure 4.21 Diffusers are shaped to convert part of the kinetic energy into pressure energy, they are the opposite if nozzles.

Example 4.8 Air enters in a diffuser at a velocity of 199 *m/s* and exists a nozzle at the same velocity. The mass flow rate through the diffuser and nozzle are constant at 10 *kg/s*. The air inlet and outlet temperatures of diffuser are 290 and 305 *K*, respectively. Both diffuser and nozzle are shown in Figure 4.22. (a) Write the mass, energy, and entropy balance equations for both diffuser and nozzle; (b) find the velocity of the air at the diffuser exit using the energy balance equation; (c) calculate the entropy generation rate for diffuser; (d) calculate the exit temperature of the air in the nozzle; and (e) find entropy generation rate in the nozzle if there is a 100 *kW* heat input supplied from a source at 600 *K*. Note that potential energy changes are negligible.

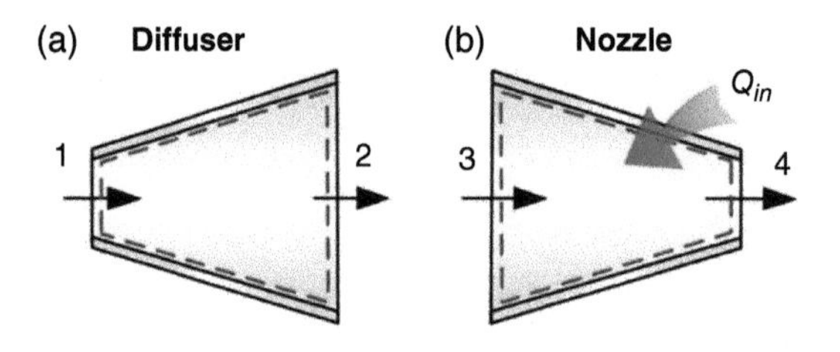

Figure 4.22 Diffuser and nozzle where the inlets and outlets states are 1 and 3, and 2 and 4, respectively.

Solution

a) Write the mass, energy, and entropy balance equations for the diffuser and nozzle.
For the diffuser:

$$\text{MBE}: \dot{m}_1 = \dot{m}_2 = \dot{m}$$

$$\text{EBE}: \dot{m}_1 h_1 + \frac{\dot{m}\, V_1^2}{2} = \dot{m}_2 h_2 + \frac{\dot{m}\, V_2^2}{2}$$

$$\text{EnBE}: \dot{m}_1 s_1 + \dot{S}_{gen} = \dot{m}_2 s_2$$

For the nozzle:

$$\text{MBE}: \dot{m}_3 = \dot{m}_4 = \dot{m}$$

$$\text{EBE}: \dot{m}_3 h_3 + \frac{\dot{m}\, V_3^2}{2} + \dot{Q}_{in} = \dot{m}_4 h_4 + \frac{\dot{m}\, V_4^2}{2}$$

$$\text{EnBE}: \dot{m}_3 s_3 + \frac{\dot{Q}_{in}}{T_s} + \dot{S}_{gen} = \dot{m}_4 s_4$$

b) Find the velocity of the air at the diffuser exit using the energy balance on the diffuser.
By applying the above given energy balance on the boundary of the nozzle, the energy balance equation can be written as:

$$\dot{m}_1 h_1 + \frac{m\, V_1^2}{2} = \dot{m}_2 h_2 + \frac{m\, V_2^2}{2}$$

Since the mass flow rate at the inlet equals that at the outlet then the above energy balance can be reduced to:

$$h_1 + \frac{V_1^2}{2} = h_2 + \frac{V_2^2}{2}$$

Assuming air as an ideal gas, the enthalpies of air at the inlet and outlet of the diffuser are found as follows:

$$T_1 = 290\ K\}\ h_1 = 290.16\,\frac{kJ}{kg}$$

$$T_2 = 305\ K\}\ h_2 = 305.22\,\frac{kJ}{kg}$$

$$290.16\,\frac{kJ}{kg} + \frac{(199\ m/s)^2}{2} \times \frac{1\ kJ/kg}{1000\ m^2/s^2} = 305.22\,\frac{kJ}{kg} + \frac{V_2^2}{2} \times \frac{1\ kJ/kg}{1000\ m^2/s^2}$$

$$\boldsymbol{V_2 = 97.4\ m/s}$$

c) Find the entropy generation rate in the diffuser with constant pressure.

$$\dot{m}_1 s_1 + \dot{S}_{gen} = \dot{m}_2 s_2$$

$$\dot{S}_{gen} = \dot{m}_2 s_2 - \dot{m}_1 s_1 = \dot{m}\left(c_p \, ln\left(\frac{T_2}{T_1}\right) - Rln\left(\frac{P_2}{P_1}\right)\right)$$

$$\dot{S}_{gen} = 10\,\frac{kg}{s} \times \left(1.005\,\frac{kJ}{kgK}\, ln\left(\frac{305}{290}\right)\right) = \mathbf{0.507}\,\frac{\mathbf{kW}}{\mathbf{K}}$$

d) From the EBE for nozzle, one calculates the exit temperature.

$$\dot{m}_3 h_3 + \frac{\dot{m}\left(V_3^2\right)}{2} + \dot{Q}_{in} = \dot{m}_4 h_4 + \frac{\dot{m}_4\left(V_4^2\right)}{2}$$

$$10\,\frac{kg}{s} \times 305.22\,\frac{kJ}{kg} + 10\,\frac{kg}{s} \times \left(\frac{\left(97.4\,\frac{m}{s}\right)^2}{2} \times \frac{1\,\frac{kJ}{kg}}{1000\,\frac{m^2}{s^2}}\right) + 100\;kW = 10\,\frac{kg}{s} \times h_4 + 10\,\frac{kg}{s}$$

$$\times \left(\frac{(199\;m/s)^2}{2} \times \frac{1\,\frac{kJ}{kg}}{1000\,\frac{m^2}{s^2}}\right)$$

$$h_4 = 300.2\,\frac{kJ}{kg}$$

$$\mathbf{T_4 = 300\;K}$$

e) Find the entropy generation rate in the nozzle with constant pressure.

$$\dot{m}_3 s_3 + \frac{\dot{Q}_{in}}{T_s} + \dot{S}_{gen} = \dot{m}_4 s_4$$

$$\dot{S}_{gen} = \dot{m}_4 s_4 - \dot{m}_3 s_3 - \frac{\dot{Q}_{in}}{T_s}$$

$$\dot{S}_{gen} = \dot{m}\left(c_p \, ln\left(\frac{T_2}{T_1}\right) - Rln\left(\frac{P_2}{P_1}\right)\right) - \frac{\dot{Q}_{in}}{T_s}$$

$$\dot{S}_{gen} = 10\,\frac{kg}{s} \times \left(1.005\,\frac{kJ}{kgK}\, ln\left(\frac{305}{290}\right)\right) - \frac{100}{600} = \mathbf{0.1735}\,\frac{\mathbf{kW}}{\mathbf{K}}$$

4.10 Exergy

As mentioned at the beginning of this chapter, thermodynamics brings two governing (so-called: constitutional type) laws, the FLT, which brings energy as a key tool, and the SLT, which brings both entropy and exergy, as shown in Figure 4.1. As a matter of fact, the FLT brings a property of energy that may be in a specific form (e in J/kg or kJ/kg) and rate form ($\dot{E}$ in W or kW). Similar to this, the SLT introduces a new thermodynamic property (so-called: exergy), which may be in a specific form (ex in J/kg or kJ/kg) and rate form ($\dot{E}x$ in W or kW).

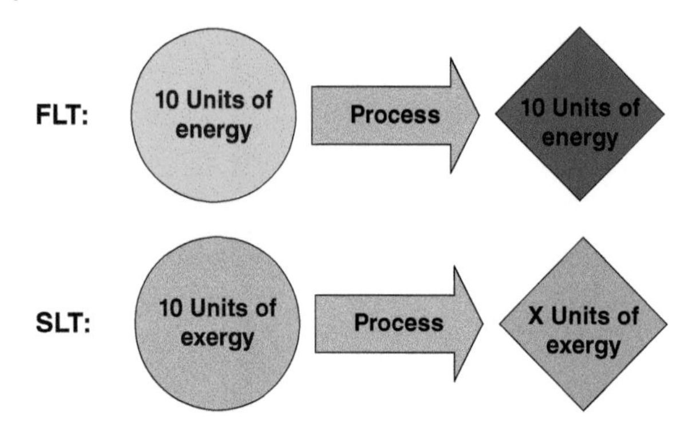

Figure 4.23 Illustration of energy conservation through the FLT and nonconservation of exergy through the SLT. It is therefore not easy to guess the value of X unless it is calculated through an exergy balance equation.

Most thermodynamicists have dwelt on entropy, rather than exergy, under the shelter of the SLT, which may not suitable to make a true comparison between these two governing laws (i.e. the FLT and SLT) through making a comparative evaluation between energy and entropy. Further to this point, one cannot compare entropy with energy but can easily and righteously do it between energy and exergy, since both properties have the same units and quantities and qualities can be compared. One can proceed to energy efficiency via energy analysis while exergy is used to go to exergy efficiency through exergy analysis. Therefore, both energy and exergy concepts should be considered for comparison purposes for design, analysis, assessment, and evaluation. Of course, this does not deny the role of entropy through the entropy balance equation, but emphasizes the limitation of entropy use.

An example is illustrated in Figure 4.23 to show a quick difference between energy and exergy concepts, obviously between the FLT and SLT. From this diagram, 10 *units* of energy is subject to a process and may result in numerous things making it up to 10 *units* of energy due to the conservation of energy principle (FLT). However, the lower section where we have 10 *units* of exergy going through the same process and ending up with X *units* of exergy in a different manner. This X may be 9 or 8 or 7 or less, but cannot make it 10 due to the nonconversation of exergy principle (SLT). Becoming less may be caused by the exergy destruction only or exergy loss or both, so it cannot be conserved.

4.11 Energy vs Exergy

Energy comes in many forms, ranging from kinetic to potential, from mechanical to chemical, from fission to fusion, and from electrical to magnetic. The role of thermodynamics is critical in each of these processes, systems, and applications in which energy transfers and energy transformations occur. The implications of thermodynamics are far-reaching and applications span the range of the human enterprise. Throughout technological history since the industrial revolution, our ability to harness energy and use it for society's needs has improved drastically. Of course, one can easily remember that the industrial revolution was fuelled by the discovery of how to exploit energy on a large scale and how to convert heat into work. Nature allows the conversion of work completely into heat, but heat cannot be entirely converted into work, and doing so requires a device (e.g. a cyclic engine). Engines attempt to optimize the conversion of heat to work.

Note that all of our daily activities one way or another involve energy transfer and energy change. The human body is a familiar example of a biological system in which the chemical energy of food or body fat is transformed into other forms of energy such as heat and work. Engineering applications of energy processes are wide ranging and include power plants to generate electricity, engines to run automobiles and aircraft, refrigeration and air conditioning systems, etc. Many examples of such systems are discussed here. In a hydroelectric power system, the potential energy of water is converted into mechanical energy through the use of an hydraulic turbine. The mechanical energy is then converted into electrical energy by an electric generator coupled to the shaft of the turbine. In a steam power generating plant, chemical or nuclear energy is converted into thermal energy in a boiler or a reactor. The energy is imparted to water, which vaporizes into steam. The energy of the steam is used to drive a steam turbine and the resulting mechanical energy is used to drive a generator to produce electric power. The steam leaving the turbine is then condensed, and the condensate is pumped back to the boiler to complete the cycle. Breeder reactors, for example, use Uranium-235 as a fuel source and can produce more fuel in the process. A solar power plant uses solar concentrators (parabolic dishes or flat mirrors) to heat a working fluid in a receiver located on a tower, where the heated fluid expands in a turbogenerator as in a conventional power plant. In a spark-ignition internal combustion engine, the chemical energy of fuel is converted into mechanical work. An air–fuel mixture is compressed and combustion is initiated by a spark device. The expansion of the combustion gases pushes against a piston, which results in the rotation of a crankshaft. Gas turbine engines, commonly used for aircraft propulsion, convert the chemical energy of fuel into thermal energy that is used to run a gas turbine. The turbine is directly coupled to a compressor that supplies the air required for combustion. The exhaust gases, upon expanding in a nozzle, create thrust. For power generation, the turbine is coupled to an electric generator and drives both the compressor and the generator. In a liquid-fuel rocket, a fuel and an oxidizer are combined, and combustion gases expand in a nozzle creating a propulsive force (thrust) to propel the rocket. A typical nuclear rocket propulsion engine offers a higher specific impulse when compared to chemical rockets. A fuel cell converts chemical energy into electrical energy directly making use of an ion-exchange membrane. When a fuel such as hydrogen is ionized, it flows from the anode through the membrane towards the cathode. The released electrons at the anode flow through an external load. In a magnetohydrodynamic generator, electricity is produced by moving high-temperature plasma through a magnetic field. A refrigeration system utilizes work supplied by an electric motor to transfer heat from a refrigerated space. Low-temperature boiling fluids such as ammonia and refrigerant-134a absorb thermal energy as they vaporize in the evaporator causing a cooling effect in the region being cooled. These are only some of the numerous engineering applications. Thermodynamics is relevant to a much wider range of processes and applications not only in engineering but also in science. A good understanding of this topic is required to improve the design and performance of energy-transfer systems.

After introducing several dimensions and applications of energy, one may look at the dimensions and applications of exergy. The exergy of a system is defined as the maximum shaft work that can be done by the composite of the system and a specified reference environment. The reference environment is assumed to be infinite, in equilibrium, and to enclose all other systems. Typically, the environment is specified by stating its temperature, pressure, and chemical composition. Note that exergy is not a conventional type thermodynamic property like other properties (energy, internal energy, etc.) but is rather a property of both a system and the reference environment.

Let's now look at its history and meanings. The term exergy comes from the Greek words ex and ergon, meaning from and work. The exergy of a system can be increased if exergy is input to it (e.g. work is done on it). The following are some terms found in the literature that are equivalent or nearly equivalent to exergy: available energy, essergy, utilizable energy, work potential, availability, and maximum obtainable work.

Exergy has the characteristic that it is conserved only when all processes occurring in a system and its environment are reversible which is practically impossible. Furthermore, exergy is destroyed whenever an irreversible process occurs. When an exergy analysis is performed on a plant such as a power station, a chemical processing plant or a refrigeration facility, the thermodynamic irreversibilities can be quantified as exergy destructions and/or exergy losses, which represent exergy consumptions (or exergy depletions) in energy quality or usefulness (e.g. wasted shaft work or wasted potential for the production of shaft work). Like energy, exergy can be transferred or transported across the boundary of a system. For each type of energy transfer or transport there is a corresponding exergy transfer or transport.

Exergy analysis takes into account the different thermodynamic values of different energy forms and quantities, e.g. work and heat. The exergy transfer associated with shaft work is equal to the shaft work. The exergy transfer associated with heat transfer, however, depends on the temperature at which it occurs in relation to the temperature of the environment.

Some important characteristics of exergy are briefly described and highlighted as follows:

- Exergy is an explicit indicator of energy quality. A brief illustration for various sources is given in Figure 4.24. The higher the temperature the higher the energy quality in regards to the applications. Work potential (or exergy content) is higher with work, such as electricity.
- If a system is in complete equilibrium with its environment, it results in a dead state and has no exergy change. More explicitly, no difference occurs in temperature, pressure, chemical composition, concentration, etc., which results in no driving force for any process.
- The exergy of a system increases more when it deviates from the reference environment. For instance, a specified quantity of hot water has a higher exergy content during the winter than on a hot summer day. A block of ice carries little exergy in winter while it can have significant exergy in summer.
- When energy loses its quality, exergy is destroyed. Exergy is technically the useful/available part of energy that really has value thermodynamically and economically.
- Exergy by definition depends not just on the state of a system or flow but also on the state of the surroundings.
- Exergy efficiency is defined a measure of approach to ideality (or reversibility). This is not necessarily true for energy efficiencies, which are often misleading.

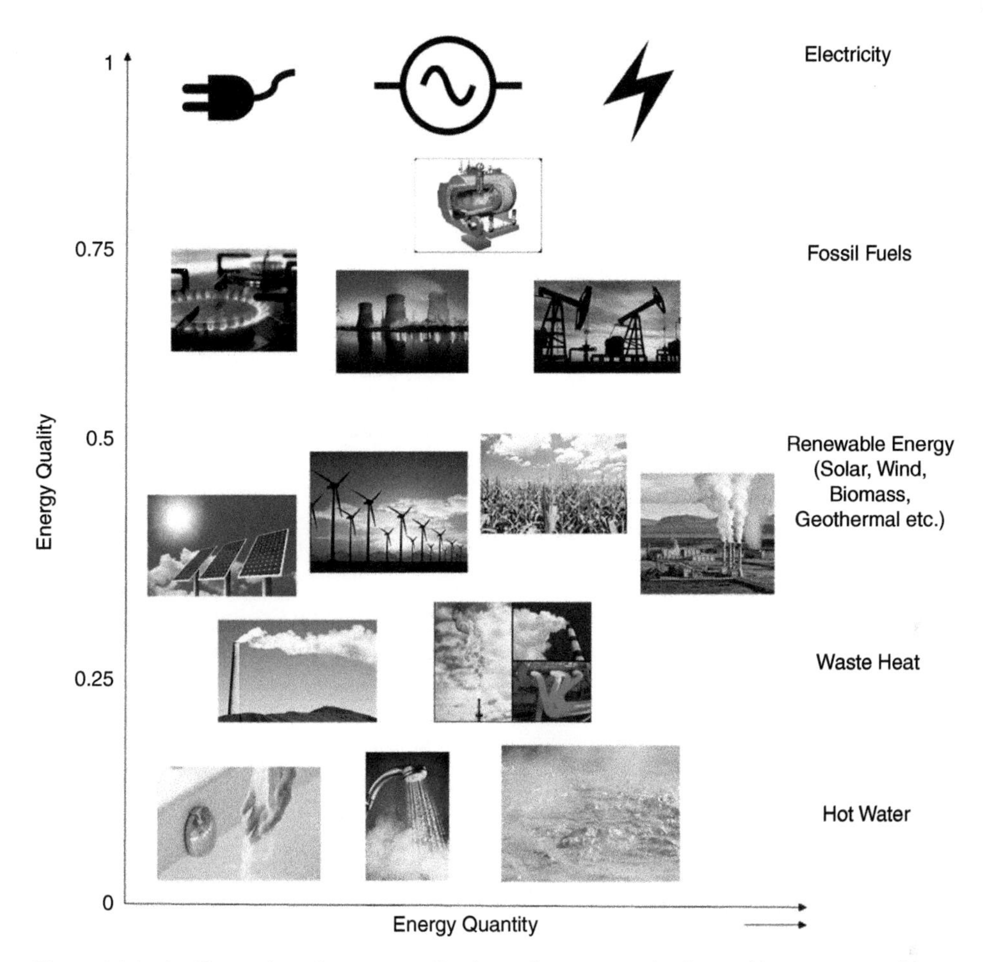

Figure 4.24 An illustration of energy quality for various sources/options with energy quantity.

Here, one can further look at linkages between energy and exergy, and the relations between exergy and both the environment and sustainable development, and can come up with some key points that highlight the importance of exergy and its utilization. Specifically, exergy analysis is recognized as an effective method and tool for:

- Combining and applying the conservation of mass and conservation of energy principles (referring to the FLT) together with the SLT for the design, analysis, assessment, and evaluation of energy systems.
- Improving the efficiencies of using energy and other resources (by identifying efficiencies that always measure the approach to ideality as well as the locations, types, and true magnitudes of wastes and losses).
- Revealing whether or not and by how much it is possible to design more efficient systems by reducing the inefficiencies.

- Addressing the impact on the environment of energy and other resource utilization, and reducing or mitigating that impact.
- Identifying whether a system contributes to achieving sustainable development or is unsustainable.

4.12 The Different Forms of Exergy

In this section the different forms of exergy are introduced, and by different forms of exergy we mean the different ways the exergy can be transferred into and out of the system and the initial and final amount of exergy the system has. So exergy can enter and leave the system in two forms for closed systems: heat transfer and work. While for open systems, exergy can enter and leave the system boundary in three different forms: following the stream of matter, heat transfer rate, and work rate. The next subsections define each form and provide the equation to calculate the form of exergy.

4.12.1 Flow Exergy

The exergy of a flowing stream of matter, which can be denoted by Ex_{flow}, is the sum of non-flow exergy and the exergy associated with the flow work of the stream (with reference to P_o), i.e.

$$Ex_{flow} = Ex_{nonflow} + (P - P_0)V \tag{4.23}$$

or Ex_{flow} is expressed in terms of physical, chemical, kinetic, and potential components as:

$$Ex_{flow} = Ex_{flow,ph} + Ex_0 + Ex_{kin} + Ex_{pot} \tag{4.24}$$

Here, $Ex_o = \sum_i (\mu_{io} - \mu_{ioo})N_i$ and $Ex_{flow,\ ph} = (H - H_0) - T_0(S - S_0)$

4.12.2 Thermal Exergy

Consider a control mass, initially at the dead state, being heated or cooled at constant volume in an interaction with some other system. The heat transfer experienced by the control mass is Q. The flow of exergy associated with the heat transfer Q is denoted by Ex_Q, and is written as:

$$Ex_Q = (1 - T_0/T)Q = \tau Q \tag{4.25}$$

where τ is called the "exergetic temperature factor."

4.12.3 Exergy of Work

The exergy associated with shaft work Ex_W is by definition W_x. The exergy transfer associated with work done by a system due to volume change is the net usable work due to the volume change, and is denoted by W_{NET}.

4.12.4 Exergy of Electricity

As for shaft work, the exergy associated with electricity which is work equals its energy.

4.13 Exergy Destruction

As highlighted in the SLT, exergy cannot be created; however, it can be destroyed, and for a process occurring in a system, the difference between the total exergy flows into and out of the system, other than the exergy that has accumulated inside the system boundary, is exergy that was destroyed and it is directly related to the entropy generated as follows:

$$Ex_d = T_0 S_{gen} \tag{4.26}$$

As we can see in the above given equation, there is direct relationship between entropy and exergy, which states "the more entropy one generates, the more exergy it destroys."

4.14 Reference Environment

Exergy is evaluated with respect to a reference environment, so the intensive properties of the reference environment partly determine the exergy of a stream or system. The reference environment is a condition where there is stable equilibrium, with all parts at rest relative to one another. No chemical reactions can occur between the environmental components. The reference environment acts as an infinite system, and is a sink and source for heat and materials. It experiences only internally reversible processes in which its intensive state remains unaltered (i.e. its temperature T_o, pressure P_o, and the chemical potentials μ_{ioo} for each of the i components present remain constant). The exergy of the reference environment is zero. The exergy of a stream or system is zero when it is in equilibrium with the reference environment. There are multiple reference environment models, some of which are listed and described here (further details are available elsewhere [1]).

4.14.1 Natural-Environment-Subsystem Models

One of the important models for the reference-environment modeling is the natural-environment-subsystem type. These types of models simulate in a realistic manner the subsystems of the natural environment. One proposed model consisted of saturated moist air and liquid water in phase equilibrium. [2] This model was later modified, [2, 3] with the modification allowing sulfur-containing material to be analyzed. The following model presents the pressure and the temperature of the reference environment shown in Table 4.1 as 25 °C and 1 atm, respectively; other properties of the model are that water (H_2O), gypsum ($CaSO_4{\cdot}2H_2O$), and limestone ($CaCO_3$) condense at 25 °C and 1 atm. It is important to know that the stable configurations of C, O, and N are taken to be CO_2, O_2, and N_2 as they are naturally available in the air.

4.15 Exergy Balance Equation for Closed Systems

This section will introduce different examples that deal with exergy balance equations of closed energy systems; a piston–cylinder device example is used to show the exergy calculation steps for closed systems. Note that in the coming examples, mass, energy, and, for specific cases, entropy balance equations are also solved. Before we start with the examples we can define the general form of the exergy balance equation as:

Table 4.1 A reference-environment model.

Temperature:	$T_o = 298.15\,K$		
Pressure:	$P_o = 1\,atm$		
Composition:	(i)	Atmospheric air saturated with H_2O at T_o and P_o, with the following composition:	
		Air constituents	Mole fraction
		N_2	0.7567
		O_2	0.2035
		H_2O	0.0303
		Ar	0.0091
		CO_2	0.0003
		H_2	0.0001
	(ii)	The following condensed phases at T_o and P_o:	
		Water (H_2O)	
		Limestone ($CaCO_3$)	
		Gypsum ($CaSO_4{\cdot}2H_2O$)	

Initial amount of exergy the system has + the amount of exergy entering the system
= Final amount of exergy the system has + the amount of exergy leaving the system
+ the amount of exergy destroyed during the process

which is formulated as

$$\Sigma Ex_{in} = \Sigma Ex_{out} + \Sigma Ex_{destroyed}$$

It is important to provide guidance to the students about how to write balance equations for some example systems which may represent some common types of problems encountered through applications, such as a rigid tank filled with substance (liquid or gas) subject to energy transfers (in terms of heat and work), heating or cooling some objects with negligible heat losses (or heat penetrations), a piston-cylinder device where there is compression work done on the system, and a piston-cylinder device where there is expansion work produced by the system. These selected example systems are clearly illustrated in Table 4.2 along with all four balance equations for mass, energy, entropy and exergy.

Table 4.2 Mass, energy, entropy and exergy balance equations written for four selected closed systems (with negligible changes in kinetic and potential energy and exergies)

Process or System	Balance Equations
A rigid tank filled with liquid subject to energy transfers 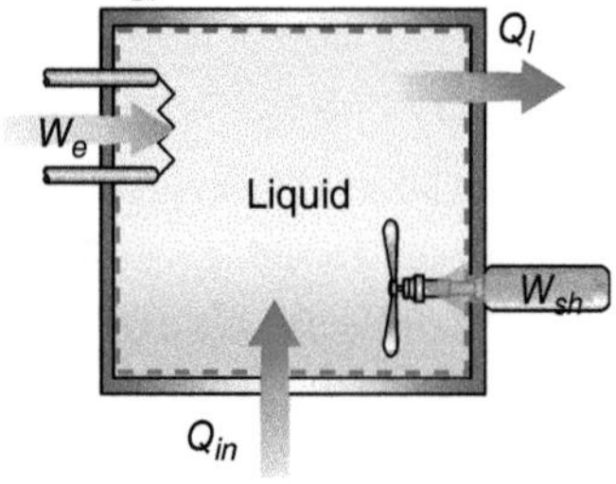	MBE: $m_1 = m_2 = m = constant$ EBE: $m_1u_1 + Q_{in} + W_e + W_{sh} = m_2u_2 + Q_l$ EnBE: $m_1s_1 + \frac{Q_{in}}{T_s} + S_{gen} = m_2s_2 + \frac{Q_l}{T_{surr}}$ ExBE: $m_1ex_1 + Ex_{Q_{in}} + W_e + W_{sh} = m_2ex_2 + Ex_{Q_l} + Ex_d$ where T_s is source temperature (*K*) and T_{surr} = immediate surrounding (boundary) temperature (*K*)

Table 4.2 (Continued)

Process or System	**Balance Equations**
Heating an oil bath with three solid metallic balls with negligible heat losses 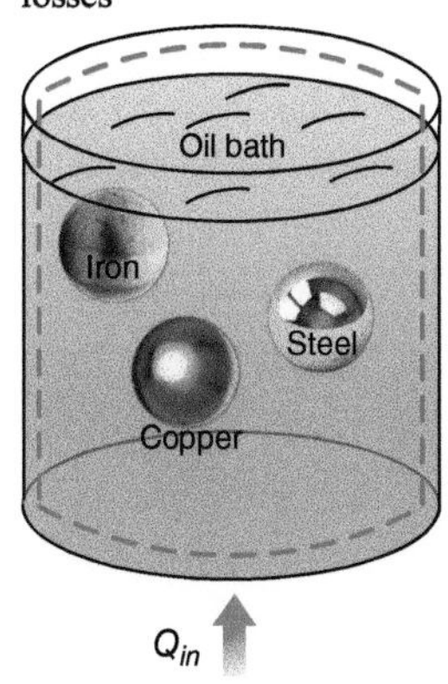	MBE: $m_{i,1} = m_{i,2} = m_i, m_{s,1} = m_{s,2} = m_s,\ m_{c,1} = m_{c,2} = m_c, and\ m_{o,1} = m_{o,2} = m_o$ EBE: $m_i u_{i,1} + m_s u_{s,1} + m_c u_{c,1} + m_o u_{o,1} + Q_{in} = m_i u_{i,2} + m_s u_{s,2} + m_c u_{c,2} + m_o u_{o,2}$ EnBE: $m_i s_{i,1} + m_s s_{s,1} + m_c s_{c,1} + m_o s_{o,1} + \frac{Q_{in}}{T_s} + S_{gen} = m_i s_{i,2} + m_s s_{s,2} + m_c s_{c,2} + m_o s_{o,2}$ ExBE: $m_i ex_{i,1} + m_s ex_{s,1} + m_c ex_{c,1} + m_o ex_{o,1} + Ex_{Q_{in}} = m_i ex_{i,2} + m_s ex_{s,2} + m_c ex_{c,2} + m_o ex_{o,2}$ where i for iron, s for steel, c for copper, and o for oil.
A compression process in a piston-cylinder device: 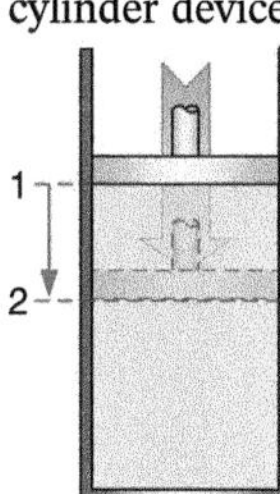	MBE: $m_1 = m_2 = m = constant$ EBE: $m_1 u_1 + W_b = m_2 u_2$ or EBE: $m_1 u_1 + m_1 P v_1 = m_2 u_2 + m_2 P v_2$ since the boundary work is defined as follows: $W_b = P\Delta V = mP\Delta v$ EnBE : $m_1 s_1 + S_{gen} = m_2 s_2$ ExBE : $m_1 ex_1 + W_b = m_2 ex_2 + Ex_d$
A closed system with variable volume (basically a piston-cylinder mechanism with boundary movement work 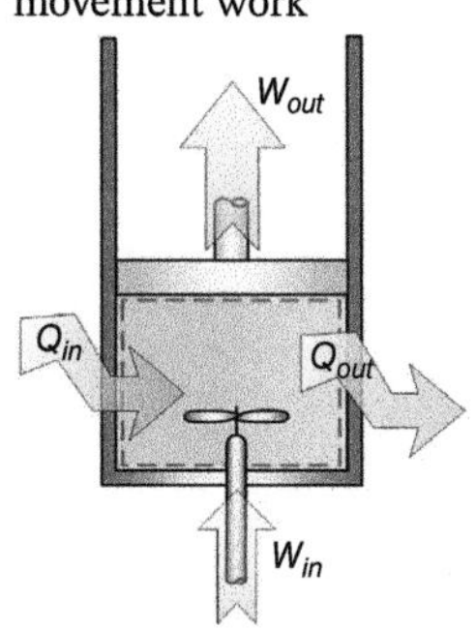	MBE: $m_1 = m_2 = m = constant$ EBE : $m_1 u_1 + Q_{in} + W_{in} = m_2 u_2 + Q_{out} + W_{out}$ where W_{out} is boundary movement work and defined as $W_b = P\Delta V = mP\Delta v$ This balance equation may also be written as EBE: $m_1 u_1 + Q_{in} + W_{in} + m_1 P v_1 = m_2 u_2 + Q_{out} + m_2 P v_2$ EnBE: $m_1 s_1 + \frac{Q_{in}}{T_s} + S_{gen} = m_2 s_2 + \frac{Q_{out}}{T_{surr}}$ ExBE: $m_1 ex_1 + Ex_{Q_{in}} + W_{in} = m_2 ex_2 + Ex_{Q_{out}} + W_b + Ex_d$

Example 4.9 Consider the closed system given Example 4.3 with Figure 4.11. In this system, there was a 1 m^3 rigid tank being heated and stirred contains air at 100 *kPa* and 20 °*C*. A 66 *kJ* of heat is transferred to the air where the 10% of the heat supplied is lost. If a 100 *kJ* mechanical work through shaft 1 and half of it by shaft 2 are provided to run the fans inside the tank. Take the source temperature for the incoming heat as 600 °*C* and immediate surrounding (boundary) temperature as 60 °*C*. The properties of air will be taken from the property tables (as given in the Appendix) or the properties software (such as EES). Take the reference temperature (ambient temperature) as 25 °*C*. (a) write all for mass balance equations for mass, energy, entropy and exergy and (b) calculate the exergy destruction.

Solution

a) Write the mass, energy, entropy balance equations:

$$\text{MBE}: m_1 = m_2 = m$$

$$\text{EBE}: m_1u_1 + W_{sh},1 + W_{sh},2 + Q_{in} = m_2u_2 + Q_l$$

$$\text{EnBE}: m_1s_1 + Q_{in}/T_s + S_{gen} = m_2s_2 + Q_l/T_{surr}$$

$$\text{ExBE}: m_1ex_1 + Ex_{Qin} + W_{sh},1 + W_{sh},2 = m_2ex_2 + Ex_{Ql} + Ex_d$$

b) Find the exergy destruction:

We had the following given, obtained from the tables (such as Appendix E1) and/or calculated earlier in Example 4.3:

$$T_o = 25\,^{\circ}C + 273 = 298\ K$$

$$T_s = 600\,^{\circ}C + 273 = 873\ K$$

$$T_{surr} = T_b = 60\,^{\circ}C + 273 = 333\ K$$

$$m = 1.189\ kg$$

$$Q_{in} = 66\ kJ \text{ and } Q_l = 66 \times 0.1 = 6.6\ kJ$$

$$W_{sh},1 = 100\ kJ \text{ and } W_{sh},2 = 50\ kJ$$

$$u_1 = 209.3\ kJ/kg \text{ and } u_2 = 385.4\ kJ/kg$$

$$(s_2 - s_1) = c_v\,ln\left(\frac{T_2}{T_1}\right) + Rln\left(\frac{v_2}{v_1}\right) = 0.718\,\frac{kJ}{kgK} \times ln\left(\frac{534.3}{293}\right) = 0.431\,\frac{kJ}{kgK}$$

$$S_{gen} = 0.457\ kJ/K$$

Here, we have two options to find the exergy destruction:

(i) To calculate directly by using the entropy generation:

$$\boldsymbol{Ex_d} = T_oS_{gen} = 298\ K \times 0.457\,\frac{kJ}{K} = \mathbf{136.1\ kJ}$$

and

(ii) To use the exergy balance equation (ExBE) written above as follows:

$$Ex_d = m(ex_1 - ex_2) + Ex_{Qin} + W_{sh,1} + W_{sh,2} - Ex_{Ql} = m(ex_1 - ex_2) + \left(1 - \frac{T_o}{T_s}\right) \times Q_{in} + W_{sh,1} + W_{sh,2} - \left(1 - \frac{T_o}{T_{surr}}\right)Q_l$$

where

$$m(ex_2 - ex_1) = m((u_2 - u_1) - T_o(s_2 - s_1)) = 1.189((176.1) - 298(0.431)) = 56.67\,kJ,$$

$$\left(1-\frac{T_o}{T_s}\right)Q_{in}=\left(1-\frac{298}{873}\right)66=43.47\,kJ \text{ and}$$

$$\left(1-\frac{T_o}{T_{surr}}\right)Q_l=\left(1-\frac{298}{333}\right)6.6=0.69\,kJ.$$

Let's substitute these into the exergy destruction equation and find the amount of exergy destruction:

$$\boldsymbol{Ex_d} = -56.67 + 43.47 + 100 + 50 - 0.69 = \mathbf{136.1\ kJ}$$

Example 4.10 A piston–cylinder device, which is being stirred, as shown in Figure 4.25, has air in it at a pressure of 400 kPa and a temperature of 17 °C. Work enters the system through the motion of a paddle wheel at an amount of 50 kJ per each kg of air inside the piston–cylinder. Heat (i.e. thermal energy) is added from a heat source with a temperature of 600 °C to the system transferring the system from state 1 to state 2 and the piston moves so that the pressure of the system remains constant at 400 kPa, though the process volume contained in the system increases to three times its original volume. (a) Write balance equations for mass, energy, entropy, and exergy for this system and (b) calculate the boundary work, the heat addition, the entropy generation, and the exergy destruction.

Solution

Before writing the balance equations and solving it for the requested items, one may need to list the assumptions that it is necessary to make.

- Air trapped inside the piston–cylinder device is treated as an ideal gas.
- The changes in the kinetic and potential energies, and hence exergies, are neglected.
- The specific heat is constant throughout the process.

a) Balance equations for mass, energy, entropy, and exergy.

Even if the problem did not specify writing the balance equations, it is always recommended to start with them as they describe all the system interactions and how it goes

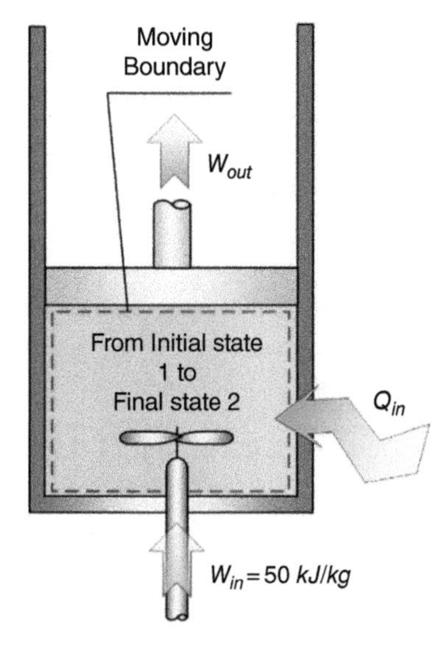

Figure 4.25 A piston–cylinder mechanism with the energies added and removed from the boundary of this device.

from one state to another. The four balance equations of the piston–cylinder device are written as:

$$\text{MBE}: m_1 = m_2 = constant$$

$$\text{EBE}: m_1u_1 + Q_{in} + W_{in,paddle} = m_2u_2 + W_{out,b}$$

$$\text{EnBE}: m_1s_1 + Q_{in}/T_s + S_{gen} = m_2s_2$$

$$\text{ExBE}: m_1ex_1 + (1-(T_o/T_s))Q_{in} + W_{in,paddle} = m_2ex_2 + W_{out} + Ex_d$$

b) Calculation of the boundary work, the heat addition, the entropy generation, and the exergy destruction.

According to ideal-gas law:

$$P_1V_1/T_1 = P_2V_2/T_2$$

with $T_2 = 870\ K$

the EBE then reduce

$$u_1 + q_{in} + w_{in,paddle} = w_{out,b} + u_2$$

Since the system is operating at a constant pressure, we use the following equation to calculate the boundary work:

$$\mathbf{w_{out,b}} = P(v_2 - v_1) = 166.5\ kJ/kg$$

The heat (thermal energy) added to the system is calculated by substituting the boundary work in the EBE:

$$207.1 + q_{in} + 50 = 166.5 + 650$$

$$\mathbf{q_{in} = 166.5 + 650 - 50 - 207.1 = 559.3\ \frac{kJ}{kg}}$$

The entropy generated during the process is calculated from the entropy balance equation:

$$m_1s_1 + Q_{in}/T_s + S_{gen} = m_2s_2$$

$$s_1 + q_{in}/T_s + s_{gen} = s_2$$

$$5.274 + \frac{559.3}{600 + 273.15} + s_{gen} = 6.417$$

$$\mathbf{s_{gen} = 0.502\ \frac{kJ}{kg\,K}}$$

The specific exergy destruction during the process is calculated based on the relationship between the exergy destruction and the entropy generation and it is utilized as follows:

$$\mathbf{ex_d} = T_o s_{gen} = 298.15 \times 0.502 = \mathbf{149.6\ \frac{kJ}{kg}}$$

Note that in this example the exergy destruction is calculated through the entropy generation. For more details on the exergy calculations for each state of the system and for those exergies entering and leaving closed systems refer to the next chapter.

4.16 Exergy Balance Equation for Open Systems

This section introduces different examples that deal with exergy balance equations of open energy systems, where again they are similar to what was introduced in the entropy balance equations of open systems. Two example are introduced in this section. The first example deals with power consuming devices, which is an air compressor, and the second example deals with a heat exchange process, which is a two fluid heat exchanger. Note that in the coming examples, mass, energy, and, for specific cases, entropy balance equations are also solved. Before we start with the examples we can define the general form of the exergy balance equation as:

The exergy rate entering the system = the exergy rate leaving the system + the amount of exergy destruction rate during the process

which is formulated as

$$\Sigma \dot{E}x_{in} = \Sigma \dot{E}x_{out} + \Sigma \dot{E}x_{destroyed}$$

Table 4.3 list a number of selected devices/components which are commonly used power generating and refrigeration units and are treated as open systems, being subject to steady-state and steady-flow processes, along with the balance equations for mass, energy, entropy and exergy.

Table 4.3 Mass, energy, entropy and exergy balance equations for some commonly used devices which are treated as open systems

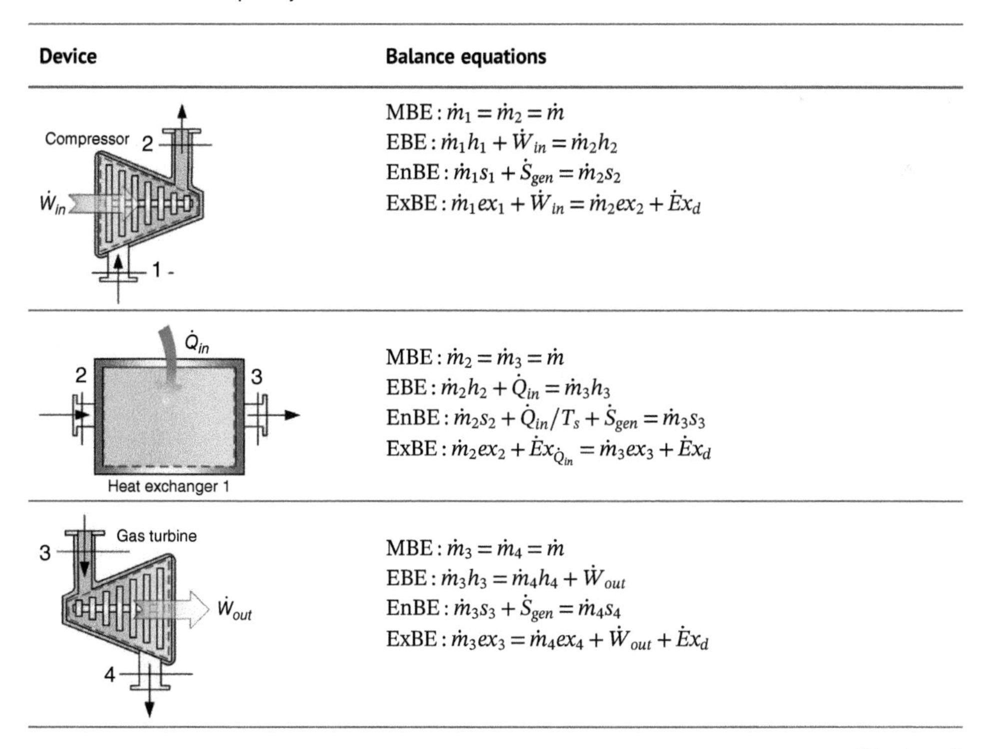

Device	Balance equations
Compressor (states 1, 2; $\dot{W}_{in}$)	MBE: $\dot{m}_1 = \dot{m}_2 = \dot{m}$ EBE: $\dot{m}_1 h_1 + \dot{W}_{in} = \dot{m}_2 h_2$ EnBE: $\dot{m}_1 s_1 + \dot{S}_{gen} = \dot{m}_2 s_2$ ExBE: $\dot{m}_1 ex_1 + \dot{W}_{in} = \dot{m}_2 ex_2 + \dot{E}x_d$
Heat exchanger 1 (states 2, 3; $\dot{Q}_{in}$)	MBE: $\dot{m}_2 = \dot{m}_3 = \dot{m}$ EBE: $\dot{m}_2 h_2 + \dot{Q}_{in} = \dot{m}_3 h_3$ EnBE: $\dot{m}_2 s_2 + \dot{Q}_{in}/T_s + \dot{S}_{gen} = \dot{m}_3 s_3$ ExBE: $\dot{m}_2 ex_2 + \dot{E}x_{\dot{Q}_{in}} = \dot{m}_3 ex_3 + \dot{E}x_d$
Gas turbine (states 3, 4; $\dot{W}_{out}$)	MBE: $\dot{m}_3 = \dot{m}_4 = \dot{m}$ EBE: $\dot{m}_3 h_3 = \dot{m}_4 h_4 + \dot{W}_{out}$ EnBE: $\dot{m}_3 s_3 + \dot{S}_{gen} = \dot{m}_4 s_4$ ExBE: $\dot{m}_3 ex_3 = \dot{m}_4 ex_4 + \dot{W}_{out} + \dot{E}x_d$

(Continued)

Table 4.3 (Continued)

Device	Balance equations
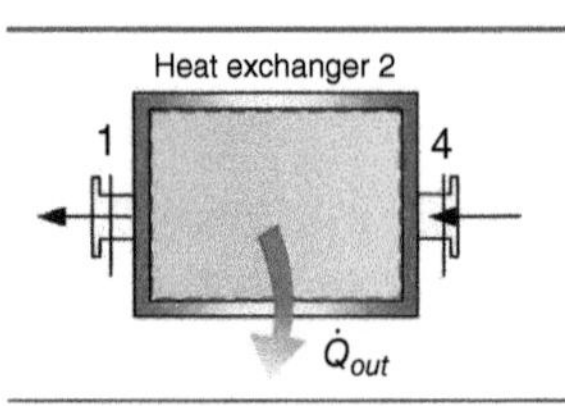	MBE : $\dot{m}_4 = \dot{m}_1 = \dot{m}$ EBE : $\dot{m}_4 h_4 = \dot{m}_1 h_1 + \dot{Q}_{out}$ EnBE : $\dot{m}_4 s_4 + \dot{S}_{gen} = \dot{m}_1 s_1 + \dot{Q}_{out}/T_b$ ExBE : $\dot{m}_4 ex_4 = \dot{m}_1 ex_1 + \dot{E}x_d + \dot{E}x_{\dot{Q}_{out}}$
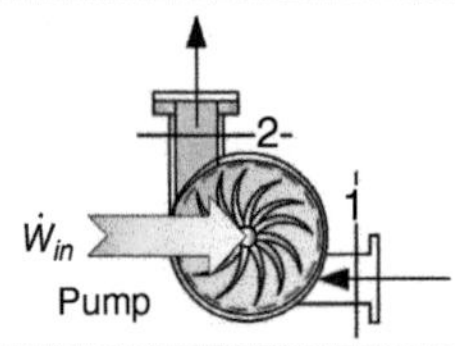	MBE : $\dot{m}_1 = \dot{m}_2 = \dot{m}$ EBE : $\dot{m}_1 h_1 + \dot{W}_{in} = \dot{m}_2 h_2$ EnBE : $\dot{m}_1 s_1 + \dot{S}_{gen},p = \dot{m}_2 s_2$ ExBE : $\dot{m}_1 ex_1 + \dot{W}_{in} = \dot{m}_2 ex_2 + \dot{E}x_{d,p}$
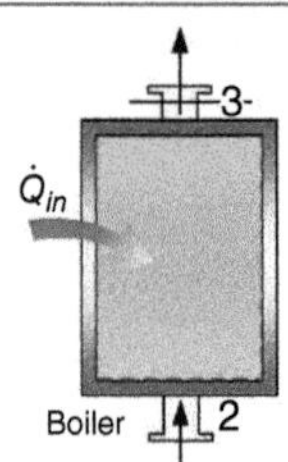	MBE : $\dot{m}_2 = \dot{m}_3 = \dot{m}$ EBE : $\dot{m}_2 h_2 + \dot{Q}_{in} = \dot{m}_3 h_3$ EnBE : $\dot{m}_2 s_2 + \dot{Q}_{in}/T_s + \dot{S}_{gen},bo = \dot{m}_3 s_3$ ExBE : $\dot{m}_2 ex_2 + \dot{E}x_{\dot{Q}_{in}} = \dot{m}_3 ex_3 + \dot{E}x_{d,bo}$
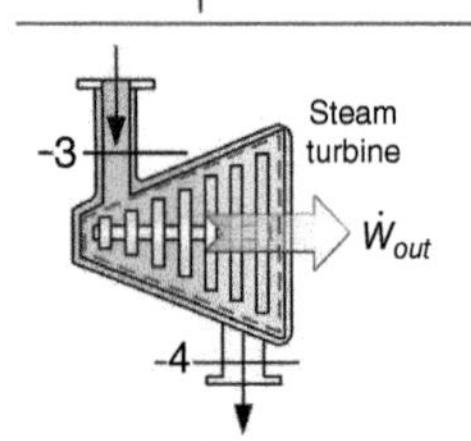	MBE : $\dot{m}_3 = \dot{m}_4 = \dot{m}$ EBE : $\dot{m}_3 h_3 = \dot{m}_4 h_4 + \dot{W}_{out}$ EnBE : $\dot{m}_3 s_3 + \dot{S}_{gen},st = \dot{m}_4 s_4$ ExBE : $\dot{m}_3 ex_3 = \dot{m}_4 ex_4 + \dot{W}_{out} + \dot{E}x_{d,st}$
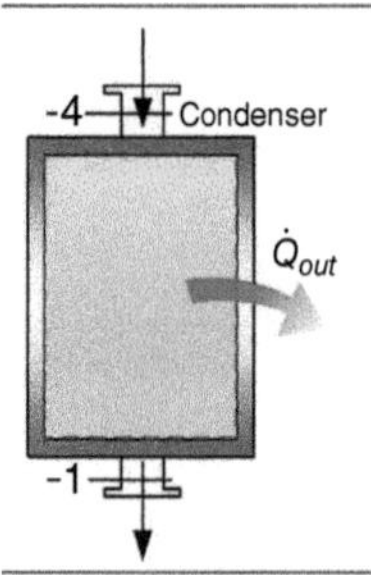	MBE : $\dot{m}_4 = \dot{m}_1 = \dot{m}$ EBE : $\dot{m}_4 h_4 = \dot{m}_1 h_1 + \dot{Q}_{co}$ EnBE : $\dot{m}_4 s_4 + \dot{S}_{gen},co = \dot{m}_1 s_1 + \dot{Q}_{out}/T_b$ ExBE : $\dot{m}_4 ex_4 = \dot{m}_1 ex_1 + \dot{E}x_{\dot{Q}_{out}} + \dot{E}x_{d,co}$
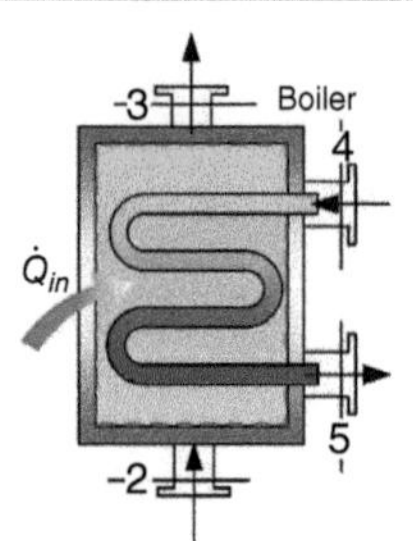	MBE : $\dot{m}_2 = \dot{m}_3 = \dot{m}$ and $\dot{m}_4 = \dot{m}_5$ EBE : $\dot{m}_2 h_2 + \dot{m}_4 h_4 + \dot{Q}_{in} = \dot{m}_3 h_3 + \dot{m}_5 h_5$ EnBE : $\dot{m}_2 s_2 + \dot{m}_4 s_4 + \frac{\dot{Q}_{in}}{T_s} + \dot{S}_{gen},bo = \dot{m}_3 s_3 + \dot{m}_5 s_5$ ExBE : $\dot{m}_2 ex_2 + \dot{m}_4 ex_4 + \dot{E}x_{\dot{Q}_{in}} = \dot{m}_3 ex_3 + \dot{m}_5 ex_5 + \dot{E}x_{d,bo}$ where $\dot{E}x_{\dot{Q}in} = \left(1 - \frac{T_o}{T_s}\right)\dot{Q}_{in}$ and $\dot{E}x_{d,bo} = T_o \times \dot{S}_{gen,bo}$
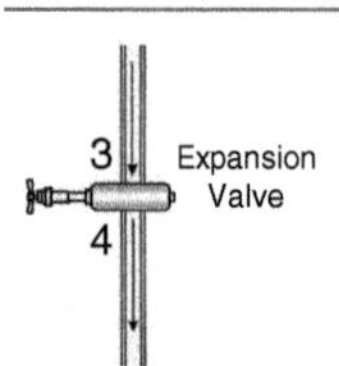	MBE : $\dot{m}_3 = \dot{m}_4 = \dot{m}$ EBE : $\dot{m}_3 h_3 = \dot{m}_4 h_4 \rightarrow h_3 = h_4$ (making the process isenthalpic) EnBE : $\dot{m}_3 s_3 + \dot{S}_{gen},3 \rightarrow 4 = \dot{m}_4 s_4$ ExBE : $\dot{m}_3 ex_3 = \dot{m}_4 ex_4 + \dot{E}x_{d,3\rightarrow 4}$

Example 4.11 The air compressor shown in Figure 4.26 compresses air that enters it at an atmospheric pressure of 100 *kPa* and a temperature of 280 *K*. After the compression process the air exits the compressor at a pressure and a temperature of 600 *kPa* and 400 *K*, respectively. The mass flow rate of the air passing through compressor is 1 *kg/s*. Note that during the compression process the compressor losses heat through its walls equivalent to 40% of the total work rate consumed by the compressor. (a) Write balance equations for mass, energy, entropy, and exergy for this system and (b) calculate the work rate consumed by the compressor and the exergy destruction rate in the compressor. Take the average temperature of 400 and 280 *K* for the boundary (immediate boundary) temperature, i.e. 340 *K*.

Solution

Before writing the balance equations and solving it for the requested items, one may need to list the assumptions that it is necessary to make.

- The heat losses occur during the compression process.
- The changes in the kinetic and potential energies and exergies are negligible.
- The air is treated as ideal gas.

a) The first step in solving this example is to write the balance equations.

$$\text{MBE}: \dot{m}_1 = \dot{m}_2$$

$$\text{EBE}: \dot{m}_1 h_1 + \dot{W}_{in} = \dot{m}_2 h_2 + \dot{Q}_{out}$$

$$\text{EnBE}: \dot{m}_1 s_1 + \dot{S}_{gen} = \dot{m}_2 s_2 + \dot{Q}_{out}/T_{surr}$$

$$\text{ExBE}: \dot{m}_1 ex_1 + \dot{W}_{in} = \dot{m}_2 ex_2 + \dot{E}x_d + \dot{E}x_{\dot{Q}_{out}}$$

b) The power consumed during the compression process needs to be calculated in this section.

Using the EBE to calculate the specific power consumed by the compressor:

$$\dot{W}_{in} = \dot{Q}_{out} + \dot{m}_2 h_2 - \dot{m}_1 h_1$$

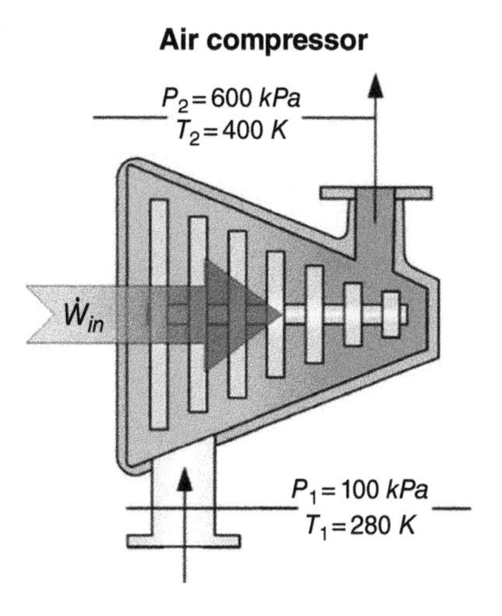

Figure 4.26 A schematic diagram of the air compressor discussed in Example 4.11.

Table 4.4 The properties of the inlet and the exit streams.

State Point	P (kPa)	T (°C)	h (kJ/kg)	s (kJ/kgK)	ex[a] (kJ/kg)
Reference state	101.3	25	298.6	5.696	0
1	100	7.0	280.5	5.637	0.509
2	600	127	401.4	5.482	166.6

[a] The reference environment temperature and pressure are taken as 25 °C and 1 atm.

Here, the properties of the stream entering and leaving the compressor are obtained from air properties tables (such as Appendix E1) or using Engineering Equation Solver (EES) software, which contains the database for the properties of most of the fluids through a large range of temperatures and pressures. The properties are tabulated in Table 4.4.

The work input needed for the compressor is calculated as:

$$\dot{m}_1 h_1 + \dot{W}_{in} = \dot{m}_2 h_2 + \dot{Q}_{out}$$

$$\dot{W}_{in} = 0.4 \times \dot{W}_{in} + \dot{m}_2 h_2 - \dot{m}_1 h_1$$

$$\dot{W}_{in} = 0.4 \times \dot{W}_{in} + 1 \times 401.4 - 1 \times 280.5$$

$$\mathbf{\dot{W}_{in} = 201.5\ kW}$$

The exergy destruction during the compression process is then calculated as follows:

$$\dot{m}_1 ex_1 + \dot{W}_{in} = \dot{m}_2 ex_2 + \dot{E}x_d + \dot{E}x_{\dot{Q}_{out}}$$

$$1\frac{kg}{s} \times +0.509\frac{kJ}{kg} + 201.5\frac{kJ}{s} = 1\frac{kg}{s} \times 166.6\frac{kJ}{kg} + \dot{E}x_d + \left(1 - \frac{298}{340}\right)\left(0.4 \times 201.5\frac{kJ}{s}\right)$$

Therefore, the exergy destruction rate is extracted and calculated as:

$$\mathbf{\dot{E}x_d = 24.43\ kW}$$

Example 4.12 Hot exhaust gases leave an internal combustion engine at a temperature and pressure of 400 °C and 150 kPa, respectively, at a rate of 0.8 kg/s. They are used to produce saturated steam at 200 °C in an insulated heat exchanger, as shown in Figure 4.27. Water enters the heat exchanger at the ambient temperature of 20 °C and the exhaust gases leave the heat exchanger at 350 °C. (a) Write balance equations for mass, energy, entropy, and exergy for this system and (b) calculate the rate of steam produced and the rate of exergy destroyed in the heat exchanger.

Solution

Note that all the subscripts are based on the numbering of the inlets and outlets shown in Figure 4.27.

It is always recommended to start with them since they describe all the system interactions and how it goes from one state to another. The four balance equations of the heat exchanger are written as:

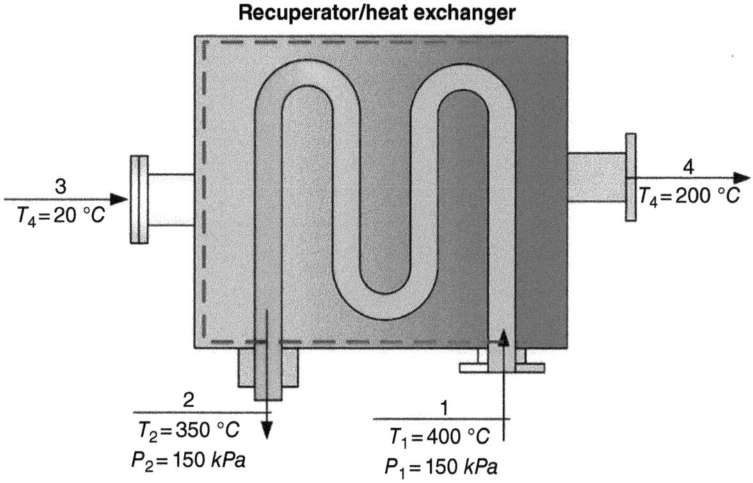

Figure 4.27 A schematic diagram of the cross-flow heat exchanger used to produce steam discussed in Example 4.12.

$$\text{MBE}: \dot{m}_1 = \dot{m}_2 = \dot{m}_w \ and \ \dot{m}_3 = \dot{m}_4 = \dot{m}_a$$

$$\text{EBE}: \dot{m}_1 h_1 + \dot{m}_3 h_3 = \dot{m}_2 h_2 + \dot{m}_4 h_4$$

$$\text{EnBE}: \dot{m}_1 s_1 + \dot{m}_3 s_3 + \dot{S}_{gen} = \dot{m}_2 s_2 + \dot{m}_4 s_4$$

$$\text{ExBE}: \dot{m}_1 ex_1 + \dot{m}_3 ex_3 = \dot{m}_2 ex_2 + \dot{m}_4 ex_4 + \dot{E}x_d$$

The properties of water are obtained from the steam tables (such as Appendix B1a) to be:

$$T_3 = 20\,^{\circ}C, \ liquid \rightarrow h_3 = 83.91\,\frac{kJ}{kg}, \ s_3 = 0.29649\,\frac{kJ}{kgK}$$

$$T_4 = 200\,^{\circ}C, \ saturated\ vapor \rightarrow h_4 = 2792.0\,\frac{kJ}{kg}, \ s_4 = 6.4302\,\frac{kJ}{kgK}$$

An energy balance on the heat exchanger gives the rate of steam production:

$$\dot{m}_1 h_1 + \dot{m}_3 h_3 = \dot{m}_2 h_2 + \dot{m}_4 h_4$$

$$\dot{m}_a h_1 + \dot{m}_w h_3 = \dot{m}_a h_2 + \dot{m}_w h_4$$

$$\dot{m}_a c_p (T_1 - T_2) = \dot{m}_w (h_4 - h_3)$$

$$0.8\,\frac{kg}{s} \times 1.063\,\frac{kJ}{kg^{o}C} \times (400 - 350)^{o}C = \dot{m}_w (2792.0 - 83.91)\,\frac{kJ}{kg}$$

$$\boldsymbol{\dot{m}_w = 0.01570\ kg/s}$$

The specific exergy changes of air that enter the heat exchanger at state 1 and leaves the exchanger at state 2:

$$\Delta ex_a = c_p\,(T_2 - T_1\,) - T_o\,(s_2 - s_1\,) = (1.063\ kJ/(kgK))(-50\ K)$$
$$-(293\ K \times -0.08206\ kJ/(kgK)) = -29.106\ kJ/kg$$

The specific exergy changes of water that enter the heat exchanger at state 3 and leaves the exchanger at state 4:

$$\Delta ex_w = (h_4 - h_3) - T_o\,(s_4 - s_3) = (2792.0 - 83.91)\ kJ/kg$$
$$-(293\ K \times (6.4302 - 0.29649)\ kJ/(kgK) = 910.913\ kJ/kg$$

The exergy destruction is determined from an exergy balance as:

$$(\dot{m}_a ex_1 + \dot{m}_w ex_3) - (\dot{m}_a ex_2 + \dot{m}_w ex_4) - \dot{E}x_d = 0$$

Rearranging and substituting, one obtains the exergy destruction rate as:

$$\dot{E}x_d = -(\dot{m}_a \Delta ex_a + \dot{m}_w \Delta ex_w) = -((0.8\ kg/s) \times (-29.106\ kJ/kg)$$
$$+ (0.01570\ kg/s) \times (910.913\ kJ/kg)) = \mathbf{8.98\ kW}$$

4.17 Exergy Efficiency

As mentioned earlier, *efficiency* is generally defined as the ratio of useful output(s) divided by the input(s) given to produce the useful output(s) as follows:

$$Efficiency = Useful\ output(s)/Input(s)$$

There are also efficiencies defined in many specific terms or specific to the process or system or application, such as thermal efficiency, combustion efficiency, motor efficiency, power plant efficiency. In this thermodynamics there are two key efficiencies to use: (i) energy efficiency coming out of the FLT and (ii) exergy efficiency coming out of the SLT.

Many use numerous efficiencies inconsistently for energy efficiency, such as thermal efficiency, which is commonly used for power plants where thermal energy (heat) is supplied to produce mechanical work (electricity). It is important for everyone to define a single efficiency, namely energy efficiency if the target is to assess the performance of a process or a system or an application energetically from the FLT point of view. When it comes to refrigeration and heat pumps, we use coefficient of performance (COP), as the values obtained out of the equation (in terms of useful output(s)/input(s)) become greater than 1, which makes us use something different from the efficiency since efficiency cannot be more than 1 or 100%. We will also specify COP into energetic-based COP (COP_{en}) and exergetic-based COP (COP_{ex}). If no subscript is given (such as COP), one should figure it out as energetic based COP.

Going back to the efficiency, we will use energy efficiency (η_{en}) only for a process or a system or an application to assess its performance energetically from the FLT point of view. So, the energy efficiency can be formulated as:

$$\eta_{en} = \frac{Useful\ energy\ output(s)}{Energy\ input(s)}$$

In order to assess the performance of a process or a system or an application from the SLT point view, we use exergy efficiency (η_{ex}), which is formulated as follows:

$$\eta_{ex} = \frac{Exergy\ of\ useful\ output(s)}{Exergy\ input(s)}$$

Here, we do not say useful exergy output since it is implicitly covered by the term of exergy (which is known as useful or available). We rather define it as exergy of useful output(s) divided by the exergy input(s). In this section, we describe the use of exergy efficiencies in assessing the utilization efficiency of energy and other resources.

In practice, many people, particularly engineers, generally use the efficiency, meaning only energy efficiency, and they know that this is something which does not make sense all the time. One may see a 100% efficiency calculated for a process or a system or an application, but realizing that the useful output is not really as much as calculated by the equation. So, they strongly feel that this energy efficiency gives something, but does not reflect the reality. This is a critical point in the discussion where we observe the need for an efficiency coming from the SLT, which is exergy efficiency.

In order to illustrate the idea of a performance parameter based on the SLT and to contrast it with an analogous energy-based efficiency, consider a control volume at steady state for which energy and exergy balances can be written, respectively, as:

$$(Energy\ in) = (Energy\ output\ in\ product) + (Energy\ emitted\ with\ waste)$$

and

$$(Exergy\ in) = (Exergy\ output\ in\ product) + (Exergy\ emitted\ with\ waste) + (Exergy\ destruction)$$

In these equations, the term product might refer to shaft work, electricity, a certain heat transfer, one or more particular exit streams, or some combination of these. The latter two terms in the exergy balance combine to constitute the exergy losses. Such losses include emissions to the surroundings such as waste heat and stack gases. The exergy destruction term in the exergy balance is caused by internal irreversibilities.

The exergy efficiency essentially gives a better understanding of performance than the energy efficiency. It stresses that both waste emissions (or external irreversibilities) and internal irreversibilities need to be dealt with to improve performance. In many cases it is the irreversibilities that are more significant and more difficult to address.

Furthermore, note that efficiency expressions each define a class of efficiencies because judgment has to be made about what is the product, what is counted as a loss, and what is the input. Different decisions about these items lead to different efficiency expressions within the class. Other SLT-based efficiency expressions also appear in the open literature. One of these is evaluated as the ratio of the sum of the exergy exiting to the sum of the exergy entering. Another class of second law efficiencies is composed of task efficiencies.

Coming back to the above formulation related point, exergy efficiency may take different forms depending on the type of the system. It is denoted by η_{ex}. Here, exergy efficiency is generally expressed as:

$$\eta_{ex} = \frac{Ex_{out}}{Ex_{in}} \tag{4.27}$$

or in rate form, it becomes:

$$\eta_{ex} = \dot{E}x_{out} / \dot{E}x_{in} \tag{4.28}$$

where

$$\dot{E}x_{in} = \dot{E}x_{out} + \dot{E}x_d \tag{4.29}$$

From the above equation, one may extract the rate of exergy output as:

$$\dot{E}x_{out} = \dot{E}x_{in} - \dot{E}x_d$$

and substitute it in Eq. (4.28), which results in:

$$\eta_{ex} = \frac{\dot{E}x_{in} - \dot{E}x_d}{\dot{E}x_{in}} = 1 - \frac{\dot{E}x_d}{\dot{E}x_{in}} \tag{4.30}$$

This can also be used to calculate the exergy efficiency, which will be same as that obtained from Eq. (4.28).

Question: If one has a system with both exergy destruction rate ($\dot{E}x_d$) and thermal exergy loss ($\dot{E}x_Q$), how should the exergy efficiency given in Eq. (4.30) be written? Here is the answer:

$$\dot{E}x_{in} = \dot{E}x_{out} + \dot{E}x_d + \dot{E}x_Q \tag{4.31}$$

where the summation of $(\dot{E}x_d + \dot{E}x_Q)$ is the so-called total exergy consumption as follows:

$$\dot{E}x_{cons} = \dot{E}x_d + \dot{E}x_Q \tag{4.32}$$

with this, the exergy efficiency has to be written as:

$$\eta_{ex} = \frac{\dot{E}x_{in} - \dot{E}x_{cons}}{\dot{E}x_{in}} = 1 - \frac{\dot{E}x_{cons}}{\dot{E}x_{in}} \tag{4.33}$$

This way the exergy efficiency becomes correct. One should check really carefully and consider both exergy destruction and exergy loss if both exist in the system.

Example 4.13 Reconsider the air compressor in Example 4.10 (Figure 4.26), which compresses air entering it at an atmospheric pressure and a temperature of 280 *K* to an exit pressure and a temperature of 600 *kPa* and 400 *K*. The mass flow rate of the air entering the compressor is around 1 *kg/s*; the compressor losses heat through its walls equivalent to 40% of the total work rate running the compressor. Calculate both the energy and exergy efficiencies of the compressor.

Solution

All four balance equations for mass, energy, entropy, and exergy are written earlier in Example 4.10 along with the calculations to find the compressor work input.

The energy and exergy efficiencies of the compressor are then calculated as follows:

$$\eta_{en} = \frac{(\dot{m}_2 h_2 - \dot{m}_1 h_1)}{\dot{W}_{in}} = \frac{1 \times (401.4 - 280.5)}{201.5} = 0.6 = \mathbf{60.0\%}$$

$$\eta_{ex} = \frac{(\dot{m}_2 ex_2 - \dot{m}_1 ex_1)}{\dot{W}_{in}} = \frac{1 \times (166.6 - (-0.509))}{201.5} = 0.8293 = \mathbf{83.0\%}$$

This shows that the exergy efficiency appears to be greater than the corresponding energy efficiency, meaning that the exergetic output of the compressor is greater than the useful energetic output of the compressor. In other forms, the exergy potential is higher than the corresponding energy potential.

4.18 Concluding Remarks

This chapter is dedicated to both entropy and exergy, which come directly from the SLT. It first introduced the entropy and exergy concepts and definitions, discussed the principles related to each of these concepts, explained why the SLT is important for practical systems and applications, compared the SLT with the FLT, and highlighted the limitations of the FLT with the fact that there is a strong need for a tool with leads to exergy. Exergy has been introduced and used for various systems. More importantly, previously written mass and energy balance equations have been extended to cover two more balance equations – entropy and exergy balance equations – making the total number of balance equations four. Writing all four balance equations correctly for thermodynamic systems, including closed and open types, has been a primary focus in this chapter. The efficiency concept has further been elaborated through energy and exergy efficiencies to assess the performances of thermodynamic systems and evaluate them for practical applications. Through the detailed example problems presented, which consider the most common types of open and closed systems, the importance of exergy is highlighted.

Study Questions/Problems

Multiple Choice Questions

1 For which of the following processes does $\dot{S}_{gen} < 0$?
 a It will never be negative, entropy generation is always positive or zero.
 b Reversible processes.
 c Irreversible processes.
 d Work producing devices.
 e Isolated systems.

2 Which of the following presents the exergy destruction?
 a $T_o S_{gen}$
 b $\left(1 - \frac{T_o}{T_b}\right) S_{gen}$
 c $T_b S_{gen}$
 d $\left(\frac{T_o}{T_b} - 1\right) S_{gen}$

3 Which of the following is the most accurate definition of thermodynamics?
  a The science of energy and exergy.
  b The science of energy and entropy.
  c The science of the FLT and SLT.
  d The science of enthalpy and internal energy.

4 Which of the following can be neglected for liquids and solids?
  a Change in density.
  b Change in entropy.
  c Change in enthalpy.
  d Change in exergy.

5 Who discovered the property entropy?
  a Clausius
  b Kelvin
  c Plank
  d Mollier
  e Carnot.

6 Arrange the following from the lowest entropy to the highest entropy (gas, solid, liquid).
  a Liquid, gas, solid
  b Solid, gas, liquid
  c Gas, liquid, solid
  d Solid, liquid, gas.

7 Write the mass, energy, entropy, and exergy balance equations for the following:

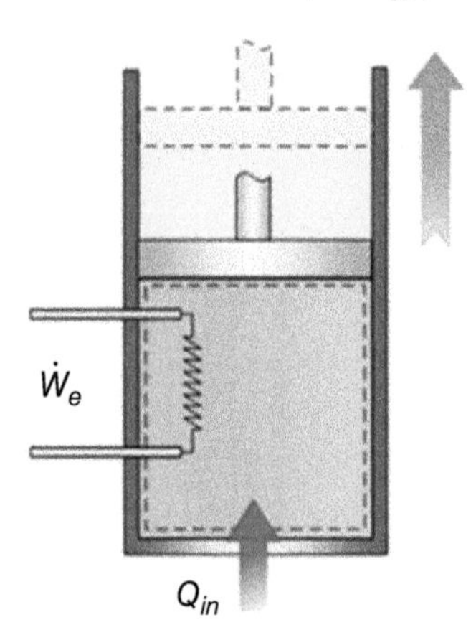

8 Which one of the following statements is true?
  a Both energy and exergy are always conserved.
  b Exergy is conserved while energy destruction takes place in real processes.
  c Energy is conserved while exergy destruction takes place in real processes.
  d None of the above.

9 Which one of the following defines the isentropic efficiency for a turbine?
   a Isentropic work/Actual work
   b Actual work/Isentropic work
   c Reversible work/Isentropic work
   d Isentropic work/Isothermal work

10 Which one of the following statements is true?
   a Entropy generation can be negative while exergy destruction is always positive
   b Exergy destruction can be negative while entropy generation is always positive
   c Both exergy destruction and entropy generation are always positive
   d None of the above

Problems

1 A piston–cylinder device, as shown in the figure, contains air that is initially at a pressure of 400 kPa. A paddle wheel is used to add work energy to the air inside the piston–cylinder device at a level of 50 kJ per each kg of air inside the piston–cylinder. Thermal energy is added to the system from a heat source with a temperature of 100 °C, transferring the system from state 1 to state 2, and the piston moves so that the temperature of the system remains constant at 17 °C. Through the process the volume contained in the system increases to three times its original volume. Calculate (a) the boundary work, (b) the thermal energy added to the system, (c) the entropy generation, and (d) the exergy destruction based on the entropy generation and by using the exergy balance equation and compare the results.

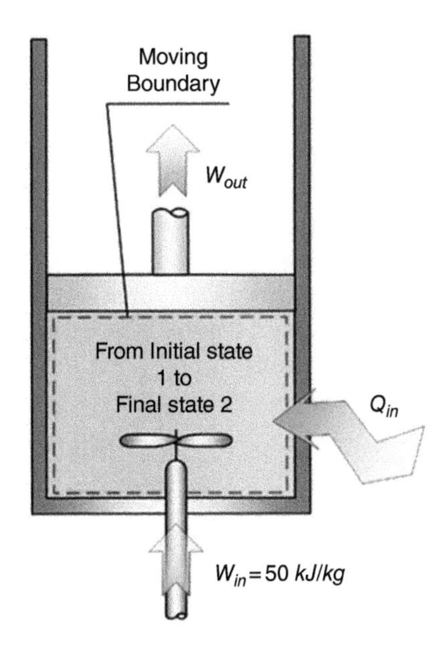

2 Consider a rigid tank filled with helium that is heated by a heat source at a temperature of 100 °*C*. The rigid tank has a volume of 1 m^3 and contains 1 *kg* of a real gas helium. Calculate (a) both the pressure and temperature of the helium in the rigid tank after 10 *kJ* of heat is transferred to the system while knowing that the system initially was at ambient temperature of 25 °*C*, (b) entropy generation, and (c) the exergy destruction based on the entropy generation and by using the exergy balance equation and compare the results.

3 A piston–cylinder device with an initial volume of 0.1 m^3, as shown in the figure, contains saturated water vapor at a temperature of 100 °*C*. The water is compressed so that the final volume is 80% of the initial volume. 40 *kJ* of heat is lost to the environment through the compression process. Write mass and energy balance equations and calculate the electrical power supplied, entropy generation, and the exergy destruction based on the entropy generation and by using the exergy balance equation and compare the results.

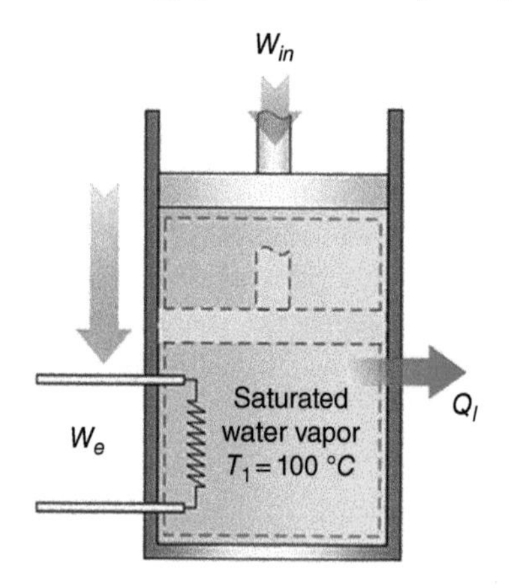

4 A 5 *l* rigid tank, as shown in the figure, contains air at 100 *kPa* and 20 °*C*. 66 *kJ* of heat is transferred to the air; 10% of the heat supplied is lost. 100 *kJ* of mechanical work through shaft 1 and 50 *kJ* of mechanical work through shaft 2 are used to run the fans inside the tank. The properties of air can be taken from the tables. Write the mass, energy, entropy, and exergy balance equations and find the final pressure and temperature of the tank, and the entropy generation and the exergy destruction based on the entropy generation and by using the exergy balance equation and compare the results.

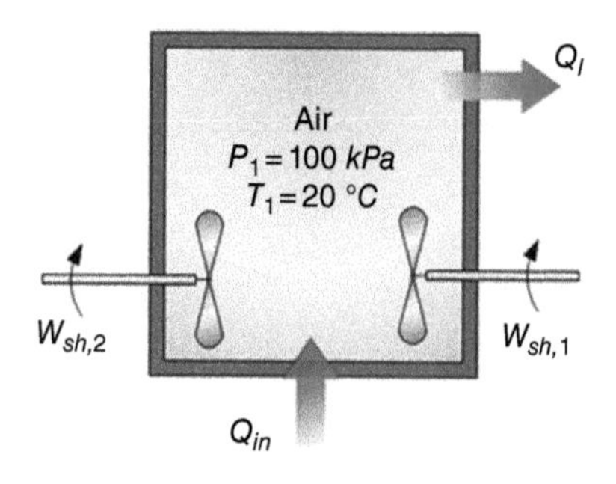

5 Water in a rigid tank, as shown in the figure, is initially at a pressure of 1 *atm*, a temperature of 25 °*C*, and has a volume of 10 *L*. 50 *kJ* of heat is transferred to the water while losing 50% of the supplied heat. An electrical heater is run on 1 *A* for 2000 *seconds* while connected to a 120 *V* source. The rotating shaft adds 1000 *kJ* of shaft work to the water in the tank while stirring it. Write the mass, energy, entropy, and exergy balance equations and calculate the final temperature of the tank.

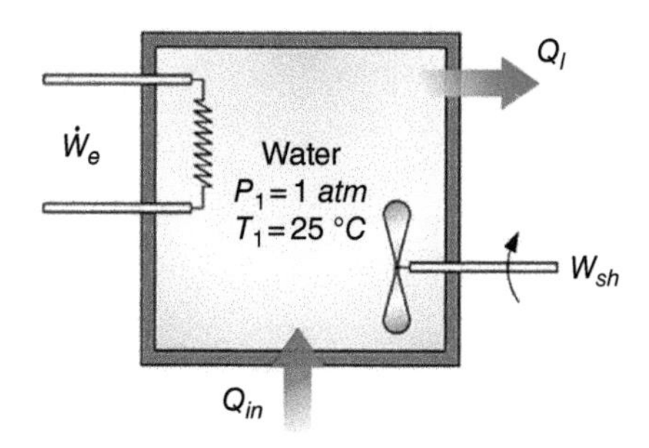

6 Air that behaves as an ideal gas is heated at a constant pressure of 200 *kPa* from a temperature of 200 to 400 *K*. Calculate the change in the internal energy of the air using (a) average specific heat capacity, (b) specific heat capacity that is a function of temperature, (c) using data from the tables, and (d) compare (a) and (b) with the answer in part (c) in terms of error difference.

7 Nitrogen that behaves as an ideal gas is heated at a constant pressure of 200 *kPa* from a temperature of 200 to 400 *K*. Calculate the change in the internal energy of the air using (a) average specific heat capacity, (b) specific heat capacity that is a function of temperature, (c) using data from the tables, and (d) compare (a) and (b) with the answer in part (c) in terms of error difference.

8 A 5 m^3 rigid tank, as shown in the figure, contains air at 100 *kPa* and 20 °*C*. 66 *kJ* of heat is transferred to the air; 10% of the heat supplied is lost. 100 *kJ* mechanical work through shaft 1 and 50 *kJ* mechanical work through shaft 2 are used to run the fans inside the tank. The properties of air can be taken from the tables. Write mass and energy balance equations and find the final pressure and temperature of the tank. Find both entropy generation and exergy destruction for this system.

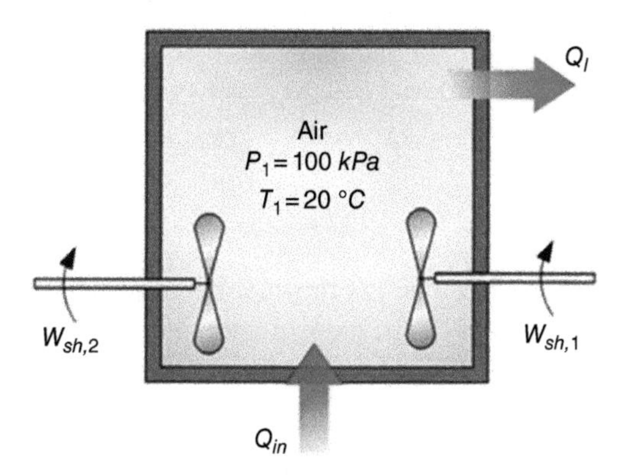

9 A piston–cylinder device, as shown in the figure, contains 2 *kg* of air that is initially at a pressure of 400 *kPa*. A paddle wheel is used to add work energy to the air inside the piston–cylinder device at a level of 50 *kJ* per each kg of air inside the piston–cylinder. Thermal energy is added to the system from a heat source with a temperature of 100 °*C*, transferring the system from state 1 to state 2. The piston moves so that the temperature of the system remains constant at 17 °*C*. Through the process the volume contained in the system increases to four times its original volume.

a Write the mass, energy, entropy, and exergy balance equations.
Calculate:
b The boundary work.
c The thermal energy added to the system.
d Entropy generation.
e Exergy destruction.

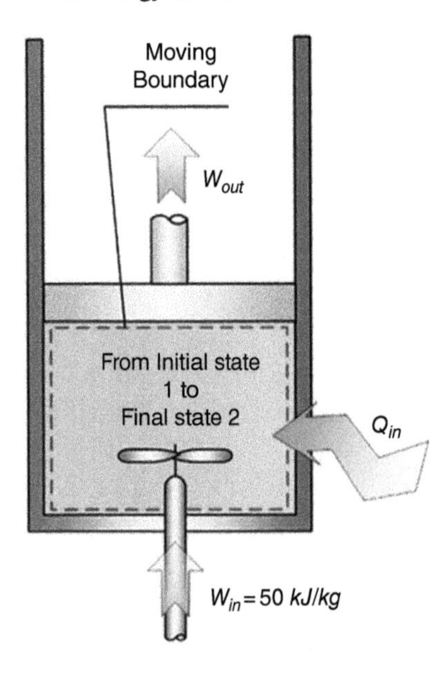

10 Consider a rigid tank filled with helium that is heated by a heat source at a temperature of 100 °*C*. The rigid tank has a volume of 1 m^3 and contains 1 *kg* of real gas helium.

a Write the mass, energy, entropy, and exergy balance equations.
b Calculate both the pressure and temperature of the helium in the rigid tank after 10 *kJ* of heat is transferred to the system while knowing that the system was initially at an ambient temperature of 25 °*C*.
c Calculate the entropy generation.
d Calculate the exergy destruction.

11 A piston–cylinder device with an initial volume of 0.1 m^3, as shown in the figure, contains saturated water vapor at a temperature of 100 °*C*. The water is compressed so that the final volume is 65% of the initial volume. 100 *kJ* of heat is lost to the environment through the compression process.

a Write mass, energy, entropy, and exergy balance equations.
b Calculate the electrical power supplied.
c Calculate the entropy generation.
d Calculate the exergy destruction.

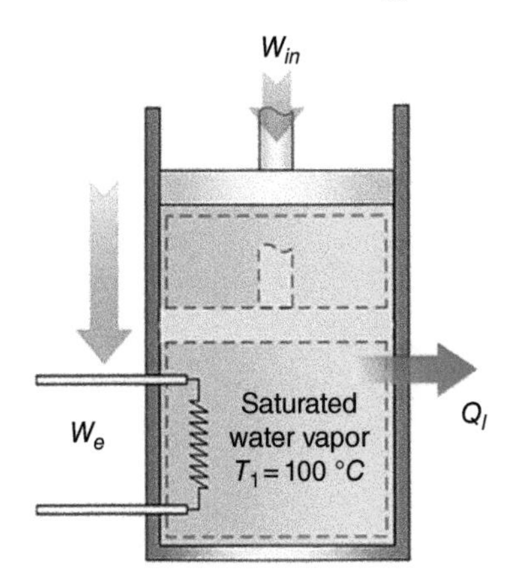

12 A water pump, as shown in the figure, raises the pressure of water from 100 *kPa* to 2.0 *MPa*. The initial temperature of water is 25 °*C*.
a Write the four balance equations.
b Calculate the work rate consumed by the pump.
c Find the temperature of the water leaving the pump.
d Calculate the rate of entropy change of water during the process.
e Calculate the entropy generation rate.
f Calculate the exergy destruction rate through the entropy generation rate and by using the exergy balance equation and compare the results.

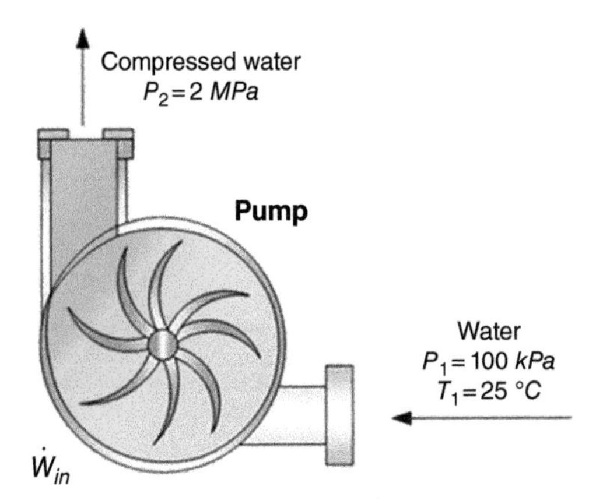

13 Consider two types of heat exchanging options as shown in the Figure below. Figure (a) shows heat transfer to the refrigerant *R*134*a* from a source temperature of 30 °*C* and Figure (b) shows the heat transfer from the refrigerant to a surrounding of 22 °*C*. During the heat gain process, the refrigerant enters the heat exchanger 1 at a temperature of −10 °*C* and exits at a temperature of −1 °*C*. The refrigerant pressure during the heat gain process is set at 10 *kPa*. During the heat loss process, the refrigerant enters heat exchanger 2 at a temperature of 30 °*C* and exits at a temperature of 25 °*C*. The refrigerant pressure during the heat loss process is set at 90 *kPa*. The mass flow rate of the refrigerant is set at 5 *kg/s*.

a Write all mass, energy, entropy and exergy balance balance equations for both heat exchangers,
b calculate the heat gained and lost by the refrigerant in heat exchangers 1 and 2 respectively,
c determine the entropy generation rate during both processes,
d find the exergy destruction rate during the both processes.

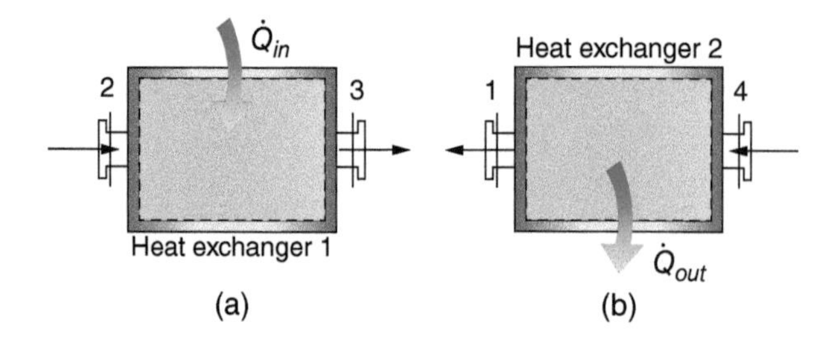

14 Water flowing at a rate of 1 *kg/s* enters a water pipe at a temperature of 100 °*C* and a pressure of 100 *kPa*, as shown in the figure. The pipe is insulated such that the pipe loses heat at a rate of 3 *kW* per meter of pipe length.

a Write the balance equations.
b Find the temperature at the outlet of the pipe if the pipe length is 10 *m*.
c Calculate the entropy generation per unit length of the pipe.
d Calculate the exergy destruction per unit length of the pipe.

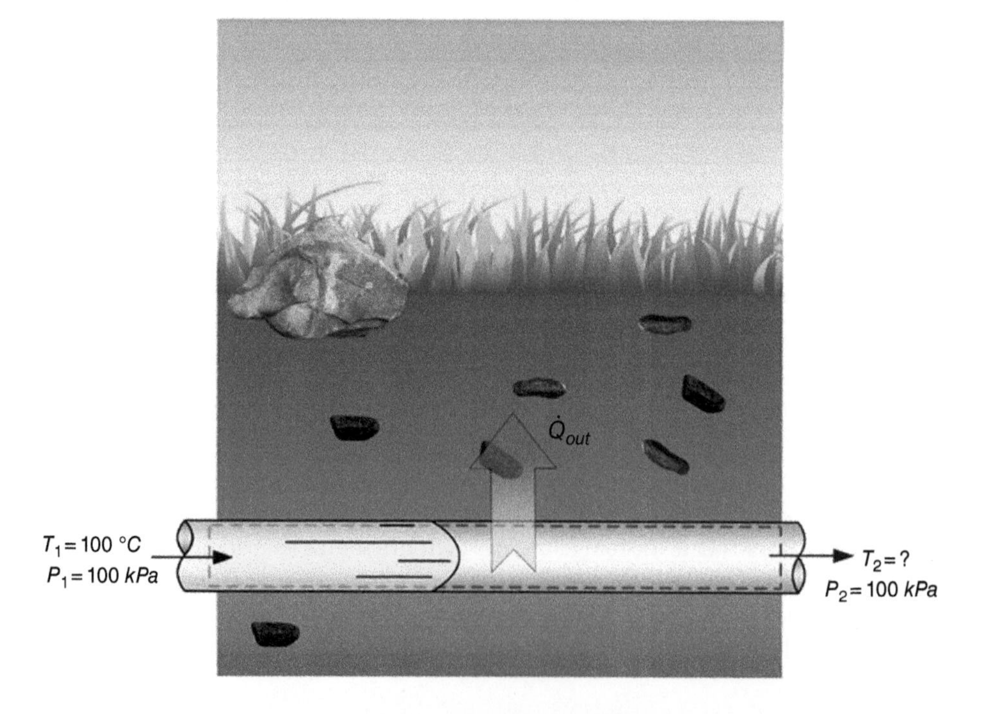

15 Air enters a nozzle steadily at a velocity of 20 *m/s*, as illustrated in Figure 3.2b. The exit area of the nozzle is 0.4 m^2. The mass flow rate through the nozzle is found to be 78.8 *kg/s*, where the thermophysical properties of air are to be used. Take the inlet and exit pressures as 600 *kPa* and 560 *kPa*, respectively.

a Write the balance equations.

b Find the velocity of the air at the nozzle exit using the energy balance on the nozzle.

c Calculate the entropy generation rate.

d Calculate the exergy destruction rate.

16 Consider a gas flow of oxygen (O_2) with a velocity of 270 *m/s* that enters an adiabatic diffuser at a pressure of 60 *kPa* and a temperature of 7 °*C* and exits the diffuser at a pressure of 70 *kPa* and a temperature of 27 °*C*. Calculate:

a The exit velocity of the fluid.

b The ratio of the inlet area to the exit area of the nozzle.

c The entropy generation.

d The exergy destruction.

17 The adiabatic steam turbine, as shown in the figure, receives 10 *kg/s* of superheated steam at a temperature of 550 °*C* and a pressure of 10 000 *kPa*. The steam turbine converts part of the energy in the steam to shaft work. The steam exits the steam turbine as a saturated vapor at a pressure of 10 *kPa* and a mixture quality of 1.

a Write the balance equations.

b Find the power output.

c Calculate the entropy generation rate.

d Calculate the exergy destruction rate.

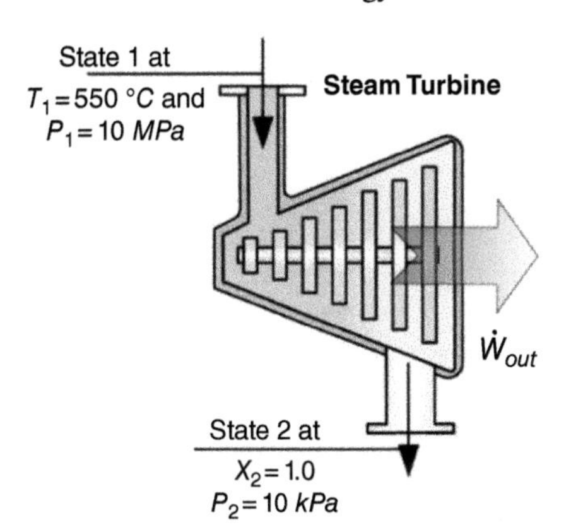

18 Consider an adiabatic steam turbine, as shown in the figure, with the following inlet and exit states: $P_1 = 12\,000$ *kPa*, $T_1 = 625$ °*C*, $P_2 = 10$ *kPa*, $x_2 = 0.95$. Taking the dead-state temperature of steam as saturated liquid at 25 °*C*:

a Write the balance equations.
b Determine work rate output by the steam turbine if the mass flow rate is known to be 2 *kg/s*.
c Calculate the entropy generation rate.
d Calculate the exergy destruction rate.

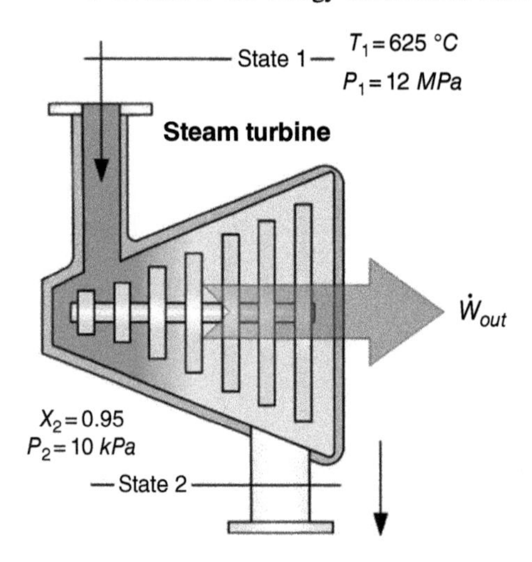

19 For the heat exchanger shown in the figure, write the mass, energy, entropy, and exergy balance equations, find the heat rate required by the heat exchanger and the entropy associated with that heat transfer if the fluid is water flowing at 0.5 *kg/s*.

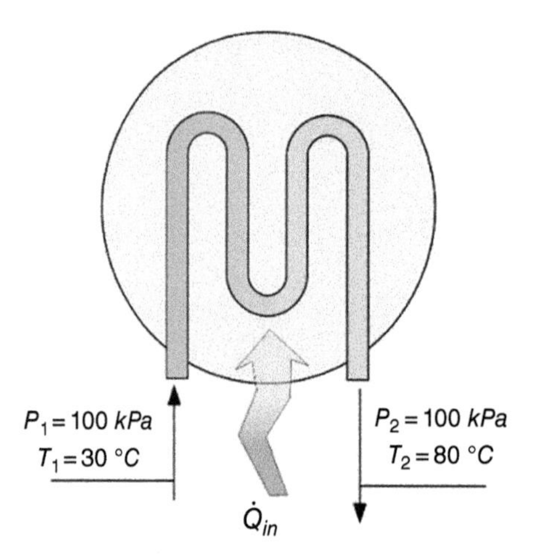

20 The water pump shown in the figure raises the pressure of water from 100 *kPa* to 2.0 *MPa*. The initial temperature of the water is 25 °*C*. Take the mass flow rate as 2.5 *kg/s*.
a Write the balance equations.
b Calculate the work rate consumed by the pump.

c Find the temperature of the water leaving the pump.
d Calculate the rate of entropy change of water during the process.
e Calculate the entropy generation rate.

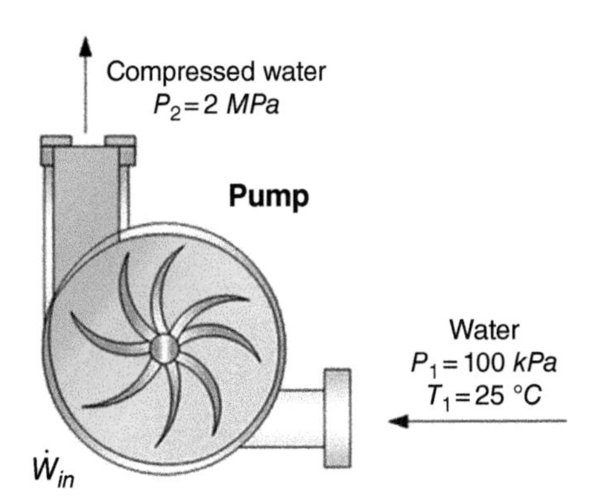

21 32 kg/s of water flow is heated at a constant pressure by $4.8 \times 10^6\ kg/h$ of exhausts gases, as shown in the figure. Assume the exhaust gasess consist only of air where you can use ideal gas equations. $\left(c_{p,air} = 1.005\, kJ/kgK\right)$.
a Write the balance equations.
b Calculate the exit temperature of the water.
c Calculate the exit temperature of the exhaust.
d Calculate the amount of heat rate transferred to the water.

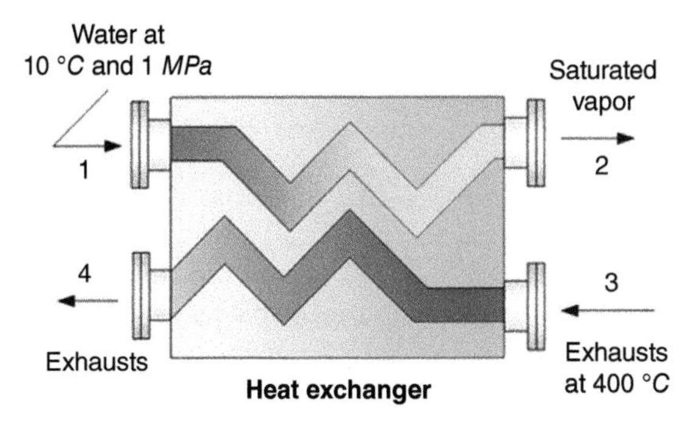

22 The pressure of water at ambient conditions is increased in a pump to 800 kPa and a temperature of 30 °C, as shown in the figure. The water entering the pump is at a temperature of 10 °C and atmospheric pressure; the mass flow rate is 1.2 kg/s.
a Write the balance equations.
b Determine the amount of power the pump consumes during the pressurization process.

c Find the entropy generation rate.
d Find the exergy content of the inlet flow.

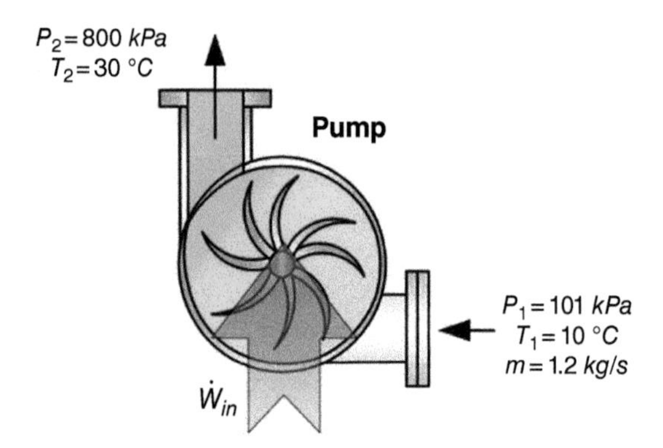

23 An air compressor compresses air from atmospheric pressure and a temperature of 280 *K* to a discharge pressure and a temperature of 600 *kPa* and 500 *K*, respectively. The mass flow rate of the air entering the compressor is 1 *kg/s*.
a Write the balance equations.
b Find the work rate consumed by the compressor.
c Find the entropy generation rate.
d Find the exergy destruction rate.

24 Consider that an adiabatic nozzle has an accelerating stream entering at a pressure of 600 *kPa*, a temperature of 500 *K*, and a velocity of 120 *m/s*. The adiabatic nozzle has a cross-sectional area ratio of 2 : 1, which will result in an exit velocity of 200 *m/s*. Assume a constant specific heat of 1.95 *kJ/kgK* and an inlet cross-sectional area of 2 m^2.
a Calculate the exit temperature.
b Calculate the exit pressure by assuming negligible change occurs in the density.
c Find the entropy generation rate.
d Find the exergy destruction rate.

25 Water at ambient pressure and a temperature of 99 °*C* is used to maintain the temperature of a building at 30 °*C* and exits the building heating system at a temperature of 35 °*C* as shown in the figure. The mass flow rate of water is 2.8 *kg/s*.
a Write the balance equations.
b Find the amount of thermal energy released from the hot water to the building per unit time.
c Find the entropy generation rate.
d Find the exergy destruction rate.

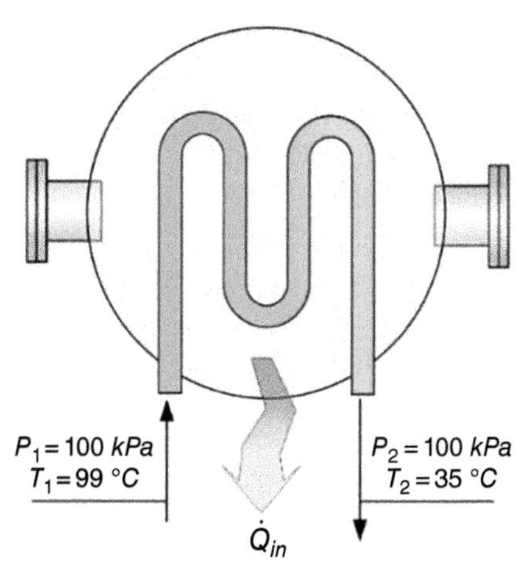

26 In an expansion valve propane enters the pipe at a pressure of 11 *bar* and a temperature of 10 °*C*. The expansion valve reduces the pressure of the propane to 3 *bar*. The mass flow rate is 1.1 *kg/s*.

a Write the balance equations.
b Determine the exit temperature of propane.
c Find the entropy generation rate.
d Find the exergy destruction rate.

27 Consider a turbine, as shown in figure (a), where there are incoming and outgoing flow exergies, work output and heat loss taking place as illustrated in the exergy flowchart (b). It is necessary to prove if the following two exergy efficiency approaches give the same results:

$$\eta_{ex} = \dot{E}x_{out}/\dot{E}x_{in} \text{ and } \eta_{ex} = 1 - \frac{\dot{E}x_{cons}}{\dot{E}x_{in}}$$

Do the necessary mathematical treatment to prove or disprove it and explain the consequences.

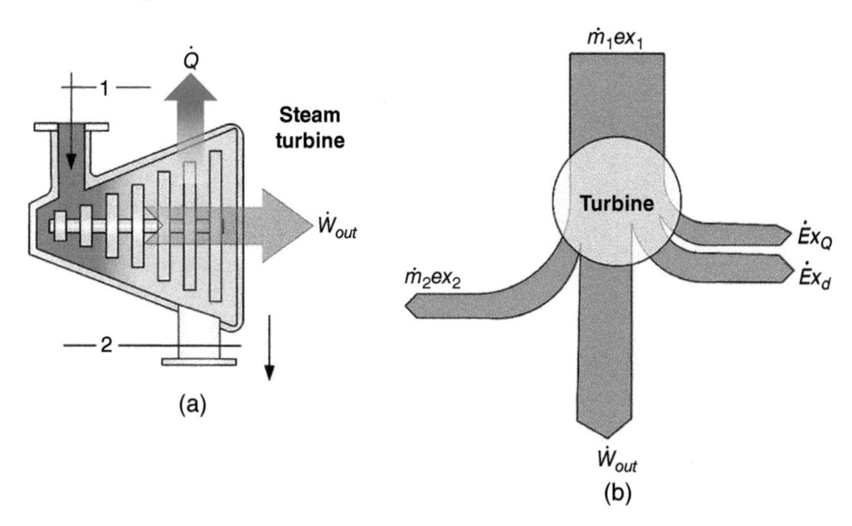

28 Nitrogen enters in a diffuser at a velocity of 299 m/s. The mass flow rate through the diffuser and nozzle are constant at 8 kg/s. The air inlet and outlet temperatures of diffuser are 300 and 315 K, respectively. Both diffuser and nozzle are shown in Figure 4.22. Take the heat input supplied from a source at 600 K.

- **a** Write the mass, energy, entropy and exergy balance equations.
- **b** Find the velocity of air at the diffuser exit using energy balance equation
- **c** Calculate the entropy generation and exergy destruction rates for diffuser
- **d** Calculate the exit temperature of air in the nozzle
- **e** Find entropy generation and exergy destruction rates in nozzle if there is a heat input supplied from a source at 600 K.

29 The air compressor shown in Figure 4.26 compresses refrigerant $R134a$ that enters at atmospheric pressure of 100 kPa and a temperature of 295 K. After the compression process, refrigerant $R134a$ exits the compressor at temperature and pressure of 405 K and 600 kPa. The mass flow rate of the refrigerant $R134a$ is 2 kg/s. Note that during the compression process, the compressor losses heat through its walls equivalent to 50% of the total work rate consumed by the compressor. Take the average temperature of 405 and 295 K for the boundary temperature, i.e. 350 K.

- **a** Write all mass, energy, entropy and exergy balance equations
- **b** Calculate the work rate consumed by the compressor
- **c** Calculate the entropy generation and exergy destruction rates in the compressor

30 Hot combustion gases enter the gas turbine as shown in the figure at state 1 and exit at state 2. The temperature at the inlet is measured to be 400 $^{\circ}C$ and pressure is set at 1 MPa. The exit pressure is set at 100 kPa and the exit temperature is measured to be 130.1 $^{\circ}C$. For a mass flow rate of 1.5 kg/s, considering the combustion gases to entail ideal gas properties for air:

- **a** Write all mass, energy, entropy and exergy balance equations
- **b** Calculate the power output from the turbine if the rate of heat loss is known to be 100 kW
- **c** Find the entropy generation rate
- **d** Calculate the exergy destruction rate.

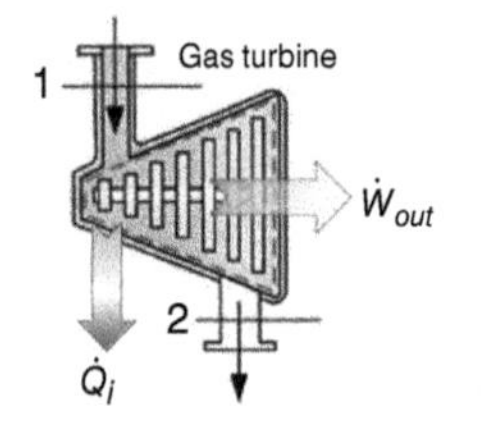

References

1 Baehr, H.D. and Schmidt, E.F. (1963). Definition und berechnung von brennstoffexergien (Definition and calculation of fuel exergy). *Brennst-Waerme-Kraft* 15: 375–381.

2 Gaggioli, R.A. and Petit, P.J. (1977). Use the second law first. *Chemtech* 7: 496–506.

3 Rodriguez, L.S.J. (1980). Calculation of available-energy quantities. In: *Thermodynamics: Second Law Analysis*, ACS Symposium Series, vol. 122, 39–60.

5

System Analysis

5.1 Introduction

Energy and the environment are recognized as two important requirements for sustaining life on earth. However, the dominant source of energy used today is fossil fuels, which negatively affect the environment. The environment is under danger due to the greenhouse gas emissions and wastes coming from human-made activities Research, development, and innovation activities have become primary in developing potential solutions to overcome the issues and develop new alternatives as well as cleaner solutions. Alternative sources of energy are being sought that can be substituted for fossil fuels, with the final goal of largely or completely replacing them. Such alternative energy sources include renewables and nuclear energy. Although these alternative energy sources can seem environmentally benign, since some of them have nearly zero carbon emissions during operation, they may have significant environmental impacts over their life cycle. This brings an important subject into consideration, that a life cycle assessment of every option is necessary. When it is done, some thermodynamic tools such as life cycle energy analysis and life cycle exergy analysis will be necessary. One may wonder why we have explained this and what the goal is. The point is that you need to plant the seeds first to get fruit later. We can represent this easily in Figure 5.1 to show how to first start planting in thermodynamics by writing the balance equations, go to energy system development, to have it ready for its fruits, which are the useful outputs ranging from power to heat/cooling, as well as other useful outputs, such as fuel and fresh water.

Further to the above mentioned points related to energy and the environment, assessment tools based on exergy are more advantageous than those based on energy for analyzing energy systems, since they provide more meaningful and realistic results. However, the environmental impact of energy systems is not evaluated through the exergy-based performance measures. Additional tools are therefore needed as the environmental impact and dimensions of the energy systems are important. Challenges, for example climate change/global warming, are attracting increasing attention recently. It is now an expectation that thermodynamics will help overcome some such challenges, particularly at the system level, by doing the right analysis, design, assessment, and improvement through the study of thermodynamics, more importantly by the tool of exergy which comes from the second law of thermodynamics (SLT).

Thermodynamics: A Smart Approach, First Edition. Ibrahim Dincer.

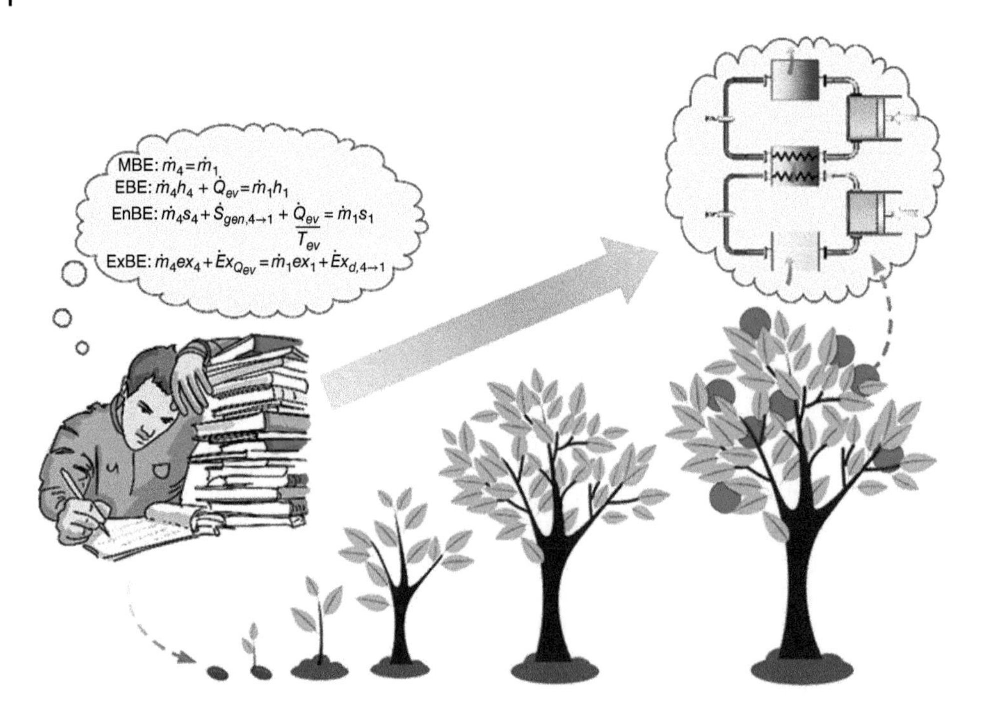

Figure 5.1 A concept of illustrating how to go from analysis to energy system development for implementation in mimicking planting to harvesting.

It is important to reemphasize that the energy spectrum covers 3S, ranging from source to service, as illustrated Figure 5.2. For everything there is a need for an energy source for energy input and services (referring to useful outputs) are necessary to produce as demanded. In between source and service there is a need to build the system to be able to produce what is needed based on the source. Between source and system, there is a need for storage, for example with solar energy as the energy source the sun does not shine all the time, so storage becomes a necessity. If after the system one produces more useful output than is needed, storage is required. This way the equation becomes 3S + 2S, which will become larger S, resulting in sustainability.

This chapter aims to make a primary focus on system analysis as a critical step in thermodynamics through the balance equations for mass, energy, entropy, and exergy for both types of closed and open systems in a more holistic manner.

5.2 Thermodynamic Laws

Thermodynamics is defined as the science of energy. The word thermodynamics is thought to be derived from the Greek words therme, which means heat, and dynamis, that means power. The origin of the word thermodynamics is descriptive of many energy conversion processes, where heat (thermal energy) is converted to work or electrical energy. Numerous thermodynamic books define thermodynamics as the science of energy and entropy. However, it is now introduced as the science of energy (which comes from the first law of

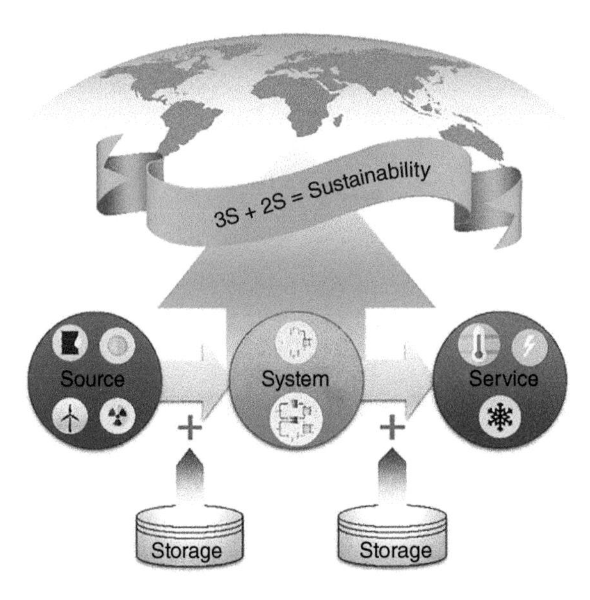

Figure 5.2 An energy spectrum illustrated for sustainable development.

thermodynamics (FLT)) and exergy (which comes from the SLT). Technically, both the FLT and SLT become really the governing laws of thermodynamics while other laws (zeroth and third) are the policy type laws, which are specific to the conditions. As mentioned earlier, the principles of thermodynamics are best presented through its four main laws, i.e. the zeroth, first, second, and third laws of thermodynamics. Although each of these laws has its own significance, the prime focus will be on the FLT and SLT.

A key weakness of the first law of thermodynamics is that it does not account meaningfully for irreversibilities, losses, and inefficiencies. Since all real systems are irreversible, there is a strong need to go beyond this law. This can be done by considering the second law of thermodynamics. The second law considers both quantity and quality of energy, and indicates that actual processes occur on their own in the direction of decreasing quality of energy. For example, water at ambient temperature and pressure is of very low quality and can be used to provide no work by interacting with the environment since it is already at ambient conditions, which are considered dead state (or reference state).

The first and second laws of thermodynamics can be combined to derive the thermodynamic quantity exergy. Many dwell on entropy, rather than exergy, in explaining the second law of thermodynamics. But dealing with exergy can facilitate comparisons between systems, and with the results of energy assessments. It is difficult to compare entropy with energy, but a comparison between energy and exergy is more straightforward since both properties have the same units. Doing so in essence compares quantities and qualities. Energy efficiencies can be seen to be less meaningful and useful than exergy efficiencies. In general, exergy and energy concepts should be considered for comparison and assessment purposes. This does not eliminate the role of entropy, but rather emphasizes a limitation in the use of entropy. Finally, before starting with the detailed analysis of different energy devices and systems, the importance of the energy and exergy analyses is highlighted in Figure 5.3, as they are linked to the environment and sustainable development. This way system analysis represents a critical step in achieving a better environment and sustainability.

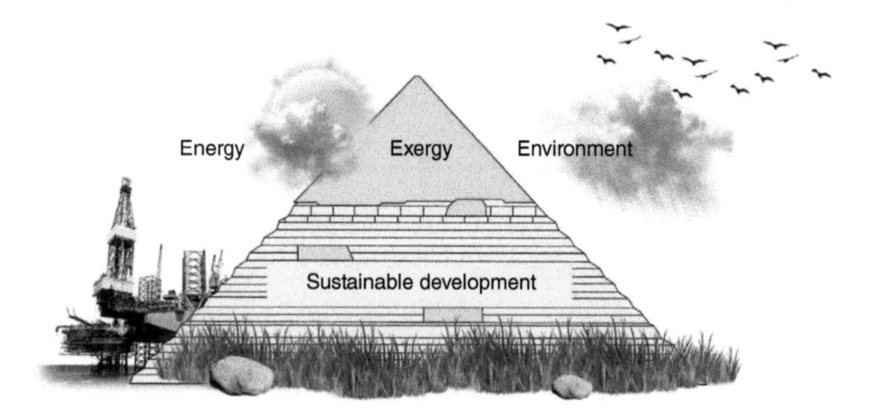

Figure 5.3 Exergy as the confluence of energy, environment, and sustainable development.

5.3 Closed Systems

In this section, the system analysis is carried out for closed systems, which include both closed systems with fixed boundary and moving boundary options to reflect the application spectrum. Table 5.1 presents the four balance equations – mass, energy, entropy, and exergy – for each of the main categories of the closed systems. The table starts with the fixed boundary closed systems, to include two examples of a rigid tank with a single material inside and another with multiple materials. It then goes to moving boundary closed systems, where two forms are presented, one that produces work and one that consumes work. Later in the chapter numerical examples are provided for each of the systems mentioned in Table 5.1.

As was shown in the earlier chapters, the closed systems can be either a fixed boundary system or moving boundary system. In this section the energy balance will be applied to various closed systems with fixed boundary.

Table 5.1 Mass, energy, entropy, and exergy balance equations for various types of closed systems.

System	Balance equations
Fixed boundary	
Q_l W_e Water W_{sh} Q_{in}	MBE: $m_1 = m_2 = constant$ EBE: $Q_{in} + W_e + W_{sh} + m_1u_1 = m_2u_2 + Q_l$ EnBE: $Q_{in}/T_s + m_1s_1 + S_{gen} = m_2s_2 + Q_l/T_b$ ExBE: $Ex_{Q_{in}} + W_e + W_{sh} + m_1ex_1 = m_2ex_2 + Ex_{Q_l} + Ex_d$ with $Ex_{Q_{in}} = \left(1 - \frac{T_o}{T_s}\right)Q_{in}$; $Ex_{Q_l} = \left(1 - \frac{T_o}{T_b}\right)Q_l$; $Ex_d = T_oS_{gen}$ where W_e is electrical work, W_{sh} is shaft work, T_s is source temperature since heat provided from a source, T_b is immediate boundary temperature since heat is leaving the system and rejected into the immediate boundary. One may also use surface temperature or average temperature of the substance ($T_{av} = (T_1+T_2)/2$) or ambient (or reference state) temperature (T_o) if T_b is not known.

Table 5.1 (Continued)

System	Balance equations
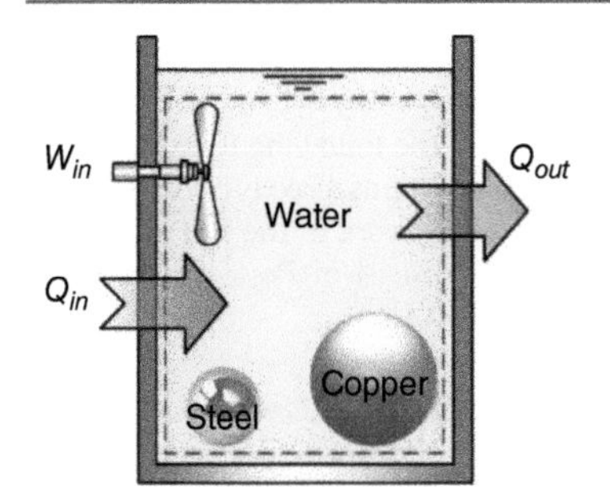	MBE: $m_{w,\,1} = m_{w,\,2} = m_w$; $m_{s,\,1} = m_{s,\,2} = m_s$; and $m_{c,\,1} = m_{c,\,2} = m_c$, EBE: $m_w u_{w,\,1} + m_s u_{s,\,1} + m_c u_{c,\,1} + W_{in} + Q_{in} = m_w u_{w,\,2} + m_s u_{s,\,2} + m_c u_{c,\,2} + Q_{out}$ EnBE: $m_w s_{w,\,1} + m_s s_{s,\,1} + m_c s_{c,\,1} + Q_{in}/T_s + S_{gen} = m_w s_{w,\,2} + m_s s_{s,\,2} + m_c s_{c,\,2} + Q_{out}/T_b$ ExBE: $m_w ex_{w,1} + m_s ex_{s,1} + m_c ex_{c,1} + W_{in} + Ex_{Q_{in}} = m_w ex_{w,2} + m_s ex_{s,2} + m_c ex_{c,2} + Ex_{Q_{out}} + Ex_d$ with $Ex_{Q_{in}} = \left(1 - \frac{T_o}{T_s}\right) Q_{in}$; $Ex_{Q_{out}} = \left(1 - \frac{T_o}{T_b}\right) Q_{out}$; $Ex_d = T_o S_{gen}$. where s denotes steel, c is copper, w is water. Also, W_{in} is shaft work, T_s is source temperature since heat provided from a source, T_b is immediate boundary temperature since Q_{out} is leaving the system and rejected into the immediate boundary. One may also use surface temperature or average temperature of the substance ($T_{av} = (T_1+T_2)/2$) or ambient (or reference state) temperature (T_o) if T_b is not known. Note that in this rigid tank there are two metallic (steel and copper) balls where their masses are included.
Moving boundary 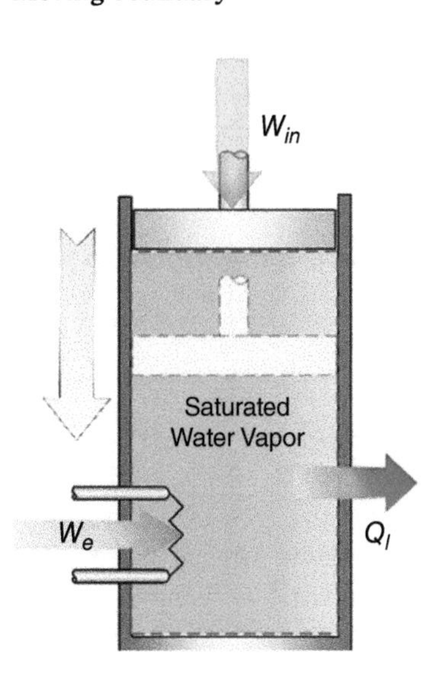	MBE: $m_1 = m_2 = constant$ EBE: $m_1 u_1 + W_e + W_{in} = m_2 u_2 + Q_l$ or one may write this EBE in the following form by considering the boundary conditions directly (where there is no need to include boundary work in the equation): $m_1 u_1 + m_1 P v_1 + W_e = m_2 u_2 + m_2 P v_2 + Q_l$ Both of these equations result in: $m h_1 + W_e = m h_2 + Q_l$ with $W_{in} = W_b = mP(v_1 - v_2)$ and $h = u + Pv$ EnBE: $m_1 s_1 + S_{gen} = m_2 s_2 + (Q_l/T_b)$ ExBE: $W_e + W_{in} + m_1 ex_1 = m_2 ex_2 + Ex_{Q_l} + Ex_d$ with $Ex_{Q_l} = \left(1 - \frac{T_o}{T_b}\right) Q_l$; $Ex_d = T_o S_{gen}$. where W_e is electrical work, W_{in} is work input (boundary work which changes the boundary) to the system to compress the fluid or gas, T_b is immediate boundary temperature since heat is leaving the system and rejected into the immediate boundary. One may also use surface temperature or average temperature of the substance ($T_{av} = (T_1+T_2)/2$) or ambient (or reference state) temperature (T_o) if T_b is not know.

(Continued)

Table 5.1 (Continued)

System	Balance equations
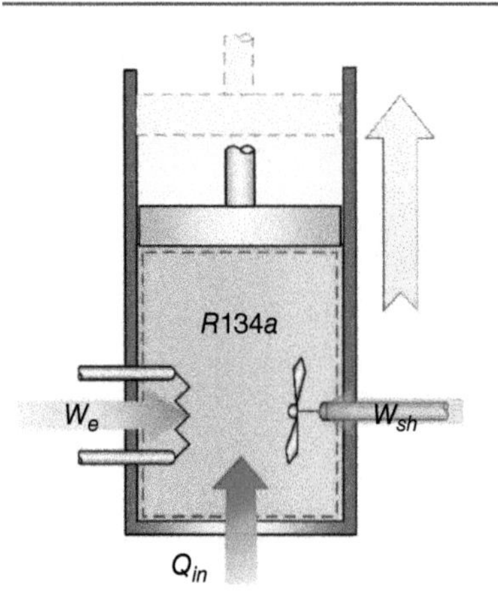	MBE: $m_1 = m_2 = \text{constant}$ EBE: $m_1u_1 + Q_{in} + W_e = m_2u_2 + W_b$ or one may write this EBE in the following form by considering the boundary conditions directly (where there is no need to include boundary work in the equation): $m_1u_1 + m_1Pv_1 + Q_{in} + W_e = m_2u_2 + m_2Pv_2$ Both of these equations result in: $mh_1 + Q_{in} + W_e = mh_2$ with $W_{out} = W_b = mP(v_2 - v_1)$ and $h = u + Pv$ EnBE: $m_1s_1 + Q_{in}/T_s + S_{gen} = m_2s_2$ ExBE: $m_1ex_1 + Ex_{Q_{in}} + W_e = m_2ex_2 + W_b + Ex_d$ with $Ex_{Q_{in}} = \left(1 - \frac{T_o}{T_s}\right)Q_{in}$; $Ex_d = T_oS_{gen}$. where W_e is electrical work, W_b is boundary movement work (which is work done by the system), T_s is source temperature. Note that no heat transfer leaving the system is shown due to the negligible heat losses.
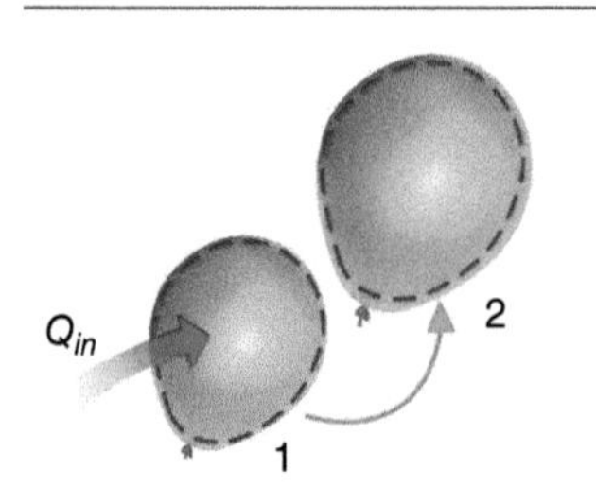	MBE: $m_1 = m_2 = \text{constant}$ EBE: $m_1u_1 + Q_{in} = m_2u_2 + W_b$ or one may write this EBE in the following form by considering the boundary conditions directly (where there is no need to include boundary work in the equation): $m_1u_1 + m_1Pv_1 + Q_{in} = m_2u_2 + m_2Pv_2$ Both of these equations result in: $mh_1 + Q_{in} = mh_2$ with $W_{out} = W_b = mP(v_2 - v_1)$ and $h = u + Pv$ EnBE: $Q_{in}/T_s + m_1s_1 + S_{gen} = m_2s_2$ ExBE: $m_1ex_1 + Ex_{Q_{in}} = m_2ex_2 + W_b + Ex_d$ with $Ex_{Q_{in}} = \left(1 - \frac{T_o}{T_s}\right)Q_{in}$; $Ex_d = T_oS_{gen}$. where W_b is boundary movement work (which is work done by the system), T_s is source temperature. Note that no heat transfer leaving the system is shown due to the negligible heat losses.

Example 5.1 Figure 5.4 shows an electrically heated and mechanically stirred rigid tank with a volume of 10 L that is initially at a pressure of 1 atm (100 kPa) and a temperature of 25 °C. A total 50 kJ of heat is transferred to the water; 10% of the supplied heat is lost. The electrical heater runns on 1 A for 10 seconds, while being connected to 120 V source. The rotating shaft adds a total of 100 kJ of shaft work to the water in the tank while stirring it. Take the source temperature as 600 °C. (a) Write the mass, energy, entropy, and exergy balance equations, (b) calculate the final temperature of the tank, (c) calculate the entropy generation, and (d) calculate the exergy destruction through the exergy balance equation.

Solution

a) Write the mass, energy, entropy, and exergy balance equations.

The changes in the kinetic energy of the system are negligible since the system velocity does not change (not accelerating nor decelerating) throughout the process. The change in the potential energy of the system can be neglected as well since the system elevation does not change throughout the process.

Based on the above mentioned assumptions the mass, energy, entropy, and exergy balance equations are written as follows:

$$\text{MBE}: m_1 = m_2 = constant$$

$$\text{EBE}: Q_{in} + \dot{W}_e \times t + W_{sh} + m_1 u_1 = m_2 u_2 + Q_l$$

(since electrical power is given, *one* needs to multiply it by time in seconds)

$$\text{EnBE}: Q_{in}/T_s + m_1 s_1 + S_{gen} = m_2 s_2 + Q_l/T_b$$

$$\text{ExBE}: Ex_{Q_{in}} + \dot{W}_e \times t + W_{sh} + m_1 ex_1 = m_2 ex_2 + Ex_{Q_l} + Ex_d$$

(since electrical power is given, one needs to multiply it by time in seconds)

b) Calculate the final temperature of the tank.

From the tables (such as Appendix B-1b) or by using a software package with built-in thermodynamic tables, such as Engineering Equation Solver EES, the saturation temperature of water at a pressure of 1 $atm = 101.325\ kPa$ is 99.61 °C. Since the temperature of water is lower than the saturation temperature, then it is in the compressed liquid state. Then one needs to go to the compressed liquid tables or another properties source to extract the properties for the subcooled water. In most published subcooled properties tables, the data sets start from a high pressure of 5 MPa, where one may use either data tables (including NIST REFPROP (National Institute of Standards and Technology REFerence fluid PROPerties)) or EES. One may even thermodynamically simplify its properties to saturated water properties and take the data for it at the given temperature, such as the saturated liquid properties at the given temperature of 25 °C:

$$u_1 = u_f @25\ ^oC = 104.7\ kJ/kg$$

$$v_1 = v_f @25\ ^oC = 0.001003\ m^3/kg$$

$$V = 10\ L \times 1 \frac{m^3}{1000\ L} = 0.01\ m^3$$

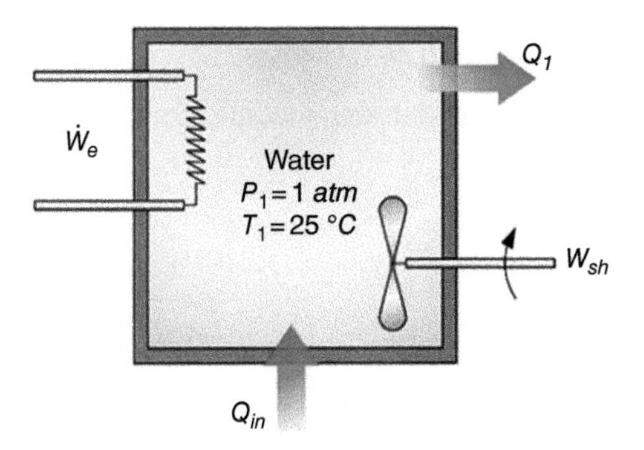

Figure 5.4 Schematic of the rigid tank discussed in Example 5.1.

$$v_1 = \frac{V_1}{m_1} \rightarrow 0.001003 = \frac{0.01}{m_1}$$

$m_1 = 9.97\ kg = m_2 = m$ since the system is a closed system where no mass enters or leaves the system.

Substituting the above obtained values in the energy balance equation is necessary for further calculations. Since the process occurred in a rigid tank then it is a constant volume process, which means we can use: $u = c_v T$

$$Q_{in} + \dot{W}_e \times \Delta t + W_{sh} + m_1 u_1 = m_2 u_2 + Q_l$$

$$Q_{in} + \dot{W}_e \times \Delta t + W_{sh} - Q_l = \Delta U = m(u_2 - u_1)$$

$$Q_{in} + \dot{W}_e \times \Delta t + W_{sh} - Q_l = mc_v(T_2 - T_1)$$

From property tables (such as Appendix A-2) or from a properties database software we can find the specific heat at a constant volume, which is taken at the initial temperature of 25 °C:

with $c_v = 4.18\,\frac{kJ}{kgK}$

$$50\ kJ + \left(1\ A \times 120\ V \times \frac{1\frac{kJ}{s}}{1000\ A.V} \times 10\ s\right) + 100\ kJ - 0.1 \times 50\ kJ$$

$$= 9.97\ kg \times 4.18\,\frac{kJ}{kgK}(T_2 - 25)$$

$\boldsymbol{T_2 = 28.51\,^{o}C}$

c) Calculate the entropy generation.

$$m_1 s_1 + \frac{Q_{in}}{T_s} + S_{gen} = m_2 s_2 + Q_l / T_b$$

$s_1 = 0.367\,\frac{kJ}{kgK}$, $s_2 = 0.416\,\frac{kJ}{kgK}$, $c_v = 4.18\,\frac{kJ}{kgK}$ from the property tables (such as Appendix B-1a) or using property database software, such as the EES, one obtains the entropy generation from the entropy balance equation as follows:

$$9.97\ kg\left(0.367\,\frac{kJ}{kgK}\right) + \frac{50\ kJ}{873} + S_{gen} = 9.97\ kg\left(0.416\,\frac{kJ}{kgK}\right) + \frac{5\ kJ}{301.5}$$

$\boldsymbol{S_{gen} = 0.45\,\frac{kJ}{K}}$

d) Calculate the exergy destruction.

$$Ex_{Q_{in}} + \dot{W}_e \times \Delta t + W_{sh} + m_1 ex_1 = m_2 ex_2 + Ex_{Q_l} + Ex_d$$

$$\left(1 - \frac{T_o}{T_s}\right) Q_{in} + \dot{W}_e \times \Delta t + W_{sh} + m_1 ex_1 = m_2 ex_2 + \left(1 - \frac{T_o}{T_b}\right) Q_l + Ex_d$$

Here, T_b is the immediate boundary temperature which is not given. It is just fine to take the average temperature.

The specific exergies for the first and the second states as well as finding or determining the source temperatures for the heat transfer rates are done as follows:

$$ex_1 = (u_1 - u_o) - T_o(s_1 - s_o)$$

$$ex_2 = (u_2 - u_o) - T_o(s_2 - s_o)$$

$$\left(1 - \frac{T_o}{T_s}\right) Q_{in} + \dot{W}_e \times t + W_{sh} + m_1((u_1 - u_o) - T_o(s_1 - s_o))$$

$$= m_2((u_2 - u_o) - T_o(s_2 - s_o)) + \left(1 - \frac{T_o}{T_b}\right) Q_l + Ex_d$$

and based on the mass balance equation, it is obtained as:

$$\left(1 - \frac{T_o}{T_s}\right) Q_{in} + \dot{W}_e \times t + W_{sh} = m((u_2 - u_1) - T_o(s_2 - s_1)) + \left(1 - \frac{T_o}{T_b}\right) Q_l + Ex_d$$

Considering T_o as 298 K and finding the values of $u_1 = 104.70\ kJ/kg$, $u_2 = 119.40\ kJ/kg$ from the property tables (such as Appendix B-1a) or using the property database software, such as the EES, the exergy destruction for this system is found as follows:

$$50\ kJ \times \left(1 - \frac{298}{873}\right) + \left(1\ A \times 120\ V \times \frac{1\ \frac{kJ}{s}}{1000\ A.V \times 10\ s}\right) + 100\ kJ = 9.97\ kg$$

$$\times ((119.40 - 104.70) - 298\,(0.416 - 0.367)) + 0.1 \times 50\ kJ \times \left(1 - \frac{298}{301.5}\right) + Ex_d$$

$$\mathbf{Ex_d = 133.09\ kJ}$$

5.3.1 Nonflow Exergy with Specific Heat Capacity

Exergy for ideal gases depends on the specific heat capacities, in this section we show how the ideal gas relations for the specific heat capacities are used to calculate the specific exergy of the gases at different conditions. As was introduced in the previous chapter, the specific exergy is defined as:

Nonflowing mass specific exergy is written as:

$$ex = (u - u_o) - T_o(s - s_o)$$

Assuming a constant specific heat capacity, we can replace the changes in the internal energy and the changes in the specific entropy with the ideal gas expression as follows:

$$ex = (c_{v,av}(T - T_O)) - T_o\left(c_{p,av} \ln\left(\frac{T}{T_o}\right) - R \ln\left(\frac{P}{P_o}\right)\right)$$

or

$$ex = (c_{v,av}(T - T_O)) - T_o\left(c_{v,av} \ln\left(\frac{T}{T_o}\right) + R \ln\left(\frac{v}{v_o}\right)\right)$$

Example 5.2 Air behaving in an ideal gas manner filling a gas container is heated at a constant pressure of 200 kPa in a piston–cylinder device from a temperature of 300 to 600 K. Calculate the change in the air specific exergy.

Solution

The change in the specific exergy of heated air in the gas container can be found as follows:

$$\Delta ex = ex_2 - ex_1$$

where the specific exergy for a nonflowing fluid is calculated as:

$$ex = (u - u_o) - T_o(s - s_o)$$

By substituting the above term in the difference in exergy for the air in the tank:

$$\Delta ex = ex_2 - ex_1 = ((u_2 - u_o) - T_o(s_2 - s_o)) - ((u_1 - u_o) - T_o(s_1 - s_o))$$

$$\Delta ex = (u_2 - u_1) - T_o(s_2 - s_1)$$

Finding the change of the internal energy by using the specific heat capacities can be done as follows:

$$du = c_v dT$$

$$\int_{u_1}^{u_2} du = \int_{T_1}^{T_2} c_v dT$$

If we assume that the specific heat capacity does not change with the variation of temperature, or that the variation of the specific heat capacity with the temperatures T_1 and T_2 is negligible, then the above integration will be solved as follows:

$$\Delta u = \int_{T_1}^{T_2} c_v dT = c_v(T_2 - T_1)$$

where c_v can be taken at the average temperature between T_1 and T_2 or it can also be taken in problems where one of the temperatures is missing and needs to be calculated at the given temperature and further iteration can also be considered to obtain a more accurate result. In this example where both the start and the end temperatures are given, then the specific heat is taken from tables (such as Appendix A-3) or from EES on the average temperature:

$$T_{av} = \frac{T_2 + T_1}{2} = \frac{300 + 600}{2} = 450\ K$$

$$\Delta u = c_{v,av}(T_2 - T_1) = 0.733\ \frac{kJ}{kgK} \times (600 - 300)K = 220\ \frac{kJ}{kg}$$

We then substitute the change in the internal energy in the change in the exergy expression:

$$\Delta ex = (c_{v,av}(T_2 - T_1)) - T_o(s_2 - s_1)$$

Then the change in the entropy of the system from one state to another is found as follows:

$$s_2 - s_1 = c_v\, ln\left(\frac{T_2}{T_1}\right) + Rln\left(\frac{v_2}{v_1}\right)$$

or

$$s_2 - s_1 = c_p \, ln\left(\frac{T_2}{T_1}\right) - Rln\left(\frac{P_2}{P_1}\right)$$

Then by substituting the entropy definition in the change in exergy since it is an ideal gas:

$$\Delta ex = (c_{v,av}(T_2 - T_1)) - T_o\left(c_p \, ln\left(\frac{T_2}{T_1}\right) - Rln\left(\frac{P_2}{P_1}\right)\right)$$

$$\Delta ex = \left(0.733 \, \frac{kJ}{kgK} \times (600 - 300)K\right) - 298 \, K\left(1.02 \, \frac{kJ}{kgK} \, ln\left(\frac{600}{300}\right) - 0.287 \, \frac{kJ}{kgK} \, ln\left(\frac{200}{200}\right)\right)$$

$$\boldsymbol{\Delta ex = (ex_2 - ex_1) = 9.211 \, \frac{kJ}{kg}}$$

which clearly shows that exergy at state point 2 increases. This is of course consistent with the SLT.

Example 5.3 A 100 kg thermal oil bath has three metallic (steel, iron, and copper) balls quenched into it as shown in Figure 5.5. Each of these ball is about 10 kg. The oil and three balls are initially at 25 °C. The bath is heated to a temperature of 575 °C at a constant pressure of 101.325 kPa. (a) Write the mass, energy, entropy, and exergy balance equations and (b) calculate the amount of heat needed for this process. Take the source temperature as 600 °C.

Solution

a) Writing balance equations is a critical step. Here, the mass, energy, entropy, and exergy balance equations for the oil bath are different than usual closed systems since this system has more than one material and object inside the selected system boundary, which is the wall of the oil bath container.

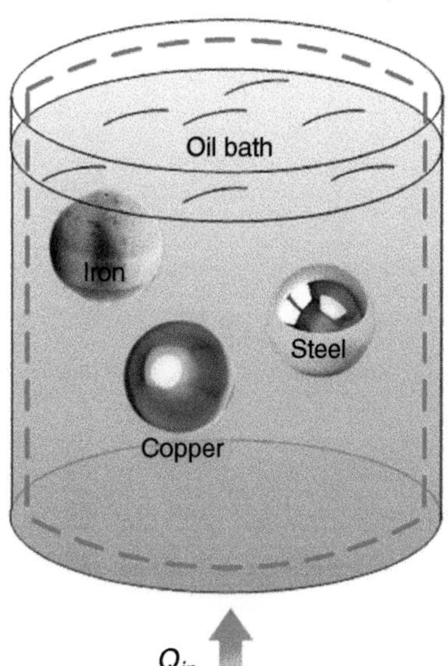

Figure 5.5 Schematic of the rigid tank discussed in Example 5.3.

$$\text{MBE}: m_{i,1} = m_{i,2} = m_i, m_{s,1} = m_{s,2} = m_s, m_{c,1} = m_{c,2} = m_c, \text{ and } m_{o,1} = m_{o,2} = m_o$$

$$\text{EBE}: m_i u_{i,1} + m_s u_{s,1} + m_c u_{c,1} + m_o u_{o,1} + Q_{in} = m_i u_{i,2} + m_s u_{s,2} + m_c u_{c,2} + m_o u_{o,2}$$

$$\text{EnBE}: m_i s_{i,1} + m_s s_{s,1} + m_c s_{c,1} + m_o s_{o,1} + Q_{in}/T_s + S_{gen} = m_i s_{i,2} + m_s s_{s,2} + m_c s_{c,2} + m_o s_{o,2}$$

$$\text{ExBE}: m_i ex_{i,1} + m_s ex_{s,1} + m_c ex_{c,1} + m_o ex_{o,1} + (1 - T_o/T_s)Q_{in} = m_i ex_{i,2} + m_s ex_{s,2} + m_c ex_{c,2} + m_o ex_{o,2} + Ex_d$$

Here, i stands for iron, s for steel, c for copper, and o for oil.

b) It is necessary to calculate the amount heat supplied, entropy generated and exergy destroyed during this process.

For amount of heat:

Given data:

$$T_0 = 25\,°C,\ T_s = 600\,°C,\ m_i = 10\ kg,\ m_s = 10\ kg,\ m_c = 10\ kg,\ m_o = 100\ kg$$

$$Cp_i = 460.5\,\frac{J}{kgK},\ Cp_s = 520.6\,\frac{J}{kgK},\ Cp_c = 376.8\,\frac{J}{kgK},\ Cp_o = 2306.1\,\frac{J}{kgK}$$

$$m_i u_{i,1} + m_s u_{s,1} + m_c u_{c,1} + m_o u_{o,1} + Q_{in} = m_i u_{i,2} + m_s u_{s,2} + m_c u_{c,2} + m_o u_{o,2}$$

$$Q_{in} = (m_i u_{i,2} - m_i u_{i,1}) + (m_s u_{s,2} - m_s u_{s,1}) + (m_c u_{c,2} - m_c u_{c,1}) + (m_o u_{o,2} - m_o u_{o,1})$$

$$Q_{in} = m_i C_{pi}(T_2 - T_1) + m_s C_{ps}(T_2 - T_1) + m_c C_{pc}(T_2 - T_1) + m_o C_{po}(T_2 - T_1)$$

$$Q_{in} = (10\ kg)\left(460.5\,\frac{J}{kgK}\right)(848 - 298)K + 10\ kg\left(520.6\,\frac{J}{kgK}\right)(848 - 298)K + 10\ kg\left(376.8\,\frac{J}{kgK}\right)(848 - 298)K + 100\ kg\left(2306.1\,\frac{J}{kgK}\right)(848 - 298)K$$

$$Q_{in} = 2532750\ J + 2863300\ J + 2072400\ J + 126835500\ J$$

$$Q_{in} = \mathbf{134{,}303{,}950\ J = 134.3\ MJ}$$

For entropy generation:

$$S_{gen} = m_i(s_{i,2} - s_{i,1}) + m_s(s_{s,2} - s_{s,1}) + m_c(s_{c,2} - s_{c,1}) + m_o(s_{o,2} - s_{o,1}) - \left(\frac{Q_{in}}{T_s}\right)$$

$$S_{gen} = (10\ kg)\left(1.079\,\frac{kJ}{kgK} - 0.488\,\frac{kJ}{kgK}\right) + (10\ kg)\left(1.158\,\frac{kJ}{kgK} - 0.582\,\frac{kJ}{kgK}\right) + (10\ kg)\left(0.942\,\frac{kJ}{kgK} - 0.527\,\frac{kJ}{kgK}\right) + (100\ kg)\left(3.147\,\frac{kJ}{kgK} - 0.364\,\frac{kJ}{kgK}\right) - \left(\frac{134303.9\ kJ}{873\ K}\right)$$

$$\mathbf{S_{gen} = 140.3\,\frac{kJ}{K}}$$

For exergy destruction:

$$m_i ex_{i,1} + m_s ex_{s,1} + m_c ex_{c,1} + m_o ex_{o,1} + Q_{in}\left(1 - \frac{T_o}{T_s}\right) = m_i ex_{i,2} + m_s ex_{s,2} + m_c ex_{c,2} + m_o ex_{o,2} + Ex_d$$

$$Ex_d = m_i(ex_{i,1} - ex_{i,2}) + m_s(ex_{s,1} - ex_{s,2}) + m_c(ex_{c,1} - ex_{c,2}) + m_o(ex_{o,1} - ex_{o,2}) + Q_{in}\left(1 - \frac{T_o}{T_s}\right)$$

$$Ex_d = m_i(u_{i,1} - u_{i,2} - T_0(s_{i,1} - s_{i,2})) + m_s(u_{s,1} - u_{s,2} - T_0(s_{s,1} - s_{s,2}))$$
$$+ m_c(u_{c,1} - u_{c,2} - T_0(s_{c,1} - s_{c,2})) + m_o(u_{o,1} - u_{o,2} - T_0(s_{o,1} - s_{o,2})) + Q_{in}\left(1 - \frac{T_o}{T_s}\right)$$

$$Ex_d = 10\ kg\left(0.4605\ \frac{kJ}{kgK}(298 - 848)K - 298\ K(0.488 - 1.079)\ \frac{kJ}{kgK}\right)$$
$$+ 10\ kg\left(0.5206\ \frac{kJ}{kgK}(298 - 848)K - 298\ K(0.582 - 1.158)\ \frac{kJ}{kgK}\right)$$
$$+ 10\ kg\left(0.3768\ \frac{kJ}{kgK}(298 - 848)K - 298\ K(0.527 - 0.942)\ \frac{kJ}{kgK}\right)$$
$$+ 100\ kg\left(2.3\ \frac{kJ}{kgK}(298 - 848)K - 298\ K(0.364 - 3.147)\ \frac{kJ}{kgK}\right)$$
$$+ 134{,}303\ kJ\left(1 - 298\ \frac{K}{873\ K}\right)$$

$$\boldsymbol{Ex_d = 40{,}786.2\ kJ = 40.7862\ MJ}$$

5.3.2 Moving Boundary Closed Systems

A closed system with a moving boundary is the other type of closed system, where movement of the boundary generates or consumes work, such as piston–cylinder mechanisms doing compression and expansion. The compression process requires work input while the expansion process produces work. In this section, examples are provided where all mass, energy, entropy, and exergy balance equations are written for system analysis and assessment.

Example 5.4 A piston–cylinder device, as shown in Figure 5.6, with an initial volume of 0.2 m^3 contains helium gas initially at 10 °C and 100 kPa. The helium is compressed through a polytropic process ($PV^n = constant$) to a pressure of 700 kPa and a temperature

Figure 5.6 A piston–cylinder device for Example 5.4.

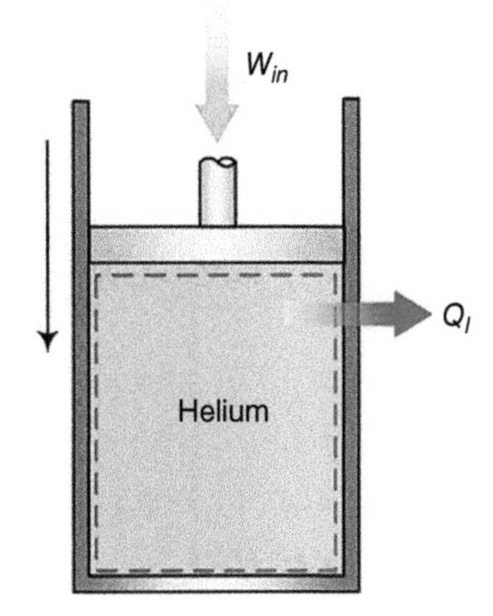

of 290 °*C*. (a) Write all mass, energy, entropy, and exergy balance equations and (b) calculate the amounts of heat lost, exergy associated with the heat loss, entropy generated and exergy destroyed. Take the polytropic exponent (n) as 1.25 for helium and the ambient temperature as 25 °*C*. The average temperature ($T_{av} = (T_1+T_2)/2$) is used for entropy and exergy calculations.

Solution

a) One first needs to write all balance equations for mass, energy, entropy, and exergy as follows:

$$\text{MBE}: m_1 = m_2 = constant$$

$$\text{EBE}: m_1u_1 + W_{in} = m_2u_2 + Q_l$$

Here, W_{in} essentially refers to boundary movement work. That's why there is no need to additionally write another boundary movement work.

$$\text{EnBE}: m_1s_1 + S_{gen} = m_2s_2 + Q_l/T_b$$

$$\text{ExBE}: m_1ex_1 + W_{in} = m_2ex_2 + Ex_{Q_l} + Ex_d$$

b) This section of the example provides calculations of the amounts of heat lost, entropy generated and exergy destroyed.

To find the necessary values needed for the question, we must first find the mass of helium in the system. To do that we take the initial conditions of the helium inside the piston–cylinder device and use the ideal gas equation as follows:

$$P_1V_1 = mRT_1$$

$$100\ kPa \times 0.2\ m^3 = m \times 2.077\ kJ/kgK \times 283.15\ K$$

$$m = \frac{100\ kPa \times 0.2\ m^3}{2.077\ kJ/kgK \times 283.15\ K} = \mathbf{0.034\ kg}$$

The work input into the system can be found by using the polytropic work relation as follows:

$$W_{in} = \frac{mR(T_2 - T_1)}{1-n}$$

$$W_{in} = \frac{0.034\ kg \times 2.077\ kJ/kgK \times (283.15\ K - 563.15\ K)}{1-1.25}$$

$$\mathbf{W_{in} = 79.09\ kJ}$$

After finding the work input to the system the heat lost (Q_l) by the system can be obtained by using the energy balance equation, as follows:

$$m_1u_1 + W_{in} = m_2u_2 + Q_l$$

$$Q_l = m_1u_1 - m_2u_2 + W_{in} \rightarrow Q_l = mC_v(T_1 - T_2) + W_{in}$$

$$Q_l = 0.034\ kg \times 3.116\ kJ/kgK(-280\ K) + 79.09\ kJ$$

$$\mathbf{Q_l = 49.426\ kJ}$$

Then one can calculate the exergy content associated with the heat loss as:

$$Ex_{Q_l} = \left(1 - \frac{T_o}{T_{av}}\right) Q_l$$

The heat lost escapes the system through its walls, which means that the temperature of the source ($T_{av} = 150 + 273 = 423\ K$) is selected to be the average temperature of the system. The systems temperature ranges from 10 to 290 °C, which will be used to calculate the average temperature as follows:

$$Ex_{Q_l} = \left(1 - \frac{298}{423}\right) \times 49.426 = \mathbf{14.61\ kJ}$$

In order to find the exergy destruction of the system, we can use the relation:

$$Ex_d = s_{gen} T_0$$

In order to find the entropy generated by the system, one can use the entropy balance equation along with the helium properties from the property tables or EES. Take $s_1 = 27.74\ \frac{kJ}{kgK}$ and $s_2 = 27.27\ \frac{kJ}{kgK}$.

$$m_1 s_1 + S_{gen} = m_2 s_2 + Q_l / T_{av}$$

$$S_{gen} = m_2 s_2 - m_1 s_1 + Q_l / T_{av}$$

$$S_{gen} = 0.034\ kg \times (27.27 - 27.74)\left(\frac{kJ}{kgK}\right) + \frac{49.426\ kJ}{423\ K}$$

$$\mathbf{S_{gen} = 0.101\ \frac{kJ}{K}}$$

One can then simply plug in the values in the exergy destruction relation as follows:

$$Ex_d = S_{gen} T_0$$

$$Ex_d = 0.101\ \frac{kJ}{K} \times 298\ K$$

$$\mathbf{Ex_d = 30.09\ kJ}$$

Example 5.5 A piston–cylinder device, as shown in Figure 5.7, has a 12 kg of refrigerant $R134a$ in the saturated vapor state at a pressure of 240 kPa. An electrical heat is turned on for 360 seconds while connected to an 110 V electrical source with 12.8 A running through the heater. In addition to the electrical heater, 300 kJ of heat are transferred to the refrigerant over the same period. (a) Write the mass, energy, entropy, and exergy balance equations, (b) calculate the final state temperature, (c) calculate the thermal exergy input to the system, (d) find the entropy generation during the process, and (e) find the exergy destruction during the process.

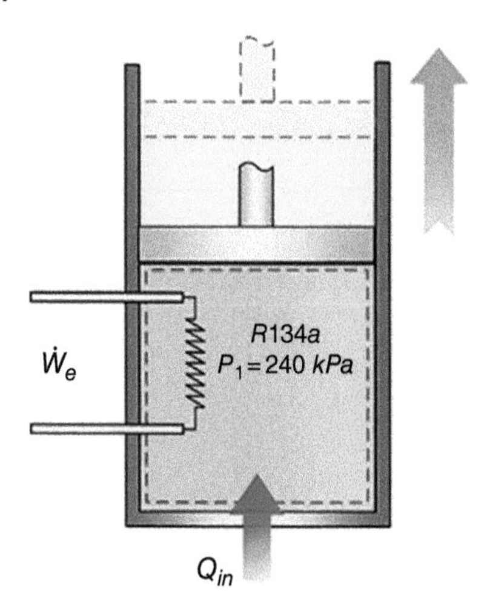

Figure 5.7 Schematic diagram of the rigid tank discussed in Example 5.5.

Solution

a) Write the mass, energy, entropy, and exergy balance equations.

The change of the kinetic energy of the system can be neglected since the system velocity does not change (not accelerating nor decelerating) throughout the process. The change in the potential energy of the system can be neglected as well since the system elevation does not change throughout the process.

Based on the above mentioned assumptions, the mass and energy balance can be written as:

$$\text{MBE}: m_1 = m_2 = constant$$

$$\text{EBE}: Q_{in} + \dot{W}_e \times \Delta t + m_1 u_1 = m_2 u_2 + W_b$$

$$\text{EnBE}: Q_{in}/T_s + m_1 s_1 + S_{gen} = m_2 s_2$$

$$\text{ExBE}: Ex_{Q_{in}} + \dot{W}_e \times \Delta t + m_1 ex_1 = m_2 ex_2 + W_b + Ex_d$$

The boundary work is done on a constant pressure process, then:

$$W_b = P \times (V_2 - V_1) = PV_2 - PV_1$$

And by using the definition of the enthalpy:

$$H = U + PV$$

then

$$Q_{in} + \dot{W}_e \times \Delta t + U_1 = U_2 + (PV_2 - PV_1)$$

$$Q_{in} + \dot{W}_e \times \Delta t = U_2 - U_1 + (PV_2 - PV_1)$$

$$= (U_2 + PV_2) - (U_1 + PV_1)$$

$$Q_{in} + \dot{W}_e \times \Delta t = H_2 - H_1$$

$$Q_{in} + \dot{W}_e \times \Delta t = m \times (h_2 - h_1)$$

b) Calculate the final state pressure and temperature.

$$Q_{in} = 300\ kJ$$

$$\dot{W}_e = IV = 12.8\ A \times 110\ V \times \frac{1\ kJ/s}{1000\ AV} = 1.408\ kJ/s$$

The $R134a$ properties at the initial state, which at saturated vapor (x = 1) and a pressure of 240 kPa, then:

$$h_1 = h_g@240\ kPa = 247.32\ \frac{kJ}{kg}$$

By substituting the values of terms in the energy balance equation:

$$300\ kJ + \left(1.408\ \frac{kJ}{s} \times 360\ s\right) = 12\ kg \times \left(h_2 - 247.32\ \frac{kJ}{kg}\right)$$

$$\boldsymbol{h_2 = 314.56\ kJ/kg}$$

Since the device is a piston–cylinder device, then the process occurs at a constant pressure.

$$\boldsymbol{P_2 = 240\ kPa}$$

In order to find the temperature of the final state, we need to define the state. However, in other cases where the properties database software is available, we enter the two independent properties and the software will provide us with the state and all other properties that we need.

By following one of the above two methods the final temperature of the piston–cylinder device becomes:

$$\boldsymbol{T_2 = 70\ ^\circ C}$$

c) Calculate the thermal exergy input to the system.

The exergy content of the heat input to the system is calculated as:

$$Ex_{Q_{in}} = \left(1 - \frac{T_o}{T_s}\right) Q_{in}$$

The heat input to the system to its walls, which means that the temperature of the source (T_s) cannot be directly selected to be the system temperature. The problem statement did not provide enough information on the source of the heat supplied to the system, which means in this case we have to make a reasonable assumption on the source temperature of the heat supplied to the piston–cylinder device. Based on the principles of heat transfer, in order to have the heat to go from one source to a destination, the source temperature must always be higher than the temperature of the destination. An acceptable approximation of the final temperature of the system is as follows:

$$Ex_{Q_{in}} = \left(1 - \frac{298}{70 + 273}\right) \times 300 = \mathbf{39.4\ kJ}$$

d) The entropy generation during the process can be found from the entropy balance equation:

$$Q_{in}/T_s + m_1 s_1 + S_{gen} = m_2 s_2$$

where s_1 can be found from the thermodynamic tables (such as Appendix B-1b) or properties database software:

$$s_1 = s_g@240\ kPa = 0.9346\ \frac{kJ}{kgK}$$

Also, s_2 can be found at $P_2 = 240\,kPa$ and $h_2 = 314.56\,kJ/kg$ as:

$$s_2 = 1.156\ \frac{kJ}{kgK}$$

The entropy generation can thus be found as:

$$S_{gen} = m_2 s_2 - m_1 s_1 - Q_{in}/T_s$$

$$S_{gen} = (12\ kg)\left(1.156\ \frac{kJ}{kgK} - 0.9346\ \frac{kJ}{kgK}\right) - \frac{300\ kJ}{343\ K}$$

$$\mathbf{S_{gen} = 1.78\ \frac{kJ}{K}}$$

e) The exergy destruction can be calculated from the exergy balance equation as follows:

$$Ex_d = Ex_{Q_{in}} + \dot{W}_e \times t + m(ex_1 - ex_2)$$

$$Ex_d = 39.4\ kJ + \left(1.408\ \frac{kJ}{s}\ 360\,s\right) + (12\ kg)\left(247.32\ \frac{kJ}{kg} - 314.6\ \frac{kJ}{kg}\right) - 298\ K$$

$$\left(0.9346\ \frac{kJ}{kgK} - 1.156\ \frac{kJ}{kgK}\right)$$

$$\mathbf{Ex_d = 530.6\ kJ}$$

Example 5.6 Consider a piston–cylinder device, as shown in Figure 5.8, which initially contains air at a pressure of $400\,kPa$ and a temperature of $17\,°C$. The air in the piston–cylinder device receives $50\,kJ$ of work per each kg of air in the device through the rotating paddle. The air in the device receives thermal energy (heat) from a source with a temperature of $600\,°C$. The piston–cylinder device operates at a constant pressure of approximately $400\ kPa$, which makes the process an *isobaric process*. The air volume in the device increases to three times its original volume. (a) Write the mass, energy, entropy, and exergy balance equations for this system, (b) find the boundary work produced by the mechanism, (c) find the heat input from the source to the system, (d) calculate the entropy generation, and (e) calculate the exergy destruction.

Solution

a) Write the mass, energy, entropy, and exergy balance equations.

Analyze the piston–cylinder device using the balance equations, which can be written as:

$$\text{MBE}: m_1 = m_2 = constant$$

$$\text{EBE}: m_1 u_1 + Q_{in} + W_{in} = m_2 u_2 + W_{out}$$

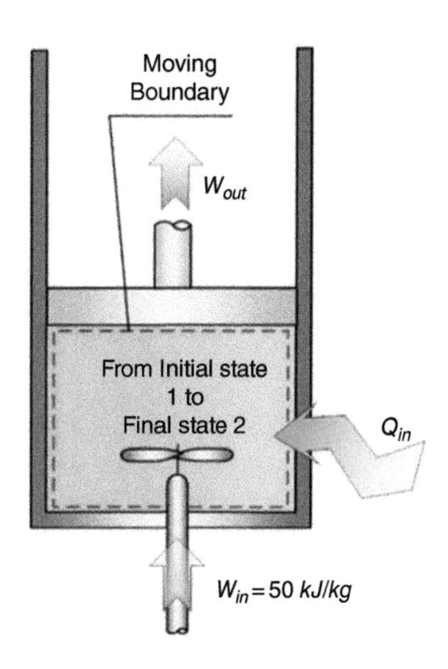

Figure 5.8 A schematic diagram of a piston–cylinder mechanism as a closed system involving heat input Q_{in}, work input W_{in}, and boundary work output W_{out}, as discussed in Example 5.6.

Here, W_{out} is the boundary movement work as produced by the system.

$\text{EnBE}: m_1s_1 + Q_{in}/T_s + S_{gen} = m_2s_2$

$\text{ExBE}: m_1ex_1 + (1-(T_o/T_s))Q_{in} + W_{in} = m_2ex_2 + W_{out} + Ex_d$

b) Calculate the boundary work produced by the system.

Note that the temperature of the air inside the device remains inconstant through the process and the pressure remains constant due to the piston motion and production of boundary work. One can write the ideal gas equation for this as

$$P_1V_1/T_1 = P_2V_2/T_2$$

and find $T_2 = 870\ K$

In addition, the EBE is reduced to

$$u_1 + q_{in} + w_{in} = u_2 + w_{out}$$

Since the system operates at a constant pressure (so-called: isobaric process), one can then use the following equation to calculate the boundary work produced by the system:

$$\mathbf{w_{out} = w_b} = P(v_2 - v_1) = \mathbf{166.5\ kJ/kg}$$

c) The heat input to the system is calculated as follows, by using the EBE in specific terms:

$$207.1\ kJ/kg + q_{in} + 50\ kJ/kg = 166.5\ kJ/kg + 650\ kJ/kg$$

$$\mathbf{q_{in} = 559.3\ kJ/kg}$$

d) The entropy generation is calculated based on the entropy balance equation written at the beginning of the solution as the first step in solving energy systems:

$$m_1s_1 + Q_{in}/T_s + S_{gen} = m_2s_2$$

By dividing the above equation by the mass of the system, which is constant since the system is a closed energy system, the division result is:

$$s_1 + q_{in}/T_s + s_{gen} = s_2$$

$$5.274 + 559.3/(600 + 273.15) + s_{gen} = 6.417$$

$$\mathbf{s_{gen} = 0.502\ kJ/(kgK)}$$

e) The exergy destruction is calculated using the ExBE as follows:

$$m_1 ex_1 + (1-(T_o/T_s))Q_{in} + W_{in} = m_2 ex_2 + W_{out} + Ex_d$$

$$Ex_d = m(ex_1 - ex_2) + (1-(T_o/T_s))Q_{in} + W_{in} - W_{out}$$

$$Ex_d = m(u_1 - u_2 - T_0(s_1 - s_2)) + (1-(T_o/T_s))Q_{in} + W_{in} - W_{out}$$

$$Ex_d = \left((207.1-650)kJ/kg - 298\ K(5.274-6.417)\frac{kJ}{kgK}\right) + \left(1-\left(\frac{298}{873.15}\right)\right)559.3\frac{kJ}{kg} + 50\frac{kJ}{kg} - 166.5\frac{kJ}{kg}$$

$$\mathbf{Ex_d = 149.6\ \frac{kJ}{kg}}$$

Example 5.7 A balloon with a 12 *kg* of refrigerant *R134a* in the saturated vapor state at a pressure of 240 *kPa*, as shown in Figure 5.9, is expanded in a process, during which a total of 300 *kJ* of heat is transferred to the refrigerant. Take the source temperature of heat as 300 °*C*. (a) Write the mass, energy, entropy, and exergy balance equations for this closed system and (b) calculate the final state temperature, entropy generation, and exergy destruction. Take $T_o = 25\,°C$.

Solution

a) Write the mass, energy, entropy, and exergy balance equations.

The change in kinetic energy of the system can be neglected since the system velocity does not change (not accelerating nor decelerating) throughout the process. The change in the potential energy of the system can be neglected as well since the system elevation does not change throughout the process.

$$\text{MBE}: m_1 = m_2 = constant$$

$$\text{EBE}: Q_{in} + m_1 u_1 = m_2 u_2 + W_b$$

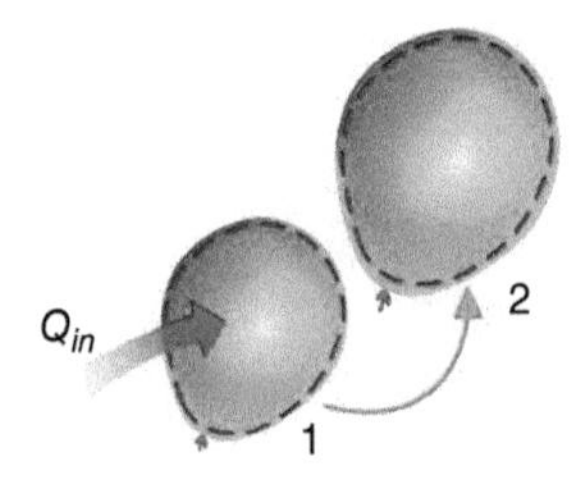

Figure 5.9 Expanding balloon due to the heat addition, as discussed in Example 5.7.

$$\text{EnBE}: Q_{in}/T_s + m_1 s_1 + S_{gen} = m_2 s_2$$

$$\text{ExBE}: Ex_{Q_{in}} + m_1 ex_1 = m_2 ex_2 + W_b + Ex_d$$

The boundary work is done on a constant pressure process, then:

$$W_b = P \times (V_2 - V_1) = PV_2 - PV_1$$

And by using the definition of the enthalpy we can rearrange the energy equation and substitute the enthalpy values, as follows:

$$h = u + PV$$

$$Q_{in} + m_1 u_1 + PV_1 = m_2 u_2 + PV_2$$

$$Q_{in} = m_2 u_2 - m_1 u_1 + (PV_2 - PV_1)$$

$$Q_{in} = (m_2 u_2 + PV_2) - (m_1 u_1 + PV_1)$$

$$Q_{in} = m_2 h_2 - m_1 h_1$$

$$Q_{in} = m \times (h_2 - h_1)$$

b) Calculate the final state temperature, entropy generation and exergy destruction.

Given the properties of $R134a$ at the initial state, which are a saturated vapor ($x = 1$) and a pressure of 240 kPa, then:

$$h_1 = h_g @ 240\ kPa = 247.32\ \frac{kJ}{kg}$$

By substituting the values of terms in the energy balance equation:

$$Q_{in} = m \times (h_2 - h_1)$$

$$300\ kJ = 12\ kg \times \left(h_2 - 247.32\ \frac{kJ}{kg}\right)$$

$$h_2 = 272.32\ kJ/kg$$

As mentioned in the problem statement, the expansion of the balloon can be assumed at a constant pressure:

$$P_1 = \mathbf{P_2 = 240\ kPa}$$

$$\left\{\begin{matrix} P_2 = 240\ kPa \\ h_2 = 272.32\ \frac{kJ}{kg} \end{matrix}\right\} \mathbf{T_2 = 23.4\ ^oC}$$

In order to find the entropy generated by the system, one can use the entropy balance equation along with the $R134a$ properties from the property tables (such as Appendix B-3c) or EES. Take $s_1 = 0.9346\ \frac{kJ}{kgK}$ and $s_2 = 1.023\ \frac{kJ}{kgK}$ accordingly.

$$m_1 s_1 + Q_{in}/T_s + S_{gen} = m_2 s_2$$

$$S_{gen} = m_2 s_2 - m_1 s_1 - Q_{in}/T_s$$

$$S_{gen} = 12\ kg \times (1.023 - 0.9346)\left(\frac{kJ}{kgK}\right) + \frac{300\ kJ}{(300 + 273)\ K}$$

$$\boldsymbol{S_{gen} = 1.58\,\frac{kJ}{K}}$$

One can then simply plug in the values in the exergy destruction relation as follows:

$$Ex_d = S_{gen}T_0$$

$$Ex_d = 1.58\,\frac{kJ}{K} \times 298.15\ K$$

$$\boldsymbol{Ex_d = 471.08\ kJ}$$

5.4 Open Systems

As mentioned in previous chapters, open systems are those where mass can enter and/or leave across the boundary of the system. The specific details of such systems with various examples were discussed earlier. Here, the prime focus is made on system analysis, because of the main goal of this chapter, and there are numerous devices/units considered as open systems, ranging from nozzles and diffusers to pumps and turbines. These devices are units that are treated as a steady state-steady flow process (SSSF), where there is no change in flow with time. However, there is another type of open system, namely unsteady-state uniform-flow (USUF), where there are charging and discharging systems (for example, filling and emptying tanks). These will be introduced through the examples where balance equations are written for mass, energy, entropy, and exergy and numerous items, such as entropy generation rate and exergy destruction rates, are calculated.

5.4.1 Steady-state Steady-flow Systems

As introduced earlier, SSSF processes are recognized in a way where the properties inside the boundary of the system do not change with time. This section will provide examples on the application of the energy balance equation on SSSF. Before going to the detailed examples on each of the SSSF system, Table 5.2 tabulates the mass, energy, entropy, and exergy balance equations for the selected types of system.

Table 5.2 The mass, energy, entropy, and exergy balance equations of each of the SSSF systems selected.

System	Balance equations
Pipe $\dot{Q}_{out}$ State 1 → Hot Fluid → State 2	MBE: $\dot{m}_1 = \dot{m}_2 = \dot{m}$ EBE: $\dot{m}_1h_1 = \dot{m}_2h_2 + \dot{Q}_{out}$ EnBE: $\dot{m}_1s_1 + \dot{S}_{gen} = \dot{m}_2s_2 + \dot{Q}_{out}/T_b$ ExBE: $\dot{m}_1ex_1 = \dot{m}_2ex_2 + \dot{E}x_{\dot{Q}_{out}} + \dot{E}x_d$ with $\dot{E}x_{Q_{out}} = \left(1-\frac{T_o}{T_b}\right)\dot{Q}_{out}$; $\dot{E}x_d = T_o\dot{S}_{gen}$

(*Continued*)

Table 5.2 (Continued)

<table>
<tr><th>System</th><th>Balance equations</th></tr>
<tr><td></td><td>where T_b is the immediate boundary temperature since heat is leaving the system and rejected into the immediate boundary. One may also use surface temperature or average temperature of the fluid ($T_{av} = (T_1+T_2)/2$) or ambient (or reference state) temperature (T_o) if the above ones are not known.</td></tr>
<tr><td>Nozzle

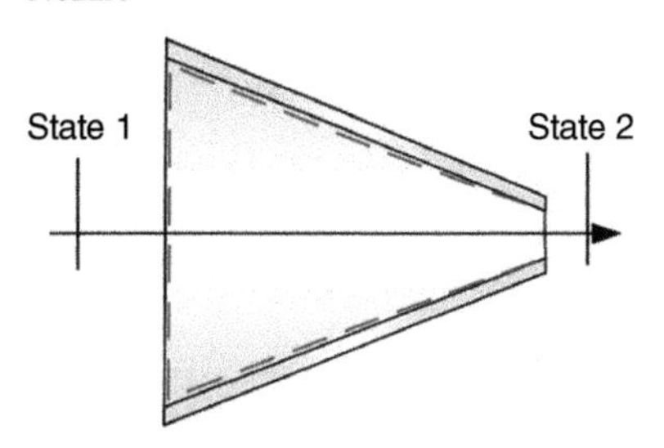

</td><td>MBE: $\dot{m}_1 = \dot{m}_2 = \dot{m}$
EBE: $\dot{m}_1 h_1 + m\frac{V_1^2}{2} = \dot{m}_2 h_2 + m\frac{V_2^2}{2}$
EnBE: $\dot{m}_1 s_1 + \dot{S}_{gen} = \dot{m}_2 s_2$
ExBE: $\dot{m}_1 ex_1 + m\frac{V_1^2}{2} = \dot{m}_2 ex_2 + m\frac{V_2^2}{2} + \dot{E}x_d$
Here, one should note that nozzles are treated as adiabatic devices and that there is no work crossing the boundary.</td></tr>
<tr><td>Diffuser

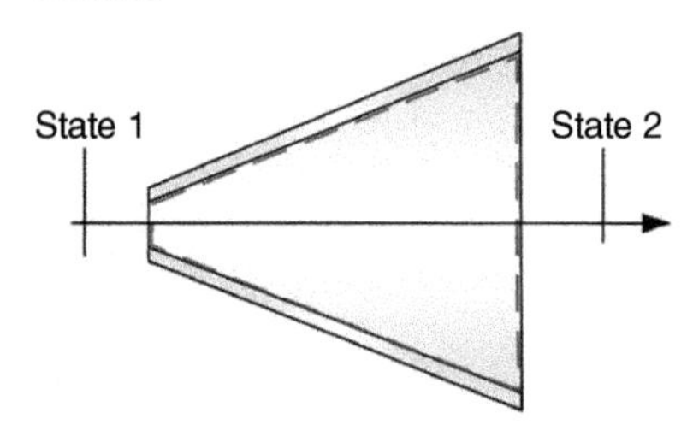

</td><td>MBE: $\dot{m}_1 = \dot{m}_2 = \dot{m}$
EBE: $\dot{m}_1 h_1 + m\frac{V_1^2}{2} = \dot{m}_2 h_2 + m\frac{V_2^2}{2}$
EnBE: $\dot{m}_1 s_1 + \dot{S}_{gen} = \dot{m}_2 s_2$
ExBE: $\dot{m}_1 ex_1 + m\frac{V_1^2}{2} = \dot{m}_2 ex_2 + m\frac{V_2^2}{2} + \dot{E}x_d$
Here, one should note that diffusers are treated as adiabatic devices and that there is no work crossing the boundary.</td></tr>
<tr><td>Turbine

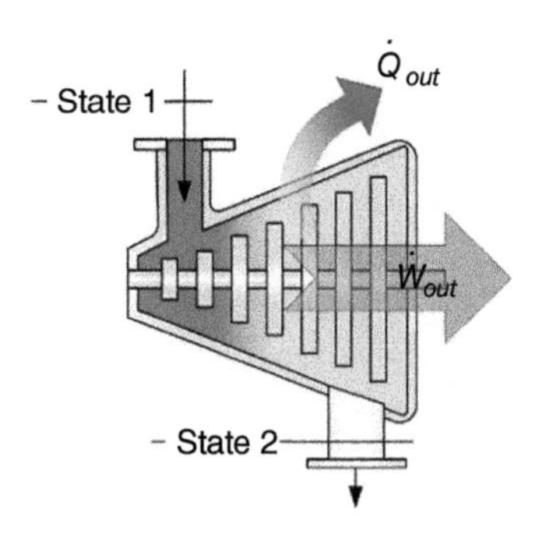

</td><td>MBE: $\dot{m}_1 = \dot{m}_2 = \dot{m}$
EBE: $\dot{m}_1 h_1 = \dot{m}_2 h_2 + \dot{Q}_{out} + \dot{W}_{out}$
EnBE: $\dot{m}_1 s_1 + \dot{S}_{gen} = \dot{m}_2 s_2 + \dot{Q}_{out}/T_b$
ExBE: $\dot{m}_1 ex_1 = \dot{m}_2 ex_2 + \dot{E}x_{\dot{Q}_{out}} + \dot{W}_{out} + \dot{E}x_d$</td></tr>
</table>

(Continued)

Table 5.2 (Continued)

System	Balance equations
Compressor 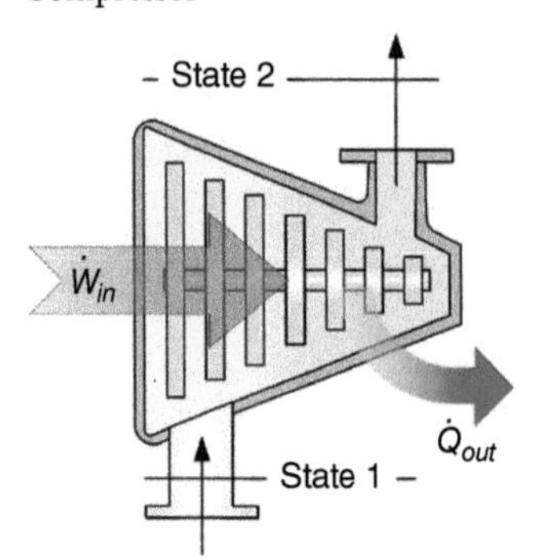	MBE: $\dot{m}_1 = \dot{m}_2$ EBE: $\dot{m}_1 h_1 + \dot{W}_{in} = \dot{m}_2 h_2 + \dot{Q}_{out}$ EnBE: $\dot{m}_1 ex_1 + \dot{W}_{in} = \dot{m}_2 ex_2 + \dot{E}x_d + \dot{E}x_{\dot{Q}_{out}}$
Pump 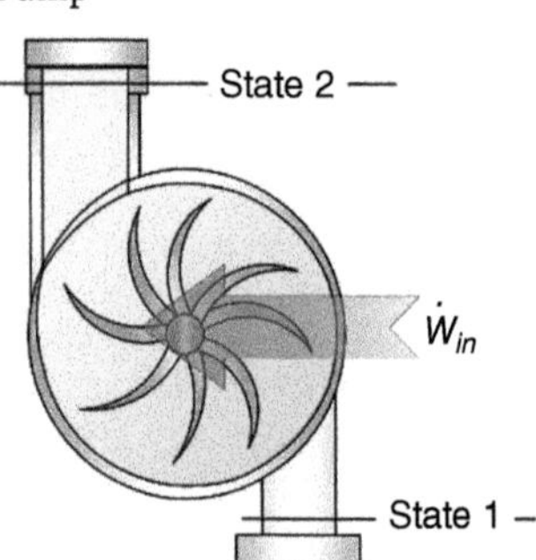	MBE: $\dot{m}_1 = \dot{m}_2 = \dot{m}$ EBE: $\dot{m}_1 h_1 + \dot{W}_{in} = \dot{m}_2 h_2$ Here, one should keep in mind that the pump work is shaft work, which is defined as $\dot{W}_{in} = \dot{m}v(P_2 - P_1)$, which truly comes out of the above written balance equation if one considers Molier's enthalpy equation: $h = u + Pv$. EnBE: $\dot{m}_1 s_1 + \dot{S}_{gen} = \dot{m}_2 s_2$ ExBE: $\dot{m}_1 ex_1 + \dot{W}_{in} = \dot{m}_2 ex_2 + \dot{E}x_d$
Expansion valve 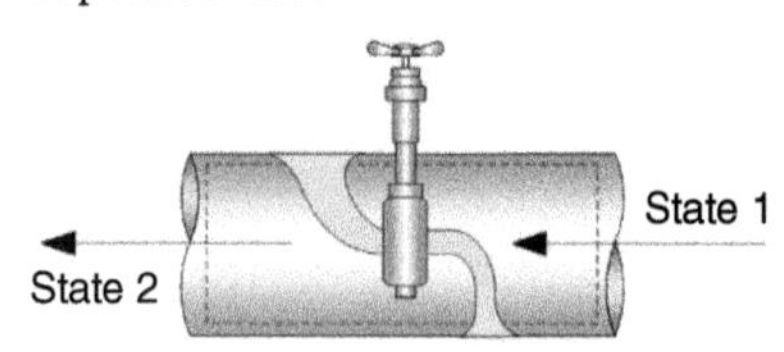	MBE: $\dot{m}_1 = \dot{m}_2 = \dot{m}$ EBE: $\dot{m}_1 h_1 = \dot{m}_2 h_2 \rightarrow h_1 = h_2$ which is a so-called isenthalpic process. This is truly confirmed by the energy balance equation. EnBE: $\dot{m}_1 s_1 + \dot{S}_{gen} = \dot{m}_2 s_2$ ExBE: $\dot{m}_1 ex_1 = \dot{m}_2 ex_2 + \dot{E}x_d$
Heat exchanger 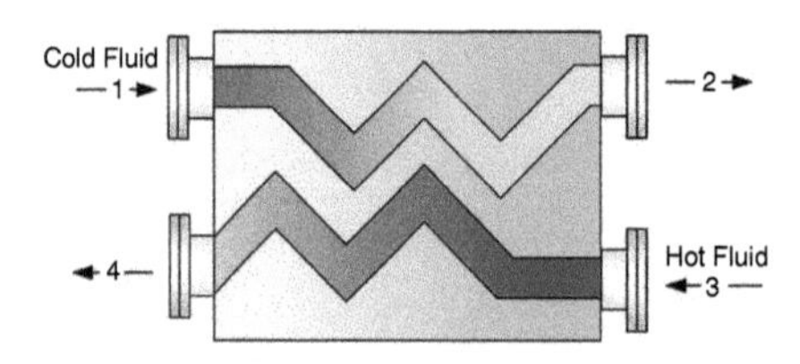	MBE: $\dot{m}_1 = \dot{m}_2 = \dot{m}_{cf}, \dot{m}_3 = \dot{m}_4 = \dot{m}_{hf}$ EBE: $\dot{m}_1 h_1 + \dot{m}_3 h_3 = \dot{m}_2 h_2 + \dot{m}_4 h_4$ EnBE: $\dot{m}_1 s_1 + \dot{m}_3 s_3 + \dot{S}_{gen} = \dot{m}_2 s_2 + \dot{m}_4 s_4$ ExBE: $\dot{m}_1 ex_1 + \dot{m}_3 ex_3 = \dot{m}_2 ex_2 + \dot{m}_4 ex_4 + \dot{E}x_d$ where *cf* is cold fluid and hf is hot fluid.

(*Continued*)

Table 5.2 (Continued)

System	Balance equations
Mixing chamber 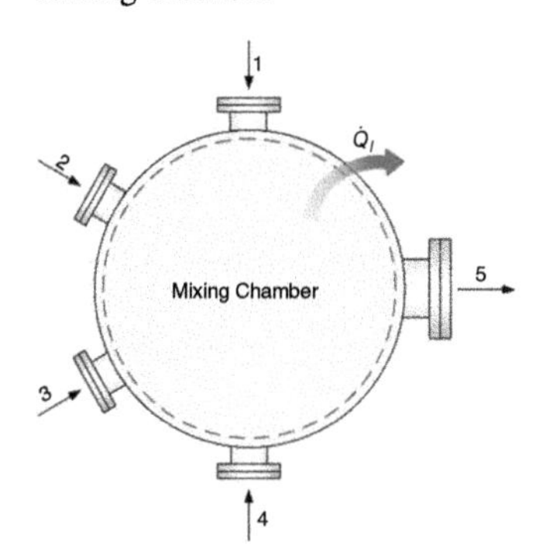	MBE: $\dot{m}_1 + \dot{m}_2 + \dot{m}_3 + \dot{m}_4 = \dot{m}_5$ EBE: $\dot{m}_1 h_1 + \dot{m}_2 h_2 + \dot{m}_3 h_3 + \dot{m}_4 h_4 = \dot{m}_5 h_5 + \dot{Q}_l$ EnBE: $\dot{m}_1 s_1 + \dot{m}_2 s_2 + \dot{m}_3 s_3 + \dot{m}_4 s_4 + S_{gen} = \dot{m}_5 s_5 + \dot{Q}_l/T_b$ ExBE: $\dot{m}_1 ex_1 + \dot{m}_2 ex_2 + \dot{m}_3 ex_3 + \dot{m}_4 ex_4 = \dot{m}_5 ex_5 + \dot{E}x_{\dot{Q}_l} + \dot{E}x_d$

Example 5.8 Hot water at a temperature of 95 °C and a pressure of 100 kPa flows in a pipe at a flow rate of 1 kg/s, as shown in Figure 5.10. The pipe loses heat at a rate of 1 kW per meter along the pipe length. (a) Write the mass, energy, entropy, and exergy balance equations, (b) find the temperature of the water leaving the pipe after 100 m of pipe length, and (c) calculate thermal exergy loss, rate entropy generation rate, and exergy destruction rate. Take the pipe surface temperature as 60 °C and reference temperature (i.e. surroundings temperature) as 25 °C.

Solution

a) Write the mass, energy, entropy, and exergy balance equations.

$$\text{MBE}: \dot{m}_1 = \dot{m}_2 = \dot{m}$$

$$\text{EBE}: \dot{m}_1 h_1 = \dot{m}_2 h_2 + \dot{Q}_{out}$$

$$\text{EnBE}: \dot{m}_1 s_1 + \dot{S}_{gen} = \dot{m}_2 s_2 + \dot{Q}_{out}/T_{su}$$

$$\text{ExBE}: \dot{m}_1 ex_1 = \dot{m}_2 ex_2 + \dot{E}x_{\dot{Q}_{out}} + \dot{E}x_d$$

b) Find the temperature of the water leaving the pipe after 100 m of pipe length.

$$\dot{Q}_{out} = 1 \frac{kW}{m} \times 100\ m$$

$$\dot{\mathbf{Q}}_{\mathbf{out}} = \mathbf{100\ kW}$$

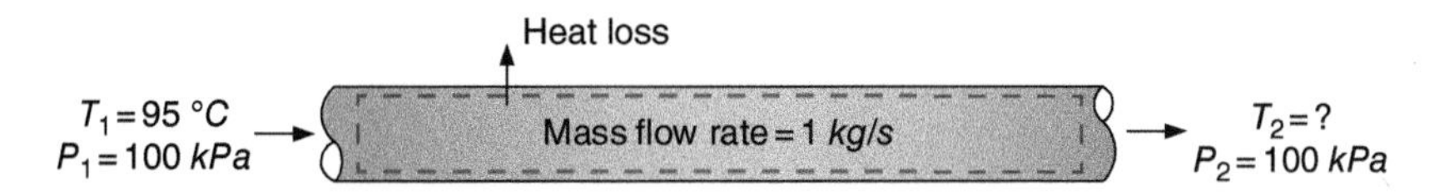

Figure 5.10 A pipe flow problem in Example 5.8.

$$\text{at} P_1 = 100\ kPa\ and\ T_1 = 95\ ^oC, h_1 = 398.1\ \frac{kJ}{kg}$$

$$\dot{m}_1 h_1 = \dot{m}_2 h_2 + \dot{Q}_{out}$$

$$1\ \frac{kg}{s} \times 398.1\ \frac{kJ}{kg} = 1\ \frac{kg}{s} \times h_2 + 1\ \frac{kW}{m} \times 100\ m$$

From the above written energy balance equation, one obtains $h_2 = 298.1\ \frac{kJ}{kg}$

By assuming no pressure loss occurs in the pipe, $P_1 = P_2 = 100\ kPa$ and $h_2 = 298.1\ \frac{kJ}{kg}$ then one can use the property tables (such as Appendix B-1d) or EES to find the temperature:

$$\boldsymbol{T_2 = 71.19\ ^oC}$$

c) Calculate the exergy associated with the heat loss rate, entropy generation rate, and exergy destruction rate.

The exergy content of the heat loss is calculated as follows:

$$\dot{Ex}_{\dot{Q}_{out}} = \left(1 - \frac{T_0}{T_{su}}\right)\dot{Q}_{out}$$

Note that the surface temperature of the pipe is taken as 60 °C, respectively to calculate the thermal exergy rate (or exergy rate of the heat losses) as follows:

$$\dot{Ex}_{\dot{Q}_{out}} = \left(1 - \frac{298}{60 + 273}\right) \times 100\ kW$$

$$\boldsymbol{\dot{Ex}_{\dot{Q}_{out}} = 10.51\ kW}$$

In order to find the entropy generated by the system, one can use the entropy balance equation along with the water inlet and exit properties from the property tables (such as Appendix B-1d) or EES. Take $s_1 = 1.25\ \frac{kJ}{kgK}$ and $s_2 = 0.9696\ \frac{kJ}{kgK}$.

$$\dot{m}_1 s_1 + \dot{S}_{gen} = \dot{m}_2 s_2 + \dot{Q}_{out}/T_{su}$$

$$\dot{S}_{gen} = \dot{m}_2 s_2 - \dot{m}_1 s_1 + \dot{Q}_{out}/T_{su}$$

$$\dot{S}_{gen} = 1\ \frac{kg}{s} \times (0.9696 - 1.25)\left(\frac{kJ}{kgK}\right) + \frac{100\ kW}{(60 + 273)\ K}$$

$$\boldsymbol{\dot{S}_{gen} = 0.020\ \frac{kW}{K}}$$

One can then simply plug in the values in the exergy destruction relation as follows:

$$\dot{Ex}_d = \dot{S}_{gen} T_0$$

$$\dot{Ex}_d = 0.020\ \frac{kW}{K} \times 298.15\ K$$

$$\boldsymbol{\dot{Ex}_d = 5.96\ kW}$$

Example 5.9 Consider a diffuser as shown in Figure 5.11a with inlet and exit pressures of 90 and 94 *kPa*, respectively. The inlet air temperature, the entrance area and the inlet air velocity are 10 °*C*, 6.4 m^2 and 94 *m/s*. Treat air as an ideal gas and use its properties. (a) Write the mass, energy, entropy, and exergy balance equations, (b) find the mass flow rate and exit temperature of the air leaving the diffuser, and (c) calculate the entropy generation rate and exergy destruction rate.

In addition, there is a nozzle, as shown in Figure 5.11b, with helium flowing through an entrance area of 0.6 m^2 at a velocity and temperature of 24 *m/s* and 25 °*C*. The exit area is measured to be 0.1 m^2 and the pressure is measured to be 90 *kPa*. The exit density is given as 0.142 kg/m^3. (a) Write the mass, energy, entropy, and exergy balance equations, (b) find the mass flow rate and exit velocity of helium, and (c) calculate the entropy generation rate and exergy destruction rate. Both of these units can be treated as adiabatic devices.

Solution for the Diffuser

a) Write the balance equations for mass, energy, entropy, and exergy.

$$\text{MBE}: \dot{m}_1 = \dot{m}_2 = \dot{m}$$

$$\text{EBE}: \dot{m}_1 h_1 + m\frac{V_1^2}{2} = \dot{m}_2 h_2 + m\frac{V_2^2}{2}$$

$$\text{EnBE}: \dot{m}_1 s_1 + \dot{S}_{gen} = \dot{m}_2 s_2$$

$$\text{ExBE}: \dot{m}_1 ex_1 + m\frac{V_1^2}{2} = \dot{m}_2 ex_2 + m\frac{V_2^2}{2} + \dot{E}x_d$$

b) Calculate the mass flow rate and exit temperature of the air leaving the diffuser.

The mass flow rate of the air can be calculated using the ideal gas equation. As the gas can be assumed to be an ideal gas:

$$Pv = RT$$

We can rearrange the specific ideal gas equation, to find the specific volume:

$$v_1 = \frac{RT_1}{P_1}$$

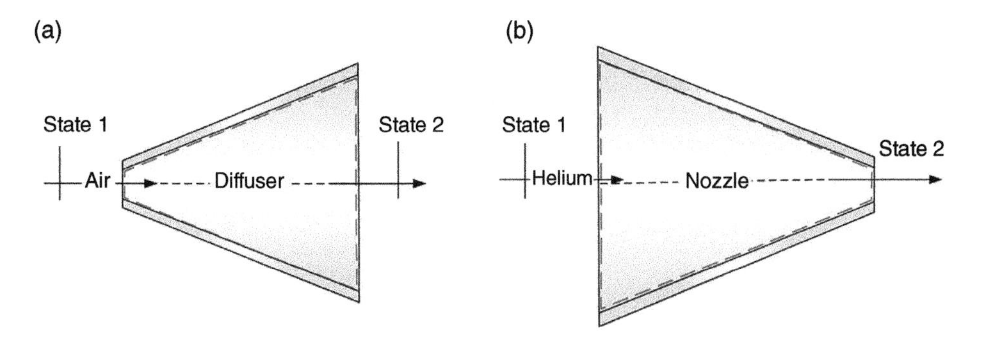

Figure 5.11 Schematic diagrams of (a) diffuser and (b) nozzle for Example 5.9.

$$v_1 = \frac{\left(0.287\ \frac{kPa\,m3}{kgK}\right)(283\ K)}{90\ kPa} = 0.902\ \frac{m^3}{kg}$$

$$\dot{m} = \frac{1}{v_1} V_1 A_1$$

$$\dot{m} = \frac{1}{0.902\ \frac{m^3}{kg}}\left(94\ \frac{m}{s}\right)(6.4\ m^2)$$

$$\mathbf{\dot{m} = 666.9\ kg/s}$$

We now need to calculate the exit temperature of the air leaving the diffuser by using the EBE:

$$\dot{m}_1 h_1 + \frac{m\,V_1^2}{2} = \dot{m}_2 h_2 + \frac{m\,V_2^2}{2}$$

$$\dot{m}_1\left(h_1 + \frac{V_1^2}{2}\right) = \dot{m}_2\left(h_2 + \frac{V_2^2}{2}\right)$$

As the mass flow rate is constant we can cancel out the inlet and outlet values, leaving us with just the enthalpy and inlet and exit velocities:

$$h_2 = h_1 - \left(\frac{V_2^2 - V_1^2}{2}\right)$$

As the diffuser is a pressure increasing device, the velocity at the exit of the diffuser can be assumed to be zero $V_1 \gg V_2$:

$$h_2 = 283.14\ \frac{kJ}{kg} - \left(\frac{0 - \left(94\ \frac{m}{s}\right)^2}{2}\left(\frac{1\ \frac{kJ}{kg}}{1000\ \frac{m^2}{s^2}}\right)\right)$$

$$h_2 = 287.56\ kJ/kg$$

Using the property tables (such as Appendix E-1) or EES, the enthalpy can then be used to find the temperature of the air at the exit, which in this case is:

$$\mathbf{T_2 = 14.02\ ^oC}$$

c) Calculate the entropy generation rate and exergy destruction rate.

The entropy generation can be calculated by isolating $\dot{S}_{gen}$:

$$\dot{m}_1 s_1 + \dot{S}_{gen} = \dot{m}_2 s_2$$

The entropy change of the system when rearranged is equal to $\dot{S}_{gen}$ as follows:

$$\dot{S}_{gen} = \dot{m}_2 s_2 - \dot{m}_1 s_1$$

$$\dot{S}_{gen} = \dot{m}(s_2 - s_1)$$

As we do not have the final entropy value, it can be assumed that under ideal gas relations that the specific heat remains constant. The relationship is:

$$s_2 - s_1 = C_p \, ln\left(\frac{T_2}{T_1}\right) - R \, ln\left(\frac{P_2}{P_1}\right)$$

$$s_2 - s_1 = 1.004 \text{ kJ}/kgK \; ln\left(\frac{287}{283}\right) - 0.287\,\frac{kPa\ m^3}{kg\ K}\, ln\left(\frac{94\ kPa}{90\ kPa}\right)$$

$$s_2 - s_1 = 0.00161\ kJ/kgK$$

$$\dot{S}_{gen} = 666.9\ kg/s \times 0.00161\ kJ/kgK$$

$$\boldsymbol{\dot{S}_{gen} = 1.073\ kW/K}$$

After the entropy generation is calculated one can proceed to calculate the exergy destruction rate using the following relationship:

$$\dot{E}x_d = \dot{S}_{gen} T_0 = 1.073\, kW/K \times 298\ K$$

$$\boldsymbol{\dot{E}x_d = 319.77\ kW}$$

Solution for the Nozzle

a) Write the balance equations for mass, energy, entropy, and exergy:

$$\text{MBE}: \dot{m}_1 = \dot{m}_2 = \dot{m}$$

$$\text{EBE}: \dot{m}_1 h_1 + \frac{m\ V_1^2}{2} = \dot{m}_2 h_2 + \frac{m\ V_2^2}{2}$$

$$\text{EnBE}: \dot{m}_1 s_1 + \dot{S}_{gen} = \dot{m}_2 s_2$$

$$\text{ExBE}: \dot{m}_1 ex_1 + \frac{m\ V_1^2}{2} = \dot{m}_2 ex_2 + \frac{m\ V_2^2}{2} + \dot{E}x_d$$

b) Find the mass flow rate and exit velocity of helium.

Helium flows through a nozzle through an entrance area of 0.6 m^2 at a speed and temperature of 24 m/s and 25 °C. The exit area is measured to be 0.1 m^2, the pressure is measured at 90 kPa. The exit density is 0.142 kg/m^3.

The mass flow rate can be defined with the following relationship:

$$\dot{m} = \rho A V$$

where ρ is the density, A is the area, and V is the velocity of the fluid.

$$\dot{m} = 0.1615\,\frac{kg}{m^3}\left(0.6\ m^2\right)\left(24\,\frac{m}{s}\right)$$

$$\boldsymbol{\dot{m} = 2.33\ kg/s}$$

The final velocity at the exit of the nozzle can be calculated based on the mass balance equation, as follows:

$$\dot{m} = \rho_1 A_1 V_1 = \rho_2 A_2 V_2$$

One can then isolate for the exit velocity as the mass flow rate stays constant.

$$2.33\,\frac{kg}{s} = 0.142\,\frac{kg}{m^3} \times 0.1\ m^2 \times V_2$$

$$\boldsymbol{V_2 = 164.08\ m/s}$$

Using the property tables or EES one can find the entropy values for the inlet and exit flows:

$$s_1 = 31.58\,\frac{kJ}{kgK} \text{ and } s_2 = 31.76\,\frac{kJ}{kgK}$$

One may obtain the following equation from the EnBE for calculating the entropy generation rate:

$$\dot{S}_{gen} = \dot{m}(s_2 - s_1)$$

$$\dot{S}_{gen} = 2.33\,\frac{kg}{s}\left(31.76\,\frac{kJ}{kgK} - 31.58\,\frac{kJ}{kgK}\right)$$

$$\boldsymbol{\dot{S}_{gen} = 0.4194\ kW/K}$$

After the entropy generation is calculated, one can proceed to calculate the exergy destruction rate using the following relationship:

$$\dot{E}x_d = \dot{S}_{gen} T_0 = 0.4194\ kW/K \times 298\ K$$

$$\boldsymbol{\dot{E}x_d = 124.98\ kW}$$

Example 5.10 Consider an adiabatic steam turbine, as shown in Figure 5.12, which receives 10 kg/s of superheated steam at a temperature of 500 °C and a pressure of 10 000 kPa. The steam turbine converts part of the energy in the steam to the shaft work

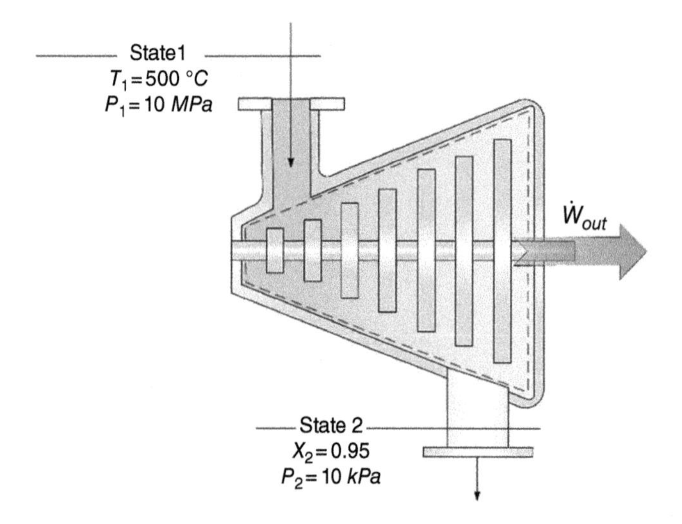

Figure 5.12 Schematic diagram of the steam turbine for Example 5.10.

output. The steam exits the steam turbine as a saturated vapor at a pressure of 10 kPa and a mixture quality of 0.95. Take the dead-state temperature of steam to be saturated liquid water at 25 °C. (a) Write the mass, energy, entropy, and exergy balance equations, (b) find the turbine work rate, and (c) calculate the entropy generation rate and exergy destruction rate.

Solution

a) Write the mass, energy, entropy, and exergy balance equations.

$$\text{MBE}: \dot{m}_1 = \dot{m}_2 = \dot{m}$$

$$\text{EBE}: \dot{m}_1 h_1 = \dot{m}_2 h_2 + \dot{W}_{out}$$

$$\text{EnBE}: \dot{m}_1 s_1 + \dot{S}_{gen} = \dot{m}_2 s_2$$

$$\text{ExBE}: \dot{m}_1 ex_1 = \dot{m}_2 ex_2 + \dot{W}_{out} + \dot{E}x_d$$

b) Find the turbine work rate.

In order to find the work rate produced by the steam turbine we use the EBE to find the work production rate.

Using P_1 and T_1, h_1 is obtained as 3374 kJ/kg from the tables (such as Appendix B-1b) or EES software. In addition, h_2 is obtained from the following equation:

$$h_2 = h_f + xh_{fg} = 191.81\,\frac{kJ}{kg} + 0.95 \times 2392.1\,\frac{kJ}{kg} = 2464\,\frac{kJ}{kg}$$

Using the EBE, we calculate the specific power produced by the steam turbine as follows:

$$\dot{m}_1 h_1 = \dot{m}_2 h_2 + \dot{W}_{out}$$

From the MBE, $\dot{m}_1 = \dot{m}_2 = \dot{m}$ and by dividing by the mass flow rate the EBE becomes: $h_1 = h_2 + w_{out}$ with the given data of temperatures and pressures, the enthalpy values are obtained and substituted to calculate the specific work:

$$3374 = 2464 + w_{out}$$

$$w_{out} = 909.8\ kJ/kg$$

and the turbine work rate becomes:

$$\dot{W}_{out} = \dot{m} \times w_{out} = 10\,\frac{kg}{s} \times 909.8\,\frac{kJ}{kg} = \mathbf{9.098\ MW}$$

c) Calculate the entropy generation rate and exergy destruction rate.

Using P_1 and T_1, s_1 is obtained as 6.5995 kJ/kgK from the tables (such as Appendix B-1b) or EES software.

$$s_2 = s_f + xs_{fg} = 0.6492\,\frac{kJ}{kgK} + 0.95 \times 7.4996\,\frac{kJ}{kgK} = 7.7738\,\frac{kJ}{kgK}$$

Using the EnBE, one may extract the following:

$$\dot{S}_{gen} = \dot{m}(s_2 - s_1) = 10(7.7738 - 6.5995) = \mathbf{11.743\ kW/K}$$

Then, using the ExBE the exergy destroyed through the expansion process is calculated as follows:

$$\dot{m}_1 ex_1 = \dot{m}_2 ex_2 + \dot{W}_{out} + \dot{E}x_d$$

From the MBE, $\dot{m}_1 = \dot{m}_2 = \dot{m}$ and by dividing by the mass flow rate, the ExBE becomes:

$$ex_1 = ex_2 + w_{out} + ex_d$$

$$ex_d = 1412 - 151.1 - 909.8 = 350.7\ kJ/kg$$

$$\dot{E}x_d = \dot{m} \times ex_d = 10\ \frac{kg}{s} \times 350.7\ \frac{kJ}{kg} = \mathbf{3.51\ MW}$$

Example 5.11 13 kg/s of steam at 873.15 K and 8000 kPa expands in a steam turbine, as shown in Figure 5.13, to saturated vapor at 300 kPa; 10% of it is taken outside the turbine and the rest is further expanded to a saturated liquid vapor mixture with a quality of 85% and pressure of 10 kPa. Neglect the changes in kinetic and potential energies and assume a total heat loss of 1000 kW. Take the surface (immediate boundary) temperature of the turbine as 77 °C. (a) Write the mass, energy, entropy, and exergy balance equations, (b) calculate the thermal exergy loss, and (c) find the entropy generation rate and exergy destruction rate accordingly.

Solution

a) Write the mass, energy, entropy, and exergy balance equations.

$$\text{MBE}: \dot{m}_1 = \dot{m}_2 + \dot{m}_3$$

where $\dot{m}_2 = 0.1 \times \dot{m}_1$ as defined in the problem.

$$\text{EBE}: \dot{m}_1 h_1 = \dot{m}_2 h_2 + \dot{m}_3 h_3 + \dot{Q}_l + \dot{W}_{out}$$

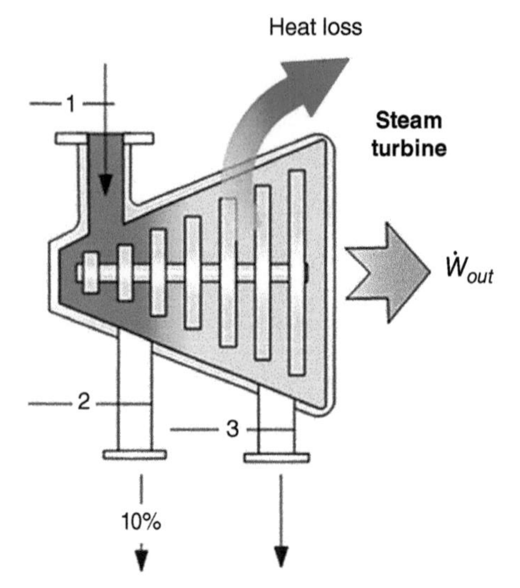

Figure 5.13 A schematic sketch of the steam turbine with two exits for Example 5.11.

where the changes in kinetic and potential energies are negligible.

$$\text{EnBE}: \dot{m}_1 s_1 + \dot{S}_{gen} = \dot{m}_2 s_2 + \dot{m}_3 s_3 + \dot{Q}_l / T_b$$

$$\text{ExBE}: \dot{m}_1 ex_1 = \dot{m}_2 ex_2 + \dot{m}_3 ex_3 + \dot{E}x_{\dot{Q}_l} + \dot{W}_{out} + \dot{E}x_d$$

b) Calculate the thermal exergy loss.

The exergy content of the heat lost from the turbine to the surroundings is calculated as follows:

$$\dot{E}x_{\dot{Q}_l} = \left(1 - \frac{T_o}{T_b}\right)\dot{Q}_l = \left(1 - \frac{298}{T_b}\right) \times 1000\ kW$$

Here, the surface (boundary) temperature for the heat loss is given at $77\,°C$ ($350\,K$). The exergy associated with the heat loss can then be calculated as follows:

$$\dot{E}x_{\dot{Q}_l} = \left(1 - \frac{298\ K}{350\ K}\right) \times 1000\ kW = \mathbf{148.57\ kW}$$

c) The entropy generation rate can be calculated from the EnBE as follows:

$$\dot{S}_{gen} = \dot{m}_2 s_2 + \dot{m}_3 s_3 - \dot{m}_1 s_1 + \frac{\dot{Q}_l}{T_b}$$

The problem has provided enough state point properties in order to obtain $s_1 = 7.022 \frac{kJ}{kgK}$, $s_2 = 6.992 \frac{kJ}{kgK}$, and $s_3 = 7.024 \frac{kJ}{kgK}$. Substituting into the above equation:

$$\dot{S}_{gen} = 1.3 \frac{kg}{s}\left(6.992 \frac{kJ}{kgK}\right) + 11.7 \frac{kg}{s}\left(7.024 \frac{kJ}{kgK}\right) - 13 \frac{kg}{s}\left(7.024 \frac{kJ}{kgK}\right) + \frac{1000\ kW}{350\ K}$$

$$\dot{S}_{gen} = \mathbf{2.838\ kW/K}$$

After the entropy generation rate is calculated, one can now calculate the exergy destruction using the following relation:

$$\dot{E}x_d = \dot{S}_{gen} T_0$$

$$\dot{E}x_d = 2.838\ kW = K \times 298\ K$$

$$\dot{E}x_d = \mathbf{845.7\ kW}$$

Example 5.12 The hot exhaust gases leaving an internal combustion engine at $400\,°C$, $150\,kPa$, and at a rate of $0.8\,kg/s$ are used to produce saturated steam at $200\,°C$ in an insulated heat exchanger, as shown in Figure 5.14. Water enters the heat exchanger at the ambient temperature of $20\,°C$ and the exhaust gases leave the heat exchanger at $350\,°C$. (a) Write the mass, energy, entropy, and exergy balance equations, (b) determine the rate of steam production, and (c) calculate the entropy generation rate and exergy destruction rate in the heat exchanger.

Solution

a) Write the mass, energy, entropy, and exergy balance equations for the heat exchanger.

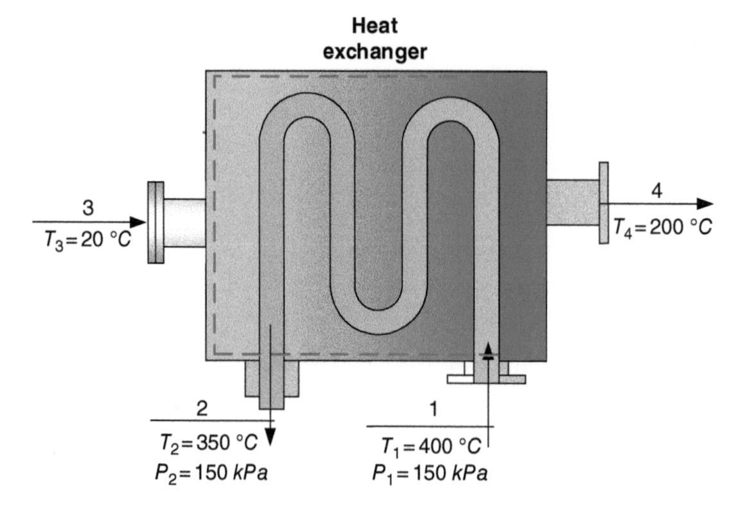

Figure 5.14 A schematic diagram of the cross-flow heat exchanger used to produce steam in Example 5.12.

$$\text{MBE}: \dot{m}_1 = \dot{m}_2 = \dot{m}_{gas}\, and\, \dot{m}_3 = \dot{m}_4 = \dot{m}_{water}$$

$$\text{EBE}: \dot{m}_1 h_1 + \dot{m}_3 h_3 = \dot{m}_2 h_2 + \dot{m}_4 h_4$$

$$\text{EnBE}: \dot{m}_1 s_1 + \dot{m}_3 s_3 + \dot{S}_{gen} = \dot{m}_2 s_2 + \dot{m}_4 s_4$$

$$\text{ExBE}: \dot{m}_1 ex_1 + \dot{m}_3 ex_3 = \dot{m}_2 ex_2 + \dot{m}_4 ex_4 + \dot{E}x_d$$

Here, we denote the inlet and exit states of exhaust gases by (1) and (2) and that of the water by (3) and (4).

b) The properties of water are obtained from the steam tables (such as Appendix B-1a) or EES database to be:

$$T_3 = 20\ ^oC, \text{liquid} \rightarrow h_3 = 83.91\ \frac{kJ}{kg},\ s_3 = 0.29649\ \frac{kJ}{kgK}$$

$$T_4 = 200\ ^oC, \text{saturated vapor} \rightarrow h_4 = 2792.0\ \frac{kJ}{kg},\ s_4 = 6.4302\ \frac{kJ}{kgK}$$

An energy balance on the heat exchanger gives the rate of steam production as follows:

$$\dot{m}_1 h_1 + \dot{m}_3 h_3 = \dot{m}_2 h_2 + \dot{m}_4 h_4$$

$$\dot{m}_a c_p (T_1 - T_2) = \dot{m}_w (h_4 - h_3)$$

$$0.8\ \frac{kg}{s} \times 1.063\ \frac{kJ}{kgK}(400 - 350) = \dot{m}_w(2792.0 - 83.91)$$

$$\mathbf{\dot{m}_w = 0.01570\ kg/s}$$

c) The entropy generation rate can be calculated by using the EnBE:

$$\dot{m}_1 s_1 + \dot{m}_3 s_3 + \dot{S}_{gen} = \dot{m}_2 s_2 + \dot{m}_4 s_4$$

The entropy change of the system when rearranged is equal to $\dot{S}_{gen}$ as follows:

$$\dot{S}_{gen} = \dot{m}_2 s_2 + \dot{m}_4 s_4 - \dot{m}_1 s_1 - \dot{m}_3 s_3$$

The question provided enough state point properties in order to obtain all the entropy values for the state points:

$$\dot{S}_{gen} = 0.01570\,\frac{kg}{s}\left(6.13\,\frac{kJ}{kgK}\right) + 0.8\,\frac{kg}{s}\left(-0.08206\,\frac{kJ}{kgK}\right)$$

$$\boldsymbol{\dot{S}_{gen} = 0.031\ kW/K}$$

The specific exergy changes of air and water streams as they flow in the heat exchanger are:

$$\begin{aligned}\Delta ex_a &= c_p\,(T_2 - T_1) - T_o\,(s_2 - s_1)\\ &= 1.063\ kJ/kgK(-50\ K) - 293\ K(-0.08206\ kJ/kgK)\\ &= -29.106\ kJ/kg\end{aligned}$$

$$\begin{aligned}\Delta ex_w &= (h_4 - h_3) - T_o\,(s_4 - s_3)\\ &= (2792.0 - 83.91)\ kJ/kg - 293\ K(6.4302 - 0.29649)\ kJ/kgK = 910.913\ kJ/kg\end{aligned}$$

The exergy destruction rate is determined from the ExBE as given below:

$$(\dot{m}_a ex_1 + \dot{m}_w ex_3) - (\dot{m}_a ex_2 + \dot{m}_w ex_4) - \dot{E}x_d = 0$$

Rearranging and substituting, one obtains:

$$\dot{E}x_d = \dot{m}_a \Delta ex_a + \dot{m}_w \Delta ex_w = \left(0.8\,\frac{kg}{s}\right)\times\left(-29.106\,\frac{kJ}{kg}\right) + \left(0.01570\,\frac{kg}{s}\right)\times\left(910.913\,\frac{kJ}{kg}\right)$$

$$\boldsymbol{\dot{E}x_d = 8.98\ kW}$$

Example 5.13 Air compressor, as shown in Figure 5.15, takes in air at atmospheric pressure and at a temperature of 280 *K*, and the air exiting the compressor leaves at a pressure and a temperature of 600 *kPa* and 400 *K*. The air mass flow rate of the air being compressed is 1 *kg/s*; the compressor loses heat through its walls equivalent to an amount of 40% of the total work rate running the compressor. (a) Write the mass, energy, entropy, and exergy balance equations, (b) calculate the compressor work input and heat rejection rate, and (c) find the entropy generation rate and exergy destruction rate.

Solution

One may start with some assumptions: (i) the heat losses occur in the compression process, (ii) the changes in the kinetic and potential energies of air are negligible, and (iii) the air is treated as an ideal gas.

a) Write the mass, energy, entropy, and exergy balance equations for the air compressor.

$$\text{MBE}: \dot{m}_1 = \dot{m}_2$$

$$\text{EBE}: \dot{m}_1 h_1 + \dot{W}_{in} = \dot{m}_2 h_2 + \dot{Q}_{out}$$

$$\text{EnBE}: \dot{m}_1 s_1 + \dot{S}_{gen} = \dot{m}_2 s_2 + \dot{Q}_{out}/T_b$$

$$\text{ExBE}: \dot{m}_1 ex_1 + \dot{W}_{in} = \dot{m}_2 ex_2 + \dot{E}x_d + \dot{E}x_{\dot{Q}_{out}}$$

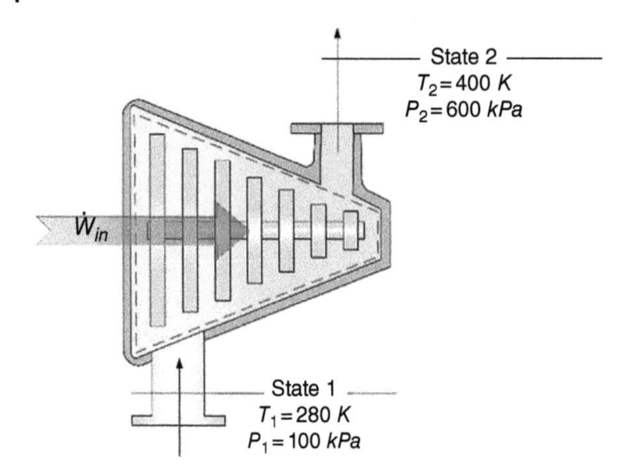

Figure 5.15 A schematic diagram of the compressor as considered in Example 5.13.

b) Calculate the compressor work input and heat rejection rate.

The power consumed during the compression process needs to be calculated in this section. Using the EBE to calculate the specific power consumed by the compressor is obtained as follows:

$$\dot{W}_{in} = \dot{Q}_{out} + \dot{m}_2 h_2 - \dot{m}_1 h_1$$

Here, the properties of the stream entering and leaving the compressor are obtained from air properties tables (such as Appendix E-1) or using the EES, which contains the database for the properties of most of the fluids through a large range of temperatures and pressures, and the properties are tabulated in Table 5.3:

Table 5.3 The properties of the inlet and exit streams.

State Point	*P* (kPa)	*T* (°C)	*h* (kJ/kg)	*s* (kJ/kgK)	*ex* (kJ/kg)
Reference state	101.3	25	298.6	5.696	0
1	100	7.0	280.5	5.637	0.509
2	600	127	401.4	5.482	166.6

The work input rate needed for the compressor is calculated as:

$$\dot{W}_{in} = 0.4 \times \dot{W}_{in} + \dot{m}_2 h_2 - \dot{m}_1 h_1$$

$$\dot{W}_{in} = 0.4 \times \dot{W}_{in} + 1 \times 401.4 - 1 \times 280.5$$

$$\mathbf{\dot{W}_{in} = 201.5\ kW}$$

Since 40% of this compressor work is lost as through heat rejection, the heat rejection rate becomes:

$$\mathbf{\dot{Q}_{out} = 0.4 \times 201.5\ kW = 80.6\ kW}$$

c) Find the entropy generation rate and exergy destruction rate.
Follows the methodology:

$$\dot{m}_1 s_1 + \dot{S}_{gen} = \dot{m}_2 s_2 + \frac{\dot{Q}_{out}}{T_b}$$

The entropy change of the system when rearranged is equal to $\dot{S}_{gen}$ as follows:

$$\dot{S}_{gen} = \dot{m}_2 s_2 - \dot{m}_1 s_1 + \frac{\dot{Q}_{out}}{T_b}$$

The question provided enough state point properties in Table 5.1 in order to obtain all the entropy values for the state points. One more thing is that the boundary or surface temperature of the turbine is not specified. Thus, it is fine to take the average temperature of the inlet and exit states, 340 K. Hence:

$$\dot{S}_{gen} = 1\,\frac{kg}{s}\left(5.482 - 5.637\,\frac{kJ}{kgK}\right) + \left(\frac{80.6\ kW}{340\ K}\right)$$

$$\boldsymbol{\dot{S}_{gen} = 0.082\ kW/K}$$

In addition, the exergy destruction rate is calculated from the ExBE as follows:

$$\dot{m}_1 ex_1 + \dot{W}_{in} = \dot{m}_2 ex_2 + \dot{E}x_d + \dot{E}x_{\dot{Q}_{out}}$$

$$1\,\frac{kg}{s} \times \left(-0.509\,\frac{kJ}{kg}\right) + 201.5\,\frac{kJ}{s} = 1\,\frac{kg}{s} \times 166.6\,\frac{kJ}{kg} + \dot{E}x_d + \left(1 - \frac{298\ K}{340\ K}\right) \times \left(80.6\,\frac{kJ}{s}\right)$$

$$\boldsymbol{\dot{E}x_d = 24.43\ kW}$$

where T_o is the ambient temperature and T_b is the boundary temperature of system, which can be approximated as the average temperature of the compressor since the heat escapes the compressor from its walls to the surrounding environment. However, the other approximation is to consider the highest temperature in the compressor. In this example, the average temperature of the compressor is taken as the sink (boundary) temperature.

Example 5.14 An electric motor–pump unit, as shown in Figure 5.16, is used to raise the pressure of water from ambient conditions to a pressure of 900 kPa and a temperature of 30 °C. The water entering the pump is at a temperature of 25 °C, at atmospheric pressure, and the mass flow rate is 0.5 kg/s. (a) Write the mass, energy, entropy, and exergy balance equations for the pump, (b) calculate the pump work rate, and (c) find the entropy generation rate and exergy destruction rate.

Solution

a) Write the mass, energy, entropy, and exergy balance equations.

$$\text{MBE}: \dot{m}_{in} = \dot{m}_{out} = \dot{m}$$

$$\text{EBE}: \dot{m}_{in} h_{in} + \dot{W}_{in} = \dot{m}_{out} h_{out}$$

$$\text{EnBE}: \dot{m}_{in} s_{in} + \dot{S}_{gen} = \dot{m}_{out} s_{out}$$

$$\text{ExBE}: \dot{m}_{in} ex_{in} + \dot{W}_{in} = \dot{m}_{out} ex_{out} + \dot{E}x_d$$

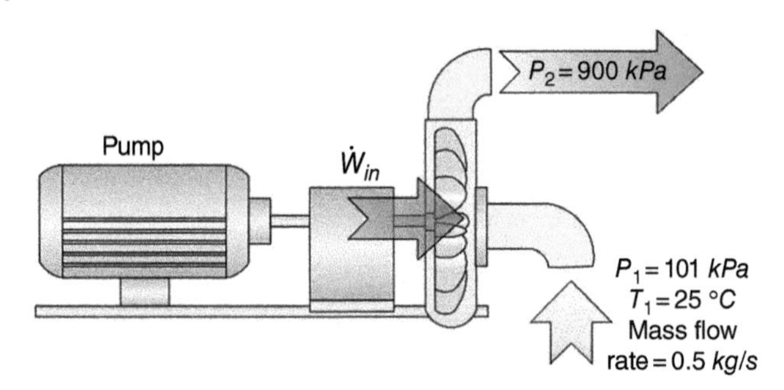

Figure 5.16 An electric motor and pump system for Example 5.14.

b) Calculate the pump work rate.

We use the EBE to calculate the specific power consumed by the compressor as follows:

$$\dot{m}_{in}h_{in} + \dot{W}_{in} = \dot{m}_{out}h_{out}$$

Here, the properties of the stream entering and leaving the pump are found from water properties tables (such as Appendix B-1d) or by using Engineering Equation Solver (EES), which contains the database for the properties of most of the fluids through a large range of temperatures and pressures. The properties of each state point are shown in Table 5.4.

The work consumption rate of the pump can be calculated by rearranging the EBE as follows:

$$\dot{W}_{in} = \dot{m}_2h_2 - \dot{m}_1h_1$$

Substituting the given mass flow rate and the specific enthalpy values from Table 5.4 into the rearranged EBE yields the following work consumption rate:

$$\dot{W}_{in} = \left(0.5 \times 126.5\,\frac{kJ}{kg}\right) - \left(0.5 \times 104.8\,\frac{kJ}{kg}\right)$$

$$\mathbf{\dot{W}_{in} = 10.82\ kW}$$

Table 5.4 The properties of the inlet and the exit streams.

Stream	*P (kPa)*	*T (°C)*	*h (kJ/kg)*	*s (kJ/kgK)*	*ex (kJ/kg)*
Reference	101.3	25	104.8	0.3669	0
State point 1	101.325	25	104.8	0.3669	0
State point 2	900	30	126.5	0.4363	0.9742

c) Determine the entropy generation and exergy destruction rates.

The entropy generation rate can be calculated by using the entropy balance equation, as follows:

$$\dot{m}_1s_1 + \dot{S}_{gen} = \dot{m}_2s_2 \rightarrow \dot{S}_{gen} = \dot{m}_2s_2 - \dot{m}_1s_1$$

The properties to their corresponding state points can be found in Table 5.4 and are used to calculate the entropy generation rate as follows:

$$\dot{S}_{gen} = 0.5\frac{kg}{s}\left(0.4363 - 0.3669\frac{kJ}{kg\ K}\right)$$

$$\mathbf{\dot{S}_{gen} = 0.0347\ kW/K}$$

The exergy destruction rate can be calculated through the exergy balance of the pump as follows:

$$\dot{m}_{in}ex_{in} + \dot{W}_{in} = \dot{m}_{out}ex_{out} + \dot{E}x_d$$

$$0.5\frac{kg}{s} \times 0\frac{kJ}{kg} + 10.82\ kW = 0.5\frac{kg}{s} \times \left(0.9742\frac{kJ}{kg}\right) + \dot{E}x_d$$

$$\mathbf{\dot{E}x_d = 10.33\ kW}$$

Example 5.15 In a vapor-compression refrigeration cycle, consider refrigerant *R134a* at a pressure of 0.8 *MPa* and in the saturated liquid state flowing through an expansion valve, as shown in Figure 5.17. The refrigerant is then throttled down to a pressure of 0.12 *MPa*. (a) Write the mass, energy, entropy, and exergy balance equations, (b) determine the temperature of the refrigerant leaving the throttle valve, and (c) calculate the specific entropy generation and specific exergy destruction during this process.

Solution

a) Write the mass, energy, entropy, and exergy balance equations.

$$\text{MBE}: \dot{m}_1 = \dot{m}_2 = \dot{m}$$

$$\text{EBE}: \dot{m}_1 h_1 = \dot{m}_2 h_2 \rightarrow h_1 = h_2$$

which makes the process isenthalpic as discussed earlier.

$$\text{EnBE}: \dot{m}_1 s_1 + \dot{S}_{gen} = \dot{m}_2 s_2$$

$$\text{ExBE}: \dot{m}_1 ex_1 = \dot{m}_2 ex_2 + \dot{E}x_d$$

b) Calculate the temperature of the refrigerant leaving the throttling valve.

By using a properties software or properties tables (such as Appendix B-3b) the enthalpy at the inlet of the throttling valve is found as follows:

$$\left.\begin{aligned} P_1 &= 0.8\ MPa \\ x_1 &= 0\ (\text{saturated liquid}) \end{aligned}\right\} h_1 = 95.5\ kJ/kg$$

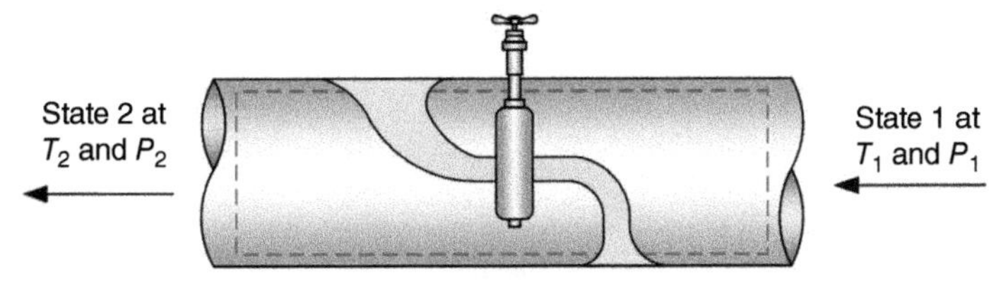

Figure 5.17 Throttling valve reducing the pressure of the refrigerant in Example 5.15.

Then by using the energy balance equation, h_2 is calculated as follows:

$$\dot{m}_1 h_1 = \dot{m}_2 h_2 \rightarrow \text{by using MBE} \rightarrow h_1 = h_2 = 95.5\ kJ/kg$$

$$\left.\begin{array}{r} P_2 = 0.12\ MPa \\ h_2 = 95.5\ \dfrac{kJ}{kg} \end{array}\right\} \mathbf{T_2 = -22.32\ ^oC}$$

c) Calculate the specific entropy generation and exergy destruction as follows:

$$\left.\begin{array}{r} P_1 = 0.8\ MPa \\ x_1 = 0\,(\text{saturated liquid}) \end{array}\right\} s_1 = 0.354\ \frac{kJ}{kgK}$$

$$\left.\begin{array}{r} P_2 = 0.12\ MPa \\ h_2 = 95.5\ \dfrac{kJ}{kg} \end{array}\right\} s_2 = 0.3838\ \frac{kJ}{kgK}$$

$$\dot{m}_1 s_1 + \dot{S}_{gen} = \dot{m}_2 s_2$$

$$\dot{S}_{gen} = \dot{m}_2 s_2 - \dot{m}_1 s_1$$

$$\mathbf{s_{gen}} = s_2 - s_1 = 0.3838 - 0.354 = \mathbf{0.0298\ \frac{kJ}{kgK}}$$

$$\mathbf{ex_d} = T_o s_{gen} = 298\ K \times 0.0298\ \frac{kJ}{kgK} = \mathbf{8.88\ \frac{kJ}{kg}}$$

Example 5.16 Propane enters a heat exchanger, as shown in Figure 5.18, at 4 °*C* at a mass flow rate of 32.6 *kg/s* to be heated with exhaust gases for later use at a constant pressure of 1 *MPa*. The exhaust gases flow at a rate of $2.484 \times 10^5\ kg/h$ and a temperature of 400 °*C*. Treat both propane and exhaust gases as ideal gases. (a) Write the mass, energy, entropy, and exergy balance equations, (b) calculate the exit temperatures of propane and exhaust air, (c) calculate the heat rate transferred to the propane, and (d) determine the entropy generation rate and exergy destruction rate.

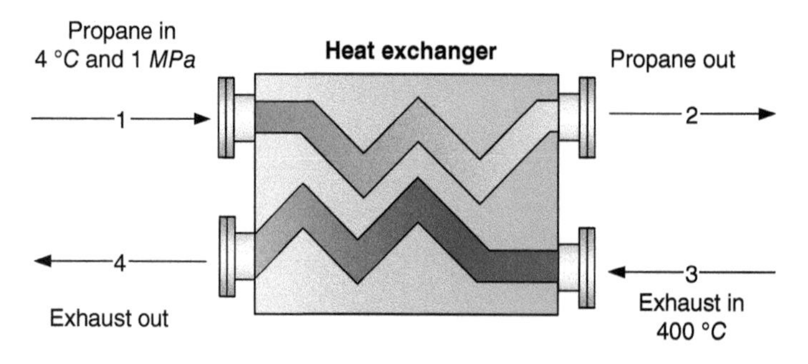

Figure 5.18 Schematic of the heat exchanger discussed in Example 5.16.

Solution

a) Write mass, energy, entropy, and exergy balance equations.

$$\text{MBE}: \dot{m}_1 = \dot{m}_2 = \dot{m}_p, \dot{m}_3 = \dot{m}_4 = \dot{m}_a$$

$$\text{EBE}: \dot{m}_1 h_1 + \dot{m}_3 h_3 = \dot{m}_2 h_2 + \dot{m}_4 h_4$$

$$\text{EnBE}: \dot{m}_1 s_1 + \dot{m}_3 s_3 + \dot{S}_{gen} = \dot{m}_2 s_2 + \dot{m}_4 s_4$$

$$\text{ExBE}: \dot{m}_1 ex_1 + \dot{m}_3 ex_3 = \dot{m}_2 ex_2 + \dot{m}_4 ex_4 + \dot{E}x_d$$

b) Calculate the exit temperatures of the propane and exhaust air.

From property tables or from property database software one can find the saturation temperature of propane.

$$\boldsymbol{T_2 = T_{sat@P=1\ MPa} = 26.4\ ^oC}$$

As for the exhaust gases one can find the temperature using the energy balance equations as follows:

$$\dot{m}_2 h_2 - \dot{m}_1 h_1 = \dot{m}_3 h_3 - \dot{m}_4 h_4$$

$$\dot{m}_p c_p (T_2 - T_1) = \dot{m}_a c_p (T_3 - T_4)$$

By using a properties software or properties tables (such as Appendix H-1) the specific heat of propane is found to be, $c_p = 2.632 \frac{kJ}{kgK}$

$$32.6 \frac{kg}{s} \times 2.632 \frac{kJ}{kgK} \times (299.4 - 277)K = 2.484 \times 10^5 \frac{kg}{h} \times \frac{1\ h}{3600\ s} \times 1.005 \frac{kJ}{kgK} (400 + 273 - T_4)\ K$$

$$\boldsymbol{T_4 = 372.24\ ^oC}$$

c) Calculate the heat rate transferred to the propane.

$$\dot{m}_3 h_3 = \dot{m}_4 h_4 + \dot{Q}_{out} \rightarrow \dot{m}_3 h_3 - \dot{m}_4 h_4 = \dot{Q}_{out} \rightarrow \dot{Q}_{out} = \dot{m}_a c_p (T_3 - T_4)$$

$$\dot{Q}_{out} = 2.484 \times 10^5 \frac{kg}{h} \times \frac{1\ h}{60\ min} \times \frac{1\ min}{60\ s} \times 1.005 \frac{kJ}{kgK} (400 - 372.24)\ K$$

$$\boldsymbol{\dot{Q}_{out} = 1922\ kW = 1.92\ MW}$$

d) Calculate the entropy generation rate and exergy destruction rate.

$$\dot{m}_1 s_1 + \dot{m}_3 s_3 + \dot{S}_{gen} = \dot{m}_2 s_2 + \dot{m}_4 s_4$$

By using a properties based software (such as EES) or properties tables, $s_1 = s_{f@T=10\,^oC} = 0.151\ kJ/kgK$ and $s_2 = s_{g\,@\,1MPa} = 6.586\ kJ/kgK$

$$\dot{S}_{gen} = \dot{m}_a \times (s_4 - s_3) + \dot{m}_p (s_2 - s_1)$$

$$\dot{S}_{gen} = \dot{m}_a \times \left(c_p\, ln\left(\frac{T_4}{T_3}\right) - R\, ln\left(\frac{P_4}{P_3}\right) \right) + \dot{m}_p (s_2 - s_1)$$

$$\dot{S}_{gen} = 2.484 \times 10^5 \frac{kg}{h} \times \frac{1\ h}{3600\ s} \times \left(1.005 \frac{kJ}{kgK} \times ln\left(\frac{372.24 + 273}{400 + 273}\right) \right)$$
$$+ 32.6 \frac{kg}{s} \times \left(2.632 \frac{kJ}{kgK} \times ln\left(\frac{26.4 + 273}{4 + 273}\right) \right) \frac{kJ}{kgK}$$

$$\boldsymbol{\dot{S}_{gen} = 3.751\ kW/K}$$

Furthermore, the exergy destruction rate for this heat exchanger is calculated as follows:

$$\dot{\boldsymbol{E}}\boldsymbol{x}_{\boldsymbol{d}} = T_o\dot{S}_{gen} = 298\ K \times 3.751\ kW/K = \mathbf{1117.89\ kW}$$

Example 5.17 Consider a two connected steam turbines with the state properties as shown in the Figure 5.19. Assume that the changes in kinetic and potential energies are negligible. (a) Write mass, energy, entropy, and exergy balance equations for each turbine separately, (b) calculate the work rates produced by both turbines, (c) find the temperature of the water leaving turbine 2, and (d) calculate the total entropy generation rate and total exergy destruction rate.

Solution

a) Write mass, energy and entropy balance equations for each turbine separately.

$$\text{MBE}: \dot{m}_1 = \dot{m}_2 = \dot{m}_3 = \dot{m}_w$$

$$\text{EBE (turbine 1)}: \dot{m}_w h_1 = \dot{m}_w h_2 + \dot{W}_{out,1} + \dot{Q}_{out,1}$$

$$\text{EBE (turbine 2)}: \dot{m}_w h_2 = \dot{m}_w h_3 + \dot{W}_{out,2} + \dot{Q}_{out,2}$$

$$\text{EBE (overall)}: \dot{m}_w h_1 = \dot{m}_w h_3 + \dot{W}_{out,t} + \dot{Q}_{out,1} + \dot{Q}_{out,2}$$

$$\text{EnBE (turbine 1)}: \dot{m}_w s_1 + \dot{S}_{gen,1} = \dot{m}_w s_2 + \dot{Q}_{out,1}/T_{av,1}$$

$$\text{EnBE (turbine 2)}: \dot{m}_w s_2 + \dot{S}_{gen,2} = \dot{m}_w s_3 + \dot{Q}_{out,2}/T_{av,2}$$

$$\text{EnBE (overall)}: \dot{m}_w s_1 + \dot{S}_{gen} = \dot{m}_w s_3 + \dot{Q}_{out,1}/T_{av,1} + \dot{Q}_{out,2}/T_{av,2}$$

$$\text{ExBE (turbine 1)}: \dot{m}_w ex_1 = \dot{m}_w ex_2 + \dot{W}_{out,1} + \dot{E}x_{\dot{Q}_{out,1}} + \dot{E}x_d$$

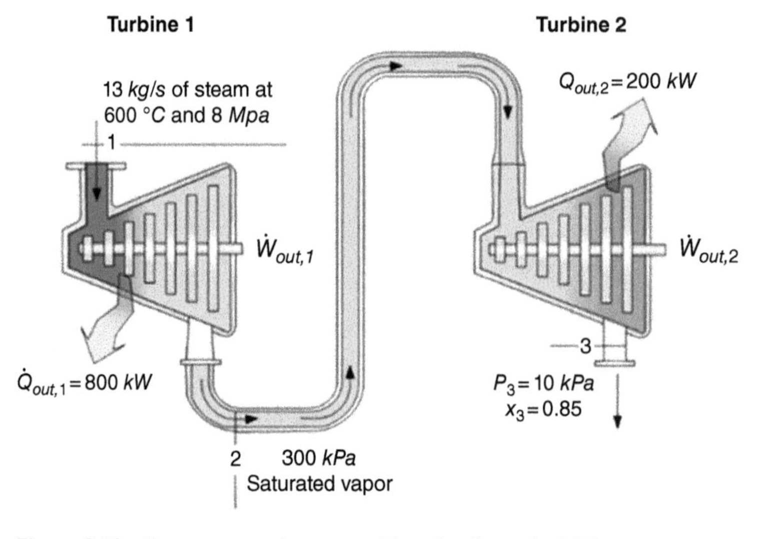

Figure 5.19 Two connected steam turbines for Example 5.17.

$$\text{ExBE (turbine 2)}: \dot{m}_w ex_2 = \dot{m}_w ex_3 + \dot{W}_{out,2} + \dot{E}x_{\dot{Q}_{out,2}} + \dot{E}x_d$$

$$\text{ExBE (overall)}: \dot{m}_w ex_1 = \dot{m}_w ex_3 + \dot{W}_{out,t} + \dot{E}x_{\dot{Q}_{out,1}} + \dot{E}x_{\dot{Q}_{out,2}} + \dot{E}x_d$$

where t stands for total and w for water.

b) Calculate the work rates produced by both turbines.

From a properties based software (such as EES) or properties tables (such as Appendix B-1c), one obtains:

$$\left.\begin{aligned} P_1 &= 8000\ kPa \\ T_1 &= 600\ ^oC \end{aligned}\right\} \begin{aligned} h_1 &= 3642.4\ kJ/kg \\ s_1 &= 7.0221\ kJ/kgK \end{aligned}$$

$$\left.\begin{aligned} P_2 &= 300\ kPa \\ x_2 &= 1 \end{aligned}\right\} h_2 = 2724.9\ kJ/kg$$

$$\dot{m}_w h_1 = \dot{m}_w h_2 + \dot{W}_{out,1} + \dot{Q}_{out,1}$$

$$\left(13\ \frac{kg}{s} \times 3642.4\ \frac{kJ}{kg}\right) = \left(13\ \frac{kg}{s} \times 2724.9\ \frac{kJ}{kg}\right) + \left(800\ \frac{kJ}{s}\right) + \dot{W}_{out,1}$$

$$\mathbf{\dot{W}_{out,1} = 11128\ kW}$$

$$h_3 = h_f + x(h_{fg})$$

$$h_3 = 191.81 + 0.85 \times (2392.1) = 2225.095\ kJ/kg$$

$$\dot{m}_w h_2 = \dot{m}_w h_3 + \dot{W}_{out,2} + \dot{Q}_{out,2}$$

$$\left(13\ \frac{kg}{s} \times 2724.9\ \frac{kJ}{kg}\right) = \left(13\ \frac{kg}{s} \times 2225.095\ \frac{kJ}{kg}\right) + \left(200\ \frac{kJ}{s}\right) + \dot{W}_{out,2}$$

$$\mathbf{\dot{W}_{out,2} = 6297\ kW}$$

The total turbine work is calculated as follows:

$$\mathbf{\dot{W}_{out,t} = \dot{W}_{out,1} + \dot{W}_{out,2} = 17425\ kW}$$

c) Find the temperature of the water leaving turbine 2.

By using a properties based software or properties tables (such as Appendix B-1b):

$$\left.\begin{aligned} P_3 &= 10\ kPa \\ x_3 &= 0.85 \end{aligned}\right\} \mathbf{T_3 = 45.82\ ^oC}$$

d) Calculate the total entropy generation rate and the total exergy destruction rate.

$$s_3 = s_f + x(s_{fg})$$

$$s_3 = 0.6492 + 0.85 \times 8.1488 = 7.57568\ kJ/kg$$

$$\dot{m}_w s_1 + \dot{S}_{gen} = \dot{m}_w s_3 + \dot{Q}_{out,1}/T_{avg,1} + \dot{Q}_{out,2}/T_{avg,2}$$

$$\dot{S}_{gen} = 13\ \frac{kg}{s} \times (7.57568 - 7.0221)\ \frac{kJ}{kgK} + \frac{800\ kW}{(873 + 406.6)/2} + \frac{200\ kW}{(318.8 + 406.6)/2}$$

$$\mathbf{\dot{S}_{gen} = 8.998\ kW/K}$$

The total exergy destruction rate then becomes

$$\mathbf{\dot{E}x_d} = T_o \dot{S}_{gen} = 298\ K \times 8.998\ kW/K = \mathbf{2681.4\ kW}$$

Example 5.18 Four different streams of water enter a mixing chamber, as shown in Figure 5.20; 10 kW of heat is rejected from this mixing chamber. Each stream comes in at a flow rate of 1 kg/s. Consider the boundary temperature as the exiting fluid temperature. (a) Write the mass, energy, entropy, and exergy balance equations, (b) identify the state of the water exiting the mixing chamber, (c) find the temperature of the water leaving the chamber, and (d) find the exergy content of the heat losses, accordingly.

Solution

a) Write the mass, energy, entropy, and exergy balance equations.

$$\text{MBE}: \dot{m}_1 + \dot{m}_2 + \dot{m}_3 + \dot{m}_4 = \dot{m}_5$$

$$\text{EBE}: \dot{m}_1 h_1 + \dot{m}_2 h_2 + \dot{m}_3 h_3 + \dot{m}_4 h_4 = \dot{m}_5 h_5 + \dot{Q}_{out}$$

$$\text{EnBE}: \dot{m}_1 s_1 + \dot{m}_2 s_2 + \dot{m}_3 s_3 + \dot{m}_4 s_4 + S_{gen} = \dot{m}_5 s_5 + \dot{Q}_{out}/T_b$$

$$\text{ExBE}: \dot{m}_1 ex_1 + \dot{m}_2 ex_2 + \dot{m}_3 ex_3 + \dot{m}_4 ex_4 = \dot{m}_5 ex_5 + \dot{E}x_{\dot{Q}_{out}} + \dot{E}x_d$$

b) Identify the state of the water exiting the mixing chamber.

$$\left.\begin{array}{l} P_1 = 100\ kPa \\ T_1 = 10\ ^oC \end{array}\right\} h_1 = 42.08\ kJ/kg$$

$$\left.\begin{array}{l} P_2 = 100\ kPa \\ T_2 = 120\ ^oC \end{array}\right\} h_2 = 2716\ kJ/kg$$

$$\left.\begin{array}{l} P_3 = 100\ kPa \\ T_3 = 300\ ^oC \end{array}\right\} h_3 = 3074\ kJ/kg$$

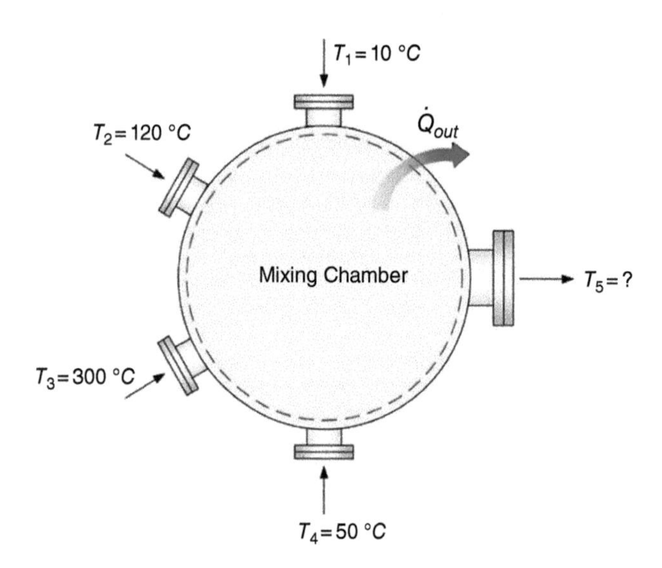

Figure 5.20 Mixing chamber with four inlets and one exit in Example 5.18.

$$\left.\begin{array}{r}P_4 = 100\ kPa\\ T_4 = 50\ ^oC\end{array}\right\} h_4 = 209.4\ kJ/kg$$

$$\left(1\frac{kg}{s}\times 42.08\frac{kJ}{kg}\right)+\left(1\frac{kg}{s}\times 2716\frac{kJ}{kg}\right)+\left(1\frac{kg}{s}\times 3074\frac{kJ}{kg}\right)+\left(1\frac{kg}{s}\times 209.4\frac{kJ}{kg}\right)$$

$$=\left(4\frac{kg}{s}\right)\times h_5 + 10\ kW$$

$$h_5 = 1507\ kJ/kg$$

$$\left.\begin{array}{r}P_5 = 100\ kPa\\ h_5 = 1507\ kJ/kg\end{array}\right\} \text{saturated liquid vapor mixture}$$

c) Find the temperature of the water leaving the chamber.

$$\left.\begin{array}{r}P_5 = 100\ kPa\\ h_5 = 1507\ kJ/kg\end{array}\right\} \boldsymbol{T_5 = 99.6\ ^oC}$$

$$h_5 = h_f + x_5\left(h_{fg}\right)$$

$$x_5 = \frac{h_5 - h_f}{h_{fg}} = \frac{1507 - 417.5}{2675 - 417.5} = 0.48$$

$$v_5 = v_f + x_5\left(v_{fg}\right)$$

$$v_5 = 0.001043 + 0.48\times(1.694 - 0.001043) = 0.814\ m^3/kg$$

d) Find the exergy content of the heat losses.

$$\dot{E}x_{\dot{Q}_{out}} = \left(1 - \frac{T_o}{T_b}\right)\dot{Q}_{out}$$

The problem states that the boundary temperature is same as the temperature of exiting stream, which then becomes:

$$T_b = T_5 = 99.6\ ^oC$$

Thus, the exergy content of the heat losses is calculated as follows:

$$\dot{E}x_{\dot{Q}_{out}} = \left(1 - \frac{298}{99.6 + 273}\right)\times 10\ kW = \mathbf{2.00\ kW}$$

5.4.2 Unsteady-state Uniform-flow Processes

USUF processes were described earlier in the previous chapter. Common examples are filling and discharging tanks, reservoirs, etc., where time dependency is important. Here, systems are presented in the forthcoming sections to illustrate system analysis and assessment through mass, energy, entropy, and exergy balance equations and efficiencies.

Example 5.19 Consider an insulated rigid tank with a volume of 1 m^3, as shown in Figure 5.21. It initially has saturated steam at a pressure of 0.5 MPa and the tank is being filled with steam from a supply line that has a temperature of 400 °C and a pressure of 9.0 MPa. The supply line fills the tank for a duration of seven minutes at a mass flow rate

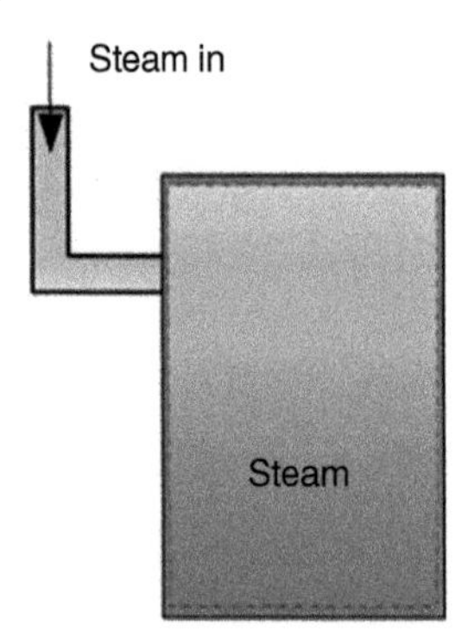

Figure 5.21 Schematic presentation of the tank being filled with steam in Example 5.19.

of 0.01 kg/s. The tank is filled with steam and reaches a pressure of 2.0 MPa. (a) Write the mass, energy, entropy, and exergy balance equations, (b) determine the tank's final temperature, and (c) calculate the entropy generation and exergy destruction of this process.

Solution

a) Write the mass, energy, entropy, and exergy balance equations.

$$\text{MBE}: m_1 + \dot{m}_{in} \times t = m_2$$

$$\text{EBE}: m_1u_1 + \dot{m}_{in}h_{in} \times t = m_2u_2$$

$$\text{EnBE}: m_1s_1 + \dot{m}_{in}s_{in} \times t + S_{gen} = m_2s_2$$

$$\text{ExBE}: m_1ex_1 + \dot{m}_{in}ex_{in} \times t = m_2ex_2 + Ex_d$$

b) Determine the tank's final temperature.

From the saturated water pressure data from tables (such as Appendix B-1b) or software databases (such as EES) we get the initial specific volume to be:

$$\left.\begin{matrix} x_1 = 1 \\ P_1 = 500\ kPa \end{matrix}\right\} v_1 = 0.37483\ \frac{m^3}{kg}$$

By knowing the tank volume, we get the initial mass as:

$$m_1 = \frac{V_1}{v_1} = \frac{1\ m^3}{0.37483\frac{m^3}{kg}} = 2.668\ kg$$

From the MBE, the final mass is calculated as:

$$m_2 = 2.668\ kg + 0.01\ \frac{kg}{s} \times 7\ min \times \frac{60\ s}{1\ min} = 6.868\ kg$$

Since the tank is rigid (constant volume), the specific volume at the final state is:

$$v_2 = \frac{V_2}{m_2} = \frac{V_1}{m_2} = \frac{1\ m^3}{6.868\ kg} = 0.1456\ \frac{m^3}{kg}$$

Now, as the final pressure and final specific volume are known and from superheated water data from the tables (such as Appendix B-1c) or software databases (such as EES), we get after interpolation:

$$\left.\begin{array}{r} v_2 = 0.1456\ \dfrac{m^3}{kg} \\ P_2 = 2\ MPa \end{array}\right\} \boldsymbol{T_2 = 377.7\ ^oC}$$

c) Calculate the entropy generation and exergy destruction of this process.
From the saturated water pressure data from tables (such as Appendix B-1b) or software databases (such as EES), we get the initial specific entropy to be:

$$\left.\begin{array}{r} x_1 = 1 \\ P_1 = 500\ kPa \end{array}\right\} s_1 = 6.8207\ \frac{kJ}{kgK}$$

Finding the superheated water data from tables (such as Appendix B-1b) or software databases (such as EES), the specific entropy for the inlet stream and the final states are:

$$\left.\begin{array}{r} T_{in} = 400\ ^oC \\ P_{in} = 9\ MPa \end{array}\right\} s_{in} = 6.2876\ \frac{kJ}{kgK}$$

$$\left.\begin{array}{r} v_2 = 0.1456\ \dfrac{m^3}{kg} \\ P_2 = 2\ MPa \end{array}\right\} s_2 = 7.053\ \frac{kJ}{kgK}$$

Substituting the above values with the masses and the inlet mass flow rate into the EnBE, the entropy generation of this process can be evaluated as:

$$S_{gen} = m_2 s_2 - m_1 s_1 - \dot{m}_{in} s_{in} \times t$$

$$S_{gen} = 6.868\ kg \times 7.053\ \frac{kJ}{kgK} - 2.668\ kg \times 6.8207\ \frac{kJ}{kgK} - 0.01\ \frac{kg}{s} \times 6.2876\ \frac{kJ}{kgK} \times 7\ \text{min} \times \frac{60\ s}{1\ \text{min}}$$

$$\boldsymbol{S_{gen} = 3.834\ \frac{kJ}{K}}$$

Assuming the environment temperature to be 298 K, the exergy destruction of this process can be calculated using the entropy generation as:

$$\boldsymbol{Ex_d} = T_o S_{gen} = 298\ K \times 3.834\ \frac{kJ}{K} = \boldsymbol{1143\ kJ}$$

Example 5.20 A 0.5 m^3 solid tank, as shown in Figure 5.22, contains saturated liquid water at 180 °C. Water is drained from the bottom of the tank when the valve at the bottom is opened. The temperature of the water is kept constant by transferring heat to the water from a source at a temperature of 330 °C. Determine the amount of transferred heat when the water is drained at a mass flow rate of 3.69 kg/s for one minute. Take the environmental temperature and pressure to be 298 K and 100 kPa, respectively. (a) Write the mass, energy, entropy, and exergy balance equations, (b) find the amount of transferred heat, and (c) determine the entropy generation of this process and find the exergy destruction of this process.

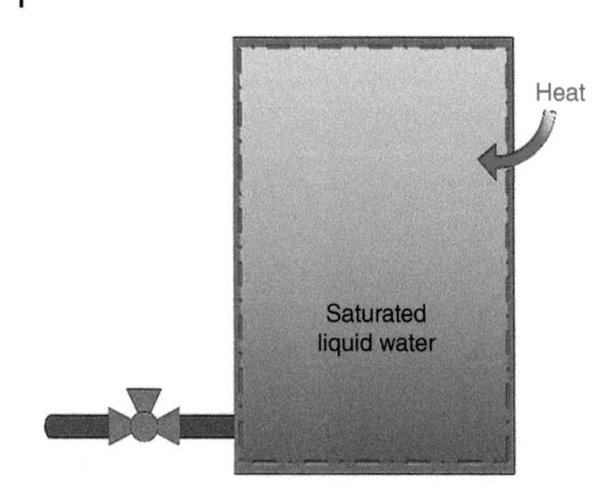

Figure 5.22 Schematic diagram of the tank being discharged for Example 5.20.

Solution

a) Write the mass, energy, entropy, and exergy balance equations.

$$\text{MBE}: m_1 = \dot{m}_{out} \times t + m_2$$

$$\text{EBE}: m_1 u_1 + Q_{in} = m_2 u_2 + \dot{m}_{out} h_{out} \times t$$

$$\text{EnBE}: m_1 s_1 + \frac{Q_{in}}{T_s} + S_{gen} = m_2 s_2 + \dot{m}_{out} s_{out} \times t$$

$$\text{ExBE}: m_1 ex_1 + Q_{in}\left(1 - \frac{T_o}{T_s}\right) = m_2 ex_2 + \dot{m}_{out} ex_{out} \times t + Ex_d$$

b) Determine the amount of transferred heat to the tank.

From saturated water temperature tables (such as Appendix B-1a) or databases, we get:

$$\left.\begin{array}{r} x_1 = 0 \\ T_1 = 180\ °C \end{array}\right\} v_1 = 0.001127\ \frac{m^3}{kg}$$

$$\left.\begin{array}{r} x_1 = 0 \\ T_1 = 180\ °C \end{array}\right\} u_1 = 761.9\ \frac{kJ}{kg}$$

$$\left.\begin{array}{r} x_{out} = 0 \\ T_{out} = 180\ °C \end{array}\right\} h_{out} = 763.1\ \frac{kJ}{kg}$$

The initial and the final masses in the tank are:

$$m_1 = \frac{V_1}{v_1} = \frac{0.5\ m^3}{0.001127\ m^3/kg} = 443.7\ kg$$

$$m_2 = m_1 - \dot{m}_{out} \times t = 443.7\ kg - 3.69\ \frac{kg}{s} \times 60\ s = 221.8\ kg$$

Now, we determine the final internal energy:

$$v_2 = \frac{V_2}{m_2} = \frac{0.5\ m^3}{221.8\ kg} = 0.002254\ \frac{m^3}{kg}$$

$$\left.\begin{array}{r} v_2 = 0.002254\,\dfrac{m^3}{kg} \\ T_2 = 180\ °C \end{array}\right\} u_2 = 772.6\,\frac{kJ}{kg}$$

Hence, the heat transfer to the tank by applying the EBE is:

$$Q_{in} = m_2u_2 + \dot{m}_{out}h_{out} \times t - m_1u_1$$

$$Q_{in} = (221.8\ kg) \times \left(772.6\,\frac{kJ}{kg}\right) + 3.69\,\frac{kg}{s} \times \left(763.1\,\frac{kJ}{kg}\right) \times 60\ s - (443.7\ kg) \times \left(761.9\,\frac{kJ}{kg}\right)$$

$$\mathbf{Q_{in} = 2563\ kJ}$$

c) Calculate the entropy generation and find the exergy destruction of this process.
First, find the specific entropy of the initial, final, and draining states as follows:

$$\left.\begin{array}{r} x_1 = 0 \\ T_1 = 180\ °C \end{array}\right\} s_1 = 2.139\,\frac{kJ}{kgK}$$

$$\left.\begin{array}{r} v_2 = 0.002254\,\dfrac{m^3}{kg} \\ T_2 = 180\ °C \end{array}\right\} s_2 = 2.165\,\frac{kJ}{kgK}$$

$$\left.\begin{array}{r} x_{out} = 0 \\ T_{out} = 180\ °C \end{array}\right\} s_{out} = 2.139\,\frac{kJ}{kgK}$$

Next, EnBE is applied and the values are substituted to give the entropy generation:

$$S_{gen} = m_2s_2 + \dot{m}_{out}s_{out} \times t - m_1s_1 - \frac{Q_{in}}{T_s}$$

$$S_{gen} = 221.8\ kg \times 2.165\,\frac{kJ}{kgK} + 3.69\,\frac{kg}{s} \times 2.139\,\frac{kJ}{kgK} \times 60\ s - 443.7\ kg \times 2.139\,\frac{kJ}{kgK} - \frac{2563\ kJ}{603\ K}$$

$$\mathbf{S_{gen} = 1.302\,\frac{kJ}{K}}$$

In addition, we need to calculate the exergy destruction of the process. Before evaluating the specific exergy for the initial, final, and draining states, the ambient state specific internal energy, enthalpy, entropy and volume of pure water must be determined.

$$\left.\begin{array}{r} P_o = 100\ kPa \\ T_o = 298\ K \end{array}\right\} u_o = 104.83\,\frac{kJ}{kg}$$

$$\left.\begin{array}{r} P_o = 100\ kPa \\ T_o = 298\ K \end{array}\right\} h_o = 104.83\,\frac{kJ}{kg}$$

$$\left.\begin{array}{r} P_o = 100\ kPa \\ T_o = 298\ K \end{array}\right\} s_o = 0.3672\,\frac{kJ}{kg}$$

$$\left.\begin{array}{l} P_o = 100\ kPa \\ T_o = 298\ K \end{array}\right\} v_o = 0.001003\ \frac{m^3}{kg}$$

So,

$$ex_1 = u_1 - u_o + P_o(v_1 - v_o) - T_o(s_1 - s_o)$$

$$ex_1 = 761.9\ \frac{kJ}{kg} - 104.83\ \frac{kJ}{kg} + 100\ kPa \times \left(0.001127\ \frac{m^3}{kg} - 0.001003\ \frac{m^3}{kg}\right)$$
$$-298\ K \times \left(2.139\ \frac{kJ}{kg\,K} - 0.3672\ \frac{kJ}{kg\,K}\right)$$

$$ex_1 = 129.09\ \frac{kJ}{kg}$$

Similarly, for the final state:

$$ex_2 = 132.15\ \frac{kJ}{kg}$$

$$ex_{out} = h_{out} - h_o - T_o(s_{out} - s_o)$$

$$ex_{out} = 763.1\ \frac{kJ}{kg} - 104.83\ \frac{kJ}{kgK} - 298K \times \left(2.139\ \frac{kJ}{kgK} - 0.3672\ \frac{kJ}{kgK}\right)$$

$$ex_{out} = 130.27\ \frac{kJ}{kg}$$

Applying the ExBE to find the exergy destruction rate:

$$Ex_d = m_1 ex_1 + Q_{in}\left(1 - \frac{T_o}{T_s}\right) - m_2 ex_2 - \dot{m}_{out} ex_{out} \times t$$

$$Ex_d = 443.7\ kg \times 129.09\ \frac{kJ}{kg} + 2563\ kJ \times \left(1 - \frac{298\ K}{603\ K}\right) - 221.8\ kg \times 132.15\ \frac{kJ}{kg}$$
$$-3.69\ \frac{kg}{s} \times 130.27\ \frac{kJ}{kg} \times 60\ s$$

$$\mathbf{Ex_d = 368.9\ kJ}$$

Example 5.21 A hot air balloon, as shown in Figure 5.23, is to be inflated to reach a volume of 100 m^3 and a pressure of 120 kPa, and it should have a temperature of 45 °C. Initially, the balloon has a volume of 30 m^3 and a pressure of 100 kPa, and is in thermal equilibrium with the environment that has a temperature of 25 °C. To inflate it, an external compressed air supply is used that supplies air at a pressure of 400 kPa and a mass flow rate of 0.01 kg/s of air at a temperature of 60 °C. Assume air to be an ideal gas. Take specific heat values at 300 K. The ambient pressure is taken as 100 kPa. (a) Write the mass, energy, entropy, and exergy balance equations, (b) calculate the time it takes to inflate the hot air balloon in minutes, (c) determine the boundary work involved with this process, (d) determine the heat loss from the balloon, (e) find the exergy destruction of this process, and (f) calculate the entropy generation of this process.

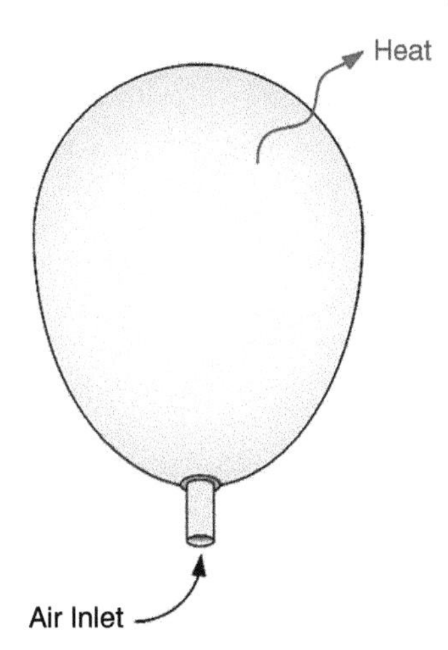

Figure 5.23 Schematic diagram of the balloon being filled with air in Example 5.21.

Solution

a) Write the mass, energy, entropy, and exergy balance equations.

$$\text{MBE}: m_1 + \dot{m}_{in} \times t = m_2$$

$$\text{EBE}: m_1u_1 + \dot{m}_{in}h_{in} \times t = m_2u_2 + W_b + Q_{loss}$$

$$\text{EnBE}: m_1s_1 + \dot{m}_{in}s_{in} \times t + S_{gen} = m_2s_2 + \frac{Q_{loss}}{T_b}$$

$$\text{ExBE}: m_1ex_1 + \dot{m}_{in}ex_{in} \times t = m_2ex_2 + W_b + Q_{loss}\left(1 - \frac{T_o}{T_b}\right) + Ex_d$$

b) Determine the time taken to inflate the hot air balloon.
This time can be found be applying the MBE along with the ideal gas law as follows:

$$m_1 = \frac{P_1V_1}{R_{air}T_1} = \frac{100\ kPa \times 30\ m^3}{0.2870\ \frac{kJ}{kgK} \times 298\ K} = 35.08\ kg$$

Similarly, for the final state, the mass is obtained as:

$$m_2 = 131.48\ kg$$

From the MBE and rearranging for time:

$$t = \frac{m_2 - m_1}{\dot{m}_{in}} = \frac{131.48\ kg - 35.08\ kg}{0.01\ \frac{kg}{s}} = 9640\ s \times \frac{1\ min}{60\ s}$$

$$\boldsymbol{t = 160.7\ min}$$

c) Calculate the boundary work involved with this process.
The boundary work for this balloon is assumed to expand linearly with pressure. This gives the boundary work as:

$$W_b = \frac{P_1 + P_2}{2}(V_2 - V_1) = \frac{100\ kPa + 120\ kPa}{2}(100\ m^3 - 30\ m^3)$$

$$\boldsymbol{W_b = 7700\ kJ}$$

d) Determine the heat loss involved with this process.

Using the EBE, the heat loss can be determined after substituting $h = c_pT$ for the inlet air and $u = c_vT$ for the air inside the balloon since air behaves as an ideal gas into it and we get

$$m_1 c_v T_1 + \dot{m}_{in} c_p T_{in} \times t = m_2 c_v T_2 + W_b + Q_{loss}$$

Rearranging for the heat loss and substituting the values in, from the respective tables (such as Appendix A-3) or databases, $c_p = 1.005\ \frac{kJ}{kgK}$, and $c_v = 0.718\ \frac{kJ}{kgK}$

$$Q_{loss} = 35.08\ kg \times 0.718 \times 298\ K + 0.01\ \frac{kg}{s} \times 1.005 \times 333\ K \times 9640\ s - 131.48\ kg \times 0.718 \times 318\ K - 7700\ kJ$$

$$\mathbf{Q_{loss} = 2048\ kJ}$$

e) To determine the exergy destruction of this process as follows:

To apply the ExBE, the T_{avg} can be taken as the average temperature between the initial and final state of air inside the hot air balloon:

$$T_{av} = \frac{T_1 + T_2}{2} = 298\ \frac{K + 318\ K}{2} = 308\ K$$

Also, the specific exergy values are determined as:

$ex_1 = 0\ \frac{kJ}{kg}$, since it has the same temperature and pressure as the environment.

$$ex_2 = u_2 - u_o + P_o(v_2 - v_o) - T_o(s_2 - s_o)$$

$$ex_2 = c_v(T_2 - T_o) + P_o\left(\frac{R_{air}\ T_2}{P_2} - \frac{R_{air}\ T_o}{P_o}\right) - T_o\left(c_p\, ln\left(\frac{T_2}{T_o}\right) - R_{air}\, ln\left(\frac{P_2}{P_o}\right)\right)$$

Then,

$$ex_2 = 1.028\ \frac{kJ}{kg}$$

$$ex_{in} = h_{in} - h_o - T_o(s_{in} - s_o)$$

$$ex_{in} = c_P(T_{in} - T_o) - T_o\left(c_p\, ln\left(\frac{T_{in}}{T_o}\right) - R_{air}\, ln\left(\frac{P_{in}}{P_o}\right)\right)$$

$$ex_{in} = 120.5\ \frac{kJ}{kg}$$

Now, the ExBE becomes when rearranging for the exergy destruction:

$$Ex_d = m_1 ex_1 + \dot{m}_{in} ex_{in} \times t - m_2 ex_2 - W_b - Q_{loss}\left(1 - \frac{T_o}{T_{av}}\right)$$

$$Ex_d = 35.08\ kg \times 0\ \frac{kJ}{kg} + 0.01\ \frac{kg}{s} \times 120.5\ \frac{kJ}{kg} \times 9640\ s - 131.48\ kg \times 1.028\ \frac{kJ}{kg} - 7700\ kJ - 2048\ kJ\left(1 - \frac{298\ K}{308\ K}\right)$$

$$\mathbf{Ex_d = 3715\ kJ}$$

f) Calculating the entropy generation of this process follows directly from the exergy destruction:

$$S_{gen} = \frac{Ex_d}{T_o} = \frac{3715\ kJ}{298\ K}$$

$$\mathbf{S_{gen} = 12.47\ \frac{kJ}{K}}$$

5.5 Exergy Efficiency

The definition of exergy efficiency is based on the second law of thermodynamics. It is also called the second law efficiency or exergy efficiency. In some sources it is also called effectiveness. Here, we use exergy efficiency and second law efficiency interchangeably. Sometimes effectiveness will be used to mean the performance of some devices, such as refrigerators and heat pumps.

Exergy efficiency may take different forms depending on the type of the system. It is denoted by η_{ex}. Here, exergy efficiency is generally expressed as:

$$\eta_{ex} = \frac{Exergy\ output}{Required\ exergy\ input} = \frac{Ex_{out}}{Ex_{in}} \tag{5.1}$$

Note that exergy output is the exergy of useful output. If there are multiple outputs, referring to cogeneration, trigeneration, etc., it will be total exergy of all useful outputs.
or in rate form:

$$\eta_{ex} = \dot{E}x_{out}/\dot{E}x_{in} \tag{5.2}$$

where

$$\dot{E}x_{in} = \dot{E}x_{out} + \dot{E}x_d \tag{5.3}$$

However, as was explained in Section 4.12 of the previous chapter, care must be taken when finding the exergy efficiency by using the following definition that is based on the exergy destruction rate:

$$\eta_{ex} = 1 - \frac{\dot{E}x_d}{\dot{E}x_{in}} \tag{5.4}$$

Equation (5.4) is not always valid as explained earlier, since it considers the exergy losses (note that exergy losses are not the same as the exergy destruction) as part of the desired exergy output. In this section, the calculation of the exergy efficiency of various energy systems, where some were introduced earlier in this chapter and others, is carried out.

Example 5.22 Consider an adiabatic steam turbine, as shown in Figure 5.24, receiving a 10 kg/s of superheated steam at a temperature of 500 °C and a pressure of 10 000 kPa. The steam turbine converts part of the energy in the steam to shaft work. The steam exits the steam turbine as a saturated vapor at a pressure of 10 kPa and a mixture quality of 0.95. Take the dead-state (reference environment) temperature of steam to be saturated liquid water at 25 °C. (Note that this example is a continuation of Example 5.10). (a) Write the mass,

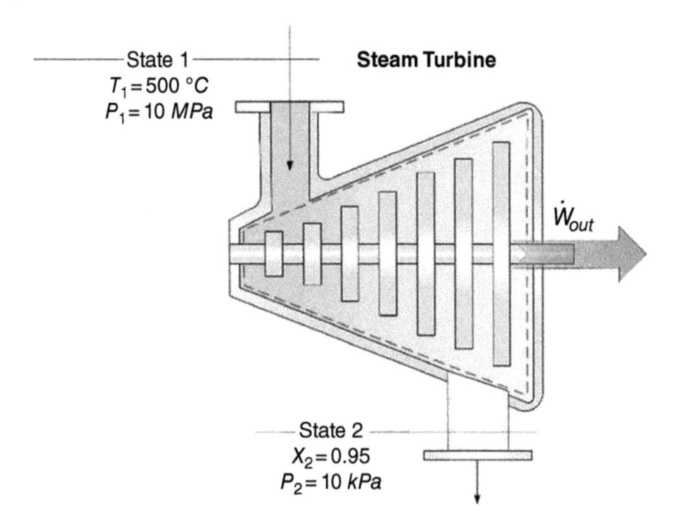

Figure 5.24 Schematic diagram of steam turbine for Example 5.22.

energy, entropy, and exergy balance equations for this turbine and (b) calculate the exergy efficiency and compare this with the energy efficiency.

Solution

a) Write the mass, energy, entropy, and exergy balance equations.

$$\text{MBE}: \dot{m}_1 = \dot{m}_2$$

$$\text{EBE}: \dot{m}_1 h_1 = \dot{m}_2 h_2 + \dot{W}_{out}$$

$$\text{EnBE}: \dot{m}_1 s_1 + \dot{S}_{gen} = \dot{m}_2 s_2$$

$$\text{ExBE}: \dot{m}_1 ex_1 = \dot{m}_2 ex_2 + \dot{W}_{out} + \dot{E}x_d$$

b) Calculate both the energy and exergy efficiencies of the turbine.

We can see that the exergy that enters the turbine is in the exergy associated with the steam entering the turbine, which is the required energy input to the system. The desired output of the turbine is the work rate, which means that the exergy efficiency of the turbine can be presented as:

$$\eta_{ex} = \frac{\dot{W}_{out}}{\dot{E}x_{in}} = \frac{\dot{W}_{out}}{\dot{m}_1 ex_1}$$

along with energy efficiency *as introduced earlier*:

$$\eta_{en} = \frac{\dot{W}_{out}}{\dot{E}n_{in}} = \frac{\dot{W}_{out}}{\dot{m}_1 h_1}$$

From Example 5.10, we have the following:

$$\dot{m} = 10\,\frac{kg}{s},\ h_1 = 3375\,\frac{kJ}{kg},\ ex_1 = 1412\,\frac{kJ}{kg},\ w_{out} = 909.8\,\frac{kJ}{kg}$$

The exergy efficiency is calculated as:

$$\eta_{ex} = \frac{10\,\frac{kg}{s} \times 909.8\,\frac{kJ}{kg}}{10\,\frac{kg}{s} \times 1412\,\frac{kJ}{kg}} = 0.644 = \mathbf{64.4\%}$$

Looking at the energy efficiency:

$$\eta_{en} = \frac{10\,\frac{kg}{s} \times 909.8\,\frac{kJ}{kg}}{10\,\frac{kg}{s} \times 3375\,\frac{kJ}{kg}} = 0.2699 = \mathbf{26.99\%}$$

One may think that exergy efficiency has always to be smaller than the corresponding energy efficiency. This is not true! When one sees in the equation that incoming flow exergy is smaller than the incoming flow energy with the same output, which is actual work, it will of course give higher exergy efficiency since its work potential is higher.

Example 5.23 An air compressor, as shown in Figure 5.25, takes in air at atmospheric pressure and a temperature of 280 *K*; the air exiting the compressor leaves at a pressure and temperature of 600 *kPa* and 400 *K*, respectively. The mass flow rate of the air being compressed is 1 *kg/s*; the compressor loses heat through its walls equivalent to 40% of the total work rate running the compressor. Calculate both energy and exergy efficiencies of the air compressor. (Note that this example is a continuation of Example 5.13)

Solution

The general definitions of energy and exergy efficiencies as are written as:

$$\eta_{en} = \frac{Desired\ energy\ output}{Total\ energy\ input} \text{ and } \eta_{ex} = \frac{Exergy\ output}{Total\ exergy\ input}$$

Use the specific enthalpy and specific exergy values (h_1 = 280.5 *kJ/kg*, h_2 = 401.4 *kJ/kg*, ex_1 = 0.5579 *kJ/kg* and ex_2 = 166.6 *kJ/kg*).

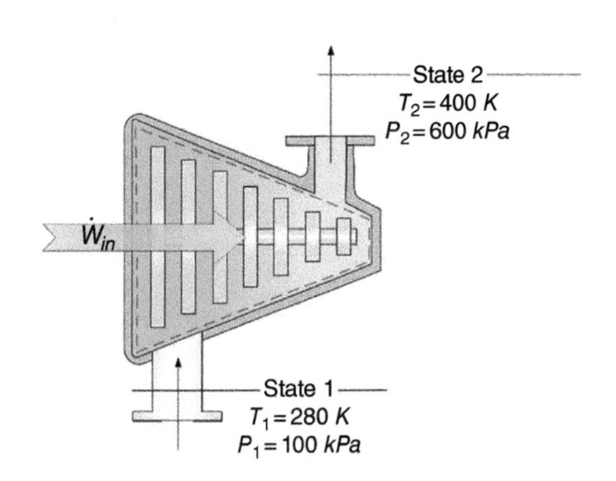

Figure 5.25 A schematic diagram of the steam turbine for Example 5.23.

The breakdown of the exergy through the compressor leads to the following definition:

$$\eta_{ex} = \frac{\dot{m}(ex_2 - ex_1)}{\dot{W}_{in}} = \frac{1\,\frac{kg}{s} \times (166.6 - 0.509)\frac{kJ}{kg}}{201.5\ kW} = 0.829 = \mathbf{82.9\%}$$

and the energy efficiency becomes:

$$\eta_{en} = \frac{\dot{m}(h_2 - h_1)}{\dot{W}_{in}} = \frac{1\,\frac{kg}{s} \times (401.4 - 280.5)\frac{kJ}{kg}}{201.5\ kW} = 0.6 = \mathbf{60.0\%}$$

One may think that exergy efficiency has always to be smaller than the corresponding energy efficiency. This is not true! When one sees in the equation that incoming flow exergy is smaller than the incoming flow energy with the same output, which is actual work, it will of course give higher exergy efficiency since its work potential is higher.

Example 5.24 This is a continuation of an earlier example problem (Example 5.16) to study its performance through energy and exergy efficiencies. 20 kg/s of water is heated at a constant pressure by 4.8×10^6 kg/h of exhaust gases, as shown in Figure 5.26. Assume the exhaust gases consist only of air where you can use ideal gas equations. Determine the exergy and energy efficiencies of the heat exchanger. Take C_p for air as 1.005 kJ/kgK. Consider a total heat loss rate of 32 MW.

Solution

The properties of the streams as well as other results are adapted from the results in Example 5.16. The energy and exergy efficiencies of the heat exchanger can be presented as:

$$\eta_{en} = \frac{Desired\ energy\ output}{Total\ energy\ input} \text{ and } \eta_{ex} = \frac{Exergy\ output}{Required\ exergy\ input}$$

which can be formulated for this heat exchanger as:

$$\eta_{ex} = \frac{\dot{m}_w \times (ex_2 - ex_1)}{\dot{m}_a \times (ex_3 - ex_4)} \text{ and } \eta_{en} = \frac{\dot{m}_w \times (h_2 - h_1)}{\dot{m}_a \times (h_3 - h_4)}$$

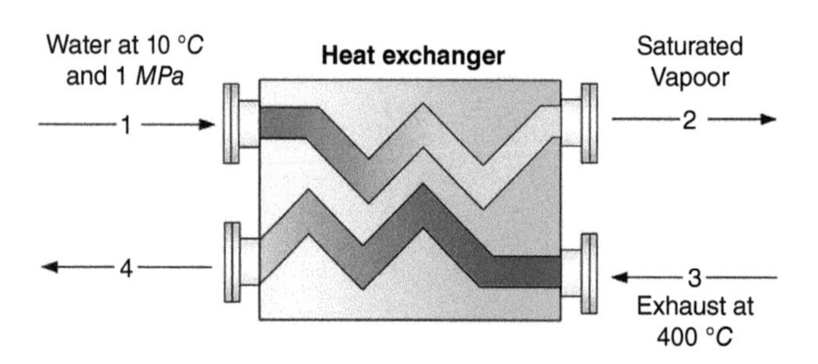

Figure 5.26 Schematic of the heat exchanger discussed in Example 5.24.

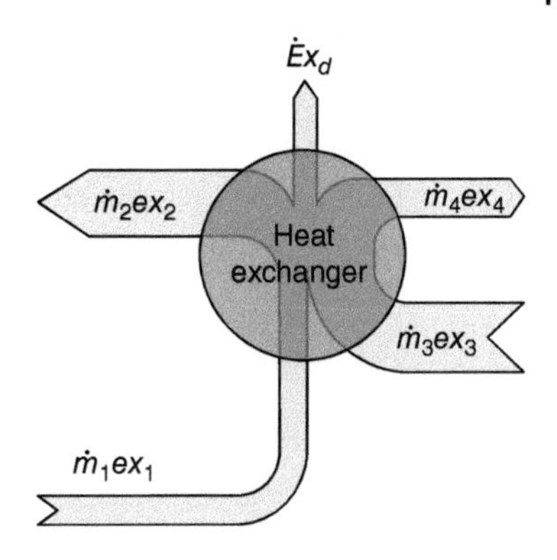

Figure 5.27 Breakdown of the exergy through the heat exchanger shown in Figure 5.26.

Based on the example description and on the breakdown of the exergy through the heat exchanger shown in Figure 5.27, we can decide that the desired output of the heat exchanger is heating the pressurized water to produce saturated vapor and the required input is the exergy supplied in the exhausts coming into the heat exchanger. Then the exergy efficiency of the heat exchanger can be referred to as:

$$\dot{m}_a(ex_3 - ex_4) = \dot{m}_a\left(c_p(T_3 - T_4) - T_o(s_3 - s_4)\right)$$

$$\dot{m}_a(ex_3 - ex_4) = 4.8 \times 10^6\ \frac{kg}{h} \times \frac{1\ h}{60\ min} \times \frac{1\ min}{60\ s}$$
$$\times \left(1.005\ \frac{kJ}{kgK}(400 - 335.2)\ K - 298\ K \times \left(1.005\ \frac{kJ}{kgK} \times \ln\left(\frac{400 + 273}{335.2 + 273}\right)\right)\right) = 46404.2\ kW$$

$$ex_2 - ex_1 = (h_2 - h_1) - T_o(s_2 - s_1)$$

$$ex_2 - ex_1 = \left(2778\ \frac{kJ}{kg} - 41.99\ \frac{kJ}{kg}\right) - 298\ K \times \left(6.586\ \frac{kJ}{kgK} - 0.151\ \frac{kJ}{kgK}\right)$$

$$ex_2 - ex_1 = 818.38\ \frac{kJ}{kg}$$

The exergy efficiency based on the specific exergies calculated above is:

$$\eta_{ex} = \frac{20\ \frac{kg}{s} \times 818.38\ \frac{kJ}{kg}}{46404.2\ kW} = 0.3527 = \mathbf{35.27\%}$$

and the corresponding energy efficiency is calculated as follows:

$$\eta_{en} = \frac{20\ kg/s\left(2778\ \frac{kJ}{kg} - 41.99\ \frac{kJ}{kg}\right)}{4.8 \times 10^6\ \frac{kg}{h} \times \frac{1\ h}{60\ min} \times \frac{1\ min}{60\ s} \times \left(1.005\ \frac{kJ}{kgK}(400 - 335.2)\ K\right)} = 0.6302 = \mathbf{63.02\%}$$

5.6 Closing Remarks

The primary focus of this chapter has been on the analysis and assessment of thermodynamic systems before getting into power generating and refrigeration cycles. In this regard, it has dwelled on writing the balance equations for mass, energy, entropy, and exergy for all types of thermodynamic systems, including closed (with no boundary change and with boundary change) and open (with SSSF and USUF separately) systems, with many examples and problems to better illustrate it and make it easier to solve practical applications. Finally, exergy efficiencies are introduced for various devices and compared with the corresponding energy efficiencies, along with some examples to illustrate the differences.

Study Questions/Problems

Concept Questions

1 Explain how to connect thermodynamics to planting and harvesting.

2 Explain why the middle S is so critical in the 3S concept given in Figure 5.2.

3 Explain the key differences between steady SSSFs and unsteady SSSFs.

4 Explain what temperature to choose in quantifying heat rejection.

5 Define and compare energy and exergy, and energy efficiency and exergy efficiency.

6 Is exergy efficiency always smaller than the corresponding energy efficiency? How do you check the correctness of the finding?

7 Explain how pressure changes in a nozzle and a diffuser and list the differences.

8 Explain how one can link exergy destruction to entropy generation.

9 What happens to thermal exergy leaving a system if the boundary temperature equals the reference temperature?

10 Which one of the following works, such as shaft work, boundary movement work and electrical work, is associated with entropy?

Problems

1 Air that behaves in ideal gas behavior is heated at a constant pressure of 200 *kPa* from a temperature of 200 to 400 *K*. Calculate the change in the internal energy, change in entropy and change in the exergy of the air using (a) average specific heat capacity, (b) specific heat capacity that is a function of temperature, (c) using data from tables (such as Appendix E-1), (d) compare (a) and (b) with the answer in part (c) in terms of error difference, and calculate both entropy generation and exergy destruction.

2 Nitrogen that behaves in ideal gas behavior is heated at a constant pressure of 200 *kPa* from a temperature of 200 to 400 *K*. Calculate the change in the internal energy, change

in entropy and change in the exergy of the air using (a) average specific heat capacity, (b) specific heat capacity that is a function of temperature, (c) using data from tables (such as Appendix E-1), (d) compare (a) and (b) with the answer in part (c) in terms of error difference, and calculate both entropy generation and exergy destruction.

3 Carbon dioxide that behaves in ideal gas behavior is heated at a constant pressure of 200 *kPa* from a temperature of 150 to 300 *K*. Calculate the change in the internal energy of the air using (a) average specific heat capacity, (b) specific heat capacity that is a function of temperature, (c) using data from the tables (such as Appendix E-3). (d) Compare (a) and (b) with the answer in part (c) in terms of error difference (e) calculate both entropy generation and exergy destruction.

4 Consider a rigid tank filled with 10 *kg* air at 100 *kPa* and 20 °*C* as shown in the figure. This tank has two shafts working at 120 *kJ* each crossing the boundary for stirring in addition to having heat supplied from a source at 600 °*C*; heat losses are up to 12% of the total heat input. The immediate boundary temperature is 45 °*C* while the final temperature of the air is 95 °*C*. (a) Write the mass, energy, entropy, and exergy balance equations, (b) calculate the amounts of heat input and heat rejection, and (c) determine the entropy generation and exergy destruction for this process. Take both the surrounding temperature and reference (dead-state) temperature as 25 °*C*.

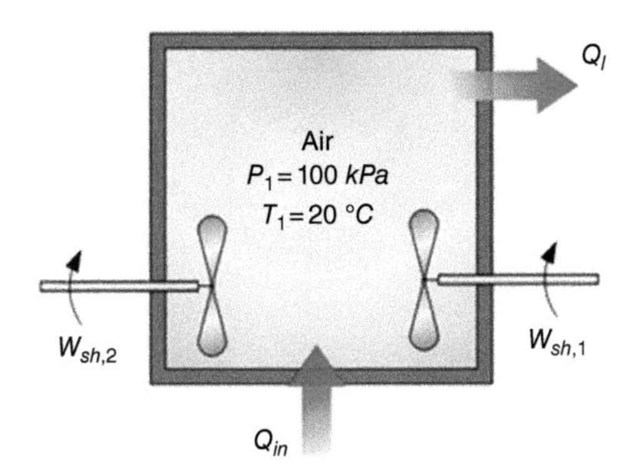

5 Consider a rigid tank filled with saturated liquid water at 1 *atm* and 25 °*C* as shown in the figure. It has a volume of 20 *L*. A total of 300 *kJ* of heat is transferred to the water, which becomes saturated vapor in the final state; 20% of the heat supplied is lost. An electrical heater runs on 1 *A* for 20 *s* while connected to 120 *V* source. The rotating shaft adds 120 *kJ* of shaft work to the water in the tank while stirring it. (a) Write the mass, energy, entropy, and exergy balance equations, (b) calculate the thermal exergy input and thermal exergy loss, and (c) determine the entropy generation and exergy destruction for this process. Take both the surrounding temperature and reference (dead-state) temperature as 50 °*C*.

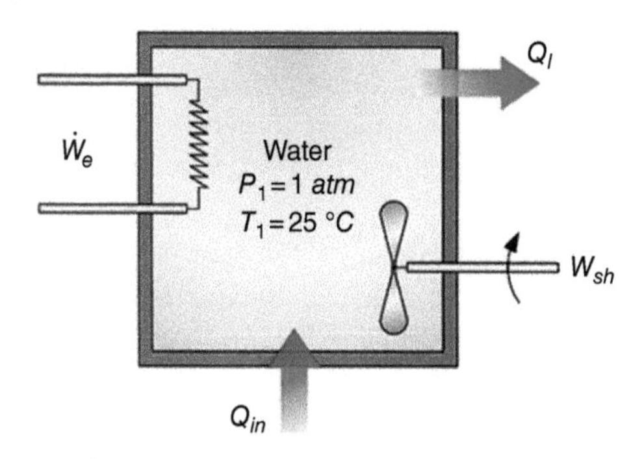

6 Consider a rigid tank filled with saturated liquid water at 1 *atm* and 20 °*C* as shown in the figure. It has a volume of 30 *L*. A total of 800 *kJ* of heat is transferred to the water where there ais a 5 *kg* steel ball and 10 kg copper ball quenched into it. 12% of the heat input is rejected. The work input required for stirring is 120 *kJ*. All (water, steel ball and copper ball) are eventually expected reach a temperature of 85 °*C* in equilibrium. (a) Write the mass, energy, entropy, and exergy balance equations, (b) calculate all heat transfer amounts, and (c) determine the entropy generation and exergy destruction for this process. Take both the surrounding temperature and reference (dead-state) temperature as 25 °*C*.

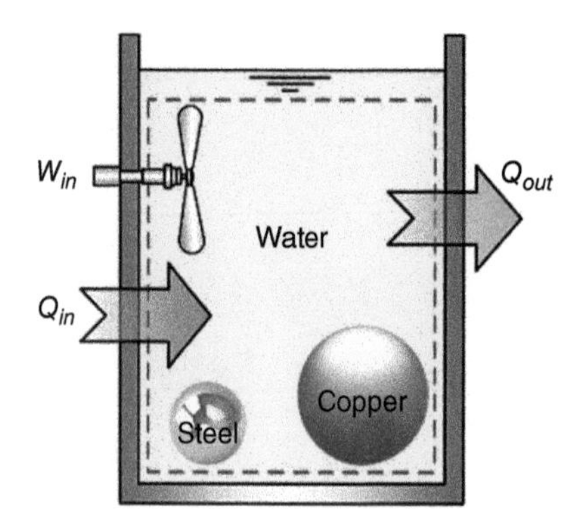

7 Refer to Problem 6 and consider thermal oil instead of water with mass of 50 *kg*. (a) Write the mass, energy, entropy and exergy balance equations (b) calculate amount of heat needed for the process (c) calculate entropy generation and (d) calculate exergy destruction.

8 Consider a rigid tank filled with helium that is heated by a heat source at a temperature of 100 °*C*. The rigid tank has a volume of 1 m^3 and contains a 1 *kg* of real gas helium. Calculate both the pressure and temperature of the helium in the rigid tank after 10 *kJ* of heat is transferred to the system while knowing that the system initially was at an ambient temperature of 25 °*C*. (a) Write the mass, energy, entropy, and exergy balance

equations for this system, (b) calculate the entropy generation, and (c) calculate the exergy destruction.

9 Consider a piston–cylinder device, as shown in the figure, which initially contains nitrogen at a pressure of 250 kPa. The gas in the piston–cylinder device receives 50 kJ of work per kg of nitrogen in the device through the rotating paddle. The gas in the device receives thermal energy (heat) from a source with a temperature of 300 °C. The piston–cylinder device operates at a constant temperature of approximately 20 °C, which makes the process an *isothermal process*. The gas volume in the device increases to four times of its original volume. (a) Write the mass, energy, entropy, and exergy balance equations for this system, (b) find the boundary work produced by the mechanism, (c) find the heat input from the source to the system, (d) calculate the entropy generation, and (e) calculate the exergy destruction.

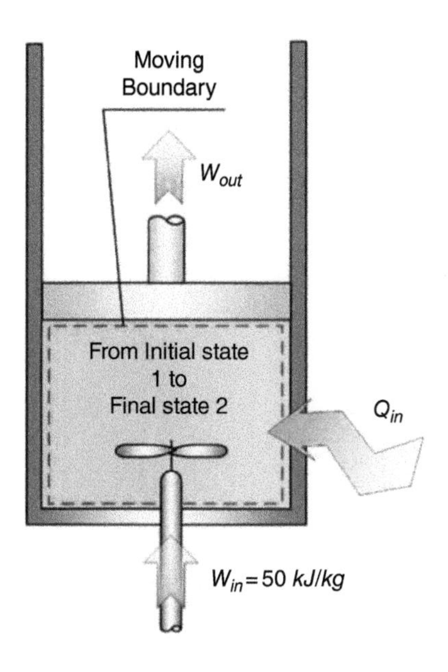

10 A piston–cylinder device with an initial volume of 0.1 m^3, as shown in the figure, contains saturated water vapor at a temperature of 100 °C. The water is compressed so that the final volume is 80% of the initial volume. 40 kJ of heat is lost to the environment through the compression process. (a) Write the mass, energy, entropy, and exergy balance equations for this system, (b) find the work input and electrical work input to the system, (c) calculate the heat rejection, (d) calculate the entropy generation, and (e) calculate the exergy destruction.

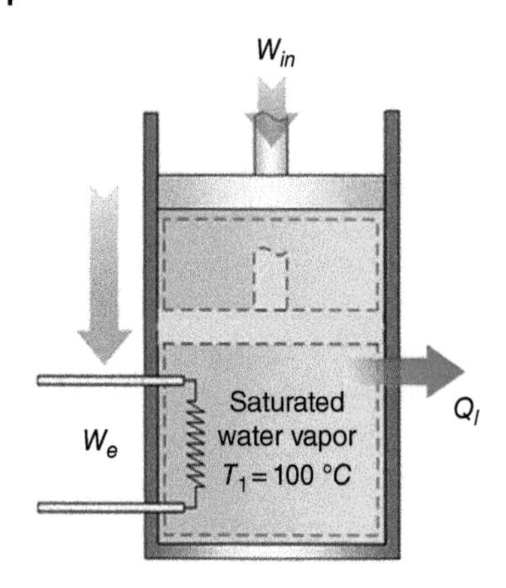

11 Consider a piston–cylinder device, as shown in the figure, which initially contains saturated liquid refrigerant $R134a$ at a temperature of $10\,°C$. The refrigerant in the piston–cylinder device receives 60 kJ of electrical work and 40 kJ of shaft work as well as heat from a source at $400\,°C$. The refrigerant is heated to become a saturated vapor at $60\,°C$ during the expansion process where the initial volume increases three times. No heat losses take place during the expansion. The surrounding temperature is the same as the reference temperature at $10\,°C$. (a) Write the mass, energy, entropy, and exergy balance equations for this system, (b) find the boundary work produced by the mechanism, (c) find the heat input from the source to the system, (d) calculate the entropy generation, and (e) calculate the exergy destruction.

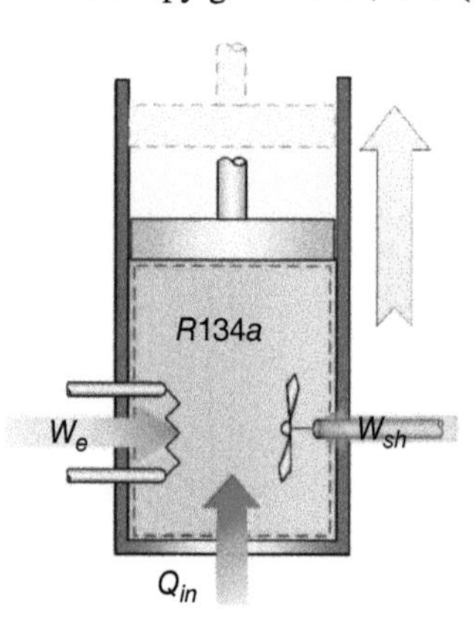

12 A balloon with a 1 kg of helium gas with a density of $0.323\,kg/m^3$ at a pressure of 200 kPa and temperature of $25\,°C$, as shown in the figure, is expanded during a process where a total of 100 kJ of heat is transferred to the gas. The final temperature of the helium becomes $65\,°C$ at constant pressure. Take the temperature of heat source as $100\,°C$ and treat helium as an ideal gas. (a) Write the mass, energy, entropy, and exergy balance equations for this expanding closed system, (b) calculate the volume increase and the amount of heat input, and (c) calculate both entropy generation and exergy destruction. Take $T_o = 25\,°C$.

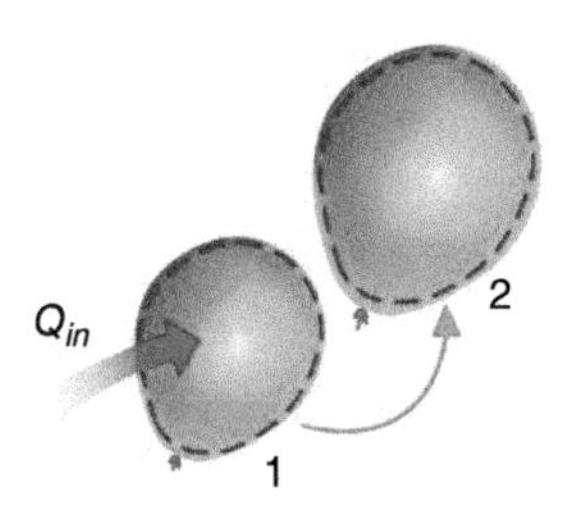

13 Reconsider Problem 12 for hydrogen gas with the density of 0.1627 kg/m^3. Write the mass, energy, entropy and exergy balance equations, (b) calculate increase in volume, (c) calculate boundary work, (d) calculate amount of heat input for the process, (e) calculate entropy generation and (f) calculate exergy destruction.

14 Consider a pipe where 1.2 kg/s of flowing hot steam enters at 200 °C, as shown in the figure. It loses some of its heat and exits at 195 °C. Assume that the steam is saturated vapor. (a) Write the mass, energy, entropy, and exergy balance equations for this pipe flow system, (b) find the amount of heat rejected, and (c) calculate the entropy generation rate and exergy destruction rate. Take the surface temperature as 100 °C and the surrounding temperature as 25 °C.

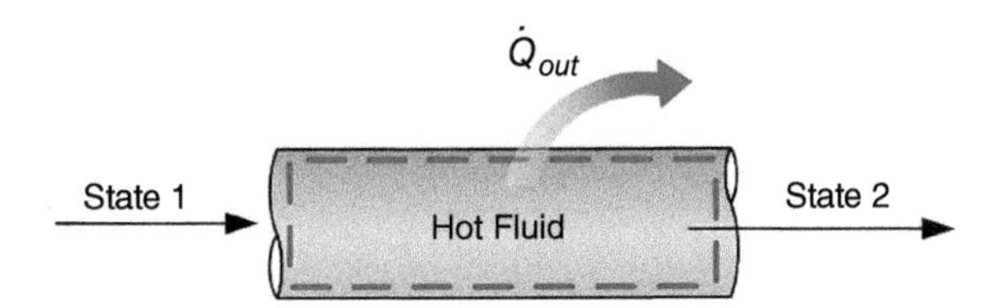

15 Consider an adiabatic nozzle with cross-sectional area of 0.4 m^2, as shown in the figure, where there is carbon dioxide entering at a velocity of 15 m/s, a temperature of 80 °C, and a pressure of 210 kPa. The gas exits at a temperature of 50 °C and a pressure of 110 kPa. (a) Write the mass, energy, entropy, and exergy balance equations for this nozzle, (b) find the exit velocity and mass flow rate of the gas, and (c) calculate the entropy generation rate and exergy destruction rate. Take the surrounding temperature as 25 °C.

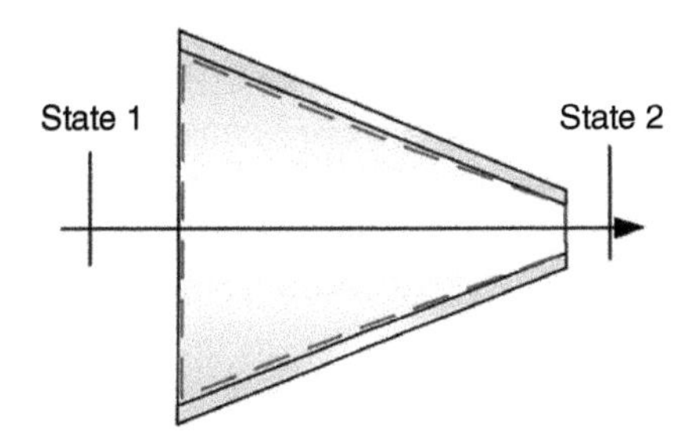

16 Consider an adiabatic diffuser with cross-sectional area of 0.1 m^2, as shown in the figure, where nitrogen enters at a velocity of 250 m/s, a temperature of 50 °C, and a pressure of 110 kPa. The gas exits at a temperature of 80 °C and a pressure of 150 kPa. (a) Write the mass, energy, entropy, and exergy balance equations for this diffuser, (b) find the exit velocity and mass flow rate of the gas, and (c) calculate the entropy generation rate and exergy destruction rate. Take the surrounding temperature as 25 °C.

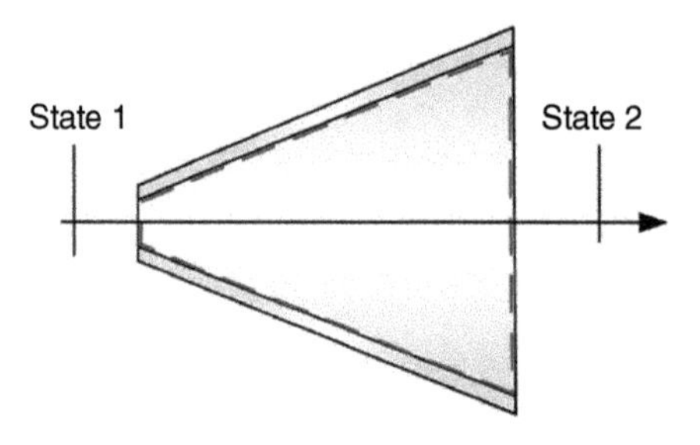

17 Consider a steam turbine, as shown in the figure, where superheated vapor enters at 500 °C and 2.5 MPa and exits at 250 kPa as saturated vapor. The mass flow rate of vapor is 10 kg/s. There is some heat rejected, accounting for about 10% of the incoming flow energy. (a) Write the mass, energy, entropy, and exergy balance equations for this turbine, (b) find the amount of heat rejected and work produced, and (c) calculate the entropy generation rate and exergy destruction rate, Take the surface temperature as 100 °C and the surrounding temperature as 25 °C.

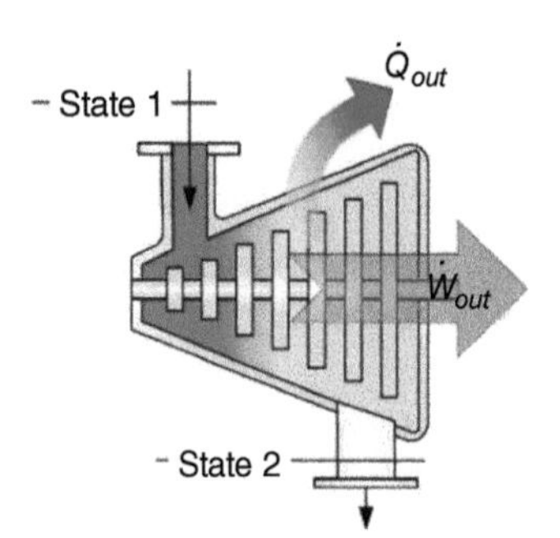

18 Consider an ammonia compressor, as shown in the figure, where the saturated ammonia vapor enters at −18 °C and 200 kPa and exits at 30 °C and 500 kPa as superheated ammonia vapor. The mass flow rate is 1 kg/s. Some heat is rejected, accounting for about 35% of the work input. (a) Write the mass, energy, entropy, and exergy balance equations for this compressor, (b) find the amount of heat rejected and work input, and (c) calculate the entropy generation rate and exergy destruction rate, Take the surface temperature as 30 °C and the surrounding temperature as 20 °C.

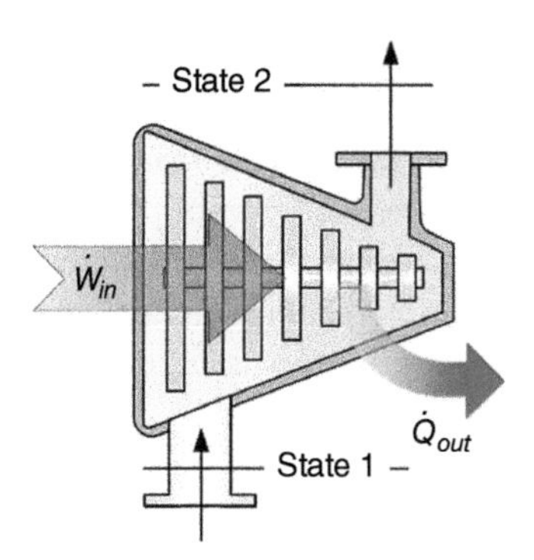

19 A water pump raises the pressure of water from 100 *kPa* to 10 *MPa*, as shown in the figure. The initial temperature of the water is 25 °*C*. (a) Write the mass, energy, entropy, and exergy balance equations for this pump, (b) find the specific work input, and (c) calculate the specific entropy generation and exergy destruction. Take the surrounding (reference) temperature as 20 °*C*.

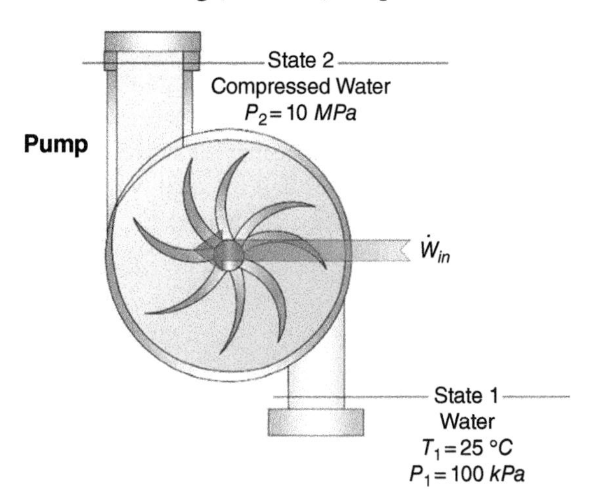

20 In a vapor-compression refrigeration cycle, consider refrigerant ammonia at a pressure of 500 *kPa* and in the saturated liquid state flowing through an expansion valve, as shown in the figure. The refrigerant is then throttled down to a pressure of 200 *kPa*. (a) Write the mass, energy, entropy, and exergy balance equations, (b) determine the temperature of the refrigerant leaving the throttle valve, and (c) calculate the specific entropy generation and specific exergy destruction during this process.

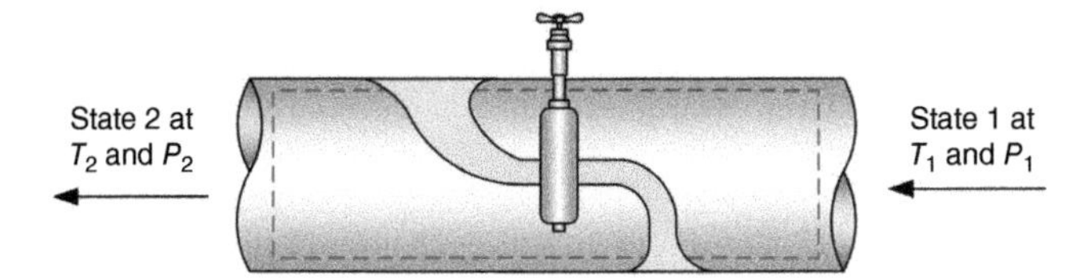

21 Consider a well-insulated heat exchanger, as shown in figure, where hot fluid (refrigerant $R134a$) enters at $40\,°C$ as a saturated vapor and leaves at the same temperature but as a saturated liquid with a mass flow rate of $1.5\,kg/s$. On the cold fluid side, tap water enters at $15\,°C$ at a flow rate of $2.1\,kg/s$. (a) Write the mass, energy, entropy, and exergy balance equations for this heat exchanger, (b) find the heat gained by the water and the exit temperature of the water, and (c) calculate the entropy generation rate and exergy destruction rate. Take the surrounding (reference) temperature as $20\,°C$.

22 Consider a well-insulated mixing chamber, as shown in figure, where a saturated water vapor stream at $110\,°C$ with a mass flow rate of $10\,kg/s$ and a saturated liquid water stream at $20\,°C$ with a flow rate of $15\,kg/s$ are mixed to give the desired output as a mixture of saturated liquid and vapor at $80\,°C$. (a) Write the mass, energy, entropy, and exergy balance equations for this mixing chamber, (b) find the quality of vapor exiting the chamber, and (c) calculate the entropy generation rate and exergy destruction rate. Take the surrounding (reference) temperature as $25\,°C$.

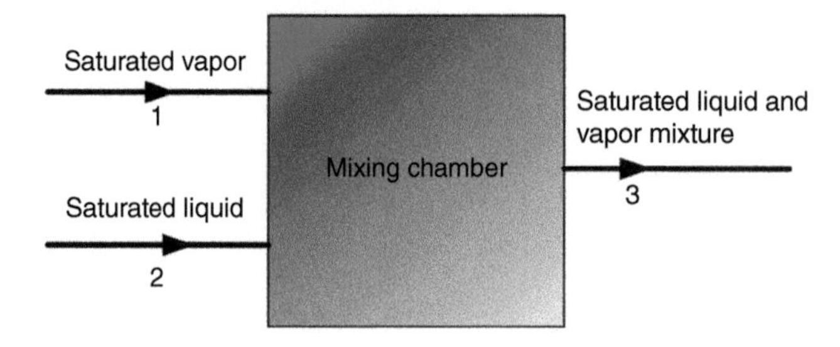

23 Reconsider problem 12 for a nozzle and calculate both the energy and exergy efficiencies.

24 Reconsider problem 13 for a diffuser and calculate both the energy and exergy efficiencies.

25 Reconsider problem 14 for a steam turbine and calculate both the energy and exergy efficiencies.

26 Reconsider problem 15 for an ammonia compressor and calculate both the energy and exergy efficiencies.

27 Reconsider problem 16 for a water pump and calculate both the energy and exergy efficiencies.

28 Reconsider problem 17 for an expansion valve and calculate both the energy and exergy efficiencies.

29 Reconsider problem 18 for a heat exchanger and calculate both the energy and exergy efficiencies.

30 Reconsider problem 19 for a mixing chamber and calculate both the energy and exergy efficiencies.

6

Power Cycles

6.1 Introduction

Energy transformation or conversion has been one of the key priorities for human beings since the industrial revolution. As everyone knows, there are energy transformations/conversions from one form to another taking place in nature with no external forces needed. The first law of thermodynamics (FLT) states that energy can be transformed from one form to another but cannot be created nor destroyed (which leads to the fact that it just changes forms). However, analyzing a process within a closed or open system is not accurate in some cases, specifically in cases where energy is being converted to work. For example, a ball could be dropped from an arbitrary height; as the shot is falling its gravitational potential energy is being transformed into kinetic energy. However, the reverse process can not naturally occur, as work needs to be done to lift the ball back to the original height. In accordance with the FLT, this process' energy is conserved therefore the law is valid. This is why the second law of thermodynamics (SLT) is needed to remove such discrepancies and false assumptions while correcting and giving a more accurate insight from a thermodynamic view. Some systems allow for a reverse process, such as a ball going back up to the height it was dropped using a system that requires an input of energy. For example, supplying electricity to a motor connected to a pulley system, which would reel the ball back to the original height. However, in the field of thermodynamics, most of the power generation systems are heat engines. Heat engines are treated as heat-driven machines that are designed for the efficient conversion of thermal or chemical energy to mechanical or electric energy. The Rankine and the Brayton cycle, or both combined, are considered to be the main power generation cycles used in industry. The Rankine cycle is a steam-driven cycle while the Brayton cycle is a gas-driven cycle. Both cycles convert thermal energy to mechanical energy and electricity. In addition to the categorization of the power cycles based on the phase of the working fluid through the cycle, the power cycles can also be categorized based whether the cycle is open or closed. Where in open cycles the working fluid is continuously renewed through each cycle, in closed cycles the working fluid is returned to its initial state to be used again in the following cycle. When the power cycle takes energy in the heat form and then converts it to power, this type of cycle is referred to as the heat engine.

This chapter discusses one of the most important applications of thermodynamics, which is power generation. Power generation is usually achieved by thermodynamics systems that run through a cycle (thermodynamic cycle) and these cycles are referred to as power cycles

Thermodynamics: A Smart Approach, First Edition. Ibrahim Dincer.

since they are used to generate power. The systems or the devices that operates through a thermodynamic cycle are referred to as engines. The power cycles can be categorized based on the phase of the working fluid; they are either gas cycles and vapor cycles. Gas cycles are those thermodynamic cycles in which the working fluid remains throughout the cycle in the gaseous phase, while in vapor cycles the working fluid changes phase between the vapor and liquid phases throughout the cycle. This chapter introduces each cycle and the balance equations of that cycle and describes how these equations are used to investigate the cycle and assess its performance using the energy and exergy efficiencies.

6.2 Carnot Concept for Power Generation

Further to our earlier discussion, the Carnot is technically a concept, although it is treated as a cycle by many; it is usually confused in many thermodynamics lectures. The Carnot concept is basically the concept of ideality and shows the maximum potential performance a power engine or a thermodynamics cycle can achieve. It essentially gives us the limits. In order to clearly illustrate how the principle of the Carnot concept is applied to power generation cycles, especially the internal combustion cycles that can be approximated as piston–cylinder devices, these cycles will be introduced one by one throughout the chapter. The result of the application of the Carnot concept on the previously mentioned power cycle is a Carnot cycle that results in the highest efficiency of any cycles of heat engines. This comes as the Carnot cycle neglects all irreversibility's such as friction, heat losses, etc. This makes the Carnot cycle a completely reversible cycle

Figure 6.1 shows multiple illustrations of the Carnot cycle using an ideal piston and its *P-v* and *T-s* diagrams as applied to heat engines in general. The Carnot cycle consists of the following processes:

- **Reversible isothermal expansion (process 1-2):** This is a process where the gas absorbs thermal energy q_{in} and expands at a constant temperature T_H. As the gas expands, work is done on its surrounding, as the cylinder head is moving. Theoretically as the air expands the gas should experience a temperature drop. However, a heat source is provided to maintain the gas's isothermal state.
- **Reversible adiabatic expansion (process 2-3):** This is a process in which the gas expands and experiences a temperature drop from T_H to T_L. The gas continues to expand however; the thermal heat source is not provided anymore but replaced by insulation. The piston in this process is assumed to be frictionless. Furthermore, the process is assumed to be in quasi-equilibrium.
- **Reversible isothermal compression (3-4):** This is a process where the piston goes back to its original position as the external forces from the surroundings are doing work on the gas. The insulation in this process is removed and a heat sink at a temperature of T_l. is provided to the cylinder's head. The heat sink ensures the gas isothermal state as its compressed.
- **Reversible adiabatic compression (process 4-1):** This is a process where insulation replaces the heat sink, and as the piston continues to compress the gas, the temperature of the gas rises to its original temperature of T_h.

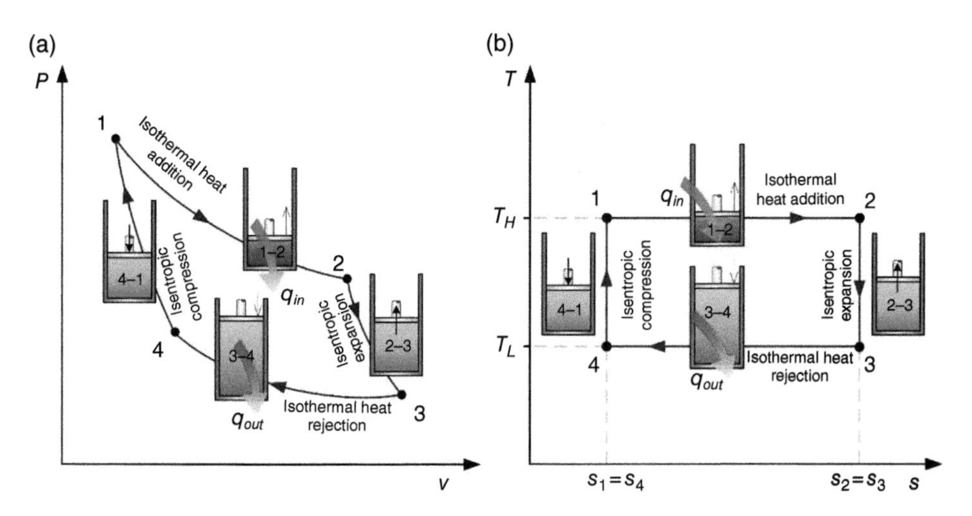

Figure 6.1 Illustrations of (a) *P-v* diagram and (b) the *T-s* diagram of the Carnot cycle applied to heat engines using an ideal piston based on a closed system.

Both compression and expansion processes of the piston as previously mentioned are isothermal in processes 1-2 and 3-4 and adiabatic in the remaining processes. The temperatures at state 1 and 2 are assumed to be equal to T_H and the temperatures at state 3 and 4 are assumed to be equal to T_L. The area inside the *P-v* diagram represents the network output of the system along with the quasi-static equilibrium boundary. During the processes 1-2-3, work is being done by the gas onto its surroundings as the gas is expanding, as seen in Figure 6.1. As for process 3-4-1, work is being done on the gas as the gas is being compressed and the correlating decreasing volume can be seen in the figure. Here, the efficiency of the Carnot cycle as mentioned previously is the highest of any other cycle, as it is assumed to consist of fully reversible processes. The thermal efficiency of a Carnot heat engine can then be defined as the heat transferred to the engine from the heat source Q_H and the heat rejected by the system to the heat sink Q_L.

In conjunction with this, it is important to state that the Carnot cycle is completely reversible, allowing all the process that it experiences to be reversed. If the processes are reversed the cycle becomes the Carnot refrigeration cycle. This is done by reversing when the heat is rejected and absorbed by the cycle.

6.3 Heat Engines

This chapter is technically about power generating cycles, which are essentially known as heat engines. In addition to the categorization of the power cycle based on the phase of the working fluid through the cycle, the power cycles can also be categorized based on whether the cycle is an open or closed system; the categorization of the power cycles is show in Figure 6.2, where in open cycles the working fluid is continuously renewed through each cycle while in closed cycles the working fluid is returned to its initial state to be used again

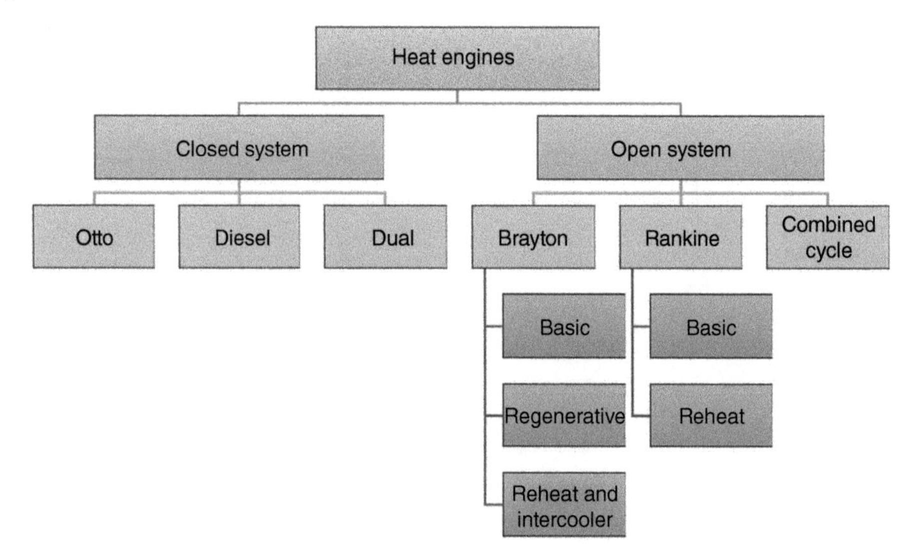

Figure 6.2 Classification of heat engines based on whether the cycle is an open or closed system.

in the following cycle. When the power cycle takes thermal energy in the form of heat and then converts it to power this type of cycle is referred to as a heat engine.

Furthermore, one can see in Figure 6.2 that heat engines can be classified based on whether the cycle components or the cycle as a whole is approximated as a closed or an open system. Those that the cycle as a whole is made up from a single system that operates in a closed system behavior, specifically as a piston–cylinder device and the cycle is such that the device undergoes different processes in sequence, are the Otto, Diesel, and the combination of both the Dual cycles. However, other cycles are made up of a set of open systems that makes up the whole cycle, in which the systems have multiple devices; each device continuously performs a specific action throughout the cycle. In this chapter, the cycles included in Figure 6.2 are covered and discussed in detail along with their *T-s* and *P-v* diagrams and applications; numerous example problems are provided to better understand the concepts and the way balance equations are written.

A heat engine, as outlined in the introduction, is a thermodynamic system that is turned into cycle. It essentially receives energy in the form of heat and converts it to power; for example, heat engines such as the Otto cycle, Diesel cycle, Rankine, and Brayton cycles are common types. If one looks at heat engines or what can also be referred to as power cycles from an input and output point of view, one can present these cycles in a simple in and out form as shown in the schematic in Figure 6.3 for both energy (Figure 6.3a) and exergy flows (Figure 6.3b).

Consider a heat engine shown in Figure 6.3, where the heat engine receives a total thermal energy of Q_H from a heat source at a temperature of T_H. Some part of this total amount of heat received by the heat engine from the heat source is converted to useful net work output (W_{net}), while the remaining part of that heat, which is denoted in Figure 6.3a as Q_L, is rejected to a heat sink at a temperature of T_L. In this regard, while Figure 6.3a presents the energy interactions within the heat engine, Figure 6.3b shows the exergy interactions within the heat engine in terms of inputs and outputs. The heat engine receives an exergy

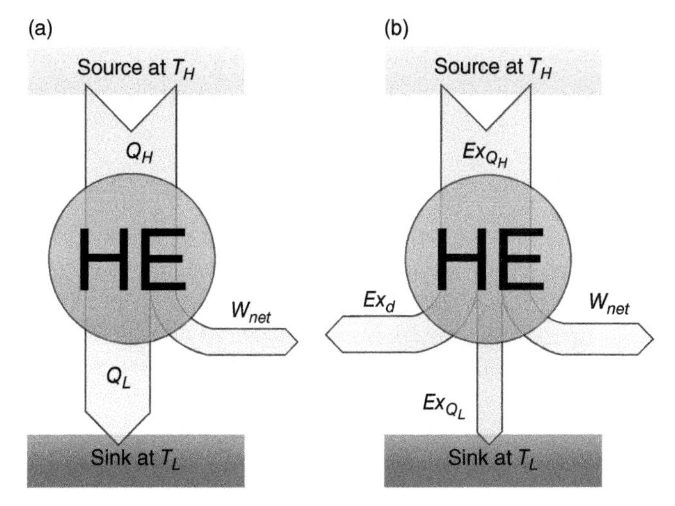

Figure 6.3 Schematic diagram of the energy and exergy interactions of a heat engine in terms of inputs and outputs.

content associated with the heat transfer with a magnitude of $\dot{E}x_{\dot{Q}_H}$; part of that exergy is converted to work output with the remaining part of it destroyed and rejected to the environment (Ex_{Q_L}).

6.3.1 Performance Assessment

The performance of heat engines is measured through the energy and exergy efficiencies as follows:

$$\eta_{en} = \frac{W_{net}}{Q_H} \tag{6.1}$$

$$\eta_{ex} = \frac{W_{net}}{Ex_{Q_H}} \tag{6.2}$$

Here, η denotes the energy efficiency, ψ presents the exergy efficiency, and the exergy content of the heat transfer from the high temperature heat source is calculated as was presented in earlier chapters:

$$Ex_{Q_H} = \left(1 - \frac{T_o}{T_H}\right) Q_H \tag{6.3}$$

The energy efficiency of a heat engine can be represented in terms of the heat transfers only through the power production process by considering the energy balance equation (EBE) for the entire cycle. In this regard, the energy efficiency (the so-called: thermal efficiency) is defined accordingly:

$$\eta_{en} = \frac{W_{net}}{Q_{in}} = \frac{Q_{in} - Q_{out}}{Q_{in}} = 1 - \frac{Q_{out}}{Q_{in}} = 1 - \frac{Q_L}{Q_H}$$

Here, $\dot{Q}_H$ is the heat that the engine receives from the high temperature source while the heat rejected to the low temperature heat reservoir is denoted by Q_L.

A heat engine that consists of all reversible processes is called a reversible heat engine or a Carnot heat engine. The thermal efficiency of a Carnot heat engine may be expressed by the temperatures of the wo reservoirs with which the heat engine exchanges heat:

$$\eta_{rev} = 1 - \frac{T_L}{T_H} \tag{6.4}$$

Here, T_H is the source temperature and T_L is the sink temperature where heat is rejected (i.e. lake, ambient air, etc.). This is the maximum thermal efficiency a heat engine operating between two reservoirs at T_H and T_L can have.

The second law (exergy) efficiency can be expressed for work-producing devices such as a turbine as the ratio of the useful work output to the maximum possible (reversible) work output:

$$\eta_{ex,wpd} = \dot{W}_{out}/\dot{W}_{rev,out} \tag{6.5}$$

Here, the subscript wpd refers to work-producing devices. This definition is more general since it can be applied to processes (in turbines, piston–cylinder devices, etc.) as well as to cycles. Note that the exergy efficiency cannot exceed 100%. We can also define an exergy efficiency for work-consuming noncyclic (such as compressors) and cyclic (such as refrigerators) devices as the ratio of the minimum (reversible) work input to the useful work input:

$$\eta_{ex,wcd} = \dot{W}_{rev,in}/\dot{W}_{in} \tag{6.6}$$

Here, the subscript wcd refers to work-consuming devices.

In the previous sections we have introduced the concept of a reversible cycle, which is derived from the concept introduced by the Carnot principle. The coming section explains in detail the Carnot cycle and how it is used to measure the best possible performance of heat engines that are operating between two specified temperatures, one being the temperature of the heat source and the second being the temperature of heat sink that the heat is rejected to.

The coming sections each introduce a power cycle and present the balance equations for each component of the power cycle followed by an example to show how the power cycle is solved and analyzed. The first sections start with power cycles that are based on closed systems, which are the Otto cycle, Diesel cycle and Dual cycle, all of which are based on a closed system with a moving boundary, the piston–cylinder device. Then these cycles are followed by the power generation cycles that are based on open systems; these are the Rankine cycle and the Brayton cycle.

Example 6.1 A heat engine, as shown in Figure 6.3, receives 100 kW heat from a at a temperature of 1000 K. The heat engine produces a net power output of 30 kW and heat is rejected to an environment with a temperature of 25 °C. (a) Write the overall mass, energy, entropy, and exergy balance equations, (b) calculate the amount of heat rejected to the environment (heat sink), (c) calculate the amount of exergy destroyed and exergy rejected to the environment, and (d) find the energy and exergy efficiencies of the heat engine, as well as (e) reversible (Carnot) efficiency.

Solution

a) Write the overall mass, energy, entropy, and exergy balance equations for the heat engine shown in Figure 6.3.

For the mass balance equation (MBE:) we have to assume whether the heat engine is operating in a closed cycle or open cycle form. In this example we are going to write the MBE for both cases.

For the case where the cycle operates in the closed cycle form, the MBE is written as:

$$\text{MBE}: m = constant$$

$$\text{EBE}: Q_{in} = Q_{out} + W_{net}$$

$$\text{EnBE}: \frac{Q_{in}}{T_H} + S_{gen} = \frac{Q_{out}}{T_L}$$

$$\text{ExBE}: Ex_{\dot{Q}_{in}} = Ex_{Q_{out}} + W_{net} + Ex_d$$

However, if the heat engine is operating in the open cycle form, then the MBE is written as:

$$\text{MBE}: \dot{m}_{in} = \dot{m}_{out} = constant$$

$$\text{EBE}: \dot{Q}_{in} = \dot{Q}_{out} + \dot{W}_{net}$$

$$\text{EnBE}: \frac{\dot{Q}_{in}}{T_H} + \dot{S}_{gen} = \frac{\dot{Q}_{out}}{T_L}$$

$$\text{ExBE}: \dot{E}x_{\dot{Q}_{in}} = \dot{E}x_{\dot{Q}_{out}} + \dot{W}_{net} + \dot{E}x_d$$

b) Calculate the amount of heat rejected to the environment (heat sink).

Use the energy balance equation (EBE) to find the amount of heat rejected to the environment as follows:

$$\dot{Q}_{in} = \dot{Q}_{out} + \dot{W}_{net}$$

$$100\ kW = \dot{Q}_{out} + 30\ kW$$

$$\boldsymbol{\dot{Q}_{out} = 70\ kW}$$

c) Calculate the amount of exergy destroyed and exergy rejected to the environment.

Since the heat is rejected to the environment, and due to the lack of details on the source temperature of the rejected heat, then based on the definition of the exergy content associated with heat transfer, the exergy associated with the rejected heat is equal to **zero**.

The amount of exergy destroyed during the power generation process is calculated using the exergy balance equation (ExBE) as follows:

$$\dot{E}x_{\dot{Q}_{in}} = \dot{E}x_{\dot{Q}_{out}} + \dot{W}_{net} + \dot{E}x_d$$

$$\left(1 - \frac{T_o}{T_H}\right) Q_{in} = 0 + \dot{W}_{net} + \dot{E}x_d$$

$$\left(1 - \frac{298\ K}{1000\ K}\right) \times 100\ kW = 0 + 30\ kW + \dot{E}x_d$$

$$\boldsymbol{\dot{E}x_d = 40.2\ kW}$$

d) Calculate the energy and exergy efficiencies of the heat engine.

The energy efficiency and the exergy efficiency of the heat engine are calculated respectively as follows:

$$\eta_{en} = \frac{\dot{W}_{net}}{\dot{Q}_{in}} = \frac{30\ kW}{100\ kW} = \mathbf{30\%}$$

$$\eta_{ex} = \frac{\dot{W}_{net}}{\dot{Ex}_{\dot{Q}_{in}}} = \frac{\dot{W}_{net,out}}{\left(1 - \frac{T_o}{T_H}\right)Q_{in}} = \frac{30\ kW}{\left(1 - \frac{298\ K}{1000\ K}\right) \times 100\ kW} = \mathbf{42.7\%}$$

e) Calculate the reversible (Carnot) efficiency.

Based on the information provided in the problem, $T_H = 1000\ K$ and $T_L = 25°C = 298\ K$, one can calculate the reversible (Carnot) efficiency as follows:

$$\eta_{rev} = 1 - \frac{T_L}{T_H} = 1 - \frac{298\ K}{1000\ K} = 0.702 = \mathbf{70.2\%}$$

This is the maximum possible (ideal) energy efficiency of a heat engine operating between source and sink (essentially between two temperatures) in a reversible manner.

It is important to mention here that the same exergy efficiency can be obtained if the energy efficiency is divided by the reversible (Carnot) efficiency as follows:

$$\eta_{ex} = \frac{\eta_{en}}{\eta_{rev}} = \frac{30}{70.2} = \mathbf{42.7\%}$$

which means that the actual or the real engine operates at an efficiency that is 42.7% of the maximum possible achievable efficiency for an engine operating between the same operating temperatures of source and sink.

Example 6.2 Consider two heat engines, both having a thermal efficiency of 30%. One of the engines (engine *A*) receives heat from a source at 600 *K*, while the other one (engine *B*) receives heat from a source at 1000 *K*. Both engines reject heat to a medium at 300 *K*. At the first glance, both engines seem to be performing equally well. When one takes a second look at these engines in light of the SLT (through the exergy concept), however, we see a totally different picture. The subject matter engines can achieve the best performance if they operate as reversible engines where the highest efficiency is possible, which will lead to the Carnot efficiency. (a) Calculate the energy efficiency of the reversible engines *A* and *B* and (b) calculate their exergy efficiencies.

Solution

a) Calculate the energy efficiencies of the reversible engine *A* and engine *B* as follows:

$$\eta_{rev,A} = \left(1 - \frac{T_o}{T_H}\right) = \left(1 - \frac{300\ K}{600\ K}\right) = \mathbf{50\%}$$

$$\eta_{rev,B} = \left(1 - \frac{T_o}{T_H}\right) = \left(1 - \frac{300\ K}{1000\ K}\right) = \mathbf{70\%}$$

Engine A has a 50% useful work potential relative to the heat provided to it, engine B has 70%. Now it is becoming apparent that engine B has a greater work potential available and thus should do a lot better than engine A. Therefore, we can say that engine B is performing poorly relative to engine A even though both have the same thermal efficiency.

b) Calculate the exergy efficiencies of engines A and B as follows:

It is obvious from this example that the first-law efficiency alone is not a realistic measure of performance of engineering devices. To overcome this deficiency, we define an exergy efficiency (or second-law efficiency) for heat engines as the ratio of the actual thermal efficiency to the maximum possible (reversible) thermal efficiency under the same conditions:

$$\eta_{ex,A} = \frac{0.30}{0.50} = 0.60 = \mathbf{60\%}$$

$$\eta_{ex,B} = \frac{0.30}{0.70} = 0.43 = \mathbf{43\%}$$

Note that engine A is converting 60% of the available work potential to useful work. This ratio is only 43% for engine B.

6.4 Otto Cycle

The Otto cycle is best suited for spark ignition engines. This cycle was established in the 1870s after Nicolaus Otto successfully demonstrated a four-stroke spark ignition engine. The Otto cycle and its processes are demonstrated in Figure 6.4 with illustrations of the piston–cylinder mechanisms on the diagrams for pressure-volume (P-v) in Figure 6.4a and temperature-entropy (T-s) in Figure 6.4b.

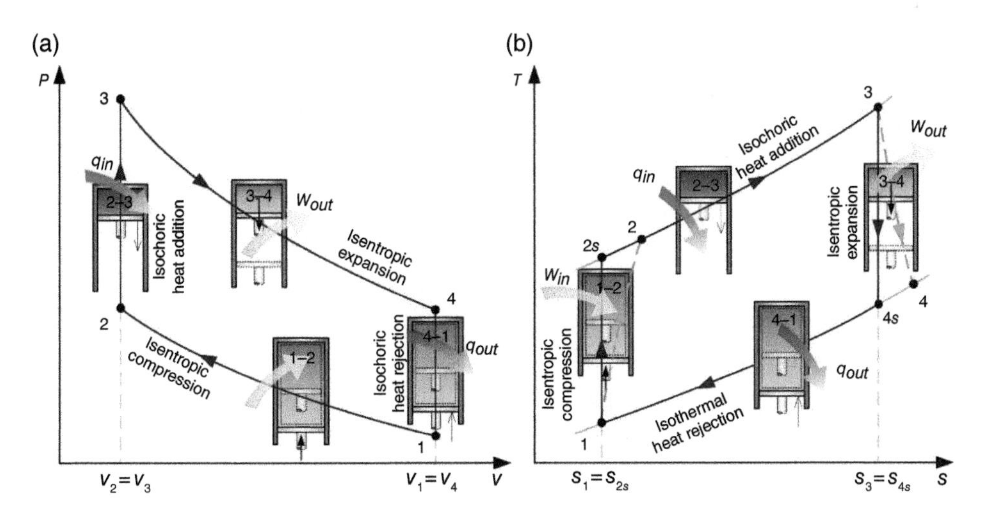

Figure 6.4 A basic, ideal Otto cycle processes along with (a) P-v and (b) T-s diagrams.

The piston–cylinder operation in the Otto cycle is demonstrated along with the *P-v* diagram. The overall cycle includes two isentropic and two isochoric processes for the working fluid (gas which is assumed to be air). The example shown in the diagram provides the following relations for volume ratio:

- Volume ratio (r_v): $r_v = \frac{v_1}{v_2}$
- Pressure ratio (r): $r = \frac{P_3}{P_1}$

In thermodynamic state one there is a gas enclosed

$$T_1 = T_0$$

There are four processes in the Otto cycle that illustrate the operation of a spark ignition (gasoline) engine as briefly explained below:

- **Process 1-2:** This process includes an isentropic compression of the working gas (air) as external work is being applied. The piston moves upwards in this process. The process is considered both isentropic and adiabatic.
- **Process 2-3:** This process includes an isochoric heat addition where the piston remains in a fixed position as the heat is being added to the working fluid. In the case of internal combustion engines, this heat would be added by the combustion process of the fuel. The internal temperature and pressure of the cylinder increase considerably during this process.
- **Process 3-4:** This process includes an isentropic expansion. During this process, the piston moves down and generates usable work. The expansion process continues until the maximum volume of the stroke is reached $v_4 = v_1$.
- **Process 4-1:** This process includes an isochoric heat rejection from the cylinder. The piston remains fixed in its final position from the last process while heat is rejected from the working gas to a heat sink in contact with the working gas (air).

The Otto cycle is considered to be externally irreversible as its heat addition and removal processes are not isothermal. The efficiency of the Otto cycle is lower than that of the Carnot cycle. The cycle efficiency can be derived under the air assumption. As the isochoric processes take place, there are no work exchanges between the cylinder and its surroundings. The heat input and output can be described by the following relations:

$$q_{in} = C_v(T_3 - T_2) \; and \; q_{out} = C_v(T_4 - T_1)$$

The isentropic equations can be derived from the equations above. The isentropic equations are $T_2 = r_v^{k-1} T_1$ and $T_3 = r_v^{k-1} T_4$, and by using these relations it can be derived that the efficiency of the internally reversible Otto cycle is:

$$\eta_{en} = 1 - \frac{q_{in}}{q_{out}} = 1 - \left(\frac{T_3 - T_2}{T_4 - T_1}\right) = 1 - r_v^{1-k}$$

The exergy efficiency of the cycle is derived as follows:

$$\eta_{ex} = \frac{w_{net}}{ex_{in}} = \frac{w_{net}}{q_{in}\left(1 - \frac{T_0}{T_s}\right)} = \frac{\eta_{en}}{\left(1 - \frac{T_0}{T_s}\right)}$$

The variation of energy and exergy efficiency of the Otto cycle in relation with the volume ratio is provided in three pressure ratios. The typical range for the volume ratios for the compression process in spark ignition engines varies between 7 and 11. The energy efficiency range for the ideal Otto cycle is within the range 40–50%. The exergy efficiency of the Otto cycle depends on additional parameters, such as source and sink temperatures, apart from r_v, which is described as the ratio between the maximum and minimum volumes of the Otto cycle. The exergy efficiency for the pressure ratio range of 7–11 and for $r = 70$–120 ranges from 44 to 68%. From the exergy efficiency range of the Otto cycle it can be seen that the magnitude of the exergy destruction is due to the heat transfer from the heat source and to the heat sink on the exergy efficiency. The ratio of the exergy destruction to the exergy output is within the range 32–66%. Finally, an important definition for the Otto cycle is the mean effective pressure (*MEP*), which is calculated as follows:

$$MEP = \frac{W_{net}}{V_{max} - V_{min}} \tag{6.7}$$

Example 6.3 Consider an Otto cycle, as shown in Figure 6.4, that has a compression ratio of 8, an isentropic compression efficiency of 85%, and an isentropic expansion efficiency of 95%. At the beginning of the compression, the air in the cylinder is at a 90 *kPa* and 15 °*C*. The maximum gas temperature is found to be 1260 °*C*. (Use constant specific heats of $c_p = 1.005$ *kJ/kgK*, $c_v = 0.7176$ *kJ/kgK*, $k = 1.4$, and $R = 0.287$ *kJ/kgK*). (a) Write the mass, energy, entropy, and exergy balance equations, (b) calculate the heat supplied to the cycle per unit mass, and (c) find energy efficiency.

Solution

a) Write the mass, energy, entropy, and exergy balance equations.
   For the four processes of the Otto cycle these are shown in Table 6.1.

b) Calculate the heat supplied to the cycle per unit mass.
   Before starting the solution to find the heat supplied during each cycle on a unit mass basis we need to define the properties given in the problem statement; they are as follows:

$$r = 8 = \frac{v_{max}}{v_{min}}, \ \ \eta_{is,c} = 0.85, \ \ \eta_{is,exp} = 0.95, \ \ T_1 = 15\,^{\circ}C \text{ and } P_1 = 90\ kPa$$

$$\boldsymbol{T_3 = 1260\,^{0}C}$$

where the subscripts is, c, and exp stand for isentropic, compression, and expansion.

All heat and work interactions occurring through the cycle are usually found from the EBEs of the cycle, except specific situations where they can be found indirectly through efficiencies and other given information.

Table 6.1 The mass, energy, entropy, and exergy balance equations for the Otto cycle of Example 6.3.

Process	Balance equations
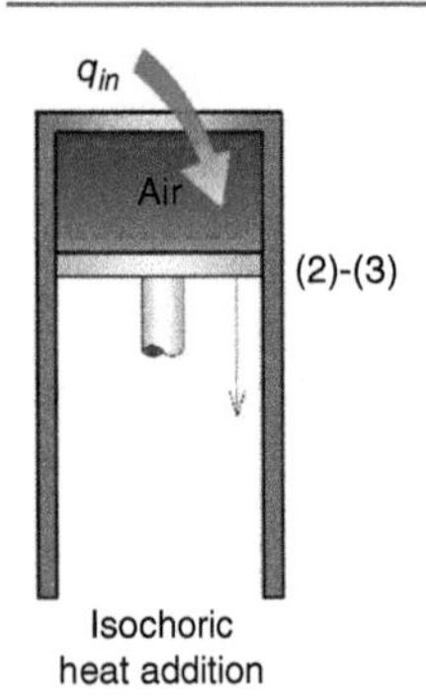	MBE: $m_1 = m_2 = m = \text{constant}$ EBE: $m_1u_1 + W_{in} = m_2u_2$ or $u_1 + w_{in} = u_2$ EnBE: $m_1s_1 + S_{gen} = m_2s_2$ or $s_1 + s_{gen} = s_2$ ExBE: $m_1ex_1 + W_{in} = m_2ex_2 + Ex_d$ or $ex_1 + w_{in} = ex_2 + ex_d$
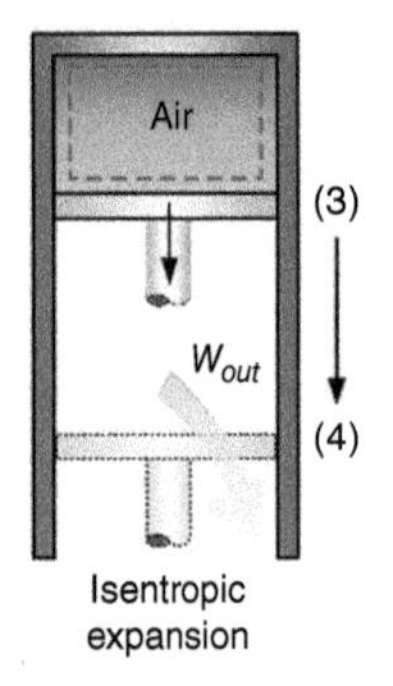	MBE: $m_2 = m_3 = m = \text{constant}$ EBE: $m_2u_2 + Q_{in} = m_3u_3$ or $u_2 + q_{in} = u_3$ EnBE: $m_2s_2 + Q_{in}/T_s + S_{gen} = m_3s_3$ or $s_2 + q_{in}/T_s + s_{gen} = s_3$ ExBE: $m_2ex_2 + \left(1-\left(\frac{T_o}{T_s}\right)\right)Q_{in} = m_3ex_3 + Ex_d$ or $ex_2 + \left(1-\left(\frac{T_o}{T_s}\right)\right)q_{in} = ex_3 + ex_d$
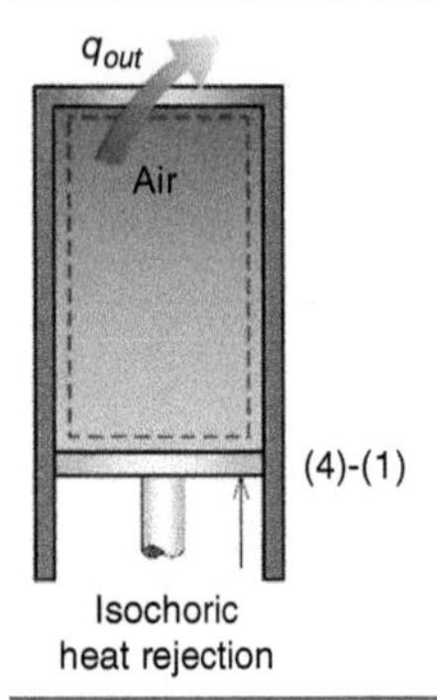	MBE: $m_3 = m_4 = m = \text{constant}$ EBE: $m_3u_3 = m_4u_4 + W_{out}$ or $u_3 = u_4 + w_{out}$ EnBE: $m_3s_3 + S_{gen} = m_4s_4$ or $s_3 + s_{gen} = s_4$ ExBE: $m_3\text{ex}_3 = m_4\text{ex}_4 + W_{out} + Ex_d$ or $ex_3 = ex_4 + w_{out} + ex_d$
	MBE: $m_4 = m_1 = m = \text{constant}$ EBE: $m_4u_4 = m_1u_1 + Q_{out}$ or $u_4 = u_1 + q_{out}$ EnBE: $m_4s_4 + S_{gen} = m_1u_1 + Q_{out}/T_b$ or $s_4 + s_{gen} = u_1 + q_{out}/T_b$ ExBE: $m_4ex_4 = m_1ex_1 + \left(1-\left(\frac{T_o}{T_b}\right)\right)Q_{out} + Ex_d$ or $ex_4 = ex_1 + \left(1-\left(\frac{T_o}{T_b}\right)\right)q_{out} + ex_d$

The process that has the heat addition step is the process from 2 to 3 and the MBE and EBE of that process are:

$$\text{MBE} : m_2 = m_3 = m = constant$$

$$\text{EBE} : m_2 u_2 + Q_{in} = m_3 u_3$$

Then from the above two equations one obtains:

$$u_2 + q_{in} = u_3$$

By assuming that the specific heat capacity does not change with the changes in temperature, including the temperature ranges the cycle goes through or the changes in the specific heat with temperature over the temperature range considered in the cycle, then the above equation can be written in terms of the temperature and the specific heats capacities, since the working gas (air) is treated as an ideal gas, as:

$$c_v T_2 + q_{in} = c_v T_3$$

$$0.7176\ \frac{kJ}{kgK} \times T_2 + q_{in} = 0.7176\ \frac{kJ}{kgK} \times (1260 + 273)\ K$$

Since the process 1-2 has an isentropic efficiency of 85% then by finding T_2 if the process was isentropic and by using the isentropic efficiency of the compression process, we can find the actual temperature for state 2, so that one can continue the solution as follows:

$$\frac{T_{2s}}{T_1} = \left(\frac{v_1}{v_2}\right)^{k-1} \rightarrow T_{2s} = T_1\left(\frac{v_1}{v_2}\right)^{k-1} \rightarrow T_{2s} = T_1\,(r)^{k-1}$$

$$T_{2s} = (15 + 273)\,(8)^{1.4-1} = \mathbf{661.6\ K}$$

$$\eta_{is,c} = \frac{h_{2s} - h_1}{h_2 - h_1} = \frac{c_p(T_{2s} - T_1)}{c_p(T_2 - T_1)} = \frac{(T_{2s} - T_1)}{(T_2 - T_1)}$$

$$0.85 = \frac{(661.6 - (15 + 273))}{(T_2 - (15 + 273))}$$

$$\mathbf{T_2 = 727.5\ K}$$

Then going back to the mass basis EBE of the heat addition process:

$$0.7176\ \frac{kJ}{kgK} \times 727.5\ K + q_{in} = 0.7176\ \frac{kJ}{kgK} \times (1260 + 273)\ K$$

$$\mathbf{q_{in} = 578\ \frac{kJ}{kg}}$$

c) The energy efficiency of the Otto cycle can be written as:

$$\eta_{en} = \frac{w_{out}}{q_{in}} = 1 - \frac{q_{out}}{q_{in}}$$

Here, we need to find the heat rejected in the process 4-1 since it is the only missing parameter in the energy efficiency equation above.

The process that has the heat rejection step is the process from 4 to 1 and the MBE and EBE of that process are:

$$\text{MBE} : m_4 = m_1 = m = const.$$

$$\text{EBE} : m_4 u_4 = m_1 u_1 + Q_{out}$$

Then from these two equations:

$$u_4 = u_1 + q_{out}$$

Similarly, by using the specific heat capacity assumption then the above equation can be written in terms of the temperature and the specific heats capacities, based on the equation, $u = c_v T$, as follows:

$$c_v T_4 = c_v T_1 + q_{out}$$

Since the process 4-1 has an isentropic efficiency of 95% then by finding T_4 if the process was isentropic and by using the isentropic efficiency of the process we can find the actual to temperature for state 4 so that we can continue the solution:

$$\frac{T_{4s}}{T_3} = \left(\frac{v_3}{v_4}\right)^{k-1} \rightarrow T_{4s} = T_3 \left(\frac{v_3}{v_4}\right)^{k-1} \rightarrow T_{4s} = T_3 \left(\frac{1}{r}\right)^{k-1}$$

$$T_{4s} = (1260 + 273) \left(\frac{1}{8}\right)^{1.4-1} = \mathbf{667.3\ K}$$

The isentropic efficiency of the expansion process is:

$$\eta_{is,exp} = \frac{h_3 - h_4}{h_3 - h_{4s}} = \frac{c_p(T_3 - T_4)}{c_p(T_3 - T_{4s})} = \frac{(T_3 - T_4)}{(T_3 - T_{4s})}$$

$$0.95 = \frac{((1260 + 273) - T_4)}{((1260 + 273) - 667.3)}$$

$$\boldsymbol{T_4 = 710.6\ K}$$

Then going back to the mass basis EBE of the heat rejection process:

$$c_v T_4 = c_v T_1 + q_{out}$$

$$0.7176\ \frac{kJ}{kgK} \times 710.6\ K = 0.7176\ \frac{kJ}{kgK} \times (15 + 273) + q_{out}$$

$$\boldsymbol{q_{out} = 303.3\ \frac{kJ}{kg}}$$

Then we go back to the energy efficiency equation of the cycle:

$$\eta_{en} = 1 - \frac{q_{out}}{q_{in}}$$

$$\eta_{en} = 1 - \frac{303.3}{578} = 0.475 = \mathbf{47.5\%}$$

Based on Example 6.3, one can note that the Otto cycle had an isentropic efficiency that is not 100% for both the compression and expansion, which means that the behavior of the Otto cycle discussed in Example 6.3 deviates from ideality. The resulting temperature-entropy diagram of the cycle is shown in Figure 6.4b where the *T-s* diagram is illustrated.

The Otto cycle, as mentioned earlier, runs on fuels that are combusted by igniting a fuel and air mixture through a spark plug. Unlike Otto engines, the Diesel engine ignites the mixture of fuel and air through the autoignition temperature, where these high temperatures are achieved by compressing the mixtures to a high pressure, as a result of which the temperature will increase as well.

Here, one should note that the ignition temperature of a fuel is the minimum temperature where it does ignite. However, the autoignition temperature is the minimum temperature required to self-ignite the combustion of the fuel without using any spark or flame. For a given fuel under the conditions selected, both ignition and autoignition temperatures become the same. This is essentially the temperature necessary to provide the activation energy required for combustion. Each fuel has a different autoignition temperature at atmospheric pressure (100 kPa), for example: 210 °C for diesel, 263 °C for gasoline, 365 °C for ethanol, 385 °C for methanol, 450 °C for hydrogen, 470 °C for propane, 537 °C for natural gas (methane), 651 °C for ammonia, etc. The autoignition temperature decreases with the increase in length of hydrocarbon chain. The higher the ignition temperature the more difficult to ignite. As a result, the flame temperature will be lower.

Example 6.4 Consider an Otto cycle, as shown in Figure 6.4, that has a compression ratio of six, a compression efficiency of 85%, and an expansion efficiency of 90%. At the beginning of the compression process, air is at 101.3 kPa and 25 °C, and the maximum temperature and pressure that occur during the cycle are $T_{max} = 1300\ K$ and $P_{max} = 4\ MPa$. Accounting for the variation of specific heats of air with temperature, assuming $T_s = 2000\ K$ and $T_0 = 293\ K$.

a) Write the mass, energy, entropy, and exergy balance equations for each process.
b) Find the net work and the heat transferred to the air during the heat addition process.
c) Calculate the exergy destruction during the expansion process.
d) Find the energy and exergy efficiencies.

Solution

The Otto cycle is illustrated on a *T-s* diagram in Figure 6.5, with the data provided to help better understand.

a) Write mass, energy, entropy, and exergy balance equations for each process.
Process 1-2: Isentropic compression

$$\text{MBE}: m_1 = m_2 = m = constant$$

$$\text{EBE}: m_1u_1 + W_{in} = m_2u_2$$

$$\text{EnBE}: m_1s_1 + S_{gen1-2} = m_2s_2$$

$$\text{ExBE}: m_1ex_1 + W_{in} = m_2ex_2 + Ex_{d1-2}$$

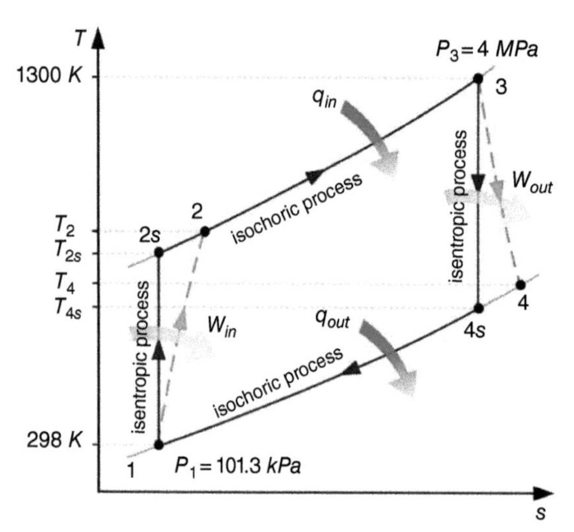

Figure 6.5 T-s diagram of the Otto cycle for Example 6.4.

Process 2-3: Isochoric heat addition

MBE: $m_2 = m_3 = m = const.$

EBE: $m_2u_2 + Q_{in} = m_3u_3$

EnBE: $m_2s_2 + Q_{in}/T_s + S_{gen2-3} = m_3s_3$

ExBE: $m_2ex_2 + \left(1 - \left(\frac{T_o}{T_s}\right)\right)Q_{in} = m_3ex_3 + Ex_{d2-3}$

Process 3-4: Isentropic expansion

MBE: $m_3 = m_4 = m = const.$

EBE: $m_3u_3 = m_4u_4 + W_{out}$

EnBE: $m_3s_3 + S_{gen3-4} = m_4s_4$

ExBE: $m_3ex_3 = m_4ex_4 + W_{out} + Ex_{d3-4}$

Process 4-1: Isochoric heat rejection

MBE: $m_4 = m_1 = m = const.$

EBE: $m_4u_4 = m_1u_1 + Q_{out}$

EnBE: $m_4s_4 + S_{gen4-1} = m_1u_1 + Q_{out}/T_b$

ExBE: $m_4ex_4 = m_1ex_1 + \left(1 - \left(\frac{T_o}{T_b}\right)\right)Q_{out} + Ex_{d4-1}$

The properties for air at 298 K are $R = 0.287$ kJ/kgK, $C_P = 1.005$ kJ/kgK, $C_v = 0.718$ kJ/kgK, and $k = 1.4$.

b) Find the heat added to the air as follows:

$T_1 = 298\ K \quad u_1 = 214.3\ kJ/kg \quad v_1 = 0.85\ m^3/kg$

$$T_{2s} = T_1\left(\frac{v_1}{v_2}\right)^{k-1} = 300\ K(6)^{1.4-1} = \mathbf{614.3\ K}$$

$$\eta_c = \frac{T_{2s} - T_1}{T_2 - T_1} = 0.85 \quad T_2 = 669.8\ K$$

$$u_2 = 489\ kJ/kg \qquad v_2 = 0.1417\ m^3/kg$$

$$\left(\frac{P_2}{P_1}\right)^{\frac{k-1}{k}} = \frac{T_{2s}}{T_1} \qquad P_2 = 1274\ kPa$$

$$T_{4s} = T_3\left(\frac{v_3}{v_4}\right)^{k-1} = 1300\ K\left(\frac{1}{6}\right)^{1.4-1} = 634.9\ K$$

$$\eta_{exp} = \frac{T_3 - T_4}{T_3 - T_{4s}} = 0.9 \qquad T_4 = 701.4\ K$$

$$q_{in} = C_v(T_3 - T_2) = 0.718\,(1300 - 669.8) = \mathbf{452.5\ kJ/kg}$$

$$w_{net} = C_v[(T_3 - T_4) - (T_2 - T_1)] = 0.718\,[(1300 - 701.4) - (669.8 - 298)] = \mathbf{164.3\ kJ/kg}$$

c) Find the exergy destruction during the isentropic expansion process.

$$ex_3 = C_v(T_3 - T_0) - T_0.(s_3 - s_0) = 570.1\ kJ/kg$$

$$ex_4 = C_v(T_4 - T_0) - T_0.(s_4 - s_0) = 133.8\ kJ/kg$$

$$u_3 = u_4 + w_{out} \qquad w_{out} = C_v[(T_3 - T_4)] = 429.8\ kJ/kg$$

$$ex_3 = ex_4 + w_{out} + ex_{d3-4}$$

$$ex_{d3-4} = \mathbf{6.5\ kJ/kg}$$

d) Calculate both the energy and exergy efficiencies.

$$\eta_{en} = \frac{w_{net}}{q_{in}} = \frac{164.3\ \frac{kJ}{kg}}{452.5\ \frac{kJ}{kg}} = \mathbf{36.31\%}$$

and

$$\eta_{ex} = \frac{w_{net}}{q_{in}\left(1 - \frac{T_0}{T_s}\right)} = \frac{164.3\ \frac{kJ}{kg}}{452.5\ \frac{kJ}{kg}\left(1 - \frac{293\ K}{2000\ K}\right)} = \mathbf{42.55\%}$$

Example 6.5 An Otto cycle, as shown in Figure 6.4, has nonisentropic compression and expansion processes; the compression ratio is six, the isentropic compression efficiency is 80%, and the isentropic expansion efficiency is 85%. At the beginning of the compression process, air is at 100 kPa and 300 K. The maximum gas temperature is 1500 K and the maximum pressure is 4200 kPa ($T_s = 2000\ K$ and $T_0 = 293\ K$).

a) Write mass, energy, entropy, and exergy balance equations for each process.
b) Find the heat addition and the net work output.
c) Calculate the exergy destruction for the expansion process.
d) Find the energy and exergy efficiencies.

Solution

First we illustrate the Otto cycle on a T-s diagram, as shown in Figure 6.6, with the data provided to help better understand.

a) The mass, energy, entropy, and exergy balance equations can be written as:
Process 1-2 Isentropic compression:

$$\text{MBE}: m_1 = m_2 = m = const.$$

$$\text{EBE}: m_1u_1 + W_{in} = m_2u_2$$

$$\text{EnBE}: m_1s_1 + S_{gen1-2} = m_2s_2$$

$$\text{ExBE}: m_1ex_1 + W_{in} = m_2ex_2 + Ex_{d1-2}$$

Process 2-3 Isochoric heat addition:

$$\text{MBE}: m_2 = m_3 = m = const.$$

$$\text{EBE}: m_2u_2 + Q_{in} = m_3u_3$$

$$\text{EnBE}: m_2s_2 + Q_{in}/T_s + S_{gen2-3} = m_3s_3$$

$$\text{ExBE}: m_2ex_2 + \left(1 - \left(\frac{T_o}{T_s}\right)\right)Q_{in} = m_3ex_3 + Ex_{d2-3}$$

Process 3-4 Isentropic expansion:

$$\text{MBE}: m_3 = m_4 = m = const.$$

$$\text{EBE}: m_3u_3 = m_4u_4 + W_{out}$$

$$\text{EnBE}: m_3s_3 + S_{gen3-4} = m_4s_4$$

$$\text{ExBE}: m_3ex_3 = m_4ex_4 + W_{out} + Ex_{d3-4}$$

Process 4-1 Isochoric heat rejection:

$$\text{MBE}: m_4 = m_1 = m = const.$$

$$\text{EBE}: m_4u_4 = m_1u_1 + Q_{out}$$

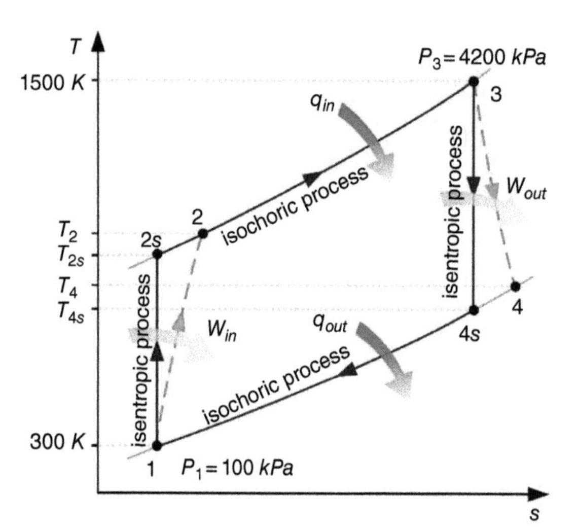

Figure 6.6 T-s diagram of the Otto cycle for Example 6.5.

$$\text{EnBE}: m_4 s_4 + S_{gen4-1} = m_1 u_1 + Q_{out}/T_b$$

$$\text{ExBE}: m_4 ex_4 = m_1 ex_1 + \left(1 - \left(\frac{T_o}{T_b}\right)\right) Q_{out} + Ex_{d4-1}$$

b) The properties for air at 300 K are $R = 0.287$ kJ/kgK, $C_P = 1.005$ kJ/kgK, $C_v = 0.718$ kJ/kgK, and $k = 1.4$.

$$T_1 = 300\ K \quad u_1 = 214.3\ kJ/kg \quad v_1 = 0.8611\ m^3/kg$$

$$T_{2s} = T_1 \left(\frac{v_1}{v_2}\right)^{k-1} = 300\ K(6)^{1.4-1} = 614.3\ K$$

$$\eta_c = \frac{T_{2s} - T_1}{T_2 - T_1} = 0.8 \qquad T_2 = 692.9\ K$$

$$u_2 = 507.1\ kJ/kg \qquad v_2 = 0.1435\ m^3/kg$$

$$\left(\frac{P_2}{P_1}\right)^{\frac{k-1}{k}} = \frac{T_{2s}}{T_1} \qquad P_2 = 1229\ kPa$$

$$T_{4s} = T_3 \left(\frac{v_3}{v_4}\right)^{k-1} = 1500\ K\left(\frac{1}{6}\right)^{1.4-1} = \mathbf{732.5\ K}$$

$$\eta_{exp} = \frac{T_3 - T_4}{T_3 - T_{4s}} = \mathbf{0.85} \qquad T_4 = 847.7\ K$$

$$q_{in} = C_v(T_3 - T_2) = 0.718\ (1500 - 692.9) = \mathbf{579.5\ kJ/kg}$$

$$w_{net} = C_v[(T_3 - T_4) - (T_2 - T_1)] = 0.718\ [(1500 - 847.7) - (692.9 - 300)] = \mathbf{186.3\ kJ/kg}$$

c) Calculate the exergy destruction during the expansion process as follows:

$$ex_3 = C_v(T_3 - T_0) - T_0.(s_3 - s_0) = 660\ kJ/kg$$

$$ex_4 = C_v(T_4 - T_0) - T_0.(s_4 - s_0) = 178.5\ kJ/kg$$

$$u_3 = u_4 + w_{out} \qquad w_{out} = C_v[(T_3 - T_4)] = 468.4\ kJ/kg$$

$$ex_3 = ex_4 + w_{out} + ex_{d3-4}$$

$$ex_{d3-4} = \mathbf{13.1\ kJ/kg}$$

d) Both the energy and exergy efficiencies are calculated for the cycle under the given conditions as follows:

$$\eta_{en} = \frac{w_{net}}{q_{in}} = \frac{186.3\ \frac{kJ}{kg}}{579.5\ \frac{kJ}{kg}} = \mathbf{32.15\%}$$

and

$$\eta_{ex} = \frac{w_{net}}{q_{in}\left(1 - \frac{T_0}{T_s}\right)} = \frac{186.3\ \frac{kJ}{kg}}{579.5\ \frac{kJ}{kg}\left(1 - \frac{293\ K}{2000\ K}\right)} = \mathbf{37.66\%}$$

6.5 Diesel Cycle

Historically, the Diesel cycle was first introduced by the German researcher Rudolf Diesel in 1895. The Diesel cycle has been a widely accepted and used cycle that has many applications, especially for industrial power generation as large engine generators can be installed. There have been many power generators that have used the Diesel cycle. Therefore, the engines are enlisted as compression ignition internal combustion engines. The combustion process occurs when the air is compressed to the autoignition temperature; pressurized fuel is then sprayed and instantaneously combusts. For the combustion process to occur the air temperature has to be over 800 *K*; the volume ratio, also known as the compression ratio for the diesel cycle, is in the range r_v = 12–24. With such a compression ratio and high temperatures a spark is no longer required to initiate the combustion process in a compression ignition engine. The rate of the reaction due to no spark plug causes the combustion to be relatively slow compared to that of the spark ignition engine. The pressure within the cylinder can be kept relatively steady contingent that the injection process of the fuel occurs at the top dead center of the cylinder.

Although the Diesel cycle looks similar to the Otto cycle due to the three common processes, there is heat addition taking place during an isobaric process. The volume during the combustion process increases and the temperature of the gas will also increase. The four processes that make up the Diesel cycle are shown in Figure 6.7 along with *P*-*v* (in Figure 6.7a) and *T*-*s* (in Figure 6.7b) diagrams. It is also equally important to explain each of these processes briefly.

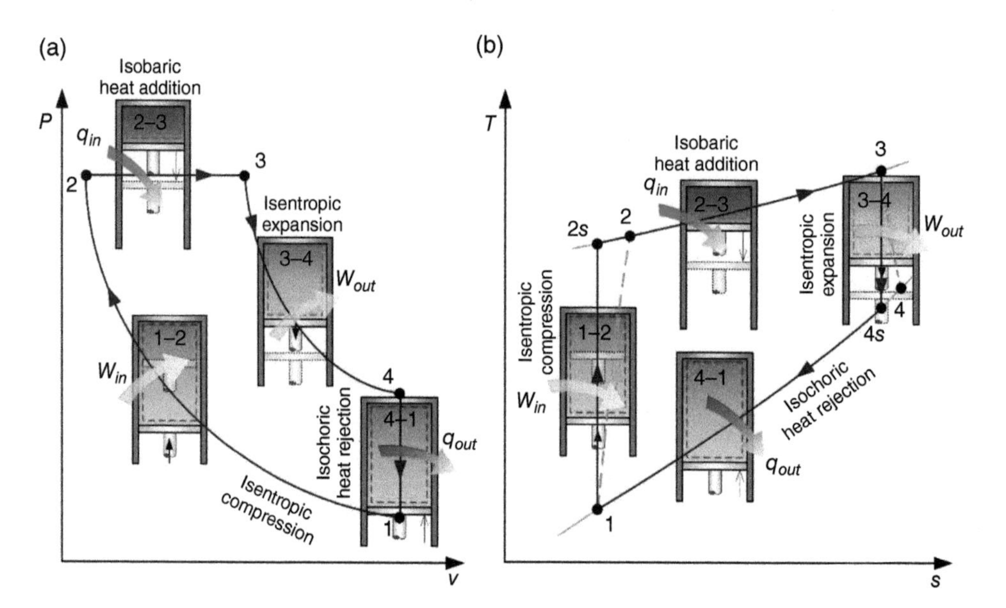

Figure 6.7 A basic, ideal Diesel cycle processes along with (a) *P*-*v* diagram and (b) *T*-*s* diagram.

- **Process 1-2:** This process includes an isentropic compression of the working gas (air) as external work is being applied. The piston moves upwards in this process. The process is considered both isentropic and adiabatic.
- **Process 2-3:** This process is different, compared to the Otto cycle, and includes an isobaric heat addition where the volume of the cylinder increases as the heat is being added. The temperature of the gas increases as the heat is being added, even though the volume is increasing.
- **Process 3-4:** This process includes an isentropic expansion. During this process, the piston moves down and generates usable work. The expansion process continues until the maximum volume of the stroke is reached $v_4 = v_1$.
- **Process 4-1:** This process includes isochoric heat rejection from the cylinder. The piston remains fixed in its final position from the last process while heat is rejected from the working gas to a heat sink in contact with the working gas (air).

The energy efficiency of the Diesel cycle under the standard air assumptions is given by the equation:

$$\eta_{en} = \frac{w_{net}}{q_{in}} = 1 - \frac{q_{out}}{q_{in}} = 1 - \frac{C_v(T_4 - T_1)}{C_p(T_3 - T_2)} = 1 - \frac{T_1\left(\frac{T_4}{T_1} - 1\right)}{kT_2\left(\frac{T_3}{T_2} - 1\right)} \tag{6.8}$$

In addition, the exergy efficiency for the Diesel cycle is, in a general form, defined as:

$$\eta_{ex} = \frac{w_{net}}{ex_{in}} = \frac{w_{net}}{q_{in}\left(1 - \frac{T_0}{T_s}\right)} = \frac{\eta_{en}}{\left(1 - \frac{T_0}{T_s}\right)} \tag{6.9}$$

In addition to those concepts introduced in the Otto cycle, for the Diesel cycle there is also the cut-off ratio, which is special for the Diesel cycles and is defined as the ratio of the volumes after and before the combustion process; it can be written as:

$$r_c = \frac{V_3}{V_2} = \frac{v_3}{v_2} \tag{6.10}$$

With the help of the definitions of the cut-off ratio and the pressure ratio we can find out that the energy efficiency of the Diesel cycle can be presented in terms of these ratios only and materials constants as:

$$\eta_{en} = 1 - \frac{1}{r^{k-1}}\left[\frac{r_c^k - 1}{k(r_c - 1)}\right] \tag{6.11}$$

Various types of crude oil and petroleum products have been used for the production of thermal energy, including gasoline, heavy duty oil, diesel, kerosene, etc. In buildings, crude oil and petroleum products are also heavily used for hot water production and space heating while in industries they are used to provide heat for various processes. Petroleum products are burnt to produce heat that is utilized to produce either hot gases or hot water to provide heating for different processes. Figure 6.8 shows the schematic of a heat engine based on diesel fuel for coproduction of steam and power. In a typical heat engine, when the

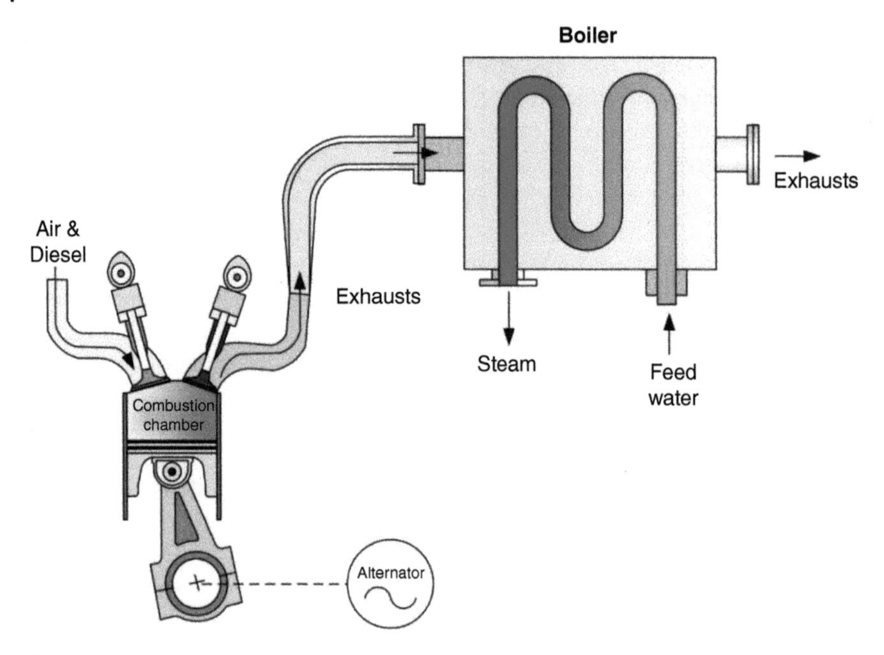

Figure 6.8 A typical diesel oil engine for co-production of steam and power.

electricity is produced the heat gets rejected at different temperature levels and can be used for the production of steam in various industries, such as food, paper, textile, etc. By using the rejected heat at different levels certain advantages are obtained: higher efficiency, lesser environmental impact per *kWh*, better cost, and operational flexibility.

Example 6.6 A power production engine that operates on an ideal Diesel cycle, as illustrated in Figure 6.7, has a compression ratio of 20 and air as the working fluid in a piston–cylinder device. The temperature and the pressure of the air at the beginning of the compression process are 20 °*C* and 95 *kPa*, respectively. The maximum temperature in the cycle reaches 2200 *K*. (Use constant specific heats of $c_p = 1.005\ kJ/(kgK)$, $c_v = 0.7176\ kJ/kgK$, $k = 1.4$, and $R = 0.287\ kJ/kgK$). Take T_o as 300 *K* and T_s as 2200 *K*.

a) Write the mass, energy, entropy, and exergy balance equations for each process of the cycle.
b) Calculate the specific heat of addition and heat rejection.
c) Calculate the net work.
d) Find both energy and exergy efficiencies.

Solution

a) Writing balance equation for the mass, energy, entropy, and exergy balance equations for the cycle processes is the most significant part. These balance equations are given in Table 6.2 to make it easier to learn and understand what really happens in the cycle.

Table 6.2 A list of the mass, energy, entropy, and exergy balance equations for the Diesel cycle of Example 6.6.

<table>
<tr><th>Process</th><th>Balance equations</th></tr>
<tr><td>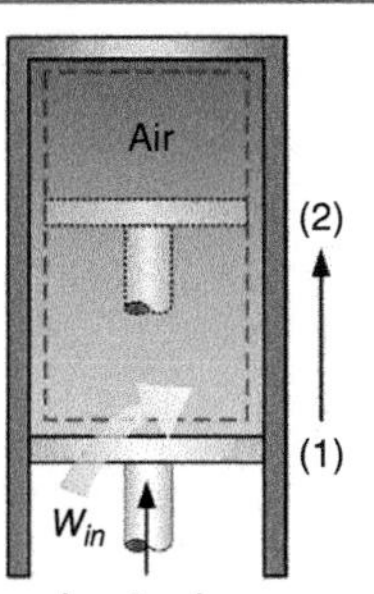

Isentropic compression</td><td>MBE: $m_1 = m_2 = m = constant$
EBE: $m_1u_1 + W_{in} = m_2u_2$
or $u_1 + w_{in} = u_2$
EnBE: $m_1s_1 + S_{gen} = m_2s_2$
or $s_1 + s_{gen} = s_2$
ExBE: $m_1ex_1 + W_{in} = m_2ex_2 + Ex_d$
or $ex_1 + w_{in} = ex_2 + ex_d$</td></tr>
<tr><td>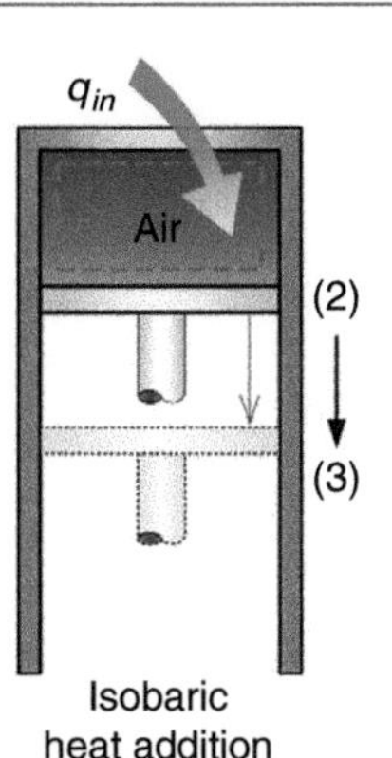

Isobaric heat addition</td><td>MBE: $m_2 = m_3 = m = constant$
EBE: $m_2u_2 + Q_{in} = m_3u_3 + W_b$
or $u_2 + q_{in} = u_3 + w_b$
EnBE: $m_2s_2 + Q_{in}/T_s + S_{gen} = m_3s_3$
or $s_2 + q_{in}/T_s + s_{gen} = s_3$
ExBE: $m_2ex_2 + \left(1 - \left(\frac{T_o}{T_s}\right)\right)Q_{in} = m_3ex_3 + W_b + Ex_d$
or $ex_2 + \left(1 - \left(\frac{T_o}{T_s}\right)\right)q_{in} = ex_3 + w_b + ex_d$</td></tr>
<tr><td>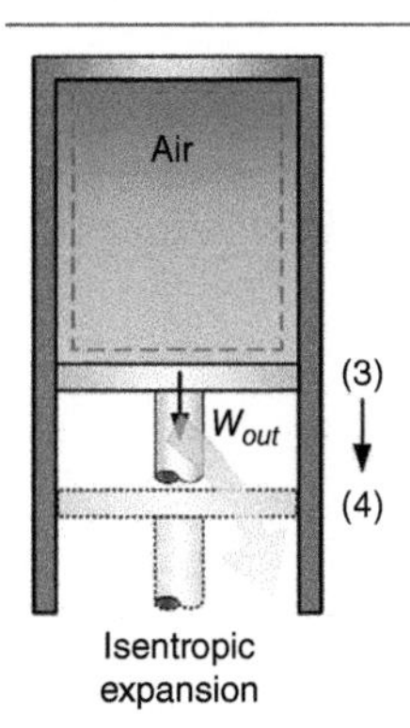

Isentropic expansion</td><td>MBE: $m_3 = m_4 = m = constant$
EBE: $m_3u_3 = m_4u_4 + W_{out}$
or $u_3 = u_4 + w_{out}$
EnBE: $m_3s_3 + S_{gen} = m_4s_4$
or $s_3 + s_{gen} = s_4$
ExBE: $m_3ex_3 = m_4ex_4 + W_{out} + Ex_d$
or $ex_3 = ex_4 + w_{out} + ex_d$</td></tr>
<tr><td>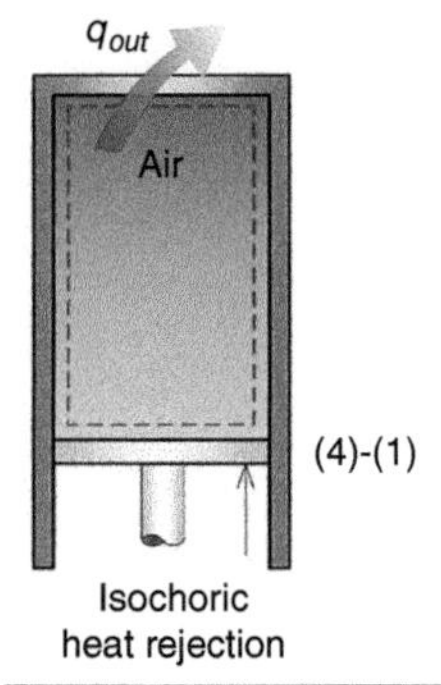

Isochoric heat rejection</td><td>MBE: $m_4 = m_1 = m = constant$
EBE: $m_4u_4 = m_1u_1 + Q_{out}$
or $u_4 = u_1 + q_{out}$
EnBE: $m_4s_4 + S_{gen} = m_1s_1 + Q_{out}/T_b$
or $s_4 + s_{gen} = s_1 + q_{out}/T_b$
ExBE: $m_4ex_4 = m_1ex_1 + \left(1 - \left(\frac{T_o}{T_b}\right)\right)Q_{out} + Ex_d$
or $ex_4 = ex_1 + \left(1 - \left(\frac{T_o}{T_b}\right)\right)q_{out} + ex_d$</td></tr>
</table>

b) Calculate both the specific heat of addition and specific heat rejection.

Before starting the solution to find the heat supplied during each cycle on a unit mass basis we need to define the properties given in the problem statement; they are:

$$r = 20 = \frac{v_{max}}{v_{min}}, \ T_1 = 20\,^{o}C \text{ and } P_1 = 95\ kPa$$

$$T_3 = 2200\ K$$

All heat and work interactions occurring through the cycle are found usually from the EBEs of the cycle, except specific situations which can be found indirectly through efficiencies and other given information.

The process that has the heat addition step is the process from 2 to 3 and its EBE is taken from Table 6.2 as follows:

$$u_2 + q_{in} = u_3 + w_b$$

From the previous chapters, we know that we can present the boundary work in terms of the pressure and the volume as follows:

$$w_b = \int_2^3 P_i dv$$

Since the piston–cylinder device in the example expands at a constant pressure, then the above equation can be represented as:

$$w_b = P\int_2^3 dv = P(v_3 - v_2)$$

$$w_b = P(v_3 - v_2)$$

Substituting the boundary work expression back into the energy balance equation that is presented in terms of mass basis gives:

$$u_2 + q_{in} = u_3 + P(v_3 - v_2)$$

$$u_2 + Pv_2 + q_{in} = u_3 + Pv_3 \rightarrow h_2 + q_{in} = h_3$$

By assuming that the specific heat capacity does not change with the changes in temperature, including the temperature ranges the cycle goes through or the changes in the specific heat with temperature over the temperature range considered in the cycle, then the above equation can be written in terms of the temperature and the specific heats capacities as:

$$c_p T_2 + q_{in} = c_p T_3$$

$$1.005\ \frac{kJ}{kgK} \times T_2 + q_{in} = 1.005\ \frac{kJ}{kgK} \times 2200\ K$$

Since the process from 1 to 2 is an isentropic process then T_2 is found as follows:

$$\frac{T_{2s}}{T_1} = \left(\frac{v_1}{v_2}\right)^{k-1} \rightarrow \ T_{2s} = T_1\left(\frac{v_1}{v_2}\right)^{k-1} \rightarrow T_{2s} = T_1\,(r)^{k-1}$$

$$T_{2s} = (20 + 273)\,(20)^{1.4-1} = 971.1\ K$$

Then, going back to the mass basis EBE of the heat addition process:

$$1.005\ \frac{kJ}{kgK} \times 971.1\ K + q_{in} = 1.005\ \frac{kJ}{kgK} \times 2200\ K$$

$$\boldsymbol{q_{in} = 1235\ \frac{kJ}{kg}}$$

For the specific heat rejection, this time we get the EBE from the table, again for the process between the state points 1 and 4, as follows:

$$u_4 = u_1 + q_{out}$$

Similarly, by using the specific heat capacity assumption, the above equation can be written in terms of the temperature and the specific heats capacities as follows:

$$c_v T_4 = c_v T_1 + q_{out}$$

$$0.7176\ \frac{kJ}{kgK} T_4 = 0.7176\ \frac{kJ}{kgK} \times (20 + 273)\ K + q_{out}$$

As the process from 2 to 3 is a constant pressure process, by using the ideal gas law the two states can be related to each other as follows:

$$\frac{P_3 v_3}{T_3} = \frac{P_2 v_2}{T_2} \rightarrow \frac{v_3}{T_3} = \frac{v_2}{T_2} \rightarrow v_3 = \frac{2200\ K}{971.1\ K} v_2 = 2.265\, v_2$$

Since the process from 4 to 1 is an isentropic process, T_4 is found as follows:

$$\frac{T_{4s}}{T_3} = \left(\frac{v_3}{v_4}\right)^{k-1} \rightarrow T_{4s} = T_3 \left(\frac{v_3}{v_4}\right)^{k-1} \rightarrow T_{4s} = T_3 \left(\frac{2.265\, v_2}{v_4}\right)^{k-1}$$

$$T_{4s} = T_3 \left(\frac{2.265}{r}\right)^{k-1} = 2200\ K \left(\frac{2.265}{20}\right)^{0.4} = \mathbf{920.6\ K}$$

Going back to the mass basis EBE of the heat addition process:

$$0.7176\ \frac{kJ}{kgK} \times 920.6\ K = 0.7176\ \frac{kJ}{kgK} \times (20 + 273)\ K + q_{out}$$

$$\boldsymbol{q_{out} = 450.6\ K}$$

c) The net work can be calculated based on the entire cycle as we write EBE for the overall cycle as follows:

$$\text{EBE}: Q_{in} = W_{net} + Q_{out} \rightarrow q_{in} = w_{net} + q_{out}$$

$$q_{in} = w_{net} + q_{out}$$

$$w_{net} = q_{in} - q_{out} = 1235 - 450.6 = \mathbf{784.4}\ \boldsymbol{\frac{kJ}{kg}}$$

d) Both energy and exergy efficiencies are calculated as follows:

$$\eta_{en} = \frac{w_{net}}{q_{in}} = \frac{784.4\ \frac{kJ}{kg}}{1235\ \frac{kJ}{kg}} = \mathbf{63.51\%}$$

and

$$\eta_{ex} = \frac{w_{net}}{q_{in}\left(1 - \frac{T_0}{T_s}\right)} = \frac{784.4\ \frac{kJ}{kg}}{1235\ \frac{kJ}{kg}\left(1 - \frac{300\ K}{2200\ K}\right)} = \mathbf{73.54\%}$$

Example 6.7 A Diesel cycle, as shown in Figure 6.7, has a compression ratio of 16 and a cutoff ratio of 2. At the beginning of the compression process, the working fluid is at 100 *kPa*, 27 °*C*, and 2 *L*. Consider a compression efficiency of 80% and an expansion efficiency of 85%; $T_s = 2000\ K$ and $T_0 = 300\ K$. Treat the working gas as air.

a) Write mass, energy, entropy, and exergy balance equations for each process.
b) Calculate the temperature and pressure of air for each process.
c) Calculate the net work output.
d) Calculate the entropy generation and exergy destruction for the heat addition process.
e) Find the energy and exergy efficiencies.

Solution

We first illustrate the Diesel cycle on a *T-s* diagram, as shown in Figure 6.9, with the data provided to help better understanding and proceed accordingly for the solution.

a) Write the mass, energy, entropy, and exergy balance equations as follows:
Process 1-2 compression:

$$\text{MBE}: m_1 = m_2 = m = constant$$

$$\text{EBE}: m_1 u_1 + W_{in} = m_2 u_2$$

$$\text{EnBE}: m_1 s_1 + S_{gen} = m_2 s_2$$

$$\text{ExBE}: m_1 ex_1 + W_{in} = m_2 ex_2 + Ex_d$$

Process 2-3 isobaric heat addition:

$$MBE: m_2 = m_3 = m = constant$$

$$\text{EBE}: m_2 u_2 + Q_{in} = m_3 u_3$$

$$\text{EnBE}: m_2 s_2 + Q_{in}/T_s + S_{gen} = m_3 s_3$$

$$\text{ExBE}: m_2 ex_2 + \left(1 - \left(\frac{T_o}{T_s}\right)\right) Q_{in} = m_3 ex_3 + Ex_d$$

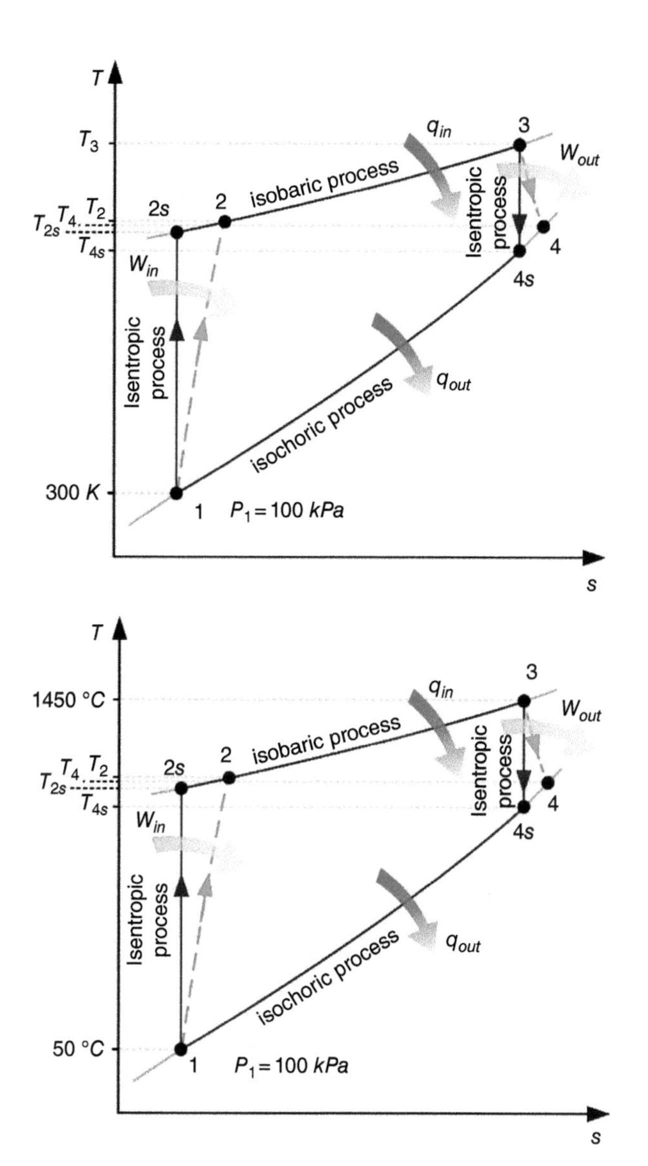

Figure 6.9 *T-s* diagram of the Diesel cycle for Examples 6.7 and 6.8.

Process 3-4 expansion:

$$\text{MBE}: m_3 = m_4 = m = constant$$

$$\text{EBE}: m_3 u_3 = m_4 u_4 + W_{out}$$

$$\text{EnBE}: m_3 s_3 + S_{gen} = m_4 s_4$$

$$\text{ExBE}: m_3 ex_3 = m_4 ex_4 + W_{out} + Ex_d$$

Process 4-1 heat rejection:

$$\text{MBE}: m_4 = m_1 = m = constant$$

$$\text{EBE}: m_4u_4 = m_1u_1 + Q_{out}$$

$$\text{EnBE}: m_4s_4 + S_{gen} = m_1b_1 + Q_{out}/T_b$$

$$\text{ExBE}: m_4ex_4 = m_1ex_1 + \left(1 - \left(\frac{T_o}{T_b}\right)\right)Q_{out} + Ex_d$$

b) Find all temperatures and pressures as follows:

$$\frac{v_1}{v_2} = r = 16 \qquad v_2 = \frac{2\ L}{16} = 0.125\ L$$

$$\frac{v_3}{v_2} = r_c = 2 \qquad v_3 = 0.25\ L$$

$$v_4 = v_1 = 2\ L$$

The properties for air at 300 K are R = 0.287 kJ/kgK, C_P = 1.005 kJ/kgK, C_v = 0.718 kJ/kgK, and k = 1.4.

$$T_{2s} = T_1\left(\frac{v_1}{v_2}\right)^{k-1} = 300\ K\left(\frac{2}{0.125}\right)^{1.4-1} = \mathbf{909.43\ K}$$

$$\eta_c = \frac{T_{2s} - T_1}{T_2 - T_1} = 0.8$$

$$\mathbf{T_2 = 1062\ K} \qquad \mathbf{s_2 = 5.925\ kJ/kgK}$$

$$P_2 = P_1\left(\frac{v_1}{v_2}\right)^{k} = 100\ kPa\left(\frac{2}{0.125}\right)^{1.4} = \mathbf{4850.29\ kPa}$$

$$P_3 = P_2 = 4850.29\ kPa$$

$$\frac{P_3v_3}{T_3} = \frac{P_2v_2}{T_2} \qquad T_3 = T_2\left(\frac{v_3}{v_2}\right) = 1062\ K \times 2 = 2124\ K \qquad s_3 = 6.763\ kJ/kgK$$

$$T_{4s} = T_3\left(\frac{v_3}{v_4}\right)^{k-1} = 2124\ K\left(\frac{0.25}{2}\right)^{1.4-1} = \mathbf{924.52\ K}$$

$$\eta_{exp} = \frac{T_3 - T_4}{T_3 - T_{4s}} = 0.85$$

$$\mathbf{T_4 = 1104.43\ K}$$

$$P_4 = P_3\left(\frac{v_3}{v_4}\right)^{k} = 4850.29\ kPa\left(\frac{0.25}{2}\right)^{1.4} = \mathbf{263.9\ kPa}$$

$$m = \frac{P_1v_1}{RT_1} = \frac{100 \times 0.002}{0.287 \times 300} = \mathbf{0.002323\ kg}$$

c) Calculate the amount of heat addition (Q_{in}) as follows:

$$m_2h_2 + Q_{in} = m_3h_3$$

$$Q_{in} = m(h_3 - h_2) = mCp(T_3 - T_2) = 0.002323\ kg \times 1.005\ kJ/kgK(2124\ K - 1062\ K)$$

$$\mathbf{Q_{in} = 2.48\ kJ}$$

Calculate the heat rejection through the EBE equation for the heat rejection process:

$$m_4u_4 = m_1u_1 + Q_{out}$$

$$Q_{out} = mCv(T_4 - T_1) = 0.002323\ kg \times 0.718\ kJ/kgK(1104.43\ K - 300\ K)$$

$$\boldsymbol{Q_{out} = 1.342\ kJ}$$

Furthermore, W_{net} is calculated as follows:

$$\boldsymbol{W_{net}} = Q_{in} - Q_{out} = 2.48\ kJ - 1.342\ kJ = \boldsymbol{1.138\ kJ}$$

d) Find the entropy generation and exergy destruction for the heat addition process as follows:

$$m_2s_2 + Q_{in}/T_s + S_{gen2-3} = m_3s_3 \qquad \boldsymbol{S_{gen2-3} = 0.0007058\ kJ/K}$$

$$m_2ex_2 + \left(1 - \left(\frac{T_o}{T_s}\right)\right)Q_{in} = m_3ex_3 + Ex_{d2-3}$$

$$ex_2 = Cp(T_2 - T_0) - T_0.(s_2 - s_0) = 699.6\ kJ/kg$$

$$ex_3 = Cp(T_3 - T_0) - T_0.(s_3 - s_0) = 1516\ kJ/kg$$

$$\boldsymbol{Ex_{d2-3} = 0.2117\ kJ}$$

e) In order to evaluate the performance of the cycle, it is necessary to study both energy and exergy efficiencies. The energy efficiency of the cycle can be evaluated as:

$$\eta_{en} = \frac{W_{net}}{Q_{in}} = \frac{1.138\ kJ}{2.48\ kJ} = \boldsymbol{45.89\%}$$

and the exergy efficiency can be calculated as:

$$\eta_{ex} = \frac{W_{net}}{Q_{in}\left(1 - \frac{T_0}{T_s}\right)} = \frac{1.138\ kJ}{2.48\ kJ\left(1 - \frac{300\ K}{2000\ K}\right)} = \boldsymbol{53.99\%}$$

Example 6.8 A Diesel cycle, as shown in Figure 6.7, with air as the working gas has a compression ratio of 16. Air is at 50 °C and 100 kPa at the beginning of the compression process and at 1450 °C at the end of the heat addition process. Take a compression efficiency of 85%, an expansion efficiency of 90%, $T_s = 2000\ K$, and $T_0 = 300\ K$

a) Write mass, energy, entropy, and exergy balance equations for each process.
b) Find the cutoff ratio.
c) Calculate the temperature and pressure for each process.
d) Calculate the entropy generation and exergy destruction for the heat addition process.
e) Find the energy and exergy efficiencies.

Solution

Before getting into details, first illustrate the Diesel cycle on a T-s diagram, as shown in Figure 6.9, with the data provided to help better understanding and proceed accordingly for the solution.

a) Write the mass, energy, entropy, and exergy balance equations for each process.

Process 1-2 compression:

$$\text{MBE}: m_1 = m_2 = m = constant$$

$$\text{EBE}: m_1u_1 + W_{in} = m_2u_2$$

$$\text{EnBE}: m_1s_1 + S_{gen1-2} = m_2s_2$$

$$\text{ExBE}: m_1ex_1 + W_{in} = m_2ex_2 + Ex_{d1-2}$$

Process 2-3 heat addition:

$$MBE: m_2 = m_3 = m = constant$$

$$\text{EBE}: m_2h_2 + Q_{in} = m_3h_3$$

$$\text{EnBE}: m_2s_2 + Q_{in}/T_s + S_{gen2-3} = m_3s_3$$

$$\text{ExBE}: m_2ex_2 + \left(1 - \left(\frac{T_o}{T_s}\right)\right)Q_{in} = m_3ex_3 + Ex_{d2-3}$$

Process 3-4 expansion:

$$\text{MBE}: m_3 = m_4 = m = constant$$

$$\text{EBE}: m_3u_3 = m_4u_4 + W_{out}$$

$$\text{EnBE}: m_3s_3 + S_{gen3-4} = m_4s_4$$

$$\text{ExBE}: m_3ex_3 = m_4ex_4 + W_{out} + Ex_{d3-4}$$

Process 4-1 heat rejection:

$$\text{MBE}: m_4 = m_1 = m = constant$$

$$\text{EBE}: m_4u_4 = m_1u_1 + Q_{out}$$

$$\text{EnBE}: m_4s_4 + S_{gen4-1} = m_1s_1 + Q_{out}/T_b$$

$$\text{ExBE}: m_4ex_4 = m_1ex_1 + \left(1 - \left(\frac{T_o}{T_b}\right)\right)Q_{out} + Ex_{d4-1}$$

b) Find the cutoff ratio.

The properties for air at 323 K are $R = 0.287$ kJ/kgK, $C_P = 1.0065$ kJ/kgK, $C_v = 0.7195$ kJ/kgK, and $k = 1.399$.

$$\frac{v_1}{v_2} = r = 16$$

$$T_{2s} = T_1\left(\frac{v_1}{v_2}\right)^{k-1} = 323\ K(16)^{1.399-1} = 976.44\ K$$

$$\eta_{is,c} = \frac{T_{2s} - T_1}{T_2 - T_1} = 0.85$$

$$T_2 = 1092\ K$$

$$\frac{P_3v_3}{T_3} = \frac{P_2v_2}{T_2} \qquad \frac{v_3}{v_2} = \frac{T_3}{T_2} = \frac{1723}{1092} = 1.578$$

c) Find all temperatures and pressures.

$$T_{4s} = T_3\left(\frac{v_3}{v_4}\right)^{k-1} = 1723\ K\left(\frac{1.578v_2}{v_4}\right)^{1.399-1} \quad : v_4 = v_1$$

$$T_{4s} = 1723\ K\left(\frac{1.578v_2}{v_1}\right)^{1.399-1} = 1723\ K\left(\frac{1.578}{16}\right)^{1.399-1} = 683.7\ K$$

$$\eta_{is,exp} = \frac{T_3 - T_4}{T_3 - T_{4s}} = 0.9$$

$$\boldsymbol{T_4 = 787.6\ K}$$

$$P_2 = P_1\left(\frac{v_1}{v_2}\right)^k = 100\ kPa(16)^{1.399} = \mathbf{4836.86\ kPa}$$

$$\boldsymbol{P_3 = P_2 = 4836.86\ kPa}$$

$$\boldsymbol{P_4} = P_3\left(\frac{v_3}{v_4}\right)^k = 4836.86\ kPa\left(\frac{1.578}{16}\right)^{1.399} = \mathbf{189.3\ kPa}$$

d) Find the specific entropy generation and specific exergy destruction for the heat addition process.

$$h_2 + q_{in} = h_3$$

$$q_{in} = (h_3 - h_2) = Cp(T_3 - T_2) = 1.0065\ kJ/kgK(1723\ K - 1092\ K)$$

$$\boldsymbol{q_{in} = 634.4\ kJ/kg}$$

$$s_2 + q_{in}/T_s + s_{gen2-3} = s_3$$

$$s_2 = 5.958\ kJ/kgK \qquad s_3 = 6.504\ kJ/kgK$$

$$\boldsymbol{s_{gen2-3} = 0.2281\ kJ/kgK}$$

$$ex_2 + \left(1 - \left(\frac{T_o}{T_s}\right)\right)q_{in} = ex_3 + ex_{d2-3}$$

$$ex_2 = C_p(T_2 - T_0) - T_0.(s_2 - s_0) = 719.9\ kJ/kg$$

$$ex_3 = C_p(T_3 - T_0) - T_0.(s_3 - s_0) = 1191\ kJ/kg$$

$$\boldsymbol{ex_{d2-3} = 68.42\ kJ/kg}$$

e) The energy efficiency of the cycle can be evaluated as:

$$w_{net} = q_{in} - q_{out}$$

One needs to find q_{out}

$$u_4 = u_1 + q_{out}$$

$$q_{out} = C_v(T_4 - T_1) = 0.7195\ kJ/kgK(787.6\ K - 323\ K)$$

$$\boldsymbol{q_{out} = 333.6\ kJ/kg}$$

To find w_{net}

$$\boldsymbol{w_{net}} = q_{in} - q_{out} = 634.4\ \frac{kJ}{kg} - 333.6\ \frac{kJ}{kg} = \mathbf{300.8\ kJ/kg}$$

The energy efficiency of the cycle is calculated as:

$$\boldsymbol{\eta_{en}} = \frac{w_{net}}{q_{in}} = \frac{300.8\ kJ/kg}{634.4\ kJ/kg} = \mathbf{47.41\%}$$

In addition, the exergy efficiency can be calculated as

$$\eta_{ex} = \frac{w_{net}}{q_{in}\left(1 - \frac{T_0}{T_s}\right)} = \frac{300.8\ kJ/kg}{634.4\ kJ/kg\left(1 - \frac{300\ K}{2000\ K}\right)} = \mathbf{55.78\%}$$

6.6 Dual Cycle

The difference between the Otto cycle and the Diesel cycle is mainly in the step of heat addition, where, as explained above, the Otto cycle heat addition step occurs at a constant volume (isochoric) process while in the Diesel cycle the heat addition process occurs at a constant pressure (isobaric) process. In modern engines some are equipped with high-speed compression ignition systems. In high-speed compression ignition (turbo) engines, the fuel is injected into the chamber earlier than in Diesel engines, where the fuel starts to ignite at later stages of the compression process, where this part of the combustion can be approximated as heat addition at a constant volume. However, in these engines the combustion does not stop at the top dead center, it continues; this results in the pressure inside the piston–cylinder device remaining high during the expansion process, which can be approximated as constant pressure heat addition in the simplified cycle. The simplified or ideal cycle of the earlier mention mode of combustion is referred to as the dual cycle. This is also recognized as another heat engine cycle, which is technically a combination of the heat addition processes of both the Otto and Diesel cycles; it is believed to have been first discovered by a Russian–German engineer Gustav Trinkler. Some still call it the Trinkler cycle. It is, however, commonly named the Dual cycle. Figure 6.10 shows a schematic of the Dual cycle, which can be considered a combination of the Otto cycle and the Diesel cycle. The key advantage of this Dual cycle is that more time is allowed for the fuel to achieve better and

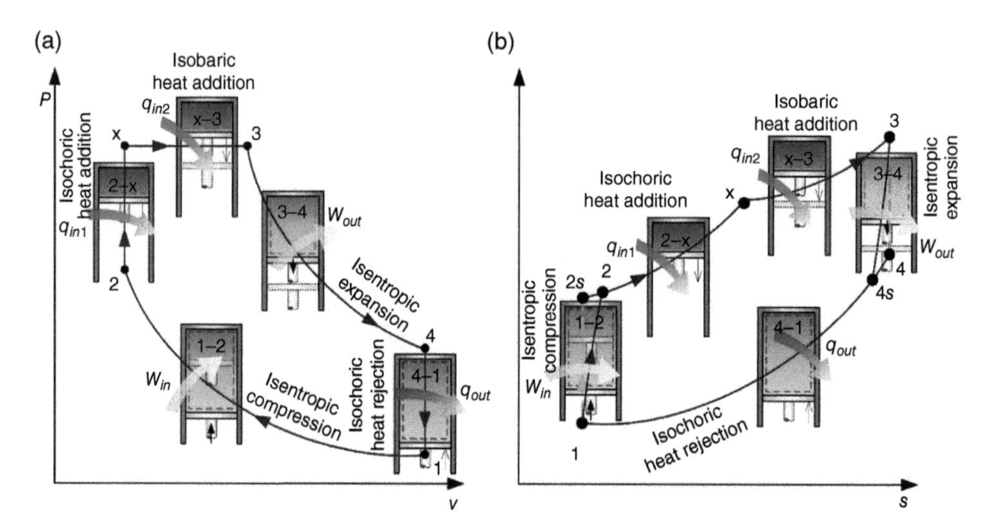

Figure 6.10 Schematic diagram of the processes of a Dual cycle along with (a) *P-v* and (b) *T-s* diagrams.

more complete combustion. The variation of pressure with specific volume of the air inside the piston–cylinder device making up the Dual cycle is plotted in Figure 6.10a. As shown, the heat addition process can be divided into two processes, the first going from state 2 to an intermediate state x is a heat addition process occurring at a constant volume, while the rest of the heat addition process occurs from state x to state 3, which occurs at a constant pressure (9 Figure 6.10a). In addition, the processes are illustrated on a temperature-entropy diagram in Figure 6.10b with distinctions made for isentropic (ideal) and nonisentropic (actual) cases.

Analyzing the Dual cycle is quite similar to the methodologies outlined in both sections of the Otto cycle and Diesel cycle where the mass, energy, entropy, and exergy balance equations were written. Balance equations for each process of the Dual cycle are listed in Table 6.3.

Table 6.3 The mass, energy, entropy, and exergy balance equations for a Dual cycle.

Process	Balance equations
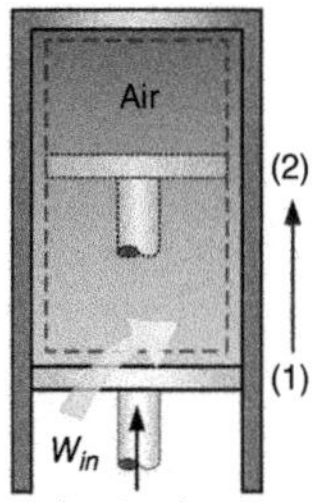 Air, (2), (1), W_{in} Isentropic compression	MBE: $m_1 = m_2 = m = constant$ EBE: $m_1u_1 + W_{in} = m_2u_2$ or $u_1 + w_{in} = u_2$ EnBE: $m_1s_1 + S_{gen} = m_2s_2$ or $s_1 + s_{gen} = s_2$ ExBE: $m_1ex_1 + W_{in} = m_2ex_2 + Ex_d$ or $ex_1 + w_{in} = ex_2 + ex_d$
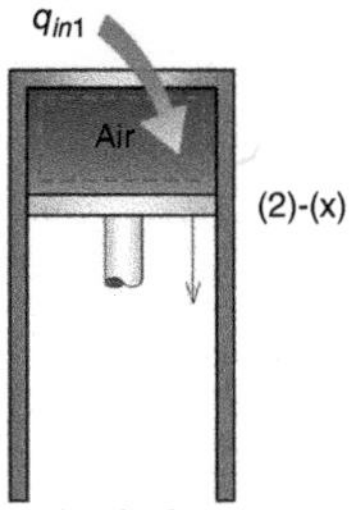 q_{in1}, Air, (2)-(x) Isochoric heat addition	MBE: $m_2 = m_x = m = constant$ EBE: $m_2u_2 + Q_{in1} = m_xu_x$ or $u_2 + q_{in1} = u_x$ EnBE: $m_2s_2 + Q_{in1}/T_s + S_{gen} = m_xs_x$ or $s_2 + q_{in1}/T_s + s_{gen} = s_x$ ExBE: $m_2ex_2 + \left(1 - \left(\frac{T_o}{T_s}\right)\right)Q_{in1} = m_xex_x + Ex_d$ or $ex_2 + \left(1 - \left(\frac{T_o}{T_s}\right)\right)q_{in1} = ex_x + ex_d$
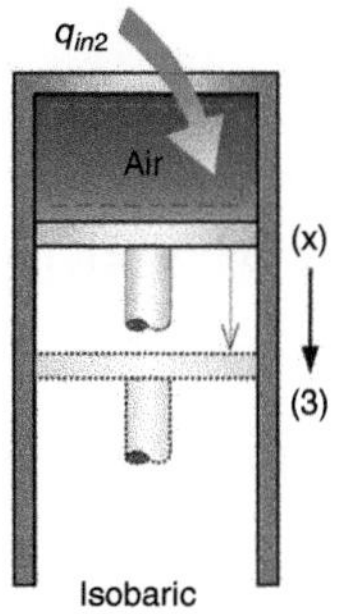 q_{in2}, Air, (x), (3) Isobaric heat addition	MBE: $m_x = m_3 = m = constant$ EBE: $m_xu_x + Q_{in2} = m_3u_3 + W_b$ or $u_x + q_{in2} = u_3 + w_b$ EnBE: $m_xs_x + Q_{in2}/T_s + S_{gen} = m_3s_3$ or $s_x + q_{in2}/T_s + s_{gen} = s_3$ ExBE: $m_xex_x + \left(1 - \left(\frac{T_o}{T_s}\right)\right)Q_{in2} = m_3ex_3 + W_b + Ex_d$ or $ex_x + \left(1 - \left(\frac{T_o}{T_s}\right)\right)q_{in2} = ex_3 + w_b + ex_d$

(Continued)

Table 6.3 (Continued)

Process	Balance equations
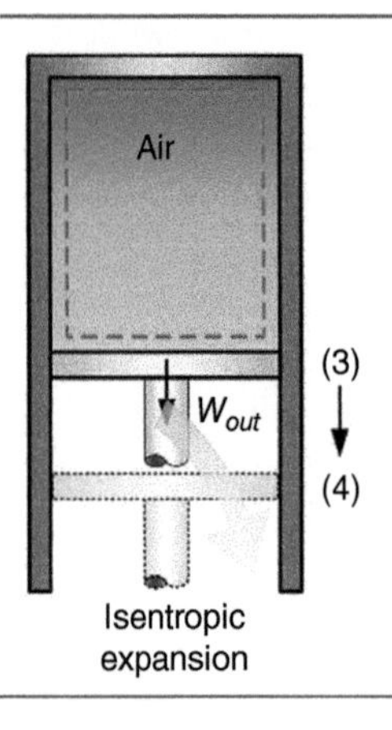	MBE: $m_3 = m_4 = m = constant$ EBE: $m_3u_3 = m_4u_4 + W_{out}$ or $u_3 = u_4 + \mathrm{w_{out}}$ EnBE: $m_3s_3 + S_{gen} = m_4s_4$ or $s_3 + s_{gen} = s_4$ ExBE: $m_3ex_3 = m_4ex_4 + W_{out} + Ex_d$ or $ex_3 = ex_4 + w_{out} + ex_d$
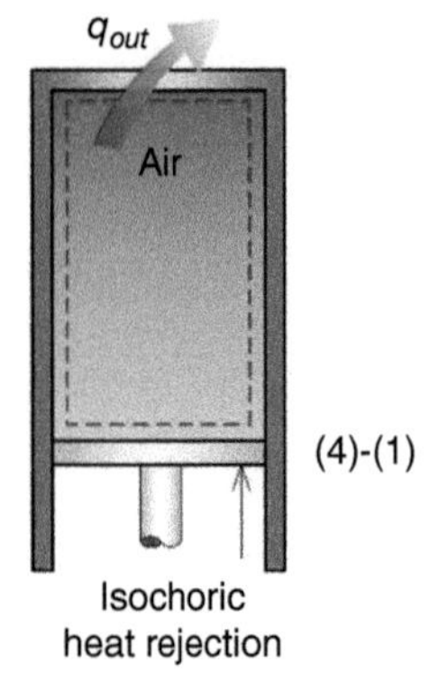	MBE: $m_4 = m_1 = m = constant$ EBE: $m_4u_4 = m_1u_1 + Q_{out}$ or $u_4 = u_1 + q_{out}$ EnBE: $m_4s_4 + S_{gen} = m_1s_1 + Q_{out}/T_b$ or $s_4 + s_{gen} = s_1 + q_{out}/T_b$ ExBE: $m_4ex_4 = m_1ex_1 + \left(1 - \left(\frac{T_o}{T_b}\right)\right)Q_{out} + Ex_d$ or $ex_4 = ex_1 + \left(1 - \left(\frac{T_o}{T_b}\right)\right)q_{out} + ex_d$

Example 6.9 A dual cycle, as shown in Figure 6.10, operates with a cutoff ratio of 2 and compression ratio of 15. In addition, during the constant volume process of heat addition, the pressure ratio is known to be 1.5. At the beginning of the compression process, the operating conditions are known to be 100 *kPa* and 25 °*C*. The isentropic efficiency of the compression and expansion processes are known to be 70 and 85%, respectively.

a) Write the mass, energy, entropy, and exergy balance equations for each process cycle.
b) Determine the amount of heat input to the cycle and find the maximum gas pressure and temperature for both ideal and actual cycles.
c) Find the entropy generation and exergy destruction during the heat and work input processes.
d) Draw the *P-v* and *T-s* diagrams for the cycle.
e) Determine the energy and exergy efficiencies of the actual cycle if the source temperature is 3000 *K*.

Solution

a) Write the mass, energy, entropy, and exergy balance equations for each process.
For the compression proess:

$$\text{MBE}: m_1 = m_2 = m = constant$$

$$\text{EBE}: m_1u_1 + W_{in} = m_2u_2$$

$$\text{EnBE}: m_1s_1 + S_{gen} = m_2s_2$$

$$\text{ExBE}: m_1ex_1 + W_{in} = m_2ex_2 + Ex_d$$

For the constant volume heat addition process:

$$\text{MBE}: m_2 = m_x = m = constant$$

$$\text{EBE}: m_2u_2 + Q_{in1} = m_xu_x$$

$$\text{EnBE}: m_2s_2 + Q_{in1}/T_s + S_{gen} = m_xs_x$$

$$\text{ExBE}: m_2ex_2 + \left(1 - \left(\frac{T_o}{T_s}\right)\right)Q_{in1} = m_xex_x + Ex_d$$

For the constant pressure heat addition process:

$$\text{MBE}: m_x = m_3 = m = constant$$

$$\text{EBE}: m_xu_x + Q_{in2} = m_3u_3 + W_b$$

$$\text{EnBE}: m_xs_x + Q_{in2}/T_s + S_{gen} = m_3s_3$$

$$\text{ExBE}: m_xex_x + \left(1 - \left(\frac{T_o}{T_s}\right)\right)Q_{in2} = m_3ex_3 + W_b + Ex_d$$

For the expansion process:

$$\text{MBE}: m_3 = m_4 = m = constant$$

$$\text{EBE}: m_3u_3 = m_4u_4 + W_{out}$$

$$\text{EnBE}: m_3s_3 + S_{gen} = m_4s_4$$

$$\text{ExBE}: m_3ex_3 = m_4ex_4 + W_{out} + Ex_d$$

For the constant volume heat rejection process:

$$\text{MBE}: m_4 = m_1 = m = constant$$

$$\text{EBE}: m_4u_4 = m_1s_1 + Q_{out}$$

$$\text{EnBE}: m_4s_4 + S_{gen} = m_1s_1 + Q_{out}/T_b$$

$$\text{ExBE}: m_4ex_4 = m_1ex_1 + \left(1 - \left(\frac{T_o}{T_b}\right)\right)Q_{out} + Ex_d$$

b) Assuming air as an ideal gas and taking air-standard assumptions, the specific volume can be found at state 1 as:

$$v_1 = \frac{RT_1}{P_1} = \frac{\left(0.287\ \frac{kPam^3}{kgK}\right)(298\ K)}{100\ kPa} = \mathbf{0.855}\ \frac{\boldsymbol{m^3}}{\boldsymbol{kg}}$$

Also, from the given compression ratio, the specific volume at state 2 can be found as:

$$v_2 = \frac{v_1}{r} = \frac{0.855\ \frac{m^3}{kg}}{15} = \mathbf{0.057\ \frac{m^3}{kg}}$$

After the compression process is completed P_2, P_3, T_2 and T_x can be evaluated for the ideal cycle as:

$$P_2 = P_1\left(\frac{v_1}{v_2}\right)^k = P_1 r^k = (100\ kPa)\,(15)^{1.4} = \mathbf{4431\ kPa}$$

$$T_{2s} = T_1\left(\frac{v_1}{v_2}\right)^{k-1} = T_1 r^{k-1} = (298\ K)(15)^{0.4} = \mathbf{880.3\ K}$$

$$P_x = P_3 = r_k P_2 = (1.5)(4431) = \mathbf{6647\ kPa}$$

$$T_{xs} = T_{2s}\left(\frac{P_3}{P_2}\right) = (880.3\ K)\left(\frac{6647\ kPa}{4431\ kPa}\right) = \mathbf{1320.5\ K}$$

Moreover, from the definition of the cutoff ratio the specific volume at state 3 (v_3) can be found as:

$$v_3 = r_c v_x = r_c v_2 = (2)\left(0.057\ \frac{m^3}{kg}\right) = \mathbf{0.114\ \frac{m^3}{kg}}$$

The maximum temperatures in the cycle at states 3 (T_3) can be found as:

$$T_{3s} = T_{xs}\left(\frac{v_3}{v_x}\right) = (1320.5\ K)\left(\frac{0.114\ \frac{m^3}{kg}}{0.057\ \frac{m^3}{kg}}\right) = \mathbf{2641\ K}$$

The temperature at state 4 (T_4) can be evaluaetd as:

$$T_{4s} = T_{3s}\left(\frac{v_3}{v_4}\right)^{k-1} = (2641\ K)\left(\frac{0.114\ \frac{m^3}{kg}}{0.855\ \frac{m^3}{kg}}\right)^{0.4} = \mathbf{1180\ K}$$

The heat added to the cycle can thus be found as:

$$Q_{in,tot} = Q_{in1} + Q_{in2}$$

where Q_{in1} and Q_{in2} can be found from the energy balance equations:

$$m_2 u_2 + Q_{in1} = m_x u_x$$

$$m_x u_x + Q_{in2} = m_3 u_3 + W_b$$

If constant specific heats are assumed, the above equations can be rewritten per unit mass and solved as follows:

$$q_{in1,s} = u_{xs} - u_{2s} = c_v(T_{xs} - T_{2s}) = \left(0.718\ \frac{kJ}{kgK}\right)(1320.5\ K - 880.3\ K) = \mathbf{316\ \frac{kJ}{kg}}$$

$$q_{in2,s} = u_3 - u_x + P_3(v_3 - v_x) = u_3 + P_3v_3 - (u_x + P_xv_x) = h_3 - h_x = c_p(T_3 - T_x)$$

$$q_{in2,s} = \left(1.005\ \frac{kJ}{kgK}\right)(2641\ K - 1320.5\ K)$$

$$= \mathbf{1327\ \frac{kJ}{kg}}$$

Thus, the total heat input to the cycle per unit mass for the ideal cycle is:

$$q_{in,total,s} = q_{in1,s} + q_{in2,s} = 316\ \frac{kJ}{kg} + 1327\ \frac{kJ}{kg} = \mathbf{1643\ \frac{kJ}{kg}}$$

For the actual cycle, state 2 properties can be found from the isentropic compression efficiency:

$$\eta_{is,C} = \frac{u_{2s} - u_1}{u_2 - u_1}$$

The above equation can be re-arranged to solve for T_{2a} as:

$$T_{2a} = T_1 + \frac{T_{2s} - T_1}{\eta_{is,C}} = T_{2a} = \mathbf{1129\ K}$$

Next, the actual temperature at state x can be found as:

$$T_x = T_2\left(\frac{P_3}{P_2}\right) = \mathbf{1694\ K}$$

Also, the actual specific volume at state 3 can be found as:

$$v_3 = r_cv_x = r_cv_2 = \mathbf{0.114\ \frac{m^3}{kg}}$$

The maximum temperature in the actual cycle T_{3a} can thus be found as:

$$T_3 = T_x\left(\frac{v_3}{v_x}\right) = (1694\ K)\left(\frac{0.114\ \frac{m^3}{kg}}{0.057\ \frac{m^3}{kg}}\right) = \mathbf{3388\ K}$$

Further, state 4 can be found from the expansion isentropic efficiency as:

$$u_4 = u_3 - \eta_{is,\,exp}\,(u_3 - u_{4s})$$

$$= 3056\ \frac{kJ}{kg} - 0.85\left(3056\ \frac{kJ}{kg} - 915.6\ \frac{kJ}{kg}\right) = \mathbf{1237\ \frac{kJ}{kg}}$$

The heat added to the actual cycle can thus be found as:

$$Q_{in,tot,a} = Q_{in1,a} + Q_{in2,a}$$

where $Q_{in1,\,a}$ and $Q_{in2,\,a}$ can be found from the energy balance equations:

$$m_2u_2 + Q_{in1} = m_xu_x$$

$$m_xu_x + Q_{in2} = m_3u_3 + W_b$$

If constant specific heats are assumed, the above equations can be rewritten per unit mass and solved as follows:

$$q_{in1,a} = u_x - u_2 = c_v(T_x - T_2) = \left(0.718\ \frac{kJ}{kgK}\right)(1694\ K - 1129\ K) = \mathbf{405.7}\ \frac{\mathbf{kJ}}{\mathbf{kg}}$$

$$q_{in2,a} = u_3 - u_x + P_3(v_3 - v_x) = u_3 + P_3v_3 - (u_x + P_xv_x) = h_3 - h_x = c_p(T_3 - T_x)$$

$$q_{in2,a} = \left(1.005\ \frac{kJ}{kgK}\right)(3388\ K - 1694\ K) = \mathbf{1702}\ \frac{\mathbf{kJ}}{\mathbf{kg}}$$

Thus, the total heat input to the cycle per unit mass for the actual cycle is:

$$q_{in,tot,a} = q_{in1,a} + q_{in2,a} = 405.7\ \frac{kJ}{kg} + 1702\ \frac{kJ}{kg} = \mathbf{2108}\ \frac{\mathbf{kJ}}{\mathbf{kg}}$$

c) The entropy generation during the heat addition processes can be found as:

$$s_{gen,2-x} = s_x - s_2 - \frac{q_{in1}}{T_s}$$

$$= 6.391\ \frac{kJ}{kgK} - 6.022\ \frac{kJ}{kgK} - \frac{405.7\ \frac{kJ}{kg}}{3000\ K} = \mathbf{0.232}\ \frac{\mathbf{kJ}}{\mathbf{kgK}}$$

$$s_{gen,x-3} = s_3 - s_x - \frac{q_{in2}}{T_s}$$

$$= 7.27\ \frac{kJ}{kgK} - 6.391\ \frac{kJ}{kgK} - \frac{1702\ \frac{kJ}{kg}}{3000\ K} = \mathbf{0.312}\ \frac{\mathbf{kJ}}{\mathbf{kgK}}$$

Next, during the work input process, the entropy generation can be found from:

$$s_{gen,1-2} = s_2 - s_1 = 6.022\ \frac{kJ}{kgK} - 5.699\ \frac{kJ}{kgK} = \mathbf{0.323}\ \frac{\mathbf{kJ}}{\mathbf{kgK}}$$

From the entropy generation values, the exergy destruction can be evaluated as:

$$ex_{d,2-x} = T_0 s_{gen,2-x} = (298\ K)\left(0.232\ \frac{kJ}{kgK}\right) = \mathbf{69.14}\ \frac{\mathbf{kJ}}{\mathbf{kg}}$$

$$ex_{d,x-3} = T_0 s_{gen,x-3} = (298\ K)\left(0.312\ \frac{kJ}{kgK}\right) = \mathbf{92.9}\ \frac{\mathbf{kJ}}{\mathbf{kg}}$$

$$ex_{d,1-2} = T_0 s_{gen,1-2} = (298\ K)\left(0.323\ \frac{kJ}{kgK}\right) = \mathbf{96.3}\ \frac{\mathbf{kJ}}{\mathbf{kg}}$$

d) Based on the data given and some other temperature, pressure, volume and specific entropy data, drawing *P*-*v* and *T*-*s* diagrams is necessary, which is done as illustrated in Figure 6.11.

e) The energy efficiency of the cycle can be evaluated as:

$$\eta_{en} = \frac{w_{net}}{q_{tot}} = \frac{(u_3 - u_4) - (u_2 - u_1)}{q_{tot}}$$

$$= \frac{\left(3056\ \frac{kJ}{kg} - 1237\ \frac{kJ}{kg}\right) - \left(870.9\ \frac{kJ}{kg} - 212.9\ \frac{kJ}{kg}\right)}{2108\ \frac{kJ}{kg}} = \mathbf{55.1\%}$$

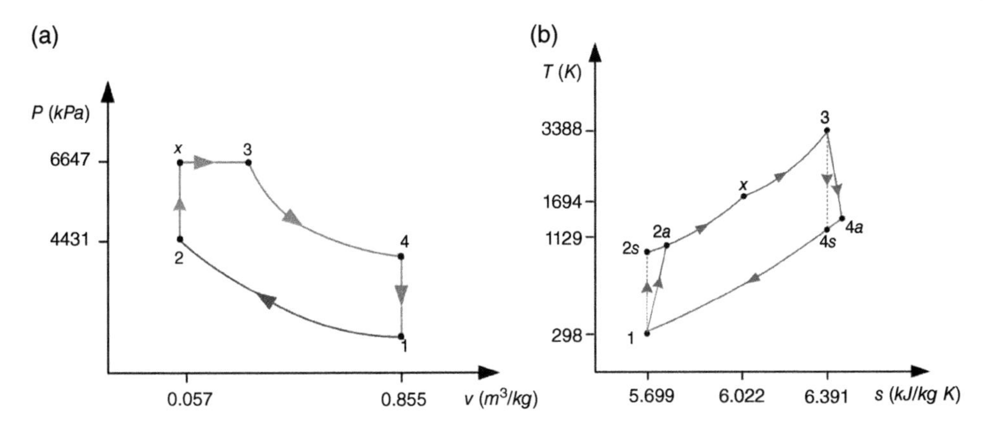

Figure 6.11 Property diagrams for Example 6.9: (a) *P-v* diagram and (b) *T-s* diagram.

Similarly, the exergy efficiency can be calculated as:

$$\eta_{ex} = \frac{w_{net}}{q_{in,tot}\left(1-\frac{T_0}{T_s}\right)} = \frac{\left(3056\ \frac{kJ}{kg} - 1237\ \frac{kJ}{kg}\right) - \left(870.9\ \frac{kJ}{kg} - 212.9\ \frac{kJ}{kg}\right)}{2108\ \frac{kJ}{kg}\left(1-\frac{298}{3000}\right)} = \mathbf{61.1\%}$$

Example 6.10 Air at 101 kPa and 27 °C enters the compression process of a dual cycle. The temperature at the end of the constant volume heat addition process is measured at 1300 K and at the end of constant pressure heat-addition process it is measured to be 2250 K. The compression ratio of the cycle is set at 15. Assume variable specific heats for air.

a) Write the mass energy, entropy, and exergy balance equations for all processes.
b) Find the fraction of constant volume heat input process as compared to the total heat input.
c) Determine the entropy generation and exergy destruction during the constant volume heat addition process if the pressure ratio during this process is 1.5.
d) Draw the *T-s* and *P-v* diagrams for the cycle.
e) Determine the energy and exergy efficiencies of the cycle if the source temperature is 4000 K.

Solution

a) Write the mass, energy, entropy, and exergy balance equations for each process
   Compression:

   MBE: $m_1 = m_2 = m = const.$

   EBE: $m_1 u_1 + W_{in} = m_2 u_2$

$$\text{EnBE}: m_1s_1 + S_{gen} = m_2s_2$$

$$\text{ExBE}: m_1ex_1 + W_{in} = m_2ex_2 + Ex_d$$

Constant volume heat addition:

$$\text{MBE}: m_2 = m_x = m = const.$$

$$\text{EBE}: m_2u_2 + Q_{in1} = m_xu_x$$

$$\text{EnBE}: m_2s_2 + Q_{in1}/T_s + S_{gen} = m_xs_x$$

$$\text{ExBE}: m_2ex_2 + \left(1-\left(\frac{T_o}{T_s}\right)\right)Q_{in1} = m_xex_x + Ex_d$$

Constant pressure heat addition:

$$\text{MBE}: m_x = m_3 = m = const.$$

$$\text{EBE}: m_xu_x + Q_{in2} = m_3u_3 + W_b$$

$$\text{EnBE}: m_xs_x + Q_{in2}/T_s + S_{gen} = m_3s_3$$

$$\text{ExBE}: m_xex_x + \left(1-\left(\frac{T_o}{T_s}\right)\right)Q_{in2} = m_3ex_3 + W_b + Ex_d$$

Expansion:

$$\text{MBE}: m_3 = m_4 = m = const.$$

$$\text{EBE}: m_3u_3 = m_4u_4 + W_{out}$$

$$\text{EnBE}: m_3s_3 + S_{gen} = m_4s_4$$

$$\text{ExBE}: m_3ex_3 = m_4ex_4 + W_{out} + Ex_d$$

Constant volume heat rejection:

$$\text{MBE}: m_4 = m_1 = m = const.$$

$$\text{EBE}: m_4u_4 = m_1u_1 + Q_{out}$$

$$\text{EnBE}: m_4s_4 + S_{gen} = m_1s_1 + Q_{out}/T_b$$

$$\text{ExBE}: m_4ex_4 = m_1ex_1 + \left(1-\left(\frac{T_o}{T_b}\right)\right)Q_{out} + Ex_d$$

b) Air-standard assumptions and negligible potential and kinetic energy changes are considered.

The temperature at state 1 is given as 300 K. The other properties at this temperature can be obtained from thermodynamic property database tables (such as Appendix E-1) or Engineering Equation Solver (EES) software.

$$u_1 = 214.07\ \frac{kJ}{kg}\ \ and\ v_{r1} = 621.2$$

From the given compression ratio of 15, the properties at state 2 can be found:

$$v_{r2} = \frac{v_2}{v_1}v_{r1} = \frac{1}{15}(621.2) = \mathbf{41.9}$$

At this value of v_{r2}, the temperature and internal energy can be found from tables (such as Appendix E-1) or the EES:

$$T_2 = 840\ K \text{ and } u_2 = 624.95\ kJ/kg$$

Next, the maximum temperature in the cycle is given as 2250 *K*. The properties at this state can be found from tables (such as Appendix E-1):

$$h_3 = 2566.4 \, \frac{kJ}{kg} \text{ and } v_{r3} = 1.864$$

The total heat input can evaluated from the energy balance:

$$q_{in,tot} = q_{in,2-x} + q_{in,x-3} = (u_x - u_2) + (h_3 - h_x)$$

$$q_{in,tot} = \left(1022.82 \, \frac{kJ}{kg} - 624.95 \, \frac{kJ}{kg}\right) + \left(2566.4 \, \frac{kJ}{kg} - 1395.97 \, \frac{kJ}{kg}\right) = \mathbf{1568 \ kJ/kg}$$

The amount of heat input during the constant volume heat input process can be found:

$$q_{in,2-x} = (u_x - u_2) = \left(1022.82 \, \frac{kJ}{kg} - 624.95 \, \frac{kJ}{kg}\right) = \mathbf{397.9} \, \frac{\mathbf{kJ}}{\mathbf{kg}}$$

The fraction of constant volume heat input process as compared to the total heat input can thus be found:

$$\frac{q_{in,2-x}}{q_{in,tot}} = \frac{397.9 \, \frac{kJ}{kg}}{1568 \ kJ/kg} = \mathbf{0.25}$$

c) The specific entropy generation during the constant volume heat addition process can be evaluated as:

$$s_{gen} = s_x - s_2 - \frac{q_{in}}{T_s} = 5.989 \, \frac{kJ}{kgK} - 5.686 \, \frac{kJ}{kgK} - \frac{397.9 \, \frac{kJ}{kg}}{4000 \ K} = \mathbf{0.2033} \, \frac{\mathbf{kJ}}{\mathbf{kgK}}$$

The specific exergy destruction can thus be found:

$$ex_d = T_0 s_{gen} = (298 \ K)\left(0.2033 \, \frac{kJ}{kgK}\right) = \mathbf{60.5} \, \frac{\mathbf{kJ}}{\mathbf{kg}}$$

d) Based on the data given and some other temperature, pressure, volume and specific entropy data, drawing *P-v* and *T-s* diagrams is necessary; this is done as illustrated in Figure 6.12.

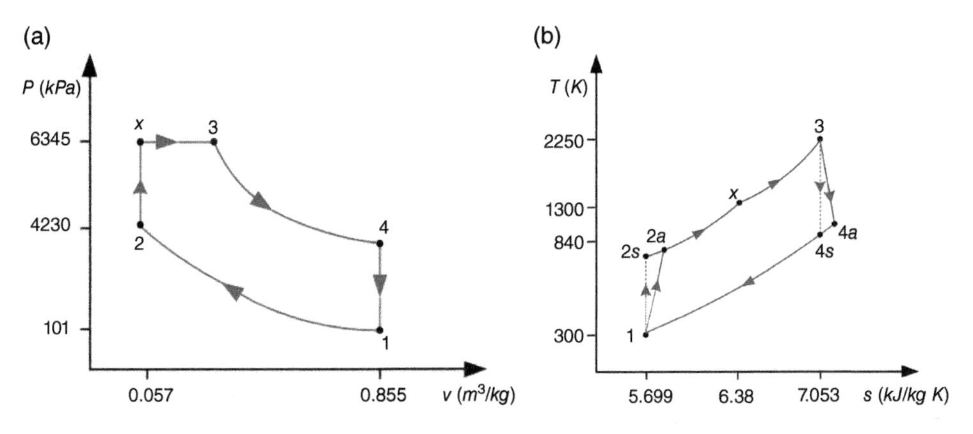

Figure 6.12 Property diagrams for Example 6.10: (a) *P-v* diagram and (b) *T-s* diagram.

e) The energy and exergy efficiencies of the overall process can be found:

$$\eta_{en} = \frac{w_{net}}{q_{in}} = \frac{(u_3 - u_4) - (u_2 - u_1)}{q_{in}} = \frac{\left(1921.3\ \frac{kJ}{kg} - 1162\ \frac{kJ}{kg}\right) - \left(624.95\ \frac{kJ}{kg} - 214.07\ \frac{kJ}{kg}\right)}{1568\ kJ/kg} = \mathbf{22.2\%}$$

and

$$\eta_{ex} = \frac{w_{net}}{q_{in}\left(1 - \frac{T_0}{T_s}\right)} = \frac{\left(1921.3\ \frac{kJ}{kg} - 1162\ \frac{kJ}{kg}\right) - \left(624.95\ \frac{kJ}{kg} - 214.07\ \frac{kJ}{kg}\right)}{1568\ kJ/kg\left(1 - \frac{298\ K}{4000\ K}\right)} = \mathbf{24.0\%}$$

6.7 Brayton Cycle

After introducing the three different cycles of power generation that are essentially based on closed systems, the forthcoming sections will cover air standard Brayton cycles (where air is the working gas throughout) and steam Rankine cycles (where water is the working fluid throughout). These Brayton and Rankine cycles contain different components that are treated as open systems and analyzed accordingly by writing the balance equations for mass, energy, entropy, and exergy, just like done earlier for the processes of the previous Otto, Diesel, and Dual cycles. First we will introduce the Brayton cycle and, since it has air as the working fluid, ideal gas equations and other approximations that were introduced earlier will also be used.

In the early 1930s, combustion-based gas-turbine power plants started to be commercially developed, based on an open type simple Brayton cycle, as shown in Figure 6.13. The simple Brayton cycle includes a turbo compressor, combustion chamber, and a gas turbine. In the 1930s the combustion turbines were thought to be the best technology to convert chemical exergy of gaseous and liquid fuels in electric power. The combustion turbine could be used with various types of fuel, such as natural gas, fuel oil, and gasified coal. The combustion turbine was further developed with changes and additions, which led to a more advanced combustion turbine power generator. The advanced combustion turbine power generators allowed for a higher efficiency as they integrated regenerators, gas reheater, compressor intercoolers, and turbine blade cooling (this allows for higher operating temperature).

The combustion turbine also had start-up times comparable to the steam Rankine power plant; in some cases start-up times were lower than one minute. The short start-up times allowed the combustion turbines to be recommended for use in regional grids to compensate for peak loads. The efficiency of a combustion turbine power plant is known to reach values over 40% when the expelled gases have elevated temperatures of over 625 *K*. If the combustion turbine is combined with a Rankine cycle power plant, the efficiencies are known to reach and surpass 55%. This cycle is known as the combined cycle.

In addition, we will make the air-standard assumption where the working fluid (i.e., fuel and air) is treated as air, which will make the Brayton cycle the air-standard Brayton cycle where air is the only working fluid throughout the cycle. This way one may use ideal gas

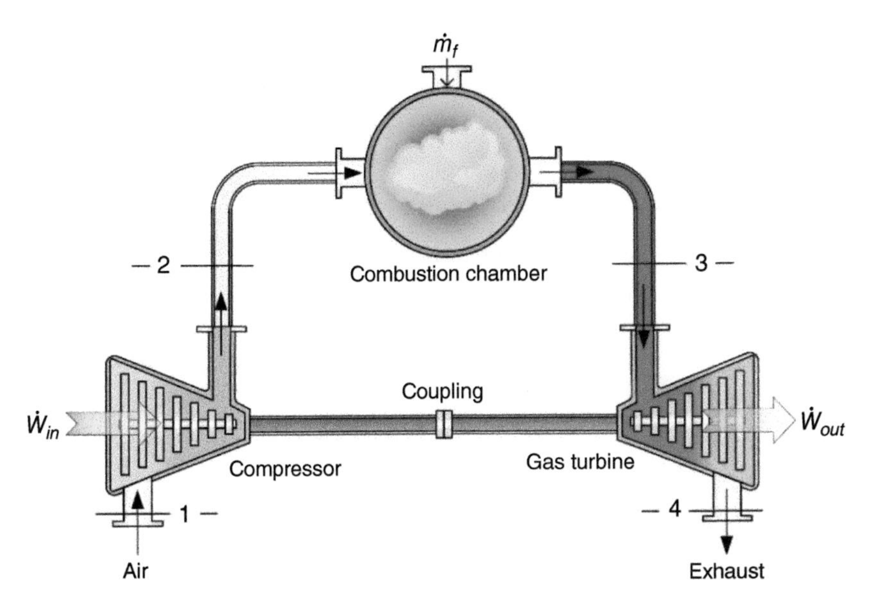

Figure 6.13 An open-cycle real gas turbine engine with a combustion chamber where fuel is fed in.

equations for air directly for thermodynamic calculations. In regards to the operation of the actual Brayton cycle, the working fluid is compressed through the compressor and the pressurized air fuel mixture is then combusted in the combustion chamber, from where the pressurized combustion gases are later delivered to the gas turbine for work production. Excess air (4–50 time that required) is provided to the combustion chamber to increase fuel utilization. Therefore, using air as a combustion gas is a good approximation and is widely accepted. When the working fluid is modeled as an ideal gas using air, with the specific heat value of 1.005 kJ/kgK at a temperature of 298 K and $\gamma = 1.4$, this combustion turbine cycle is named the air standard Brayton cycle. If the cycle compromised of internally reversible processes, then the cycle is denoted as an ideal standard air Brayton cycle. The simple Brayton cycle is compromised of four processes, which include isentropic compression, isobaric heat addition, isentropic expansion, and isobaric heat rejection. On a mechanical level, the processes are performed by the following components, the compressor, heater, turbine, and cooler. The Brayton cycle has isobaric processes for heat addition and rejection, instead of isothermal processes. This makes the Brayton cycle an externally irreversible cycle.

As introduced earlier in Figure 6.13, the gas turbine is really the heart of the Brayton cycle where the work rate is generated by expanding the compressed combustion exhausts in an open type system. There is also a common approach when we idealize and solve the problems, as shown in Figure 6.14, where we assume that air comes into the compressor and moves to the heat exchanger, after being compressed, to get heated up to the desired temperature before it is delivered into a gas turbine where the work rate is generated. The expanded air leaves the cycle apparently.

The majority of combustion turbine power plants operates based on the open Brayton cycle shown in Figure 6.13. The open Brayton cycle is directly connected to the atmosphere.

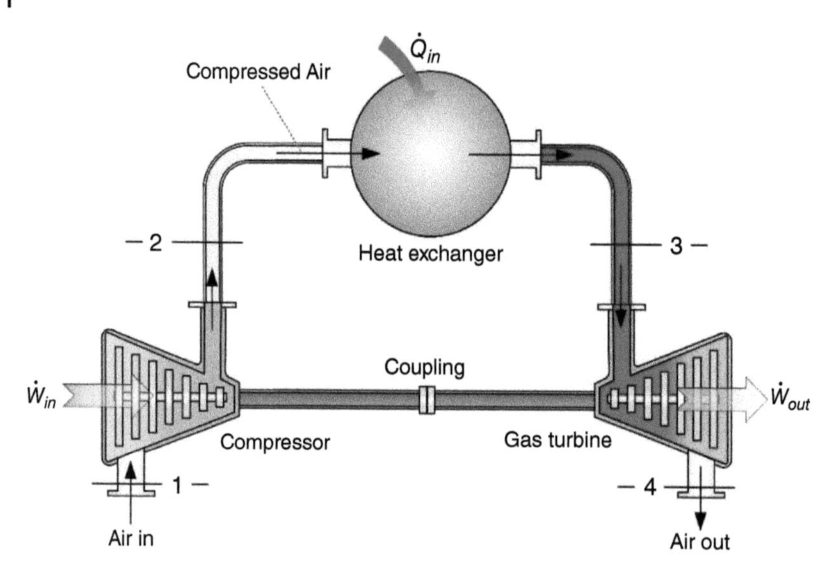

Figure 6.14 An open-cycle ideal gas turbine engine (air-standard Brayton cycle).

Therefore, its intake and exhaust pressures are equal to atmospheric pressure. These constraints represent a limitation regarding working conditions. This constraint can be overcome with the utilization of a heat exchanger, which takes the role of the cooler. With the use of a heat exchanger, the turbine could be made to discharge in a vacuum, increasing the power output and the efficiency. The turbine discharge pressure could be adjusted by the means of changing the heat sink temperature. However, the closed Brayton cycle must operate with an inert working fluid that could include air and helium, etc. The heat addition process can be achieved from external combustion by means of heat transfer to the heater (process 2-3). The working fluid and the combustion gases are never in contact throughout the process, and with these changes the Brayton cycle becomes the external combustion engine. The key point is that both the open and closed Brayton cycles are equivalent thermodynamically. However, the open cycle uses the atmosphere as its cooler whereas the closed Brayton cycle uses the heat exchanger as its cooler.

As mentioned earlier, the closed configuration of the ideal Brayton cycle is where the air that exits the turbine is recirculated to be used again through the cycle; this is done by adding a heat exchanger that cools down the air to the inlet temperature of the compressor. Figure 6.15 shows a schematic diagram of the closed-cycle configuration of the ideal Brayton cycle. The property diagrams for the closed-type air-standard Brayton cycle are illustrated in Figure 6.16a for pressure-specific volume (P-v) and in Figure 6.16b for temperature-specific entropy (T-s). Figure 6.16b also illustrates ideal and actual compression and expansion processes. Considering the cycle and its P-v and T-s diagrams, there are four processes occurring, as given in Table 6.4.

Note that the ideal gas equations that were presented earlier and in the previous chapters that apply to isentropic processes with constant specific heat apply here as well.

These equations are presented here as well where they will also be used to drive an energy efficiency expression that depends on properties of the working fluid and the main

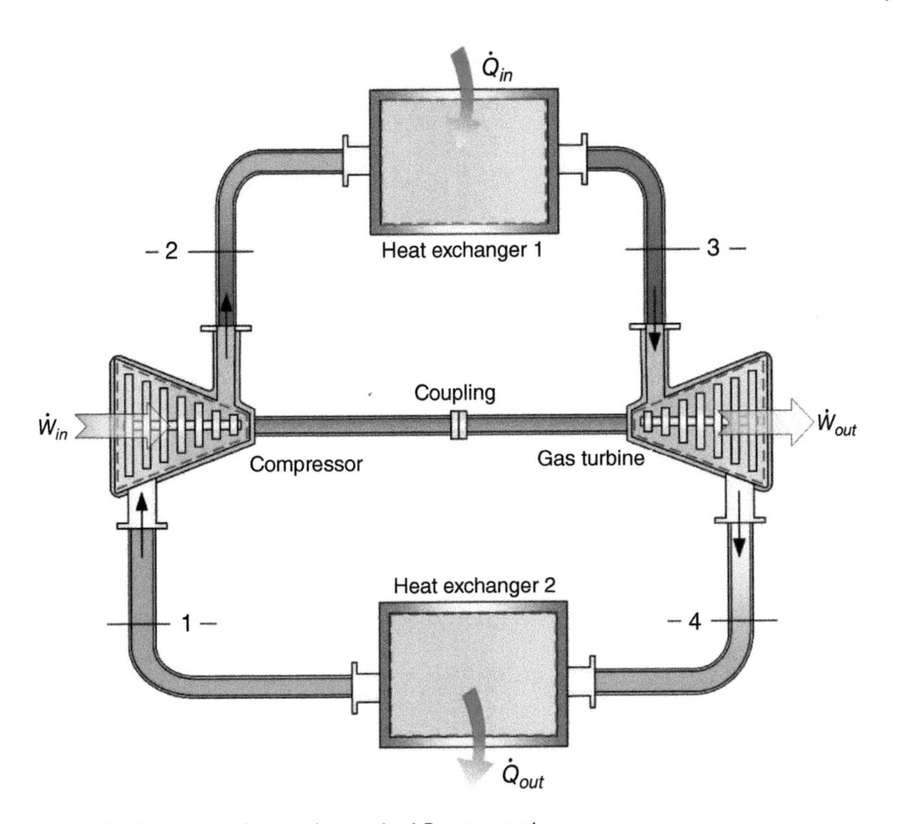

Figure 6.15 A closed-type air-standard Brayton cycle.

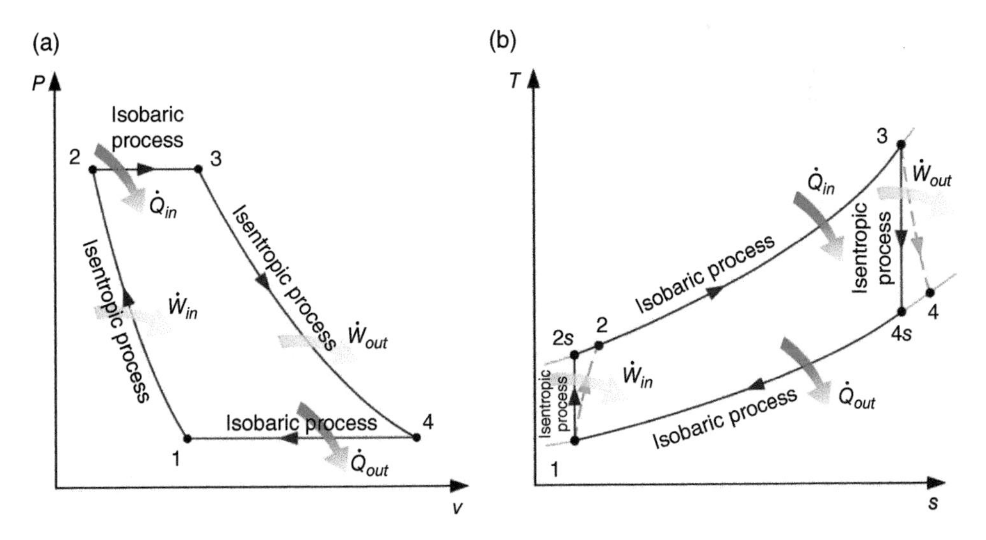

Figure 6.16 A closed-cycle gas-turbine engine (a) temperature versus the specific entropy and (b) pressure versus the specific volume.

Table 6.4 A list of processes of ideal and actual air-standard Brayton cycle.

Process	Ideal cycle	Actual cycle
Compression	Isentropic (1-2*s*)	Nonisentropic (1-2)
Isobaric	Heat addition (2*s*-3)	Heat addition (2-3)
Expansion	Isentropic (3-4*s*)	Nonisentropic (3-4)
Isobaric	Heat rejection (4*s*-1)	Heat rejection (4-1)

operating parameters of the cycle. For the isentropic processes in the cycle, which are from 1-2 and 3-4, the following expression are true:

$$\frac{T_2}{T_1} = \left(\frac{P_2}{P_1}\right)^{(k-1)/k} \quad and \quad \frac{T_3}{T_4} = \left(\frac{P_3}{P_4}\right)^{(k-1)/k}$$

An important parameter of the Brayton cycle is the pressure ratio of the cycle, which is equal to the highest pressure in the cycle over the lowest pressure of cycle, which can be written as:

$$r_p = \frac{Highest\ pressure}{Lowest\ pressure} = \frac{P_2}{P_1} = \frac{P_3}{P_4} = \frac{P_2}{P_4} = \frac{P_3}{P_1} \tag{6.12}$$

Here, r_P is the pressure ratio of the Brayton cycle. The energy efficiency of the Brayton cycle can be written as all other power generation cycle as:

$$\eta_{Br} = \frac{w_{net,out}}{q_{in}} = 1 - \frac{q_{out}}{q_{in}} = 1 - \frac{c_v(T_4 - T_1)}{c_p(T_3 - T_2)} = 1 - \frac{T_1((T_4/T_1) - 1)}{T_2((T_3/T_2) - 1)}$$

By considering the above arrangement of the temperature to find the Brayton cycle efficiency, when we substitute the ideal gas isentropic relations the energy efficiency will be presented in terms of the pressure ratio and air properties as follows:

$$\eta_{Br} = 1 - \frac{1}{r_p^{(k-1)/k}} \tag{6.13}$$

Here, k is the specific heat ratio of the working fluid. Using the definition of the energy efficiency of the Brayton cycle in Eq. (6.13) we can investigate the effect of the pressure ratio on the energy efficiency of the Brayton cycle that is operating between two specified maximum and minimum temperatures in the cycle; the results of this investigation are presented in Figure 6.17 in a temperature-entropy diagram.

Figure 6.17 shows how the temperature-entropy diagram of the Brayton cycle changes with the changing pressure ratio of the cycle that is operating between two specified temperatures. The area enclosed inside the *T-s* diagram equal to the net work produced by the cycle, which means as the area increases the more power the cycle generates. One can see in Figure 6.17 that the pressure ratio of 8.2 resulted in the largest amount of work produced compared to increasing or decreasing the pressure ratio to 15 or 2.

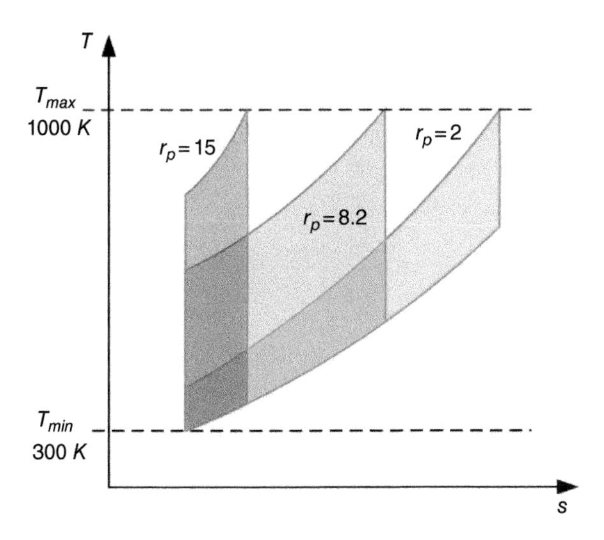

Figure 6.17 Variation of the *T-s* diagram of the Brayton cycle with the variation of the pressure ratio.

Finally, for the Brayton cycle an important concept is the back work ratio, which is the ratio of the work consumed by the compressor (w_c) over the work produced by the turbine (w_t) in the cycle. This can be written as:

$$r_{bw} = \frac{w_c}{w_t} = \frac{(h_2 - h_1)}{(h_3 - h_4)} \tag{6.14a}$$

Considering the working gas as an ideal gas, one can write:

$$r_{bw} = \frac{(T_2 - T_1)}{(T_3 - T_4)} \tag{6.14b}$$

Here, r_{bw} is the back work ratio.

Furthermore, it is important to evaluate the performance of air-standard Brayton cycles based on the general definition of the efficiency:

$$Efficiency = Net\ work\ output/Heat\ input$$

which is commonly used for all heat engines (and power generating cycles).

In this regard, the energy and exergy efficiencies are defined as:

$$\eta_{en} = \frac{\dot{W}_{net}}{\dot{Q}_{in}} \tag{6.15}$$

and

$$\eta_{ex} = \frac{\dot{W}_{net}}{\dot{E}x_{Qin}} \tag{6.16}$$

Before closing this section, the mass, energy, entropy, and exergy balance equations for each component of the cycle shown in Figure 6.15 are shown in Table 6.5.

Table 6.5 The mass, energy, entropy, and exergy balance equations for each component of the Brayton cycle.

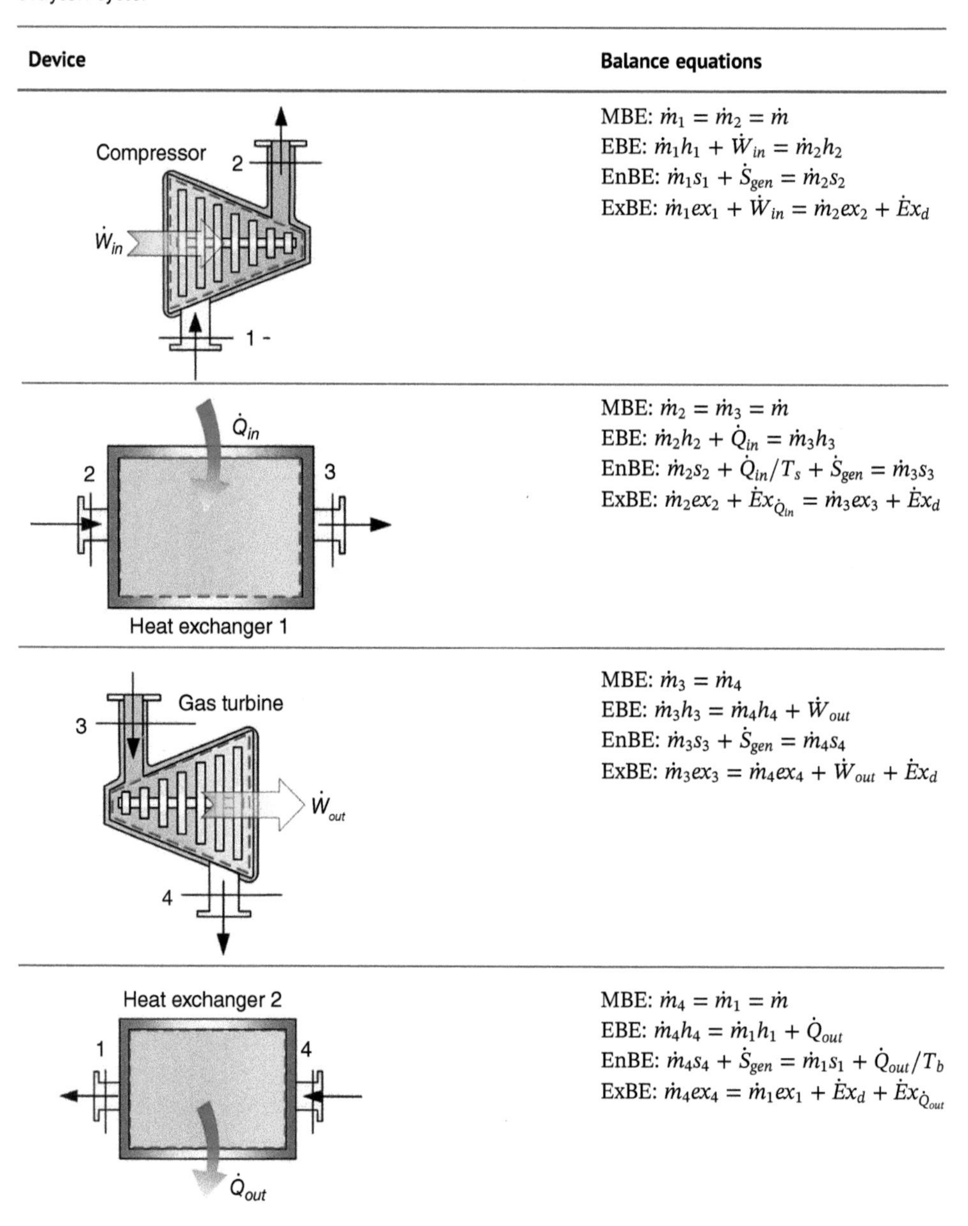

Device	Balance equations
Compressor	MBE: $\dot{m}_1 = \dot{m}_2 = \dot{m}$ EBE: $\dot{m}_1 h_1 + \dot{W}_{in} = \dot{m}_2 h_2$ EnBE: $\dot{m}_1 s_1 + \dot{S}_{gen} = \dot{m}_2 s_2$ ExBE: $\dot{m}_1 ex_1 + \dot{W}_{in} = \dot{m}_2 ex_2 + \dot{E}x_d$
Heat exchanger 1	MBE: $\dot{m}_2 = \dot{m}_3 = \dot{m}$ EBE: $\dot{m}_2 h_2 + \dot{Q}_{in} = \dot{m}_3 h_3$ EnBE: $\dot{m}_2 s_2 + \dot{Q}_{in}/T_s + \dot{S}_{gen} = \dot{m}_3 s_3$ ExBE: $\dot{m}_2 ex_2 + \dot{E}x_{\dot{Q}_{in}} = \dot{m}_3 ex_3 + \dot{E}x_d$
Gas turbine	MBE: $\dot{m}_3 = \dot{m}_4$ EBE: $\dot{m}_3 h_3 = \dot{m}_4 h_4 + \dot{W}_{out}$ EnBE: $\dot{m}_3 s_3 + \dot{S}_{gen} = \dot{m}_4 s_4$ ExBE: $\dot{m}_3 ex_3 = \dot{m}_4 ex_4 + \dot{W}_{out} + \dot{E}x_d$
Heat exchanger 2	MBE: $\dot{m}_4 = \dot{m}_1 = \dot{m}$ EBE: $\dot{m}_4 h_4 = \dot{m}_1 h_1 + \dot{Q}_{out}$ EnBE: $\dot{m}_4 s_4 + \dot{S}_{gen} = \dot{m}_1 s_1 + \dot{Q}_{out}/T_b$ ExBE: $\dot{m}_4 ex_4 = \dot{m}_1 ex_1 + \dot{E}x_d + \dot{E}x_{\dot{Q}_{out}}$

Example 6.11 Consider a closed-type ideal Brayton cycle, as shown in Figure 6.15, operating at a pressure ratio of 12 with the air leaving the heat addition process at a temperature of 600 °C. The air is at a pressure and a temperature of 100 kPa and 15 °C, respectively, when it enters the air compressor in the cycle. Note that both the turbine and the compressor operate isentropically. Treat air as an ideal gas throughout the cycle. (a) Write the mass, energy, entropy, and exergy balance equations of each component in the cycle, (b) find the specific compressor work, and (c) calculate the specific heat addition.

Solution

a) The balance equations for this particular system as shown in Figure 6.15 are the same as given in Table 6.5.

b) Find the compressor work consumption per unit mass of air.

All heat and work interactions occurring through the cycle are found usually from the EBEs of the cycle, except for specific situations where they can be found indirectly through efficiencies and other given information.

The process that has the work consumption process is the compression process 1-2 and the MBE and EBE of that process are:

$$\text{MBE}: \dot{m}_1 = \dot{m}_2 = \dot{m}$$

$$\text{EBE}: \dot{m}_1 h_1 + \dot{W}_{in} = \dot{m}_2 h_2$$

From the above two equations, one has:

$$h_1 + w_{in} = h_2$$

By assuming that the specific heat capacity does not change with the changes in temperature, including the temperature ranges the cycle goes through or the changes in the specific heat with temperature over the temperature range considered in the cycle, then the above equation can be written in terms of the temperature and the specific heats capacities as follows:

$$c_p T_1 + w_{in} = c_p T_2$$

$$1.005\ \frac{kJ}{kgK} \times (15 + 273)\ K + w_{in} = 1.005\ \frac{kJ}{kgK} \times T_2$$

Since the process 1-2 is an isentropic process, then $T_2 = T_{2s}$ is found as follows:

$$\frac{T_{2s}}{T_1} = \left(\frac{P_2}{P_1}\right)^{(k-1)/k} \rightarrow T_{2s} = T_1 \left(\frac{P_2}{P_1}\right)^{(k-1)/k} \rightarrow T_{2s} = T_1\left(r_p\right)^{(k-1)/k}$$

$$T_{2s} = (15 + 273)\,(12)^{(1.4-1)/1.4} = \mathbf{585.8\ K}$$

Going back to the mass basis EBE of the heat addition process:

$$1.005\ \frac{kJ}{kgK} \times (15 + 273)\ K + w_{in} = 1.005\ \frac{kJ}{kgK} \times 585.8\ K$$

$$w_{in} = \mathbf{299.3\ \frac{kJ}{kg}}$$

c) Calculate the heat added to the air in the heat addition step per unit mass of air.

The process that has the heat addition step is the process 2-3 and the MBE and EBE of that process are:

$$\text{MBE}: \dot{m}_2 = \dot{m}_3 = \dot{m}$$

$$\text{EBE}: \dot{m}_2 h_2 + \dot{Q}_{in} = \dot{m}_3 h_3$$

Then from the above two equations:

$$h_2 + q_{in} = h_3$$

By assuming that the specific heat capacity does not change with the changes in temperature, including the temperature ranges the cycle goes through or the changes in the specific heat with temperature over the temperature range considered in the cycle, then the above equation can be written in terms of the temperature and the specific heats capacities as follows, based on $h = c_pT$:

$$c_pT_2 + q_{in} = c_pT_3$$

$$1.005\ \frac{kJ}{kgK} \times 585.8\ K + q_{in} = 1.005\ \frac{kJ}{kgK} \times (600 + 273)$$

$$\mathbf{q_{in} = 288.6\ \frac{kJ}{kg}}$$

Example 6.12 An open-type gas-turbine power plant operating on a Brayton cycle, as shown in Figure 6.18a, has a pressure ratio of 12. Air enters to the compressor at 305 K and 100 kPa with mass flow rate 15 kg/s; it leaves the compressor at 680 K. Air enters the turbine at 1410 K and exits at 853 K. $T_s = 2000\ K$, $T_0 = 293\ K$, and $P_0 = 100\ kPa$.

a) Write mass, energy, entropy, and exergy balance equations for each process.
b) Find the back work ratio.
c) Calculate the isentropic efficiency for the compressor and the turbine.
d) Calculate exergy destruction and the entropy generation for the compressor and turbine.
e) Calculate the energy and exergy efficiencies.

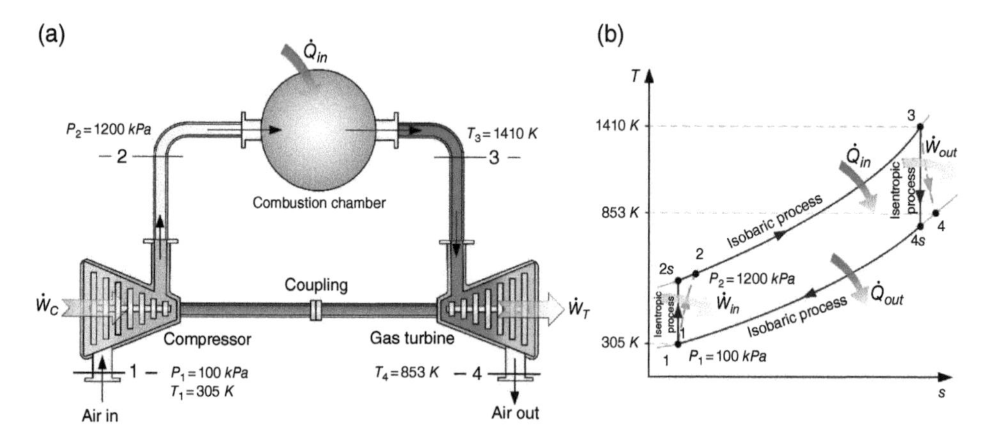

Figure 6.18 (a) An open-type air-standard Brayton cycle and (b) its T-s diagram for Example 6.12.

Solution

It is first important to sketch the schematic diagram of the problem cycle (as shown in Figure 6.18a) along with its *T-s* diagram (as shown in Figure 6.18b) with the information and property data provided. This is necessary before beginning to write the balance equations for mass, energy, entropy, and exergy of each component of the cycle.

a) The mass, energy, entropy, and exergy balance equations can be written as follows:

For the compressor:

$$\text{MBE}: \dot{m}_1 = \dot{m}_2$$

$$\text{EBE}: \dot{m}_1 h_1 + \dot{W}_C = \dot{m}_2 h_2$$

$$\text{EnBE}: \dot{m}_1 s_1 + \dot{S}_{gen,C} = \dot{m}_2 s_2$$

$$\text{ExBE}: \dot{m}_1 ex_1 + \dot{W}_C = \dot{m}_2 ex_2 + \dot{E}x_{d,C}$$

For the combustion chamber:

$$\text{MBE}: \dot{m}_2 = \dot{m}_3$$

$$\text{EBE}: \dot{m}_2 h_2 + \dot{Q}_{CC} = \dot{m}_3 h_3$$

$$\text{EnBE}: \dot{m}_2 s_2 + \frac{\dot{Q}_{CC}}{T_s} + \dot{S}_{gen,Cc} = \dot{m}_3 s_3$$

$$\text{ExBE}: \dot{m}_2 ex_2 + \dot{Q}_{CC}\left(1 - \frac{T_0}{T_s}\right) = \dot{m}_3 ex_3 + \dot{E}x_{d,CC}$$

For the turbine:

$$\text{MBE}: \dot{m}_3 = \dot{m}_4$$

$$\text{EBE}: \dot{m}_3 h_3 = \dot{W}_T + \dot{m}_4 h_4$$

$$\text{EnBE}: \dot{m}_3 s_3 + \dot{S}_{gen,T} = \dot{m}_4 s_4$$

$$\text{ExBE}: \dot{m}_3 ex_3 = \dot{W}_T + \dot{m}_4 ex_4 + \dot{E}x_{d,T}$$

It is now necessary to find T_2 and T_4 as required.

$T_1 = 305\ K,$	$P_1 = 100\ kPa$	$h_1 = 305.5\ kJ/kg$	$s_1 = 5.722\ kJ/kgK$
$\frac{P_2}{P_1} = r = 12$	$P_2 = 1200\ kPa$	$\mathbf{T_2 = 680\ K}$	$s_2 = 5.832\ kJ/kgK$
		$h_2 = 692.1\ kJ/kg$	
$T_3 = 1410\ K$	$P_3 = P_2$	$h_3 = 1528\ kJ/k$	$s_3 = 6.66\ kJ/kgK$
$\mathbf{T_4 = 853\ K}$	$P_4 = P_1$	$h_4 = 880.8\ kJ/kg$	$s_4 = 6.792\ kJ/kgK$

b) Find the back work ratio.

$$\dot{W}_C = \dot{m}_1(h_2 - h_1) = 15\,(692.1\ kJ/kg - 305.5\ kJ/kg) = \mathbf{5800\ kW}$$

$$\dot{W}_T = \dot{m}_1(h_3 - h_4) = 15\,(1528\ kJ/kg - 880.8\ kJ/kg) = \mathbf{9701\ kW}$$

$$\boldsymbol{r_{bw}} = \dot{W}_C/\dot{W}_t = \mathbf{0.5979}$$

c) Calculate the isentropic efficiencies for both the compressor and turbine,
These are calculated based on the enthalpy values obtained:

$P_2 = 1200\ kPa \qquad s_{2s} = s_1 = 5.722\ kJ/kgK \qquad h_{2s} = 621.4\ kJ/kg$

$$\eta_{is,c} = \frac{h_{2s} - h_1}{h_2 - h_1} = \mathbf{81.7\%}$$

$P_4 = 100\ kPa \qquad s_{4s} = s_3 = 6.66\ kJ/kgK \qquad h_{4s} = 774.7\ kJ/kg$

$$\eta_{is,t} = \frac{h_3 - h_4}{h_3 - h_{4s}} = \mathbf{85.91\%}$$

d) Determine the exergy destruction and entropy generation rates for the compressor and the turbine.

$$ex_1 = (h_1 - h_0) - T_0.(s_1 - s_0) = \mathbf{0.231\ kJ/kg}$$

$$ex_2 = (h_2 - h_0) - T_0.(s_2 - s_0) = \mathbf{354.8\ kJ/kg}$$

$$\dot{m}_1 s_1 + \dot{S}_{gen,C} = \dot{m}_2 s_2 \qquad \dot{S}_{gen,C} = \mathbf{1.643\ kW/K}$$

$$\dot{m}_1 ex_1 + \dot{W}_C = \dot{m}_2 ex_2 + \dot{E}x_{d,C} \qquad \dot{E}x_{d,C} = \mathbf{481.3\ kW}$$

$$ex_3 = (h_3 - h_0) - T_0.(s_3 - s_0) = \mathbf{947.5\ kJ/kg}$$

$$ex_4 = (h_4 - h_0) - T_0.(s_4 - s_0) = \mathbf{262.1\ kJ/kg}$$

$$\dot{m}_3 s_3 + \dot{S}_{gen,T} = \dot{m}_4 s_4 \qquad \dot{S}_{gen,T} = \mathbf{1.979\ kW/K}$$

$$\dot{m}_3 ex_3 = \dot{W}_T + \dot{m}_4 ex_4 + \dot{E}x_{d,T} \qquad \dot{E}x_{d,T} = \mathbf{579.8\ kW}$$

e) Find the energy and exergy efficiencies of the cycle.

$$\dot{m}_2 h_2 + \dot{Q}_{CC} = \dot{m}_3 h_3 \qquad \dot{Q}_{CC} = 12531\ kW$$

$$\dot{W}_{net} = \dot{W}_T - \dot{W}_C = 3901\ kW$$

$$\eta_{en} = \frac{\dot{W}_{net}}{\dot{Q}_{in}} = \frac{3901}{12531} = \mathbf{31.13\%}$$

and

$$\eta_{ex} = \frac{\dot{W}_{net}}{\dot{Q}_{in}\left(1 - \frac{T_0}{T_s}\right)} = \frac{3901}{12531\left(1 - \frac{293}{2000}\right)} = \mathbf{36.48\%}$$

Example 6.13 A closed-type gas-turbine power plant, as shown in Figure 6.19a, operating on the simple Brayton cycle has a pressure ratio of 15. The air enters the compressor at atmospheric pressure and a temperature of 30 °C at a rate of 14 m^3/s; it leaves the turbine at 630 °C. The gas exits the combustion chamber at 1327 °C. Assume a compressor isentropic efficiency of 80%. Consider variable specific heats and the source temperature as 2000 K. Take T_o as 293 K and P_o as 100 kPa.

a) Write mass, energy, entropy, and exergy balance equations for each component.
b) Calculate the back work ratio and the net power output.

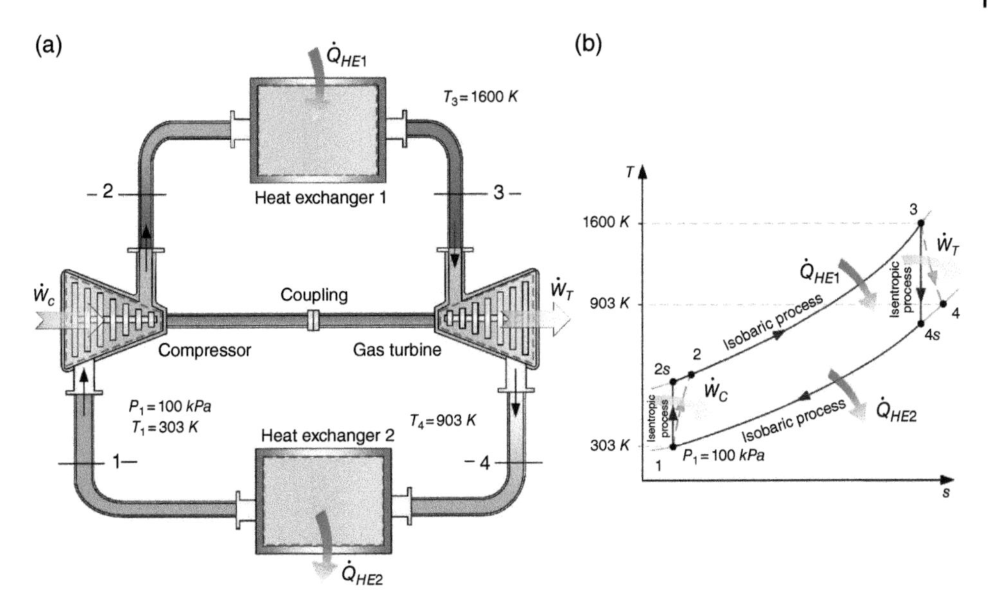

Figure 6.19 (a) A closed-type air-standard Brayton cycle and (b) its *T-s* diagram for Example 6.13.

c) Calculate the isentropic efficiency of the turbine.
d) Calculate exergy destruction for the turbine.
e) Find the energy and exergy efficiencies.

Solution

It is first important to draw the schematic diagram of the problem cycle (as shown in Figure 6.19a) along with its *T-s* diagram (as shown in Figure 6.19b) with the information and property data provided. This is necessary before beginning to write the balance equations for mass, energy, entropy, and exergy of each component of the cycle.

a) Write the mass, energy, entropy, and exergy balance equations for each component of the cycle:

For the compressor:

$$\text{MBE}: \dot{m}_1 = \dot{m}_2$$

$$\text{EBE}: \dot{m}_1 h_1 + \dot{W}_C = \dot{m}_2 h_2$$

$$\text{EnBE}: \dot{m}_1 s_1 + \dot{S}_{gen,C} = \dot{m}_2 s_2$$

$$\text{ExBE}: \dot{m}_1 ex_1 + \dot{W}_C = \dot{m}_2 ex_2 + \dot{E}x_{d,C}$$

For heat exchanger 1:

$$\text{MBE}: \dot{m}_2 = \dot{m}_3$$

$$\text{EBE}: \dot{m}_2 h_2 + \dot{Q}_{HE1} = \dot{m}_3 h_3$$

$$\text{EnBE: } \dot{m}_2 s_2 + \frac{\dot{Q}_{HE1}}{T_s} + \dot{S}_{gen,HE1} = \dot{m}_3 s_3$$

$$\text{ExBE: } \dot{m}_2 ex_2 + \dot{Q}_{HE1}\left(1 - \frac{T_0}{T_s}\right) = \dot{m}_3 ex_3 + \dot{E}x_{d,HE1}$$

For the turbine:

$$\text{MBE: } \dot{m}_3 = \dot{m}_4$$

$$\text{EBE: } \dot{m}_3 h_3 = \dot{W}_T + \dot{m}_4 h_4$$

$$\text{EnBE: } \dot{m}_3 s_3 + \dot{S}_{gen,T} = \dot{m}_4 s_4$$

$$\text{ExBE: } \dot{m}_3 ex_3 = \dot{W}_T + \dot{m}_4 ex_4 + \dot{E}x_{d,T}$$

For heat exchanger 2:

$$\text{MBE: } \dot{m}_4 = \dot{m}_1$$

$$\text{EBE: } \dot{m}_4 h_4 = \dot{m}_1 h_1 + \dot{Q}_{HE2}$$

$$\text{EnBE: } \dot{m}_4 s_4 + \dot{S}_{gen,HE2} = \dot{m}_1 s_1 + \dot{Q}_{HE2}/T_b$$

$$\text{ExBE: } \dot{m}_4 ex_4 = \dot{m}_1 ex_1 + \left(1 - \left(\frac{T_o}{T_b}\right)\right)\dot{Q}_{HE2} + \dot{E}x_{d,HE2}$$

b) Find the back work ratio and the net power output.

$T_1 = 303\ K, P_1 = 100\ kPa$ $\quad h_1 = 303.4\ kJ/kg$ $\quad s_1 = 5.715\ kJ/kgK$

$\frac{P_2}{P_1} = r = 15 \quad P_2 = 1500\ kPa$ $\quad s_{2s} = s_1 = 5.715\ kJ/kgK$ $\quad h_{2s} = 657.6\ kJ/kg$

$\eta_c = \frac{h_{2s} - h_1}{h_2 - h_1} = 0.8$ $\quad h_2 = 746.1\ kJ/kg$

$T_3 = 1600\ K$ $\quad h_3 = 1758\ kJ/kg$ $\quad s_3 = 6.749\ kJ/kgK$

$$\dot{m}_1 = \frac{P_1 V}{RT_1} = \frac{100 \times 14}{0.287 \times 303} = \mathbf{16.1\ kg/s}$$

$T_4 = 903\ K$ $\quad h_4 = 936.6\ kJ/kg$ $\quad s_4 = 6.855\ kJ/kgK$

$$\dot{W}_C = \dot{m}_2(h_2 - h_1) = \mathbf{7127\ kW}$$

$$\dot{W}_T = \dot{m}_3(h_3 - h_4) = \mathbf{13220\ kW}$$

$$\boldsymbol{r_{bw}} = \dot{W}_C/\dot{W}_T = \mathbf{0.5391}$$

$$\dot{Q}_{HE1} = \dot{m}_3 h_3 - \dot{m}_2 h_2 = \mathbf{16286\ kW}$$

$$\dot{Q}_{HE2} = \dot{m}_4 h_4 - \dot{m}_1 h_1 = \mathbf{10193\ kW}$$

$$\boldsymbol{\dot{W}_{net}} = \dot{W}_T - \dot{W}_C = \dot{Q}_{HE1} - \dot{Q}_{HE2} = \mathbf{6093\ kW}$$

c) Find the isentropic efficiency of the turbine.

$s_{4s} = s_3 = 6.749\ kJ/kgK \quad h_{4s} = 844.4\ kJ/kg$

$$\boldsymbol{\eta_{is,t}} = \frac{h_3 - h_4}{h_3 - h_{4s}} = \mathbf{89.95\%}$$

d) Calculate the entropy generation rate and exergy destruction rate of the turbine.

$ex_3 = (h_3 - h_0) - T_0.(s_3 - s_0)$ $\quad ex_4 = (h_4 - h_0) - T_0.(s_4 - s_0)$

$\dot{m}_3 s_3 + \dot{S}_{gen,T} = \dot{m}_4 s_4$ $\quad \mathbf{\dot{S}_{gen,T} = 1.716\ kW/K}$

$\dot{m}_3 ex_3 = \dot{W}_T + \dot{m}_4 ex_4 + \dot{E}x_{d,T}$ $\quad \mathbf{\dot{E}x_{d,T} = 502.7\ kW}$

e) Calculate both the energy and exergy efficiencies of the cycle:

$$\eta_{en} = \frac{\dot{W}_{net}}{\dot{Q}_{HE1}} = \frac{6093\ kW}{16286\ kW} = \mathbf{37.41\%}$$

and

$$\eta_{ex} = \frac{\dot{W}_{net}}{\dot{Q}_{HE1}\left(1 - \frac{T_0}{T_s}\right)} = \frac{6093\ kW}{16286\ kW\left(1 - \frac{293\ K}{2000\ K}\right)} = \mathbf{43.83\%}$$

Example 6.14 A closed-type gas-turbine power plant, as shown in Figure 6.20a, operating on the simple Brayton cycle has a pressure ratio of 16. Air enters the compressor at atmospheric pressure and with a temperature of 27 °*C* at a rate of 10 *kg/s*; it leaves the turbine at 700 °*C*. The gas exits the combustion chamber at 1800 *K*. Assume a compressor isentropic efficiency of 85%. Consider variable specific heats and take $T_s = 2500\ K$. Take T_o as 293 *K* and P_o as 100 *kPa*.

a) Write mass, energy, entropy, and exergy balance equations for each process.
b) Calculate the back work ratio and the net power output.
c) Calculate the isentropic efficiency of the turbine.

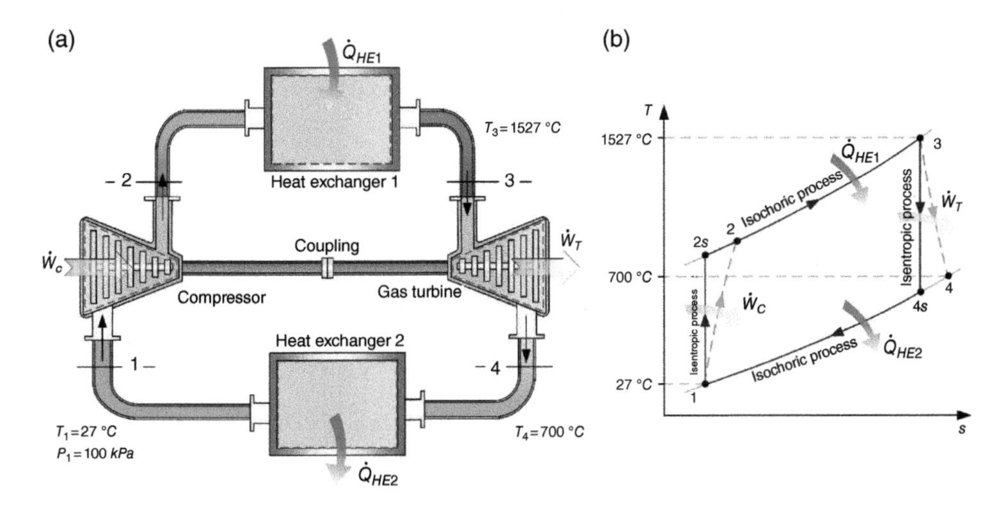

Figure 6.20 (a) A closed-type air-standard Brayton cycle and (b) its *T-s* diagram for Example 6.14.

d) Calculate the entropy generation and exergy destruction rates for the turbine and heat exchanger 1.
e) Find both the energy and exergy efficiencies of the cycle.

Solution

It is first important to draw the schematic diagram of the problem cycle (as shown in Figure 6.20a) along with its *T-s* diagram (as shown in Figure 6.20b) with the information and property data provided. This is necessary before beginning to write the balance equations for mass, energy, entropy, and exergy of each component of the cycle.

a) Write the mass, energy, entropy, and exergy balance equations for each component of the cycle.
For the compressor:

$$\text{MBE}: \dot{m}_1 = \dot{m}_2$$

$$\text{EBE}: \dot{m}_1 h_1 + \dot{W}_C = \dot{m}_2 h_2$$

$$\text{EnBE}: \dot{m}_1 s_1 + \dot{S}_{gen,C} = \dot{m}_2 s_2$$

$$\text{ExBE}: \dot{m}_1 ex_1 + \dot{W}_C = \dot{m}_2 ex_2 + \dot{E}x_{d,C}$$

For heat exchanger 1:

$$\text{MBE}: \dot{m}_2 = \dot{m}_3$$

$$\text{EBE}: \dot{m}_2 h_2 + \dot{Q}_{HE1} = \dot{m}_3 h_3$$

$$\text{EnBE}: \dot{m}_2 s_2 + \frac{\dot{Q}_{HE1}}{T_s} + \dot{S}_{gen,HE1} = \dot{m}_3 s_3$$

$$\text{ExBE}: \dot{m}_2 ex_2 + \dot{Q}_{HE1}\left(1 - \frac{T_0}{T_s}\right) = \dot{m}_3 ex_3 + \dot{E}x_{d,HE1}$$

For the turbine:

$$\text{MBE}: \dot{m}_3 = \dot{m}_4$$

$$\text{EBE}: \dot{m}_3 h_3 = \dot{W}_T + \dot{m}_4 h_4$$

$$\text{EnBE}: \dot{m}_3 s_3 + \dot{S}_{gen,T} = \dot{m}_4 s_4$$

$$\text{ExBE}: \dot{m}_3 ex_3 = \dot{W}_T + \dot{m}_4 ex_4 + \dot{E}x_{d,T}$$

For heat exchanger 2:

$$\text{MBE}: \dot{m}_4 = \dot{m}_1$$

$$\text{EBE}: \dot{m}_4 h_4 = \dot{m}_1 h_1 + \dot{Q}_{HE2}$$

$$\text{EnBE}: \dot{m}_4 s_4 + \dot{S}_{gen,HE2} = \dot{m}_1 s_1 + \dot{Q}_{HE2}/T_b$$

$$\text{ExBE}: \dot{m}_4 ex_4 = \dot{m}_1 ex_1 + \left(1 - \left(\frac{T_o}{T_b}\right)\right)\dot{Q}_{HE2} + \dot{E}x_{d,HE2}$$

b) Find the back work ratio and the net power output.

$T_1 = 300\ K$, $\quad P_1 = 100\ kPa$ $\quad h_1 = 300.4\ kJ/kg$ $\quad s_1 = 5.705\ kJ/kgK$

$\frac{P_2}{P_1} = r = 16 \qquad P_2 = 1600\ kPa \qquad s_{2s} = s_1 = 5.705\ kJ/kgK \qquad h_{2s} = 663.1\ kJ/kg$

$\eta_{is,c} = \frac{h_{2s} - h_1}{h_2 - h_1} = 0.85 \qquad h_2 = 727.1\ kJ/kg$

$T_3 = 1800\ K \qquad h_3 = 2003\ kJ/kg \qquad s_3 = 6.875\ kJ/kgK$

$T_4 = 973\ K \qquad h_4 = 1016\ kJ/kg \qquad s_4 = 6.94\ kJ/kgK$

$\dot{W}_C = \dot{m}_2(h_2 - h_1) = \mathbf{4267\ kW}$

$\dot{W}_T = \dot{m}_3(h_3 - h_4) = \mathbf{9879\ kW}$

$\boldsymbol{r_{bw}} = \dot{W}_C/\dot{W}_T = \mathbf{0.4319}$

$\dot{Q}_{HE1} = \dot{m}_3 h_3 - \dot{m}_2 h_2 = \mathbf{12763\ kW}$

$\dot{Q}_{HE2} = \dot{m}_4 h_4 - \dot{m}_1 h_1 = \mathbf{7151\ kW}$

$\dot{\boldsymbol{W}}_{\boldsymbol{net}} = \dot{W}_T - \dot{W}_C = \dot{Q}_{HE1} - \dot{Q}_{HE2} = \mathbf{5612\ kW}$

c) Find the isentropic efficiency of the turbine.

$s_{4s} = s_3 = 6.875\ kJ/kgK \qquad h_{4s} = 954.4\ kJ/kg$

$\boldsymbol{\eta_{is,t}} = \frac{h_3 - h_4}{h_3 - h_{4s}} = \mathbf{94.17\%}$

d) Find the entropy generation rate and exergy destruction rate for heat exchanger 1 and the turbine.

$ex_2 = (h_2 - h_0) - T_0(s_2 - s_0)$

$ex_3 = (h_3 - h_0) - T_0(s_3 - s_0)$

$ex_4 = (h_4 - h_0) - T_0(s_4 - s_0)$

$\dot{m}_2 s_2 + \frac{\dot{Q}_{HE1}}{T_s} + \dot{S}_{gen,HE1} = \dot{m}_3 s_3 \qquad \dot{\boldsymbol{S}}_{\boldsymbol{gen,HE1}} = \mathbf{5.65\ kW/K}$

$\dot{m}_2 ex_2 + \dot{Q}_{HE1}\left(1 - \frac{T_0}{T_s}\right) = \dot{m}_3 ex_3 + \dot{E}x_{d,HE1} \qquad \dot{\boldsymbol{E}}\boldsymbol{x}_{\boldsymbol{d,HE1}} = \mathbf{1660\ kW}$

$\dot{m}_3 s_3 + \dot{S}_{gen,T} = \dot{m}_4 s_4 \qquad \dot{\boldsymbol{S}}_{\boldsymbol{gen,T}} = \mathbf{0.6468\ kW/K}$

$\dot{m}_3 ex_3 = \dot{W}_T + \dot{m}_4 ex_4 + \dot{E}x_{d,T} \qquad \dot{\boldsymbol{E}}\boldsymbol{x}_{\boldsymbol{d,T}} = \mathbf{189.5\ kW}$

e) Calculate both the energy and exergy efficiencies of the cycle.

$$\boldsymbol{\eta_{en}} = \frac{\dot{W}_{net}}{\dot{Q}_{HE1}} = \frac{5612\ kW}{12763\ kW} = \mathbf{43.97\%}$$

and

$$\boldsymbol{\eta_{ex}} = \frac{\dot{W}_{net}}{\dot{Q}_{HE1}\left(1 - \frac{T_0}{T_s}\right)} = \frac{5612\ kW}{12763\ kW\left(1 - \frac{293\ K}{2500\ K}\right)} = \mathbf{49.81\%}$$

6.7.1 Regenerative Brayton Cycle

The efficiency of the basic Brayton cycle can be increased with some modification, where the external irreversibilities experienced due to the finite temperature differences at the heat sink and source are reduced. The improvements can be achieved by the following modifications:

- Regeneration could be utilized by preheating air after it exits the compressor using the heat recovered from the hot gas exhausted by the turbine. Therefore, the heat input required is reduced by a similar amount of net work generated from the cycle, which raises the efficiency.
- Reheating in this case is performed by installing an interstage reheater in between two or more turbines. In an actual combustion turbine, typically a high and low pressure turbine are installed on the same shaft, with a secondary combustion chamber installed between them to supply heat to the working gas. Therefore, the working fluid is reheated before entering the low pressure turbine. The reheating increases the network output, which increases the cycle efficiency.

Intercooling is performed by carrying out the air compression process in stages with interstage cooling. The intercooling decreases the work consumed by the compressor, therefore decreasing the back work ratio, which increases the overall cycle efficiency.

Two types of air-standard Brayton cycle with a regeneration option are illustrated in Figure 6.21, one for an open and one for a closed type system. The regenerator is used to preheat the working gas (air) coming from the compressor (2) prior to entering the combustion chamber (as seen in Figure 6.21a) or the main heat exchanger (3) (as seen in Figure 6.21b) to get the necessary heat before entering in the gas turbine (4). After the expansion and work rate production, the stream of the hot low-pressure gas exhausted by the turbine (5) is further cooled down by the regenerator after transferring its heat to heat the working gas (air). After the regenerator the stream is either exhausted to the environment (6) as shown in Figure 6.21a or recycled by the means of a heat exchanger (heat exchanger 2), which is part of a closed type cycle. The temperature at the temperature outlet must be higher than that at the compressor exit, as the heat transfer in the regenerator is performed from stream 5-6 to 2-3. Therefore, the constraints of the cycle are $T_5 > T_2$. Furthermore, both *T-s* and *P-v* diagrams of these two regenerative Brayton cycle are shown in Figure 6.22. where it is clearly seen that that the reduction in the heat input requirement is quite noticeable.

One can derive a simple expression for the exergy efficiency of the Brayton cycle with a regenerator, using the air standard and no internal irreversibility assumptions. The no internal irreversibility assumption comes from the implication that the regenerator is well suited to match the cold and hot stream profiles. The outlet temperature of the regenerator at state 3 reaches a similar temperature of the turbine outlet at state 5. Therefore, it could be assumed that $T_3 = T_5$ due to the standard air assumption with a constant specific heat. The energy balance for the regenerator unit includes the following relation:

$$h_3 - h_2 = h_5 - h_6 \rightarrow T_3 - T_2 = T_5 - T_6 \rightarrow T_2 = T_6 \tag{6.17}$$

The assumption of no internal irreversibility's derives from the implication that the cycle experiences isentropic expansion and compression processes. These processes can be

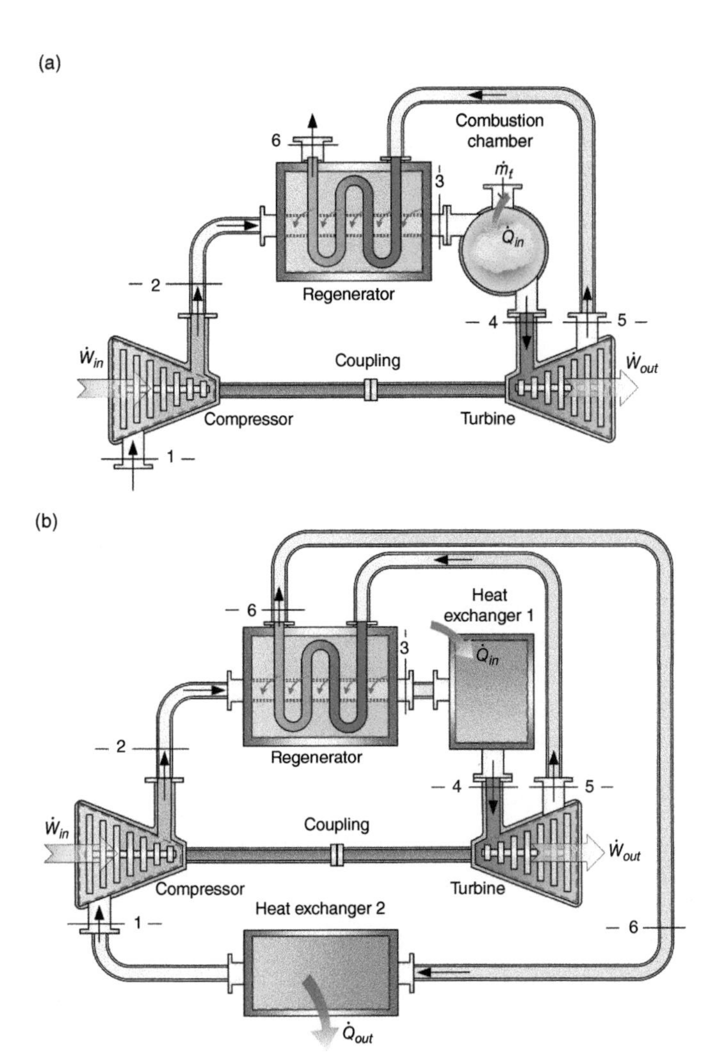

Figure 6.21 Illustration of two types of regenerative Brayton cycle: (a) open type and (b) closed type.

modeled based on the ideal gas equation of state, similar to the simple configuration indicated above.

$$\frac{T_2}{T_1} = \left(\frac{P_2}{P_1}\right)^k = \frac{T_4}{T_5} = \left(\frac{P_4}{P_5}\right)^k \tag{6.18}$$

The energy efficiency is derived from the overall cycle energy balance, which includes that the net work output has to be equal to the heat input minus the heat rejected. Therefore, the energy efficiency is considered to be one minus the heat rejected by the heat input. Using the two equations the energy efficiency becomes:

$$\eta = 1 - \frac{c_p(T_6 - T_1)}{c_p(T_4 - T_3)} = 1 - \frac{(T_1T_2)1 - \left(\frac{T_1}{T_2}\right)}{(T_1T_2)1 - \left(\frac{T_1}{T_2}\right)} = 1 - \tau r^k, where\ \tau = \frac{T_1}{T_4} \tag{6.19}$$

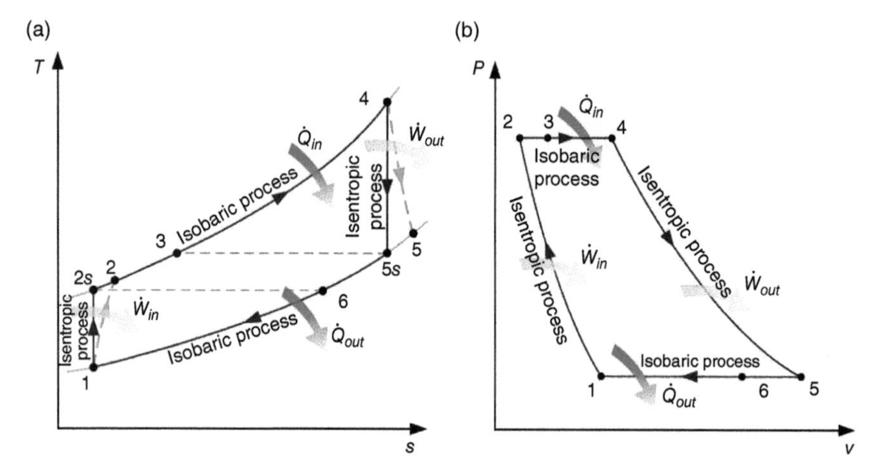

Figure 6.22 Property diagrams of regenerative Brayton cycles: (a) *T-s* diagram and (b) *P-v* diagram.

The above energy efficiency equation includes the ratio m, which represents the relation between the highest and lowest temperatures in the cycles. Given that $T_4 > T_2 = r^k T_1$, one should note that when $m < r^{-k}$ the energy efficiency reduces to zero. Furthermore, if one predefines the condition that the overall cycle efficiency with the inclusion of regeneration must be higher than that without the modification, it must satisfy the following inequality:

$$1 - r^{-k} < 1 - \mathrm{T}r^k \text{ or } \mathrm{T} < r^{2k} \tag{6.20}$$

This more restrictive than the previous condition. It should also be noted that the cycle efficiency reaches the Carnot efficiency regardless of the pressure ratio r, if the heat source temperature is infinity, as $m = 0$ and hence $\eta = 1$. Furthermore, one can use the Efficiency Improvement Factor (EIF) specifically for the regenerative Brayton cycle in comparison to the simple Brayton cycle meaning.

$$EIF = \frac{1 - \mathrm{T}r^k}{1 - r^{-k}} = \frac{\eta_{regeneration}}{\eta_{basic}} \tag{6.21}$$

The above equation illustrates the *EIFs* dependence on the heat sink and source temperature ratio (m) and various pressure ratios r for the standard air assumption. Low pressure ratios and high heat source temperatures produce a better *EIF*.

In this regard, the energy and exergy efficiencies are defined as:

$$\eta_{en} = \frac{\dot{W}_{net}}{\dot{Q}_{in}} \tag{6.22}$$

and

$$\eta_{ex} = \frac{\dot{W}_{net}}{\dot{E}x_{Qin}} \tag{6.23}$$

One should note that the heat input due to regeneration will be less than what is normally needed for a basic air-standard Brayton cycle.

Example 6.15 An industrial plant is equipped with a regenerative gas turbine cycle, as shown in Figure 6.23a. The air inlet conditions at the compressor are 101.325 *kPa* and 27 °*C*. The compressor pressurizes air to 10 times the inlet pressure. In addition, the maximum air temperature in the gas turbine is measured as 790 °*C*. Also, a temperature difference of 10 °*C* is measured between the cold air stream leaving the regenerator and the hot air stream entering the regenerator. The net mechanical power output is 106 *kW* and the heat input source temperature is 2500 *K*. Also, the compressor and turbine operate with isentropic efficiencies of 87% and 85%, respectively.

a) Write the mass energy, entropy, and exergy balance equations for all system components.
b) Determine the heat rejection and addition rate for both the actual and ideal cycles.
c) Determine the entropy generation rate in the combustion chamber.
d) Find the exergy destruction rate in the combustion chamber.
e) Determine the energy and exergy efficiencies of the cycle.

Solution

It is first important to draw the schematic diagram of the problem cycle (as shown in Figure 6.23a) along with its *T-s* diagram (as shown in Figure 6.23b) with the information and property data provided. This is necessary before beginning to write the balance equations for mass, energy, entropy, and exergy of each component of the cycle.

a) Write the mass energy, entropy, and exergy balance equations for all system components.

For the compressor:

$$\text{MBE}: \dot{m}_1 = \dot{m}_2$$

$$\text{EBE}: \dot{m}_1 h_1 + \dot{W}_C = \dot{m}_2 h_2$$

$$\text{EnBE}: \dot{m}_1 s_1 + \dot{S}_{gen,C} = \dot{m}_2 s_2$$

$$\text{ExBE}: \dot{m}_1 ex_1 + \dot{W}_C = \dot{m}_2 ex_2 + \dot{E}x_{d,C}$$

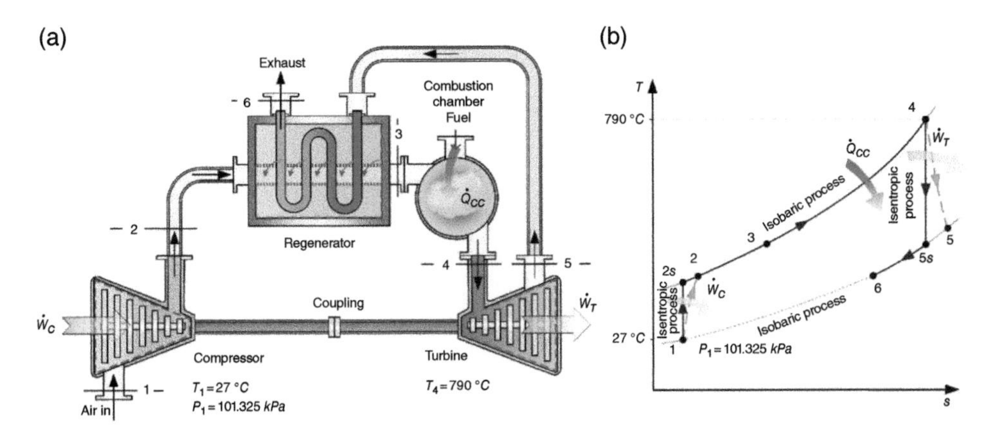

Figure 6.23 (a) A regenerative Brayton cycle and (b) its *T-s* diagram for Example 6.15.

For the regenerator:

MBE: $\dot{m}_2 = \dot{m}_3$ and $\dot{m}_5 = \dot{m}_6$

EBE: $\dot{m}_2 h_2 + \dot{m}_5 h_5 = \dot{m}_3 h_3 + \dot{m}_6 h_6$

EnBE: $\dot{m}_2 s_2 + \dot{m}_5 s_5 + \dot{S}_{gen,reg} = \dot{m}_3 s_3 + \dot{m}_6 s_6$

ExBE: $\dot{m}_2 ex_2 + \dot{m}_5 ex_5 = \dot{m}_3 ex_3 + \dot{m}_6 ex_6 + \dot{E}x_{d,reg}$

For the combustion chamber:

MBE: $\dot{m}_3 = \dot{m}_4$

EBE: $\dot{m}_3 h_3 + \dot{Q}_{CC} = \dot{m}_4 h_4$

EnBE: $\dot{m}_3 s_3 + \dfrac{\dot{Q}_{CC}}{T_s} + \dot{S}_{gen,Cc} = \dot{m}_4 s_4$

ExBE: $\dot{m}_3 ex_3 + \dot{Q}_{CC}\left(1 - \dfrac{T_0}{T_s}\right) = \dot{m}_4 ex_4 + \dot{E}x_{d,CC}$

For the turbine:

MBE: $\dot{m}_4 = \dot{m}_5$

EBE: $\dot{m}_4 h_4 = \dot{W}_T + \dot{m}_5 h_5$

EnBE: $\dot{m}_4 s_4 + \dot{S}_{gen,T} = \dot{m}_5 s_5$

ExBE: $\dot{m}_4 ex_4 = \dot{W}_T + \dot{m}_5 ex_5 + \dot{E}x_{d,T}$

b) Air-standard assumptions are considered along with the assumption of negligible potential and kinetic energy changes. For the ideal cycle, the temperature at state 2 can be found as:

$$T_{2s} = T_1 (c_r)^{\frac{k-1}{k}} = (300\ K)(10)^{\frac{0.4}{1.4}} = \mathbf{579.2\ K}$$

Also, for the actual cycle, T_2 can be found as:

$$T_{2a} = T_1 + \frac{T_{2s} - T_1}{\eta_{is,C}} = 300\ K + \frac{579.2\ K - 300\ K}{0.87} = \mathbf{620.9\ K}$$

Similarly, the isentropic temperature at state 5 can evaluated as:

$$T_{5s} = T_4 \left(\frac{1}{c_r}\right)^{\frac{k-1}{k}} = (1063\ K)\left(\frac{1}{10}\right)^{\frac{0.4}{1.4}} = \mathbf{550.6\ K}$$

Also, for the actual cycle, T_{5a} can be found as:

$$T_5 = T_4 - \eta_{is,exp}(T_4 - T_{5s}) = 1063\ K - (0.85)(1063\ K - 550.6\ K) = \mathbf{627.4\ K}$$

The energy balance of the regenerator can be written as:

$$\dot{m}_2 h_2 + \dot{m}_5 h_5 = \dot{m}_3 h_3 + \dot{m}_6 h_6$$

Assuming constant specific heats, the above equation can be re-arranged as:

$$(T_2 - T_3) = (T_6 - T_5)$$

The relation between T_3 and T_5 is given as:

$$T_3 = T_5 - 10 = 627.4\ K - 10 = \mathbf{617.4\ K}$$

Substituting this in the above equation gives:

$$T_6 = (T_2 - T_3) + T_5 = (620.9\ K - 617.4\ K) + 627.4\ K = \mathbf{631\ K}$$

The net power output is known to be 106 kW. This can be used to find the mass flow rate required for both the actual and ideal cycles:

$$\dot{m}_a = \frac{\dot{W}_{net}}{w_{net,a}} = \frac{106\ kW}{c_p((T_4 - T_5) - (T_2 - T_1))} =$$

$$\frac{106\ kW}{\left(1\ \frac{kJ}{kgK}\right)((1063\ K - 627.4\ K) - (620.9\ K - 300\ K))} = \mathbf{0.9241\ \frac{kg}{s}}$$

$$\dot{m}_{ideal} = \frac{\dot{W}_{net}}{w_{net,ideal}} = \frac{106\ kW}{c_p((T_4 - T_{5s}) - (T_{2s} - T_1))}$$

$$= \frac{106\ kW}{\left(1\ \frac{kJ}{kgK}\right)((1063\ K - 550.6\ K) - (579.2\ K - 300\ K))} = \mathbf{0.455\ \frac{kg}{s}}$$

Further, the heat input rate for both actual and ideal cases can be found from the energy balance of the combustion chamber:

$$\dot{Q}_{CC,a} = \dot{m}_4 h_4 - \dot{m}_3 h_3 = \dot{m}_a c_p (T_4 - T_3) = 0.9241\ \frac{kg}{s}\left(1.005\ \frac{kJ}{kgK}(1063\ K - 617.4\ K)\right) = \mathbf{413.8\ kW}$$

$$\dot{Q}_{CC,ideal} = \dot{m}_4 h_4 - \dot{m}_3 h_3 = \dot{m}_{ideal} c_p (T_4 - T_3) = 0.455\ \frac{kg}{s}\left(1.005\ \frac{kJ}{kgK}(1063\ K - 617.4\ K)\right)$$

$$= \mathbf{203.8\ kW}$$

The heat rejection rate can be evaluated as:

$$\dot{Q}_{rej,a} = \dot{m}_6 h_6 - \dot{m}_1 h_1 = \dot{m}_a (c_p (T_6 - T_1))$$

$$= 0.9241\ \frac{kg}{s}\left(1.005\ \frac{kJ}{kgK}(631\ K - 300\ K)\right) = \mathbf{307.4\ kW}$$

$$\dot{Q}_{rej,ideal} = \dot{m}_6 h_6 - \dot{m}_1 h_1 = \dot{m}_{ideal} (c_p (T_6 - T_1))$$

$$= 0.45\ \frac{kg}{s}\left(1.005\ \frac{kJ}{kgK}(631\ K - 300\ K)\right)$$

$$= \mathbf{149.7\ kW}$$

c) Find the entropy generation rate from the entropy balance of the combustion chamber.

$$\dot{S}_{gen,CC} = \dot{m}_4 s_4 - \dot{m}_3 s_3 - \frac{\dot{Q}_{CC}}{T_s}$$

where $s_4 = 6.376\ \frac{kJ}{kgK}$ at $P_4 = 1013.25\ kPa$ and $T_4 = 790\ ^\circ C$. Also, $s_3 = 6.175\ \frac{kJ}{kgK}$ can be found at $P_3 = 1010.325\ kPa$ and $T_3 = 617.4\ ^\circ C$

Substituting the above values gives:

$$\dot{S}_{gen,CC} = 0.9241\ \frac{kg}{s}\left(6.376\ \frac{kJ}{kgK} - 6.175\ \frac{kJ}{kgK}\right) - \frac{413.8\ kW}{2500\ K} = \mathbf{0.02\ \frac{kW}{K}}$$

d) Find the exergy destruction rate from the exergy balance of the combustion chamber.

$$\dot{m}_3 ex_3 + \dot{Q}_{CC}\left(1 - \frac{T_0}{T_s}\right) = \dot{m}_4 ex_4 + \dot{E}x_{d,CC}$$

where the specific exergy at states 3 and 4 can be found as:

$$\begin{aligned} ex_3 &= (h_3 - h_0) - T_0(s_3 - s_0) \\ &= \left(922.5\ \frac{kJ}{kg} - 300.6\ \frac{kJ}{kg}\right) - 300\ K\left(6.175\ \frac{kJ}{kgK} - 5.697\ \frac{kJ}{kgK}\right) \\ &= \mathbf{480\ kJ/kg} \end{aligned}$$

$$\begin{aligned} ex_4 &= (h_4 - h_0) - T_0(s_4 - s_0) \\ &= \left(1119\ \frac{kJ}{kg} - 300.6\ \frac{kJ}{kg}\right) - 300\ K\left(6.378\ \frac{kJ}{kgK} - 5.697\ \frac{kJ}{kgK}\right) \\ &= \mathbf{615.8\ kJ/kg} \end{aligned}$$

Substituting these values in the ExBE to find the exergy destruction rate:

$$\dot{E}x_{d,CC} = \dot{m}_3 ex_3 + \dot{Q}_{CC}\left(1 - \frac{T_0}{T_s}\right) - \dot{m}_4 ex_4$$

$$= \left(0.9241\ \frac{kg}{s}\right)\left(480\ \frac{kJ}{kg}\right) + 413.8\ kW\left(1 - \frac{300\ K}{2500\ K}\right) - \left(0.9241\ \frac{kg}{s}\right)(615.8\ kJ/kg)$$

$$= \mathbf{238\ kW}$$

e) Find the energy efficiency of the overall cycle.

$$\eta_{en} = \frac{\dot{W}_{net,a}}{\dot{Q}_{CC,a}} = \frac{106\ kW}{413.8\ kW} = \mathbf{25.6\%}$$

The exergy efficiency of the overall cycle is:

$$\eta_{ex} = \frac{\dot{W}_{net}}{\dot{Q}_{CC}\left(1 - \frac{T_0}{T_s}\right)} = \frac{106\ kW}{413.8\ kW\left(1 - \frac{300\ K}{2500\ K}\right)} = \mathbf{29.1\%}$$

Example 6.16 Consider an air standard regenerative Brayton cycle, as shown in Figure 6.24, with a pressure ratio of seven. The minimum temperature in the cycle is measured as 37 °C and the maximum temperature is 877 °C. The isentropic efficiency of the compressor used is specified as 70%. Also, the turbine is specified to have an isentropic efficiency of 85%. The regenerator effectiveness is known to be 65%.

a) Write the mass energy, entropy, and exergy balance equations for all system components.

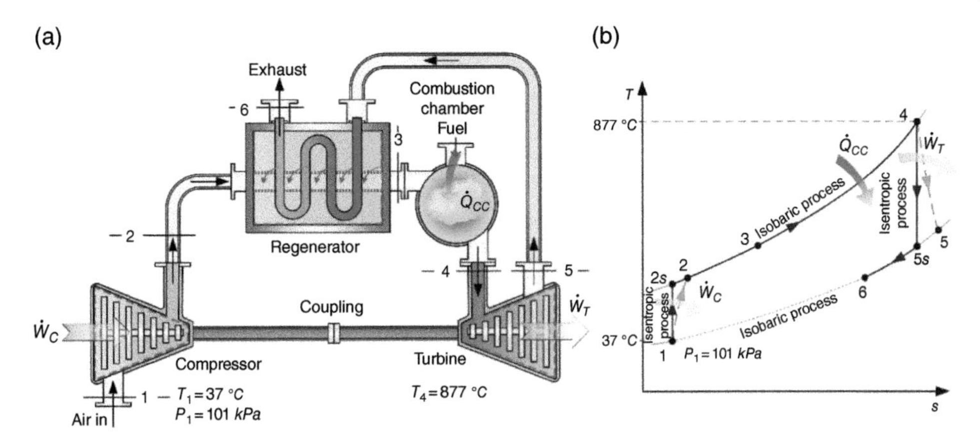

Figure 6.24 (a) A regenerative Brayton cycle and (b) its *T-s* diagram for Example 6.16.

b) Determine the air temperature at the turbine exit.
c) Find the net power output.
d) Determine the entropy generation rates in the compressor if the mass flow rate of air is 2.3 *kg/s*.
e) Find the exergy destruction rate in the turbine with the mass flow rate of 2.3 *kg/s*.
f) Determine the energy and exergy efficiencies of the cycle if the source temperature is 2500 *K*.

Solution

It is first important to draw the schematic diagram of the problem cycle (as shown in Figure 6.24a) along with its *T-s* diagram (as shown in Figure 6.24b) with the information and property data provided. This is necessary before beginning to write the balance equations for mass, energy, entropy, and exergy of each component of the cycle.

a) Write the mass energy, entropy, and exergy balance equations for all system components.

For the compressor:

$$\text{MBE}: \dot{m}_1 = \dot{m}_2$$

$$\text{EBE}: \dot{m}_1 h_1 + \dot{W}_C = \dot{m}_2 h_2$$

$$\text{EnBE}: \dot{m}_1 s_1 + \dot{S}_{gen,C} = \dot{m}_2 s_2$$

$$\text{ExBE}: \dot{m}_1 ex_1 + \dot{W}_C = \dot{m}_2 ex_2 + \dot{E}x_{d,C}$$

For the regenerator:

$$\text{MBE}: \dot{m}_2 = \dot{m}_3 \text{ and } \dot{m}_5 = \dot{m}_6$$

$$\text{EBE}: \dot{m}_2 h_2 + \dot{m}_5 h_5 = \dot{m}_3 h_3 + \dot{m}_6 h_6$$

$$\text{EnBE: } \dot{m}_2 s_2 + \dot{m}_5 s_5 + \dot{S}_{gen,reg} = \dot{m}_3 s_3 + \dot{m}_6 s_6$$

$$\text{ExBE: } \dot{m}_2 ex_2 + \dot{m}_5 ex_5 = \dot{m}_3 ex_3 + \dot{m}_6 ex_6 + \dot{E}x_{d,reg}$$

For the combustion chamber:

$$\text{MBE: } \dot{m}_3 = \dot{m}_4$$

$$\text{EBE: } \dot{m}_3 h_3 + \dot{Q}_{CC} = \dot{m}_4 h_4$$

$$\text{EnBE: } \dot{m}_3 s_3 + \frac{\dot{Q}_{CC}}{T_s} + \dot{S}_{gen,Cc} = \dot{m}_4 s_4$$

$$\text{ExBE: } \dot{m}_3 ex_3 + \dot{Q}_{CC}\left(1 - \frac{T_0}{T_s}\right) = \dot{m}_4 ex_4 + \dot{E}x_{d,CC}$$

For the turbine:

$$\text{MBE: } \dot{m}_4 = \dot{m}_5$$

$$\text{EBE: } \dot{m}_4 h_4 = \dot{W}_T + \dot{m}_5 h_5$$

$$\text{EnBE: } \dot{m}_4 s_4 + \dot{S}_{gen,T} = \dot{m}_5 s_5$$

$$\text{ExBE: } \dot{m}_4 ex_4 = \dot{W}_T + \dot{m}_5 ex_5 + \dot{E}x_{d,T}$$

b) The temperature at state 1 is given as $37\,°C = 310\,K$. The properties at this temperature can be found from tables (such as Appendix E-1) as:

$$h_1 = 310.24\ \frac{kJ}{kg} \text{ and } P_{r1} = 1.555$$

Next, the pressure ratio can be used to find P_{r2} as:

$$\frac{P_{r2}}{P_{r1}} = \frac{P_2}{P_1}$$

$$P_{r2} = P_{r1}\left(\frac{P_2}{P_1}\right) = 1.555\,(7) = \mathbf{10.885}$$

Also, the isentropic enthalpy can be found from the thermodynamics properties tables (such as Appendix E-1) as (such as Appendix E-1) P_{r2} as $h_{2s} = 541.26\ kJ/kg$.

The isentropic efficiency of the compressor is given as 70%. This can be used to find the enthalpy at the compressor exit:

$$\eta_{is,C} = \frac{h_{2s} - h_1}{h_2 - h_1}$$

$$h_2 = h_1 + \frac{h_{2s} - h_1}{\eta_{is,C}} = \left(310.24\ \frac{kJ}{kg}\right) + \frac{541.26\ \frac{kJ}{kg} - 310.24\ \frac{kJ}{kg}}{0.7} = \mathbf{640.3\ \frac{kJ}{kg}}$$

Next, since the highest temperature in cycle occurs at state 4, $T_4 = 877\,°C = 1150\,K$. Other properties at state 4 can be found from tables (such as Appendix E-1) or the *EES* as:

$$h_4 = 1219.25\ \frac{kJ}{kg} \text{ and } P_{r4} = 200.15$$

The pressure ratio can also be used to find P_{r5}:

$$P_{r5} = P_{r4}\left(\frac{P_5}{P_4}\right) = 200.15\left(\frac{1}{7}\right) = \mathbf{28.59}$$

The isentropic enthalpy h_{5s} can be found at P_{r5} as $h_{5s} = 711.8\ \frac{kJ}{kg}$.

The isentropic efficiency of the turbine can be used to find the actual enthalpy at state 4:

$$h_5 = h_4 - \eta_{is,T}(h_4 - h_{5s}) = 1219.25\ \frac{kJ}{kg} - 0.85\left(1219.25\ \frac{kJ}{kg} - 711.8\ \frac{kJ}{kg}\right) = \mathbf{787.9\ \frac{kJ}{kg}}$$

Using the above value of h_5, the temperature at turbine exit can be found from tables (such as Appendix E-1) or the *EES* as:

$$\boldsymbol{T_5 = 495.4\ ^oC}$$

c) Find the net power output from the cycle.
This can be found by subtracting the compressor work from the turbine work:

$$w_{net} = w_t - w_c = (h_4 - h_5) - (h_2 - h_1)$$

$$= \left(1219.25\ \frac{kJ}{kg} - 787.9\ \frac{kJ}{kg}\right) - \left(640.3\ \frac{kJ}{kg} - 310.24\ \frac{kJ}{kg}\right) = \mathbf{101.3\ \frac{kJ}{kg}}$$

d) Determine the entropy generation rate in the compressor.
This can be evaluated from the entropy balance as:

$$\dot{S}_{gen,C} = \dot{m}_2(s_2 - s_1)$$

The entropy at state 2 (s_2) can be found from P_2 and h_2 from a thermodynamic property database considering air enters at atmospheric pressure at state 1 (101 *kPa*) and is compressed with a pressure ratio of 7–707 *kPa*:

$$s_2 = 5.904\ \frac{kJ}{kgK}$$

Similarly, the entropy at state 1 can be found from $P_1 = 101\ kPa$ and $T_1 = 37\ °C$ as:

$$s_1 = 5.736\ \frac{kJ}{kgK}$$

Substituting these in the above equation gives:

$$\dot{S}_{gen,C} = 2.3\ \frac{kg}{s}\left(5.904\ \frac{kJ}{kgK} - 5.736\ \frac{kJ}{kgK}\right) = \mathbf{0.3864\ \frac{kW}{K}}$$

e) Find the exergy destruction rate in the turbine.
This can be found by applying the ExBE:

$$\dot{Ex}_{d,T} = \dot{m}_4(ex_4 - ex_5) - \dot{W}_T$$

where ex_4 and ex_5 can be found as:

$$ex_4 = (h_4 - h_0) - T_0(s_4 - s_0)$$
$$= \left(1219.25\ \frac{kJ}{kg} - 300.6\ \frac{kJ}{kg}\right) - 303\ K\left(6.571\ \frac{kJ}{kgK} - 5.697\ \frac{kJ}{kgK}\right)$$
$$= \mathbf{654.2\ kJ/kg}$$

$$ex_5 = (h_5 - h_0) - T_0(s_5 - s_0)$$
$$= \left(787.9\ \frac{kJ}{kg} - 300.6\ \frac{kJ}{kg}\right) - 303\ K\left(6.674\ \frac{kJ}{kgK} - 5.697\ \frac{kJ}{kgK}\right)$$
$$= \mathbf{191\ kJ/kg}$$

Substituting these specific exergy values gives:

$$\dot{E}x_{d,T} = 2.3\ \frac{kg}{s}\left(654.2\ \frac{kJ}{kg} - 191\ \frac{kJ}{kg}\right) - 2.3\ \frac{kg}{s}\left(1219.25\ \frac{kJ}{kg} - 787.9\ \frac{kJ}{kg}\right) = \mathbf{73.3\ kW}$$

f) Find the energy efficiency of the cycle.

$$\eta_{en} = \frac{\dot{W}_{net}}{\dot{Q}_{in}}$$

The effectiveness of the regenerator is:

$$\epsilon = \frac{h_3 - h_2}{h_5 - h_2}$$

The above equation can be rearranged to solve for h_3:

$$h_3 = \epsilon(h_5 - h_2) + h_2 = 0.65\left(787.9\ \frac{kJ}{kg} - 640.3\ \frac{kJ}{kg}\right) + 640.3\ \frac{kJ}{kg} = \mathbf{736.2\ \frac{kJ}{kg}}$$

The energy efficiency of the overall regenerative Brayton cycle is defined as:

$$\eta_{en} = \frac{\dot{W}_{net}}{\dot{Q}_{in}} = \frac{\left(2.3\ \frac{kg}{s}\right)\left(101.3\ \frac{kJ}{kg}\right)}{\left(2.3\ \frac{kg}{s}\right)\left(1219.25\ \frac{kJ}{kg} - 736.2\ \frac{kJ}{kg}\right)} = \mathbf{20.9\%}$$

The overall exergy efficiency is evaluated as:

$$\eta_{ex} = \frac{\dot{W}_{net}}{\dot{Q}_{in}\left(1 - \frac{T_0}{T_s}\right)} = \frac{\left(2.3\ \frac{kg}{s}\right)\left(101.3\ \frac{kJ}{kg}\right)}{(1111\ kW)\left(1 - \frac{303\ K}{2500\ K}\right)} = \mathbf{23.9\%}$$

6.8 Rankine Cycle

In the previous Brayton cycle, the working gas was air, leading to the name the air-standard Brayton cycle. It is sometimes called a gas turbine cycle. In this section, the focus is on another very significant heat engine cycle, which was discovered by in 1859 by a Scottish engineer, William J.M. Rankine, the so-called Rankine cycle. This cycle is also recognized as a steam engine cycle, steam turbine cycle, steam Rankine cycle, etc. In the cycle, the working fluid is water, which goes through phase changes to be steam at high temperature and high pressure with high work potential (and energy content) and generate work in a steam turbine. This is a heat-driven cycle and what makes it the heat engine cycle. It has been the most widely used cycle since the industrial revolution.

The Rankine cycle consists of four key components, namely a pump, a boiler (or a heat exchanger depending on the heat source), a steam turbine, and a condenser (Figure 6.25a). In addition, a list of processes of the ideal and actual steam Rankine cycles is shown in Table 6.6; these processes are also clearly seen in Figure 6.25b on a T-s diagram.

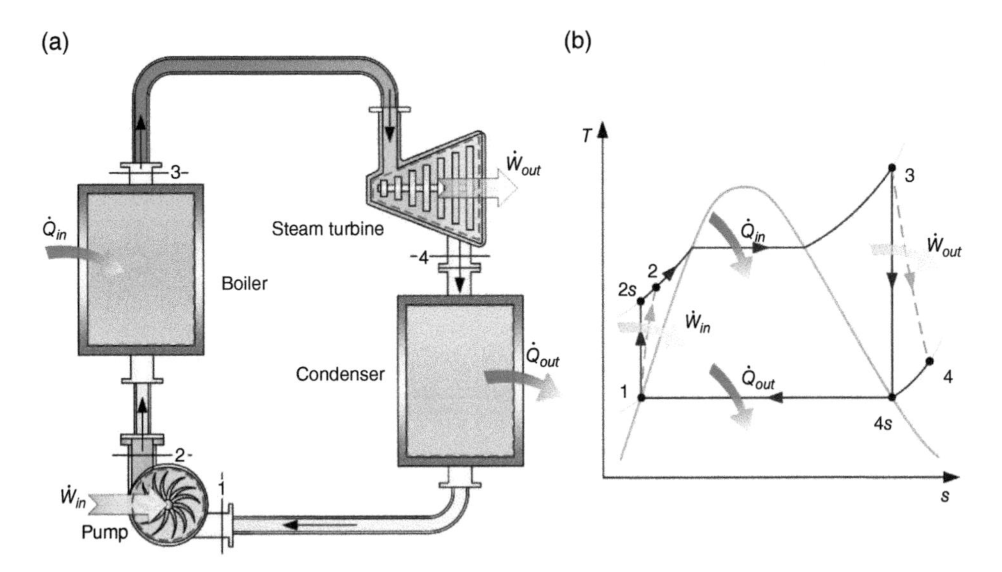

Figure 6.25 (a) A schematic diagram of a simple Rankine cycle and (b) its T-s diagram showing both ideal and actual behaviors for the pump and steam turbine.

Table 6.6 A list of processes of the ideal and actual steam Rankine cycles.

Process	Ideal cycle	Actual cycle
Adiabatic pumping (compression)	Isentropic (1-2*s*)	Nonisentropic (1-2)
Isobaric	Heat addition (2*s*-3)	Heat addition (2-3)
Expansion	Isentropic (3-4*s*)	Nonisentropic (3-4)
Isobaric	Heat rejection (4*s*-1)	Heat rejection (4-1)

Table 6.7 The mass, energy, entropy, and exergy balance equations for each component of a simple Rankine cycle.

Device	Balance equation
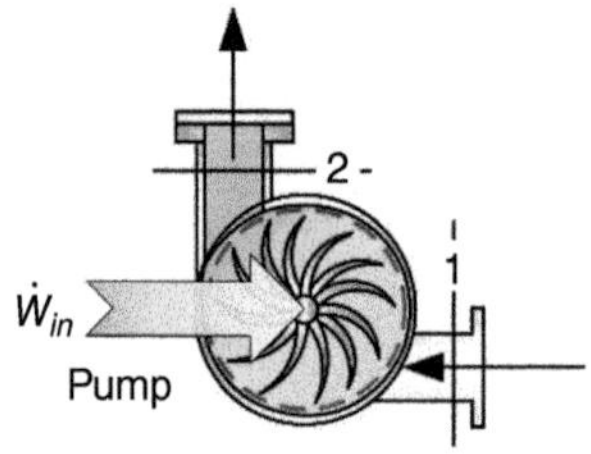 2 1 $\dot{W}_{in}$ Pump	MBE: $\dot{m}_1 = \dot{m}_2 = \dot{m}$ EBE: $\dot{m}_1 h_1 + \dot{W}_{in} = \dot{m}_2 h_2$ EnBE: $\dot{m}_1 s_1 + \dot{S}_{gen,p} = \dot{m}_2 s_2$ ExBE: $\dot{m}_1 ex_1 + \dot{W}_{in} = \dot{m}_2 ex_2 + \dot{E}x_{d,p}$
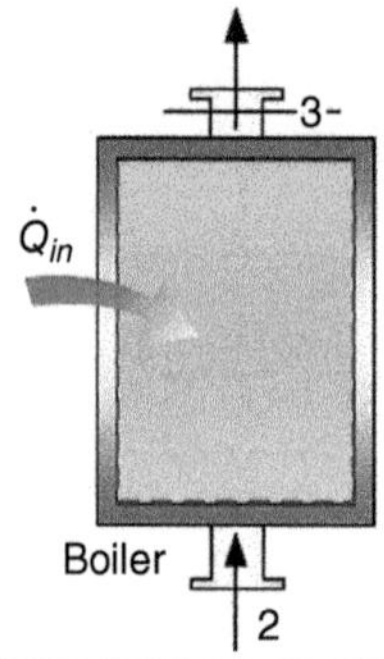 3 $\dot{Q}_{in}$ Boiler 2	MBE: $\dot{m}_2 = \dot{m}_3 = \dot{m}$ EBE: $\dot{m}_2 h_2 + \dot{Q}_{in} = \dot{m}_3 h_3$ EnBE: $\dot{m}_2 s_2 + \dot{Q}_{in}/T_s + \dot{S}_{gen,bo} = \dot{m}_3 s_3$ ExBE: $\dot{m}_2 ex_2 + \dot{E}x_{\dot{Q}_{in}} = \dot{m}_3 ex_3 + \dot{E}x_{d,bo}$
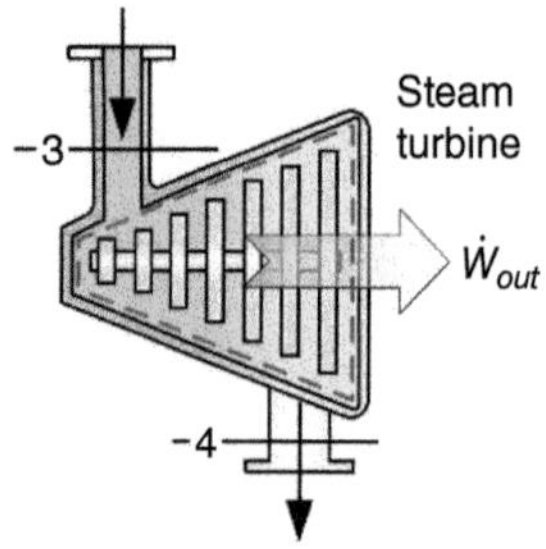 Steam turbine 3 $\dot{W}_{out}$ 4	MBE: $\dot{m}_3 = \dot{m}_4 = \dot{m}$ EBE: $\dot{m}_3 h_3 = \dot{m}_4 h_4 + \dot{W}_{out}$ EnBE: $\dot{m}_3 s_3 + \dot{S}_{gen,st} = \dot{m}_4 s_4$ ExBE: $\dot{m}_3 ex_3 = \dot{m}_4 ex_4 + \dot{W}_{out} + \dot{E}x_{d,st}$
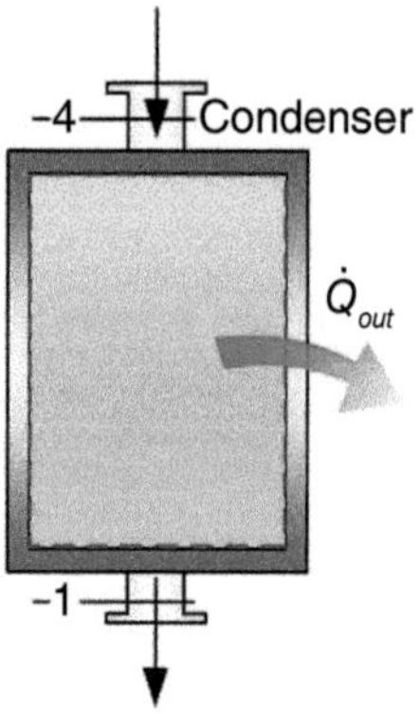 4 Condenser $\dot{Q}_{out}$ 1	MBE: $\dot{m}_4 = \dot{m}_1 = \dot{m}$ EBE: $\dot{m}_4 h_4 = \dot{m}_1 h_1 + \dot{Q}_{co}$ EnBE: $\dot{m}_4 s_4 + \dot{S}_{gen,co} = \dot{m}_1 s_1 + \dot{Q}_{out}/T_b$ ExBE: $\dot{m}_4 ex_4 = \dot{m}_1 ex_1 + \dot{E}x_{\dot{Q}_{out}} + \dot{E}x_{d,co}$

When one looks at the operating principle of the steam Rankine cycle, it has four components along with the four processes listed in Table 6.6. This is a closed cycle and there is no equivalent open version. In its operation, water, as saturated liquid water, is sent to a pump, where its pressure is raised, before being sent to a boiler. The boiler is responsible for converting the water into high pressure and high temperature steam using thermal energy (heat) supplied from an external source. Then high pressure and high temperature steam exiting the boiler enters the steam turbine, which converts part of the thermal and pressure energy stored in the steam to shaft work. The shaft work is used to rotate the shaft of the electric generator to produce electricity. The water exiting the steam turbine, usually a mixture of liquid and vapor, is sent to the condenser, where it is converted to the liquid phase by rejecting heat. Changing the orientation or the location of any one of the devices in the cycle will result in damaging the overall system and the Rankine cycle will no longer be capable of generating electricity. For a better understanding of how these cycles work, the following example considers a Rankine cycle and analyzes it using energy and exergy analyses. The variation in temperature of the water (the working fluid of the cycle) with the variation of the specific entropy of the cycle is shown in Figure 6.25a. In this regard, the mass, energy, entropy, and exergy balance equations for each device and component of the cycle are shown in Table 6.7.

With regard to the performance of steam Rankine cycle, both the energy and exergy efficiencies of the steam Rankine cycle will be used, as defined as:

$$\eta_{en} = \frac{\dot{W}_{net}}{\dot{Q}_{in}} = 1 - \frac{\dot{Q}_{out}}{\dot{Q}_{in}}$$

and

$$\eta_{ex} = \frac{\dot{W}_{net}}{\dot{E}x_{\dot{Q}_{in}}} = \frac{\dot{W}_{net}}{\left(1 - \frac{T_0}{T_s}\right)\dot{Q}_{in}}$$

Example 6.17 Consider an actual Rankine cycle, as shown in Figure 6.26a, with an inlet pressure to the turbine of 7 *MPa* and a temperature of 500 °*C*; the turbine exit pressure is 100 *kPa*. The isentropic efficiency of the turbine is 94%; the pump, condenser, and boiler pressure losses are neglected. The temperature of the water leaving the condenser is 50 °*C* and the mass flow rate in the cycle is 20 *kg/s* of water.

a) Write the mass, energy, entropy, and exergy balance equations of each component in the cycle.
b) Calculate the heat addition to the boiler.
c) Calculate the net power produced.
d) Calculate the energy and exergy efficiencies of the cycle.

Solution

It is first important to draw the schematic diagram of the problem cycle (as shown in Figure 6.26a) along with its *T-s* diagram (as shown in Figure 6.26b) with the information

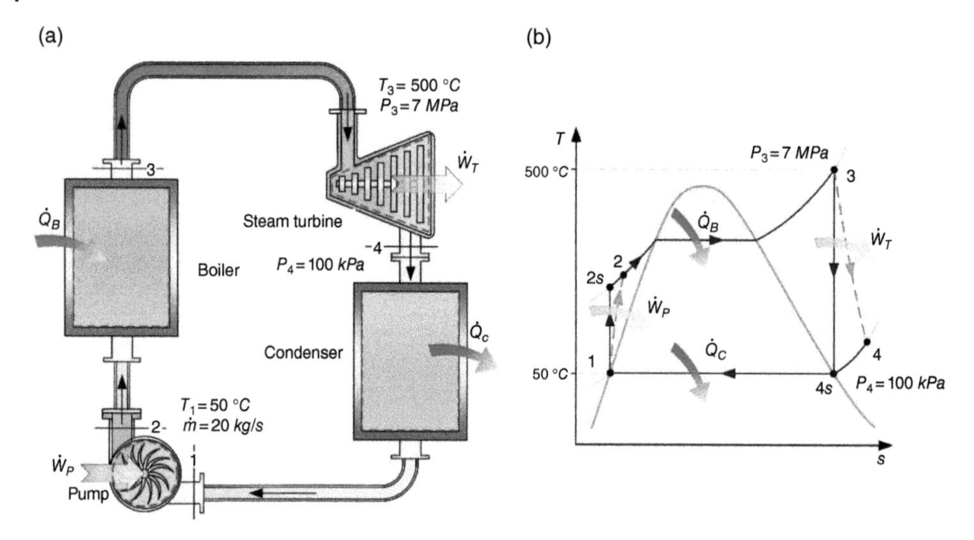

Figure 6.26 (a) Schematic diagram and (b) *T-s* diagram of the steam Rankine cycle discussed in Example 6.17.

and property data provided. This is necessary before beginning to write the balance equations for mass, energy, entropy, and exergy of each component of the cycle.

a) Write the balance equations for each component of the cycle for mass, energy, entropy, and exergy.

For the pump:

$$\text{MBE}: \dot{m}_1 = \dot{m}_2 = \dot{m}$$

$$\text{EBE}: \dot{m}_1 h_1 + \dot{W}_p = \dot{m}_2 h_2$$

$$\text{EnBE}: \dot{m}_1 s_1 + \dot{S}_{gen,p} = \dot{m}_2 s_2$$

$$\text{ExBE}: \dot{m}_1 ex_1 + \dot{W}_p = \dot{m}_2 ex_2 + \dot{E}x_{d,P}$$

For the boiler:

$$\text{MBE}: \dot{m}_2 = \dot{m}_3 = \dot{m}$$

$$\text{EBE}: \dot{m}_2 h_2 + \dot{Q}_B = \dot{m}_3 h_3$$

$$\text{EnBE}: \dot{m}_2 s_2 + \dot{Q}_B/T_s + \dot{S}_{gen,B} = \dot{m}_3 s_3$$

$$\text{ExBE}: \dot{m}_2 ex_2 + \dot{E}x_{\dot{Q}_{in}} = \dot{m}_3 ex_3 + \dot{E}x_{d,B}$$

For the turbine:

$$\text{MBE}: \dot{m}_3 = \dot{m}_4 = \dot{m}$$

$$\text{EBE}: \dot{m}_3 h_3 = \dot{m}_4 h_4 + \dot{W}_T$$

$$\text{EnBE}: \dot{m}_3 s_3 + \dot{S}_{gen,T} = \dot{m}_4 s_4$$

$$\text{ExBE}: \dot{m}_3 ex_3 = \dot{m}_4 ex_4 + \dot{W}_T + \dot{E}x_{d,T}$$

For the condenser:

$$\text{MBE}: \dot{m}_4 = \dot{m}_1 = \dot{m}$$

$$\text{EBE}: \dot{m}_4 h_4 = \dot{m}_1 h_1 + \dot{Q}_C$$

$$\text{EnBE}: \dot{m}_4 s_4 + \dot{S}_{gen,C} = \dot{m}_1 s_1 + \dot{Q}_C / T_b$$

$$\text{ExBE}: \dot{m}_4 ex_4 = \dot{m}_1 ex_1 + \dot{E}x_{\dot{Q}C} + \dot{E}x_{d,C}$$

b) Using the energy balance equation and going back to the EBE of the boiler, we can see that in order to calculate the thermal energy that the boiler receives from an external source we first need to find the enthalpy entering the boiler. The enthalpy of the stream entering the boiler is the same the enthalpy of the stream exiting the pump:

$$w_p = v_1(P_2 - P_1) = 0.001012 \times (7000 - 100) = \mathbf{6.984\ kJ/kg}$$

where v_1 is taken as the specific volume of the saturated liquid (so it is v_f at T_1 or P_1 given).

We now proceed to calculate: $w_p = h_2 - h_1 \rightarrow 6.984 = \text{h}_2 - 209.4$

$$h_2 = 216.4\ kJ/kg$$

$$q_B = h_3 - h_2 = 3411 - 216.4 = \mathbf{3194\ kJ/kg}$$

$$\dot{Q}_B = \dot{m} \times q_B = 20 \times 3194 = \mathbf{63883\ kW}$$

c) Calculate the net power produced.

In the cycle there is the pump that consumes power and the steam turbine that produces power. The power produced by the turbine is calculated based on the energy balance over the turbine:

$$w_{is,T} = h_3 - h_{4s} = 3411 - 2675 = \mathbf{736\ kJ/kg}$$

$$w_T = w_{is,T}/\eta_{is,T} = 736/0.94 = \mathbf{783\ kJ/kg}$$

$$w_{net} = w_T - w_p = 783 - 6.984 = \mathbf{776.016\ kJ/kg}$$

$$\dot{W}_{net} = \dot{m} \times w_{net} = 20 \times 776.016 = \mathbf{15520.32\ kW}$$

d) Calculate the energy and the exergy efficiency equations for the overall cycle.

Since no mass enters or leaves the overall Rankine cycle, it is treated as a closed energy system and based on that the efficiencies are derived. Therefore, the energy efficiency of the cycle is obtained as:

$$\eta_{en} = \frac{\dot{W}_{net}}{\dot{Q}_B} = 15520.32/63883 = \mathbf{24.29\%}$$

The exergy efficiency is then obtained as

$$\eta_{ex} = \frac{\dot{W}_{net}}{\dot{E}x_{\dot{Q}_B}} = 15520.32/42076 = \mathbf{36.89\%}$$

Example 6.18 In a Rankine cycle using water as working fluid, as shown in Figure 6.27a, steam enters the turbine at a temperature of 400 °C and at 3 MPa pressure. Assume the turbine isentropic efficiency to be 85%. The turbine outlet is condensed at 50 kPa pressure. The source temperature is 600 °C.

a) Write the mass, energy, entropy, and exergy balance equations.
b) Calculate the net work generated by the both actual and isentropic cycles.
c) Determine input heat in the boiler.
d) Calculate the efficiencies for both the actual and isentropic cycles.

Solution

It is first important to draw the schematic diagram of the problem cycle (as shown in Figure 6.27a) along with its *T-s* diagram (as shown in Figure 6.27b) with the information and property data provided. This is necessary before beginning to write the balance equations for mass, energy, entropy, and exergy of each component of the cycle.

a) The changes in the kinetic and potential energies of the system are negligible since the system velocity and elevation do not change throughout the process.

   Based on the above mentioned assumptions, the mass, energy, entropy, and exergy balance for each component are written as:

   For the pump:

$$\text{MBE}: \dot{m}_1 = \dot{m}_2 = \dot{m}$$

$$\text{EBE}: \dot{m}_1 h_1 + \dot{W}_{in} = \dot{m}_2 h_2$$

$$\text{EnBE}: \dot{m}_1 s_1 + \dot{S}_{gen,P} = \dot{m}_2 s_2$$

$$\text{ExBE}: \dot{m}_1 ex_1 + \dot{W}_{in} = \dot{m}_2 ex_2 + \dot{E}x_{d,P}$$

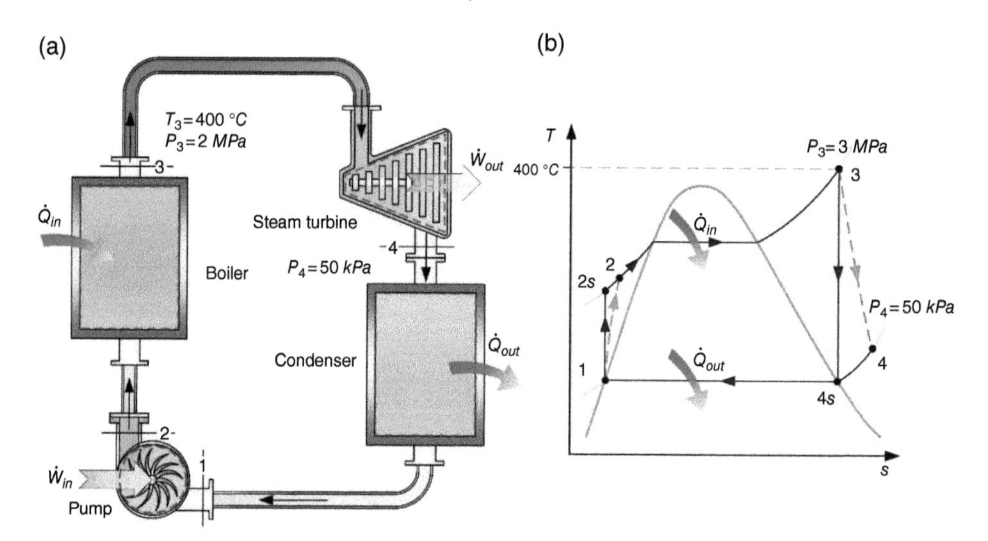

Figure 6.27 (a) Schematic diagram and (b) *T-s* diagram of the steam Rankine cycle discussed in Example 6.18.

For the boiler:

$$\text{MBE}: \dot{m}_2 = \dot{m}_3 = \dot{m}$$

$$\text{EBE}: \dot{m}_2 h_2 + \dot{Q}_{in} = \dot{m}_3 h_3$$

$$\text{EnBE}: \dot{m}_2 s_2 + \frac{\dot{Q}_{in}}{T_s} + \dot{S}_{gen,B} = \dot{m}_3 s_3$$

$$\text{ExBE}: \dot{m}_2 ex_2 + \dot{Q}_{in}\left(1 - \frac{T_o}{T_s}\right) = \dot{m}_3 ex_3 + \dot{E}x_{d,B}$$

For the turbine:

$$\text{MBE}: \dot{m}_3 = \dot{m}_4 = \dot{m}$$

$$\text{EBE}: \dot{m}_3 h_3 = \dot{m}_4 h_4 + \dot{W}_{out}$$

$$\text{EnBE}: \dot{m}_3 s_3 + \dot{S}_{gen,T} = \dot{m}_4 s_4$$

$$\text{ExBE}: \dot{m}_3 ex_3 = \dot{m}_4 ex_4 + \dot{W}_{out} + \dot{E}x_{d,T}$$

For the condenser:

$$\text{MBE}: \dot{m}_4 = \dot{m}_1 = \dot{m}$$

$$\text{EBE}: \dot{m}_4 h_4 = \dot{m}_1 h_1 + \dot{Q}_{out}$$

$$\text{EnBE}: \dot{m}_4 s_4 + \dot{S}_{gen,C} = \dot{m}_1 s_1 + \frac{\dot{Q}_{out}}{T_b}$$

$$\text{ExBE}: \dot{m}_4 ex_4 = \dot{m}_1 ex_1 + \dot{Q}_{out}\left(1 - \frac{T_o}{T_b}\right) + \dot{E}x_{d,C}$$

b) Calculate the net work generated by the both actual and isentropic cycles.
State 1: $P_1 = 50\ kPa$
Sat. liquid

$$h_1 = 340.5\ \frac{kJ}{kg}$$

$$v_1 = 0.00103\ \frac{m^3}{kg}$$

State 2: $P_2 = 3\ MPa$

$$s_2 = s_1$$

$$w_{in} = v_1(P_2 - P_1) = \left(0.00103\ \frac{m^3}{kg}\right)(3000 - 50)\ kPa = \mathbf{3.038}\ \frac{\mathbf{kJ}}{\mathbf{kg}}$$

$$h_2 = h_1 + w_{in} = (340.5 + 3.038)\ \frac{kJ}{kg} = \mathbf{343.5}\ \frac{\mathbf{kJ}}{\mathbf{kg}}$$

State 3: $P_3 = 3\ MPa$

$$T_3 = 400\ °C$$

$$h_3 = 3231.7\ \frac{kJ}{kg}$$

$$s_3 = 6.923 \frac{kJ}{kgK}$$

State 4: $P_4 = 50\ kPa$

$$s_{4s} = s_3$$

$$x_4 = \frac{s_{4s} - s_f}{s_{fg}}$$

$$x_4 = 0.897$$

$$h_{4s} = h_f + x_4 h_{fg} = \mathbf{2407.9} \frac{\mathbf{kJ}}{\mathbf{kg}}$$

$$w_{is,out} = h_3 - h_{4s} = 3231.7 - 2407.9 = \mathbf{823.8} \frac{\mathbf{kJ}}{\mathbf{kg}}$$

$$w_{is,net} = w_{is,out} - w_{is,in} = 823.8 - 3.038 = \mathbf{820.76} \frac{\mathbf{kJ}}{\mathbf{kg}}$$

$$\eta_{is,T} = \frac{h_3 - h_{4a}}{h_3 - h_{4s}}$$

$$h_{4a} = 2531 \frac{kJ}{kg}$$

$$w_{out} = h_3 - h_{4a} = 3231.7 - 2531 = \mathbf{700.2} \frac{\mathbf{kJ}}{\mathbf{kg}}$$

$$w_{net} = w_{out} - w_{in} = 700.2 - 3.038 = \mathbf{697.2} \frac{\mathbf{kJ}}{\mathbf{kg}}$$

c) Calculate the specific heat input and specific heat rejection values.

$$q_{in} = h_3 - h_2 = 3231.7 - 343.5 = \mathbf{2887} \frac{\mathbf{kJ}}{\mathbf{kg}}$$

$$q_{out} = h_{4s} - h_1 = 2407.9 - 340.5 = \mathbf{2067.4} \frac{\mathbf{kJ}}{\mathbf{kg}}$$

d) Calculate the energy efficiency of this cycle in ideal case (isentropic for pump and turbine).

$$\eta_{en_{is}} = \frac{w_{net_{is}}}{q_{in}} = \frac{820.76}{2887.7} = \mathbf{28.4\%}$$

Both energy and exergy efficiencies of the actual Rankine cycle result in:

$$\eta_{en} = \frac{w_{net_a}}{q_{in}} = \frac{697.2}{2887.7} = \mathbf{24.2\%}$$

and

$$\eta_{ex} = \frac{w_{net_a}}{q_{in}\left(1 - \frac{T_o}{T_s}\right)} = \frac{697.2}{2887.7\left(1 - \frac{298}{873}\right)} = \mathbf{36.7\%}$$

Example 6.19 In a steam Rankine cycle, as shown in Figure 6.28, the pump compresses the working fluid to 5 *MPa* and the condenser operates at 75 *kPa*. The turbine inlet temperature and source temperature are 500 °*C*. The mass flow rate of the working fluid is 10 *kg/s* and the isentropic efficiency of the turbine is 90%. Assume turbine isentropic efficiency to be 80%. The losses are assumed negligible and the water after the condenser becomes a saturated liquid at the saturation temperature.

a) Write the mass, energy, entropy, and exergy balance equations.
b) Calculate the heat input rate of the boiler and the net work rate generated by the turbine.
c) Calculate the entropy generation rate and exergy destruction rate of the boiler.
d) Calculate the cycle energy and exergy efficiencies.

Solution

It is now important to draw the schematic diagram of the problem cycle (as shown in Figure 6.28a) along with its *T-s* diagram (as shown in Figure 6.28b) with the information and property data provided accordingly. This is necessary before beginning to write the balance equations for mass, energy, entropy, and exergy of each component of the cycle.

a) Calculate the mass, energy, entropy, and exergy balances.

   The changes of the kinetic and potential energy of the system are negligible since the system velocity and elevation does not change throughout the process.

   Based on these assumptions, the mass, energy, entropy, and exergy balance for each component are:

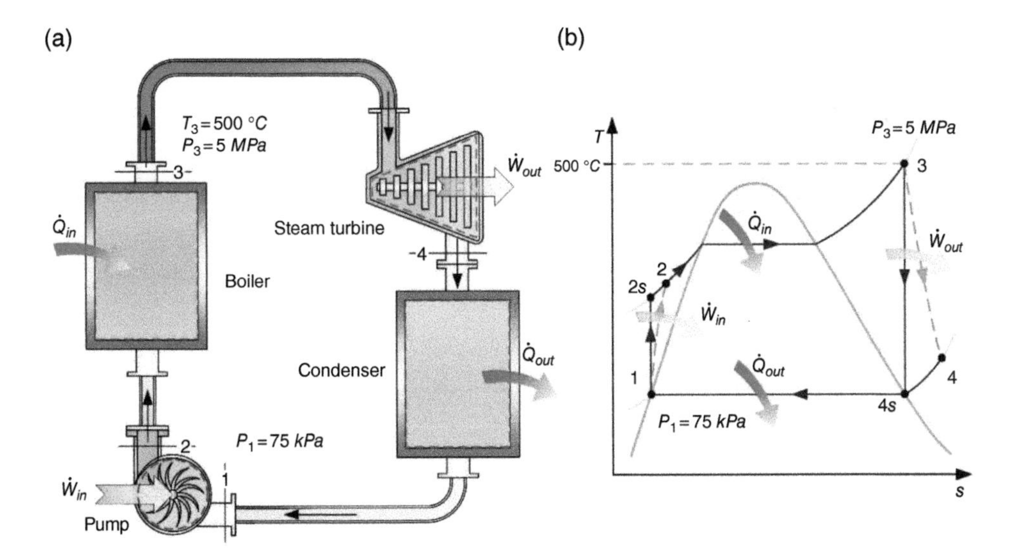

Figure 6.28 (a) Schematic diagram and (b) *T-s* diagram of the steam Rankine cycle discussed in Example 6.19.

For the pump:

$$\text{MBE}: \dot{m}_1 = \dot{m}_2 = \dot{m}$$

$$\text{EBE}: \dot{m}_1 h_1 + \dot{W}_{in} = \dot{m}_2 h_2$$

$$\text{EnBE}: \dot{m}_1 s_1 + \dot{S}_{gen,P} = \dot{m}_2 s_2$$

$$\text{ExBE}: \dot{m}_1 ex_1 + \dot{W}_{in} = \dot{m}_2 ex_2 + \dot{E}x_{d,P}$$

For the boiler:

$$\text{MBE}: \dot{m}_2 = \dot{m}_3 = \dot{m}$$

$$\text{EBE}: \dot{m}_2 h_2 + \dot{Q}_{in} = \dot{m}_3 h_3$$

$$\text{EnBE}: \dot{m}_2 s_2 + \frac{\dot{Q}_{in}}{T_s} + \dot{S}_{gen,B} = \dot{m}_3 s_3$$

$$\text{ExBE}: \dot{m}_2 ex_2 + \dot{Q}_{in}\left(1 - \frac{T_o}{T_s}\right) = \dot{m}_3 ex_3 + \dot{E}x_{d,B}$$

For the turbine:

$$\text{MBE}: \dot{m}_3 = \dot{m}_4 = \dot{m}$$

$$\text{EBE}: \dot{m}_3 h_3 = \dot{m}_4 h_4 + \dot{W}_{out}$$

$$\text{EnBE}: \dot{m}_3 s_3 + \dot{S}_{gen,T} = \dot{m}_4 s_4$$

$$\text{ExBE}: \dot{m}_3 ex_3 = \dot{m}_4 ex_4 + \dot{W}_{out} + \dot{E}x_{d,T}$$

For the condenser:

$$\text{MBE}: \dot{m}_4 = \dot{m}_1 = \dot{m}$$

$$\text{EBE}: \dot{m}_4 h_4 = \dot{m}_1 h_1 + \dot{Q}_{out}$$

$$\text{EnBE}: \dot{m}_4 s_4 + \dot{S}_{gen,C} = \dot{m}_1 s_1 + \frac{\dot{Q}_{out}}{T_b}$$

$$\text{ExBE}: \dot{m}_4 ex_4 = \dot{m}_1 ex_1 + \dot{Q}_{out}\left(1 - \frac{T_o}{T_b}\right) + \dot{E}x_{d,C}$$

b) Calculate the heat input rate of the boiler and the net work rate generated by the turbine.

It is first necessary to get the required properties either from tables (such as Appendix B-1b) or the EES to proceed with the calculations:

State 1: $P_1 = 75\ kPa$

$$T_1 = T_{sat@75\ kPa} = 91.76\ °C$$

$$h_1 = 384.4\ \frac{kJ}{kg}$$

$$v_1 = 0.001037\ \frac{m^3}{kg}$$

State 2: $w_{in} = v_1(P_2 - P_1) = \left(0.001037\ \frac{m^3}{kg}\right)(5000 - 75)\ kPa = \mathbf{5.108}\ \frac{\mathbf{kJ}}{\mathbf{kg}}$

$$\dot{W}_{in} = \dot{m} \times w_{in} = 10\ \frac{kg}{s} \times 5.108\ \frac{kJ}{kg} = \mathbf{51.08}\ \boldsymbol{kW}$$

$$h_2 = h_1 + w_{in} = (384.4 + 5.108)\ \frac{kJ}{kg} = \mathbf{389.5}\ \boldsymbol{\frac{kJ}{kg}}$$

State 3: $P_3 = 5\ MPa$

$$T_3 = 500\,°C$$

$$h_3 = 3434.7\ \frac{kJ}{kg}$$

$$s_3 = 6.978\ \frac{kJ}{kgK}$$

State 4: $P_4 = 75\ kPa$

$$s_4 = s_3$$

$$x_4 = \frac{s_4 - s_f}{s_{fg}}$$

$$x_4 = 0.9235$$

$$h_{4s} = h_f + x_4 h_{fg} = \mathbf{2488.1}\ \boldsymbol{\frac{kJ}{kg}}$$

$$\dot{W}_{out,is} = \dot{m}(h_3 - h_{4s}) = 10\ \frac{kg}{s}\ (3434.7 - 2488.1)\ \frac{kJ}{kg} = \mathbf{9465.6}\ \boldsymbol{kW}$$

$$\dot{W}_{net_{is}} = \dot{W}_{out,is} - \dot{W}_{in,is} = 9465.6 - 51.08 = \mathbf{9414}\ \boldsymbol{kW}$$

$$\eta_{is} = \frac{h_3 - h_4}{h_3 - h_{4s}}$$

$$h_4 = 2677\ \frac{kJ}{kg}$$

$$w_{out} = h_3 - h_4 = 3434.7 - 2677 = \mathbf{757.7}\ \boldsymbol{\frac{kJ}{kg}}$$

$$w_{net} = w_{out} - w_{in} = 757.7 - 51.08/10 = \mathbf{752.6}\ \boldsymbol{\frac{kJ}{kg}}$$

$$\dot{Q}_{in} = \dot{m}(h_3 - h_2) = 10\ kg/s\,(3434.7 - 389.5)\ \frac{kJ}{kg} = \mathbf{30452}\ \boldsymbol{kW}$$

c) Calculate the entropy generation rate and exergy destruction rate for the boiler.

$$\dot{m}_2 s_2 + \frac{\dot{Q}_{in}}{T_s} + \dot{S}_{gen,B} = \dot{m}_3 s_3$$

$$\dot{S}_{gen,B} = 10\ \frac{kg}{s}\ (6.978 - 1.217)\ \frac{kJ}{kgK} - \frac{30452}{773}\ \frac{kJ}{s.K} = \mathbf{18.21}\ \boldsymbol{\frac{kW}{K}}$$

$$\dot{m}_2 ex_2 + \dot{Q}_{in}\left(1 - \frac{T_o}{T_s}\right) = \dot{m}_3 ex_3 + \dot{Ex}_{d,B}$$

$$\dot{E}x_{d,B} = 10\ \frac{kg}{s}\ (32.76 - 1359.7)\ \frac{kJ}{kg} + 30452\ kW\left(1 - \frac{298}{773}\right) = \mathbf{5443\ kW}$$

d) Calculate the energy efficiency for the ideal case where the pump and turbine are isentropic devices.

$$\eta_{en,_{is}} = \frac{\dot{W}_{net_{is}}}{\dot{Q}_{in}} = \frac{9414}{30452} = \mathbf{30.9\%}$$

Overall cycle energy and exergy efficiencies can be written as:

$$\eta_{en} = \frac{\dot{W}_{net}}{\dot{Q}_{in}} = \frac{7526}{30452} = \mathbf{24.71\%}$$

and

$$\eta_{ex} = \frac{\dot{W}_{net}}{\dot{Q}_{in}\left(1 - \frac{T_o}{T_s}\right)} = \frac{7526}{30452\left(1 - \frac{298}{773}\right)} = \mathbf{40.21\%}$$

6.8.1 Ideal Reheat Rankine Cycle

There are various ways and methods – lowering the condenser pressure (P_1), superheating the steam to increase its temperature (T_3), and increasing boiler pressure (P_2) – that can be followed to improve the overall performance of a Rankine cycle. In addition, there is a general modification made to the Rankine cycle, known as the "Reheat Rankine Cycle or Reheat Cycle," where two steam turbines (high- and low-temperature ones) are used. It aims to discharge the steam in two stages of steam turbine with a reheat step between the two stages; this reheat step increases the temperature of the steam before it enters the second stage steam turbine, as shown in Figure 6.29a. It is also important to draw the T-s diagram to clearly see the processes taking place in the cycle components (Figure 6.29b). Here, ideal and actual behaviors are illustrated for the pump and turbines.

In modern power plants that are based on the Rankine cycle, incorporating the cycle with reheat results in an increase in the overall energy efficiency by 4–5%, due to the increase in the average temperature at which heat is transferred to the steam. Note that the average temperature of the reheat process is a determining factor of the improvement of the Rankine cycle since it increases the overall power produce by the cycle (the enclosed area by the temperature-entropy plot). The average temperature of the reheat process is calculated as follows, noting that the T_5 is equal to T_3:

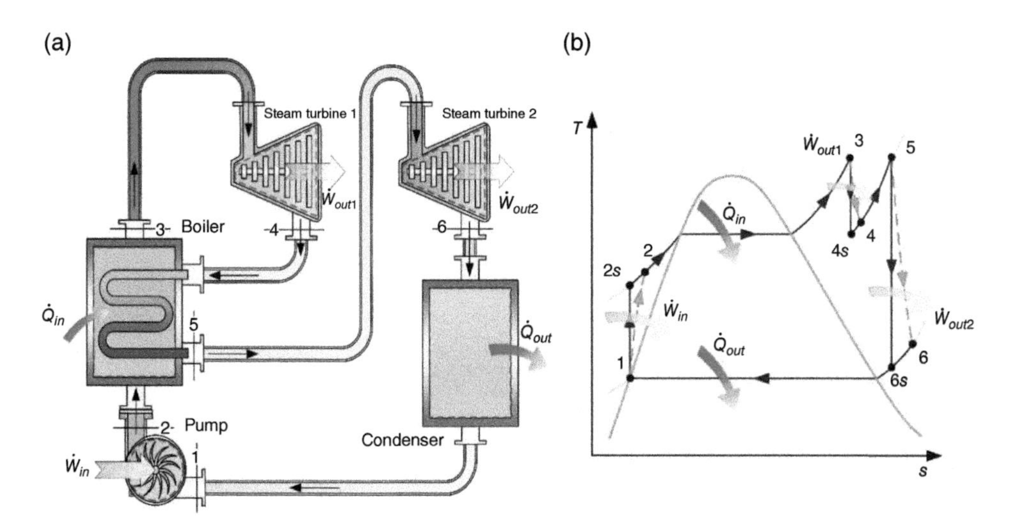

Figure 6.29 (a) A schematic diagram of a reheat Rankine cycle and (b) its *T*-*s* cycle covering both ideal and actual cases for both the pump and steam turbines.

$$T_{avg,reheat} = \frac{T_3 + T_4}{2} \tag{6.24}$$

Here, avg refers to the average. Note that the average temperature of the reheat can be increased by increasing the number of the reheat stages and the expansion stages. However, it was found that increasing the number of reheat and expansion stages more than two stages is not practical.

The variation in temperature of the water as the working fluid of the cycle with the variation of the specific entropy throughout the cycle is shown in Figure 6.29b. The mass, energy, entropy, and exergy balance equations for each device and component of the cycle are shown in Table 6.8.

With regard to the performance of reheat Rankine cycles, we will use both energy and exergy efficiencies of the reheat Rankine cycle defined as:

$$\eta_{en} = \frac{\dot{W}_{net}}{\dot{Q}_{in}} = 1 - \frac{\dot{Q}_{out}}{\dot{Q}_{in}}$$

and

$$\eta_{ex} = \frac{\dot{W}_{net}}{\dot{E}x_{\dot{Q}_{in}}} = \frac{\dot{W}_{net}}{\left(1 - \frac{T_0}{T_s}\right)\dot{Q}_{in}}$$

Table 6.8 The mass, energy, entropy, and exergy balance equations for each component of a reheat Rankine cycle.

<table>
<tr><th>Device</th><th>Balance equation</th></tr>
<tr><td>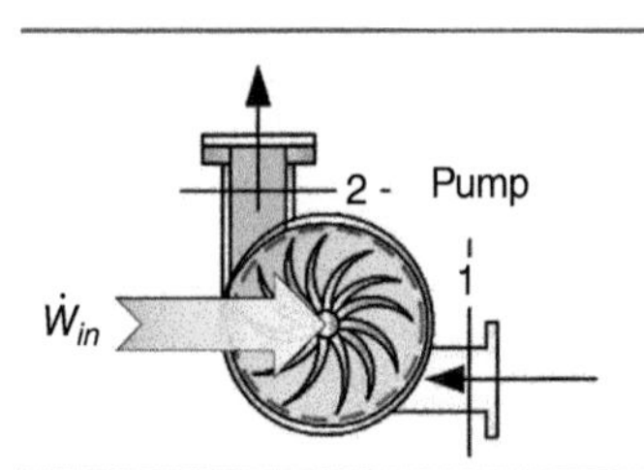</td><td>MBE: $\dot{m}_1 = \dot{m}_2 = \dot{m}$
EBE: $\dot{m}_1 h_1 + \dot{W}_{in} = \dot{m}_2 h_2$
EnBE: $\dot{m}_1 s_1 + \dot{S}_{gen,p} = \dot{m}_2 s_2$
ExBE: $\dot{m}_1 ex_1 + \dot{W}_{in} = \dot{m}_2 ex_2 + \dot{E}x_{d,p}$</td></tr>
<tr><td>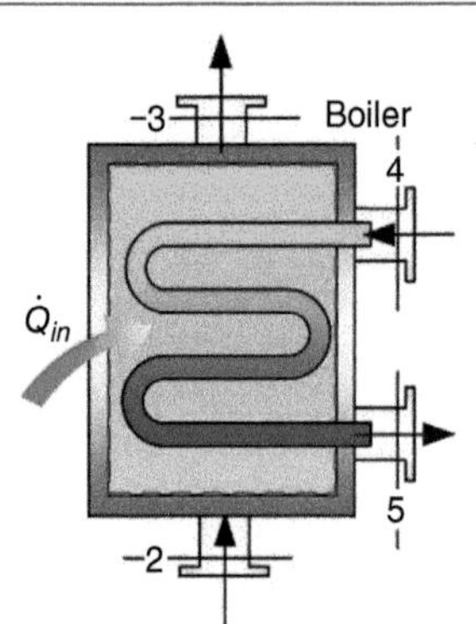</td><td>MBE: $\dot{m}_2 = \dot{m}_3 = \dot{m}$; $\dot{m}_4 = \dot{m}_5$
EBE: $\dot{m}_2 h_2 + \dot{m}_4 h_4 + \dot{Q}_{in} = \dot{m}_3 h_3 + \dot{m}_5 h_5$
EnBE: $\dot{m}_2 s_2 + \dot{m}_4 s_4 + \frac{\dot{Q}_{in}}{T_s} + \dot{S}_{gen,bo} = \dot{m}_3 s_3 + \dot{m}_5 s_5$
ExBE: $\dot{m}_2 ex_2 + \dot{m}_4 ex_4 + \dot{E}x_{\dot{Q}_{in}} = \dot{m}_3 ex_3 + \dot{m}_5 ex_5 + \dot{E}x_{d,bo}$
where
$\dot{E}x_{\dot{Q}_{in}} = \left(1 - \frac{T_o}{T_s}\right)\dot{Q}_{in}$ and $\dot{E}x_{d,bo} = T_o \times \dot{S}_{gen,bo}$</td></tr>
<tr><td>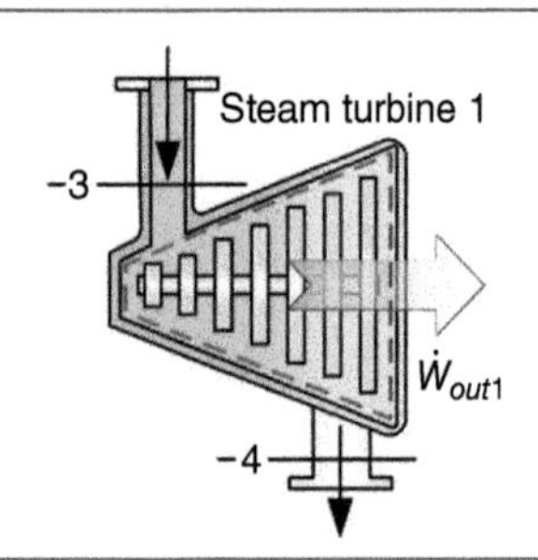</td><td>MBE: $\dot{m}_3 = \dot{m}_4 = \dot{m}$
EBE: $\dot{m}_3 h_3 = \dot{m}_4 h_4 + \dot{W}_{out1}$
EnBE: $\dot{m}_3 s_3 + \dot{S}_{gen,st} = \dot{m}_4 s_4$
ExBE: $\dot{m}_3 ex_3 = \dot{m}_4 ex_4 + \dot{W}_{out1} + \dot{E}x_{d,st1}$</td></tr>
<tr><td>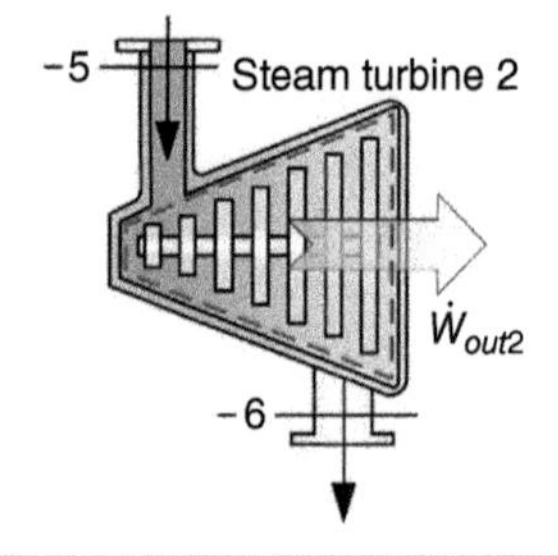</td><td>MBE: $\dot{m}_5 = \dot{m}_6 = \dot{m}$
EBE: $\dot{m}_5 h_5 = \dot{m}_6 h_6 + \dot{W}_{out2}$
EnBE: $\dot{m}_5 s_5 + \dot{S}_{gen,st1} = \dot{m}_6 s_6$
ExBE: $\dot{m}_5 ex_5 = \dot{m}_6 ex_6 + \dot{W}_{out2} + \dot{E}x_{d,st2}$</td></tr>
<tr><td>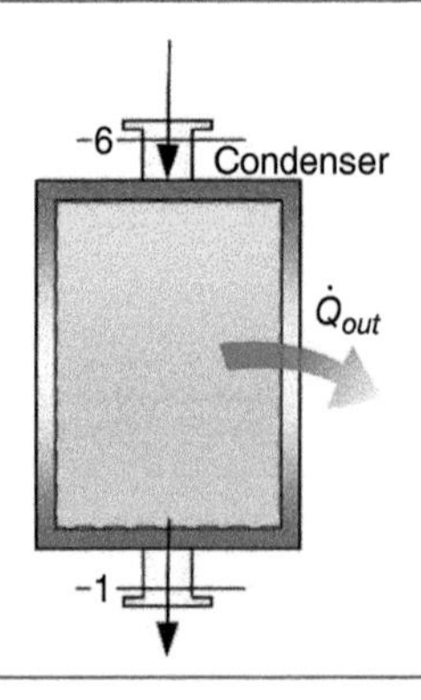</td><td>MBE: $\dot{m}_6 = \dot{m}_1 = \dot{m}$
EBE: $\dot{m}_6 h_6 = \dot{m}_1 h_1 + \dot{Q}_{out}$
EnBE: $\dot{m}_6 s_6 + \dot{S}_{gen,co} = \dot{m}_1 s_1 + \dot{Q}_{out}/T_b$
ExBE: $\dot{m}_6 ex_6 = \dot{m}_1 ex_1 + \dot{E}x_{\dot{Q}_{out}} + \dot{E}x_{d,co}$
where
$\dot{E}x_{\dot{Q}_{out}} = \left(1 - \frac{T_o}{T_b}\right)\dot{Q}_{out}$ and $\dot{E}x_{d,co} = T_o \times \dot{S}_{gen,co}$</td></tr>
</table>

Example 6.20 Consider an ideal reheat Rankine cycle, as shown in Figure 6.30a, that has a boiler pressure of 5000 *kPa*; the reheat pipes that runs through the boiler are at a pressure of 1200 *kPa* and the operating pressure of the condenser is 20 *kPa*. The turbines are designed to produce a 0.96 quality mixture at their exit. The reference temperature is 27 °*C* and the source temperature is 727 °*C*.

a) Write the mass, energy, entropy, and exergy balance equations of the boiler with the reheat.
b) Find the inlet temperature of the first and the second turbines.
c) Find the energy and exergy efficiencies of the ideal reheat Rankine cycle.

Solution

It is now important to draw the schematic diagram of the problem cycle (as shown in Figure 6.30a) along with its *T-s* diagram (as shown in Figure 6.30b) with the information and property data provided accordingly. This is necessary before beginning to write the balance equations for mass, energy, entropy, and exergy of each component of the cycle.

a) Write the mass, energy, entropy, and exergy balance equations of the boiler with the reheat option.

   Since the only difference between the ideal Rankine cycle and the ideal reheat Rankine cycle is the boiler with the reheat piping, its four balance equations are written as:

$$\text{MBE}: \dot{m}_2 = \dot{m}_3 \; and \; \dot{m}_4 = \dot{m}_5$$

$$\text{EBE}: \dot{m}_2 h_2 + \dot{m}_4 h_4 + \dot{Q}_{in} = \dot{m}_3 h_3 + \dot{m}_5 h_5$$

$$\text{EnBE}: \dot{m}_2 s_2 + \dot{m}_4 s_4 + \dot{Q}_{in}/T_s + \dot{S}_{gen,B} = \dot{m}_3 s_3 + \dot{m}_5 s_5$$

$$\text{ExBE}: \dot{m}_2 ex_2 + \dot{m}_4 ex_4 + \dot{E}x_{\dot{Q}_{in}} = \dot{m}_3 ex_3 + \dot{m}_5 ex_5 + \dot{E}x_{d,B}$$

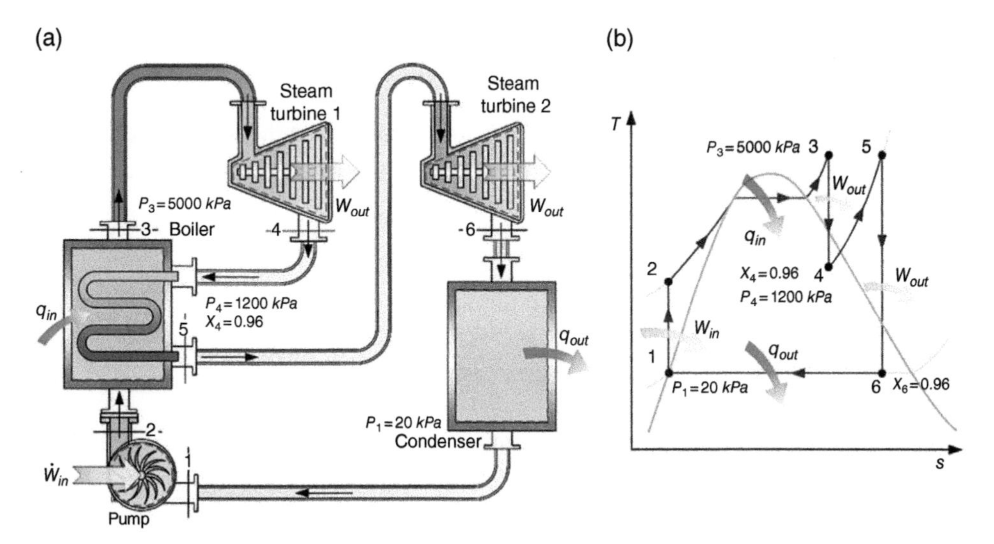

Figure 6.30 (a) Schematic diagram of a reheat Rankine cycle and (b) its *T-s* cycle in ideal form.

b) Find the inlet temperatures of the first and the second turbines.

Here, T_3 is the inlet temperature of the steam entering the first turbine and T_5 is temperature of the steam entering the second turbine.

The properties of the different states throughout the cycle are taken from property tables (such as Appendix B-1b) or from software with built-in thermophysical properties within its database, such as the EES.

The output from the condenser is always maintained as a saturated water to avoid further cooling that will later requires additional heating by the boiler. The enthalpy and the specific volume at the exit of the condenser and inlet to the pump are found as follows:

$$\left.\begin{array}{r} P_1 = 20\ kPa \\ x_1 = 0\,(sat.liq.) \end{array}\right\} \begin{array}{l} h_1 = 251.42\ kJ/kg \\ v_1 = 0.0010172\ m^3/kg \end{array}$$

$$w_p = v_1(P_2 - P_1) = 0.0010172\ \frac{m^3}{kg} \times (5000 - 20)\ kPa \times \frac{1\ kJ}{1\ kPa \times m^3}$$

$$w_p = \mathbf{5.065\ kJ/kg}$$

Then from the MBE and EBE of the pump, which are written as follows:

$$\text{MBE}: \dot{m}_1 = \dot{m}_2 = \dot{m} = constant$$

$$\text{EBE}: \dot{m}_1 h_1 + \dot{W}_{in} = \dot{m}_2 h_2$$

$$h_1 + w_{in} = h_2$$

$$251.42\ \frac{kJ}{kg} + 5.065\ \frac{kJ}{kg} = h_2$$

$$h_2 = 256.49\ \frac{kJ}{kg}$$

For state 4, which is the exit of the first turbine:

$$\left.\begin{array}{r} P_4 = 1200\ kPa \\ x_4 = 0.96 \end{array}\right\} \begin{array}{l} h_4 = h_f + x_4 h_{fg} = 798.33 + 0.96 \times 1985.4 = \mathbf{2704.3\ kJ/kg} \\ s_4 = s_f + x_4 s_{fg} = 2.2159 + 0.96 \times 4.3058 = \mathbf{6.3495\ kJ/kgK} \end{array}$$

For state 3, since the turbine is isentropic, which means that the isentropic efficiency is 100%, then:

$$\left.\begin{array}{r} P_3 = 5000\ kPa \\ s_3 = s_4 \end{array}\right\} \begin{array}{l} h_3 = 3006.9\ kJ/kg \\ \mathbf{T_3 = 327.2\,^\circ C} \end{array}$$

For state 6, the exit state of the second turbine:

$$\left.\begin{array}{r} P_6 = 20\ kPa \\ x_6 = 0.96 \end{array}\right\} \begin{array}{l} h_6 = 2514.6\ kJ/kg \\ s_6 = 7.6242\ kJ/kgK \end{array}$$

For state 5, since the turbine is isentropic, which means that it has an isentropic efficiency of 100%, then:

$$\left.\begin{array}{r} P_5 = 1200\ kPa \\ s_5 = s_6 \end{array}\right\} \begin{array}{l} h_5 = 3436.0\ kJ/kg \\ \mathbf{T_5 = 481.1\,^\circ C} \end{array}$$

c) Find the energy efficiency of the Rankine cycle.

$$\eta = \frac{w_{net}}{q_{in}} = \frac{q_{in} - q_{out}}{q_{in}}$$

From the MBE and EBE of the boiler:

$$h_2 + h_4 + q_{in} = h_3 + h_5$$

$$q_{in} = h_3 + h_5 - h_2 - h_4 = 3006.9\ \frac{kJ}{kg} + 3436.0\ \frac{kJ}{kg} - 256.49\ \frac{kJ}{kg} - 2704.3\ \frac{kJ}{kg}$$

$$= \mathbf{3482.0}\ \frac{\mathbf{kJ}}{\mathbf{kg}}$$

From the MBE and EBE of the condenser:

$$h_6 = h_1 + q_{out}$$

$$q_{out} = h_6 - h_1 = 2514.6\ \frac{kJ}{kg} - 251.42\ \frac{kJ}{kg} = \mathbf{2263.2}\ \frac{\mathbf{kJ}}{\mathbf{kg}}$$

Here, $w_{net} = q_{in} - q_{out} = 3482.0 - 2263.2 = \mathbf{1218.8}\ \mathbf{kJ/kg}$

The overall energy efficiency of the reheat cycle is:

$$\eta_{en} = \frac{w_{net}}{q_{in}} = \frac{1218.8}{3482.0} = 0.35 = \mathbf{35\%}$$

The overall exergy efficiency of the reheat cycle for a source temperature of 1000 K and a reference temperature of 300 K becomes:

$$\eta_{ex} = \frac{w_{net}}{\left(1 - \frac{T}{T_s}\right) q_{in}} = \frac{1218.8}{\left(1 - \frac{300}{1000}\right) 3482.0} = 0.50 = \mathbf{50\%}$$

Example 6.21 Consider a reheat Rankine cycle, as shown in Figure 6.31a, with the pump operating at the pressure of 10 MPa and the boiler heating the steam at 500 °C. The condenser is set to operate at the pressure of 75 kPa and the source temperature is 500 °C. The isentropic efficiencies of high-pressure and low-pressure turbines are assumed to be 75%. The moisture content in the exit stream from the low-pressure turbine is 9.124% and consider that that steam is reheated to the same temperature as the high-pressure turbine inlet temperature.

a) Write the mass, energy, entropy, and exergy balance equations for all components of the reheat cycle.
b) Determine the pressure at the inlet of the low-pressure turbine.
c) Calculate the net work generated by the low- and high-pressure turbines.
d) Calculate the entropy generation and exergy destruction of the boiler.
e) Calculate the isentropic efficiency of the cycle.
f) Calculate the overall energy and exergy efficiencies.

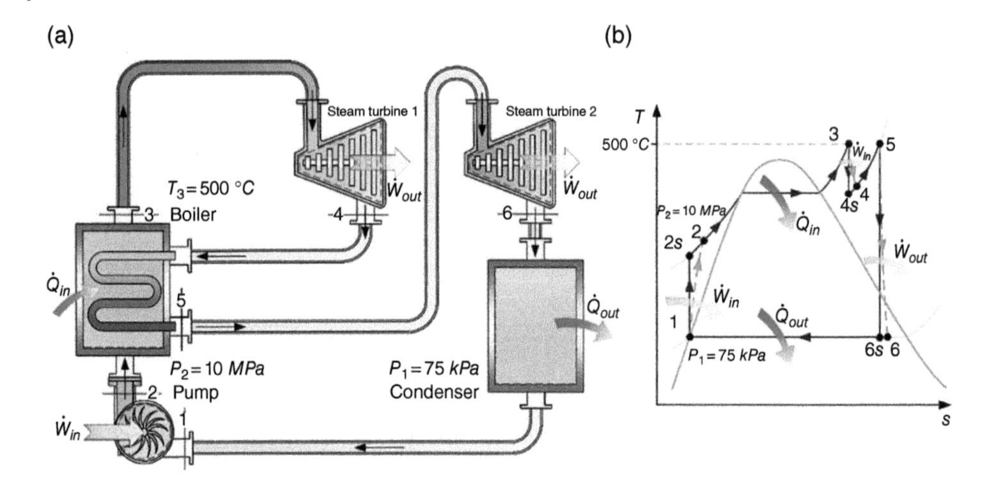

Figure 6.31 (a) A schematic diagram of a reheat Rankine cycle and (b) its *T-s* cycle covering both ideal and actual cases for both pump and steam turbines.

Solution

It is now important to draw the schematic diagram of the problem cycle (as shown in Figure 6.31a) along with its *T-s* diagram (as shown in Figure 6.31b) with the information and property data provided accordingly. This is necessary before beginning to write the balance equations for mass, energy, entropy, and exergy of each component of the cycle.

a) Write the mass, energy, entropy, and exergy balance for each component.

For the pump:

$$\text{MBE}: \dot{m}_1 = \dot{m}_2 = \dot{m}$$

$$\text{EBE}: \dot{m}_1 h_1 + \dot{W}_{in} = \dot{m}_2 h_2$$

$$\text{EnBE}: \dot{m}_1 s_1 + \dot{S}_{gen,P} = \dot{m}_2 s_2$$

$$\text{ExBE}: \dot{m}_1 ex_1 + \dot{W}_{in} = \dot{m}_2 ex_2 + \dot{E}x_{d,P}$$

For the boiler:

$$\text{MBE}: \dot{m}_2 + \dot{m}_4 = \dot{m}_3 + \dot{m}_5$$

$$\text{EBE}: \dot{m}_2 h_2 + \dot{m}_4 h_4 + \dot{Q}_{in} = \dot{m}_3 h_3 + \dot{m}_5 h_5$$

$$\text{EnBE}: \dot{m}_2 s_2 + \dot{m}_4 s_4 + \frac{\dot{Q}_{in}}{T_s} + \dot{S}_{gen,B} = \dot{m}_3 s_3 + \dot{m}_5 s_5$$

$$\text{ExBE}: \dot{m}_2 ex_2 + \dot{m}_4 ex_4 + \dot{Q}_{in}\left(1 - \frac{T_o}{T_s}\right) = \dot{m}_3 ex_3 + \dot{m}_5 ex_5 + \dot{E}x_{d,B}$$

For high-pressure turbine (1):

$$\text{MBE}: \dot{m}_3 = \dot{m}_4 = \dot{m}$$

$$\text{EBE}: \dot{m}_3 h_3 = \dot{m}_4 h_4 + \dot{W}_{out1}$$

$$\text{EnBE}: \dot{m}_3 s_3 + \dot{S}_{gen,T1} = \dot{m}_4 s_4$$

$$\text{ExBE}: \dot{m}_3 ex_3 = \dot{m}_4 ex_4 + \dot{W}_{out,1} + \dot{E}x_{d,T1}$$

For low-pressure turbine (2):

$$\text{MBE}: \dot{m}_5 = \dot{m}_6 = \dot{m}$$

$$\text{EBE}: \dot{m}_5 h_5 = \dot{m}_6 h_6 + \dot{W}_{out2}$$

$$\text{EnBE}: \dot{m}_5 s_5 + \dot{S}_{gen,T2} = \dot{m}_6 s_6$$

$$\text{ExBE}: \dot{m}_5 ex_5 = \dot{m}_6 ex_6 + \dot{W}_{out,2} + \dot{E}x_d$$

For the condenser:

$$\text{MBE}: \dot{m}_6 = \dot{m}_1 = \dot{m}$$

$$\text{EBE}: \dot{m}_6 h_6 = \dot{m}_1 h_1 + \dot{Q}_{out}$$

$$\text{EnBE}: \dot{m}_6 s_6 + \dot{S}_{gen,C} = \dot{m}_1 s_1 + \frac{\dot{Q}_{out}}{T_b}$$

$$\text{ExBE}: \dot{m}_6 ex_6 = \dot{m}_1 ex_1 + \dot{Q}_{out}\left(1 - \frac{T_o}{T_b}\right) + \dot{E}x_{d,C}$$

b) Determine the pressure at the inlet of the low-pressure turbine.
State 6: $P_6 = P_1 = 75\ kPa$

$$x_6 = 1 - 0.09124 = 0.90876$$

$$s_{6s} = s_f + x_6 s_{fg} = \mathbf{6.888}\ \frac{\mathbf{kJ}}{\mathbf{kgK}}$$

$$h_{6s} = h_f + x_6 h_{fg} = \mathbf{2455}\ \frac{\mathbf{kJ}}{\mathbf{kg}}$$

State 5: $T_5 = 500\ °C$

$$s_5 = s_{6s}$$

$$h_5 = 3423\ \frac{kJ}{kg}$$

$$P_5 = 5.9\ MPa$$

Thus, the steam should be reheated at 5.9 MPa pressure to avoid a moisture content above 9.124%. The enthalpies of each state need to be determined to calculate the cycle efficiencies.

c) Calculate the net work generated by the low- and high-pressure turbines.
State 1: $P_1 = 75\ kPa$
Saturated liquid

$$h_1 = 384.4\ \frac{kJ}{kg}$$

$$v_1 = 0.001037\ \frac{m^3}{kg}$$

State 2: $P_2 = 10\,000\ kPa$

$$s_2 = s_1$$

$$w_{in} = v_1(P_2 - P_1) = \left(0.001037\ \frac{m^3}{kg}\right)(10000 - 75)\ kPa = \mathbf{10.29}\ \frac{\mathbf{kJ}}{\mathbf{kg}}$$

$$h_2 = h_1 + w_{in} = (384.4 + 10.29)\ \frac{kJ}{kg} = \mathbf{394.7}\ \frac{\mathbf{kJ}}{\mathbf{kg}}$$

State 3: $P_1 = 10\ MPa$

$$T_3 = 500\ ^\circ C$$

$$h_3 = 3214\ \frac{kJ}{kg}$$

$$s_3 = 6.599\ \frac{kJ}{kgK}$$

State 4: $P_4 = 5.9\ MPa$

$$s_{4s} = s_3$$

$$T_4 = 414\,^\circ C$$

$$h_{4s} = 3214\ \frac{kJ}{kg}$$

$$w_{out1,is} = h_3 - h_{4s} = 3375 - 3214 = \mathbf{161}\ \frac{\mathbf{kJ}}{\mathbf{kg}}$$

$$w_{out2,is} = h_5 - h_{6s} = 3423 - 2455 = \mathbf{968}\ \frac{\mathbf{kJ}}{\mathbf{kg}}$$

$$w_{net_{is}} = w_{out1,is} + w_{out2,is} - w_{in,is} = \mathbf{1129}\ \frac{\mathbf{kJ}}{\mathbf{kg}}$$

For the actual turbine work:

$$\eta_{T1,is} = \frac{h_3 - h_4}{h_3 - h_{4s}}$$

$$h_4 = 3254\ \frac{kJ}{kg}$$

$$w_{out1} = h_3 - h_4 = 3375 - 3254 = \mathbf{121}\ \frac{\mathbf{kJ}}{\mathbf{kg}}$$

$$\eta_{T2,is} = \frac{h_5 - h_6}{h_5 - h_{6s}}$$

$$h_6 = 2697\ \frac{kJ}{kg}$$

$$w_{out2} = h_5 - h_6 = 3423 - 2697 = \mathbf{726}\ \frac{\mathbf{kJ}}{\mathbf{kg}}$$

$$w_{net} = w_{out1} + w_{out2} - w_{in} = \mathbf{836.7}\ \frac{\mathbf{kJ}}{\mathbf{kg}}$$

For the input heat:

$$q_{in} = (h_3 - h_2) + (h_5 - h_4) = (3375 - 394.7) + (3423 - 3254) = \mathbf{3149}\ \frac{\mathbf{kJ}}{\mathbf{kg}}$$

d) Calculate the entropy generation and exergy destruction of the boiler.
The source temperature for the boiler is taken as the output temperature, which is 500 °*C*. For entropy generation and exergy destruction

$$s_2 + s_4 + \frac{q_{in}}{T_s} + s_{gen,B} = s_3 + s_5$$

$$s_{gen,B} = (6.599 - 1.213)\ \frac{kJ}{kgK} + (6.888 - 6.599)\ \frac{kJ}{kgK} - \frac{3149}{773}\ \frac{kJ}{kgK} = \mathbf{1.601}\ \frac{\mathbf{kJ}}{\mathbf{kgK}}$$

$$ex_2 + ex_4 + q_{in}\left(1 - \frac{T_o}{T_s}\right) = ex_3 + ex_5 + ex_{d,B}$$

$$ex_{d,B} = (37.69 - 1412.9)\ \frac{kJ}{kg} + (1042.6 - 1133)\ \frac{kJ}{kg} + 2980\ \frac{kJ}{kg}\left(1 - \frac{298}{773}\right) = \mathbf{474.2}\ \frac{\mathbf{kJ}}{\mathbf{kg}}$$

e) Calculate the isentropic efficiency of the cycle.
The energy efficiency of the isentropic (ideal) cycle is then found as

$$\eta_{en_{is}} = \frac{w_{net_{is}}}{q_{in}} = \frac{1129}{3149} = \mathbf{35.8\%}$$

f) Calculate the overall energy and exergy efficiencies.
The overall energy and exergy efficiencies of the actual reheat cycle are:

$$\eta_{en} = \frac{w_{net}}{q_{in}} = \frac{836.7}{3149} = \mathbf{26.57\%}$$

and

$$\eta_{ex} = \frac{w_{net}}{q_{in}\left(1 - \frac{T_o}{T_s}\right)} = \frac{836.7}{3149\left(1 - \frac{298}{773}\right)} = \mathbf{43.24\%}$$

Example 6.22 In a slightly different type (regenerative) of Rankine cycle with an open-type feedwater heater as shown in Figure 6.32, the boiler heats the steam to 650 °*C* and the steam reaches the turbine at 12 *MPa*. The mass fraction from the turbine to the feedwater heater is 0.3. The isentropic efficiency of the turbine can be assumed to be 76% while the pumps are 100% efficient isentropically. The condenser operates at 50 *kPa* pressure and steam enters the feedwater heater at 1 *MPa*; the high-pressure pump operates at 1 *MPa*.

a) Write the mass, energy, entropy, and exergy balance equations.
b) Calculate the net work of the turbine.
c) Calculate the entropy generation and exergy destruction of the condenser.
d) Calculate the energy and exergy efficiencies.

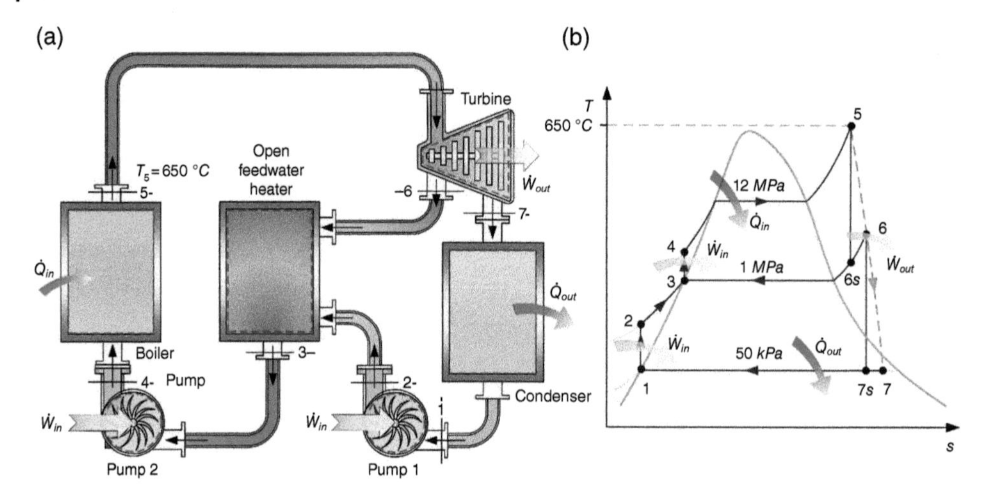

Figure 6.32 (a) Schematic diagram of a regenerative-type Rankine cycle and (b) its *T-s* cycle covering both ideal and actual cases for both pump and steam turbines.

Solution

It is now important to draw the schematic diagram of the problem cycle (as shown in Figure 6.32a) along with its *T-s* diagram (as shown in Figure 6.32b) with the information and property data provided accordingly. This is necessary before beginning to write the balance equations for mass, energy, entropy, and exergy of each component of the cycle. Consider $s_5 = s_{6s} = s_{7s}$.

a) Write the mass, energy, entropy, and exergy balance for each component.

For pump 1:

$$\text{MBE}: \dot{m}_1 = \dot{m}_2 = \dot{m}$$

$$\text{EBE}: \dot{m}_1 h_1 + \dot{W}_{in} = \dot{m}_2 h_2$$

$$\text{EnBE}: \dot{m}_1 s_1 + \dot{S}_{gen} = \dot{m}_2 s_2$$

$$\text{ExBE}: \dot{m}_1 ex_1 + \dot{W}_{in} = \dot{m}_2 ex_2 + \dot{E}x_d$$

For the open feedwater heater:

$$\text{MBE}: \dot{m}_2 + \dot{m}_6 = \dot{m}_3$$

$$\text{EBE}: \dot{m}_2 h_2 + \dot{m}_6 h_6 = \dot{m}_3 h_3$$

$$\text{EnBE}: \dot{m}_2 s_2 + \dot{m}_6 s_6 + \dot{S}_{gen} = \dot{m}_3 s_3$$

$$\text{ExBE}: \dot{m}_2 ex_2 + \dot{m}_6 ex_6 = \dot{m}_3 ex_3 + \dot{E}x_d$$

For pump 2:

$$\text{MBE}: \dot{m}_3 = \dot{m}_4 = \dot{m}$$

$$\text{EBE}: \dot{m}_3 h_3 + \dot{W}_{in} = \dot{m}_4 h_4$$

$$\text{EnBE}: \dot{m}_3 s_3 + \dot{S}_{gen} = \dot{m}_4 s_4$$

$$\text{ExBE}: \dot{m}_3 ex_3 + \dot{W}_{in} = \dot{m}_4 ex_4 + \dot{E}x_d$$

For the boiler:

$$\text{MBE}: \dot{m}_4 = \dot{m}_5 = \dot{m}$$

$$\text{EBE}: \dot{m}_4 h_4 + \dot{Q}_{in} = \dot{m}_5 h_5$$

$$\text{EnBE}: \dot{m}_4 s_4 + \frac{\dot{Q}_{in}}{T_s} + \dot{S}_{gen} = \dot{m}_5 s_5$$

$$\text{ExBE}: \dot{m}_4 ex_4 + \dot{Q}_{in}\left(1 - \frac{T_o}{T_s}\right) = \dot{m}_5 ex_5 + \dot{E}x_d$$

For the turbine:

$$\text{MBE}: \dot{m}_5 = \dot{m}_6 + \dot{m}_7$$

$$\text{EBE}: \dot{m}_5 h_5 = \dot{m}_6 h_6 + \dot{m}_7 h_7 + \dot{W}_{out}$$

$$\text{EnBE}: \dot{m}_5 s_5 + \dot{S}_{gen} = \dot{m}_6 s_6 + \dot{m}_7 s_7$$

$$\text{ExBE}: \dot{m}_5 ex_5 = \dot{m}_6 ex_6 + \dot{m}_7 ex_7 + \dot{W}_{out} + \dot{E}x_d$$

For the condenser:

$$\text{MBE}: \dot{m}_7 = \dot{m}_1 = \dot{m}$$

$$\text{EBE}: \dot{m}_7 h_7 = \dot{m}_1 h_1 + \dot{Q}_{out}$$

$$\text{EnBE}: \dot{m}_7 s_7 + \dot{S}_{gen} = \dot{m}_1 s_1 + \frac{\dot{Q}_{out}}{T_b}$$

$$\text{ExBE}: \dot{m}_7 ex_7 = \dot{m}_1 ex_1 + \dot{Q}_{out}\left(1 - \frac{T_o}{T_b}\right) + \dot{E}x_d$$

b) Calculate the net work of the turbine.
State 1: $P_1 = 50\ kPa$
Saturated liquid

$$h_1 = 340.5\ \frac{kJ}{kg}$$

$$v_1 = 0.00103\ \frac{m^3}{kg}$$

State 2: $P_2 = 1\ MPa$

$$s_2 = s_1$$

$$w_{pump1} = v_1(P_2 - P_1) = \left(0.00103\ \frac{m^3}{kg}\right)(1000 - 50)\ kPa = \mathbf{0.98}\ \frac{\mathbf{kJ}}{\mathbf{kg}}$$

$$h_2 = h_1 + w_{pump} = (340.5 + 0.98)\ \frac{kJ}{kg} = \mathbf{341.5}\ \frac{\mathbf{kJ}}{\mathbf{kg}}$$

State 3: $P_1 = 1\ MPa$
Saturated liquid

$$h_3 = 762.6\ \frac{kJ}{kg}$$

$$v_3 = 0.001127 \ \frac{m^3}{kg}$$

State 4: $P_4 = 12 \ MPa$

$$s_4 = s_3$$

$$w_{pump2} = v_3(P_4 - P_3) = \left(0.001127 \ \frac{m^3}{kg}\right)(12000 - 1000) \ kPa = \mathbf{12.4} \ \frac{\mathbf{kJ}}{\mathbf{kg}}$$

$$h_4 = h_3 + w_{pump} = (762.6 + 12.4) \ \frac{kJ}{kg} = \mathbf{774.9} \ \frac{\mathbf{kJ}}{\mathbf{kg}}$$

State 5: $T_5 = 650\,°C$

$$P_5 = 12 \, MPa$$

$$h_5 = 3734 \ \frac{kJ}{kg}$$

$$s_5 = 6.945 \ \frac{kJ}{kgK}$$

State 6: $P_6 = 1 \ MPa$

$$s_{6s} = s_5$$

$$h_{6s} = 2952 \ \frac{kJ}{kg}$$

$$T_6 = \mathbf{254.3}\,°\mathbf{C}$$

State 7: $P_7 = 50 \ kPa$

$$s_{7s} = s_5$$

$$s_{7s} = s_f + x_7 s_{fg}$$

$$x_7 = 0.9$$

$$h_{7s} = h_f + x_7 h_{fg} = \mathbf{2416} \ \frac{\mathbf{kJ}}{\mathbf{kg}}$$

For input heat:

$$q_{in} = (h_5 - h_4) = (3734 - 774.9) = \mathbf{2959} \ \frac{\mathbf{kJ}}{\mathbf{kg}}$$

$$w_{out,is} = 1h_5 - 0.3h_{6s} - 0.7h_{7s} = (1 \times 3734) - (0.3 \times 2953) - (0.7 \times 2416)$$

$$w_{out,is} = \mathbf{1157} \ \frac{\mathbf{kJ}}{\mathbf{kg}}$$

$$w_{net_{is}} = w_{turb_{is}} - w_{pump1} - w_{pump2} = \mathbf{1144} \ \frac{\mathbf{kJ}}{\mathbf{kg}}$$

For actual turbine work:

$$\eta_{is} = \frac{h_5 - h_6}{h_5 - h_{6s}}$$

$$h_6 = 3140\ \frac{kJ}{kg}$$

$$\eta_{is} = \frac{h_5 - h_7}{h_5 - h_{7s}}$$

$$h_7 = 2732\ \frac{kJ}{kg}$$

$$w_{out} = 1h_5 - 0.3h_6 - 0.7h_7 = (1 \times 3734) - (0.3 \times 3140) - (0.7 \times 2854)$$

$$w_{out} = \mathbf{880}\ \frac{\mathbf{kJ}}{\mathbf{kg}}$$

$$w_{net} = w_{out} - w_{in} = \mathbf{866.6}\ \frac{\mathbf{kJ}}{\mathbf{kg}}$$

c) Calculate the entropy generation and exergy destruction of the condenser.
The boundary temperature for the condenser is either given or taken as the average of the input and output temperatures, which is 53 °C in this problem. For entropy generation and exergy destruction of condenser, one obtains:

$$s_7 + s_{gen,C} = s_1 + \frac{q_{out}}{T_b}$$

$$s_{gen,C} = (1.091 - 7.823) + \frac{2392}{326} = \mathbf{0.605\ kJ/kgK}$$

$$ex_7 = ex_1 + q_{out}\left(1 - \frac{T_o}{T_b}\right) + ex_{d,C}$$

$$ex_{d,C} = (405.1 - 19.87)\ \frac{kJ}{kg} - 2392\ \frac{kJ}{kg}\left(1 - \frac{298}{326}\right) = \mathbf{180}\ \frac{\mathbf{kJ}}{\mathbf{kg}}$$

d) Calculate the energy and exergy efficiencies.
The energy efficiency of the ideal cycle can be found as:

$$\eta_{en_{is}} = \frac{w_{net_{is}}}{q_{in}} = \frac{1144}{2959} = \mathbf{38.7\%}$$

The overall energy and exergy efficiencies of the actual cycle can be found as:

$$\eta_{en} = \frac{w_{net}}{q_{in}} = \frac{780.6}{2959} = \mathbf{26.38\%}$$

$$\eta_{ex} = \frac{w_{net}}{q_{in}\left(1 - \frac{T_o}{T_s}\right)} = \frac{866.6}{2959\left(1 - \frac{298}{923}\right)} = \mathbf{43.25\%}$$

6.8.2 Cogeneration Rankine Cycle

The main use of the Rankine cycles is power generation as was introduced earlier in the chapter, which means that the main goal for a Rankine cycle is to convert as much as it can from the delivered thermal energy to work, which is the most valuable form of energy.

The remaining part of the delivered heat that was not converted to work is rejected through the condenser such that the working fluid can be returned to its starting point so that the cycle can run again. This rejected heat is often rejected to a heat sink that can be considered as a heat reservoir where, no matter how much heat you reject to it, its temperature will remain constant, such as a river, lake, sea, or ocean. The ultimate goal is to make use of this rejected heat and have it as a useful commodity that may be used for heating, hot water or cooling. In this way the system produces two useful outputs, namely work (and hence power) and heat with the same input to the boiler.

In many industries, such as paper, oil production, refining, and others, process heat is required to operate. Process heat is delivered usually from a source with a temperature that varies from 150 to 200 °C, although it can also be at lower temperatures than that range depending on the type of industry and specific application. Such industries and plants also consumes power in the form of electricity, which means that with specific parameters and operating conditions we can use the usual Rankine cycle as a power generation unit as well as a heat production system, where the process heat comes from the condenser that might be operating at a higher pressure or receive a very hot steam, as shown in Figure 6.33a along with its *T-s* diagram in Figure 6.33b.

In Figure 6.33, a basic Rankine cycle is used in a cogeneration plant, where the heat being rejected by the condenser is used as process heat for various purposes:

$$\dot{Q}_{out} = \dot{Q}_p = \dot{m}_a(h_b - h_a) \tag{6.25}$$

where the subscript p refers to process heat.

In regards to the performance of a cogeneration Rankine cycle (as shown in Figure 6.33a), we will define both energy and exergy efficiencies of the cogeneration Rankine cycle as follows:

$$\eta_{en} = \frac{\dot{W}_{net} + (\dot{m} \times \Delta h_{ab})}{\dot{Q}_{in}}$$

where Δh_{ab} is the amount of flow energy gained from the condenser.

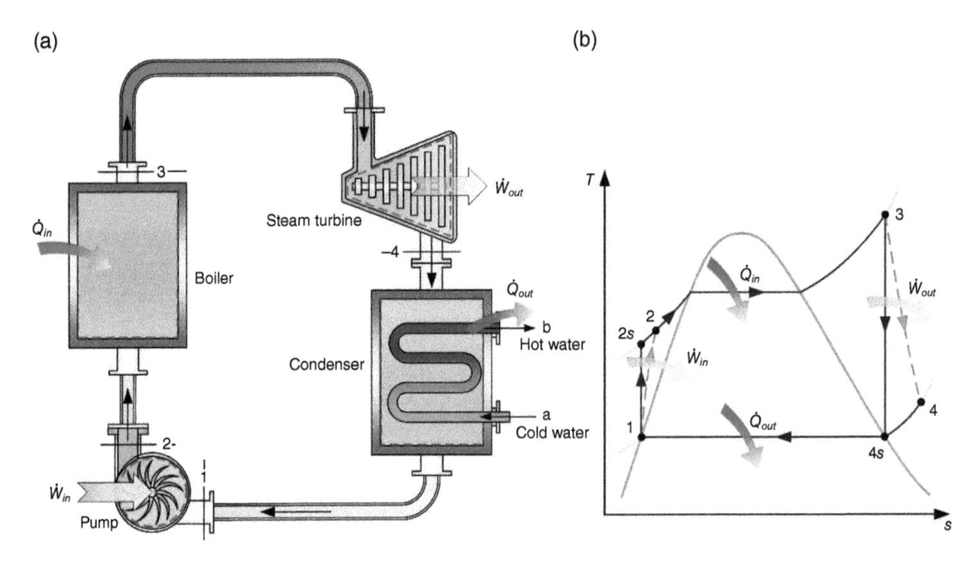

Figure 6.33 A basic Rankine cycle with a useful heat output to a cogeneration plant.

It may also be defined in the following form if the heat is directly transferred:

$$\eta_{en} = \frac{\dot{W}_{net} + \dot{Q}_p}{\dot{Q}_{in}}$$

and

$$\eta_{ex} = \frac{\dot{W}_{net} + \dot{E}x_{ab}}{\dot{E}x_{\dot{Q}_{in}}} = \frac{\dot{W}_{net} + (\dot{m} \times \Delta ex_{ab})}{\left(1 - \frac{T_0}{T_s}\right)\dot{Q}_{in}}$$

based on the closed type arrangement given in Figure 6.33a, where Δex_{ab} is the amount of flow exergy gained from the condenser.

The exergy efficiency may also be defined in the following form if the heat is directly transferred:

$$\eta_{ex} = \frac{\dot{W}_{net} + + \dot{E}x_{\dot{Q}_p}}{\dot{E}x_{\dot{Q}_{in}}} = \frac{\dot{W}_{net} + \left(1 - \frac{T_0}{T_b}\right)\dot{Q}_p}{\left(1 - \frac{T_0}{T_s}\right)\dot{Q}_{in}}$$

Of course, there are possibilities to design different options with the Rankine cycle for cogenerational purposes. Figure 6.34 shows another design of the Rankine cycle that can have an adjustable load between process heat at high temperatures and power production in addition to having heat recovered from the condenser for useful purposes.

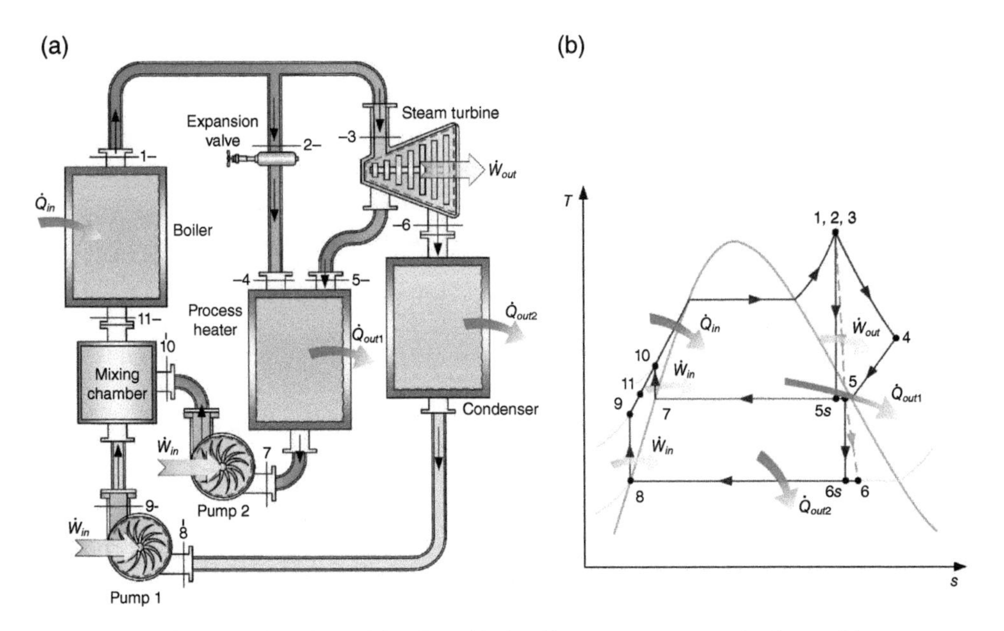

Figure 6.34 Cogeneration plant with adjustable load between power production and heat processing.

As shown in Figure 6.34 the cogeneration loads are controlled by how much of the steam exits the turbine to be sent to the heater that provides the process heat. Analyzing the cogeneration plant shown in Figure 6.34 is done through the balance equations, which are used to derive the energy and exergy efficiency definitions as well as the utilization factor in energy and exergy aspects. The four balance equations for the overall cogeneration plant shown in Figure 6.34 can be written as:

$$\text{MBE}: \dot{m} = constant$$

$$\text{EBE}: \dot{Q}_{in} + \dot{W}_{in1} + \dot{W}_{in1} = \dot{W}_{out} + \dot{Q}_{out,1} + \dot{Q}_{out,2}$$

$$\text{EnBE}: \dot{Q}_{in}/T_s + \dot{S}_{gen} = \dot{Q}_{out,1}/T_{b1} + \dot{Q}_{out,2}/T_{b2}$$

$$\text{EBE}: \dot{E}x_{\dot{Q}_{in}} + \dot{W}_{in1} + \dot{W}_{in1} = \dot{W}_{out} + \dot{E}x_{\dot{Q}_{out,1}} + \dot{E}x_{\dot{Q}_{out,2}} + \dot{E}x_d$$

The energy and exergy efficiencies of the overall cogeneration Rankine cycle are defined as:

$$\eta = \frac{\dot{W}_{net} + \dot{Q}_{out,t}}{\dot{Q}_{in}} \tag{6.26}$$

where $\dot{Q}_{out,t} = (\dot{Q}_{out,1} + \dot{Q}_{out,2})$ which is total useful heat output.

$$\psi = \frac{\dot{W}_{out} + \dot{E}x_{\dot{Q}_{out,t}}}{\dot{E}x_{\dot{Q}_{in}}} \tag{6.27}$$

where $\dot{E}x_{\dot{Q}_{out,t}} = (\dot{E}x_{\dot{Q}_{out,1}} + \dot{E}x_{\dot{Q}_{out,2}})$ which is total useful thermal exergy output.

Example 6.23 A cogeneration plant, as shown in Figure 6.35, has steam with a mass flow rate of 13 *kg/s* entering the turbine at a pressure of 8000 *kPa* and a temperature of 600 °*C*, where 10% of the steam leaves the turbine from the first exit when the steam in the turbine reaches a pressure of 300 *kPa* as a saturated vapor and the remaining steam is expanded further to a pressure of 10 *kPa* and a quality of 85%. The boiler receives a net amount of heat of 60 *MW*. Consider 2000 *kW* of heat recovered from the condenser.

a) Write the mass, energy, entropy, and exergy balance equations for the boiler and the steam turbine.
b) Find the energy efficiency of the cycle for single generation (consider pump works negligible).
c) Find the energy efficiency of the cycle for generation (consider pump works negligible).

Solution

a) Write the mass, energy, entropy, and exergy balance equations for the boiler and the steam turbine.
   The boiler balance equations can be written as:

$$\text{MBE}: \dot{m}_2 + \dot{m}_7 = \dot{m}_4$$

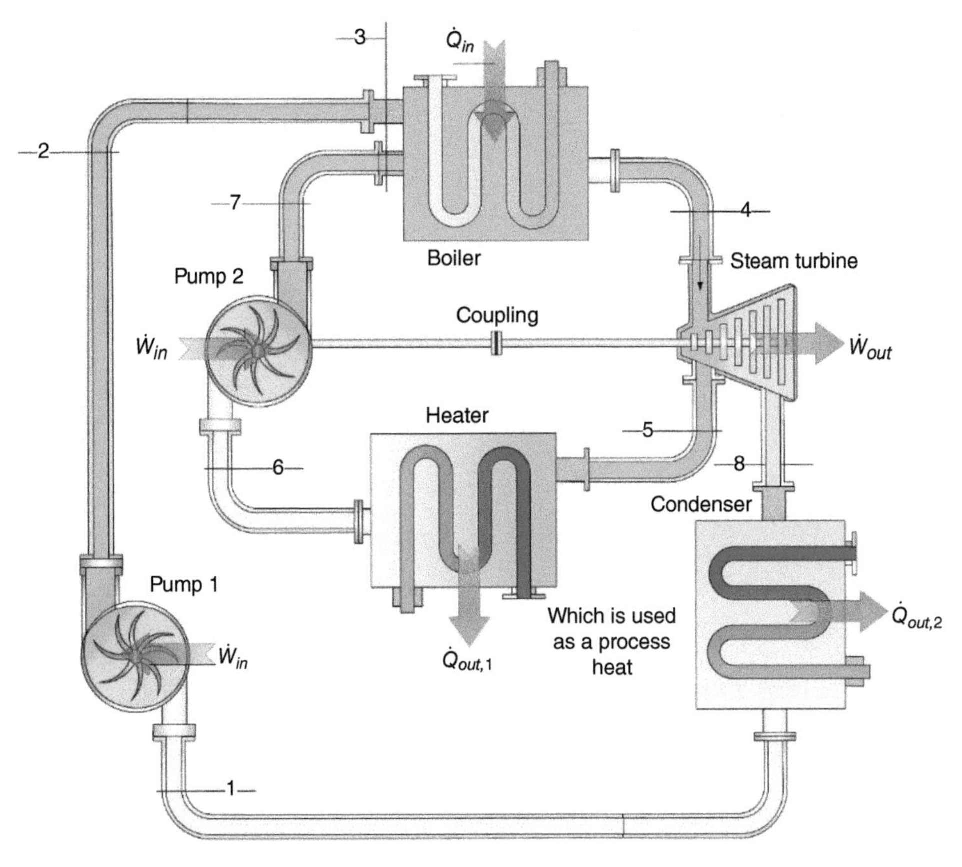

Figure 6.35 Cogeneration plant discussed in Example 6.23.

$$\text{EBE}: \dot{m}_2 h_2 + \dot{m}_7 h_7 + \dot{Q}_{in} = \dot{m}_4 h_4$$

$$\text{EnBE}: \dot{m}_2 s_2 + \dot{m}_7 s_7 + \dot{Q}_{in}/T_s + \dot{S}_{gen} = \dot{m}_4 s_4$$

$$\text{ExBE}: \dot{m}_2 ex_2 + \dot{m}_7 ex_7 + \dot{E}x_{\dot{Q}_{in}} = \dot{m}_4 ex_4 + \dot{E}x_d$$

The steam turbine balance equations can be written as:

$$\text{MBE}: \dot{m}_4 = \dot{m}_5 + \dot{m}_8 \text{ and } \dot{m}_5 = 0.1\,\dot{m}_4 \text{ and } \dot{m}_8 = 0.9\,\dot{m}_4$$

$$\text{EBE}: \dot{m}_4 h_4 = \dot{m}_5 h_5 + \dot{m}_8 h_8 + \dot{W}_{out}$$

$$\text{EnBE}: \dot{m}_4 s_4 + \dot{S}_{gen} = \dot{m}_5 s_5 + \dot{m}_8 s_8$$

$$\text{ExBE}: \dot{m}_4 ex_4 = \dot{m}_5 ex_5 + \dot{m}_8 ex_8 + \dot{W}_{out} + \dot{E}x_d$$

b) Find the energy efficiency of the cycle.

The energy efficiency of the cogeneration plant can be written as follows since we can neglect the work rate consumed by the pumps:

$$\eta = \frac{\dot{W}_{out}}{\dot{Q}_{in}}$$

In order to find the work produced by the cogeneration plant we need to solve for $\dot{W}_{out}$ in the EBE of the steam turbine, which is done as follows:

$$\dot{m}_4 h_4 = \dot{m}_5 h_5 + \dot{m}_8 h_8 + \dot{W}_{out}$$

$$\dot{W}_{out} = \dot{m}_4 h_4 - \dot{m}_5 h_5 - \dot{m}_8 h_8$$

$$\dot{m}_4 = 13\ \frac{kg}{s},\quad \dot{m}_5 = 0.1 \times \dot{m}_4 = 1.3\ \frac{kg}{s},\quad \dot{m}_8 = 0.9 \times \dot{m}_4 = 11.7\ \frac{kg}{s}$$

By using a properties based software or properties tables (such as Appendix B-1c):

$$\left.\begin{matrix} P_4 = 8000\ kPa \\ T_4 = 600\ ^oC \end{matrix}\right\} \begin{matrix} h_4 = 3642.4\ kJ/kg \\ s_4 = 7.0221\ kJ/kgK \end{matrix}$$

$$\left.\begin{matrix} P_5 = 300\ kPa \\ x_5 = 1 \end{matrix}\right\} h_5 = 2724.9\ kJ/kg$$

$$\left.\begin{matrix} P_8 = 10\ kPa \\ x_8 = 0.85 \end{matrix}\right\} h_8 = h_f + x(h_{fg}) = 191.81 + 0.85 \times (2392.1) = \mathbf{2225.095\ kJ/kg}$$

$$\dot{W}_{out} = 13\ \frac{kg}{s} \times 3642.4\ \frac{kJ}{kg} - 1.3\ \frac{kg}{s} \times 2724.9\ \frac{kJ}{kg} - 11.7\ \frac{kg}{s} \times 2225.095\ \frac{kJ}{kg}$$

$$\dot{W}_{out} = \mathbf{17775.2\ kW}$$

$$\eta_{en,sg} = \frac{\dot{W}_{out}}{\dot{Q}_{in}} = \frac{17775.2\ kW}{60000\ kW} = \mathbf{29.60\%} \text{ for single generation}$$

where sg stands for single generation.

c) Find the cogeneration Rankine cycle efficiency this time with negligible pump works:

$$\eta_{en,cg} = \frac{\dot{W}_{out} + \dot{Q}_{out,t}}{\dot{Q}_{in}}$$

In order to find the cogeneration efficiency, we need to find the process heat the cogeneration plant produces from the heater shown in Figure 6.35, which is found by solving the EBE of the heater and can be written as:

$$\dot{m}_5 h_5 = \dot{m}_6 h_6 + \dot{Q}_{out,1}$$

$$\left.\begin{matrix} P_6 = 300\ kPa \\ x_6 = 0 \end{matrix}\right\} h_6 = 561.4\ kJ/kg$$

$$\dot{Q}_{out,1} = \dot{m}_5 h_5 - \dot{m}_6 h_6 = 1.3\ \frac{kg}{s} \times 2724.9\ \frac{kJ}{kg} - 1.3\ \frac{kg}{s} \times 561.4\ \frac{kJ}{kg}$$

$$\dot{Q}_{out,1} = \mathbf{2812.55\ kW}$$

$\dot{Q}_{out,2} = 2000\ kW$ which is given in the problem.

$$\dot{Q}_{out,t} = (2812.55 + 2000) = \mathbf{4812.55\ kW}$$

Then the cogeneration efficiency is calculated as follows:

$$\eta_{en,cg} = \frac{\dot{W}_{out} + \dot{Q}_{out,t}}{\dot{Q}_{in}} = \frac{17775.2\ kW + 4812.55\ kW}{60000\ kW} = \mathbf{37.64\%}$$

which shows that the generation cycle is 8.04% more efficient than the single generation system.

Example 6.24 The turbine of a cogeneration Rankine plant, as shown in Figure 6.36a, operates at a temperature and pressure of 600 °*C* and 5 *MPa*. 20% of the steam is extracted from the turbine at 400 *kPa* and the remaining 80% is expanded to 50 *kPa*. In order to meet the high thermal energy requirements, 20% of the steam from the boiler is throttled to 400 *kPa*. The steam exits the process heater at 400 *kPa* as saturated liquid. The mass flow rate of the steam fed to the boiler is 10 *kg/s*. Assume turbine isentropic efficiency to be 80%. Assume the turbine and pump to be adiabatic and neglect the losses and pressure drops.

a) Write the mass, energy, entropy, and exergy balance equations for each component of the cogeneration Rankine cycle.
b) Calculate the enthalpies for each state.
c) Calculate the work rate of the turbine.
d) Determine the process heat rate.
e) Calculate the entropy generation rate and exergy destruction rate of the boiler.

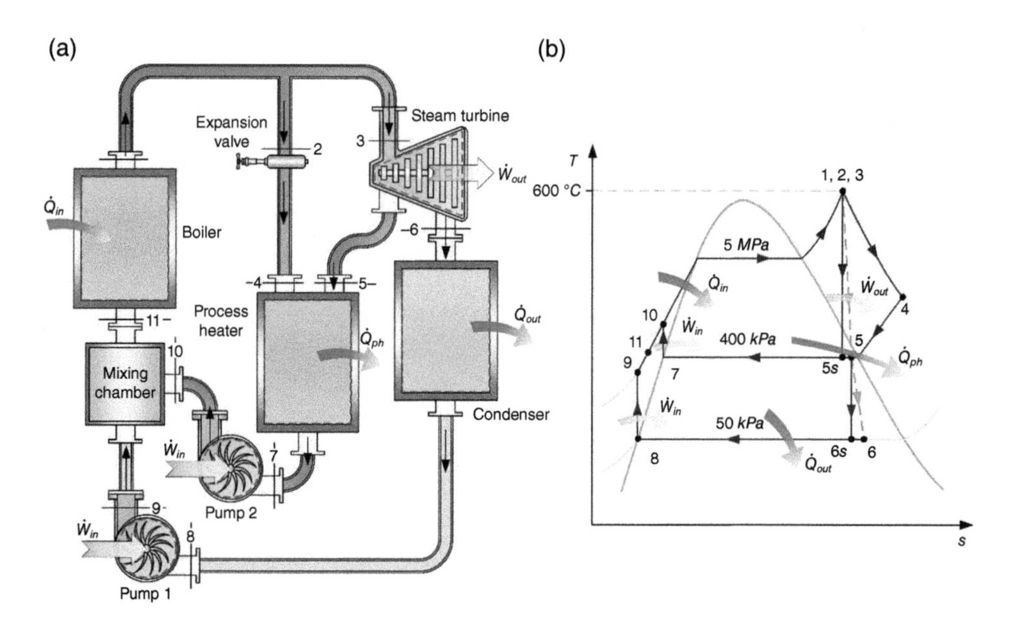

Figure 6.36 (a) A cogeneration Rankine plant with (b) its *T-s* diagram as discussed in Example 6.24.

f) Calculate the energy and exergy efficiencies of the cogeneration Rankine cycle. Also, find the respective efficiencies by considering cases where there is only process heat received and where both process heat and condenser heat are used as useful outputs.

Solution

It is now important to draw the schematic diagram of the problem cycle (as shown in Figure 6.36a) along with its *T-s* diagram (as shown in Figure 6.36b) with the information and property data provided accordingly. This is necessary before beginning to write the balance equations for mass, energy, entropy, and exergy of each component of the cycle.

a) Write the mass, energy, entropy, and exergy balance for each component.

For pump 1:

$$\text{MBE}: \dot{m}_8 = \dot{m}_9 = \dot{m}$$

$$\text{EBE}: \dot{m}_8 h_8 + \dot{W}_{in1} = \dot{m}_9 h_9$$

$$\text{EnBE}: \dot{m}_8 s_8 + \dot{S}_{gen,P1} = \dot{m}_9 s_9$$

$$\text{ExBE}: \dot{m}_8 ex_8 + \dot{W}_{in1} = \dot{m}_9 ex_9 + \dot{E}x_{d,P1}$$

For the mixing chamber:

$$\text{MBE}: \dot{m}_9 + \dot{m}_{10} = \dot{m}_{11}$$

$$\text{EBE}: \dot{m}_9 h_9 + \dot{m}_{10} h_{10} = \dot{m}_{11} h_{11}$$

$$\text{EnBE}: \dot{m}_9 s_9 + \dot{m}_{10} s_{10} + \dot{S}_{gen,P} = \dot{m}_{11} s_{11}$$

$$\text{ExBE}: \dot{m}_9 ex_9 + \dot{m}_{10} ex_{10} = \dot{m}_{11} ex_{11} + \dot{E}x_{d,MC}$$

For pump 2:

$$\text{MBE}: \dot{m}_7 = \dot{m}_{10} = \dot{m}$$

$$\text{EBE}: \dot{m}_7 h_7 + \dot{W}_{in2} = \dot{m}_{10} h_{10}$$

$$\text{EnBE}: \dot{m}_7 s_7 + \dot{S}_{gen,P2} = \dot{m}_{10} s_{10}$$

$$\text{ExBE}: \dot{m}_7 ex_7 + \dot{W}_{in2} = \dot{m}_{10} ex_{10} + \dot{E}x_{d,P2}$$

For the boiler:

$$\text{MBE}: \dot{m}_{11} = \dot{m}_1 = \dot{m}$$

$$\text{EBE}: \dot{m}_{11} h_{11} + \dot{Q}_{in} = \dot{m}_1 h_1$$

$$\text{EnBE}: \dot{m}_{11} s_{11} + \frac{\dot{Q}_{in}}{T_s} + \dot{S}_{gen,B} = \dot{m}_1 s_1$$

$$\text{ExBE}: \dot{m}_{11} ex_{11} + \dot{Q}_{in}\left(1 - \frac{T_o}{T_s}\right) = \dot{m}_1 ex_1 + \dot{E}x_{d,B}$$

For the process heater:

$$\text{MBE}: \dot{m}_4 + \dot{m}_5 = \dot{m}_7$$

$$\text{EBE}: \dot{m}_4 h_4 + \dot{m}_5 h_5 = \dot{m}_7 h_7 + \dot{Q}_{PH}$$

$$\text{EnBE}: \dot{m}_4 s_4 + \dot{m}_5 s_5 + \dot{S}_{gen,PH} = \dot{m}_7 s_7 + \frac{\dot{Q}_{PH}}{T_b}$$

$$\text{ExBE}: \dot{m}_4 ex_4 + \dot{m}_5 ex_5 = \dot{m}_7 ex_7 + \dot{Q}_{PH}\left(1 - \frac{T_o}{T_b}\right) + \dot{E}x_{d,PH}$$

For the turbine:

$$\text{MBE}: \dot{m}_3 = \dot{m}_6 + \dot{m}_5$$

$$\text{EBE}: \dot{m}_3 h_3 = \dot{m}_6 h_6 + \dot{m}_5 h_5 + \dot{W}_{out}$$

$$\text{EnBE}: \dot{m}_3 s_3 + \dot{S}_{gen,T} = \dot{m}_6 s_6 + \dot{m}_5 s_5$$

$$\text{ExBE}: \dot{m}_3 ex_3 = \dot{m}_6 ex_6 + \dot{m}_5 ex_5 + \dot{W}_{out} + \dot{E}x_{d,T}$$

For the condenser:

$$\text{MBE}: \dot{m}_6 = \dot{m}_8 = \dot{m}$$

$$\text{EBE}: \dot{m}_6 h_6 = \dot{m}_8 h_8 + \dot{Q}_{out}$$

$$\text{EnBE}: \dot{m}_6 s_6 + \dot{S}_{gen,C} = \dot{m}_8 s_8 + \frac{\dot{Q}_{out}}{T_b}$$

$$\text{ExBE}: \dot{m}_6 ex_6 = \dot{m}_8 ex_8 + \dot{Q}_{out}\left(1 - \frac{T_o}{T_b}\right) + \dot{E}x_{d,C}$$

For the expansion valve:

$$\text{MBE}: \dot{m}_2 = \dot{m}_4 = \dot{m}$$

$$\text{EBE}: \dot{m}_2 h_2 = \dot{m}_4 h_4 \rightarrow h_2 = h_4 \text{ which isenthalpic process}$$

$$\text{EnBE}: \dot{m}_2 s_2 + \dot{S}_{gen,EV} = \dot{m}_4 s_4$$

$$\text{ExBE}: \dot{m}_2 ex_2 = \dot{m}_4 ex_4 + \dot{E}x_{d,EV}$$

b) Calculate the enthalpies for each state.

State 1: $P_1 = 5\ MPa$

$T_1 = 600\ °C$

$$h_1 = 3667\ \frac{kJ}{kg}$$

$$s_1 = 7.261\ \frac{kJ}{kgK}$$

State 2: $T_2 = 600\ °C$

$$s_2 = s_1$$

$$h_2 = 3667\ \frac{kJ}{kg}$$

State 3: $T_3 = 600\ °C$

$$h_3 = h_2$$

State 4: $P_4 = 400\ kPa$

$$h_2 = h_4$$

$$s_4 = 8.414\ \frac{kJ}{kgK}$$

$T_4 = 583.4\ °C$

State 5: $P_5 = 400\ kPa$

$s_3 = s_{5s}$

$h_{5s} = 2904\ \frac{kJ}{kg}$

State 6: $P_6 = 50\ kPa$

$s_{6s} = s_{5s}$

$h_{6s} = 2527\ \frac{kJ}{kg}$

$T_6 = 81.32\ °C$

State 7: $P_7 = 400\ kPa$

$x_7 = 0$

$h_7 = 604.7\ \frac{kJ}{kg}$

$v_7 = 0.001084\ \frac{m^3}{kg}$

State 8: $P_8 = 50\ kPa$

$x_8 = 0$

$h_8 = 340.6\ \frac{kJ}{kg}$

$v_8 = 0.001030\ \frac{m^3}{kg}$

State 9: $P_9 = 5\ MPa$

$s_9 = s_8$

$$w_{in1} = v_8(P_9 - P_8) = \left(0.001030\ \frac{m^3}{kg}\right)(5000 - 50)\ kPa = \mathbf{5.09\ \frac{kJ}{kg}}$$

$$\dot{W}_{in1} = \dot{m}w_{in1} = 6.4\ \frac{kg}{s}\left(5.08\ \frac{kJ}{kg}\right) = \mathbf{32.57\ kW}$$

$$h_9 = h_8 + \dot{W}_{in1} = (340.6 + 32.57) = \mathbf{308\ \frac{kJ}{kg}}$$

State 10: $P_{10} = 5\ MPa$

$$w_{in2} = v_7(P_{10} - P_7) = \left(0.001084\ \frac{m^3}{kg}\right)(5000 - 400)\ kPa = \mathbf{4.99\ kJ/kg}$$

$$\dot{W}_{in2} = \dot{m}w_{in2} = 3.6\ \frac{kg}{s}\left(4.99\ \frac{kJ}{kg}\right) = \mathbf{17.7\ kW}$$

$$h_{10} = h_7 + w_{in2} = (604.7 + 4.99) = \mathbf{607.7\ \frac{kJ}{kg}}$$

State 11: $P_{11} = 5\ MPa$

Form the EBE for mixing chamber, one obtains

$$h_{11} = 415.9\ \frac{kJ}{kg}$$

c) Calculate the work rate of the turbine.

$$\dot{Q}_{in} = \dot{m}_1(h_1 - h_{11}) = 10\ \frac{kg}{s}(3667 - 415.9)\ \frac{kJ}{kg} = \mathbf{32511\ kW}$$

$$\dot{W}_{out,is} = \dot{m}_3 h_3 - \dot{m}_5 h_{5s} - \dot{m}_6 h_{6s} = (8 \times 3667) - (1.6 \times 2904) - (6.4 \times 2527)\ \frac{kJ}{kg}$$

$$= \mathbf{8517\ kW}$$

$$\dot{W}_{net_{is}} = \dot{W}_{out,is} - \dot{W}_{in1} - \dot{W}_{in2} = 8517 - 32.57 - 17.7 = \mathbf{8467\ kW}$$

For actual turbine work:

$$\eta_{is} = \frac{h_1 - h_5}{h_1 - h_{5s}}$$

$$h_5 = 3056\ \frac{kJ}{kg}$$

$$\eta_{is} = \frac{h_1 - h_6}{h_1 - h_{6s}}$$

$$h_6 = 2755\ \frac{kJ}{kg}$$

$$\dot{W}_{out} = \dot{m}_3 h_3 - \dot{m}_5 h_5 - \dot{m}_6 h_6 = (8 \times 3667) - (1.6 \times 3056) - (6.4 \times 2755)\ kW = \mathbf{6814\ kW}$$

$$\dot{W}_{net} = \dot{W}_{out} - \dot{W}_{in1} - \dot{W}_{in2} = 6814 - 32.57 - 17.7 = \mathbf{6764\ kW}$$

d) Determine the process heat rate.

For the output heat, we follow the following methodology by using the energy balance equation:

$$\dot{Q}_p = \dot{m}_4 h_4 + \dot{m}_5 h_5 - \dot{m}_7 h_7$$

$$\dot{m}_4 = (0.2)\left(10\ \frac{kg}{s}\right) = 2\ \frac{kg}{s}$$

$$\dot{m}_5 = \left(8 - 0.8(8)\ \frac{kg}{s}\right) = 1.6\ \frac{kg}{s}$$

$$\dot{m}_7 = \left((2 + 1.6)\ \frac{kg}{s}\right) = 3.6\ \frac{kg}{s}$$

$$\dot{Q}_p = \dot{m}_4 h_4 + \dot{m}_5 h_5 - \dot{m}_7 h_7 = \mathbf{10047\ kW}$$

$$\dot{Q}_{cond} = \dot{m}_6(h_6 - h_8) = \mathbf{13993\ kW}$$

e) Calculate the entropy generation rate and exergy destruction rate of the boiler.

The source temperature for the boiler is taken as the output temperature, which is $600\,°C$. For entropy generation and exergy destruction:

$$\dot{m}_{11} s_{11} + \frac{\dot{Q}_{in}}{T_s} + \dot{S}_{gen} = \dot{m}_1 s_1$$

$$\dot{S}_{gen} = 10\ \frac{kg}{s}(7.261 - 1.285)\ \frac{kJ}{kgK} - \frac{32511}{873}\ \frac{kW}{K} = \mathbf{22.52}\ \frac{\mathbf{kW}}{\mathbf{K}}$$

$$\dot{m}_{11}ex_{11} + \dot{Q}_{in}\left(1 - \frac{T_o}{T_{source}}\right) = \dot{m}_1 ex_1 + \dot{E}x_d$$

$$\dot{E}x_d = 10\ \frac{kg}{s}(37.57 - 1508)\ \frac{kJ}{kg} + 32511\ kW\left(1 - \frac{298}{873}\right) = \mathbf{6709}\ \boldsymbol{kW}$$

f) Calculate both energy and exergy efficiencies of the overall cycle.

The energy efficiency of the cycle under isentropic processes for pump and turbine are found to be:

$$\eta_{en_{is}} = \frac{\dot{W}_{net_{is}} + \dot{Q}_p}{\dot{Q}_{in}} = \frac{8467 + 10047}{32511} = \mathbf{56.9\%}$$

The actual energy and exergy efficiencies of the overall cycle are:

$$\eta_{en} = \frac{\dot{W}_{net_a} + \dot{Q}_p}{\dot{Q}_{in}} = \frac{6764 + 10047}{32511} = \mathbf{51.7\%}$$

$$\eta_{ex} = \frac{\dot{W}_{net_a} + \dot{Q}_p\left(1 - \frac{T_o}{T_b}\right)}{\dot{Q}_{in}\left(1 - \frac{T_o}{T_s}\right)} = \frac{6764 + 10047\left(1 - \frac{298}{416}\right)}{32511\left(1 - \frac{298}{873}\right)} = \mathbf{44.9\%}$$

The energy and exergy efficiencies of the overall cycle by considering 50% of the condenser heat ($\dot{Q}_{out}$) as a useful output are written as:

$$\dot{Q}_{tot} = \dot{Q}_p + \left(0.5 \times \dot{Q}_{out}\right) = 10047 + 0.5(13993) = \mathbf{17044}\ \boldsymbol{kW}$$

$$\eta_{en} = \frac{\dot{W}_{net} + \dot{Q}_{tot}}{\dot{Q}_{in}} = \frac{6764 + 17044}{32511} = \mathbf{73.2\%}$$

$$\eta_{ex} = \frac{\dot{W}_{net} + \dot{Q}_{tot}\left(1 - \frac{T_o}{T_b}\right)}{\dot{Q}_{in}\left(1 - \frac{T_o}{T_s}\right)} = \frac{6764 + 17044\left(1 - \frac{298}{416}\right)}{32511\left(1 - \frac{298}{873}\right)} = \mathbf{54\%}$$

The energy and exergy efficiencies of the overall cycle by considering 100% of the condenser heat ($\dot{Q}_{out}$) as a useful output are written as follows:

$$\dot{Q}_{tot} = \dot{Q}_p + \dot{Q}_{out} = 10047 + 13993 = \mathbf{24040}\ \boldsymbol{kW}$$

$$\eta_{en} = \frac{\dot{W}_{net} + \dot{Q}_{tot}}{\dot{Q}_{in}} = \frac{6764 + 24040}{32511} = \mathbf{94.7\%}$$

$$\eta_{ex} = \frac{\dot{W}_{net} + \dot{Q}_{tot}\left(1 - \frac{T_o}{T_b}\right)}{\dot{Q}_{in}\left(1 - \frac{T_o}{T_s}\right)} = \frac{6764 + 24040\left(1 - \frac{298}{416}\right)}{32511\left(1 - \frac{298}{873}\right)} = \mathbf{63.4\%}$$

Example 6.25 In a cogeneration Rankine plant, as shown in Figure 6.37, the boiler heats the steam to 550 °*C* and turbine operates at 3 *MPa*. 15% of the steam is extracted from the turbine at 1 *MPa* and the rest is expanded to 50 *kPa*. Assume the turbine isentropic efficiency to be 72%. The steam extracted from the turbine is mixed with feedwater after being condensed. The input steam flow rate to the turbine is 25 *kg/s*.

a) Write the mass, energy, entropy, and exergy balance equations.
b) Determine the boiler heat input rate.
c) Calculate the work rate of the turbine.
d) Calculate the entropy generation and exergy destruction of the condenser.
e) Calculate the energy and exergy efficiencies.

Solution

It is now important to draw the schematic diagram of the problem cycle (as shown in Figure 6.37a) along with its *T-s* diagram (as shown in Figure 6.37b) with the information and property data provided accordingly. This is necessary before beginning to write the balance equations for mass, energy, entropy, and exergy of each component of the cycle. Consider $s_6 = s_{7s} = s_{8s}$.

a) Write the mass, energy, entropy, and exergy balance equations.
For pump 1:

$$\text{MBE}: \dot{m}_1 = \dot{m}_2 = \dot{m}$$

$$\text{EBE}: \dot{m}_1 h_1 + \dot{W}_{in1} = \dot{m}_2 h_2$$

$$\text{EnBE}: \dot{m}_1 s_1 + \dot{S}_{gen,P1} = \dot{m}_2 s_2$$

$$\text{ExBE}: \dot{m}_1 ex_1 + \dot{W}_{in1} = \dot{m}_2 ex_2 + \dot{E}x_{d,P1}$$

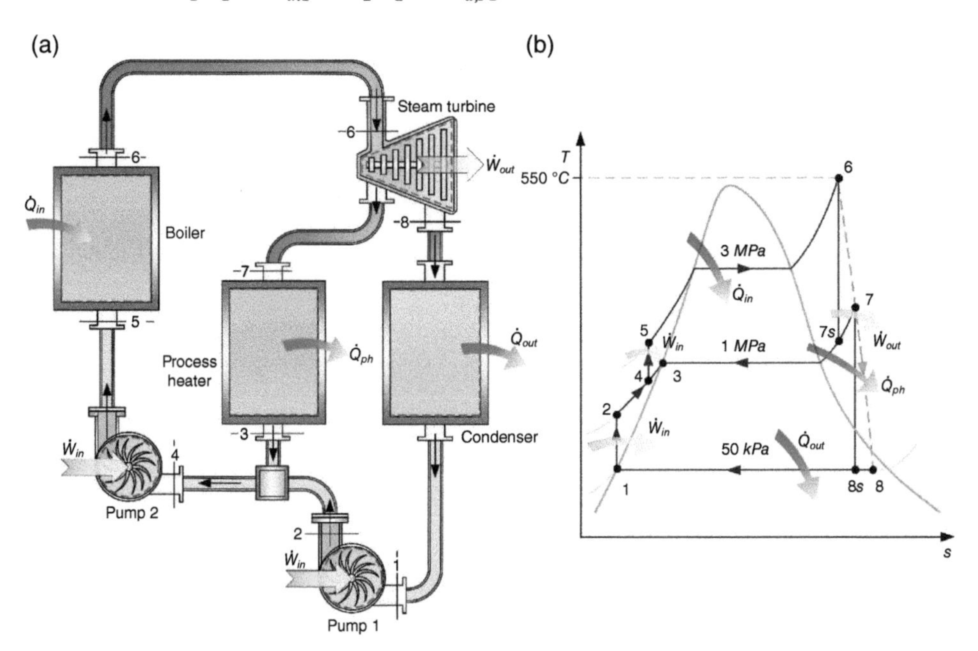

Figure 6.37 (a) A cogeneration Rankine plant with (b) its *T-s* diagram as discussed in Example 6.25.

For the mixing chamber:

$$\text{MBE}: \dot{m}_2 + \dot{m}_3 = \dot{m}_4$$

$$\text{EBE}: \dot{m}_2 h_2 + \dot{m}_3 h_3 = \dot{m}_4 h_4$$

$$\text{EnBE}: \dot{m}_2 s_2 + \dot{m}_3 s_3 + \dot{S}_{gen,MC} = \dot{m}_4 s_4$$

$$\text{ExBE}: \dot{m}_2 ex_2 + \dot{m}_3 ex_3 = \dot{m}_4 ex_4 + \dot{E}x_{d,MC}$$

For pump 2:

$$\text{MBE}: \dot{m}_4 = \dot{m}_5 = \dot{m}$$

$$\text{EBE}: \dot{m}_4 h_4 + \dot{W}_{in2} = \dot{m}_5 h_5$$

$$\text{EnBE}: \dot{m}_4 s_4 + \dot{S}_{gen} = \dot{m}_5 s_5$$

$$\text{ExBE}: \dot{m}_4 ex_4 + \dot{W}_{in2} = \dot{m}_5 ex_5 + \dot{E}x_d$$

For the boiler:

$$\text{MBE}: \dot{m}_5 = \dot{m}_6 = \dot{m}$$

$$\text{EBE}: \dot{m}_5 h_5 + \dot{Q}_{in} = \dot{m}_6 h_6$$

$$\text{EnBE}: \dot{m}_5 s_5 + \frac{\dot{Q}_{in}}{T_s} + \dot{S}_{gen,B} = \dot{m}_6 s_6$$

$$\text{ExBE}: \dot{m}_5 ex_5 + \dot{Q}_{in}\left(1 - \frac{T_o}{T_s}\right) = \dot{m}_6 ex_6 + \dot{E}x_{d,B}$$

For the turbine:

$$\text{MBE}: \dot{m}_6 = \dot{m}_7 + \dot{m}_8$$

$$\text{EBE}: \dot{m}_6 h_6 = \dot{m}_7 h_7 + \dot{m}_8 h_8 + \dot{W}_{out}$$

$$\text{EnBE}: \dot{m}_6 s_6 + \dot{S}_{gen,T} = \dot{m}_7 s_7 + \dot{m}_8 s_8$$

$$\text{ExBE}: \dot{m}_6 ex_6 = \dot{m}_7 ex_7 + \dot{m}_8 ex_8 + \dot{W}_{out} + \dot{E}x_{d,T}$$

For the process heater:

$$\text{MBE}: \dot{m}_7 = \dot{m}_3 = \dot{m}$$

$$\text{EBE}: \dot{m}_7 h_7 = \dot{m}_3 h_3 + \dot{Q}_{ph}$$

$$\text{EnBE}: \dot{m}_7 s_7 + \dot{S}_{gen,PH} = \dot{m}_3 s_3 + \frac{\dot{Q}_{ph}}{T_b}$$

$$\text{ExBE}: \dot{m}_7 ex_7 = \dot{m}_3 ex_3 + \dot{Q}_{ph}\left(1 - \frac{T_o}{T_b}\right) + \dot{E}x_{d,PH}$$

For the condenser:

$$\text{MBE}: \dot{m}_8 = \dot{m}_1 = \dot{m}$$

$$\text{EBE}: \dot{m}_8 h_8 = \dot{m}_1 h_1 + \dot{Q}_{out}$$

$$\text{EnBE}: \dot{m}_8 s_8 + \dot{S}_{gen,C} = \dot{m}_1 s_1 + \frac{\dot{Q}_{out}}{T_b}$$

$$\text{ExBE}: \dot{m}_8 ex_8 = \dot{m}_1 ex_1 + \dot{Q}_{out}\left(1 - \frac{T_o}{T_b}\right) + \dot{E}x_{d,C}$$

b) Determine the boiler heat input rate. / c) Calculate the work rate of the turbine.

State 1: $P_1 = 50\ kPa$

$$x_1 = 0$$

State 2: $P_2 = 1\ MPa$

$$h_1 = h_{f@50\ kPa} = 340.5\ \frac{kJ}{kg}$$

$$v_1 = 0.00103\ \frac{m^3}{kg}$$

$$w_{in1} = v_1(P_2 - P_1) = \left(0.00103\ \frac{m^3}{kg}\right)(1000 - 50)\ kPa = \mathbf{0.9784}\ \frac{\mathbf{kJ}}{\mathbf{kg}}$$

$$h_2 = h_1 + w_{in1} = (340.5 + 0.9784)\ \frac{kJ}{kg} = \mathbf{341.5}\ \frac{\mathbf{kJ}}{\mathbf{kg}}$$

$$\dot{W}_{in1} = \dot{m} \times w_{in1} = 21.25\ kg/s \times 0.9784\ kJ/kg = \mathbf{20.8\ kW}$$

State 3: $P_3 = 1\ MPa$

$$x_3 = 0$$

$$h_3 = h_{f@1\ MPa} = \mathbf{762.5}\ \frac{\mathbf{kJ}}{\mathbf{kg}}$$

State 4: $P_4 = 1\ MPa$

$$P_5 = 3\ MPa$$

$$\dot{m}_2 h_2 + \dot{m}_3 h_3 = \dot{m}_4 h_4$$

$$\dot{m}_2 = (0.85)\left(25\ \frac{kg}{s}\right) = \mathbf{21.25}\ \frac{\mathbf{kg}}{\mathbf{s}}$$

$$\dot{m}_3 = (0.15)\left(25\ \frac{kg}{s}\right) = \mathbf{3.75}\ \frac{\mathbf{kg}}{\mathbf{s}}$$

$$h_4 = \frac{(\dot{m}_2 h_2 + \dot{m}_3 h_3)}{\dot{m}_4} = \frac{\left(21.25\ \frac{kg}{s} \times h_2\right) + \left(3.75\ \frac{kg}{s} \times h_3\right)}{25\ \frac{kg}{s}} = 404.7\ \frac{kJ}{kg}$$

$$v_4 = v_{@h_4} = 0.00104\ \frac{m^3}{kg}$$

$$w_{in2} = v_4(P_5 - P_4) = \left(0.00104\ \frac{m^3}{kg}\right)(3000 - 1000)\ kPa = \mathbf{2.08}\ \frac{\mathbf{kJ}}{\mathbf{kg}}$$

$$h_5 = h_4 + w_{in2} = (404.7 + 2.08)\ \frac{kJ}{kg} = \mathbf{406.7}\ \frac{\mathbf{kJ}}{\mathbf{kg}}$$

$$\dot{W}_{in2} = \dot{m} \times w_{in2} = 21.25\ kg/s \times 2.08\ kJ/kg = \mathbf{52.01\ kW}$$

State 6: $P_6 = 3\ MPa$

$$T_6 = 550\ °C$$

$$h_6 = 3570\ \frac{kJ}{kg}$$

$$s_4 = 7.375\ \frac{kJ}{kgK}$$

State 7: $P_7 = 1\ MPa$

$$s_{7s} = s_6$$

$$h_{7s} = 3205\ \frac{kJ}{kg}$$

State 8: $P_8 = 50\ kPa$

$$s_{8s} = s_6$$

$$x_8 = \frac{s_{8s} - s_f}{s_{fg}} = \mathbf{0.9667}$$

$$h_{8s} = h_f + x_8 h_{fg} = \mathbf{2569}\ \boldsymbol{\frac{kJ}{kg}}$$

For process heat:

$$\dot{m}_7 = 0.15 \times 25\ \frac{kg}{s} = \mathbf{3.75}\ \boldsymbol{\frac{kg}{s}}$$

$$\dot{Q}_{ph,is} = \dot{m}_7(h_{7s} - h_3) = 3.75\ \frac{kg}{s}(3205 - 762.5)\ \frac{kJ}{kg} = \mathbf{9160}\ \boldsymbol{kW}$$

Then,

$$\dot{W}_{out,is} = \dot{m}_6 h_6 - \dot{m}_7 h_{7s} - \dot{m}_8 h_{8s}$$

$$\dot{W}_{out,is} = \left((25)\ \frac{kg}{s} \times h_6\right) - \left((0.15 \times 25)\ \frac{kg}{s} \times h_{7s}\right) - \left((0.85 \times 25)\ \frac{kg}{s} \times h_{8s}\right)$$

$$\dot{W}_{out,is} = \mathbf{22641}\ \boldsymbol{kW}$$

$$\dot{W}_{net_{is}} = \dot{W}_{out,is} - \dot{W}_{in1} - \dot{W}_{in2} = 22641 - 20.8 - 52.01 = \mathbf{22567}\ \boldsymbol{kW}$$

$$\eta_{is} = \frac{h_6 - h_7}{h_6 - h_{7s}}$$

$$h_7 = 3307\ \frac{kJ}{kg}$$

$$\eta_{is} = \frac{h_6 - h_8}{h_6 - h_{8s}}$$

$$h_8 = 2849\ \frac{kJ}{kg}$$

$$\dot{Q}_{ph} = \dot{m}_7(h_7 - h_3) = 3.75\ \frac{kg}{s}(3307 - 762.5)\ \frac{kJ}{kg} = \mathbf{9543}\ \boldsymbol{kW}$$

Then,

$$\dot{W}_{out} = \dot{m}_6 h_6 - \dot{m}_7 h_7 - \dot{m}_8 h_8$$

$$\dot{W}_{out} = \left((25)\ \frac{kg}{s} \times h_6\right) - \left((0.15 \times 25)\ \frac{kg}{s} \times h_7\right) - \left((0.85 \times 25)\ \frac{kg}{s} \times h_8\right)$$

$$\dot{W}_{out} = \mathbf{16308\ kW}$$

$$\dot{W}_{net} = \dot{W}_{out} - \dot{W}_{in1} - \dot{W}_{in2} = 16308 - 20.8 - 52.01 = \mathbf{16235\ kW}$$

For input heat:

$$\dot{Q}_{in} = \dot{m}_5(h_6 - h_5) = 25\ \frac{kg}{s}(3570 - 406.7)\ \frac{kJ}{kg} = \mathbf{79083\ kW}$$

d) Calculate the entropy generation and exergy destruction of the condenser.

The immediate boundary temperature for the condenser may directly be given or taken as the average of the input and output temperatures, which is 53 °C in this case. For the entropy generation and exergy destruction of the condenser:

$$\dot{Q}_{out} = \dot{m}_8(h_8 - h_1) = 21.25\ \frac{kg}{s}(2849 - 340.5)\ \frac{kJ}{kg} = \mathbf{53305\ kW}$$

$$\dot{m}_8 s_8 + \dot{S}_{gen,C} = \dot{m}_1 s_1 + \frac{\dot{Q}_{out}}{T_b}$$

$$\dot{S}_{gen,C} = 21.25\ \frac{kg}{s}(1.091 - 8.097)\ \frac{kJ}{kgK} + \frac{53305}{326}\ \frac{kW}{K} = \mathbf{14.63}\ \frac{\mathbf{kW}}{\mathbf{K}}$$

$$\dot{m}_8 ex_8 = \dot{m}_1 ex_1 + \dot{Q}_{out}\left(1 - \frac{T_o}{T_b}\right) + \dot{E}x_{d,C}$$

$$Ex_{d,C} = 21.25\ \frac{kg}{s}(440.7 - 19.87)\ \frac{kJ}{kg} - 53305\ kW\left(1 - \frac{298}{326}\right) = \mathbf{4364\ kW}$$

e) Calculate the energy and exergy efficiencies.

Here, we consider the process heat as the only useful heat output. In this regard, both energy and exergy efficiencies become:

$$\eta_{en} = \frac{\dot{W}_{net} + \dot{Q}_{ph}}{\dot{Q}_{in}} = \frac{16235 + 9543}{79083} = \mathbf{32.6\%}$$

and

$$\eta_{ex} = \frac{\dot{W}_{net} + \dot{Q}_{ph}\left(1 - \frac{T_o}{T_b}\right)}{\dot{Q}_{in}\left(1 - \frac{T_o}{T_s}\right)} = \frac{16235 + 9543\left(1 - \frac{298}{326}\right)}{79083\left(1 - \frac{298}{823}\right)} = \mathbf{33.8\%}$$

6.8.3 Combined Brayton–Rankine Cycles

The continuous efforts by researchers and engineers in industry to improve existing power plants and achieve higher efficiencies have lead them to integrate the gas turbine (air-standard Brayton) cycle with the Rankine cycle, where the connection between the two cycles is the boiler and the resulting integrated energy system is called a combined cycle. In the combined cycle the Brayton cycle tops the bottoming Rankine cycle and the resulting efficiency is higher than each of the cycles individually. There are potentially two arrangements of this kind of combined cycle: one with an open-type Brayton cycle, as shown in Figure 6.38a, and one with a closed-type of Brayton cycle, as shown in Figure 6.38b.

In the combined cycles, if we compare between the topping Brayton cycle and the bottoming Rankine cycle, the gas turbine operates at very high temperatures compared to the steam turbine. Typical inlet temperatures of the gas entering the gas turbine are usually around 1500 °*C* and over the temperature for the burner exit of turbojet engines. However, for steam turbines the maximum steam temperature entering the turbine is around 620 °*C*. Although the high temperature inlet to the gas turbine leads to the higher energy efficiency of gas turbines compared to steam turbines, the gases leave the gas turbine at a temperature of around 500 °*C*, which results in large losses. The fact that the exhaust temperature of the gas leaving the turbine is high made it possible to integrate the Rankine cycle as a bottoming cycle to recover the heat from gas turbine exhausts and use it to generate steam for the Rankine cycle, as shown in Figure 6.38a. The property plot of temperature and entropy for all of the nine states through the combined cycle is shown in Figure 6.39a.

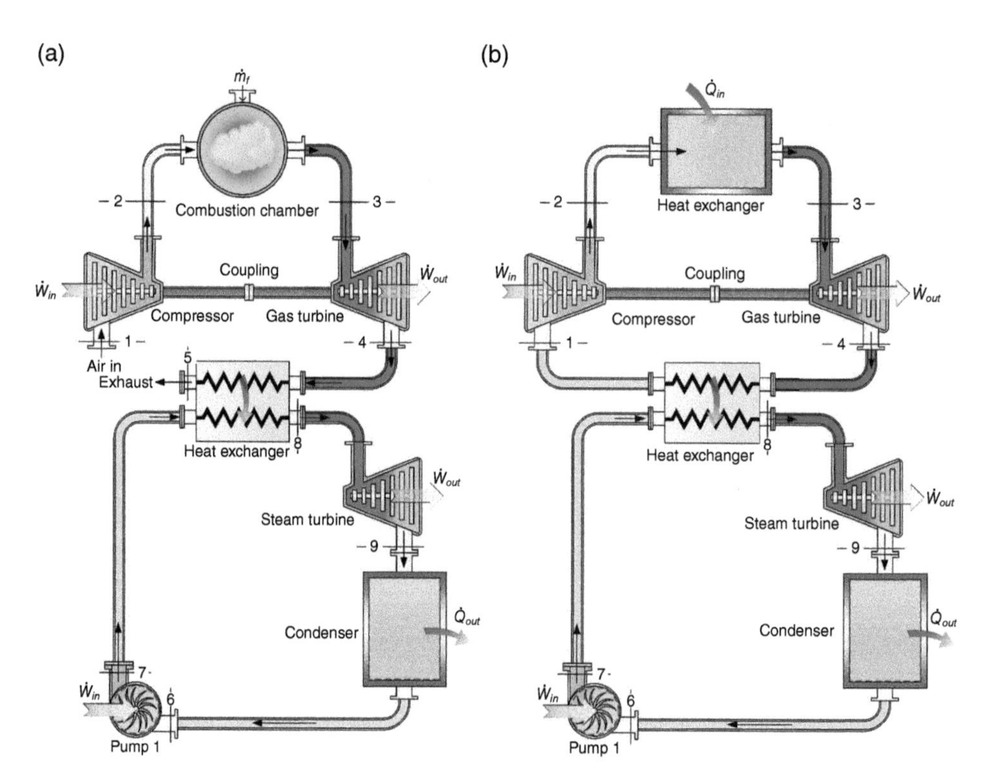

Figure 6.38 Schematic diagrams of two combined cycles: (a) with an open-type Brayton cycle and (b) with a closed-type Brayton cycle.

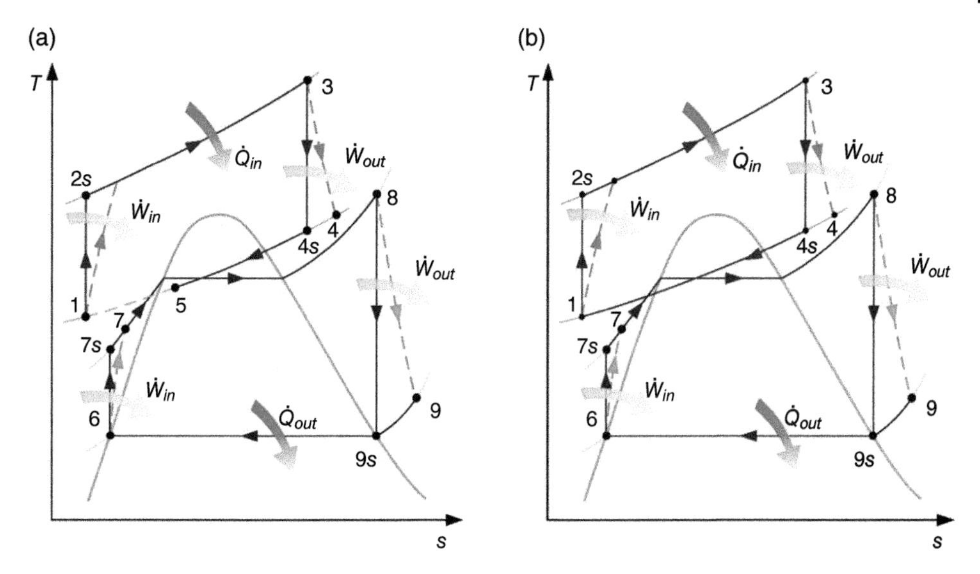

Figure 6.39 *T-s* diagrams of the combined cycles: (a) with an open-type Brayton cycle and (b) with a closed-type Brayton cycle.

Example 6.26 A combined cycle, as shown in Figure 6.38a along with its *T-s* diagram in Figure 6.39a, has the Brayton cycle topping and the Rankine cycle as the bottoming cycle. The air in the ideal Brayton cycle receives 130 MW of heat from a heat source that is at a temperature of 1500 °C, and the Brayton cycle generates a net amount of power equal to 55.4 *MW*. The gas leaving the gas turbine is at a temperature of 650 °*C* and has an energy equal to 75 *MW*. The boiler of the bottoming Rankine cycle is able to recover 80% of that energy and generate superheat steam that makes the Rankine cycle generate a net amount of power equal to 21.7 *MW*.

a) Write the mass, energy, entropy, and exergy balance equations for the overall topping Brayton cycle, bottoming Rankine cycle, and the integrated combined cycle.
b) Calculate the energy and exergy efficiencies of the topping Brayton cycle.
c) Calculate the energy and exergy efficiencies of the bottoming Rankine cycle.
d) Calculate the energy and exergy efficiencies of the combined cycle.

Solution

a) Write the mass, energy, entropy, and exergy balance equations for the overall topping Brayton cycle, bottoming Rankine cycle, and the integrated combined cycle
For the Brayton cycle:

$$\text{MBE}: \dot{m} = constant$$

$$\text{EBE}: \dot{Q}_{in} + \dot{W}_{in} = \dot{Q}_{out} + \dot{W}_{out}$$

$$\text{EnBE}: \dot{Q}_{in}/T_s + \dot{S}_{gen} = \dot{Q}_{out}/T_b$$

$$\text{ExBE}: \dot{E}x_{\dot{Q}_{in}} + \dot{W}_{in} = \dot{E}x_{\dot{Q}_{out}} + \dot{W}_{out} + \dot{E}x_d$$

For the Rankine cycle:

$$\text{MBE}: \dot{m} = constant$$

$$\text{EBE}: \dot{Q}_{in} + \dot{W}_{in} = \dot{Q}_{out} + \dot{W}_{out}$$

$$\text{EnBE}: \dot{Q}_{in}/T_s + \dot{S}_{gen} = \dot{Q}_{out}/T_b$$

$$\text{ExBE}: \dot{E}x_{\dot{Q}_{in}} + \dot{W}_{in} = \dot{E}x_{\dot{Q}_{out}} + \dot{W}_{out} + \dot{E}x_d$$

For the combined cycle:

$$\text{MBE}: \dot{m} = constant$$

$$\text{EBE}: \dot{Q}_{in} + \dot{W}_{in} = \dot{Q}_{out} + \dot{W}_{out}$$

$$\text{EnBE}: \dot{Q}_{in}/T_s + \dot{S}_{gen} = \dot{Q}_{out}/T_b$$

$$\text{ExBE}: \dot{E}x_{\dot{Q}_{in}} + \dot{W}_{in} = \dot{E}x_{\dot{Q}_{out}} + \dot{W}_{out} + \dot{E}x_d$$

b) Calculate the energy and exergy efficiencies of the topping Brayton cycle.

The energy and exergy efficiencies of the topping Brayton cycle are calculated respectively as follows:

$$\eta_{en} = \frac{\dot{W}_{net,out}}{\dot{Q}_{in}} = \frac{55.4\ MW}{130\ MW} = \mathbf{42.6\%}$$

$$\eta_{ex} = \frac{\dot{W}_{net,out}}{\dot{E}x_{\dot{Q}_{in}}} = \frac{\dot{W}_{net,out}}{\left(1 - \frac{T_o}{T_s}\right)\dot{Q}_{in}} = \frac{55.4\ MW}{\left(1 - \frac{298\ K}{1500 + 273}\right) \times 130\ MW} = \mathbf{51.2\%}$$

c) Calculate the energy and exergy efficiencies of the bottoming Rankine cycle.

The energy and exergy efficiencies of the bottoming Rankine cycle are calculated respectively as follows:

$$\eta_{en} = \frac{\dot{W}_{net,out}}{\dot{Q}_{in}} = \frac{21.7\ MW}{0.8 \times 75\ MW} = \mathbf{36.2\%}$$

$$\eta_{ex} = \frac{\dot{W}_{net,out}}{\dot{E}x_{\dot{Q}_{in}}} = \frac{\dot{W}_{net,out}}{\left(1 - \frac{T_o}{T_s}\right)\dot{Q}_{in}} = \frac{21.7\ MW}{\left(1 - \frac{298\ K}{650 + 273}\right) \times 0.8 \times 75\ MW} = \mathbf{53.4\%}$$

d) Calculate the energy and exergy efficiencies of the combined cycle.

The energy and exergy efficiencies of the combined cycle are calculated respectively as follows:

$$\eta_{en} = \frac{\dot{W}_{net,out}}{\dot{Q}_{in}} = \frac{\dot{W}_{net,out,BC} + \dot{W}_{net,out,RC}}{\dot{Q}_{in,BC}} = \frac{55.4\ MW + 21.7\ MW}{130\ MW} = \mathbf{59.3\%}$$

$$\eta_{ex} = \frac{\dot{W}_{net,out}}{\dot{E}x_{\dot{Q}_{in}}} = \frac{\dot{W}_{net,out,BC} + \dot{W}_{net,out,RC}}{\left(1 - \frac{T_o}{T_s}\right)\dot{Q}_{in}} = \frac{55.4\ MW + 21.7\ MW}{\left(1 - \frac{298\ K}{1500 + 273}\right) \times 130\ MW} = \mathbf{71.3\%}$$

The results clearly show that the combined cycle gives better performance with higher energy and exergy efficiencies.

Example 6.27 A combined cycle comprising of gas and steam power cycles, as shown in Figure 6.40, is used for power generation. The gas power cycle comprises an air-standard Brayton cycle, which works with a pressure ratio of 9. In this cycle, the compressor inlet state is at a temperature of 298 K and the combustion chamber exit temperature is measured to be 1351 K. The compressor, pump, and turbines operate with an isentropic efficiency of 80%. The waste heat in the gas turbine exhaust is utilized to operate a steam Rankine cycle. The steam Rankine cycle operates with the highest cycle pressure of 8 MPa and lowest cycle pressure of 10 kPa. The temperature at the steam turbine inlet is measured to be 500 °C and the exhaust gases are found to exit the heat exchanger at a temperature of 177 °C.

a) Write the mass energy, entropy, and exergy balance equations for all system components.
b) Find the ratio of the mass flow rates where the combustion gases and steam pass through the heat exchanger.
c) Determine the net work output.
d) Determine the entropy generation in both the gas and steam turbines.
e) Find the exergy destruction in both the gas and steam turbines.

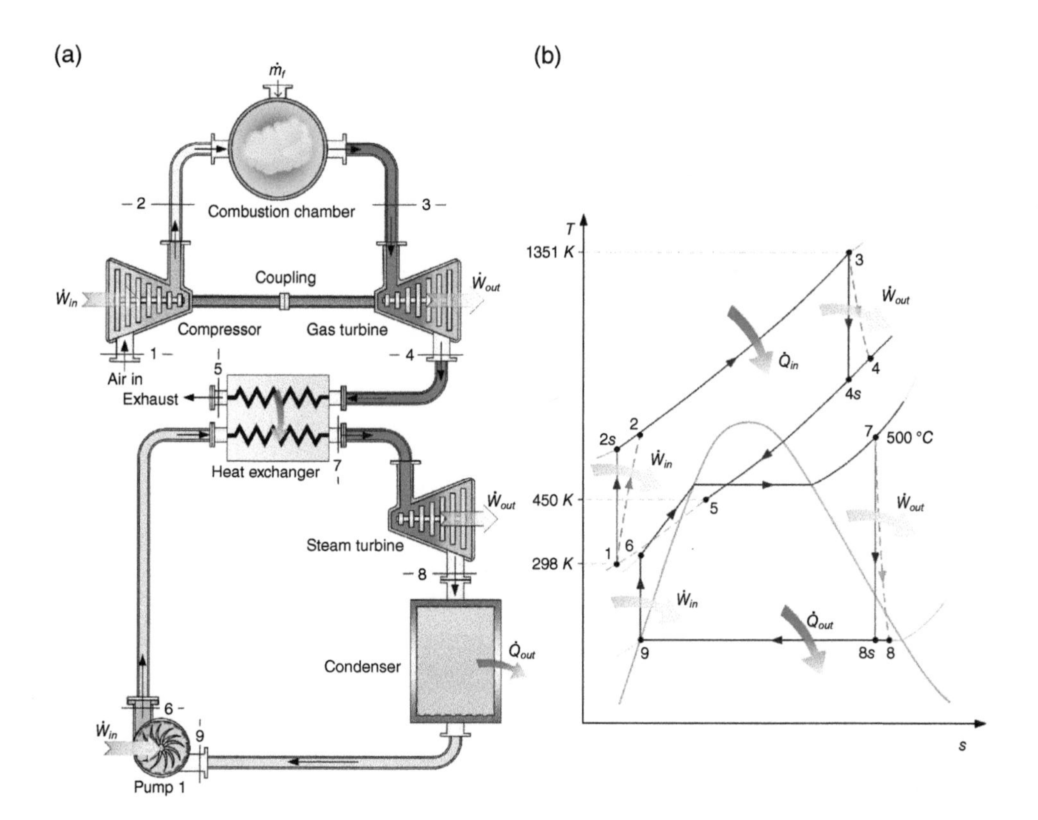

Figure 6.40 (a) A cogeneration Rankine plant with (b) its T-s diagram as discussed in Example 6.27.

f) Determine the energy and exergy efficiencies of the cycle if the source temperature is 2500 K and the mass flow rates of air and steam are 10 and 3 kg/s, respectively.

Solution

It is now important to draw the schematic diagram of the problem cycle (as shown in Figure 6.40a) along with its T-s diagram (as shown in Figure 6.40b) with the information and property data provided accordingly. This is necessary before beginning to write the balance equations for mass, energy, entropy, and exergy of each component of the cycle.

a) Write the mass energy, entropy, and exergy balance equations for all system components.

For the compressor:

$$\text{MBE}: \dot{m}_1 = \dot{m}_2$$

$$\text{EBE}: \dot{m}_1 h_1 + \dot{W}_C = \dot{m}_2 h_2$$

$$\text{EnBE}: \dot{m}_1 s_1 + \dot{S}_{gen,C} = \dot{m}_2 s_2$$

$$\text{ExBE}: \dot{m}_1 ex_1 + \dot{W}_C = \dot{m}_2 ex_2 + \dot{E}x_{d,C}$$

For the combustion chamber:

$$\text{MBE}: \dot{m}_2 = \dot{m}_3$$

$$\text{EBE}: \dot{m}_2 h_2 + \dot{Q}_{CC} = \dot{m}_3 h_3$$

$$\text{EnBE}: \dot{m}_2 s_2 + \frac{\dot{Q}_{CC}}{T_s} + \dot{S}_{gen,CC} = \dot{m}_3 s_3$$

$$\text{ExBE}: \dot{m}_2 ex_2 + \dot{Q}_{CC}\left(1 - \frac{T_0}{T_s}\right) = \dot{m}_3 ex_3 + \dot{E}x_{d,CC}$$

For the gas turbine:

$$\text{MBE}: \dot{m}_3 = \dot{m}_4$$

$$\text{EBE}: \dot{m}_3 h_3 = \dot{W}_{GT} + \dot{m}_4 h_4$$

$$\text{EnBE}: \dot{m}_3 s_3 + \dot{S}_{gen,GT} = \dot{m}_4 s_4$$

$$\text{ExBE}: \dot{m}_3 ex_3 = \dot{W}_{GT} + \dot{m}_4 ex_4 + \dot{E}x_{d,GT}$$

For the heat exchanger:

$$\text{MBE}: \dot{m}_4 = \dot{m}_5 \text{ and } \dot{m}_6 = \dot{m}_7$$

$$\text{EBE}: \dot{m}_4 h_4 + \dot{m}_6 h_6 = \dot{m}_5 h_5 + \dot{m}_7 h_7$$

$$\text{EnBE}: \dot{m}_4 s_4 + \dot{m}_6 s_6 + \dot{S}_{gen,HX} = \dot{m}_5 s_5 + \dot{m}_7 s_7$$

$$\text{ExBE}: \dot{m}_4 ex_4 + \dot{m}_6 ex_6 = \dot{m}_5 ex_5 + \dot{m}_7 ex_7 + \dot{E}x_{d,HX}$$

For the steam turbine:

$$\text{MBE}: \dot{m}_7 = \dot{m}_8$$

$$\text{EBE}: \dot{m}_7 h_7 = \dot{W}_{ST} + \dot{m}_8 h_8$$

$$\text{EnBE}: \dot{m}_7 s_7 + \dot{S}_{gen,ST} = \dot{m}_8 s_8$$

$$\text{ExBE}: \dot{m}_7 ex_7 = \dot{W}_{ST} + \dot{m}_8 ex_8 + \dot{E}x_{d,T}$$

For the condenser:

$$\text{MBE}: \dot{m}_8 = \dot{m}_9$$

$$\text{EBE}: \dot{m}_8 h_8 = \dot{Q}_{con} + \dot{m}_9 h_9$$

$$\text{EnBE}: \dot{m}_8 s_8 + \dot{S}_{gen,con} = \frac{\dot{Q}_{con}}{T_b} + \dot{m}_9 s_9$$

$$\text{ExBE}: \dot{m}_8 ex_8 = \dot{Q}_{con}\left(1 - \frac{T_0}{T_b}\right) + \dot{m}_9 ex_9 + \dot{E}x_{d,con}$$

For the pump:

$$\text{MBE}: \dot{m}_9 = \dot{m}_6$$

$$\text{EBE}: \dot{m}_9 h_9 + \dot{W}_p = \dot{m}_6 h_6$$

$$\text{EnBE}: \dot{m}_9 s_9 + \dot{S}_{gen,P} = \dot{m}_6 s_6$$

$$\text{ExBE}: \dot{m}_9 ex_9 + \dot{W}_P = \dot{m}_6 ex_6 + \dot{E}x_{d,P}$$

b) The state point properties are first obtained from the given information. At the compressor inlet, the air enters at a temperature of 298 K and a pressure of 101 kPa. Other properties at state 1 can be obtained from the tables (such as Appendix E-1) or properties database software (such as EES) as:

$$h_1 = 298.6\ \frac{kJ}{kg} \text{ and } s_1 = 5.697\ \frac{kJ}{kgK}$$

The pressure ratio of the compressor is given as 9, hence $P_2 = 9\ (P_1) = 909\ kPa$. Also, the compressor isentropic efficiency is known to be 80%. This can be used to find h_2:

$$h_2 = h_1 + \frac{h_{2s} - h_1}{\eta_{is,c}}$$

where h_{2s} is found at $P_2 = 909\ kPa$ and $s_{2s} = s_1 = 5.697\ \frac{kJ}{kgK}$ as

$$h_{2s} = 559.9\ \frac{kJ}{kg}$$

$$h_2 = 298.6\ \frac{kJ}{kg} + \frac{559.9\ \frac{kJ}{kg} - 298.6\ \frac{kJ}{kg}}{0.8} = \mathbf{625.2\ \frac{kJ}{kg}}$$

From h_2 and P_2, the specific entropy can also be found as $s_2 = 5.808\ \frac{kJ}{kgK}$.

Next, the temperature at state 3 is given as $T_3 = 1351\ K$. Also the pressure can be assumed to be constant in the combustion chamber ($P_2 = P_3 = 909\ kPa$). Thus, the specific enthalpy and entropy can be found from thermodynamic tables (such as Appendix E-1) or a software database as:

$$h_3 = 1457\ \frac{kJ}{kg} \text{ and } s_3 = 6.689\ \frac{kJ}{kgK}$$

Further, the turbine isentropic efficiency is also given as 80%. This can be used to find h_4:

$$h_4 = h_3 - \eta_{is,T}(h_3 - h_{4s})$$

where h_{4s} is found at $P_4 = 101\ kPa$ and $s_{4s} = s_3 = 6.689\ \frac{kJ}{kgK}$ as

$$h_{4s} = 798.9\ \frac{kJ}{kg}$$

$$h_4 = 1457\ \frac{kJ}{kg} - 0.8\left(1457\ \frac{kJ}{kg} - 798.9\ \frac{kJ}{kg}\right) = \mathbf{930.5}\ \frac{\mathbf{kJ}}{\mathbf{kg}}$$

$$h_2 = \mathbf{625.2}\ \frac{\mathbf{kJ}}{\mathbf{kg}}$$

Next, the temperature at state 5 is known to be $T_5 = 450\ K$. This can be used with $P_5 = P_4 = 101\ kPa$ to find h_5 and s_5 as:

$$h_5 = 452.3\ \frac{kJ}{kg}\ \text{and}\ s_5 = 6.113\ \frac{kJ}{kgK}$$

The temperature and pressure at state 7 are known to be $P_7 = 8000\ kPa$ and $T_7 = 500\ ^{\circ}C$. These can be used to find:

$$h_7 = 3399\ \frac{kJ}{kg}\ \text{and}\ s_7 = 6.727\ \frac{kJ}{kgK}$$

The properties at state 8 can be found from the isentropic efficiency of the turbine as:

$$h_8 = h_7 - \eta_{is,T}(h_7 - h_{8s})$$

$$= 3399\ \frac{kJ}{kg} - 0.8\left(3399\ \frac{kJ}{kg} - 2130\ \frac{kJ}{kg}\right) = \mathbf{2384}\ \frac{\mathbf{kJ}}{\mathbf{kg}}$$

The pressure state 9 is known to be $P_9 = 10\ kPa$ and the properties at state 1 can be found assuming the properties of saturated liquid at this pressure:

$$h_9 \approx h_{f,10\ kPa} = 191.8\ \frac{kJ}{kg}\ \text{and}\ s_9 \approx s_{f,10kPa} = 0.6492\ \frac{kJ}{kgK}$$

The properties at state 6 can then be found from the pump isentropic efficiency as:

$$h_6 = h_9 + \frac{h_{6s} - h_9}{\eta_{is,p}}$$

$$= 191.8\ \frac{kJ}{kg} + \frac{199.9\ \frac{kJ}{kg} - 191.8\ \frac{kJ}{kg}}{0.8} = \mathbf{201.9}\ \frac{\mathbf{kJ}}{\mathbf{kg}}$$

Lastly, the energy balance of the heat exchanger can be re-arranged as:

$$\dot{m}_4 h_4 - \dot{m}_5 h_5 = \dot{m}_7 h_7 - \dot{m}_6 h_6$$

where $\dot{m}_4 = \dot{m}_5 = \dot{m}_g$ and $\dot{m}_6 = \dot{m}_7 = \dot{m}_w$

$$\dot{m}_g(h_4 - h_5) = \dot{m}_w(h_7 - h_6)$$

$$\frac{\dot{m}_w}{\dot{m}_g} = \frac{h_4 - h_5}{h_7 - h_6} = \frac{930.5\,\frac{kJ}{kg} - 452.3\,\frac{kJ}{kg}}{3399\,\frac{kJ}{kg} - 201.9\,\frac{kJ}{kg}} = 0.149$$

c) The net power output can be found from the summation of the net power output of the Brayton cycle and Rankine cycle as:

$$w_{net} = w_{net,BC} + w_{net,RC}$$

Neglecting pump work, the net work output can be evaluated as:

$$w_{net} = (h_3 - h_4) - (h_2 - h_1) + (h_7 - h_8)$$

$$= \left(1457\,\frac{kJ}{kg} - 930.5\,\frac{kJ}{kg}\right) - \left(625.2\,\frac{kJ}{kg} - 298.6\,\frac{kJ}{kg}\right) + \left(3399\,\frac{kJ}{kg} - 2384\,\frac{kJ}{kg}\right) = \mathbf{1215}\,\frac{\mathbf{kJ}}{\mathbf{kg}}$$

d) The entropy generations occurred in the gas and steam turbines can be found from the entropy balance as:

$$s_{gen,GT} = s_4 - s_3 = \mathbf{0.1573}\,\frac{\mathbf{kJ}}{\mathbf{kgK}}$$

$$s_{gen,ST} = s_8 - s_7 = \mathbf{0.796}\,\frac{\mathbf{kJ}}{\mathbf{kgK}}$$

e) The exergy destructions occurred in gas and steam turbines can be found from the ExBE as:

$$ex_{d,GT} = ex_3 - w_{GT} - ex_4$$

where the specific exergy at states 3 and 4 can be evaluated as:

$$ex_3 = (h_3 - h_0) - T_0(s_3 - s_0)$$

$$= \left(1457\,\frac{kJ}{kg} - 298.6\,\frac{kJ}{kg}\right) - 298\ K\left(6.689\,\frac{kJ}{kgK} - 5.696\,\frac{kJ}{kgK}\right)$$

$$= \mathbf{862.4\ kJ/kg}$$

$$ex_4 = (h_4 - h_0) - T_0(s_4 - s_0)$$

$$= \left(930.5\,\frac{kJ}{kg} - 298.6\,\frac{kJ}{kg}\right) - 298\ K\left(6.846\,\frac{kJ}{kgK} - 5.696\,\frac{kJ}{kgK}\right)$$

$$= \mathbf{289.1\ kJ/kg}$$

substituting these values in the ExBE gives:

$$ex_{d,GT} = 862.4\,\frac{kJ}{kg} - 526.5\,\frac{kJ}{kg} - 289.1\,\frac{kJ}{kg} = \mathbf{46.87}\,\frac{\mathbf{kJ}}{\mathbf{kg}}$$

Similarly, the exergy destruction in the steam turbine can also be found as:

$$ex_{d,ST} = ex_7 - w_{ST} - ex_8$$

where the specific exergy at states 7 and 8 can be evaluated as:

$$ex_7 = (h_7 - h_0) - T_0(s_7 - s_0)$$
$$= \left(3399\ \frac{kJ}{kg} - 104.9\ \frac{kJ}{kg}\right) - 298\ K\left(6.727\ \frac{kJ}{kgK} - 0.3672\ \frac{kJ}{kgK}\right)$$
$$= \mathbf{1399\ kJ/kg}$$

$$ex_8 = (h_8 - h_0) - T_0(s_8 - s_0)$$
$$= \left(2384\ \frac{kJ}{kg} - 104.9\ \frac{kJ}{kg}\right) - 298\ K\left(7.522\ \frac{kJ}{kgK} - 0.3672\ \frac{kJ}{kgK}\right)$$
$$= \mathbf{146.9\ kJ/kg}$$

$$ex_{d,ST} = 1399\ \frac{kJ}{kg} - 1015\ \frac{kJ}{kg} - 146.9\ \frac{kJ}{kg} = \mathbf{237.2\ \frac{kJ}{kg}}$$

f) The energy efficiency can be found from the following equation:

$$\eta_{en} = \frac{w_{net}}{q_{in,CC}}$$

where $q_{in,\ CC}$ can be found from the energy balance equation of the combustion chamber as

$$q_{in,cc} = h_3 - h_2$$

$$q_{in,cc} = 1457\ \frac{kJ}{kg} - 625.2\ \frac{kJ}{kg} = \mathbf{831.7\ \frac{kJ}{kg}}$$

$$\eta_{en} = \frac{w_{net}}{q_{in,CC}} = \frac{\left(10\ \frac{kg}{s}\right)\left[\left(1457\ \frac{kJ}{kg} - 930.5\ \frac{kJ}{kg}\right) - \left(625.2\ \frac{kJ}{kg} - 298.6\ \frac{kJ}{kg}\right)\right] + \left(3\ \frac{kg}{s}\right)\left(3399\ \frac{kJ}{kg} - 2384\ \frac{kJ}{kg}\right)}{\left(10\ \frac{kg}{s}\right)\left(831.7\ \frac{kJ}{kg}\right)}$$

$= 60.7\%$

and

$$\eta_{ex} = \frac{w_{net}}{q_{in,CC}\left(1 - \frac{T_0}{T_s}\right)}$$

$$\eta_{ex} = \frac{\left(10\ \frac{kg}{s}\right)\left[\left(1457\ \frac{kJ}{kg} - 930.5\ \frac{kJ}{kg}\right) - \left(625.2\ \frac{kJ}{kg} - 298.6\ \frac{kJ}{kg}\right)\right] + \left(3\ \frac{kg}{s}\right)\left(3399\ \frac{kJ}{kg} - 2384\ \frac{kJ}{kg}\right)}{\left(10\ \frac{kg}{s}\right)\left(831.7\ \frac{kJ}{kg}\right)\left(1 - \frac{298}{2500}\right)}$$

$= 68.9\%$

Example 6.28 The combined power generation cycle shown in Figure 6.41a includes a reheat Rankine cycle along with a Brayton cycle. The compressor of the Brayton cycle entails a pressure ratio of 7. The compressor inlet comprises an ambient air input at 17 °C. The mass flow rate of air through the compressor is known to be 40 kg/s. The temperature of the combustion gases is measured to be 1078 °C. The reheat Rankine cycle operates at the highest pressure of 5 MPa and the lowest pressure of 8 kPa. The mass flow rate of steam in the Rankine cycle is 4.6 kg/s. The exhaust gas temperature at the exit of the heat exchanger is measured to be 200 °C and the steam temperature at the inlet of the

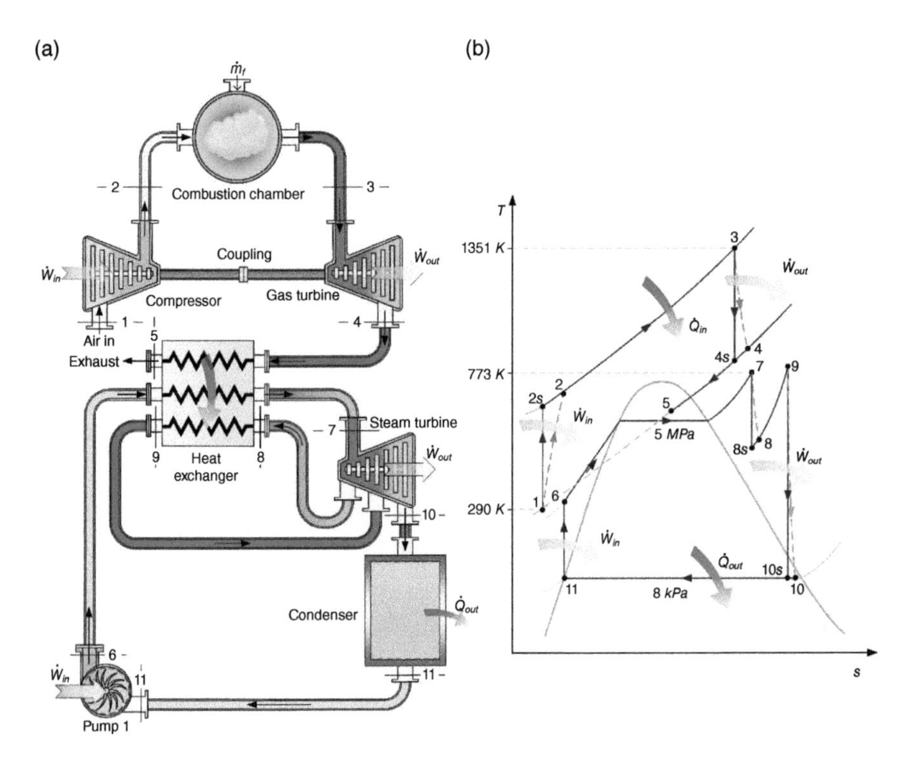

Figure 6.41 (a) A cogeneration Rankine plant with (b) its T-s diagram as discussed in Example 6.28.

high-pressure turbine is 500 °C. The exit pressure of steam from the high-pressure turbine is 1000 kPa and the reheat temperature is set as 400 °C. The isentropic efficiency of the pump, compressor, and turbines can be assumed to be 85%. The heat source temperature can be taken as 2485 °C.

a) Write the mass, energy, entropy, and exergy balance equations for all system components.
b) Find the vapor quality at the exit of the low-pressure turbine.
c) Determine the net power output from the cycle.
d) Determine the entropy generation and exergy destruction rates in the gas turbine.
e) Determine the energy and exergy efficiencies of the cycle.

Solution

It is now important to draw the schematic diagram of the problem cycle (as shown in Figure 6.41a) along with its T-s diagram (as shown in Figure 6.41b) with the information and property data provided accordingly. This is necessary before beginning to write the balance equations for mass, energy, entropy, and exergy of each component of the cycle.

a) Write the mass, energy, entropy, and exergy balance equations for all system components

For the compressor:

$$\text{MBE}: \dot{m}_1 = \dot{m}_2$$

$$\text{EBE}: \dot{m}_1 h_1 + \dot{W}_C = \dot{m}_2 h_2$$

$$\text{EnBE}: \dot{m}_1 s_1 + \dot{S}_{gen,C} = \dot{m}_2 s_2$$

$$\text{ExBE}: \dot{m}_1 ex_1 + \dot{W}_C = \dot{m}_2 ex_2 + \dot{E}x_{d,C}$$

For the combustion chamber:

$$\text{MBE}: \dot{m}_2 = \dot{m}_3$$

$$\text{EBE}: \dot{m}_2 h_2 + \dot{Q}_{CC} = \dot{m}_3 h_3$$

$$\text{EnBE}: \dot{m}_2 s_2 + \frac{\dot{Q}_{CC}}{T_s} + \dot{S}_{gen,CC} = \dot{m}_3 s_3$$

$$\text{ExBE}: \dot{m}_2 ex_2 + \dot{Q}_{CC}\left(1 - \frac{T_0}{T_s}\right) = \dot{m}_3 ex_3 + \dot{E}x_{d,CC}$$

For the gas turbine:

$$\text{MBE}: \dot{m}_3 = \dot{m}_4$$

$$\text{EBE}: \dot{m}_3 h_3 = \dot{W}_{GT} + \dot{m}_4 h_4$$

$$\text{EnBE}: \dot{m}_3 s_3 + \dot{S}_{gen,GT} = \dot{m}_4 s_4$$

$$\text{ExBE}: \dot{m}_3 ex_3 = \dot{W}_{GT} + \dot{m}_4 ex_4 + \dot{E}x_{d,GT}$$

For the heat exchanger:

$$\text{MBE}: \dot{m}_4 = \dot{m}_5 \text{ and } \dot{m}_6 = \dot{m}_7 = \dot{m}_8 = \dot{m}_9$$

$$\text{EBE}: \dot{m}_4 h_4 + \dot{m}_6 h_6 + \dot{m}_8 h_8 = \dot{m}_5 h_5 + \dot{m}_7 h_7 + \dot{m}_9 h_9$$

$$\text{EnBE}: \dot{m}_4 s_4 + \dot{m}_6 s_6 + \dot{m}_8 s_8 + \dot{S}_{gen,HX} = \dot{m}_5 s_5 + \dot{m}_7 s_7 + \dot{m}_9 s_9$$

$$\text{ExBE}: \dot{m}_4 ex_4 + \dot{m}_6 ex_6 + \dot{m}_8 ex_8 = \dot{m}_5 ex_5 + \dot{m}_7 ex_7 + \dot{m}_9 ex_9 + \dot{E}x_{d,HX}$$

For the steam turbine:

$$\text{MBE}: \dot{m}_7 = \dot{m}_8 = \dot{m}_9 = \dot{m}_{10}$$

$$\text{EBE}: \dot{m}_7 h_7 + \dot{m}_9 h_9 = \dot{W}_{ST} + \dot{m}_8 h_8 + \dot{m}_{10} h_{10}$$

$$\text{EnBE}: \dot{m}_7 s_7 + \dot{m}_9 s_9 + \dot{S}_{gen,ST} = \dot{m}_8 s_8 + \dot{m}_{10} s_{10}$$

$$\text{ExBE}: \dot{m}_7 ex_7 + \dot{m}_9 ex_9 = \dot{W}_{ST} + \dot{m}_8 ex_8 + \dot{m}_{10} ex_{10} + \dot{E}x_{d,T}$$

For the condenser:

$$\text{MBE}: \dot{m}_{10} = \dot{m}_{11}$$

$$\text{EBE}: \dot{m}_{10} h_{10} = \dot{Q}_{con} + \dot{m}_{11} h_{11}$$

$$\text{EnBE}: \dot{m}_{10} s_{10} + \dot{S}_{gen,con} = \frac{\dot{Q}_{con}}{T_b} + \dot{m}_{11} s_{11}$$

$$\text{ExBE}: \dot{m}_{10} ex_{10} = \dot{Q}_{con}\left(1 - \frac{T_0}{T_b}\right) + \dot{m}_{11} ex_{11} + \dot{E}x_{d,con}$$

For the pump:

$$\text{MBE}: \dot{m}_{11} = \dot{m}_6$$

$$\text{EBE}: \dot{m}_{11}h_{11} + \dot{W}_p = \dot{m}_6 h_6$$

$$\text{EnBE}: \dot{m}_{11}s_{11} + \dot{S}_{gen,P} = \dot{m}_6 s_6$$

$$\text{ExBE}: \dot{m}_{11}ex_{11} + \dot{W}_P = \dot{m}_6 ex_6 + \dot{E}x_{d,P}$$

b) Find the vapor quality at the exit of the low-pressure turbine.

The temperature at the compressor inlet is known to be 17 °C, the enthalpy of air at this temperature can be found from a software database or tables (such as Appendix E-1) and the entropy can also be found at this temperature and ambient pressure of 101 kPa as $h_1 = 290.6\ \frac{kJ}{kg}$ and $s_1 = 5.669\ \frac{kJ}{kgK}$

Also, since the pressure ratio is 7, $P_2 = 707\ kPa$. The other properties at state 2 can be found from the isentropic efficiency relationship:

$$h_2 = h_1 + \frac{h_{2s} - h_1}{\eta_{is,c}}$$

where $h_{2s} = 507.2\ \frac{kJ}{kg}$ at $P_2 = 707\ kPa$ and $s_{2s} = s_1 = 5.669\ \frac{kJ}{kgK}$

$$h_2 = 290.6\ \frac{kJ}{kg} + \frac{507.2\ \frac{kJ}{kg} - 290.6\ \frac{kJ}{kg}}{0.85} = \mathbf{545.5\ \frac{kJ}{kg}}$$

Next, from the given temperature at state 3 $T_3 = 1078\ ^\circ C$ and pressure $P_3 = 707\ kPa$, the enthalpy and entropy can be found as $h_3 = 1457\ kJ/kg$ and $s_3 = 6.761\ \frac{kJ}{kgK}$

The gas turbine exit state can be determined as:

$$h_4 = h_3 - \eta_{is,T}(h_3 - h_{4s})$$

$$= 1457\ \frac{kJ}{kg} - 0.85\left(1457\ \frac{kJ}{kg} - 856.9\ \frac{kJ}{kg}\right) = \mathbf{949.9\ \frac{kJ}{kg}}$$

s_4 can be found from h_4 and P_4 as $s_4 = 6.864\ \frac{kJ}{kgK}$.

Furthermore, the temperature of the exhaust gases at the heat exchanger exit is specified as T_5= 200 °C. This can be used along with the pressure of 101 kPa to find $h_5 = 475.8\ \frac{kJ}{kg}$ and $s_5 = 6.164\ \frac{kJ}{kgK}$.

Next, the thermodynamic properties at different state points of the steam Rankine cycle are found. The lowest pressure of 8 kPa is at state points 10 and 6. The enthalpy at state 11 can be found as $h_{11} \approx h_{f,8\ kPa} = 173.8\ \frac{kJ}{kg}$ and $s_{11} \approx s_{f,8kPa} = 0.5925\ \frac{kJ}{kgK}$.

The properties at state 6 can then be found from the pump isentropic efficiency as:

$$h_6 = h_{11} + \frac{h_{6s} - h_{11}}{\eta_{is,p}}$$

$$= 173.8\ \frac{kJ}{kg} + \frac{178.9\ \frac{kJ}{kg} - 173.8\ \frac{kJ}{kg}}{0.85} = \mathbf{180.1\ \frac{kJ}{kg}}$$

Moreover, from $P_9 = 1000\ kPa$ and $T_9 = 400\,^{\circ}C$, $h_9 = 3264\ \frac{kJ}{kg}$ and $s_9 = 7.467\ \frac{kJ}{kgK}$.

The enthalpy at state 10 can be found from the isentropic efficiency as:

$$h_{10} = h_9 - \eta_{is,T}(h_9 - h_{10s})$$

$$h_{10} = 3264\ \frac{kJ}{kg} - 0.85\left(3264\ \frac{kJ}{kg} - 2337\ \frac{kJ}{kg}\right) = \mathbf{2476\ \frac{kJ}{kg}}$$

From h_{10} and P_{10}, $s_{10} = 7.909\ \frac{kJ}{kgK}$

Lastly, from h_{10} and P_{10}, the vapor quality is found as:

$$\mathbf{x_{10} = 0.958}$$

c) Determine the net power output from the cycle.
The properties of state 7 can be found from $T_7 = 500\,^{\circ}C$ and $P_7 = 5000\ kPa$ as $h_7 = 3435\ \frac{kJ}{kg}$ and $s_7 = 6.978\ \frac{kJ}{kgK}$. Also, state 8 can be found as:

$$h_8 = h_7 - \eta_{is,T}(h_7 - h_{8s})$$

$$= 3435\ \frac{kJ}{kg} - (0.85)\left(3435\ \frac{kJ}{kg} - 2970\ \frac{kJ}{kg}\right) = \mathbf{3040\ \frac{kJ}{kg}}$$

Neglecting pump work, the net power output of the cycle can be found as:

$$\dot{W}_{net} = \dot{W}_{GT} - \dot{W}_c + \dot{W}_{ST}$$

$$\dot{W}_{GT} = \dot{m}_g(h_3 - h_4) = \left(40\ \frac{kg}{s}\right)\left(1457\ \frac{kJ}{kg} - 949.9\ \frac{kJ}{kg}\right) = \mathbf{20,284\ kW}$$

$$\dot{W}_c = \dot{m}_g(h_2 - h_1) = \left(40\ \frac{kg}{s}\right)\left(545.5\ \frac{kJ}{kg} - 290.6\ \frac{kJ}{kg}\right) = \mathbf{10,196\ kW}$$

$$\dot{W}_{ST} = \dot{m}_w(h_7 - h_8 + h_9 - h_{10})$$

$$= \left(4.6\ \frac{kg}{s}\right)\left(3435\ \frac{kJ}{kg} - 3040\ \frac{kJ}{kg} + 3264\ \frac{kJ}{kg} - 2476\ \frac{kJ}{kg}\right)$$

$$= \mathbf{5442\ kW}$$

Thus, $\dot{W}_{net} = 20,284\ kW - 10,196\ kW + 5442\ kW = \mathbf{15,530\ kW}$

d) Determine the entropy generation and exergy destruction rates in the gas turbine.
The entropy generation rate in the gas turbine can be evaluated as:

$$\dot{S}_{gen,GT} = \dot{m}_4 s_4 - \dot{m}_3 s_3$$

$$\dot{S}_{gen,GT} = \left(40\ \frac{kg}{s}\right)\left(6.864\ \frac{kJ}{kgK} - 6.761\ \frac{kJ}{kgK}\right) = \mathbf{4.12\ \frac{kW}{K}}$$

The entropy generation rate of the steam turbine is determined as:

$$\dot{S}_{gen,ST} = \dot{m}_8 s_8 + \dot{m}_{10} s_{10} - \dot{m}_7 s_7 - \dot{m}_9 s_9$$

$$= \dot{m}_w (s_8 + s_{10} - s_7 - s_9)$$

$$= \left(4.6\ \frac{kg}{s}\right)\left(7.104\ \frac{kJ}{kgK} + 7.909\ \frac{kJ}{kgK} - 6.978\ \frac{kJ}{kgK} - 7.467\ \frac{kJ}{kgK}\right) = \mathbf{2.613\ \frac{kW}{K}}$$

The specific exergy at state 3 can be found as:

$$ex_3 = (h_3 - h_0) - T_0(s_3 - s_0)$$

$$= \left(1457\ \frac{kJ}{kg} - 298.6\ \frac{kJ}{kg}\right) - 298\ K\left(6.761\ \frac{kJ}{kgK} - 5.696\ \frac{kJ}{kgK}\right) = \mathbf{841\ kJ/kg}$$

Similarly, the specific exergy at state 4 is determined as:

$$ex_4 = (h_4 - h_0) - T_0(s_4 - s_0)$$

$$= \left(946.9\ \frac{kJ}{kg} - 298.6\ \frac{kJ}{kg}\right) - 298\ K\left(6.864\ \frac{kJ}{kgK} - 5.696\ \frac{kJ}{kgK}\right) = \mathbf{300.2\ kJ/kg}$$

$$\dot{E}x_{d,GT} = \dot{m}_3 ex_3 - \dot{m}_4 ex_4 - \dot{W}_{GT} = \left(40\ \frac{kg}{s}\right)\left(841\ \frac{kJ}{kg} - 300.2\ \frac{kJ}{kg}\right) - 20,284\ kW = \mathbf{1348\ kW}$$

e) Determine the energy and exergy efficiencies of the cycle.
The energy efficiency of the overall cycle can be found as:

$$\eta_{en} = \frac{\dot{W}_{net}}{\dot{Q}_{in,CC}}$$

The heat input rate can be calculated as:

$$\dot{Q}_{CC} = \dot{m}_3 h_3 - \dot{m}_2 h_2 = \left(40\ \frac{kg}{s}\right)\left(1457\ \frac{kJ}{kg} - 545.5\ \frac{kJ}{kg}\right) = \mathbf{36460\ kW}$$

$$\eta_{en} = \frac{\dot{W}_{net}}{\dot{Q}_{in,CC}} = \frac{15,530\ kW}{36460\ kW} = \mathbf{42.6\%}$$

The exergy efficiency is evaluated as:

$$\eta_{ex} = \frac{\dot{W}_{net}}{\dot{Q}_{in,CC}\left(1 - \frac{T_0}{T_s}\right)} = \frac{15,530\ kW}{36,460\ kW\left(1 - \frac{298\ K}{2758\ K}\right)} = \mathbf{47.1\%}$$

6.9 Concluding Remarks

This chapter has focused on power generating cycles, such as the Otto, Diesel, Dual, open-type air-standard Brayton, closed-type air-standard Brayton, regenerative open-type air-standard Brayton, regenerative closed-type air-standard Brayton, steam Rankine, reheat Rankine, cogeneration Rankine, and combined Brayton–Rankine, and presented the methodologies to thermodynamically analyze them by writing the mass, energy, entropy, and exergy balance equations. Analyzing their performances through energy and exergy efficiencies has also been one of the key goals. Furthermore, many illustrative examples have been provided for each cycle to illustrate and explain better.

Study Questions and Problems

Questions

1 Explain how one can link power cycle to a heat engine.

2 Repeat the concept of the heat engine and develop the performance relationship about a power cycle.

3 Explain the Carnot concept with the process about a power cycle.

4 Draw the *P-v* and *T-s* diagrams of the Carnot cycle for a power generating cycle in general.

5 Classify the power cycles and explain each separately.

6 Incorporate the exergy approach into a heat engine concept.

7 Derive both energy and exergy efficiencies for power cycles in general and explain what differentiates them.

8 List potential sources of entropy generation in a power cycle.

9 List potential sources of exergy destruction in a power cycle and explain how exergy destruction is related to the entropy generation.

10 Explain what the Otto cycle is, what it does, and what processes take place. Draw the process *T-s* and *P-v* diagrams.

11 Write all mass, energy, entropy, and exergy balance equations for an actual Otto cycle.

12 Explain what the Diesel cycle is, what it does, and what processes take place. Draw the process *T-s* and *P-v* diagrams.

13 Write all mass, energy, entropy, and exergy balance equations for an actual Diesel cycle.

14 Explain what the Dual cycle is, what it does, and what processes take place. Draw the process *T-s* and *P-v* diagrams.

15 Write all mass, energy, entropy, and exergy balance equations for an actual Dual cycle.

16 Explain what the Brayton cycle is, what it does and what processes place. Draw the process *T-s* and *P-v* diagrams.

17 Explain the difference between open and closed types of the Brayton cycle.

18 Write all mass, energy, entropy, and exergy balance equations for actual open and closed types of the Brayton cycle.

19 Explain what the Rankine cycle is, what it does, and what processes take place. Draw the process *T-s* and *P-v* diagrams.

20 Write all mass, energy, entropy, and exergy balance equations for actual open and closed types of Rankine cycle.

21 Explain what the reheat Rankine cycle is, what it does, and what processes take place. Draw the process *T-s* and *P-v* diagrams.

22 Write all mass, energy, entropy, and exergy balance equations for actual open and closed types of reheat Rankine cycle.

23 Explain what the regenerative Rankine cycle is, what it does, and what processes take place. Draw the process *T-s* and *P-v* diagrams.

24 Write all mass, energy, entropy, and exergy balance equations for actual open and closed types of regenerative Rankine cycle.

25 Explain what the cogeneration Rankine cycle is, what it does, and what processes take place. Draw the process *T-s* and *P-v* diagrams.

26 Write all mass, energy, entropy, and exergy balance equations for actual open and closed types of cogeneration Rankine cycle.

27 Explain what the combined Brayton and Rankine cycle is, what it does, and what processes take place. Draw the process *T-s* and *P-v* diagrams.

28 Write all mass, energy, entropy, and exergy balance equations for actual open and closed types of the combined Brayton- Rankine cycle.

29 Define energy and exergy efficiencies for power cycles by making specific distinctions to each in term inputs and outputs.

30 Explain what back work ratio is and describe it mathematically.

Problems

1 A heat engine receives 130 kW heat from a heat source at a temperature of 1000 K, where the heat engine produces a net power output of 50 kW and the rejected heat is rejected to an environment temperature of 25 °C.
 a Write the overall mass, energy, entropy, and exergy balance equations.
 b Calculate the amount of heat rejected to the environment (heat sink).
 c Calculate the amount of exergy destroyed and exergy rejected to the environment.
 d Calculate the energy and exergy efficiencies of the heat engine.

2 Consider two heat engines, both having a thermal efficiency of 40%. One of the engines (engine A) receives heat from a source at 600 K and the other one (engine B) from a source at 1000 K. Both engines reject heat to a medium at 300 K. At the first glance, both engines seem to be performing equally well. When we take a second look at these engines in light of the SLT (exergy concept), however, we see a totally different picture. As stated earlier, the subject matter engines can achieve the best performance if they operate as reversible engines where the highest efficiency is possible, which will lead to the Carnot efficiency.
 a Calculate the energy efficiency of the reversible engine A and engine B.
 b Calculate the exergy efficiency of engines A and B.
 c Discuss the results of part (b).

3 An Otto cycle has a compression ratio of 10, an isentropic compression efficiency of 85%, and an isentropic expansion efficiency of 85%. At the beginning of the compression, the air in the cylinder is at 90 kPa and 15 °C. The maximum gas temperature is found to be 1260 °C. (Use constant specific heats of $c_p = 1.005$ kJ/kgK, $c_v = 0.7176$ kJ/kgK, $k = 1.4$, and $R = 0.287$ kJ/kgK.)
 a Write the mass, energy, entropy, and exergy balance equations.
 b Calculate the heat supplied to the cycle per unit mass.
 c Find both energy and exergy efficiencies.

4 An Otto cycle operates at a compression ratio of 9, as illustrated in the figure. Air at the beginning of the compression process is at a pressure of 95 kPa and a temperature of 295 K with the amount of heat transferred to the air during the constant volume heat addition process to be 720 kJ/kg. The isentropic efficiencies for the compression and the expansion processes can be assumed to be 80 and 85%. The engine has four cylinders with a total displacement volume of 1.25 l and is running at a speed of 4500 rpm. Assume an isentropic efficiency of 80% for both compression and expansion processes. Considering specific heat changes of air with temperature, you are required to do the following:

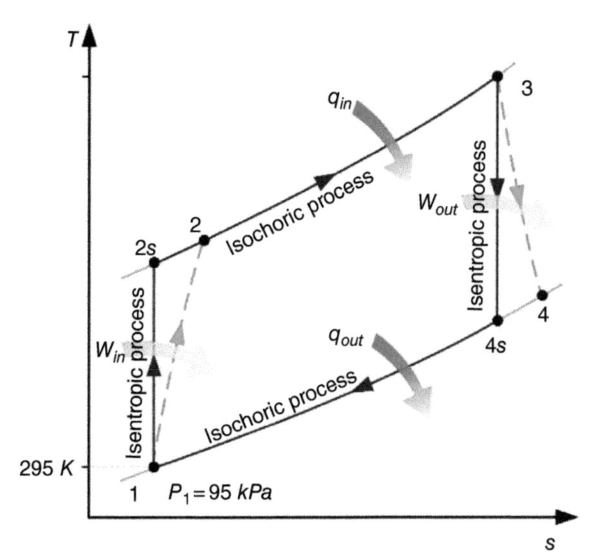

a Write the mass, energy, entropy, and exergy balance equations for each component of the system.
b Calculate the net power output and the maximum pressure and temperature during the cycle.
c Find the total entropy generation rate and exergy destruction rate for the system.
d Determine the energy and exergy efficiencies of the system.

5 A power production engine, as illustrated in the figure with its *P-v* diagram, operates on an ideal Diesel cycle that operates on a compression ratio of 20 and has air as the working fluid in the piston–cylinder device. The temperature and the pressure of the air at the beginning of the compression process are 30 °*C* and 95 *kPa*. The maximum temperature in the cycle reaches 1727 °*C*. (Use constant specific heats of $c_p = 1.005$ *kJ/(kgK)*, $c_v = 0.7176$ *kJ/kgK*, $k = 1.4$, and $R = 0.287$ *kJ/kgK*.)

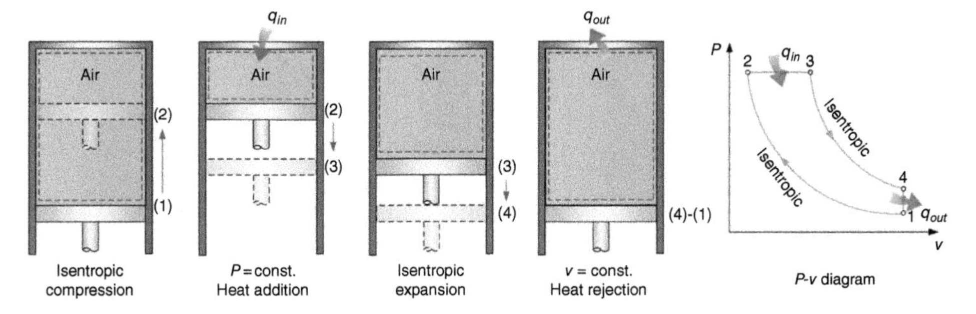

a Write the mass, energy, entropy, and exergy balance equations.
b Calculate the heat added and the heat removed in unit mass basis.
c Find the net work output.
d Calculate both energy and exergy efficiencies of the cycle.

6 An air-standard Diesel cycle operates at a compression ratio of 15 and a cutoff ratio of 1.4, as illustrated in the *T-s* diagram. Air is at a pressure and temperature of 100 *kPa* and 298 *K* with a 2 *L* of volume at the beginning of the compression process. Considering the specific heat variations with temperature:

a Write the mass, energy, entropy, and exergy balance equations for each component of the system.
b Calculate the *MEP* and temperature after heat addition process.
c Find the total entropy generation rate and exergy destruction rate for the system.
d Determine the energy and exergy efficiencies of the system.

7 An ideal Dual cycle operates at a compression ratio of 17 and a cutoff ratio of 1.8. During the constant-volume heat addition process, the pressure ratio is 1.6. Air is at a pressure and temperature of 90 *kPa* and 293 *K* at the beginning of the compression process. Considering the specific heat variations with temperature:

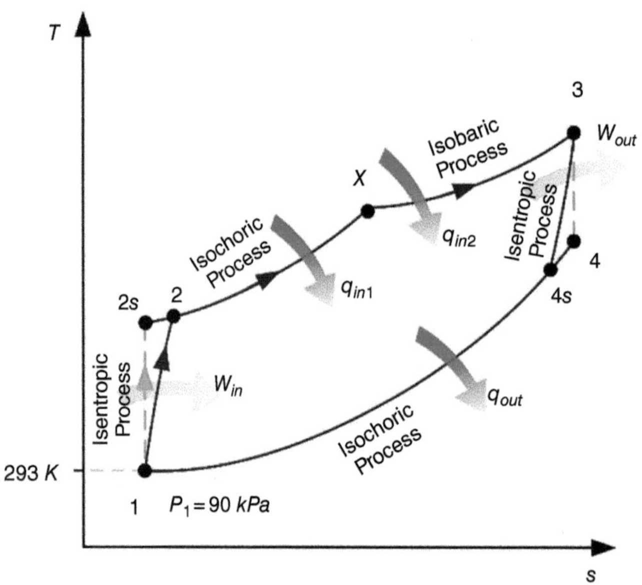

a Write the mass, energy, entropy, and exergy balance equations for all processes.
b Calculate the amount of heat addition and the maximum pressure and temperature for both ideal and actual cycles.

c Find the total entropy generation rate and exergy destruction rate for the system.
d Determine the energy and exergy efficiencies of the cycle.

8 A gas turbine power plant, as shown in the figure, operates on an actual Brayton cycle with a pressure ratio of 11. The gas enters the compressor at an ambient temperature of 298 *K* and its temperature at the turbine inlet is 1450 *K*. The work input for the compressor under ideal conditions could be considered as 260 *kJ/kg*. Assume the turbine and compressor isentropic efficiencies to be 87% and 78%, respectively.

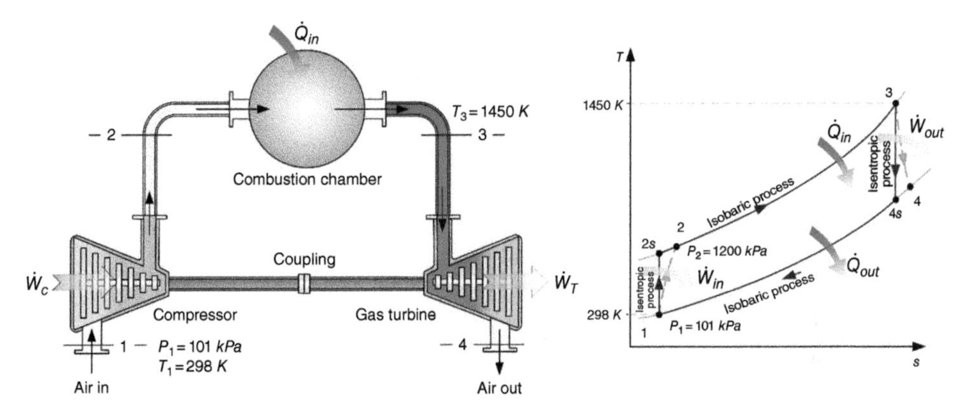

a Write the mass, energy, entropy, and exergy balance equations for each component of the system.
b Calculate the (i) rate of work done by the turbine, (ii) rate of heat rejection by the turbine, and (iii) turbine exit temperature.
c Find the total entropy generation rate and exergy destruction rate for the system.
d Determine the energy and exergy efficiencies of the system.

9 Consider an actual Brayton cycle operating at a pressure ratio of 14, as shown in the figure with its *T-s* diagram. The cycle operates at the maximum and minimum pressures of 1400 and 101.3 *kPa*, respectively. Air, at a rate of 20 *kg/s*, enters the compressor at an ambient temperature of 300 *K* and exits the turbine at a temperature of 950 *K*. Assume turbine and compressor isentropic efficiencies to be 90% and 83%, respectively.:

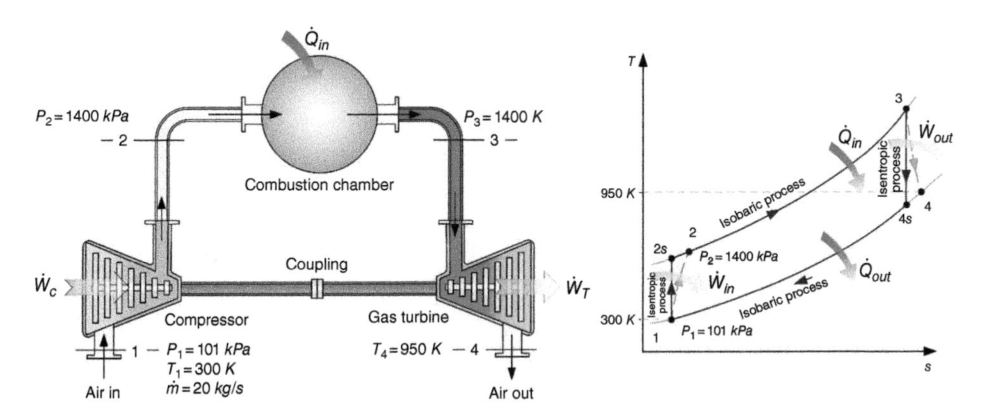

a Write the mass, energy, entropy, and exergy balance equations for each component of the cycle.

b Calculate the net power output and compressor work input rate.
c Find the total entropy generation rate and exergy destruction rate for the system.
d Determine the energy and exergy efficiencies of the cycle.

10 Re-consider the Brayton cycle shown in Problem 9. The maximum pressure is now increased to 2000 *kPa*, and the mass flow rate is decreased to 15 *kg/s*. If the ambient temperature is changed to 290 *K*, the turbine exit temperature is measured to be 600 *K* and the source temperature at the combustion chamber is 2000 *K*.
a Write the mass, energy, entropy and exergy balance equations for all cycle components.
b Find the power input required by the compressor and the net power output.
c Determine the total entropy generation and exergy destruction rates.
d Find both energy and exergy efficiencies of the cycle.

11 An closed-type Brayton cycle, as shown in the figure, operates at a pressure ratio of 10 and the air leaves the heat addition process with a temperature of 600 °*C*. The air is at a pressure and a temperature of 100 *kPa* and 45 °*C* when it enters the air compressor in the cycle. The turbine and the compressor operate in a non-isentropic manner with isentropic efficiencies of 85% and 80%, respectively.

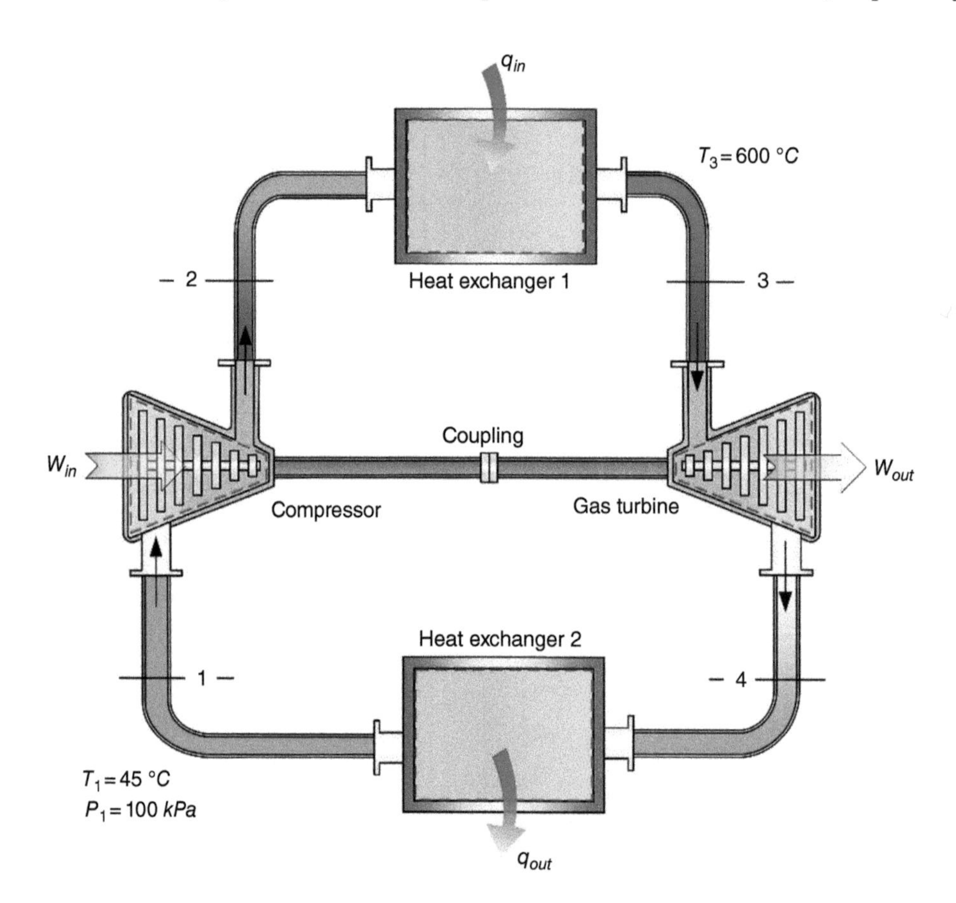

a Write the mass, energy, entropy, and exergy balance equations of each component in the cycle.

b Find the compressor work consumption per unit mass of air.
c Calculate the heat added to the air in the heat addition step per unit mass of air.
d Find both energy and exergy efficiencies of the cycle.

12 A closed type actual Brayton cycle operates at a pressure ratio of 10, as shown in the figure with its *T-s* diagram. The highest temperature of the cycle is 950 *K*. The temperature and pressure at the inlet of the compressor are 290 *K* and 120 *kPa*. Take the turbine and compressor isentropic efficiencies to be 85% and 82%, respectively.

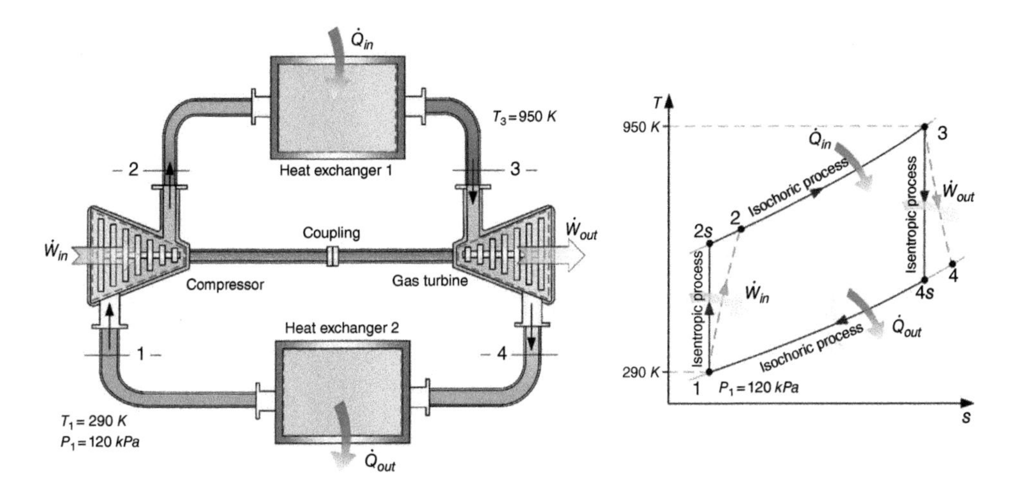

a Write the mass, energy, entropy, and exergy balance equations for each component of the system.
b Calculate the turbine exit temperature and heat output rate for the turbine.
c Find the total entropy generation rate and exergy destruction rate for the system.
d Determine the energy and exergy efficiencies of the system.

13 Consider a closed-type Brayton cycle operating at a pressure ratio of 8, as shown in the figure with its *T-s* diagram. The temperature at the inlet of the turbine is 1125 *K*. Air at a temperature and pressure of 285 *K* and 105 *kPa* enters the compressor. Take the turbine and compressor isentropic efficiencies to be 88% and 85% respectively.

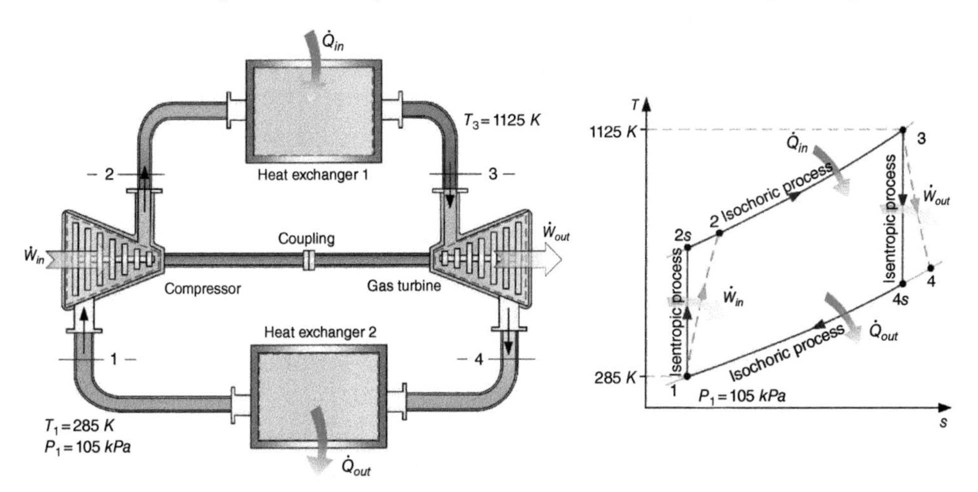

a Write the mass, energy, entropy, and exergy balance equations for each component of the cycle.
b Calculate the net power output and heat rejection rate for the turbine.
c Find the total entropy generation rate and exergy destruction rate for the system.
d Determine the energy and exergy efficiencies of the cycle.

14 A regenerative Brayton cycle with air as the working fluid operates at a pressure ratio of 8, as shown in the figure with its *T*-*s* diagram. The minimum and maximum temperatures in the cycle are 298 and 1400 *K*. Assume an isentropic efficiency of 90% for the compressor, 86% for the turbine, and a regenerator effectiveness of 75%. Take the mass flow rate of air as 1.2 *kg/s*.

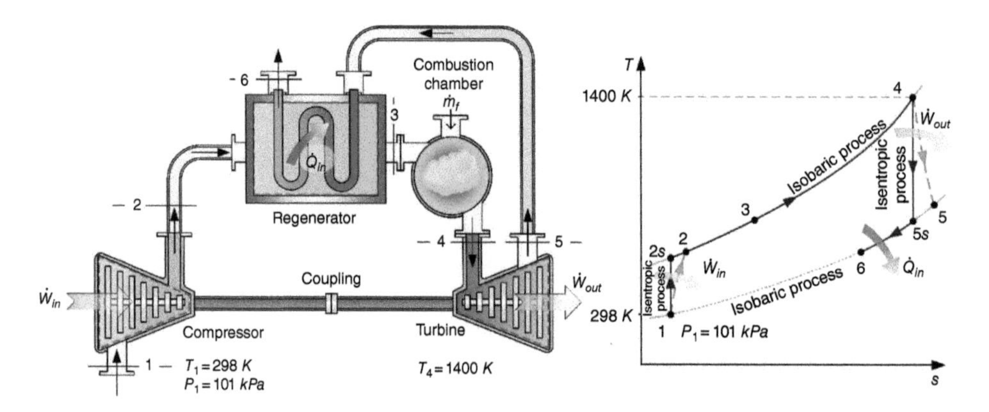

a Write the mass, energy, entropy, and exergy balance equations for each component of the cycle.
b Calculate the (i) net power output of the system, (ii) combustion chamber inlet temperature, and (iii) turbine outlet temperature.
c Find the total entropy generation rate and exergy destruction rate for the system.
d Determine the energy and exergy efficiencies of the cycle.

15 Reconsider the regenerative Brayton cycle given in Problem 14. The pressure ratio is to be changed to 15 and the ambient temperature is known to be 280 *K*. If the mass flow rate is increased to 5 *kg/s*, the following are needed.
a Write the mass, energy, entropy and exergy balance equations for all cycle components.
b Find the net power output, compressor exit temperature and turbine exit temperature.
c Determine the total rates of entropy generation and exergy destruction in the cycle.
d Find the energy and exergy efficiencies of the cycle.

16 A gas turbine cycle with regeneration operates at a pressure ratio of 11, as shown in the figure with its *T*-*s* diagram. Air enters the compressor at 305 *K* and 101.3 *kPa*. The maximum temperature within the cycle is 1200 *K* and the temperature difference between the hot air entering and cold air leaving the regenerator is 290 K. The cycle has a net power output of 150 kW. Assume an isentropic efficiency of 85% for the compressor and 90% for the turbine.

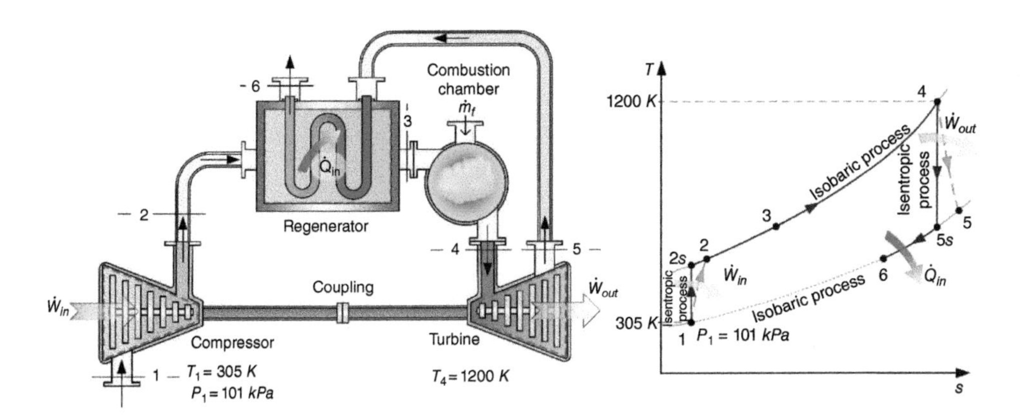

a Write the mass, energy, entropy, and ExBE balance equations for each component of the cycle.

b Calculate the (i) heat addition and rejection for the cycle and (ii) Turbine outlet temperature.

c Find the total entropy generation rate and exergy destruction rate for the system.

d Determine the energy and exergy efficiencies of the cycle.

17 An actual Rankine cycle, as shown in Figure 6.28, with an inlet pressure to the turbine of 7 MPa and a temperature of 600 °C, the turbine exit pressure is 100 kPa. The isentropic efficiency of the turbine is 94%, and the pump, condenser and the boiler pressure losses are neglected. If the temperature of the water leaving the condenser is 50 °C and the mas flow rate is 20 kg/s of water in the cycle. Take the source temperature as 1200 K.

a Write the mass, energy, entropy, and exergy balance equations of each component in the cycle.

b Calculate the thermal energy (heat) sent to the boiler.

c Calculate the net power produced.

d Calculate the energy and exergy efficiencies of the cycle.

18 A steam power plant operates on an actual Rankine cycle, as shown in the figure with its T-s diagram. The superheated steam enters the turbine at a pressure of 2.5 MPa and a temperature of 650 K. The saturated liquid vapor mixture exiting the turbine is at a pressure of 100 kPa. Consider the pump and the turbine isentropic efficiencies to be 80% and 88% respectively and mass flow rate 10 kg/s. Take source temperature as 1200 K.

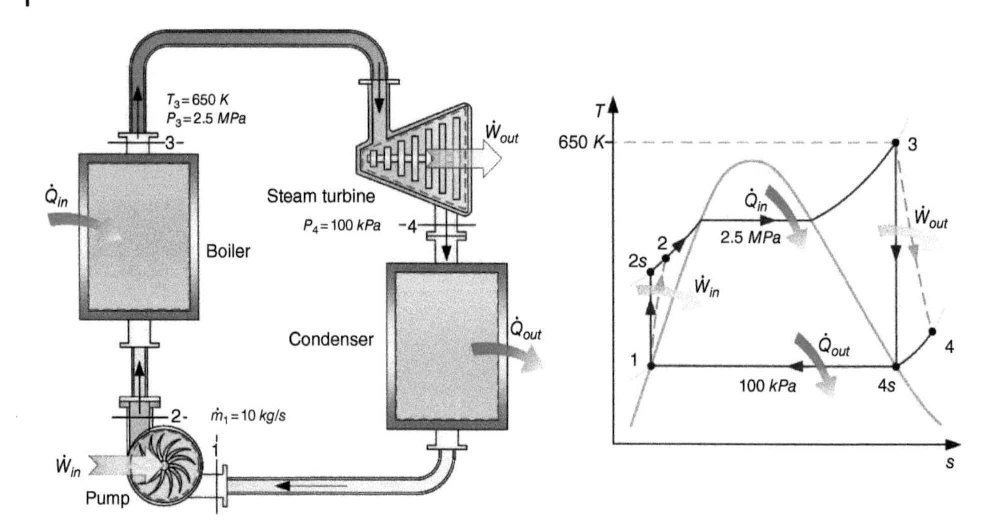

a Write the mass, energy, entropy, and exergy balance equations for each component of the cycle.
b Calculate the net power output of the plant and rate of heat input to the boiler.
c Find the total entropy generation rate and exergy destruction rate for the system.
d Determine the energy and exergy efficiencies of the cycle.

19 Consider an actual Rankine cycle operating between a maximum pressure of 4.5 *MPa* and a minimum pressure of 75 *kPa*, as shown in the figure with its *T-s* diagram. The superheated steam enters the turbine at a temperature of 600 *K* at a rate of 40 *kg/s*. Consider the pump and the turbine isentropic efficiencies to be 82% and 90%, respectively. Take source temperature as 1200 *K*.

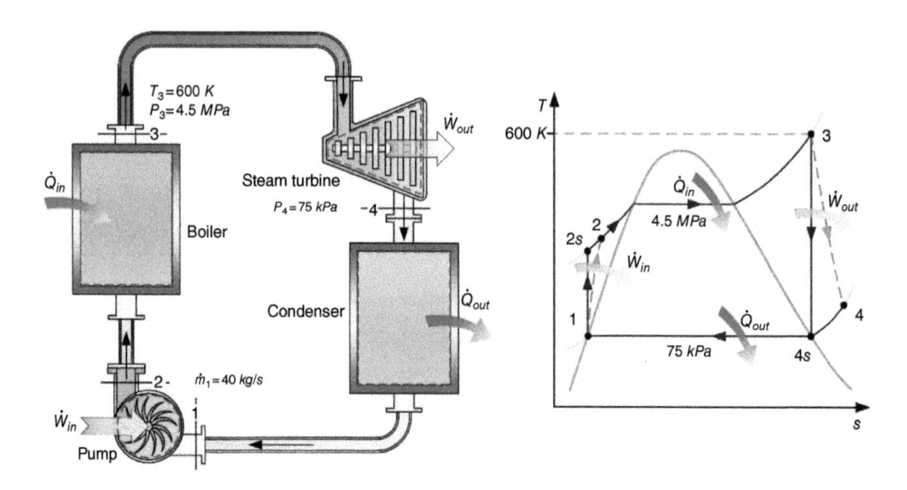

a Write the mass, energy, entropy, and exergy balance equations for each component of the cycle.

b Calculate the net power output of the plant and turbine exit temperature.
c Find the total entropy generation rate and exergy destruction rate for the system.
d Determine the energy and exergy efficiencies of the cycle.

20 Consider a Rankine cycle as shown in Figure 6.31, using water as working fluid, steam enters the turbine at a temperature of 500 °*C* and 3.5 *MPa* pressure and turbine output temperature of 50 °*C*. Assume turbine isentropic efficiency to be 88%. The turbine outlet is condensed at 50 *kPa* pressure. The source temperature is 600 °*C*.

a Write the mass, energy, entropy and exergy balance equations for all cycle components.
b Calculate the net work generated by the both actual and isentropic cycles.
c Determine input heat in the boiler.
d Calculate the efficiencies for both actual and isentropic versions of the cycle.

21 Consider a steam power plant operating on an actual reheat Rankine cycle, as shown in the figure with its *T*-*s* diagram. The superheated steam enters the high-pressure turbine at a pressure and temperature of 10 *MPa* and 750 *K*. The steam enters the low-pressure turbine at a pressure of 1.6 *MPa* after being reheated in the boiler to a temperature of 615 *K*. The mass flow rate of the steam is 5.25 *kg/s*. Assume the pump and turbine to have isentropic efficiencies of 82 and 85%, respectively, and the moisture content in the steam to be not more than 7% at the turbine outlet. Take the source temperature as 1200 *K*. The enthalpy at the exit of the low pressure turbine is 2416 *kJ/kg*.

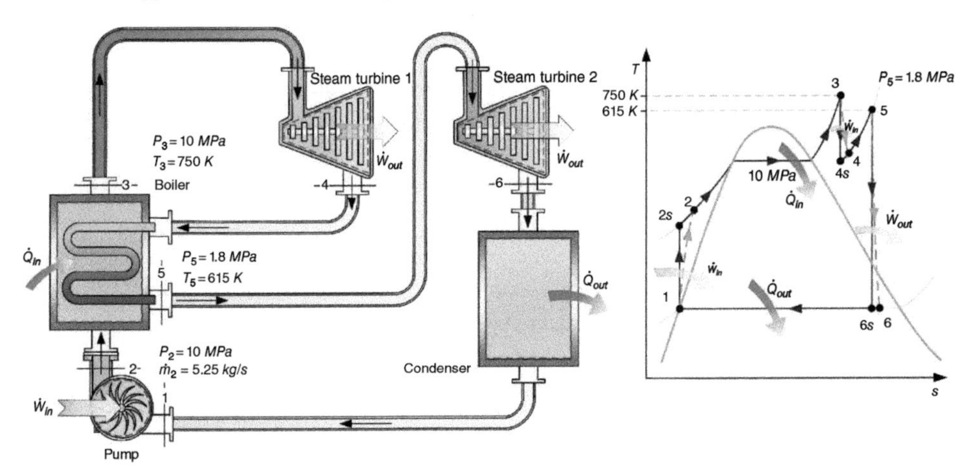

a Write the mass, energy, entropy, and exergy balance equations for each component of the cycle.
b Calculate the net power output of the plant and pressure of the condenser.
c Find the total entropy generation rate and exergy destruction rate for the system.
d Determine the energy and exergy efficiencies of the cycle.

22 A steam power plant operating on a reheat Rankine cycle, as shown in the figure with its *T*-*s* diagram, has a net power output of 60 *MW*. Steam enters the high-pressure turbine at a temperature of 810 K and a pressure of 12.5 MPa. For the low-pressure turbine, the inlet conditions are 1.25 MPa and 810 K. Assume the pump and turbine to have isentropic efficiencies of 90% and 82%, respectively. The pressure at the exit of the low pressure turbine is 75 kPa, and the mass flow rate in the cycle is 5 kg/s.

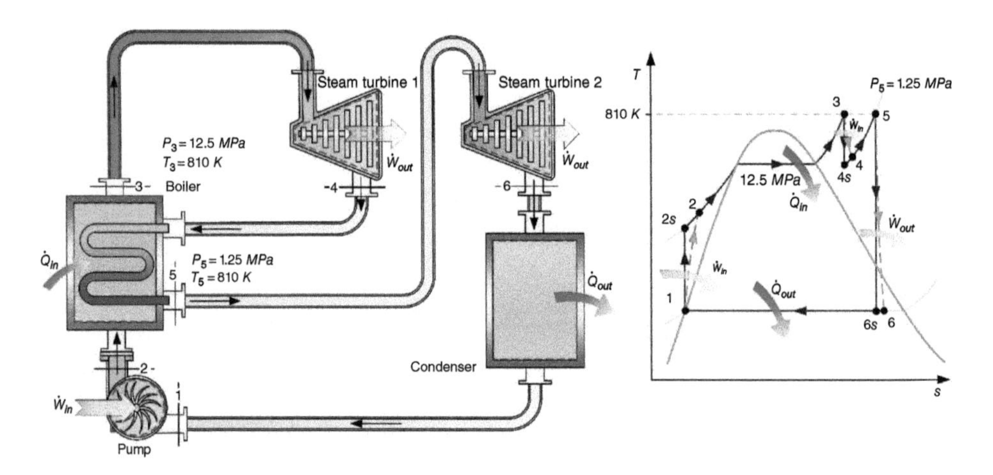

a Write the mass, energy, entropy, and exergy balance equations for each component of the system.
b Show the cycle on a T-s diagram with respect to saturation lines.
c Determine the net power output of the plant and the quality of steam at the turbine exit.
d Find the total entropy generation rate and exergy destruction rate for the system.
e Determine the energy and exergy efficiencies of the system.

23 Consider a reheat Rankine cycle as shown in Figure 6.31, operating at the turbine inlet pressure of 12 MPa and boiler heats up the steam at 450 °C. The condenser is set to operate at the pressure of 50 kPa and source temperature is 1000 °C. The isentropic efficiencies of HPT and LPT are assumed to be 75%. The moisture content in the exit stream from low-pressure turbine is 10% and consider that steam is reheated to the same temperature as the high-pressure turbine inlet temperature.
a Write the mass, energy, entropy and exergy balance equations for all system components.
b Calculate the net work generated by the low- and high-pressure turbines.
c Calculate the entropy generation and exergy destruction rates of the boiler.
d Calculate the efficiency of the cycle considering ideal operation.
e Calculate the actual energy and exergy efficiencies.

24 Consider a cogeneration plant where steam enters the turbine at a pressure of 6 MPa and a temperature of 800 K, as shown in the figure with the T-s diagram. One-third of the steam is extracted from the turbine at a pressure of 1.8 MPa for process heating.

The remaining steam is expanded to 12 *kPa*. The fraction of the steam extracted from the turbine at 1.8 *MPa* undergoes condensation and mixing with the feedwater at constant pressure, and is pumped to the pressure of the boiler at 6 *MPa*. The mass flow rate of steam through the boiler is 40 *kg/s*. Consider the turbine and the pump isentropic efficiencies to be 87% and 80%, respectively. Pressure drops and heat losses in the piping are assumed insignificant and can be neglected. Take the source temperature as 1200 *K*.

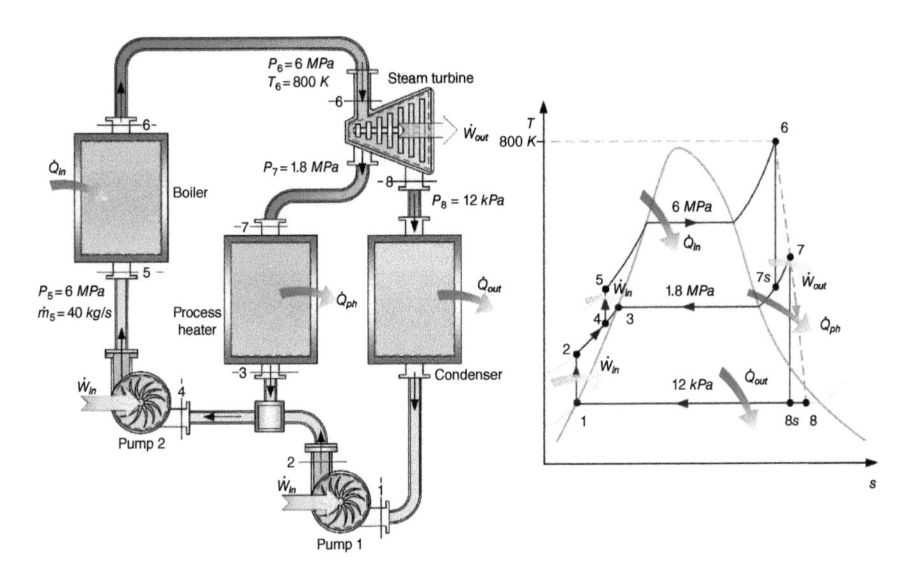

a Write the mass, energy, entropy, and exergy balance equations for each component of the system.
b Calculate the net power output of the system.
c Find the total entropy generation rate and exergy destruction rate for the system.
d Determine the energy and exergy efficiencies of the system.

25 A cogeneration plant operates at a maximum pressure of 8.5 *MPa* and a maximum temperature of 900 *K*, as shown in the figure with the *T-s* diagram. One-fifth of the steam is extracted from the turbine at a pressure of 2.5 *MPa* for process heating. The rest of the steam is expanded to a pressure of 17 *kPa*. The portion of the steam extracted from the turbine at 2.5 *MPa* is allowed to condense and mix with the feedwater at a constant pressure before it is pumped to the boiler at 8.5 *MPa*. This steam is condensed at a constant pressure and pressurized through a pump to 8.5 *MPa*, which is equal to the boiler pressure. The mass flow rate of steam through the boiler is 50 *kg/s*. Consider the turbine and the pump isentropic efficiencies to be 80% and 88%, respectively. Pressure drops and heat losses in the piping are assumed insignificant and can be neglected.

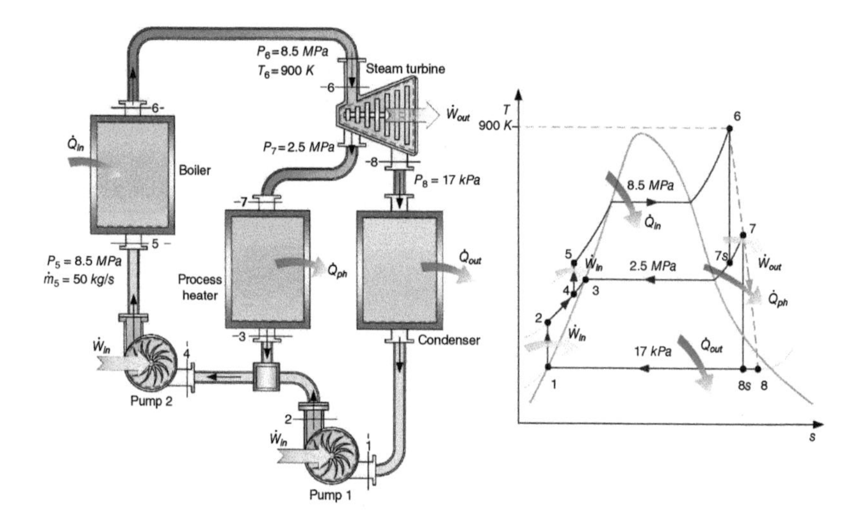

a Write the mass, energy, entropy, and exergy balance equations for each component of the system.
b Calculate the net power output of the system and the utilization factor of the plant.
c Find the total entropy generation rate and exergy destruction rate for the system.
d Determine the energy and exergy efficiencies of the system.

26 Consider a cogeneration Rankine plant as shown in Figure 6.36, operating at the boiler output temperature and pressure of 550 °*C* and 4 *MPa*. 30 percent of the steam is extracted from the turbine at 400 *kPa* and remaining 70% expands to 50 *kPa*. In order to meet the high thermal energy requirements, 15 percent of the steam from the boiler is throttled at 400 *kPa*. The steam exits the process heater at 400 *kPa* as saturated liquid. The mass flow rate of the steam fed to the boiler is 10 *kg/s*. Assume turbine isentropic efficiency to be 85%. By assuming turbine and pump to be isentropic and neglecting the losses and pressure drops.
a Write the mass, energy, entropy and exergy balance equations for all system components.
b Calculate the enthalpies on each state and rate of process heating.
c Calculate the work rate of the turbine.
d Calculate the entropy generation and exergy destruction rates of the boiler.
e Calculate both energy and exergy efficiencies of the system.

27 A combined cycle has the Brayton cycle topping and the Rankine cycle as the bottoming cycle, as shown in Figure 6.40. The air in the ideal Brayton cycle receives an amount of heat of 130 *MW* from a heat source that is at a temperature of 1700 °*C* and the Brayton cycle generates a net amount of power equal to 55.4 *MW*. The gas leaving the gas turbine is at a temperature of 750 °*C* and has an energy equal to 75 *MW*. If the boiler of the bottoming Rankine cycle was able to recover 70% of that energy and generate superheat steam, it means that the Rankine cycle generates a net amount of power equal to 21.7 *MW*.

a Write the mass, energy, entropy, and exergy balance equations for the overall topping Brayton cycle, bottoming Rankine cycle, and the integrated combined cycle.
b Calculate the energy and exergy efficiency of the topping Brayton cycle.
c Calculate the energy and exergy efficiency of the bottoming Rankine cycle.
d Calculate the energy and exergy efficiency of the combined cycle.

28 A combined gas–steam power plant is considered in this problem. The topping cycle is an actual Brayton cycle operating at a pressure ratio of 8, as shown in the figure with its *T-s* diagram. Air enters the compressor at 298 *K* with a mass flow rate of 35 *kg/s* and the gas turbine at 1275 *K*. A reheat Rankine cycle acts as the bottoming cycle operating at maximum and minimum pressures of 7.5 *MPa* and 50 *kPa*, respectively. The rate of the steam heated in a heat exchanger via waste heat recovery from the exhaust gases exiting the gas turbine is 4 *kg/s*. The exhaust gases leave the heat exchanger at a temperature of 495 *K*. Steam, after experiencing a pressure drop, exits the high-pressure turbine at a pressure of 1.25 *MPa* and is reheated to 700 *K* before being expanded in the low-pressure turbine. Consider the isentropic efficiencies for the pump, compressor, and turbines to be 85%. Take the source temperature of combustion chamber as 2000 *K*.

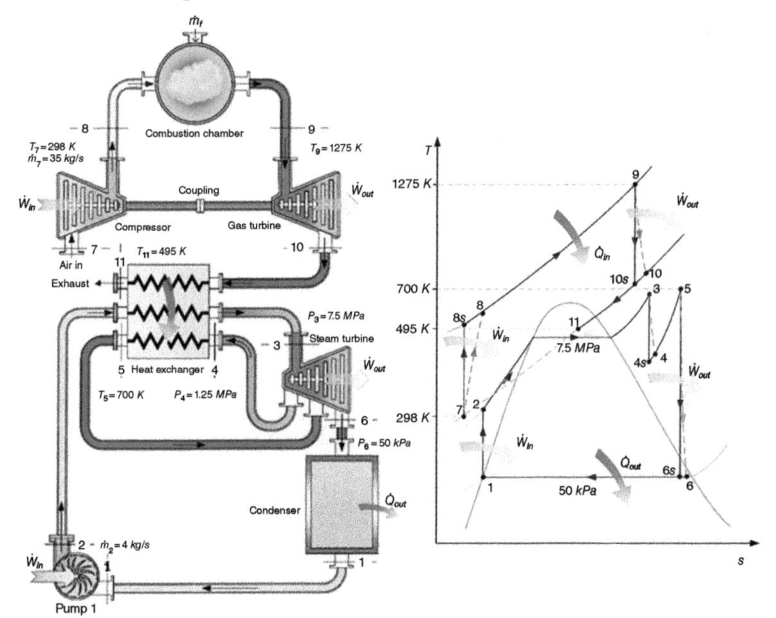

a Write the mass, energy, entropy, and exergy balance equations for each component of the system.
b Calculate the net power output of the plant and moisture content at the low-pressure turbine outlet.
c Find the total entropy generation rate and exergy destruction rate for the system.
d Determine the energy and exergy efficiencies of the system.

29 The gas turbine portion of a combined gas–steam power plant has a pressure ratio of 13. Air enters the compressor at an ambient temperature of 298 *K* at a rate of 10 *kg/s* and is heated to 1350 *K* inside the combustion chamber, as shown in the figure with the *T*-*s* diagram. The exhaust gases exiting the gas turbine are used to heat the steam to 650 *K* at 8 *MPa* in a heat exchanger. The temperature of the exhaust gases at the heat exchanger outlet is 385 *K*. The steam leaving the turbine is condensed at 12 *kPa*. Consider the turbine, compressor and pump isentropic efficiencies to be 85%, 87%, and 90%, respectively. Take the source temperature of combustion chamber as 2000 *K*.

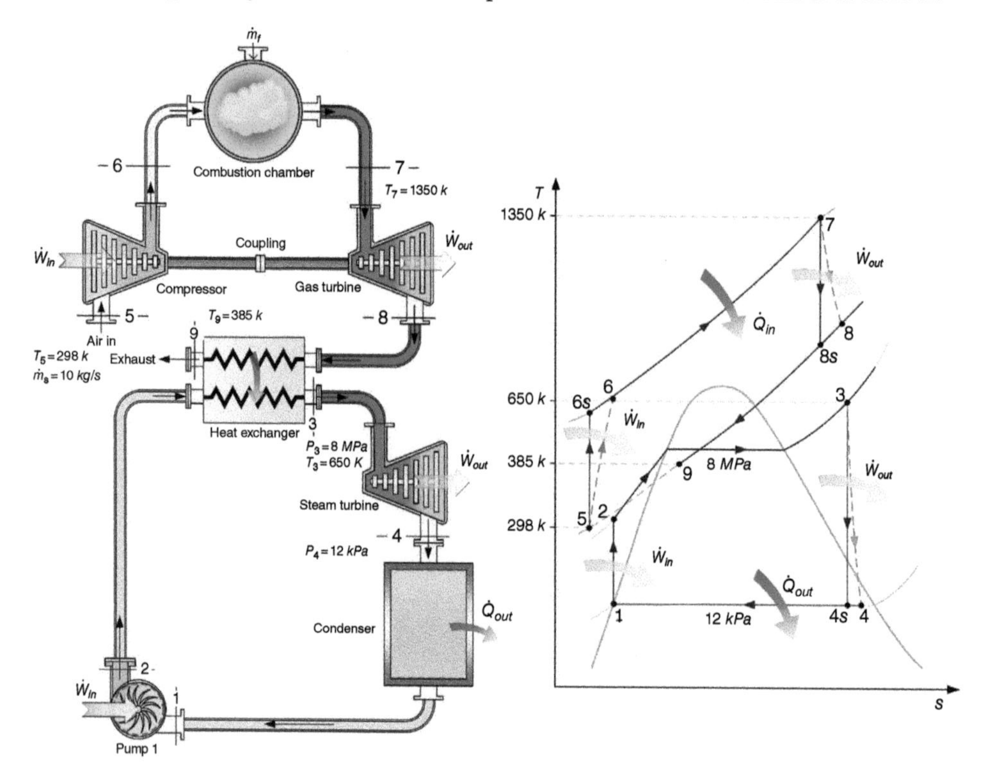

a Write the mass, energy, entropy, and exergy balance equations for each component of the system.

b Calculate the net power output of the plant and mass flow rate of the steam.

c Find the total entropy generation rate and exergy destruction rate for the system.

d Determine the energy and exergy efficiencies of the system.

30 Consider the combined cycle shown in Problem 29 with a pressure ratio of 15 for the topping Brayton cycle. Also, the steam Rankine cycle operates with a minimum pressure level of 50 *kPa*. Consider the source temperature of the combustion chamber as 2000 *K*.

a Write the mass, energy, entropy and exergy balance equations for all system components.

b Calculate net power output of the plant and mass flow rate of the steam.

c Calculate the total entropy generation rate and exergy destruction rate for the system.

d Find the energy and exergy efficiencies of the system.

7

Refrigeration and Heat Pump Cycles

7.1 Introduction

It is well-known that refrigeration has a diverse nature and finds application possibilities in many sectors with a large number of processes ranging from cooling to air conditioning and from food refrigeration to human comfort. The design, analysis, assessment, evaluation, and improvement of these systems are of great significance and come down to thermodynamics. The better one knows thermodynamics the better systems he/she can design for practical applications. As Figure 7.1 illustrates, one person who really knows thermodynamics well will design happy refrigeration and heat pump systems. Here, happy systems mean that they are well-designed, -analyzed, -assessed, -evaluated and -improved. So, learning and applying things right to the refrigeration and heat pump systems are really indispensable. That's why this present chapter is important for better systems and applications.

It is important to be reminded a couple of conceptual things. When an engineer or an engineering student undertakes the analysis of a refrigeration system and/or a heat pump system, he or she should deal with several basic aspects first, depending upon the type of the problem being studied, that may concern thermodynamics. In conjunction with this, there is a need to introduce several definitions and concepts before moving into refrigeration systems and applications in depth. Furthermore, units are of importance in the analysis of such systems and applications. One should make sure that the units used are consistently to reach the correct result. This means that there are several introductory factors to be taken into consideration to avoid getting lost. While the information in some situations is limited, it is desirable that the reader comprehend these processes. Despite assuming that the reader, if he or she is a student, has completed necessary subjects in thermodynamics, there is still a need for him or her to review, and for those who are practicing refrigeration engineers, the need is much stronger to understand the physical phenomena and practical aspects, along with a knowledge of the basic laws, principles, concepts, and definitions. The bottom line here is that thermodynamics is really key behind everything, including all refrigeration and heat pump systems and applications. As outlined in Chapter 5, the way one plants things relates to the way he/she will harvest accordingly. The way systems analysis is

Thermodynamics: A Smart Approach, First Edition. Ibrahim Dincer.

Figure 7.1 An illustration of how a better thermodynamicist can design happier systems.

introduced in Chapter 5, is the way this methodology will be employed to analyze refrigeration and heat pump systems.

As a quick reference to an earlier statement made under the second law of thermodynamics (SLT): heat travels from a high temperature source to a low temperature one, and it is possible to give several examples, such as cooling foodstuff to cooling a car engine. Let's look at a cup of hot coffee sitting on a table and losing heat to its surroundings which are at a lower temperature, as shown in Figure 7.2. If the coffee is left there long enough, it will naturally cool down and reach the thermal equilibrium with the surroundings.

However, if one wants to develop cooling and heating processes, not naturally, but forcefully to achieve desired conditions, there is a need to design refrigeration and heat pump systems to specifically desired conditions and environments as illustrated in Figure 7.3 for refrigeration (Figure 7.3a) and for heat pumps (Figure 7.3b). From this figure once can understand that the objective of the refrigerator (*R*) is to remove Q_L from the space at the low temperature (T_L), and in order to carry this out that the refrigerator consumes a work input of W_{in}. In refrigeration, the goal is to provide the desired cooling at T_L. However, for a heat pump (*HP*), work input is required in a similar fashion but with the difference that the aim is now to provide heating and achieve Q_H at T_H.

After some introductory information and a few quick illustrations, it is necessary to say that this chapter aims to focus on refrigeration and heat pump systems and provide analysis methodologies for various types of refrigeration and heat pump systems, along with examples, and evaluate their performances.

Figure 7.2 An illustration of heat being lost from a high temperature source (coffee) to low temperature surroundings.

7.2 Refrigerants

Refrigerants are the working fluids used in refrigeration and heat pump systems due to their heat absorbing capabilities during the evaporation process, which makes them suitable for practical applications. The refrigerants, which provide a cooling effect during the phase change from liquid to vapor, are commonly used in refrigeration, air conditioning, and heat pump systems, as well as chemical processes. The most natural refrigerants that humankind used in the past were water and air, and these are still widely used. However, more effective artificial refrigerants were developed in the eighteenth century to provide faster cooling. So too were respective devices to make them suitable for various applications in various sectors. In the literature, there are various ways of grouping refrigerants based on various aspects such as chemistry, application, functionality, etc. (even artificial to natural). However, there are five primary groups of refrigerants considered, namely: halocarbons, hydrocarbons, inorganic compounds, azeotropic mixtures, and nonazeotropic mixtures. Some of the most widely used refrigerants are shown in Figure 7.4.

The first group, halocarbons, contain one or more of the three halogens chlorine, fluorine, or bromine and were widely used in refrigeration and air conditioning systems as refrigerants. In this group, the halocarbons consisting of chlorine, fluorine, and carbon were the

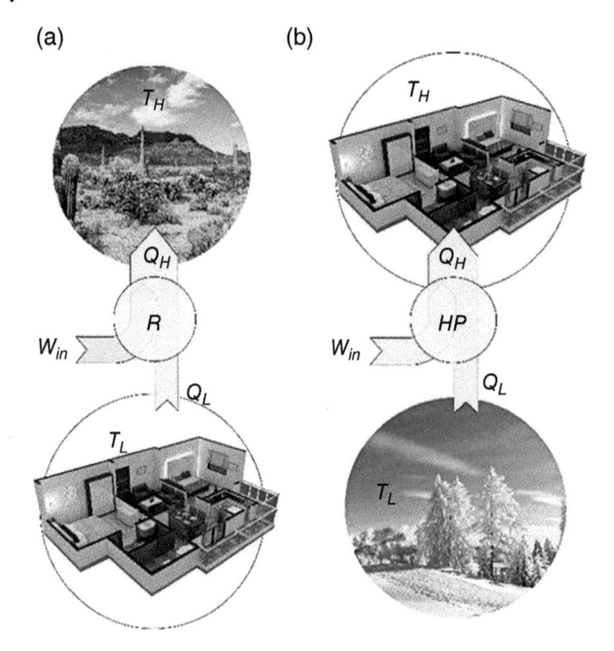

Figure 7.3 Both a refrigerator (*R*) and a heat pump (*HP*) are the same devices with a little difference, for example (a) a refrigerator is cooling a house and (b) a heat pump is heating a house.

Figure 7.4 Some of the most widely used refrigerants.

most commonly used refrigerants (so-called chlorofluorocarbons, *CFCs*). *CFCs* were commonly used as refrigerants, solvents, and foam blowing agents. The most common *CFCs* have been *CFC*-11 or *R*11, *CFC*-12 or *R*12, *CFC*-113 or *R*113, *CFC*-114 or *R*114, and *CFC*-115 or *R*115 (here R stands for refrigerant). Although *CFCs*, such as *R*11, *R*12, *R*22, *R*113, and *R*114, were very common refrigerants employed in refrigeration and air-conditioning equipment, they were used in other applications as aerosols, foams, solvents, etc. Their use was rapidly decreased, due to their environmental impact, namely stratospheric ozone depletion, and all chlorine containing ones were banned.

The second group, hydrocarbons (*HCs*) are chemical compounds that mainly consist of carbon and hydrogen. *HCs* include methane, ethane, propane, cyclopropane, butane, and cyclopentane. Although HCs are highly flammable, they offer some advantages as potential refrigerants due to the fact that they are inexpensive to produce and they have zero ozone layer depletion potential, very low global warming potential, and low toxicity.

Most of the third group of refrigerants, despite their early invention, are still used in numerous refrigeration, air conditioning, and heat pump applications as refrigerants. Some examples are ammonia (NH_3), water (H_2O), air (in a composition of $0.21O_2 + 0.78N_2 + 0.01Ar$), carbon dioxide ($CO_2$), and sulfur dioxide ($SO_2$). Among these compounds, ammonia has received the greatest attention for practical applications and even today there is more interest in.

The fourth group, azeotropic refrigerants (so-called: mixtures), consists of two substances having different properties but behaving as a single substance. The two substances cannot be separated by distillation. The most common azeotropic refrigerant is $R502$, which contains 48.8% $R22$ and 51.2% $R115$.

The last group of refrigerants, nonazeotropic mixtures, consists of multiple components of different volatility that, when used in refrigeration cycles, change composition during evaporation (boiling) or condensation. Nonazeotropic mixtures were also called zeotropic mixtures or blends. Nonazeotropic mixtures (e.g. $R11 + R12$, $R12 + R22$, $R12 + R114$, $R13B1 + R152a$, $R22 + R114$ and $R114 + R152a$, etc.) received great attention at the beginning of the twentieth century and were used up until 2000 heavily until most of the compounds ($R11$, $R12$, $R22$, etc.) were banned. There are still three of these being used, $R12 + R114$, $R22 + R114$, and $R13B1 + R152a$. It is clear that the heat transfer phenomena during the phase change of nonazeotropic mixtures are more complicated than with single-component refrigerants.

7.3 Basic Refrigeration Cycle

Refrigeration is the process of removing heat from matter, which may be a solid, liquid, or gas, and cooling it by lowering its temperature through a sensible heat transfer (or, in some cases, by lowering its specific enthalpy through a latent heat transfer, which is a phase change process). There are a number of ways of lowering temperature, some of which are of historical interest only. In some older methods, lowering of temperature was accomplished by the rapid expansion of gases under reduced pressures. Thus, cooling may be brought about by compressing air, removing the excess heat produced in compressing it, and then permitting it to expand. A lowering of temperatures is also produced by adding certain salts, such as sodium nitrate, sodium thiosulfate (hypo), and sodium sulfite, to water. The same effect is produced, but to a lesser extent, by dissolving common salt or calcium chloride in water. There are of course many more to add to these examples.

Note that two common methods of refrigeration are natural and mechanical. In natural refrigeration, there are no artificially made processes and/or mechanical devices employed to achieve the task of cooling. If one looks at the historical examples, ice has been used in refrigeration since ancient times, and it is still widely used in various locations of the world for cooling foodstuff, even for seasonal cooling purposes. In this natural technique of refrigeration, air is passed through the ice blocks to be cooled down for air conditioning

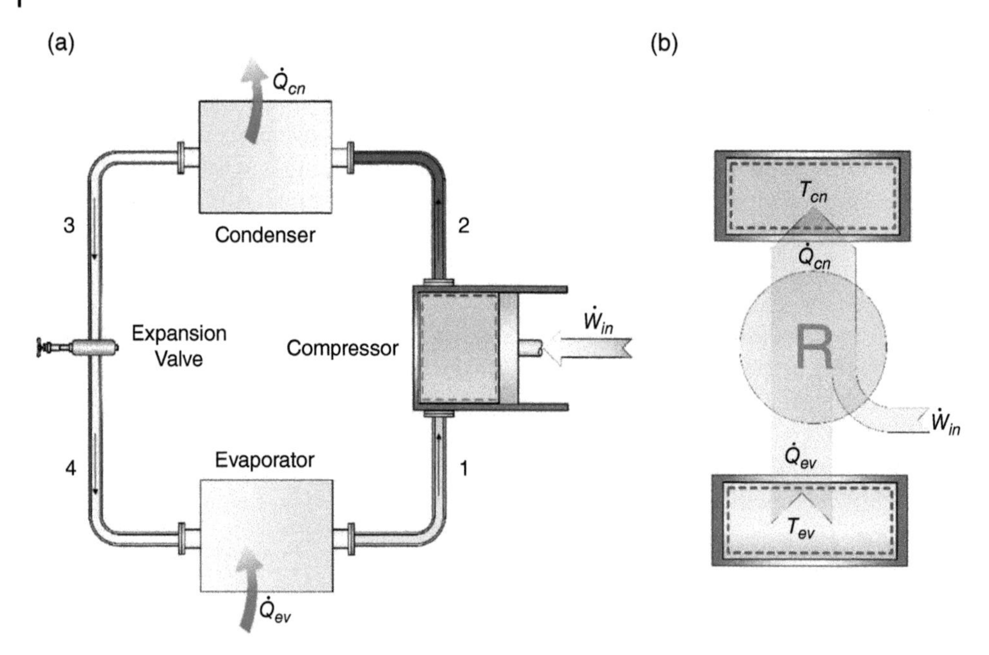

Figure 7.5 (a) Schematic diagram of a basic vapor-compression refrigeration cycle and (b) a representation of the overall refrigeration process with all inputs and outputs.

applications, and foodstuff are stored in ice for cooling and preservation purposes. One may consider circulating the air through the ice and cooling foodstuff thereafter.

It is important to look at basic refrigeration and how it operates, and how many process exist in a basic vapor compression refrigeration cycle. For this we consider the schematic diagram shown in Figure 7.5. This basic refrigeration system consists of four main components, namely evaporator, compressor, condenser, and expansion (throttling) valve. Just like other cycles, there are four key processes taking place.

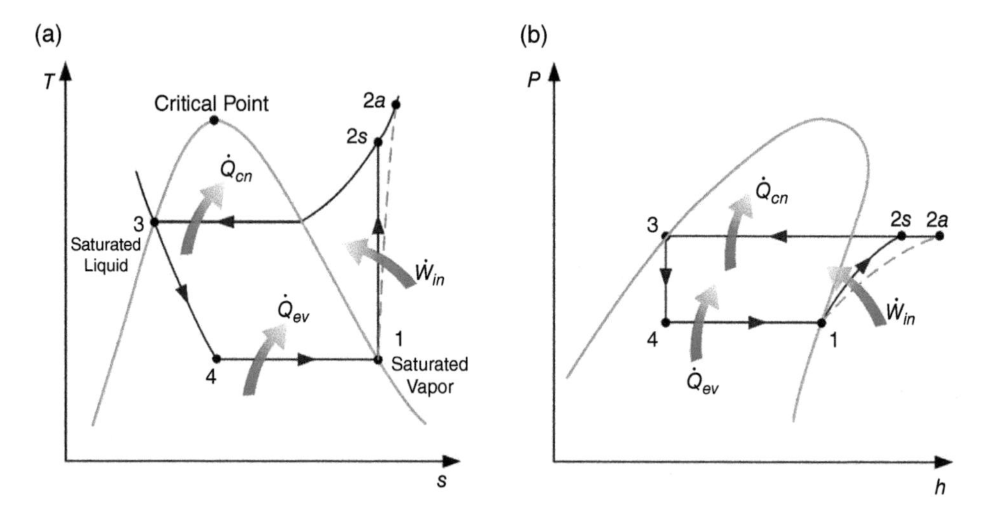

Figure 7.6 Property diagrams for the basic refrigeration cycle: (a) T-s diagram and (b) P-h diagram.

1-2 compression: (which may be isentropic compression under ideal conditions where entropy remains constant) where there is work input supplied externally.
2-3 isobaric heat rejection: where the heat is extracted from the refrigerant vapor and cooled down to a desired temperature after being condensed.
3-4 isenthalpic expansion: where the liquid refrigerant is expanded isenthalpically to a desired pressure to become a saturated liquid and vapor mixture.
4-1 isobaric heat addition: where the liquid gets into the evaporator to provide the necessary cooling by absorbing heat.

The above given basic refrigeration cycle is further illustrated on two most common property diagrams, namely *T-s* diagram and *P-h* diagram (Figure 7.6), which are important to use and illustrate the processes with the correct state points and their properties. These seem to be a confirmation of what is learned through the cycle. One should notice that there are two state points given for state 2 (as 2*s* which refers to isentropic (constant-entropy case) where the isentropic efficiency is 100% and 2*a* which refers to the actual case where isentropic efficiency is less than 100%).

Note that household fridges are most widely used versions of these basic types of vapor-compression refrigeration system, which are a critical part of daily life where we need to preserve foodstuffs and increase their shelf-life. Figure 7.7 shows a fridge with some representative foodstuffs to illustrate how it looks. In the lower section of the fridge the temperature is set to 3–4 $°C$ to minimize biological activity in foodstuffs to prevent their spoilage and increase their life. The top part is the freezer section, where foodstuffs are frozen at

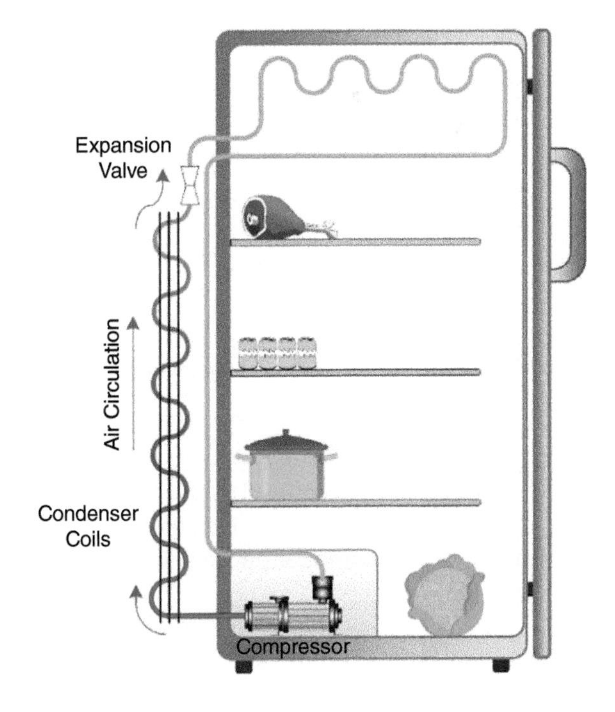

Figure 7.7 Schematic diagram of a household fridge.

$-18\,°C$ and all biological and chemical activity is stopped to preserve the foodstuffs for longer periods of time, ranging from weeks to years.

Performance Evaluation:
It is important to define the performance of any refrigeration through the energetic based coefficient of performance (COP) (COP_{en}), which is under the first law of thermodynamics (FLT), and the exergetic based COP (COP_{ex}), which is under the second law of thermodynamics (SLT) as follows:

$$COP_{R,en} = \frac{\dot{Q}_{ev}}{\dot{W}_{in}} \tag{7.1}$$

and

$$COP_{R,ex} = \frac{\dot{E}x_{\dot{Q}_{ev}}}{\dot{W}_{in}} = \frac{\left(\frac{T_o}{T_{ev}} - 1\right)\dot{Q}_{ev}}{\dot{W}_{in}} \tag{7.2}$$

where the subscripts ev and in stand for evaporator and input. T_o is the reference environment temperature, which is mainly taken as the surrounding temperature.

Example 7.1 A basic ideal refrigeration cycle, as shown in Figure 7.5, is used to refrigerate the freezer section of a fridge at $-15\,°C$ in a kitchen at $25\,°C$. The total heat gained by the refrigerant is 1500 kJ/h and the heat rejection from the condenser is 2600 kJ/h. (a) Write the mass, energy, entropy, and exergy balance equations for the overall refrigeration system as a single unit, (b) find the work input to the compressor in kW, (c) calculate the COP of the refrigerator, and (d) find the minimum work input to the compressor if a reversible refrigerator is used.

Solution

a) Write the mass, energy, entropy, and exergy balance equations for the overall refrigeration system as a single unit.

$$\text{MBE}: \dot{m}_1 = \dot{m}_2 = \dot{m}_3 = \dot{m}_4 = \dot{m}$$

$$\text{EBE}: \dot{Q}_{ev} + \dot{W}_{in} = \dot{Q}_{cn}$$

$$\text{EnBE}: \dot{Q}_{ev}/T_{ev} + \dot{S}_{gen} = \dot{Q}_{cn}/T_{cn}$$

$$\text{ExBE}: \dot{E}x_{\dot{Q}_{ev}} + \dot{W}_{in} = \dot{E}x_{\dot{Q}_{cn}} + \dot{E}x_d$$

Here, both $\dot{E}x_{\dot{Q}_{ev}}$ and $\dot{E}x_{\dot{Q}_{cn}}$ are calculated as follows:

$$\dot{E}x_{\dot{Q}_{ev}} = \left(\frac{T_o}{T_{ev}} - 1\right)\dot{Q}_{ev} \text{ and } \dot{E}x_{\dot{Q}_{cn}} = \left(1 - \frac{T_o}{T_{cn}}\right)\dot{Q}_{cn}$$

b) The work input is determined from an energy balance on the refrigeration cycle.
By using the EBE that is defined for the refrigeration system as a single unit we can rearrange it to find the consumed power by the refrigeration unit as follows:

$$\dot{Q}_{ev} + \dot{W}_{in} = \dot{Q}_{cn}$$

$$\dot{W}_{in} = \dot{Q}_{cn} - \dot{Q}_{ev} = 2600\,\frac{kJ}{h} - 1500\,\frac{kJ}{h} = 1100\,\frac{kJ}{h} \times \left(\frac{1\ kW}{3600\frac{kJ}{h}}\right) = \mathbf{0.306\ kW}$$

c) The *COP* of the refrigerator (energetic based *COP*) is calculated as:

$$COP_{R,en} = \frac{\dot{Q}_{ev}}{\dot{W}_{in}} = \frac{1500\ \frac{kJ}{h} \times \left(\frac{1\ kW}{3600\ \frac{kJ}{h}}\right)}{0.306\ kW} = \mathbf{1.36}$$

d) The maximum *COP* of the cycle is only possible with a Carnot refrigerator, which is a reversible refrigerator. The Carnot *COP* is then be used to find the minimum work input to the compressor, which leads to the highest *COP*, ideally:

$$COP_{R,en,rev} = \frac{T_{ev}}{T_{cn} - T_{ev}} = \frac{258}{298 - 258} = 6.45$$

$$\dot{W}_{min} = \frac{\dot{Q}_L}{COP_{en,R,rev}} = \frac{1500 \div 3600}{6.45} = \mathbf{0.065\ kW}$$

It is evident that the ideal work input to the compressor is 4.74 times less than the work required; it is also clearly seen from the *COPs* that COP_R is 4.74 is less than the $COP_{R,rev}$ (which is Carnot COP_R).

Example 7.2 A refrigeration system, as shown in Figure 7.8, using $R134a$ as the refrigerant (working fluid of the cycle) is employed to keep the targeted space refrigerated at a temperature of $-10\ °C$ by rejecting heat to the ambient air, which is at a temperature of $22\ °C$. $R134a$ enters the compressor at $140\ kPa$ as a saturated vapor at a flow rate of $375\ L/min$.

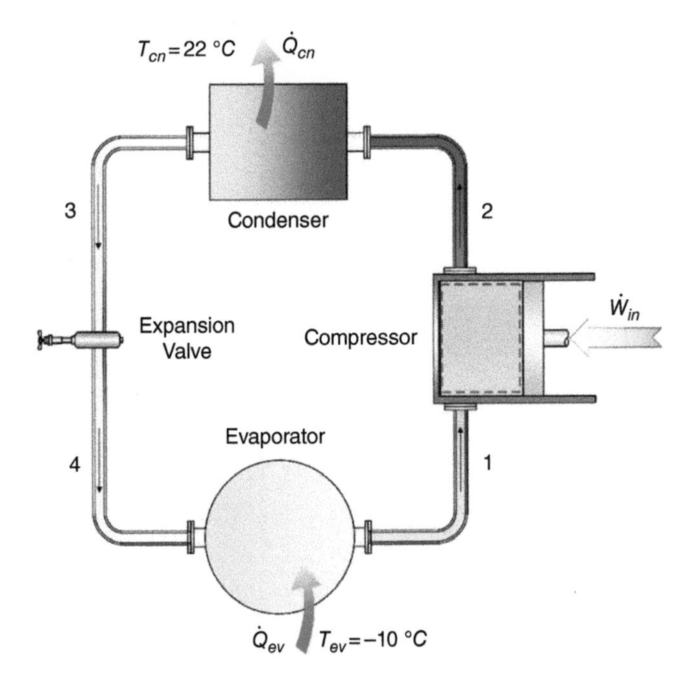

Figure 7.8 Graphical illustration of the refrigeration unit discussed in Example 7.2.

The isentropic efficiency of the compressor is 80%. The refrigerant leaves the condenser at 46.3 °C as a saturated liquid. (a) Write the mass, energy, entropy, and exergy balance equations for each component of the cycle, (b) find the rate of cooling provided by the system, (c) calculate the energetic *COP* of the system, (d) find the exergy destruction in each component of the cycle, (e) find the exergetic *COP* of the system, and (f) find the total exergy destruction in the cycle.

Solution

a) Write the mass, energy, entropy, and exergy balance equation for each component in the cycle.

In this regard, the first step in solving any thermodynamics problem is to write the four balance equations for each component of the cycle; these balance equations are presented in Table 7.1.

The overall cycle balance equations are:

$$\text{MBE}: \dot{m}_1 = \dot{m}_2 = \dot{m}_3 = \dot{m}_4 = \dot{m}$$

$$\text{EBE}: \dot{Q}_{ev} + \dot{W}_{in} = \dot{Q}_{cn}$$

$$\text{EnBE}: \dot{Q}_{ev}/T_{ev} + \dot{S}_{gen,ov} = \dot{Q}_{cn}/T_{cn}$$

$$\text{ExBE}: \dot{E}x_{\dot{Q}_{ev}} + \dot{W}_{in} = \dot{E}x_{\dot{Q}_{cn}} + \dot{E}x_{d,ov}$$

where subscript ov stands for overall.

Note that in the exergy balance equation, the condenser heat is not considered since it is equal to zero as the source temperature of the heat is the same as the ambient temperature and will result in a zero thermal exergy transfer.

b) Calculate the rate of cooling provided by the system.

One can find the properties of *R*134*a* either from properties tables (such as Appendix B-3b) or from the EES software with the properties database)

$$\left.\begin{array}{r} P_1 = 140\ kPa \\ x_1 = 1 \end{array}\right\} \begin{array}{l} h_1 = 239.17\ kJ/kg \\ s_1 = 0.9446\ kJ/(kgK) \\ v_1 = 0.1402\ m^3/kg \end{array}$$

$$\left.\begin{array}{r} P_3 = P_{sat}@46.3\,^{\circ}C \\ P_2 = P_3 = 1200\ kPa \\ s_{2s} = s_1 = 0.9446\ kJ/kg.K \end{array}\right\} h_{2s} = 284.09\ kJ/kg$$

$$\left.\begin{array}{r} P_3 = 1200\ kPa \\ x_3 = 0 \end{array}\right\} \begin{array}{l} h_3 = 117.77\ kJ/kg \\ s_3 = 0.4244\ kJ/kgK \end{array}$$

From the MBE and EBE of the expansion valve we can conclude that:

$$h_3 = h_4 = 117.77\ kJ/kg$$

$$\left.\begin{array}{r} P_4 = 140\ kPa \\ h_4 = 117.77\ kJ/kg \end{array}\right\} s_4 = 0.4674\ kJ/kgK$$

Table 7.1 The mass, energy, entropy, and exergy balance equations for each component of the refrigeration unit discussed in Example 7.2.

Component	Balance equations
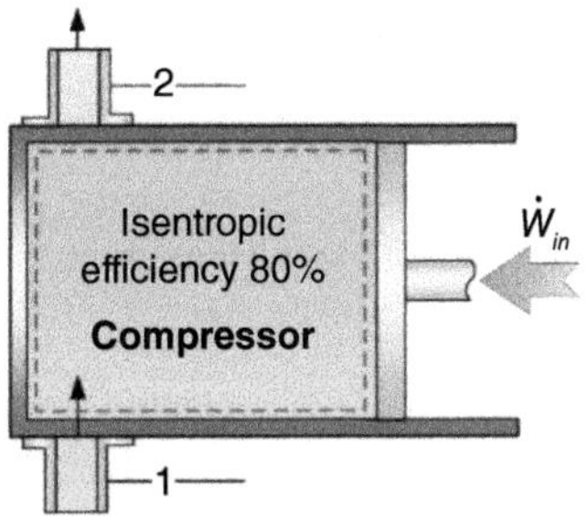 2, 1, $\dot{W}_{in}$, Isentropic efficiency 80%, **Compressor**	MBE: $\dot{m}_1 = \dot{m}_2 = \dot{m}$ EBE: $\dot{m}_1 h_1 + \dot{W}_{\text{in}} = \dot{m}_2 h_2$ EnBE: $\dot{m}_1 s_1 + \dot{S}_{gen,1\to 2} = \dot{m}_2 s_2$ ExBE: $\dot{m}_1 ex_1 + \dot{W}_{in} = \dot{m}_2 ex_2 + \dot{E}x_{d,1\to 2}$
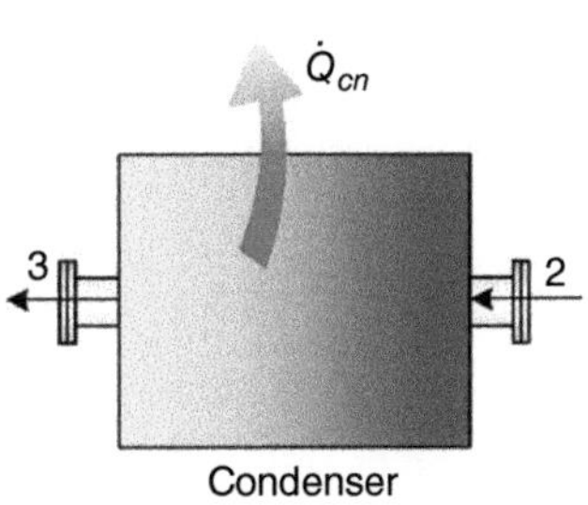 $\dot{Q}_{cn}$, 3, 2, Condenser	MBE: $\dot{m}_2 = \dot{m}_3 = \dot{m}$ EBE: $\dot{m}_2 h_2 = \dot{m}_3 h_3 + \dot{Q}_{cn}$ EnBE: $\dot{m}_2 s_2 + \dot{S}_{gen,2\to 3} = \dot{m}_3 s_3 + \dfrac{\dot{Q}_{cn}}{T_{cn}}$ ExBE: $\dot{m}_2 ex_2 = \dot{m}_3 ex_3 + \dot{E}x_{\dot{Q}_{cond}} + \dot{E}x_{d,2\to 3}$
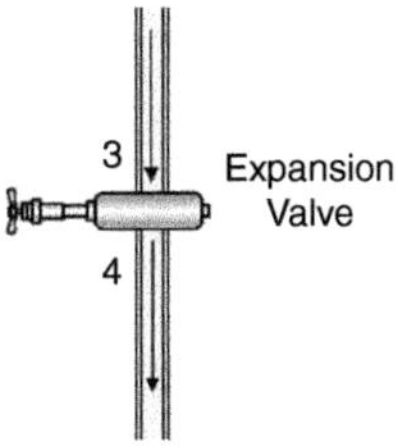 3, 4, Expansion Valve	MBE: $\dot{m}_3 = \dot{m}_4 = \dot{m}$ EBE: $\dot{m}_3 h_3 = \dot{m}_4 h_4 \rightarrow h_3 = h_4$ EnBE: $\dot{m}_3 s_3 + \dot{S}_{gen,3\to 4} = \dot{m}_4 s_4$ ExBE: $\dot{m}_3 ex_3 = \dot{m}_4 ex_4 + \dot{E}x_{d,3\to 4}$
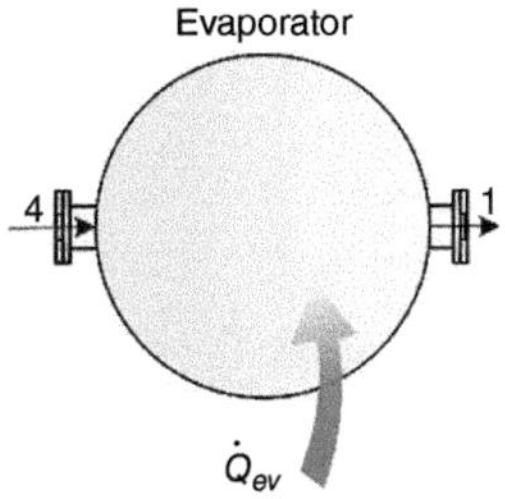 Evaporator, 4, 1, $\dot{Q}_{ev}$	MBE: $\dot{m}_4 = \dot{m}_1 = \dot{m}$ EBE: $\dot{m}_4 h_4 + \dot{Q}_{ev} = \dot{m}_1 h_1$ EnBE: $\dot{m}_4 s_4 + \dot{S}_{gen,4\to 1} + \dfrac{\dot{Q}_{ev}}{T_{ev}} = \dot{m}_1 s_1$ ExBE: $\dot{m}_4 ex_4 + \dot{E}x_{\dot{Q}_{ev}} = \dot{m}_1 ex_1 + \dot{E}x_{d,4\to 1}$

Then by using the isentropic efficiency of the compressor to find the actual enthalpy exiting the compressor (h_2):

$$\eta_{is,c} = \frac{h_{2s} - h_1}{h_2 - h_1}$$

$$0.80 = \frac{284.09 - 239.17}{h_2 - 239.17} \rightarrow h_2 = 295.32\ \frac{kJ}{kg}$$

Thus, we can use the pressure leaving the compressor and the actual enthalpy to find the actual entropy as follows:

$$\left.\begin{aligned} P_2 &= 1200\ kPa \\ h_2 &= 295.32\ kJkg \end{aligned}\right\} s_2 = 0.9783\ kJ/kgK$$

The mass flow rate of the refrigerant is:

$$\dot{m} = \frac{\dot{V}_1}{v_1} = \frac{375\ \dfrac{L}{min} \times \dfrac{1\ min}{60\ s} \times \dfrac{1\ m^3}{1000\ L}}{0.1402\ \dfrac{m^3}{kg}} = \mathbf{0.04458\ kg/s}$$

The refrigeration load accomplished in the evaporator, the rate of heat rejected from the condenser, and the compressor work input are calculated as follows from the balance equations accordingly:

$$\dot{m}_4 h_4 + \dot{Q}_{ev} = \dot{m}_1 h_1$$

$$\dot{Q}_{ev} = \dot{m}_1 h_1 - \dot{m}_4 h_4 = 0.04458\ \frac{kg}{s} \times (239.17 - 117.77)\frac{kJ}{kg} = \mathbf{5.41\ kW}$$

c) Calculate the energetic based *COP* of the cycle.

$$COP_{R,en} = \frac{\dot{Q}_{ev}}{\dot{W}_{in}}$$

We need the work consumed by the compressor, which is found by solving for $\dot{W}_{in}$ in the EBE of the compressor as follows:

$$\dot{m}_1 h_1 + \dot{W}_{in} = \dot{m}_2 h_2$$

$$\dot{W}_{comp} = \dot{m}_2 h_2 - \dot{m}_1 h_1 = \dot{m}(h_2 - h_1) = 0.04458\ \frac{kg}{s} \times (295.32 - 239.17)\frac{kJ}{kg} = 2.50\ kW$$

$$COP_{R,en} = \frac{\dot{Q}_{ev}}{\dot{W}_{in}} = \frac{5.41}{2.50} = \mathbf{2.16}$$

d) Find the exergy destruction rate of each component in the cycle.

Noting that the dead-state temperature is $T_{cn} = T_o = 295\ K$, the exergy destruction in each component of the cycle is determined as follows through the use of the entropy generation rate of each component; where each is found then the exergy destruction rate is calculated as follows:

For the compressor:

$$\dot{m}_1 s_1 + \dot{S}_{gen,1\to 2} = \dot{m}_2 s_2$$

$$\dot{S}_{gen,1\to 2} = \dot{m}_2 s_2 - \dot{m}_1 s_1 = \dot{m}(s_2 - s_1) = 0.04458\ \frac{kg}{s} \times (0.9783 - 0.9446)\frac{kJ}{kgK}$$

$$= \mathbf{0.001502\ \frac{kW}{K}}$$

$$\dot{E}x_{d,1\to 2} = T_o \dot{S}_{gen,1\to 2} = 295 \times 0.001502 = \mathbf{0.4432\ kW}$$

For the condenser:

$$\dot{m}_2 s_2 + \dot{S}_{gen,2\to 3} = \dot{m}_3 s_3 + \frac{\dot{Q}_{cond}}{T_{cn}}$$

$$\dot{S}_{gen,2\to3} = \dot{m}_3 s_3 + \frac{\dot{Q}_{cn}}{T_{cn}} - \dot{m}_2 s_2 = 0.04458\ \frac{kg}{s} \times 0.4244\ \frac{kJ}{kgK} + \frac{7.915\ kW}{295\ K} - 0.04458\ \frac{kg}{s} \times 0.9783\ \frac{kJ}{kgK}$$

$$\dot{S}_{gen,2\to3} = \mathbf{0.002138}\frac{\mathbf{kW}}{\mathbf{K}}$$

$$\dot{E}x_{d,2\to3} = T_o \dot{S}_{gen,2\to3} = 295 \times 0.002138 = \mathbf{0.6307\ kW}$$

For the expansion valve:

$$\dot{m}_3 s_3 + \dot{S}_{gen,3\to4} = \dot{m}_4 s_4$$

$$\dot{S}_{gen,3\to4} = \dot{m}_4 s_4 - \dot{m}_3 s_3 = \dot{m}(s_4 - s_3) = 0.04458\ \frac{kg}{s} \times (0.4674 - 0.4244)\frac{kJ}{kgK}$$

$$\dot{S}_{gen,3\to4} = \mathbf{0.001916\ kW/K}$$

$$\dot{E}x_{d,3\to4} = T_o \dot{S}_{gen,3\to4} = 295 \times 0.001916 = \mathbf{0.5652\ kW}$$

For the evaporator:

$$\dot{m}_4 s_4 + \dot{S}_{gen,4\to1} + \frac{\dot{Q}_{ev}}{T_{ev}} = \dot{m}_1 s_1$$

$$\dot{S}_{gen,4\to1} = \dot{m}_1 s_1 - \dot{m}_4 s_4 - \frac{\dot{Q}_{ev}}{T_{ev}}$$

$$\dot{S}_{gen,4\to1} = 0.04458\ \frac{kg}{s} \times (0.9446 - 0.4674)\frac{kJ}{kgK} - \frac{5.41\ kW}{263\ K} = \mathbf{0.000703\ kW/K}$$

$$\dot{E}x_{d,4\to1} = T_o \dot{S}_{gen,4\to1} = 295 \times 0.0006964 = \mathbf{0.2074\ kW}$$

e) The thermal exergy transferred from the low-temperature medium is:

$$\dot{E}x_{\dot{Q}_{ev}} = \dot{Q}\left(\frac{T_o}{T_{ev}} - 1\right) = 5.41 \times \left(\frac{295}{263} - 1\right) = \mathbf{0.6583\ kW}$$

The exergetic *COP* of the cycle is calculated as:

$$COP_{R,ex} = \frac{\dot{E}x_{\dot{Q}_{ev}}}{\dot{W}_{in}} = \frac{0.6583}{2.503} = 0.263$$

Let's look at this by also employing the second law efficiency or second law *COP* which may also be determined from:

$$COP_{R,ex} = COP_{R,en}/COP_{R,rev}$$

where

$$COP_{R,rev} = \frac{T_L}{T_H - T_L} = \frac{-10 + 273}{(22 - (-10))} = \mathbf{8.22}$$

Substituting,

$$COP_{R,ex} = \frac{COP_{R,en}}{COP_{R,en,rev}} = \frac{2.16}{8.22} = \mathbf{0.263}$$

These results are identical, as expected.

f) The total exergy destruction in the cycle is the difference between the exergy supplied (power input) and the exergy recovered (the exergy of the heat transferred from the low-temperature medium):

$$\dot{E}x_{d,ov} = \dot{W}_{in} - \dot{E}x_{\dot{Q}_{ev}} = (2.503 - 0.6583) = \mathbf{1.845\ kW}$$

The total exergy destruction can also be determined by adding exergy destructions in each component:

$$\dot{E}x_{d,ov} = \dot{E}x_{d,1\rightarrow2} + \dot{E}x_{d,2\rightarrow3} + \dot{E}x_{d,3\rightarrow4} + \dot{E}x_{d}$$
$$= 0.4432 + 0.6307 + 0.5652 + 0.2074 = \mathbf{1.845\ kW}$$

The results are identical, as expected. The exergy input to the cycle is equal to the actual work input, which is 2.503 *kW*. The same cooling load could have been accomplished by only 26.3% of this power (0.3163 *kW*) if a reversible system were used. The difference between the two is the exergy destructed in the cycle (as 1.845 *kW*). It can be shown that increasing the evaporating temperature and decreasing the condensing temperature would also decrease the exergy destructions in these components.

Example 7.3 A refrigeration system, as shown in Figure 7.9, operating with Refrigerant 134*a* (*R*134*a*) which enters a condenser at 1200 *kPa* at a rate of 0.051 *kg/s*. The expansion valve expands the refrigerant to 198 *kPa*. The isentropic efficiency of the compressor is 85%. The quality of refrigerant entering the evaporator is 0.20 and it leaves as a saturated vapor.

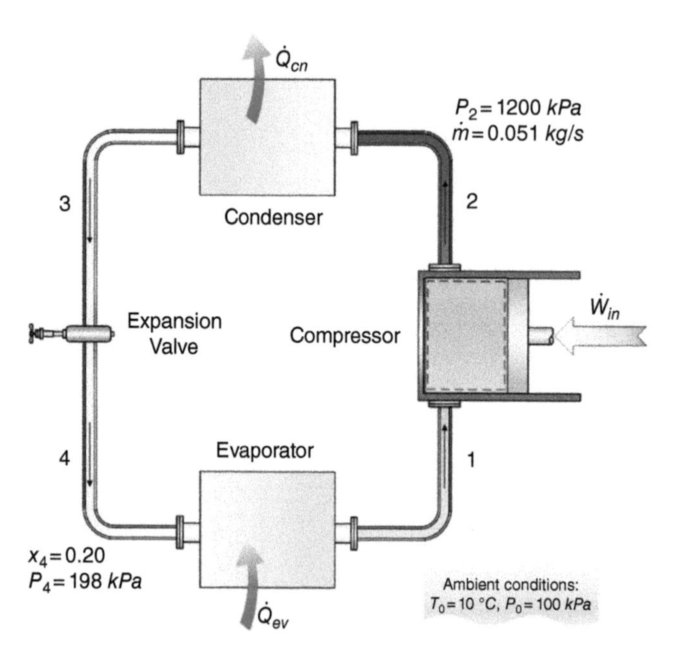

Figure 7.9 Graphical illustration of the refrigeration unit discussed in Example 7.3.

This refrigeration cycle is being used to keep a space at a temperature of $-6\,°C$. Any pressure drops and heat transfers at the connections between components can be disregarded. Take the ambient temperature and pressure as $10\,°C$ and $100\,kPa$, respectively. (a) Write the mass, energy, entropy, and exergy balance equations, (b) determine the rate of heat removal by the refrigerator and the power input to the compressor, (c) find the entropy generation and the exergy destruction rates for each component, and (d) calculate both the energetic and exergetic coefficients of performance (COP_{en}, COP_{ex}).

Solution

a) Write mass, energy, entropy, and exergy balance equations for each component in the cycle under the assumptions of steady state and steady flow operation and negligible kinetic and potential energy changes.

For the condenser:

MBE: $\dot{m}_2 = \dot{m}_3 = \dot{m}$

EBE: $\dot{m}_2 h_2 = \dot{m}_3 h_3 + \dot{Q}_{cn}$

EnBE: $\dot{m}_2 s_2 + \dot{S}_{gen,2\to3} = \dot{m}_3 s_3 + \dfrac{\dot{Q}_{cn}}{T_{cn}}$

ExBE: $\dot{m}_2 ex_2 = \dot{m}_3 ex_3 + \dot{E}x_{\dot{Q}_{cn}} + \dot{E}x_{d,2\to3}$

For the evaporator:

MBE: $\dot{m}_4 = \dot{m}_1 = \dot{m}$

EBE: $\dot{m}_4 h_4 + \dot{Q}_{ev} = \dot{m}_1 h_1$

EnBE: $\dot{m}_4 s_4 + \dot{S}_{gen,4\to1} + \dfrac{\dot{Q}_{ev}}{T_{ev}} = \dot{m}_1 s_1$

ExBE: $\dot{m}_4 ex_4 + \dot{E}x_{\dot{Q}_{ev}} = \dot{m}_1 ex_1 + \dot{E}x_{d,4\to1}$

For the compressor:

MBE: $\dot{m}_1 = \dot{m}_2 = \dot{m}$

EBE: $\dot{m}_1 h_1 + \dot{W}_{in} = \dot{m}_2 h_2$

EnBE: $\dot{m}_1 s_1 + \dot{S}_{gen,1\to2} = \dot{m}_2 s_2$

ExBE: $\dot{m}_1 ex_1 + \dot{W}_{in} = \dot{m}_2 ex_2 + \dot{E}x_{d,1\to2}$

For the expansion valve:

MBE: $\dot{m}_3 = \dot{m}_4 = \dot{m}$

EBE: $\dot{m}_3 h_3 = \dot{m}_4 h_4 \quad \to \quad h_3 = h_4$ (Isenthalpic process)

EnBE: $\dot{m}_3 s_3 + \dot{S}_{gen,3\to4} = \dot{m}_4 s_4$

ExBE: $\dot{m}_3 ex_3 = \dot{m}_4 ex_4 + \dot{E}x_{d,3\to4}$

The overall cycle balance equations are:

MBE: $\dot{m}_1 = \dot{m}_2 = \dot{m}_3 = \dot{m}_4 = \dot{m}$

$$\text{EBE: } \dot{Q}_{ev} + \dot{W}_{in} = \dot{Q}_{cn}$$

$$\text{EnBE: } \frac{\dot{Q}_{ev}}{T_{ev}} + \dot{S}_{gen,ov} = \frac{\dot{Q}_{cn}}{T_{cn}}$$

$$\text{ExBE: } \dot{E}x_{\dot{Q}_{ev}} + \dot{W}_{in} = \dot{E}x_{\dot{Q}_{cn}} + \dot{E}x_{d,ov}$$

where the subscript "ov" stands for overall.

b) Determine the rate of heat removal by the refrigerator and the power input to the compressor.

The properties of R134a are found either from properties tables (such as Appendix B-3b) or from the EES software with the properties database)

$$\left.\begin{matrix} P_1 = 198\ kPa \\ x_1 = 1 \end{matrix}\right\} \begin{matrix} h_1 = 244.6\ \frac{kJ}{kg} \\ s_1 = 0.939\ \frac{kJ}{kg\ K} \end{matrix}$$

$$\left.\begin{matrix} P_2 = 1200\ kPa \\ s_{2s} = s_1 = 0.939\ \frac{kJ}{kg\ K} \end{matrix}\right\} h_{2s} = 282.3\ \frac{kJ}{kg}$$

Then by using the isentropic efficiency of the compressor (η_c) to find the actual enthalpy exiting the compressor:

$$\text{Isentropic efficiency: } \eta_c = \frac{h_{2s} - h_1}{h_2 - h_1} = > h_2 = h_1 + \left(\frac{(h_{2s} - h_1)}{\eta_c}\right)$$

$$h_2 = 244.6 + \left(\frac{(282.3 - 244.6)}{0.85}\right) \rightarrow h_2 = 288.9\ \frac{kJ}{kg}$$

$$\left.\begin{matrix} P_2 = 1200\ kPa \\ h_2 = 288.9\ \frac{kJ}{kg} \end{matrix}\right\} s_2 = 0.9592\ \frac{kJ}{kgK}$$

$$\left.\begin{matrix} P_4 = 198\ kPa \\ x_4 = 0.20 \end{matrix}\right\} \begin{matrix} s_4 = 0.3104\ \frac{kJ}{kgK} \\ h_4 = 79.38\ \frac{kJ}{kg} \end{matrix}$$

$$\left.\begin{matrix} P_3 = 1200\ kPa \\ h_3 = h_4 = 79.38\ \frac{kJ}{kg} \end{matrix}\right\} \begin{matrix} s_3 = 0.2991\ \frac{kJ}{kgK} \\ T_3 = 20.01^{\circ}C \end{matrix}$$

The compressor work and the rate of heat removal by the refrigerator are calculated as follows:

$$\dot{Q}_{cn} = \dot{m}\ (h_2 - h_3) = 0.051\ \frac{kg}{s} \times \left[(288.9 - 79.38)\ \frac{kJ}{kg}\right] = \mathbf{10.68\ kW}$$

$$\dot{W}_{in} = \dot{m}\ (h_2 - h_1) = 0.051\ \frac{kg}{s} \times \left[(288.9 - 244.6)\ \frac{kJ}{kg}\right] = \mathbf{2.26\ kW}$$

c) Determine the entropy generation and the exergy destruction rates for each component.

For the condenser:

The EnBE is applied and the values are substituted to give the entropy generation:

$$\dot{m}_2 s_2 + \dot{S}_{gen} = \dot{m}_3 s_3 + \frac{\dot{Q}_{cn}}{T_{cn}}$$

$$\dot{S}_{gen} = \dot{m}(s_3 - s_2) + \frac{\dot{Q}_{cn}}{T_{cn}}$$

$$\dot{S}_{gen} = 0.051\ \frac{kg}{s} \times \left[(0.2991 - 0.9592)\ \frac{kJ}{kgK}\right] + \frac{10.68\ kW}{283\ K}$$

$$\mathbf{\dot{S}_{gen} = 4.1 \times 10^{-3}\ \frac{kW}{K}}$$

Before evaluating the specific exergies, the ambient state specific enthalpy and entropy are determined.

$$\left.\begin{matrix} P_o = 100\ kPa \\ T_o = 10°C \end{matrix}\right\}\ h_0 = 263.8\ \frac{kJ}{kg}, s_o = 1.063\ \frac{kJ}{kgK}$$

So,

$$ex_2 = (h_2 - h_0) - T_o \times (s_2 - s_0)$$

$$ex_2 = \left[(288.9 - 263.8)\frac{kJ}{kg}\right] - 283\left[(0.9592 - 1.063)\frac{kJ}{kgK}\right]$$

$$\mathbf{ex_2 = 54.47\ \frac{kJ}{kg}}$$

$$ex_3 = (h_3 - h_0) - T_o \times (s_3 - s_0)$$

$$ex_3 = \left[(79.38 - 263.8)\frac{kJ}{kg}\right] - 283\left[(0.2991 - 1.063)\frac{kJ}{kgK}\right]$$

$$\mathbf{ex_3 = 31.76\ \frac{kJ}{kg}}$$

Applying the ExBE to find the exergy destruction rate:

$$\dot{E}x_{d,2\rightarrow3} = m_2 ex_2 - \left[m_3 ex_3 + \left(1 - \frac{T_o}{T_{cn}}\right)\dot{Q}_{cn}\right]$$

$$\dot{E}x_{d,2\rightarrow3} = 0.051\ \frac{kg}{s} \times 54.47\ \frac{kJ}{kg} - \left[\left(0.051\ \frac{kg}{s} \times 31.76\ \frac{kJ}{kg}\right) + \left(1 - \frac{283\ K}{283\ K}\right) \times 10.68\ kW\right]$$

$$\mathbf{\dot{E}x_{d,2\rightarrow3} = 1.16\ kW}$$

For the evaporator:

The EBE is applied and the values are substituted to give $\dot{Q}_{ev}$:

$$\dot{Q}_{ev} = \dot{m}(h_1 - h_4)$$

$$\dot{Q}_{ev} = 0.051\ \frac{kg}{s} \times \left[(244.6 - 79.38)\frac{kJ}{kg}\right]$$

$$\mathbf{\dot{Q}_{ev} = 8.4262\ kW}$$

Here, the EnBE is applied and the values are substituted to give the entropy generation:

$$\dot{m}_4 s_4 + \frac{\dot{Q}_{ev}}{T_{ev}} + \dot{S}_{gen,4\to1} = \dot{m}_1 s_1$$

$$\dot{S}_{gen,4\to1} = \dot{m}(s_1 - s_4) - \frac{\dot{Q}_{ev}}{T_{ev}}$$

$$\dot{S}_{gen,4\to1} = 0.051\ \frac{kg}{s} \times \left[(0.939 - 0.3104)\frac{kJ}{kgK}\right] - \frac{8.4262\ kJ}{267\ K}$$

$$\mathbf{\dot{S}_{gen,4\to1} = 5 \times 10^{-4}\ \frac{kW}{K}}$$

$$ex_1 = (h_1 - h_0) - T_o \times (s_1 - s_0)$$

$$ex_1 = \left[(244.6 - 263.8)\frac{kJ}{kg}\right] - 283\left[(0.939 - 1.063)\frac{kJ}{kgK}\right]$$

$$\mathbf{ex_1 = 15.892\ \frac{kJ}{kg}}$$

$$ex_4 = (h_4 - h_0) - T_o \times (s_4 - s_0)$$

$$ex_4 = \left[(79.38 - 263.8)\frac{kJ}{kg}\right] - 283\left[(0.3104 - 1.063)\frac{kJ}{kgK}\right]$$

$$\mathbf{ex_4 = 28.56\ \frac{kJ}{kg}}$$

Applying the ExBE to find the exergy destruction rate:

$$\dot{E}x_{d,4\to1} = \left[\ \dot{m}_4 ex_4 + \left(\frac{T_o}{T_{ev}} - 1\right)\dot{Q}_{ev}\right] - \dot{m}_1 ex_1$$

$$\dot{E}x_{d,4\to1} = \left[\left(0.051\ \frac{kg}{s} \times 28.56\ \frac{kJ}{kg}\right) + \left(\frac{283\ K}{267\ K} - 1\right) \times 8.426\ kW\right] - 0.051\ \frac{kg}{s} \times 15.892\ \frac{kJ}{kg}$$

$$\mathbf{\dot{E}x_{d,4\to1} = 1.150\ kW}$$

For the compressor:

The EnBE is applied and the values are substituted to give the entropy generation:

$$\dot{m}_1 s_1 + \dot{S}_{gen,1\to2} = \dot{m}_2 s_2$$

$$\dot{S}_{gen,1\to2} = \dot{m}(s_2 - s_1)$$

$$\dot{S}_{gen,1\to2} = 0.051\ \frac{kg}{s} \times \left[(0.9592 - 0.939)\frac{kJ}{kgK}\right]$$

$$\mathbf{\dot{S}_{gen,1\to2} = 1 \times 10^{-3}\ \frac{kW}{K}}$$

Applying the ExBE to find the exergy destruction rate:

$$\dot{m}_1 ex_1 + \dot{W}_{in} = \dot{m}_2 ex_2 + \dot{E}x_d$$

$$\dot{E}x_{d,1\to2} = \dot{W}_{in} + \dot{m}(ex_1 - ex_2)$$

$$\dot{E}x_{d,1\to 2} = 2.26\ kW + 0.051\ \frac{kg}{s} \times \left[(15.892 - 54.47)\ \frac{kJ}{kg}\right]$$

$$\mathbf{\dot{E}x_{d,1\to 2} = 0.2925\ kW}$$

For the expansion valve:

The EnBE is applied and the values given and obtained are substituted to give the entropy generation

$$\dot{m}_3 s_3 + \dot{S}_{gen,3\to 4} = \dot{m}_4 s_4$$

$$\dot{S}_{gen,3\to 4} = \dot{m} \times (s_4 - s_3)$$

$$\dot{S}_{gen,3\to 4} = 0.051\ \frac{kg}{s} \times \left[(0.3104 - 0.2991)\ \frac{kJ}{kgK}\right]$$

$$\mathbf{\dot{S}_{gen,3\to 4} = 5.8 \times 10^{-4}\ \frac{kW}{K}}$$

Applying the ExBE to find the exergy destruction rate:

$$\dot{m}_3 ex_3 = \dot{m}_4 ex_4 + \dot{E}x_{d,3\to 4}$$

$$\dot{E}x_{d,3\to 4} = \dot{m} \times (ex_3 - ex_4)$$

$$\dot{E}x_{d,3\to 4} = 0.051\ \frac{kg}{s} \times \left[(31.76 - 28.56)\frac{kJ}{kg}\right]$$

$$\mathbf{\dot{E}x_{d,3\to 4} = 0.1632\ kW}$$

d) Determine the energetic and exergetic coefficients of performance (COP_{en}, COP_{ex}).

$$COP_{en} = \frac{\dot{Q}_{ev}}{\dot{W}_{in}}$$

$$COP_{en} = \frac{8.4262}{2.2593\ kW}$$

$$\mathbf{COP_{en} = 3.729}$$

$$COP_{ex} = \frac{\dot{E}x_{Q_{in}}}{\dot{W}_{in}}$$

$$COP_{ex} = \frac{\left(\frac{283}{267} - 1\right) \times 8.426\ kW}{2.26\ kW}$$

$$\mathbf{COP_{ex} = 0.2234}$$

7.4 Air-Standard Refrigeration Systems

There is another type of refrigeration system, so called: gas refrigeration or air refrigeration or air-standard refrigeration system. These air-standard refrigeration cycles are also known as reverse Brayton cycles. In such systems, refrigeration is accomplished by means of a noncondensing gas (commonly air) cycle rather than a refrigerant-based vapor compression cycle.

While the refrigeration load per kilogram of refrigerant circulated in a vapor-compression cycle is equal to a large fraction of the enthalpy of vaporization, in an air cycle it is only the product of the temperature rise of the gas in the low-side heat exchanger and the specific heat of the gas. Therefore, a large refrigeration load requires a large mass rate of circulation. In order to keep the equipment size smaller, the complete unit may be under pressure, which requires a closed cycle. The throttling valve used for the expansion process in a vapor-compression refrigeration cycle is usually replaced by an expansion engine (e.g. expander) for an air-cycle refrigeration system. The work required for the refrigeration effect is provided by the gas refrigerant. These systems are of great interest in applications where the weight of the refrigerating unit must be kept to a minimum, for example, in aircraft cabin cooling.

A schematic arrangement of a basic air-standard refrigeration cycle and its *T*-*s* diagram are shown in Figure 7.10. Here, two cases identified, namely isentropic (ideal) and actual, where there are both isentropic and actual compression and both isentropic and actual expansion. This basic system has four processes taking place.

- 1-2 compression process: where a compressor raises the pressure of the gas (air) from its lowest to its highest value. There are two compression processes indicated in the *T*-*s* diagram, 1-2*s* is isentropic compression and 1-2*a* is actual compression.
- 2-3 isobaric heat rejection process: where the gas (air) is cooled down to the desired temperature at constant pressure and the heat is rejected.
- 3-4 expansion process: where an expander reduces the pressure of the gas (air) from its highest value to its lowest one. There are two expansion processes indicated in the *T*-*s* diagram, 3-4*s* is isentropic expansion and 3-4*a* is actual expansion.
- 4-1 isobaric heat input process: where the gas (air) at constant pressure raises its temperature by gaining heat and providing cooling at the desired temperature. This input is known as refrigeration load.

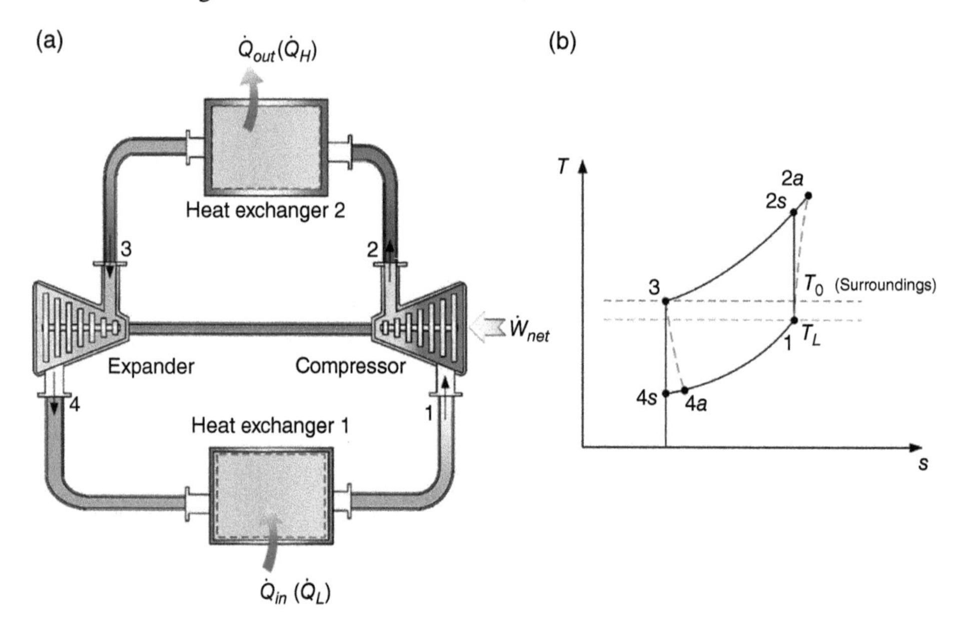

Figure 7.10 (a) Schematic diagram of the air-standard refrigeration cycle and (b) its *T*-*s* diagram for both ideal and actual compression and expansion processes.

Example 7.4 Consider an air-standard refrigeration system, as shown in Figure 7.11, where a space is refrigerated to a temperature of $-5\,°C$ by losing a total of 21.0 kW heat. The compressor receives air at a temperature of $-20\,°C$ and increases its pressure by a factor of four, while the expander receives air at a temperature of $16\,°C$. The isentropic efficiencies of the compressor and turbine are 82 and 84%, respectively. Take the environment temperature and pressure to be 283 K and 100 kPa, respectively. Assume air to behave as an ideal gas with a specific heat ratio of 1.4 and constant pressure specific heat of 1.005 $kJ/kg.K$. (a) Write the mass, energy, entropy, and exergy balance equations for each component in the refrigeration cycle and the entire cycle, (b) determine the input power needed to run this cycle, (c) find both the COP_{en} and COP_{ex} of this refrigeration cycle, (d) find the entropy generation rate of this cycle, and (e) calculate the exergy destruction rate of this cycle.

Solution

a) Write the mass, energy, entropy, and exergy balance equations for each component in the refrigeration cycle and the entire cycle.

For the compressor:

$$\text{MBE}: \dot{m}_1 = \dot{m}_2$$

$$\text{EBE}: \dot{m}_1 h_1 + \dot{W}_c = \dot{m}_2 h_2$$

$$\text{EnBE}: \dot{m}_1 s_1 + \dot{S}_{gen,c} = \dot{m}_2 s_2$$

$$\text{ExBE}: \dot{m}_1 ex_1 + \dot{W}_c = \dot{m}_2 ex_2 + \dot{E}x_{d,c}$$

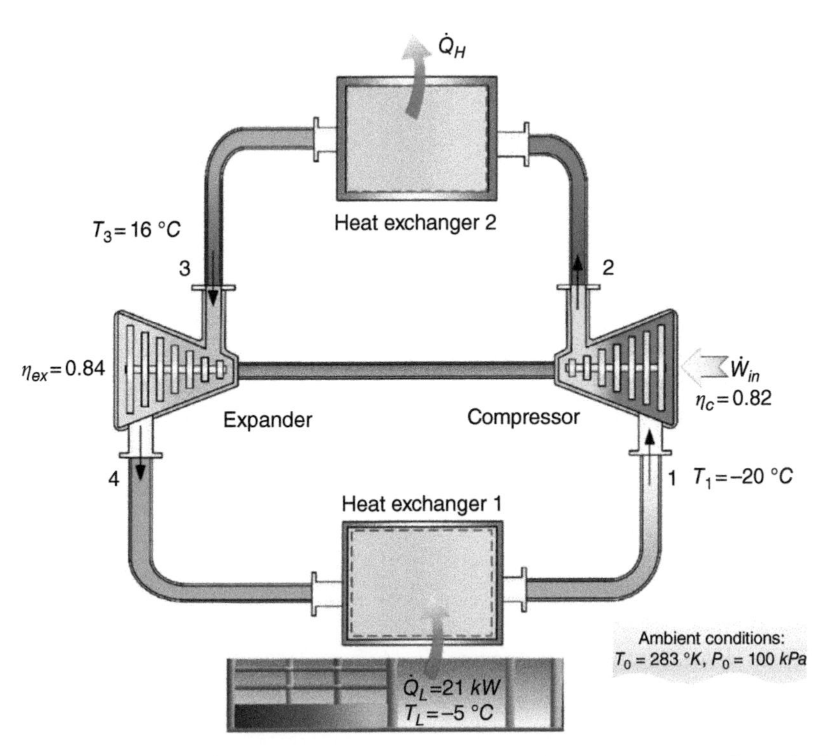

Figure 7.11 Schematic diagram of the air-standard refrigeration cycle for Example 7.4.

For heat exchanger 2:

$$\text{MBE}: \dot{m}_2 = \dot{m}_3$$

$$\text{EBE}: \dot{m}_2 h_2 = \dot{m}_3 h_3 + \dot{Q}_H$$

$$\text{EnBE}: \dot{m}_2 s_2 + \dot{S}_{gen,H} = \dot{m}_3 s_3 + \frac{\dot{Q}_H}{T_{b,H}}$$

$$\text{ExBE}: \dot{m}_2 ex_2 = \dot{m}_3 ex_3 + \dot{Q}_H\left(1 - \frac{T_o}{T_{b,H}}\right) + \dot{E}x_{d,H}$$

For the expander:

$$\text{MBE}: \dot{m}_3 = \dot{m}_4$$

$$\text{EBE}: \dot{m}_3 h_3 = \dot{m}_4 h_4 + \dot{W}_{ex}$$

$$\text{EnBE}: \dot{m}_3 s_3 + \dot{S}_{gen,ex} = \dot{m}_4 s_4$$

$$\text{ExBE}: \dot{m}_3 ex_3 = \dot{m}_4 ex_4 + \dot{W}_{ex} + \dot{E}x_{d,ex}$$

For the heat exchanger 1:

$$\text{MBE}: \dot{m}_4 = \dot{m}_1$$

$$\text{EBE}: \dot{m}_4 h_4 + \dot{Q}_L = \dot{m}_1 h_1$$

$$\text{EnBE}: \dot{m}_4 s_4 + \frac{\dot{Q}_L}{T_{s,L}} + \dot{S}_{gen,H} = \dot{m}_1 s_1$$

$$\text{ExBE}: \dot{m}_4 ex_4 + \dot{Q}_L\left(\frac{T_o}{T_{s,L}} - 1\right) = \dot{m}_1 ex_1 + \dot{E}x_{d,H}$$

For the overall cycle:

$$\text{EBE}: \dot{Q}_L + \dot{W}_{\text{in}} = \dot{Q}_H$$

$$\text{EnBE}: \frac{\dot{Q}_L}{T_{s,L}} + \dot{S}_{gen,ov} = \frac{\dot{Q}_H}{T_{b,H}}$$

$$\text{ExBE}: \dot{Q}_L\left(\frac{T_o}{T_{s,L}} - 1\right) + \dot{W}_{\text{in}} = \dot{Q}_H\left(1 - \frac{T_o}{T_{b,H}}\right) + \dot{E}x_{d,ov}$$

b) Determine the input power needed to run this cycle.

One needs to determine the input power for this refrigeration cycle, the compressor work as well as the expander work are to be found first. Starting with the compressor,

$$\frac{T_{2s}}{T_1} = \left(\frac{P_2}{P_1}\right)^{\frac{k-1}{k}}$$

Rearranging and solving for the isentropic temperature, we get:

$$T_{2s} = 253\ K \times (4)^{\frac{0.4}{1.4}} = \mathbf{375.9\ K}$$

Using the compressor isentropic efficiency:

$$\eta_c = \frac{T_{2s} - T_1}{T_2 - T_1} = 0.82$$

Solving for state 2 temperature, we get:

$$T_2 = 402.9\ K$$

For the expander, state 4 temperature is:

$$\frac{T_{4s}}{T_3} = \left(\frac{P_4}{P_3}\right)^{\frac{k-1}{k}}$$

$$T_{4s} = 289\ K \times \left(\frac{1}{4}\right)^{\frac{0.4}{1.4}} = \mathbf{194.5\ K}$$

Using the expander isentropic efficiency:

$$\eta_{ex} = \frac{T_3 - T_4}{T_3 - T_{4s}} = 0.84$$

Solving for state 4 temperature, we get:

$$T_4 = 209.6\ K$$

Before finding the compressor and expander work values, the mass flow rate of the cycle needs to be found, too. We use the MBE and EnBE for heat exchanger 1 as follows:

$$\dot{m}_4 = \dot{m}_1$$

$$\dot{m}_4 h_4 + \dot{Q}_L = \dot{m}_1 h_1$$

By knowing $h = c_p T$ for air since it is an ideal gas:

$$\dot{m}_1 = \frac{\dot{Q}_L}{c_p(T_1 - T_4)} = \frac{21.0\ kW}{1.005\ \frac{kJ}{kgK} \times (253\ K - 209.6\ K)} = \mathbf{0.481\ \frac{kg}{s}}$$

The input power is:

$$\dot{W}_{\text{in}} = \dot{W}_c - \dot{W}_{exp} = \dot{m}_1 \left[c_p(T_2 - T_1) - c_p(T_3 - T_4)\right]$$

$$\dot{W}_{\text{in}} = 0.481\ \frac{kg}{s} \times \left[1.005\ \frac{kJ}{kgK} \times (402.9\ K - 253\ K) - 1.005\ \frac{kJ}{kgK} \times (289\ K - 209.6\ K)\right]$$

$$\boldsymbol{\dot{W}_{in} = 34.08\ kW}$$

c) Calculate both COP_{en} and COP_{ex} of this refrigeration cycle.
To start with the energy COP for this refrigeration cycle:

$$\boldsymbol{COP_{en}} = \frac{\dot{Q}_L}{\dot{W}_{\text{in}}} = \frac{21.0\ kW}{34.08\ kW} = \mathbf{0.616}$$

For the exergy COP:

$$COP_{ex} = \frac{\dot{Q}_L\left(\frac{T_o}{T_{s,L}} - 1\right)}{\dot{W}_{\text{in}}} = \frac{21.0\ kW \times \left(\frac{283\ K}{268\ K} - 1\right)}{34.08\ kW}$$

$$\boldsymbol{COP_{ex} = 0.0345}$$

d) Determine the entropy generation rate of this cycle.

From the EBE for the overall cycle, the heat released to the environment is:

$$\dot{Q}_H = \dot{Q}_L + \dot{W}_{in} = 21.0\ kW + 34.08\ kW = \mathbf{55.08\ kW}$$

The boundary temperature for heat exchanger 2 will be the average temperature of states 2 and 3:

$$T_{b,H} = \frac{T_2 + T_3}{2} = \frac{402.9\ K + 289\ K}{2} = \mathbf{345.95\ K}$$

The EnBE for the overall cycle is:

$$\frac{\dot{Q}_L}{T_{s,L}} + \dot{S}_{gen,ov} = \frac{\dot{Q}_H}{T_{b,H}}$$

Solving for the entropy generation rate of the refrigeration cycle:

$$\dot{S}_{gen,ov} = \frac{\dot{Q}_H}{T_{b,H}} - \frac{\dot{Q}_L}{T_{s,L}}$$

$$\dot{S}_{gen,ov} = \frac{55.08\ kW}{345.95} - \frac{21.0\ kW}{268\ K}$$

$$\dot{\mathbf{S}}_{\mathbf{gen,overall}} = \mathbf{0.0809\ \frac{kW}{K}}$$

e) Determine the exergy destruction rate of this cycle.

Similarly, we use the ExBE for the overall cycle:

$$\dot{Q}_L\left(\frac{T_o}{T_{s,L}} - 1\right) + \dot{W}_{\text{in}} = \dot{Q}_H\left(1 - \frac{T_o}{T_{b,H}}\right) + \dot{E}x_{d,ov}$$

Solving for the exergy destruction rate and substituting the values:

$$\dot{E}x_{d,ov} = \dot{Q}_L\left(\frac{T_o}{T_{s,L}} - 1\right) + \dot{W}_{\text{in}} - \dot{Q}_H\left(1 - \frac{T_o}{T_{b,H}}\right)$$

$$\dot{E}x_{d,ov} = 21.0\ kW \times \left(\frac{283\ K}{268\ K} - 1\right) + 34.08\ kW - 55.08\ kW \times \left(1 - \frac{283\ K}{345.95\ K}\right)$$

$$\dot{\mathbf{E}}\mathbf{x}_{\mathbf{d,ov}} = \mathbf{25.23\ kW}$$

Compare this result with the overall exergy destruction rate when the overall entropy generation rate and ambient temperature are used:

$$\dot{E}x_{d,overall} = T_o\dot{S}_{gen,overall} = 283\ K \times 0.0809\ \frac{kW}{K} = \mathbf{22.89\ kW}$$

7.5 Cascade Refrigeration Systems

One should note that in various industrial applications there is a need for relatively low evaporative temperatures with a considerably large temperature and pressure difference that single vapor-compression refrigeration cycles are practically unable to achieve. One of the solutions for such cases is to perform the refrigeration in two or more stages (i.e. two or more cycles) operating in series. These refrigeration cycles are called *cascade refrigeration cycles*. Cascade systems are employed to obtain high temperature differentials between the heat source and heat sink and are applied for temperatures ranging from −70 to 100 °*C*. Application of a three-stage compression system for evaporating temperatures below −70 °*C* is limited, due to difficulties with refrigerants reaching their freezing temperatures. Impropriety of three-stage vapor compression systems can be avoided by applying a cascade vapor-compression refrigeration system.

Cascade refrigeration systems are commonly used in the liquefaction of various gases, including natural gas and some petroleum gases. There is one important point to consider to get the best possible performance for the lowest temperature targeted by cascading, that is to determine the most suitable cycles with the most suitable refrigerants. So, the cascading process technically becomes a thermodynamic problem. The appealing advantage of these cascade systems is that refrigerants with appropriate properties can be chosen, avoiding large size system components. In these systems multiple evaporators can be utilized in any one stage of compression. Refrigerants used in each stage may be different and are selected for optimum performance at the given evaporator and condenser temperatures.

A two-stage cascade system, as shown in Figure 7.12, employs two vapor-compression systems technically, but working separately with different refrigerants and interconnected in such a way that the evaporator of one system is used to serve as condenser to a lower temperature system (i.e. the evaporator from the first unit cools the condenser of the second unit). In practice, an alternative arrangement may be considered with a common condenser with a booster circuit to provide two separate evaporator temperatures.

In closing, a cascade arrangement is also known as binary arrangement and allows one of the systems to be operated at a lower temperature and pressure than would otherwise be possible with the same type and size of single-stage system. It also allows two different refrigerants to be used, and it may even allow temperatures below −150 °*C* to be generated. Figure 7.12 shows a two-stage cascade refrigeration system connected through a heat exchanger, which is the most crucial part of the cascade system, where the evaporator of the upper system (A) will cool the condenser of the lower system (B). The processes in this cascade system are summarized briefly here:

1-2 compression: (which may be isentropic compression under ideal conditions where entropy remains constant) where there is work input supplied externally. As indicated in Figure 7.12b, 1-2*s* shows isentropic compression while 1-2*a* shows actual compression.

2-3 isobaric heat rejection: where heat is extracted from the refrigerant vapor and cooled down by the evaporator of the upper system (A) to a desired temperature after being condensed.

3-4 isenthalpic expansion: where liquid refrigerant is expanded isenthalpically to a desired pressure to become saturated liquid and vapor mixture.

4-1 isobaric heat addition: where liquid gets into the evaporator to provide the necessary cooling by absorbing heat.

5-6 compression: (which may be isentropic compression under ideal conditions where entropy remains constant) where there is work input supplied externally. As indicated in Figure 7.12b, 5-6*s* shows isentropic compression while 5-6*a* shows actual compression.

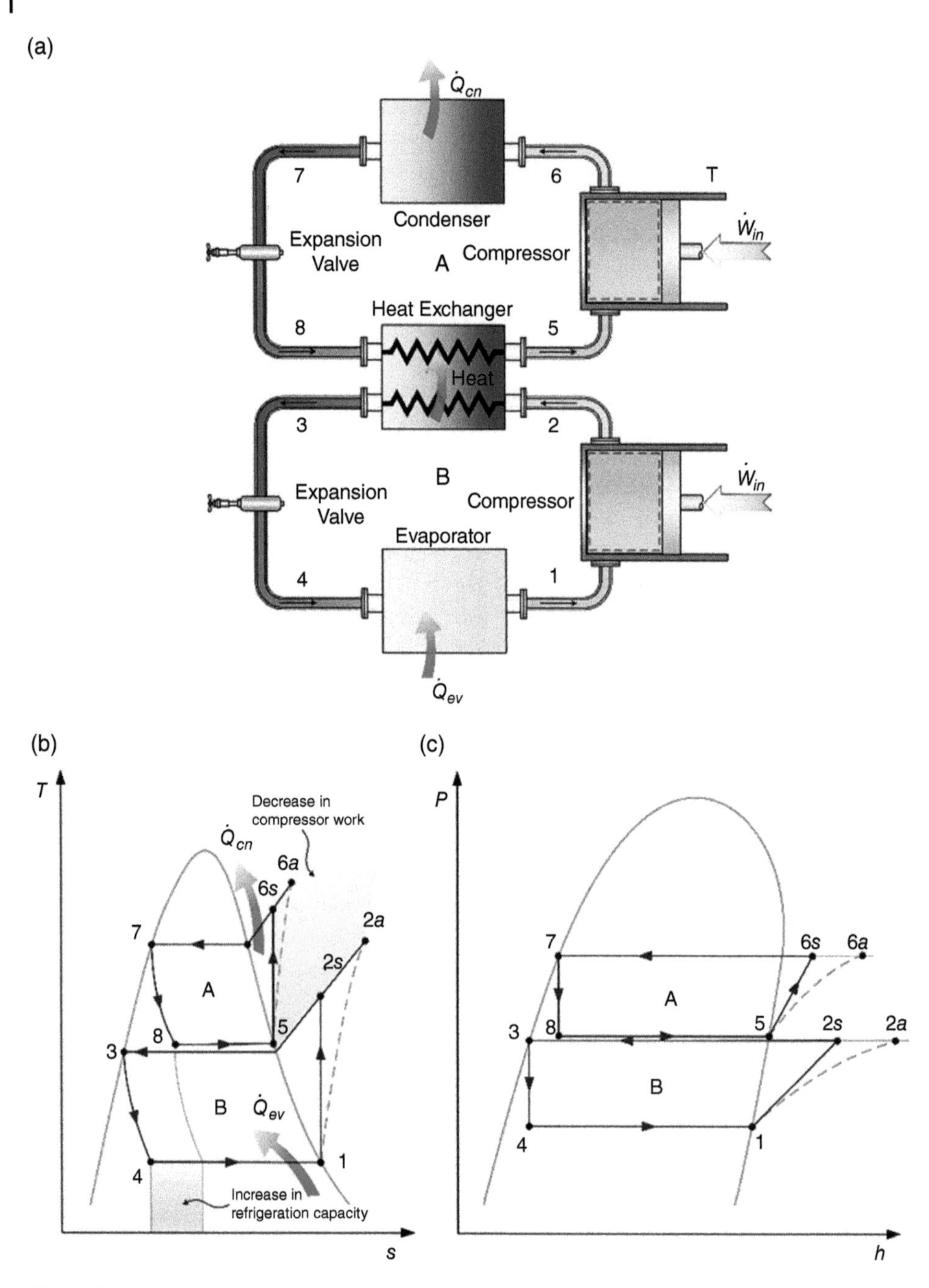

Figure 7.12 (a) Schematic diagram of the cascade refrigeration cycle, (b) its *T-s* diagram for both ideal and actual compression, and (c) its *P-h* diagram for both ideal and actual compression.

6-7 isobaric heat rejection: where heat is extracted from the refrigerant vapor and cooled down by the condenser of the upper system (A) to a desired temperature after being condensed.

7-8 isenthalpic expansion: where liquid refrigerant is expanded isenthalpically to a desired pressure to become a saturated liquid and vapor mixture.

8-5 isobaric heat addition: where liquid refrigerant gets into the evaporator to provide the necessary cooling to the refrigerant of the lower system (B) by absorbing its heat.

Example 7.5 A two-stage cascade refrigeration system, similar to the one shown in Figure 7.12, using *R*123 as refrigerant operates on an ideal vapor-compression cycle within the pressure limits of 1.2 and 0.18 *MPa*. Heat is transferred in an adiabatic counterflow heat exchanger from the lower cycle stream with a pressure of 0.5 *MPa* to the upper cycle stream with a pressure of 0.4 *MPa*. The compression processes in both cycles take place with 80% efficiency. Assume the mass flow rate of 0.1 *kg/s* in the upper cycle. (a) Write the mass, energy, entropy, and exergy balance equations for this cascade system, (b) determine the mass flow rate of refrigerant in the lower cycle (B), (c) find the rate of heat removal from the refrigerated space, and (d) calculate the energetic and exergetic coefficients of performance for this cascade refrigeration cycle.

Solution

a) Write the mass, energy, entropy, and exergy balance equations.

The upper cycle is labeled A and the lower cycle is labeled B.

For compressor B:

$$\text{MBE}: \dot{m}_1 = \dot{m}_2$$

$$\text{EBE}: \dot{m}_1 h_1 + \dot{W}_{CB} = \dot{m}_2 h_2$$

$$\text{EnBE}: \dot{m}_1 s_1 + \dot{S}_{gen,CB} = \dot{m}_2 s_2$$

$$\text{ExBE}: \dot{m}_1 ex_1 + \dot{W}_{CB} = \dot{m}_2 ex_2 + \dot{E}x_{d,CB}$$

For evaporator B:

$$\text{MBE}: \dot{m}_4 = \dot{m}_1$$

$$\text{EBE}: \dot{m}_4 h_4 + \dot{Q}_L = \dot{m}_1 h_1$$

$$\text{EnBE}: \dot{m}_4 s_4 + \frac{\dot{Q}_L}{T_{L,evaB}} + \dot{S}_{gen,evaB} = \dot{m}_1 s_1$$

$$\text{ExBE}: \dot{m}_4 ex_4 + \dot{Q}_L\left(\frac{T_0}{T_{L,eva}} - 1\right) = \dot{m}_1 ex_1 + \dot{E}x_{d,evaB}$$

For expansion valve B:

$$\text{MBE}: \dot{m}_3 = \dot{m}_4$$

$$\text{EBE}: \dot{m}_3 h_3 = \dot{m}_4 h_4$$

$$\text{EnBE}: \dot{m}_3 s_3 + \dot{S}_{gen,evB} = \dot{m}_4 s_4$$

$$\text{ExBE}: \dot{m}_3 ex_3 = \dot{m}_4 ex_4 + \dot{E}x_{d,evB}$$

For compressor A:

$$\text{MBE}: \dot{m}_5 = \dot{m}_6$$

$$\text{EBE}: \dot{m}_5 h_5 + \dot{W}_{CA} = \dot{m}_6 h_6$$

$$\text{EnBE}: \dot{m}_5 s_5 + \dot{S}_{gen,CA} = \dot{m}_6 s_6$$

$$\text{ExBE}: \dot{m}_5 ex_5 + \dot{W}_{CA} = \dot{m}_6 ex_6 + \dot{E}x_{d,CA}$$

For expansion valve A:

$$\text{MBE}: \dot{m}_7 = \dot{m}_8$$

$$\text{EBE}: \dot{m}_7 h_7 = \dot{m}_8 h_8$$

$$\text{EnBE}: \dot{m}_7 s_7 + \dot{S}_{gen,evA} = \dot{m}_8 s_8$$

$$\text{ExBE}: \dot{m}_7 ex_7 = \dot{m}_8 ex_8 + \dot{E}x_{d,evA}$$

For condenser A:

$$\text{MBE}: \dot{m}_6 = \dot{m}_7$$

$$\text{EBE}: \dot{m}_6 h_6 = \dot{Q}_H + \dot{m}_7 h_7$$

$$\text{EnBE}: \dot{m}_6 s_6 + \dot{S}_{gen,con} = \frac{\dot{Q}_H}{T_H} + \dot{m}_7 s_7$$

$$\text{ExBE}: \dot{m}_6 ex_6 = \dot{Q}_H \left(1 - \frac{T_0}{T_{H,con}}\right) + \dot{m}_7 ex_7 + \dot{E}x_{d,con}$$

For the heat exchanger:

$$\text{MBE}: \dot{m}_2 = \dot{m}_3, \dot{m}_5 = \dot{m}_8$$

$$\text{EBE}: \dot{m}_2 h_2 + \dot{m}_8 h_8 = \dot{m}_3 h_3 + \dot{m}_5 h_5$$

$$\text{EnBE}: \dot{m}_2 s_2 + \dot{m}_8 s_8 + \dot{S}_{gen,hex} = \dot{m}_3 s_3 + \dot{m}_5 s_5$$

$$\text{ExBE}: \dot{m}_2 ex_2 + \dot{m}_8 ex_8 = \dot{m}_3 ex_3 + \dot{m}_5 ex_5 + \dot{E}x_{d,hex}$$

b) The mass flow rate relation can be written using the EBE for heat exchanger as follows:

$$\dot{m}_A(h_5 - h_8) = \dot{m}_B(h_2 - h_3)$$

The specific enthalpies for each state can be determined using the property table for $R123$ refrigerant:

$$\left.\begin{array}{r} x_1 = 1 \\ P_1 = 0.18\ MPa \end{array}\right\} h_1 = 410\ \frac{kJ}{kg}$$

$$\left.\begin{array}{r} x_1 = 1 \\ P_1 = 0.18\ MPa \end{array}\right\} s_1 = 1.671\ \frac{kJ}{kgK}$$

$$s_1 = s_{2s}$$

$$\left.\begin{aligned} s_{2s} &= 1.671\ \frac{kJ}{kgK} \\ P_2 &= 0.5\ MPa \end{aligned}\right\} h_{2s} = 426.8\ \frac{kJ}{kg}$$

The actual enthalpy of state 2 can be calculated through the isentropic efficiency definition for the compressor B as follows:

$$\eta_{CB} = \frac{h_{2s} - h_1}{h_2 - h_1} = 0.8$$

$$h_2 = \frac{h_{2s} - h_1}{0.8} + h_1 = 431\ \frac{kJ}{kg}$$

$$\left.\begin{aligned} x_3 &= 0 \\ P_3 &= 0.5\ MPa \end{aligned}\right\} h_3 = 286.5\ \frac{kJ}{kg}$$

Writing the EBE for expansion valve B gives:

$$h_4 = h_3 = 286.5\ \frac{kJ}{kg}$$

$$\left.\begin{aligned} x_5 &= 1 \\ P_5 &= 0.4\ MPa \end{aligned}\right\} h_5 = 426.5\ \frac{kJ}{kg}$$

$$\left.\begin{aligned} x_5 &= 1 \\ P_5 &= 0.4\ MPa \end{aligned}\right\} s_5 = 1.681\ \frac{kJ}{kgK}$$

$$s_5 = s_{6s}$$

$$\left.\begin{aligned} s_{6s} &= 1.681\ \frac{kJ}{kgK} \\ P_6 &= 1.2\ MPa \end{aligned}\right\} h_{6s} = 444.6\ \frac{kJ}{kg}$$

The actual enthalpy of state 6 can be calculated through the isentropic efficiency definition for the compressor A as follows:

$$\eta_{CA} = \frac{h_{6s} - h_5}{h_6 - h_5} = 0.8$$

$$h_6 = \frac{h_{6s} - h_5}{0.8} + h_5 = 449.2\ \frac{kJ}{kg}$$

$$\left.\begin{aligned} x_7 &= 0 \\ P_7 &= 1.2\ MPa \end{aligned}\right\} h_7 = 332.7\ \frac{kJ}{kg}$$

Writing the EBE for expansion valve B gives:

$$h_8 = h_7 = 332.7\ \frac{kJ}{kg}$$

The mass flow rate for cycle A can now be calculated as:

$$\dot{m}_A(h_5 - h_8) = \dot{m}_B(h_2 - h_3)$$

$$\dot{\boldsymbol{m}}_{\boldsymbol{A}} = \dot{m}_B \times \frac{(h_2 - h_3)}{(h_5 - h_8)} = 0.1\ \frac{kg}{s} \times \frac{\left(431\ \frac{kJ}{kg} - 286.5\ \frac{kJ}{kg}\right)}{\left(426.5\ \frac{kJ}{kg} - 332.7\ \frac{kJ}{kg}\right)} = \mathbf{0.154\ \frac{kg}{s}}$$

c) The rate of heat removal from the refrigerated space can be calculated from EBE for evaporator B:

$$\dot{Q}_L = \dot{m}_B(h_1 - h_4)$$

$$\dot{\boldsymbol{Q}}_{\boldsymbol{L}} = 0.1\,\frac{kg}{s} \times \left(410\,\frac{kJ}{kg} - 286.5\,\frac{kJ}{kg}\right) = \mathbf{12.35}\,\frac{\boldsymbol{kJ}}{\boldsymbol{s}}$$

$$\dot{W}_{net,in} = \dot{W}_{CA} + \dot{W}_{CB} = \dot{m}_A(h_6 - h_5) + \dot{m}_B(h_2 - h_1)$$

$$\dot{\boldsymbol{W}}_{\boldsymbol{net,in}} = 0.154\,\frac{kg}{s} \times \left(449.2\,\frac{kJ}{kg} - 426.5\,\frac{kJ}{kg}\right) + 0.1\,\frac{kg}{s} \times \left(431\,\frac{kJ}{kg} - 410\,\frac{kJ}{kg}\right) = \mathbf{5.59}\ \boldsymbol{kW}$$

d) For this cascade cycle, the energetic *COP* is defined as the ratio of the heat removed from the refrigerated space to the net total work input where:

$$\boldsymbol{COP_{en}} = \left(\frac{\dot{Q}_L}{\dot{W}_{net,in}}\right) = \frac{12.35\,\frac{kJ}{s}}{5.59\,\frac{kJ}{s}} = \mathbf{2.21}$$

Moreover, assuming $T_L = \frac{T_1 + T_4}{2} = 262.75\ K$, the exergetic *COP* can be expressed as:

$$COP_{ex} = \left(\frac{\dot{E}x_{Q_L}}{\dot{W}_{net,in}}\right) = \frac{\dot{Q}_L\left(\frac{T_0}{T_L} - 1\right)}{\dot{W}_{net,in}}$$

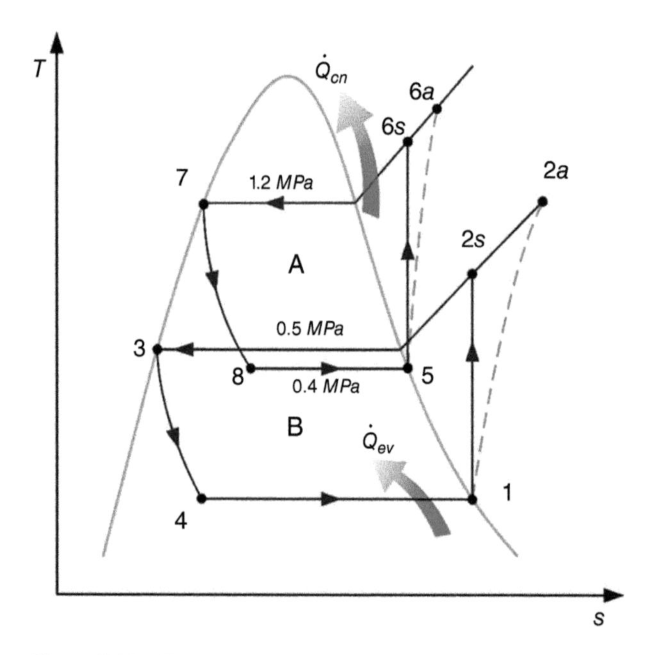

Figure 7.13 T-s diagram for the cascade system solved in Example 7.5.

$$COP_{ex} = \frac{12.35\,\frac{kJ}{s}\left(\frac{298.15\ K}{262.75\ K} - 1\right)}{5.59\,\frac{kJ}{s}} = \mathbf{0.29}$$

The T-s diagram of this example is illustrated in Figure 7.13 with isentropic and actual cases.

7.6 Heat Pumps

Heat pump cycles appear to be same as refrigeration cycles with the difference that the main purpose of running it is heating as opposed to cooling in refrigeration cycles. The principle of using a heat engine in a reverse mode as a heat pump, as shown in Figure 7.14, was initially proposed by Lord Kelvin in the nineteenth century, but it was only in the twentieth century where practical machines came into common use, particularly for refrigeration purposes. As Figure 7.14 shows, for a heat engine heat is supplied from a source to use and generate work output while some of it is rejected to the sink (surroundings). However, for the heat pump this is reversed, as heat is received from a lower temperature sink, upgraded, and transferred back to source with external work input. So, these devices can be defined as heat upgrading devices or heat upgrading cycles.

Furthermore, the basic objective of heat pumping is exactly the same as the objective of refrigeration: the heat is removed at a low temperature and rejected at a higher temperature. Most heat pumps in use today operate on a vapor-compression cycle. In this respect, the components of a vapor-compression heat pump system are exactly the same as those of

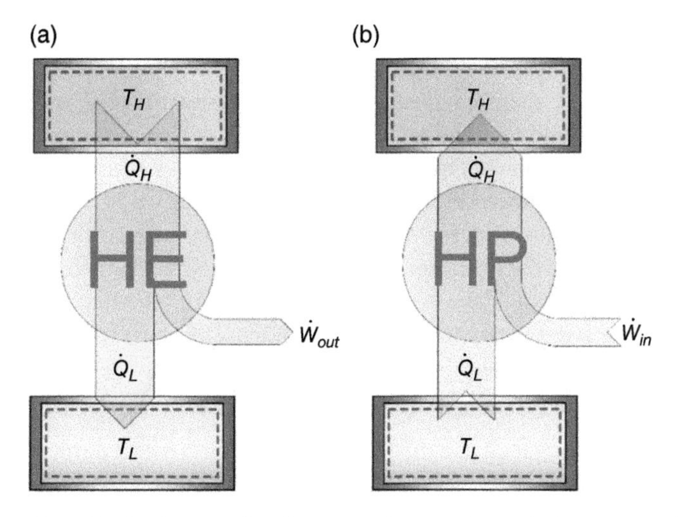

Figure 7.14 A comparison between a heat engine and a heat pump which was originally proposed by Lord Kelvin as a reverse heat engine.

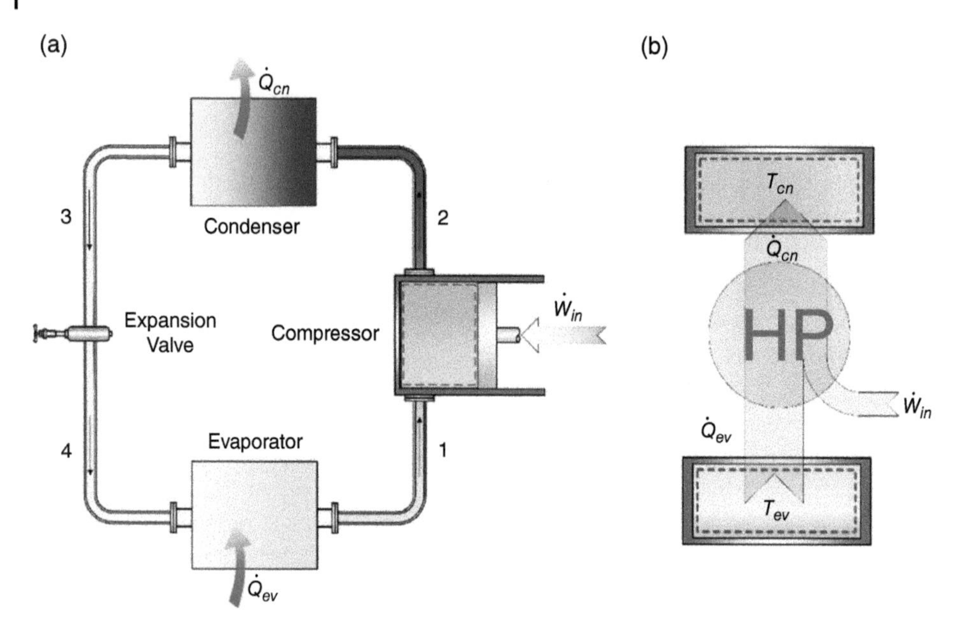

Figure 7.15 (a) Schematic diagram of the conventional heat pump showing the main components, the direction of the mass flow, and the direction of the work interactions. (b) The working principle of the heat pump where the targeted location is the medium at temperature T_{cn}.

a vapor-compression refrigeration system, namely, compressor, condenser, evaporator, and expansion device. The difference between these two systems is that a refrigeration system generally transfers heat from a low temperature to the ambient temperature, whereas a heat pump transfers heat from the ambient temperature to a higher temperature, for example, from a low-temperature heat source (e.g. air, water, or ground) to a higher temperature heat sink (e.g. air, water, or ground). For this reason, heat pump systems are identified as *reverse-cycle refrigeration systems*. One of the most common heat sources for a heat pump is air, although water is also used. Recently, there has been increasing interest in using soil (ground) as the heat source for heating and cooling applications. Ground source heat pumps thus have a good market share, along with low-temperature geothermal sources. It is important to highlight that by utilizing low-temperature resources their performance is highly improved. The conventional mechanical heat pump consisting of only one stage is shown in Figure 7.15, in which the direction of the flows and the direction of the energy interactions are shown clearly. Heat pumps usually have a higher capital cost than other heating systems. However, the heat pump more economic through its operational life, over which it will cover its initial cost and start saving money.

Performance Evaluation

For a heat pump the only difference compared to a refrigeration cycle is the useful output, which is heat for heating purposes rather than cooling load for cooling purposes in refrigeration. It is important to define the performance of any heat pump cycle/system through the energetic-based *COP* (COP_{en}), under the FLT, and the exergetic-based *COP* (COP_{ex}), under the SLT as follows:

$$COP_{HP,en} = \frac{\dot{Q}_{cn}}{\dot{W}_{\text{in}}} \tag{7.3}$$

and

$$COP_{HP,ex} = \frac{\dot{E}x_{\dot{Q}_{cn}}}{\dot{W}_{in}} = \frac{\left(1 - \frac{T_o}{T_{cn}}\right)\dot{Q}_{cn}}{\dot{W}_{in}} \tag{7.4}$$

where subscripts cn and in stand for condenser and input. T_o is the reference environment temperature which is mainly taken as surrounding temperature.

Here, we take $\dot{Q}_H = \dot{Q}_{cn}$, $\dot{Q}_L = \dot{Q}_{ev}$, $T_H = T_{cn}$, and $T_L = T_{ev}$ and, similarly, for the heat pump cycle we write

$$COP_{en.HP} = \frac{\dot{Q}_H}{\dot{W}_{\text{in}}} \tag{7.5}$$

and by using the reversible (Carnot) energy *COP* of a heat pump cycle as it has been derived earlier in the Carnot section of the book:

$$COP_{ex,HP} = \frac{COP_{en,HP}}{COP_{en,HP,rev}} = \frac{\frac{\dot{Q}_H}{\dot{W}_{in}}}{\left(\frac{1}{1 - \frac{T_L}{T_H}}\right)} = \frac{\dot{Q}_H\left(1 - \frac{T_L}{T_H}\right)}{\dot{W}_{in}}$$

One can write the following relationship between the energy and exergy *COPs* for the heat pump cycle as:

$$COP_{ex,HP} = COP_{en.HP}\left(1 - \frac{T_L}{T_H}\right) \tag{7.6}$$

After establishing *COP* relations for both the simple refrigeration cycle and the heat pump cycle, it is possible to relate these *COPs* energetically and exergetically. From the general diagram of both cycles, the EBE for both is the same:

$$\dot{Q}_H = \dot{Q}_L + \dot{W}_{in}$$

Dividing both sides of this equation by $\dot{W}_{in}$, it becomes:

$$\frac{\dot{Q}_H}{\dot{W}_{in}} = \frac{\dot{Q}_L}{\dot{W}_{in}} + \frac{\dot{W}_{in}}{\dot{W}_{in}}$$

Substituting the energy *COP* expressions derived above into it, one gets the following relation:

$$\boldsymbol{COP_{en.HP} = COP_{en.R} + 1} \tag{7.7}$$

In order to find the exergetic *COP* relationship between the two cycles, substitute Eqs. (7.5) and (7.6) into Eq. (7.7):

$$\frac{COP_{ex.HP}}{\left(1 - \frac{T_L}{T_H}\right)} = \frac{COP_{ex.R}}{\left(\frac{T_H}{T_L} - 1\right)} + 1$$

With some mathematical rearrangement, one obtains the following final equation:

$$COP_{ex.HP} = 1 - \frac{T_L}{T_H}(1 - COP_{ex.R}) \quad (7.8)$$

Note that air-source heat pumps generally have *COPs* of 2–4; they deliver 2–4 times more energy than they consume. Water and ground source heat pumps normally have *COPs* of 3–5. The *COP* of air-source heat pumps decreases as the outside temperature drops. When comparing *COPs*, make sure ratings are based on the same outside air temperature. *COPs* for ground- and water-source heat pumps do not vary as much because ground and water temperatures are more constant than air temperatures.

Reversed Operation Heat Pump

The heat pump shown in Figure 7.15 is the single-stage, mechanical heat pump, which means that the heat pump is of the simplest design. However, there are many other different designs of heat pumps. For example, in a cascaded heat pump two conventional heat pumps are connected in such a way that the condenser of the bottoming cycle rejects heat to the boiler of the upper cycle. The reason behind the cascaded heat pump is to avoid having an excessively large pressure difference between the condenser and the boiler of the heat pump. Excessively large pressure difference will result in a high work requirement on the compressor and sometimes a single compressor is not sufficient to produce that increase in the pressure, which will force the designer to have multistage compression, which will have high capital cost plus high operational cost. However, cascading heat pumps will reduce the operational cost and, thus, will result in savings. Another type of heat pump system is shown in Figure 7.16, where the same system can operate in two modes, refrigeration and heat pumping. The trick is to switch the pipe connection between the exit and the inlet of the compressor.

The forthcoming examples show how to analyze heat pump systems through energy and exergy analyses in a similar way to that in which the refrigeration cycle was analyzed. In the first example an overall cycle analysis is performed treating the heat pump as a single unit; in the second example the cycle will be analyzed in more detail.

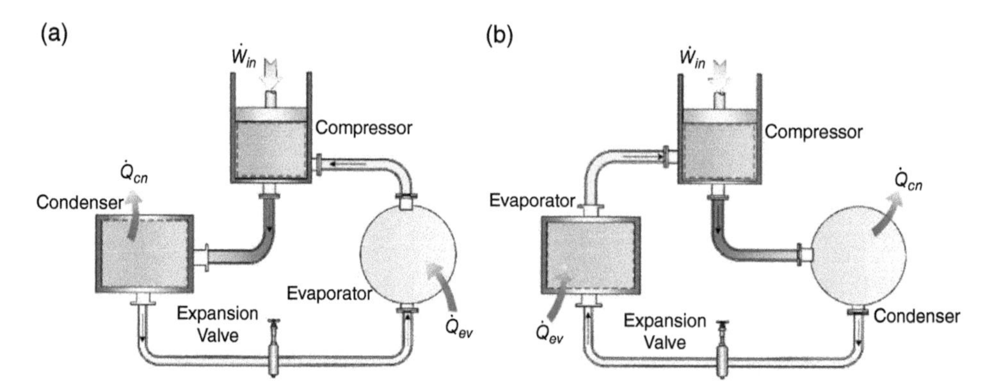

Figure 7.16 Schematic diagram of a system that can be switched between (a) a heat pump operation mode and (b) a refrigeration operation mode.

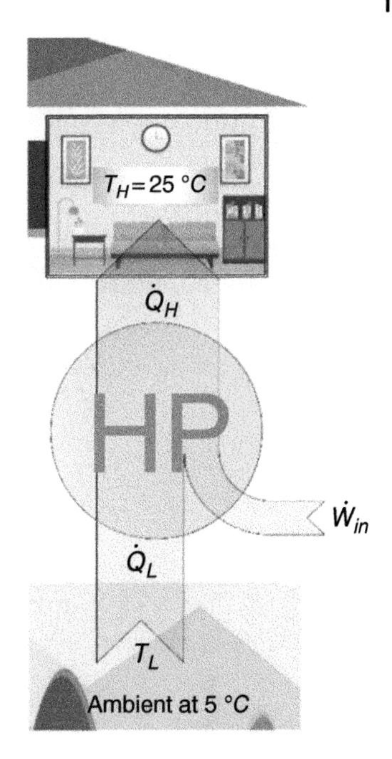

Figure 7.17 Heat pump keeping a house at a constant temperature by supplying the heat the room loses to the surrounding environment as discussed in Example 7.6.

Example 7.6 A heat pump unit, as shown in Figure 7.17, is operated to keep a room at 25 °*C* by taking up heat from an environment at 5 °*C*. The heat pump provides an amount of heat to the room equal to the total heat loss from the room to the environment, which is estimated to be 45 000 *kJ/h*. The power input to the compressor is 4.5 *kW*. (a) Write the mass, energy, entropy, and exergy balance equations for the heat pump as a single unit, (b) determine the rate of heat absorbed from the environment in *kJ/h*, (c) calculate the COP_{en} of the heat pump, (d) find the maximum rate of heat supply to the room for the given power input, (e) calculate the COP_{ex} of the heat pump, and (f) determine the minimum power input for the same heating load and the exergy destruction of the cycle.

Solution

a) Write the mass, energy, entropy, and exergy balance equation for the heat pump as a single unit:

$$\text{MBE} : \dot{m} = constant$$

$$\text{EBE} : \dot{Q}_L + \dot{W}_{in} = \dot{Q}_H$$

$$\text{EnBE} : \dot{Q}_L/T_L + \dot{S}_{gen} = \dot{Q}_H/T_H$$

$$\text{ExBE} : \dot{E}x_{\dot{Q}_L} + \dot{W}_{in} = \dot{E}x_{\dot{Q}_H} + \dot{E}x_d$$

b) Determine the rate of heat absorbed from the environment in *kJ/h*:

From the EBE of the heat pump as a single unit we can find the amount of heat taken from the environment to keep the house at a constant temperature, which is carried out as follows:

$$\dot{Q}_L + \dot{W}_{in} = \dot{Q}_H$$

$$\dot{Q}_L = \dot{Q}_H - \dot{W}_{in} = 45000\frac{kJ}{h} - \left(4.5\ kW \times \frac{3600\frac{kJ}{h}}{1\ kW}\right) = \mathbf{28800\frac{kJ}{h}}$$

c) Calculate the COP_{en} of the heat pump.

$$COP_{HP,en} = \frac{\dot{Q}_H}{\dot{W}_{in}} = \frac{\frac{45000\ kJ/h}{3600\ (kJ/h)/\ kW}}{4.5} = \mathbf{2.78}$$

d) The *COP* of the Carnot cycle operating between the same temperature limits and the maximum rate of heat supply to the room for the given power input are:

$$COP_{HP,C} = \frac{T_H}{T_H - T_L} = \frac{298}{298 - 278} = \mathbf{14.9}$$

The maximum rate of heat supplied to the room for the given power input is calculated through the use of the maximum achievable *COP*, which is the Carnot heat pump *COP* calculated as follows:

$$COP_{carnot,HP} = \frac{\dot{Q}_{H,max}}{\dot{W}_{in}}$$

$$\dot{Q}_{H,max} = COP_{carnot,HP} \times \dot{W}_{in} = 14.9 \times 4.5\ kW \times \frac{3600\ kJ/h}{1\ kW} = \mathbf{241380\ \frac{kJ}{h}}$$

e) The exergetic *COP* (COP_{ex}) of the heat pump cycle is:

$$COP_{HP,ex} = \frac{COP_{en}}{COP_{carnot}} = \frac{2.78}{14.9} = 0.\mathbf{186}$$

f) Determine the minimum power input for the same heating load and the exergy destruction of the cycle.

$$\dot{W}_{min} = \dot{E}x_{\dot{Q}_H} = \dot{Q}_H\left(1 - \frac{T_o}{T_H}\right) = 45000 \times \left(1 - \frac{278}{298}\right) = 3020\ kJ/h$$

$$\dot{E}x_d = \dot{W}_{actual} - \dot{W}_{min} = 4.5 \times 3600 - 3020 = \mathbf{13180\ kJ/h}$$

The COP_{ex} of the heat pump cycle may alternatively be determined from:

$$COP_{HP,ex} = \frac{\dot{W}_{min}}{\dot{W}_{actual}} = \frac{3020}{4.5 \times 3600} = \mathbf{0.186}$$

One can also use the earlier given equation for exergetic *COP* and calculate it for the above given heat pump:

$$COP_{HP,ex} = \frac{\dot{E}x_{\dot{Q}_H}}{\dot{W}_{in}} = \frac{\dot{Q}_H\left(1 - \frac{T}{T_H}\right)}{\dot{W}_{in}} = \frac{45000 \times \left(1 - \frac{278}{298}\right)}{4.5 \times 3600} = \mathbf{0.186}$$

The results appear to be the same as expected.

Example 7.7 A heat pump, as shown in Figure 7.18, using $R134a$ as the refrigerant of the cycle, which is the working fluid of the heat pump, takes up heat from an environment that is at a temperature of $-10\,°C$ to keep the temperature of a room at $22\,°C$. $R134a$ enters the compressor at $140\,kPa$ at a flow rate of $375\,L/min$ as a saturated vapor. The isentropic efficiency of the compressor is 80%. The refrigerant leaves the condenser at $46.3\,°C$ as a saturated liquid. (a) Write the mass, energy, entropy, and exergy balance equations of each component in the cycle, (b) find the rate of heating provided by the system, (c) find the energetic and exergetic *COPs* of the heat pump system, (d) find the exergy destruction in each component of the cycle, and (e) draw the temperature variation of the refrigerant with its entropy throughout the cycle.

Solution

a) Write the mass, energy, entropy, and exergy balance equations of each component in the cycle.

For the compressor:

$$\text{MBE}: \dot{m}_1 = \dot{m}_2$$

$$\text{EBE}: \dot{m}_1 h_1 + \dot{W}_{\text{in}} = \dot{m}_2 h_2$$

$$\text{EnBE}: \dot{m}_1 s_1 + \dot{S}_{gen,1\to 2} = \dot{m}_2 s_2$$

$$\text{ExBE}: \dot{m}_1 ex_1 + \dot{W}_{in} = \dot{m}_2 ex_2 + \dot{E}_{d,1\to 2}$$

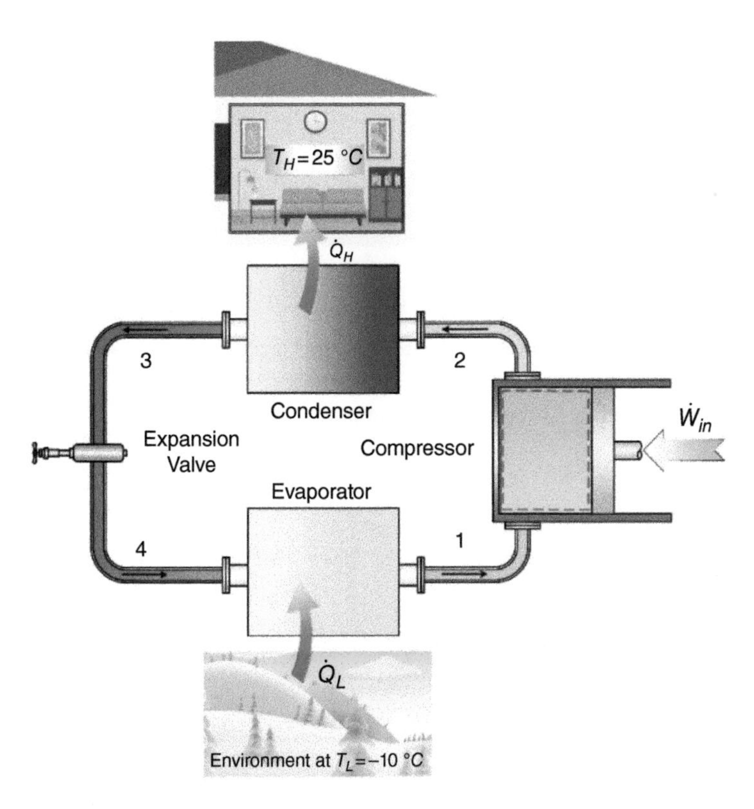

Figure 7.18 Heat pump discussed in Example 7.7.

For the condenser:

$$\text{MBE}: \dot{m}_2 = \dot{m}_3$$

$$\text{EBE}: \dot{m}_2 h_2 = \dot{m}_3 h_3 + \dot{Q}_H$$

$$\text{EnBE}: \dot{m}_2 s_2 + \dot{S}_{gen,2\to3} = \dot{m}_3 s_3 + \frac{\dot{Q}_H}{T_H}$$

$$\text{ExBE}: \dot{m}_2 ex_2 = \dot{m}_3 ex_3 + \dot{E}x_{\dot{Q}_H} + \dot{E}_{d,2\to3}$$

For the expansion valve:

$$\text{MBE}: \dot{m}_3 = \dot{m}_4$$

$$\text{EBE}: \dot{m}_3 h_3 = \dot{m}_4 h_4$$

$$\text{EnBE}: \dot{m}_3 s_3 + \dot{S}_{gen,3\to4} = \dot{m}_4 s_4$$

$$\text{ExBE}: \dot{m}_3 ex_3 = \dot{m}_4 ex_4 + \dot{E}_{d,3\to4}$$

For the evaporator:

$$\text{MBE}: \dot{m}_4 = \dot{m}_1$$

$$\text{EBE}: \dot{m}_4 h_4 + \dot{Q}_L = \dot{m}_1 h_1$$

$$\text{EnBE}: \dot{m}_4 s_4 + \dot{S}_{gen} + \frac{\dot{Q}_L}{T_L} = \dot{m}_1 s_1$$

$$\text{ExBE}: \dot{m}_4 ex_4 + \dot{E}x_{\dot{Q}_L} = \dot{m}_1 ex_1 + \dot{E}_{d,4\to1}$$

The overall cycle balance equation is:

$$\text{MBE}: \dot{m} = constant$$

$$\text{EBE}: \dot{Q}_L + \dot{W}_{\text{in}} = \dot{Q}_H$$

$$\text{EnBE}: \dot{Q}_L/T_L + \dot{S}_{gen} = \dot{Q}_H/T_H$$

$$\text{ExBE}: \dot{E}x_{\dot{Q}_L} + \dot{W}_{comp} = \dot{E}x_{d,ov} + \dot{E}x_{\dot{Q}_H}$$

b) Find the rate of heating provided by the system.

Before starting to solve the requirements of heat rates, the properties of $R134a$ are found either using property tables (such as Appendix B-3b) or through software with thermophysical properties stored in its database. These properties are:

$$\left.\begin{array}{l} P_1 = 140\ kPa \\ x_1 = 1 \end{array}\right\} \begin{array}{l} h_1 = 239.17\ kJ/kg \\ s_1 = 0.9446\ kJ/(kgK) \\ v_1 = 0.1402\ m^3/kg \end{array}$$

$$\left.\begin{array}{l} P_3 = P_{sat}@46.3\,^{\circ}C \\ P_3 = P_2 = 1200\ kPa \\ s_{2s} = s_1 = 0.9446\ kJ/kgK \end{array}\right\} h_{2s} = 284.09\ kJ/kg$$

$$\left.\begin{array}{l} P_3 = 1200\ kPa \\ x_3 = 0 \end{array}\right\} \begin{array}{l} h_3 = 117.77\ kJ/kg \\ s_3 = 0.4244\ kJ/kgK \end{array}$$

From the MBE and EBE of the expansion valve we can conclude that:

$$h_3 = h_4 = 117.77\ kJ/kg$$

$$\left.\begin{aligned} P_4 &= 140\ kPa \\ h_4 &= 117.77\ kJ/kg \end{aligned}\right\} s_4 = 0.4674\ kJ/kgK$$

Then by using the isentropic efficiency of the compressor to find the actual enthalpy exiting the compressor (h_2):

$$\eta_{is,c} = \frac{h_{2s} - h_1}{h_2 - h_1}$$

$$0.80 = \frac{284.09 - 239.17}{h_2 - 239.17} \rightarrow h_2 = 295.32\ \frac{kJ}{kg}$$

where now we can use the pressure leaving the compressor and the actual enthalpy to find the actual entropy as follows:

$$\left.\begin{aligned} P_2 &= 1200\ kPa \\ h_2 &= 295.32\ kJkg \end{aligned}\right\} s_2 = 0.9783\ kJ/kgK$$

The mass flow rate of the refrigerant is:

$$\dot{m} = \frac{\dot{V}_1}{v_1} = \frac{375 \frac{L}{min} \times \frac{1\ min}{60\ s} \times \frac{1\ m^3}{1000\ L}}{0.1402\ \frac{m^3}{kg}} = \mathbf{0.04458\ kg/s}$$

From the EBE of the boiler or evaporator:

$$\dot{m}_4 h_4 + \dot{Q}_{evaporator} = \dot{m}_1 h_1$$

$$\dot{Q}_{evaporator} = \dot{m}_1 h_1 - \dot{m}_4 h_4 = 0.04458\ \frac{kg}{s} \times (239.17 - 117.77) \frac{kJ}{kg} = \mathbf{5.41\ kW}$$

c) Find the energetic and exergetic *COPs* of the heat pump system.
The *COP* of the heat pump cycle is calculated as follows:

$$COP_{HP} = \frac{\dot{Q}_H}{\dot{W}_{in}}$$

We need the work consumed by the compressor and the heat supplied to the room. These are found by solving for $\dot{W}_{in}$ in the EBE of the compressor and solving for $\dot{Q}_H$, which is the heat rejected by the condenser, in the EBE of the condenser as follows:

$$\dot{m}_1 h_1 + \dot{W}_{comp} = \dot{m}_2 h_2$$

$$\dot{W}_{comp} = \dot{m}_2 h_2 - \dot{m}_1 h_1 = \dot{m}(h_2 - h_1) = 0.04458\ \frac{kg}{s} \times (295.32 - 239.17) \frac{kJ}{kg} = \mathbf{2.50\ kW}$$

$$\dot{m}_2 h_2 = \dot{m}_3 h_3 + \dot{Q}_{cond}$$

$$\dot{Q}_{cond} = \dot{m}_2 h_2 - \dot{m}_3 h_3 = 0.04458\ \frac{kg}{s} \times (295.32 - 117.77) \frac{kJ}{kg} = \mathbf{7.92\ kW}$$

$$COP_{HP} = \frac{\dot{Q}_H}{\dot{W}_{in}} = \frac{7.92}{2.50} = \mathbf{3.16}$$

d) Find the exergy destruction rate of each component in the cycle.

Noting that the dead-state temperature is $T_o = T_H = 295\,K$, the exergy destruction in each component of the cycle is determined as follows through the use of the entropy generation rate of each component. When each is found then the exergy destruction rate is calculated as follows:

Compressor:

$$\dot{m}_1 s_1 + \dot{S}_{gen,1\rightarrow2} = \dot{m}_2 s_2$$

$$\dot{S}_{gen,1\rightarrow2} = \dot{m}_2 s_2 - \dot{m}_1 s_1 = \dot{m}(s_2 - s_1) = 0.04458\,\frac{kg}{s} \times (0.9783 - 0.9446)\frac{kJ}{kgK} = 0.001502\,\frac{kW}{K}$$

$$\dot{E}x_{d,1\rightarrow2} = T_o \dot{S}_{gen,1\rightarrow2} = 295 \times 0.001502 = \mathbf{0.4432\ kW}$$

Condenser:

$$\dot{m}_2 s_2 + \dot{S}_{gen,2\rightarrow3} = \dot{m}_3 s_3 + \frac{\dot{Q}_{cond}}{T_H}$$

$$\dot{S}_{gen,2\rightarrow3} = \dot{m}_3 s_3 + \frac{\dot{Q}_{cond}}{T_H} - \dot{m}_2 s_2 = 0.04458\,\frac{kg}{s} \times 0.4244\,\frac{kJ}{kgK} + \frac{7.92\ kW}{295\ K} - 0.04458\,\frac{kg}{s} \times 0.9783\,\frac{kJ}{kgK}$$

$$\dot{S}_{gen,2\rightarrow3} = \mathbf{0.002138\,\frac{kW}{K}}$$

$$\dot{E}x_{d,2\rightarrow3} = T_o \dot{S}_{gen,2\rightarrow3} = 295 \times 0.002138 = \mathbf{0.6308\ kW}$$

Expansion valve:

$$\dot{m}_3 s_3 + \dot{S}_{gen,3\rightarrow4} = \dot{m}_4 s_4$$

$$\dot{S}_{gen,3\rightarrow4} = \dot{m}_4 s_4 - \dot{m}_3 s_3 = \dot{m}(s_4 - s_3) = 0.04458\,\frac{kg}{s} \times (0.4674 - 0.4244)\frac{kJ}{kgK}$$

$$\dot{S}_{gen,3\rightarrow4} = \mathbf{0.001916\ kW/K}$$

$$\dot{E}x_{d,3\rightarrow4} = T_o \dot{S}_{gen,3\rightarrow4} = 295 \times 0.001916 = \mathbf{0.5651\ kW}$$

Evaporator:

$$\dot{m}_4 s_4 + \dot{S}_{gen,4\rightarrow1} + \frac{\dot{Q}_{evaporator}}{T_L} = \dot{m}_1 s_1$$

$$\dot{S}_{gen,4\rightarrow1} = \dot{m}_1 s_1 - \dot{m}_4 s_4 - \frac{\dot{Q}_{evaporator}}{T_L}$$

$$\dot{S}_{gen,4\rightarrow1} = 0.04458\,\frac{kg}{s} \times (0.9446 - 0.4674)\frac{kJ}{kgK} - \frac{5.41\ kW}{263\ K} = \mathbf{0.0006964\ kW/K}$$

$$\dot{E}x_{d,4\rightarrow1} = T_o \dot{S}_{gen,4\rightarrow1} = 295 \times 0.0006964 = \mathbf{0.2054\ kW}$$

e) The temperature versus entropy diagram of the heat pump is plotted in Figure 7.19.

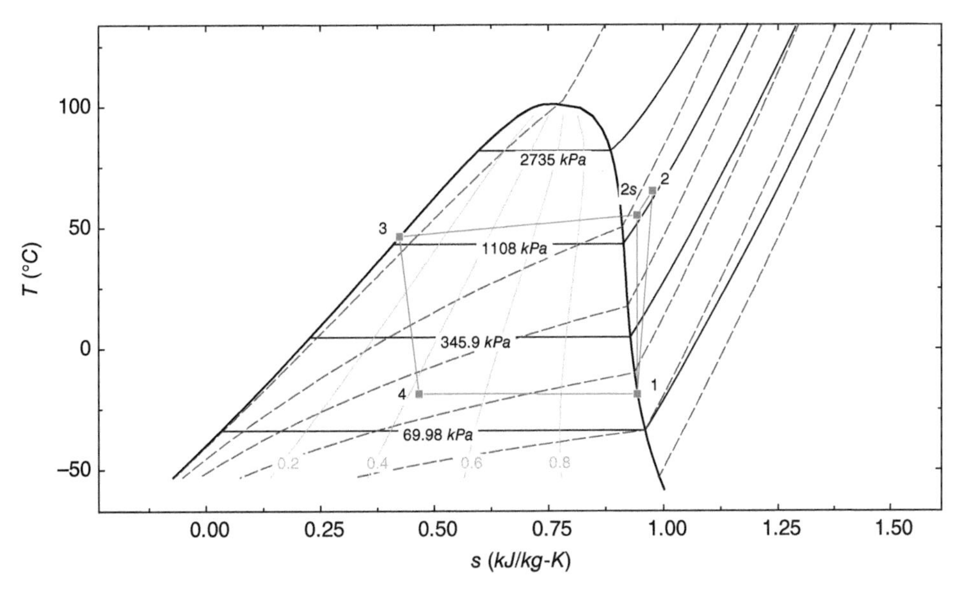

Figure 7.19 Temperature-entropy diagram of the vapor-compression heat pump cycle considered in Example 7.7.

Example 7.8 A heat pump system, as shown in Figure 7.20, that works with R134a is employed to maintain the temperature of a house at 23 °*C* where 50 *kW is* needed. The evaporator uses underground water in saturated liquid condition as the heat source. Water enters the evaporator at 65 °*C* and leaves at 45 °*C*. The saturated liquid refrigerant with a flow rate of 0.4 *kg/s* leaves the evaporator at 22 °*C* and 608.27 *kPa*. The compressor pressurizes the refrigerant to 1600 *kPa* with 81% efficiency. (a) Write the mass, energy, entropy, and exergy balance equations for each component, (b) find values for each state point, (c) determine the mass flow rate of the underground water, (d) calculate the power input to the compressor, (e) calculate the energetic and exergetic *COPs* for the system. Assume that the changes in kinetic and potential energies are negligible and steady-state operating conditions exist. $T_{cn} = 50\,°C$ and $T_o = 10\,°C$

Solution

a) Write the mass, energy, entropy, and exergy balance equation for each component in the cycle.

For the evaporator,

$$\text{MBE}: \dot{m}_{w1} = \dot{m}_{w2}, \dot{m}_4 = \dot{m}_1 = \dot{m}$$

$$\text{EBE}: \dot{m}_4 h_4 + \dot{m}_{w1} h_{w1} = \dot{m}_1 h_1 + \dot{m}_{w2} h_{w2}$$

$$\text{EnBE}: \dot{m}_4 s_4 + \dot{m}_{w1} s_{w1} + \dot{S}_{gen,4\to1} = \dot{m}_1 s_1 + \dot{m}_{w2} s_{w2}$$

$$\text{ExBE}: \dot{m}_4 ex_4 + \dot{m}_{w1} ex_{w1} = \dot{m}_1 ex_1 + \dot{m}_{w2} ex_{w2} + \dot{E}x_{d,4\to1}$$

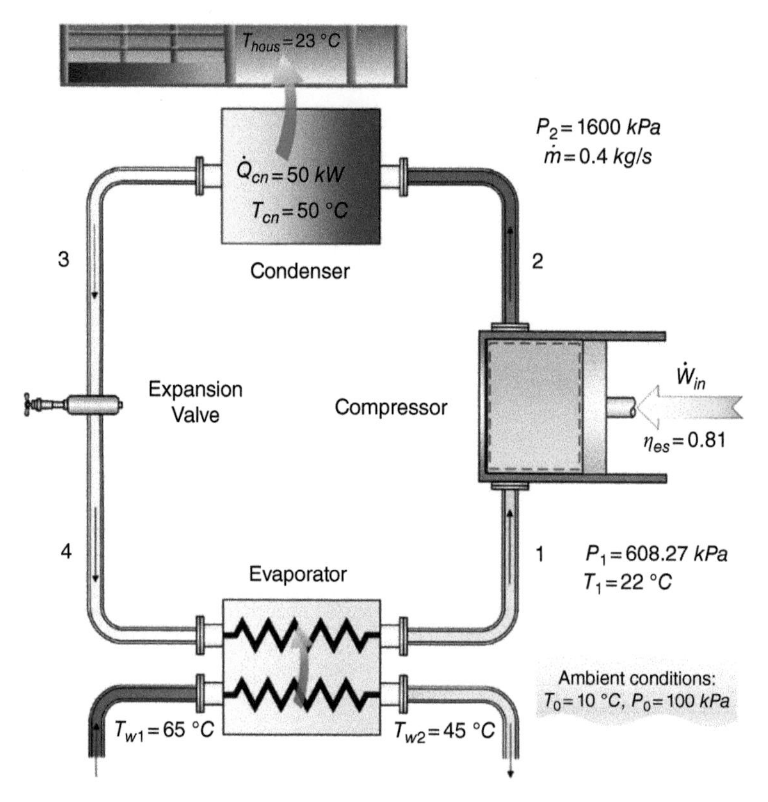

Figure 7.20 Graphical illustration of the refrigeration unit discussed in Example 7.8.

For the compressor,

$$\text{MBE}: \dot{m}_1 = \dot{m}_2 = \dot{m}$$

$$\text{EBE}: \dot{m}_1 h_1 + \dot{W}_{\text{in}} = \dot{m}_2 h_2$$

$$\text{EnBE}: \dot{m}_1 s_1 + \dot{S}_{gen,1\to 2} = \dot{m}_2 s_2$$

$$\text{ExBE}: \dot{m}_1 ex_1 + \dot{W}_{\text{in}} = \dot{m}_2 ex_2 + \dot{E}x_{d,1\to 2}$$

For the condenser,

$$\text{MBE}: \dot{m}_2 = \dot{m}_3 = \dot{m}$$

$$\text{EBE}: \dot{m}_2 h_2 = \dot{m}_3 h_3 + \dot{Q}_{cn}$$

$$\text{EnBE}: \dot{m}_2 s_2 + \dot{S}_{gen,2\to 3} = \dot{m}_3 s_3 + \frac{\dot{Q}_{cn}}{T_{cn}}$$

$$\text{ExBE}: \dot{m}_2 ex_2 = \dot{m}_3 ex_3 + \dot{EX}_{\dot{Q}_{cn}} + \dot{E}x_{d,2\to 3}$$

For the expansion valve,

$$\text{MBE}: \dot{m}_3 = \dot{m}_4 = \dot{m}$$

$$\text{EBE}: \dot{m}_3h_3 = \dot{m}_4h_4$$

$$\text{EnBE}: \dot{m}_3s_3 + \dot{S}_{gen,3\to4} = \dot{m}_4s_4$$

$$\text{ExBE}: \dot{m}_3ex_3 = \dot{m}_4ex_4 + \dot{E}x_{d,3\to4}$$

The overall cycle balance equation is:

$$\text{MBE}: \dot{m}_{w1} = \dot{m}_{w2},\ \ \dot{m}_1 = \dot{m}_2 = \dot{m}_3 = \dot{m}_4$$

$$\text{EBE}: \dot{W}_{\text{in}} = \dot{Q}_{cn}$$

$$\text{EnBE}: \dot{S}_{gen,ov} = \frac{\dot{Q}_{cn}}{T_{cn}}$$

$$\text{ExBE}: \dot{W}_{\text{in}} = \dot{E}X_{\dot{Q}_{cn}} + \dot{E}x_{d,ov}$$

where subscript ov stands for overall.

b) Find values for each state point.

From properties-based software (such as EES) or property tables (such as Appendix B-3a), one obtains:

$$\left.\begin{matrix} T_1 = 22\,°C \\ x_1 = 1 \end{matrix}\right\} \begin{matrix} h_1 = 262.6\ kJ/kg \\ s_1 = 0.922\ kJ/kgK \end{matrix}$$

$$\left.\begin{matrix} T_1 = 22\,°C \\ x_1 = 1 \end{matrix}\right\} P_1 = 608.27\ kPa$$

$$s_1 = s_{2s} = 0.922\ kJ/kgK$$

$$\left.\begin{matrix} P_2 = 1600\ kPa \\ s_{2s} = s_1 \end{matrix}\right\} h_{2s} = 282.4\ kJ/kg$$

The actual enthalpy of state 2 can be calculated through the isentropic efficiency definition for the compressor as follows:

$$\eta_c = \frac{h_{2s} - h_1}{h_2 - h_1} = 0.81$$

$$h_2 = \frac{h_{2s} - h_1}{0.81} + h_1 = \frac{282.4\ kJ/kg - 262.6\ kJ/kg}{0.81} + 262.6\ kJ/kg = \mathbf{287.04\ kJ/kg}$$

$$\mathbf{h_2} = 287.04\ kJ/kg$$

$$\left.\begin{matrix} P_2 = 1600\ kPa \\ h_2 = 287.04\ kJ/kg \end{matrix}\right\} \begin{matrix} T_2 = 65\ °C \\ s_2 = 0.935\ kJ/kgK \end{matrix}$$

Writing the EBE for the condenser gives:

$$\dot{m}_2h_2 = \dot{m}_3h_3 + \dot{Q}_{cn}$$

$$h_3 = \frac{-\dot{Q}_{cn} + \dot{m}_2h_2}{\dot{m}_3}$$

$$h_3 = \frac{(0.4\ kg/s \times 287.04\ kJ/kg - 50\ kW)}{0.4\ kg/s}$$

$$\mathbf{h_3 = 162.04\ kJ/kg}$$

$$\left.\begin{array}{l} P_3 = 1600\ kPa \\ h_3 = 162.04\ kJ/kg \end{array}\right\} \begin{array}{l} T_3 = 57.88\,°C \\ s_3 = 0.558\ kJ/kgK \end{array}$$

$$\left.\begin{array}{l} P_4 = 608.27\ kPa \\ h_4 = h_3 \end{array}\right\} \begin{array}{l} T_4 = 22\,°C \\ s_4 = 0.58\ kJ/kgK \end{array}$$

$$\left.\begin{array}{r} T_{w1} = 65\,°C \\ x_{w1} = 0\ \text{(saturated liquid)} \end{array}\right\} \begin{array}{l} h_{w1} = 272.1\ kJ/kg \\ s_{w1} = 0.894\ kJ/kgK \end{array}$$

$$\left.\begin{array}{r} T_{w2} = 45\,°C \\ x_{w2} = 0\ \text{(saturated liquid)} \end{array}\right\} \begin{array}{l} h_{w2} = 188.4\ kJ/kg \\ s_{w2} = 0.64\ kJ/kgK \end{array}$$

c) Determine the mass flow rate of the underground water.
Writing the MBE and EBE for the evaporator gives:

$$\dot{m}_w = \dot{m}_{w1} = \dot{m}_{w2}$$

$$\dot{m}_4 h_4 + \dot{m}_w h_{w1} = \dot{m}_1 h_1 + \dot{m}_w h_{w2}$$

$$\dot{m}_w = \frac{\dot{m}_1 h_1 - \dot{m}_4 h_4}{h_{w1} - h_{w2}} = \frac{(0.4\ kg/s \times 262.6\ kJ/kg) - (0.4\ kg/s \times 162.04\ kJ/kg)}{(272.1\ kJ/kg - 188.4\ kJ/kg)}$$

$$\mathbf{\dot{m}_w = 0.48\ kg/s}$$

d) The total net work input can be calculated through EBE for compressor:

$$\dot{m}_1 h_1 + \dot{W}_{\text{in}} = \dot{m}_2 h_2$$

$$\dot{W}_{\text{in}} = \dot{m}_2 h_2 - \dot{m}_1 h_1 = 0.4\ kg/s \times 287.04\ kJ/kg - 0.4\ kg/s \times 262.6\ kJ/kg$$

$$\mathbf{\dot{W}_{in} = 9.76\ kW}$$

e) Calculate the energetic and exergetic *COPs* for the system.
For this heat pump, energetic *COP*, as a measure for energetic performance of the system, can be defined as the ratio of the rejected heat from refrigerant to the net total work input:

$$COP_{en} = \frac{\dot{Q}_{cn}}{\dot{W}_{\text{in}}}$$

$$\boldsymbol{COP_{en}} = \frac{50\ kW}{9.76\ kW} = 5.12$$

The exergetic *COP* can be written as:

$$COP_{ex} = \frac{\dot{EX}_{\dot{Q}_{cn}}}{\dot{W}_{in}} = \frac{\dot{Q}_{cn}\left(1 - \dfrac{T_o}{T_{cn}}\right)}{\dot{W}_{in}}$$

$$\boldsymbol{COP_{ex}} = \frac{50\ kW \times \left(1 - \dfrac{283.15\ K}{323.15\ K}\right)}{9.76\ kW} = \mathbf{0.63}$$

Example 7.9 A reverse heat pump system, as shown in Figure 7.21, that works with *R134a* is used for both cooling and heating purposes. The compressor works with 86% efficiency and pressurizes the refrigerant from 240 to 1200 kPa. The environment temperature

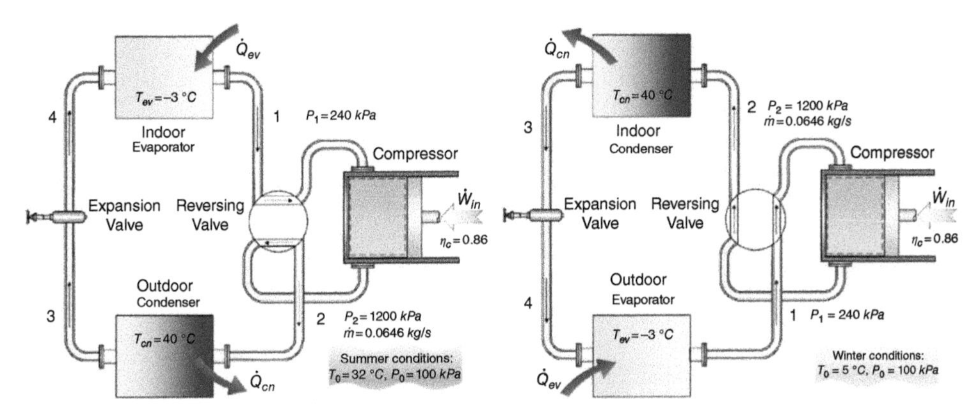

Figure 7.21 Graphical illustration of the heat pump with reversing valve unit discussed in Example 7.9.

is assumed as 32 °*C* in summer and 5 °*C* in winter when the desired room temperature is 23 °*C* for both seasons. The summer setting is effective for five months and the winter setting for seven months. The flow rate of the refrigerant is 0.0646 *kg/s*. (a) Write the mass, energy, entropy, and exergy balance equations for each component, (b) find values for each state point, (c) calculate the power input to the compressor and the heat transfer rates for condenser and evaporator, and (d) calculate the energetic and exergetic *COPs* for both settings and compare and calculate their weighted averages for a year (COP_{en}, COP_{ex}). Assume that the changes in kinetic and potential energies are negligible and steady-state operating conditions exist. $T_{cn} = 40\,°C$, $T_{ev} = -3\,°C$

Solution

a) Write the mass, energy, entropy, and exergy balance equations for each component.

For the evaporator:

$$\text{MBE}: \dot{m}_4 = \dot{m}_1$$

$$\text{EBE}: \dot{m}_4 h_4 + \dot{Q}_{ev} = \dot{m}_1 h_1$$

$$\text{EnBE}: \dot{m}_4 s_4 + \dot{S}_{gen,4\to1} + \frac{\dot{Q}_{ev}}{T_{ev}} = \dot{m}_1 s_1$$

$$\text{ExBE}: \dot{m}_4 ex_4 + \dot{E}x_{\dot{Q}_{ev}} = \dot{m}_1 ex_1 + \dot{E}x_{d,4\to1}$$

For the compressor:

$$\text{MBE}: \dot{m}_1 = \dot{m}_2 = \dot{m}$$

$$\text{EBE}: \dot{m}_1 h_1 + \dot{W}_{in} = \dot{m}_2 h_2$$

$$\text{EnBE}: \dot{m}_1 s_1 + \dot{S}_{gen,1\to2} = \dot{m}_2 s_2$$

$$\text{ExBE}: \dot{m}_1 ex_1 + \dot{W}_{in} = \dot{m}_2 ex_2 + \dot{E}x_{d,1\to2}$$

For the condenser:

$$\text{MBE}: \dot{m}_2 = \dot{m}_3 = \dot{m}$$

$$\text{EBE}: \dot{m}_2 h_2 = \dot{m}_3 h_3 + \dot{Q}_{cn}$$

$$\text{EnBE}: \dot{m}_2 s_2 + \dot{S}_{gen,2\to3} = \dot{m}_3 s_3 + \frac{\dot{Q}_{cn}}{T_{cn}}$$

$$\text{ExBE}: \dot{m}_2 ex_2 = \dot{m}_3 ex_3 + \dot{E}x_{d,2\to3} + \dot{E}x_{\dot{Q}_{cn}}$$

For the expansion valve:

$$\text{MBE}: \dot{m}_3 = \dot{m}_4 = \dot{m}$$

$$\text{EBE}: \dot{m}_3 h_3 = \dot{m}_4 h_4$$

$$\text{EnBE}: \dot{m}_3 s_3 + \dot{S}_{gen,3\to4} = \dot{m}_4 s_4$$

$$\text{ExBE}: \dot{m}_3 ex_3 = \dot{m}_4 ex_4 + \dot{E}x_{d,3\to4}$$

The overall cycle balance equation is:

$$\text{MBE}: \dot{m}_{w1} = \dot{m}_{w2} \text{ and within the cycle } \dot{m}_1 = \dot{m}_2 = \dot{m}_3 = \dot{m}_4$$

$$\text{EBE}: \dot{W}_{in} + \dot{Q}_{ev} = \dot{Q}_{cn}$$

$$\text{EnBE}: \dot{S}_{gen,ov} + \dot{Q}_{ev}/T_{ev} = \frac{\dot{Q}_{cn}}{T_{cn}}$$

$$\text{ExBE}: \dot{W}_{in} + \dot{E}x_{\dot{Q}_{ev}} = \dot{E}x_{\dot{Q}_{cn}} + \dot{E}x_{d,ov}$$

where subscript *ov* stands for overall.

b) Find values for each state point.

From the properties-based software (such as EES) or property tables (such as Appendix B-3b), one obtains:

$$\left.\begin{array}{r} P_1 = 240\ kPa \\ x_1 = 1 \end{array}\right\} \begin{array}{l} h_1 = 247.32\ kJ/kg \\ s_1 = 0.9347\ kJ/kgK \end{array}$$

$$s_1 = s_{2s} = 0.9347\ kJ/kgK$$

$$\left.\begin{array}{r} P_2 = 1200\ kPa \\ s_{2s} = s_1 \end{array}\right\} h_{2s} = 280.9\ kJ/kg$$

The actual enthalpy of state 2 can be calculated through the isentropic efficiency definition for the compressor as follows

$$\eta_c = \frac{h_{2s} - h_1}{h_2 - h_1} = 0.86$$

$$h_2 = \frac{h_{2s} - h_1}{0.86} + h_1 = \frac{280.9\ kJ/kg - 247.32\ kJ/kg}{0.86} + 247.32\ kJ/kg = \mathbf{286.37\ kJ/kg}$$

$$\mathbf{h_2} = 286.37\ kJ/kg$$

$$\left.\begin{array}{r} P_2 = 1200\ kPa \\ h_2 = 286.37\ kJ/kg \end{array}\right\} \begin{array}{l} T_2 = 57.2\,^\circ C \\ s_2 = 0.9515\ kJ/kgK \end{array}$$

$$\left.\begin{array}{r} P_3 = 1200\ kPa \\ x_3 = 0 \end{array}\right\} \begin{array}{l} h_3 = 117.8\ kJ/kg \\ s_3 = 0.4244\ kJ/kgK \end{array}$$

Writing the EBE for the condenser gives:

$$\dot{m}_2 h_2 = \dot{m}_3 h_3 + \dot{Q}_{cn}$$

$$\dot{Q}_{cn} = 0.0646\ kg/s \times 286.37\ kJ/kg - 0.0646\ kg/s \times 117.8\ kJ/kg$$

$$\boldsymbol{\dot{Q}_{cn} = 10.89\ kW}$$

$$\left.\begin{array}{l} P_4 = P_1 \\ h_4 = h_3 \end{array}\right\} s_4 = 0.4511\ kJ/kgK$$

c) Writing the EBE for the evaporator gives:

$$\dot{m}_4 h_4 + \dot{Q}_{ev} = \dot{m}_1 h_1$$

$$\dot{Q}_{ev} = \dot{m}_1 h_1 - \dot{m}_4 h_4 = (0.0646\ kg/s \times 247.32\ kJ/kg) - (0.0646\ kg/s \times 117.8\ kJ/kg)$$

$$\boldsymbol{\dot{Q}_{ev} = 8.37\ kW}$$

d) The total net work input can be calculated through the EBE for the compressor:

$$\dot{m}_1 h_1 + \dot{W}_{in} = \dot{m}_2 h_2$$

$$\dot{W}_{in} = \dot{m}_2 h_2 - \dot{m}_1 h_1 = 0.0646\ kg/s \times 286.37\ kJ/kg - 0.0646\ kg/s \times 247.32\ kJ/kg$$

$$\boldsymbol{\dot{W}_{in} = 2.52\ kW}$$

e) For the winter setting, the energetic *COP*, as a measure for the energetic performance of the system, can be defined as the ratio of the heat rejected from the refrigerant to the net total work input:

$$COP_{en,winter} = \frac{\dot{Q}_{cn}}{\dot{W}_{in}}$$

$$\boldsymbol{COP_{en,winter}} = \frac{10.89\ kW}{2.52\ kW} = \boldsymbol{4.32}$$

The exergetic *COP* can be written as:

$$COP_{ex,winter} = \frac{\dot{E}x_{\dot{Q}_{cn}}}{\dot{W}_{in}} = \frac{\dot{Q}_{cn}\left(1 - \frac{T_{o,winter}}{T_{cn}}\right)}{\dot{W}_{in}}$$

$$\boldsymbol{COP_{ex,winter}} = \frac{10.89\ kW \times \left(1 - \frac{278.15\ K}{313.15\ K}\right)}{2.52\ kW} = \boldsymbol{0.48}$$

For the summer setting, the energetic *COP*, as a measure for the energetic performance of the system can be defined as the ratio of the heat rejected from the refrigerant to the net total work input:

$$COP_{en,summer} = \frac{\dot{Q}_{ev}}{\dot{W}_{in}}$$

$$\boldsymbol{COP_{en,summer}} = \frac{8.37\ kW}{2.52\ kW} = \boldsymbol{3.32}$$

The exergetic *COP* can be written as:

$$COP_{ex,summer} = \frac{\dot{E}x_{\dot{Q}_{ev}}}{\dot{W}_{in}} = \frac{\dot{Q}_{ev}\left(\frac{T_{o,summer}}{T_{ev}} - 1\right)}{\dot{W}_{in}}$$

$$\boldsymbol{COP_{ex,summer}} = \frac{8.37\ kW \times \left(\frac{305.15\ K}{270.15\ K} - 1\right)}{2.52\ kW} = \mathbf{0.43}$$

For the weighted average of energetic *COP*, durations during the year will are considered in the following:

$$COP_{en,average} = \frac{T_{summer} \times COP_{en,summer} + T_{winter} \times COP_{en,winter}}{T_{total}}$$

$$\boldsymbol{COP_{en,average}} = \frac{5\ months \times 3.32 + 7\ months \times 4.32}{12\ months} = \mathbf{3.9}$$

The weighted average exergetic *COP* can be written as:

$$COP_{ex,average} = \frac{T_{summer} \times COP_{ex,summer} + T_{winter} \times COP_{ex,winter}}{T_{total}}$$

$$\boldsymbol{COP_{ex,average}} = \frac{5\ months \times 0.43 + 7\ months \times 0.48}{12\ months} = \mathbf{0.46}$$

7.7 Absorption Refrigeration Cycles

Absorption refrigeration cycles are known as thermal refrigeration cycles since they use heat as the input energy, rather than electricity. The principle of the absorption refrigeration cycle is not new and has been known since the early 1800s; the first absorption refrigeration machine was invented by French engineer Ferdinand P.E. Carre in 1860 as an intermittent crude ammonia absorption apparatus based on the chemical affinity of ammonia for water, basically using ammonia as the refrigerant and water as absorbent. These systems employ two working fluids, namely a refrigerant and an absorbing agent. After Carre's unit produced up to 100 *kg* of ice per hour, many large systems were thereafter manufactured for various sectors, ranging from the chemical to petroleum industries. Such systems from small to large scale applications continued their popularity until early 1900s. After vapor-compression refrigeration systems found large acceptance in the sectors, they lost their momentum due to the mechanical refrigeration (vapor compression based). After 1970s, due to the oil crisis, there was increasing interest in using absorption refrigeration systems driven by renewable heat (particularly solar and geothermal); the earlier absorption refrigeration systems were energized by steam or hot water generated from natural gas, oil-fired boilers, and electrical heaters. There is today still some interest in such systems as part of integrated energy systems where there is heat available for cooling purposes. These offer some advantages such as quite operation, high reliability, suitability for renewables, easy maintenance, etc.

Furthermore, it is considered an interesting refrigeration system that is based on a different mode of operation and has become economically attractive for those cases where there is an inexpensive source of thermal energy coming from a source with a temperature of 100–200 °*C*. These temperature sources, mentioned earlier, can come from various sources, such as geothermal, solar, and waste heat from any plants, including cogeneration systems.

An absorption refrigeration cycle uses a heat source in place of power input to a compressor (Figure 7.19). The system involves a power input to a pump and heat input in a generator. Therefore, the *COP* of the system can be expressed as:

$$COP_{AR} = \frac{\dot{Q}_L}{\dot{W}_P + \dot{Q}_{ge}} \tag{7.9}$$

where $\dot{Q}_L$ is the rate of cooling in the evaporator, $\dot{Q}_{ge}$ is the rate of heat supplied to the generator, and $\dot{W}_P$ is the power input to the pump. The power input is usually neglected as its value is small compared to heat input. The second-law efficiency may be expressed as:

$$COP_{ex,AR} = \frac{\dot{E}x_{\dot{Q}_L}}{\dot{W}_P + \dot{E}x_{\dot{Q}_{ge}}} = -\frac{\dot{Q}_L\left(1 - \frac{T_o}{T_L}\right)}{\dot{W}_P + \dot{Q}_{ge}\left(1 - \frac{T_o}{T_s}\right)} \tag{7.10}$$

where T_o, T_L, and T_s are the temperatures of dead state, cooled space, and heat source, respectively.

In order to develop a relation for the maximum (reversible) *COP* of an absorption refrigeration system, we consider a reversible heat engine and a reversible refrigerator as shown in Figure 7.22. Heat is absorbed from a source at T_s by a reversible heat engine and the waste heat is rejected to an environment T_o. Work output from the heat engine is used as the work input in the reversible refrigerator, which keeps a refrigerated space at T_L while rejecting heat to the environment at T_o.

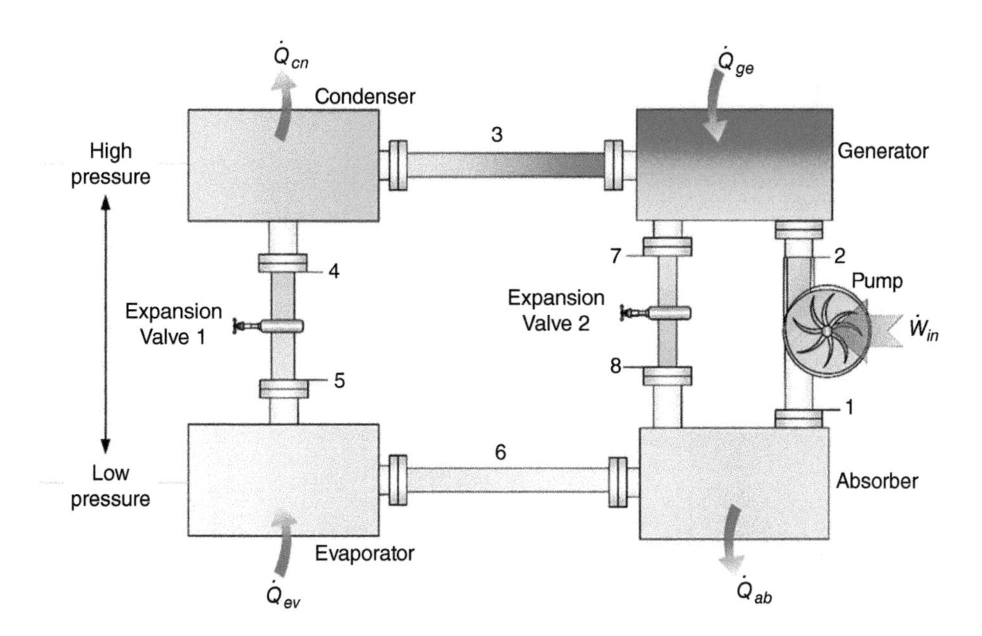

Figure 7.22 A basic absorption refrigeration system.

Starting first with, as mentioned earlier, the definition of the *COP* of the absorption refrigeration, if we assume that the work consumed by the pump is very small compared to the heat supplied to the generator then the *COP* of the absorption refrigerator can be written as:

$$COP_{AR} = \frac{\dot{Q}_L}{\dot{Q}_{ge}} = \frac{\dot{Q}_L}{\dot{Q}_{ge}} \times \frac{\dot{W}}{\dot{W}} = \frac{\dot{Q}_L}{\dot{W}} \times \frac{\dot{W}}{\dot{Q}_{ge}}$$

The above manipulation of the *COP* of the absorption refrigerator is presented graphically in Figure 7.23 as a heat engine and a refrigerator.

Since:

$$\eta = \frac{\dot{W}}{\dot{Q}_{ge}} \text{ and } COP_R = \frac{\dot{Q}_L}{\dot{W}}$$

Then the *COP* of the absorption refrigerator can be written in terms of the energy efficiency of heat engines and the *COP* of refrigerators as:

$$COP_{AR} = \eta \times COP_R$$

Then we can find the definition for the reversible absorption refrigerator as follows:

$$COP_{AR,rev} = \eta_{rev} \times COP_{R,rev}$$

$$COP_{AR,rev} = \left(1 - \frac{T_o}{T_s}\right) \times \left(\frac{T_L}{T_o - T_L}\right) \tag{7.11}$$

Using this result, the second-law (exergy) efficiency of an absorption refrigeration system can also be expressed as:

$$COP_{ex,AR} = COP_{AR}/COP_{AR,rev} \tag{7.12}$$

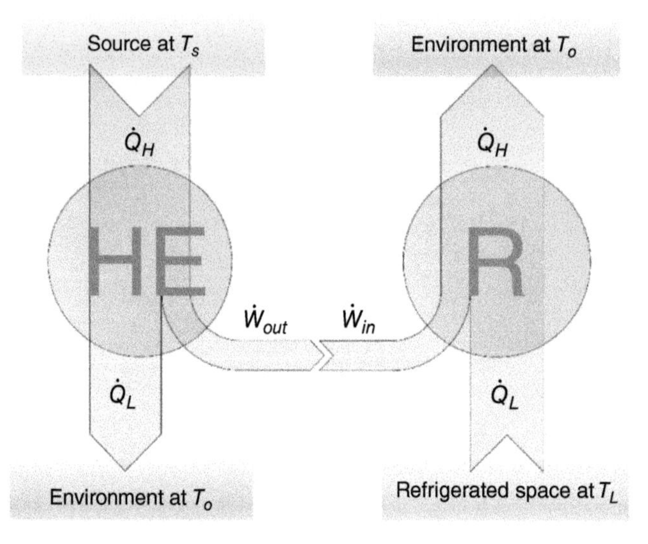

Figure 7.23 The system used to develop reversible *COP* of an absorption refrigeration system.

Example 7.10 An absorption refrigeration unit, as shown in Figure 7.22, that is used to cool a room at 20 °*C*, while the environment temperature is 30 °*C* by having the generator receives thermal energy from a source at a temperature of 200 °*C*. If the refrigeration unit has a *COP* of 4. (a) Write the mass, energy, entropy, and exergy balance equations for each component in the system, (b) calculate the maximum *COP* the absorption refrigerator can achieve operating between the above given temperatures, and (c) calculate both energetic and exergetic *COPs* of the absorption refrigerator.

Solution

a) To write the mass, energy, entropy, and exergy balance equations for each component in the system is necessary as follows;

For circulation pump:

$$\text{MBE}: \dot{m}_1 = \dot{m}_2$$

$$\text{EBE}: \dot{m}_1 h_1 + \dot{W}_{in} = \dot{m}_2 h_2$$

$$\text{EnBE}: \dot{m}_1 s_1 + \dot{S}_{gen} = \dot{m}_2 s_2$$

$$ExBE: \dot{m}_1 ex_1 + \dot{W}_{in} = \dot{m}_2 ex_2 + \dot{E}x_d$$

where $\dot{W}_{in}$ is pump work which may be shown as $\dot{W}_P$.

For generator:

$$\text{MBE}: \dot{m}_2 = \dot{m}_3 + \dot{m}_7$$

$$\text{EBE}: \dot{m}_2 h_2 + \dot{Q}_{ge} = \dot{m}_3 h_3 + \dot{m}_7 h_7$$

$$\text{EnBE}: \dot{m}_2 s_2 + \dot{Q}_{ge}/T_s + \dot{S}_{gen} = \dot{m}_3 s_3 + \dot{m}_7 s_7$$

$$\text{ExBE}: \dot{m}_2 ex_2 + \dot{E}x_{\dot{Q}_{ge}} = \dot{m}_3 ex_3 + \dot{m}_7 ex_7 + \dot{E}x_d$$

For condenser:

$$\text{MBE}: \dot{m}_3 = \dot{m}_4$$

$$\text{EBE}: \dot{m}_3 h_3 = \dot{m}_4 h_4 + \dot{Q}_{cn}$$

$$\text{EnBE}: \dot{m}_3 s_3 + \dot{S}_{gen} = \dot{m}_4 s_4 + \dot{Q}_{cn}/T_{cn}$$

$$\text{ExBE}: \dot{m}_3 ex_3 = \dot{m}_4 ex_4 + \dot{E}x_{\dot{Q}_{cn}} + \dot{E}x_d$$

For expansion valve 1:

$$\text{MBE}: \dot{m}_4 = \dot{m}_5$$

$$\text{EBE}: \dot{m}_4 h_4 = \dot{m}_5 h_5$$

$$\text{EnBE}: \dot{m}_4 s_4 + \dot{S}_{gen} = \dot{m}_5 s_5$$

$$\text{ExBE}: \dot{m}_4 ex_4 = \dot{m}_5 ex_5 + \dot{E}x_d$$

For expansion valve 2:

$$\text{MBE}: \dot{m}_7 = \dot{m}_8$$

$$\text{EBE}: \dot{m}_7 h_7 = \dot{m}_8 h_8$$

$$\text{EnBE}: \dot{m}_7 s_7 + \dot{S}_{gen} = \dot{m}_8 s_8$$

$$\text{ExBE}: \dot{m}_7 ex_7 = \dot{m}_8 ex_8 + \dot{E}x_d$$

For evaporator:

$$\text{MBE}: \dot{m}_5 = \dot{m}_6$$

$$\text{EBE}: \dot{m}_5 h_5 + \dot{Q}_{ev} = \dot{m}_6 h_6$$

$$\text{EnBE}: \dot{m}_5 s_5 + \dot{Q}_{ev}/T_{ev} + \dot{S}_{gen} = \dot{m}_6 s_6$$

$$\text{ExBE}: \dot{m}_5 ex_5 + \dot{E}x_{\dot{Q}_{ev}} = \dot{m}_6 ex_6 + \dot{E}x_d$$

For absorber:

$$\text{MBE}: \dot{m}_8 + \dot{m}_6 = \dot{m}_1$$

$$\text{EBE}: \dot{m}_8 h_8 + \dot{m}_6 h_6 = \dot{m}_1 h_1 + \dot{Q}_{ab}$$

$$\text{EnBE}: \dot{m}_8 s_8 + \dot{m}_6 s_6 + \dot{S}_{gen} = \dot{m}_1 s_1 + \dot{Q}_{ab}/T_{ab}$$

$$\text{ExBE}: \dot{m}_8 ex_8 + \dot{m}_6 ex_6 = \dot{m}_1 ex_1 + \dot{E}x_{\dot{Q}_{ab}} + \dot{E}x_d$$

b) To calculate the maximum *COP* the absorption refrigerator can achieve operating between the above given temperatures as follows:

$$COP_{AR,rev} = \left(1 - \frac{T_o}{T_s}\right) \times \left(\frac{T_L}{T_o - T_L}\right) = \left(1 - \frac{30 + 273}{200 + 273}\right) \times \left(\frac{20 + 273}{(30 + 273) - (20 + 273)}\right)$$

$$\mathbf{COP_{AR,rev} = 10.5}$$

c) To calculate the exergetic *COP* of the absorption refrigerator as follows:

$$COP_{ex,AR} = \frac{COP_{AR}}{COP_{AR,rev}} = \frac{4}{10.5} = \mathbf{0.38}$$

Example 7.11 The generator of an ammonia-water absorption chiller, as shown in Figure 7.24, receives heat of 84.2 kW from a 400 K heat source to keep a space cooled at 290 K using an evaporator that evaporates ammonia-water mixture at a pressure of 170 kPa. Both the absorber and the condenser release a total heat rate of 111.3 kW. The working fluid enters the evaporator with a quality of 0.165 and ammonia mass ratio of 0.909, and leaves with a temperature of 272 K and a quality of 0.783. Take the environment temperature and pressure to be 298 K and 100 kPa, respectively.

a) Write the mass, energy, entropy, and exergy balance equations for each component in the absorption chiller and the entire cycle.
b) Determine the heat transfer rate that takes place at the evaporator.
c) Determine the COP_{en} and COP_{ex} of this refrigeration cycle.
d) Determine the mass flow rate going through the evaporator.
e) Determine the entropy generation rate of this evaporator.
f) Determine the exergy destruction rate of this evaporator.

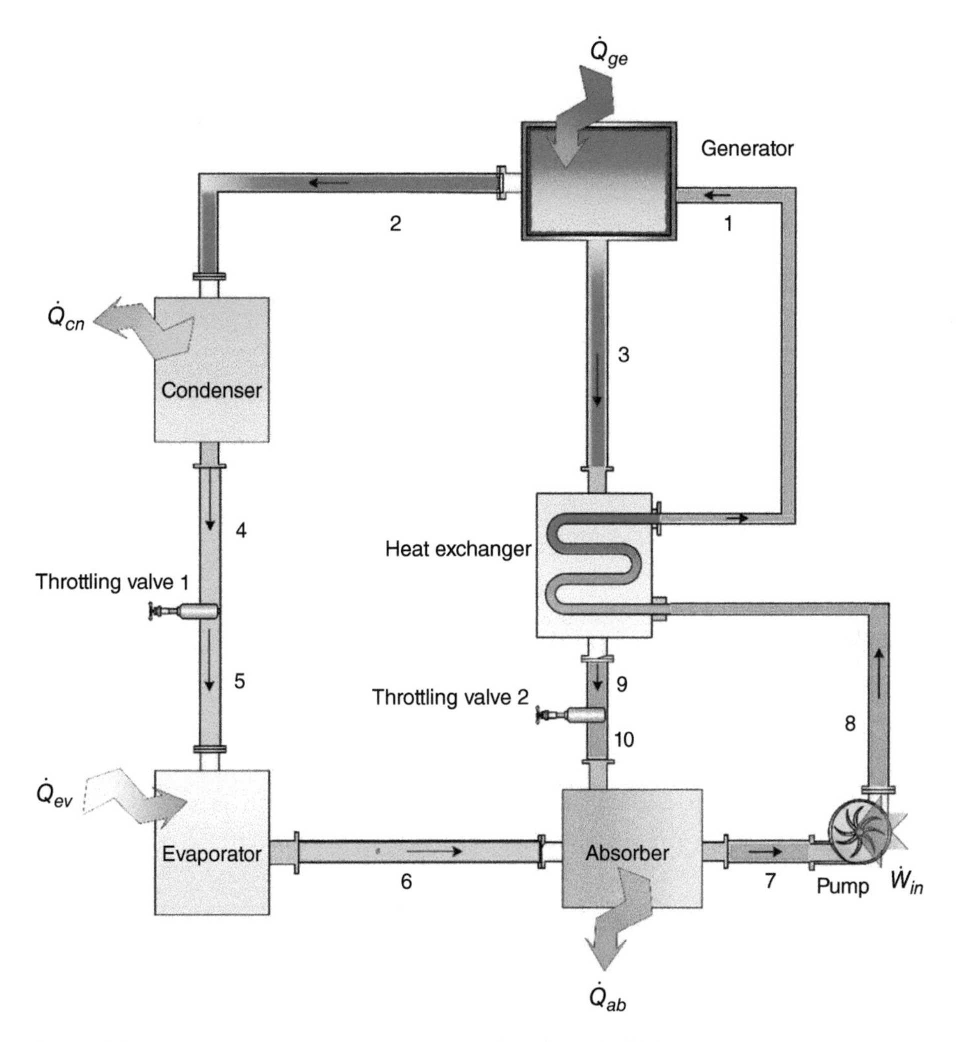

Figure 7.24 The absorption cooling system used in Example 7.11.

Solution

a) Write the mass, energy, entropy, and exergy balance equations for each component in the absorption chiller and the entire cycle.

For the generator:

$$\text{MBE}: \dot{m}_1 = \dot{m}_2 + \dot{m}_3$$

$$\text{EBE}: \dot{m}_1 h_1 + \dot{Q}_{ge} = \dot{m}_2 h_2 + \dot{m}_3 h_3$$

$$\text{EnBE}: \dot{m}_1 s_1 + \frac{\dot{Q}_{ge}}{T_{s,ge}} + \dot{S}_{gen,ge} = \dot{m}_2 s_2 + \dot{m}_3 s_3$$

$$\text{ExBE}: \dot{m}_1 ex_1 + \dot{Q}_{ge}\left(1 - \frac{T_o}{T_{s,ge}}\right) = \dot{m}_2 ex_2 + \dot{m}_3 ex_3 + \dot{E}x_{d,ge}$$

For the condenser:

$$\text{MBE}: \dot{m}_2 = \dot{m}_4$$

$$\text{EBE}: \dot{m}_2 h_2 = \dot{m}_4 h_4 + \dot{Q}_{cn}$$

$$\text{EnBE}: \dot{m}_2 s_2 + \dot{S}_{gen,cn} = \dot{m}_4 s_4 + \frac{\dot{Q}_{cn}}{T_{b,cn}}$$

$$\text{ExBE}: \dot{m}_2 ex_2 = \dot{m}_4 ex_4 + \dot{Q}_{cn}\left(1 - \frac{T_o}{T_{b,cn}}\right) + \dot{E}x_{d,cn}$$

For throttling valve 1:

$$\text{MBE}: \dot{m}_4 = \dot{m}_5$$

$$\text{EBE}: \dot{m}_4 h_4 = \dot{m}_5 h_5$$

$$\text{EnBE}: \dot{m}_4 s_4 + \dot{S}_{gen,tv1} = \dot{m}_5 s_5$$

$$\text{ExBE}: \dot{m}_4 ex_4 = \dot{m}_5 ex_5 + \dot{E}x_{d,tv1}$$

For the evaporator:

$$\text{MBE}: \dot{m}_5 = \dot{m}_6$$

$$\text{EBE}: \dot{m}_5 h_5 + \dot{Q}_{ev} = \dot{m}_6 h_6$$

$$\text{EnBE}: \dot{m}_5 s_5 + \frac{\dot{Q}_{ev}}{T_{s,ev}} + \dot{S}_{gen,ev} = \dot{m}_6 s_6$$

$$\text{ExBE}: \dot{m}_5 ex_5 + \dot{Q}_{ev}\left(\frac{T_o}{T_{s,ev}} - 1\right) = \dot{m}_6 ex_6 + \dot{E}x_{d,ev}$$

For the absorber:

$$\text{MBE}: \dot{m}_6 + \dot{m}_{10} = \dot{m}_7$$

$$\text{EBE}: \dot{m}_6 h_6 + \dot{m}_{10} h_{10} = \dot{m}_7 h_7 + \dot{Q}_{ab}$$

$$\text{EnBE}: \dot{m}_6 s_6 + \dot{m}_{10} s_{10} + \dot{S}_{gen,ab} = \dot{m}_7 s_7 + \frac{\dot{Q}_{ab}}{T_{b,ab}}$$

$$\text{ExBE}: \dot{m}_6 ex_6 + \dot{m}_{10} ex_{10} = \dot{m}_7 ex_7 + \dot{Q}_{ab}\left(1 - \frac{T_o}{T_{b,ab}}\right) + \dot{E}x_{d,ab}$$

For the pump:

$$\text{MBE}: \dot{m}_7 = \dot{m}_8$$

$$\text{EBE}: \dot{m}_7 h_7 + \dot{W}_p = \dot{m}_8 h_8$$

$$\text{EnBE}: \dot{m}_7 s_7 + \dot{S}_{gen,p} = \dot{m}_8 s_8$$

$$\text{ExBE}: \dot{m}_7 ex_7 + \dot{W}_p = \dot{m}_8 ex_8 + \dot{E}x_{d,p}$$

For throttling valve 2:

$$\text{MBE}: \dot{m}_9 = \dot{m}_{10}$$

$$\text{EBE}: \dot{m}_9 h_9 = \dot{m}_{10} h_{10}$$

$$\text{EnBE}: \dot{m}_9 s_9 + \dot{S}_{gen,tv2} = \dot{m}_{10} s_{10}$$

$$\text{ExBE}: \dot{m}_9 ex_9 = \dot{m}_{10} ex_{10} + \dot{E}x_{d,tv2}$$

For the heat exchanger:

$$\text{MBE}: \dot{m}_3 + \dot{m}_8 = \dot{m}_1 + \dot{m}_9$$

$$\text{EBE}: \dot{m}_3 h_3 + \dot{m}_8 h_8 = \dot{m}_1 h_1 + \dot{m}_9 h_9$$

$$\text{EnBE}: \dot{m}_3 s_3 + \dot{m}_8 s_8 + \dot{S}_{gen,he} = \dot{m}_1 s_1 + \dot{m}_9 s_9$$

$$\text{ExBE}: \dot{m}_3 ex_3 + \dot{m}_8 ex_8 = \dot{m}_1 ex_1 + \dot{m}_9 ex_9 + \dot{E}x_{d,he}$$

For the overall cycle:

$$\text{EBE}: \dot{Q}_{ge} + \dot{Q}_{ev} + \dot{W}_p = \dot{Q}_{ab} + \dot{Q}_{cn}$$

$$\text{EnBE}: \frac{\dot{Q}_{ge}}{T_{s,ge}} + \frac{\dot{Q}_{ev}}{T_{s,ev}} + \dot{S}_{gen,ov} = \frac{\dot{Q}_{ab}}{T_{b,ab}} + \frac{\dot{Q}_{cn}}{T_{b,cn}}$$

$$\text{ExBE}: \dot{Q}_{ge}\left(1 - \frac{T_o}{T_{s,ge}}\right) + \dot{Q}_{ev}\left(\frac{T_o}{T_{s,ev}} - 1\right) + \dot{W}_p = \dot{Q}_{ab}\left(1 - \frac{T_o}{T_{b,ab}}\right)$$

$$+ \dot{Q}_{cn}\left(1 - \frac{T_o}{T_{b,cn}}\right) + \dot{E}x_{d,ov}$$

b) Determine the heat transfer that takes place at the evaporator.
By applying the EBE for the entire cycle and ignoring the pump work, we can get the heat transfer rate of the evaporator:

$$\dot{Q}_{ev} = \dot{Q}_{ab} + \dot{Q}_{cn} - \dot{Q}_{ge} - \dot{W}_p$$

$$\dot{Q}_{ev} = 111.3\ kW - 84.2\ kW - 0\ kW$$

$$\mathbf{\dot{Q}_{ev} = 27.1\ kW}$$

c) Determine the *COP* of this refrigeration cycle.
The definition of the energy *COP* for this refrigeration cycle is:

$$\boldsymbol{COP_{en}} = \frac{\dot{Q}_{ev}}{\dot{Q}_{ge}} = \frac{27.1\ kW}{84.2\ kW} = \mathbf{0.322}$$

while the exergy *COP* is:

$$\boldsymbol{COP_{ex}} = \frac{\dot{Q}_{ev}\left(\frac{T_o}{T_{s,ev}} - 1\right)}{\dot{Q}_{ge}\left(1 - \frac{T_o}{T_{s,ge}}\right)} = \frac{27.1\ kW\left(\frac{298\ K}{290\ K} - 1\right)}{84.2\ kW\left(1 - \frac{298\ K}{400\ K}\right)} = \mathbf{0.0348}$$

d) Determine the mass flow rate going through the evaporator.
From the EES software, the specific enthalpy properties of the working mixture are

$$\left.\begin{aligned} c_{NH_3,5} &= 0.909 \\ x_5 &= 0.165 \\ P_5 &= 170\ kPa \end{aligned}\right\} h_5 = 60.4\ \frac{kJ}{kg}$$

$$\left.\begin{array}{r} c_{NH_3,6} = 0.909 \\ x_6 = 0.783 \\ P_6 = 170\ kPa \end{array}\right\} h_6 = 958.3\ \frac{kJ}{kg}$$

Now, using the MBE and EBE for the evaporator, the mass flow rate is calculated as follows:

$$\dot{m}_5 = \dot{m}_6$$

$$\dot{m}_5 h_5 + \dot{Q}_{ev} = \dot{m}_6 h_6$$

$$\dot{m}_5 = \frac{\dot{Q}_{ev}}{h_6 - h_5} = \frac{27.1\ kW}{958.3\ \frac{kJ}{kg} - 60.4\ \frac{kJ}{kg}}$$

$$\boldsymbol{\dot{m}_5 = 0.0302\ \frac{kg}{s}}$$

e) Determine the entropy generation rate of this evaporator.

Firstly, find the specific entropy values for streams 5 and 6 from EES as:

$$\left.\begin{array}{r} c_{NH_3,5} = 0.909 \\ x_5 = 0.165 \\ P_5 = 170\ kPa \end{array}\right\} s_5 = 0.506\ \frac{kJ}{kgK}$$

$$\left.\begin{array}{r} c_{NH_3,6} = 0.909 \\ x_6 = 0.783 \\ P_6 = 170\ kPa \end{array}\right\} s_6 = 3.985\ \frac{kJ}{kgK}$$

Secondly, apply the EnBE and rearrange for the entropy generation rate:

$$\dot{S}_{gen,ev} = \dot{m}_6 s_6 - \dot{m}_5 s_5 - \frac{\dot{Q}_{ev}}{T_{s,ev}}$$

$$\dot{S}_{gen,ev} = 0.0302\ \frac{kg}{s} \times 3.985\ \frac{kJ}{kgK} - 0.0302\ \frac{kg}{s} \times 0.506\ \frac{kJ}{kgK} - \frac{27.1\ kW}{290\ K}$$

$$\boldsymbol{\dot{S}_{gen,ev} = 0.0115\ \frac{kW}{K}}$$

f) Determine the exergy destruction rate of this evaporator.

Using the ExBE and rearranging for the exergy destruction rate:

$$\dot{E}x_{d,ev} = \dot{m}_5(ex_5 - ex_6) + \dot{Q}_{ev}\left(\frac{T_o}{T_{s,ev}} - 1\right)$$

Expressing the difference in specific exergy in terms of specific enthalpies and entropies:

$$\dot{E}x_{d,ev} = \dot{m}_5(h_5 - h_6 - T_o(s_5 - s_6)) + \dot{Q}_{ev}\left(\frac{T_o}{T_{s,ev}} - 1\right)$$

$$\dot{E}x_{d,ev} = 0.0302\ \frac{kg}{s} \times \left(60.4\ \frac{kJ}{kg} - 958.3\ \frac{kJ}{kg} - 298\ K\left(0.506\ \frac{kJ}{kgK} - 3.985\ \frac{kJ}{kgK}\right)\right)$$
$$+ 27.1\ kW\left(\frac{298\ K}{290\ K} - 1\right)$$

$$\dot{E}x_{d,ev} = 4.94\ kW$$

7.8 Closing Remarks

In this chapter the focus has been on various types of refrigeration and heat pump systems that are recognized as the most widely used systems worldwide. Everyone of us touches a fridge at home and almost everyone uses an air conditioner or heat pump at home. So, basically everyone has a refrigeration or heat pump system in his/her daily life. This clearly confirms how important this subject is. In this regard, this chapter has introduced the fundamentals related to refrigeration and heat pump cycles, refrigerants and their roles, basic refrigeration and heat pump systems, air-standard (gas) refrigeration systems, cascade refrigeration systems, reverse heat pump systems and absorption refrigeration systems, and thermodynamic analyses of these systems by writing all the balance equations for mass, energy, entropy, and exergy and assessment of these systems through energy and exergy based *COPs*. Examples have been provided to help better understand the subject and analysis and performance assessment of these systems.

Study Questions and Problems

Concept Questions

1 Give three examples from the daily life to emphasize the importance of refrigeration and heat pump systems.

2 Name three different Chloro Flouro Carbon based refrigerants.

3 Why was the use of CFC refrigerants banned?

4 Name four hydrocarbon based refrigerants.

5 What are the key advantages of hydrocarbon refrigerants?

6 Which refrigerants become azeotropic? Why? and How?

7 Classify the refrigerants, give one example of each one, and discuss their environmental impacts if any.

8 Draw *T-s* and *P-h* diagrams of the basic refrigeration system and indicate isentropic (ideal) and actual cases where there are differences. Discuss them with examples.

9 Explain why household fridges are used.

10 Explain what the roles of the compressor, condenser, expansion valve, and evaporator are.

11 Explain why air-standard (gas) refrigeration systems are used? What benefits do they bring?

12 Describe the key differences between vapor-compression and air-standard refrigeration cycles.

13 Which parameters play significant role in air-standard refrigeration cycle?

14 Explain the role of heat exchangers in air-standard cycle.

15 Describe the roles of compressor and expander in an air-standard refrigeration cycle?

16 What is the role net work in the air-standard (gas) refrigeration system?

17 Explain why cascading is necessary for refrigeration.

18 Compare both energetic and exergetic *COPs* and list the differences.

19 Explain why exergetic *COP* is needed.

20 Explain what the heat pump is and what it is used for.

21 Explain where heat pumps can potentially be used.

22 Discuss the sources of heat pumps with examples and list pros and cons.

23 Explain why reversed operation of a heat pump is necessary.

24 Is *COP* for a heat pump greater than the corresponding *COP* of a refrigerator? Why? How? By how much?

25 Discuss the historical importance of absorption refrigeration systems and explain what they lost momentum.

26 Define both energetic and exergetic *COPs* for an absorption refrigeration system.

27 Explain why a compressor is not used in an absorption refrigeration system? What is then used instead?

28 Explain the role of the generator in an absorption refrigeration system.

29 What are the desired sources of heat for absorption refrigeration systems? Explain why?

30 Why are two working fluids used in an absorption refrigeration system?

Problems

1 A refrigeration cycle is used to keep a food compartment at $-20\,°C$ in an environment at $25\,°C$. The total heat gain from the food comppartment is estimated to be $1500\,kJ/h$ and the heat rejection in the condenser is $2600\,kJ/h$.

- **a** Write the mass, energy, entropy, and exergy balance equations for the refrigeration system as a single unit.
- **b** Calculate the power input to the compressor in kW.
- **c** Calculate both the energetic and exergetic *COPs* of the refrigerator.
- **d** Calculate the minimum power input to the compressor if a reversible refrigerator was used.

2 A refrigeration unit that has *R*134*a* as the refrigerant (the working fluid of the cycle) is used to keep the targeted space cool at a temperature of $-10\,°C$ by rejecting heat to ambient air that is at a temperature of $22\,°C$. *R*134*a* enters the compressor at $142\,kPa$ at a flow rate of $400\,L/min$ as a saturated vapor. The isentropic efficiency of the compressor is 80%. The refrigerant leaves the condenser at $46.3\,°C$ as a saturated liquid.

- **a** Write the mass, energy, entropy, and exergy balance equation for each component in the cycle.
- **b** Find the rate of cooling provided by the system.
- **c** Find the *COP* of the refrigeration unit.
- **d** Find the entropy generation and exergy destruction in each component of the cycle.
- **e** Find the second-law efficiency of the cycle.
- **f** Find the total exergy destruction in the cycle.

3 Refrigerant 134*a* (*R*134*a*) enters a condenser at $1500\,kPa$ at a rate of $0.27\,kg/s$, as shown in the figure. A throttling valve expands the refrigerant to $450\,kPa$. The isentropic efficiency of the compressor is 85%. The quality of refrigerant entering the evaporator is 0.14 and it leaves as a saturated vapor. This refrigeration cycle is to keep a space at a temperature of $21\,°C$. Any pressure drops and heat transfers at the connections between components can be disregarded. Take the ambient temperature and pressure as $30\,°C$ and $100\,kPa$, respectively.

a Write the mass, energy, entropy, and exergy balance equations for each component.
b Determine the rate of heat removal by the refrigerator and the power input to the compressor.
c Determine the entropy generation and the exergy destruction rates for each component.
d Determine energetic and exergetic *COPs* (COP_{en}, COP_{ex}).

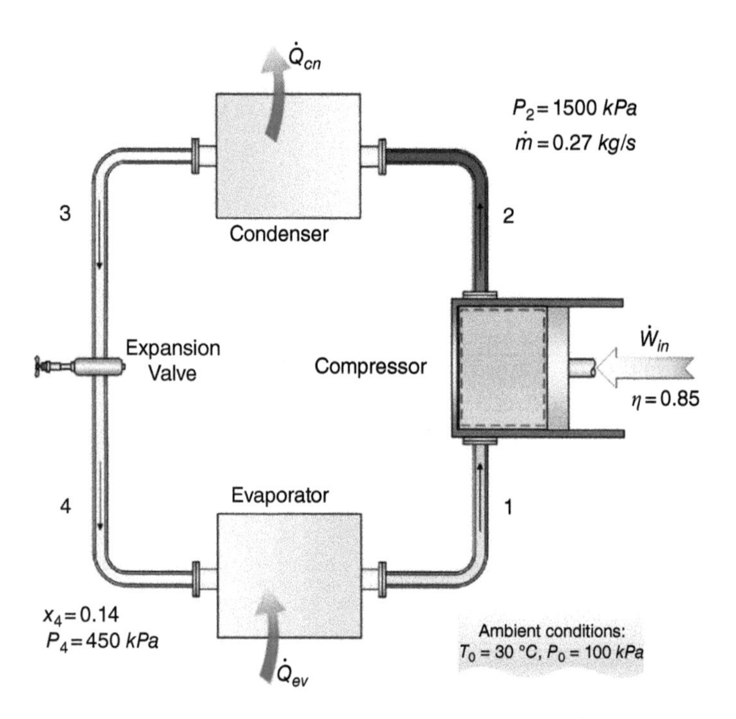

4 Refrigerant 113 (*R*113) enters a condenser at 92.9 *kPa* at a rate of 0.27 *kg/s*, as shown in the figure. A throttling valve expands the refrigerant to 2.834 *kPa*. The isentropic efficiency of the compressor is 85% and the refrigerant leaving the condenser is a saturated liquid. Also, the quality of refrigerant entering the evaporator is 0.18. This refrigeration cycle is to keep a space at a temperature of −15 °*C*. The refrigerant leaves the evaporator as a saturated vapor. Any pressure drops and heat transfers at the connections between components can be disregarded. Take the ambient temperature and pressure as 25 °*C* and 100 *kPa*, respectively.

a Write the mass, energy, entropy, and exergy balance equations.
b determine the rate of heat removal by the refrigerator and the power input to the compressor.
c determine the entropy generation and the exergy destruction rates for each component.
d determine the energetic and exergetic *COPs* (COP_{en}, COP_{ex}).

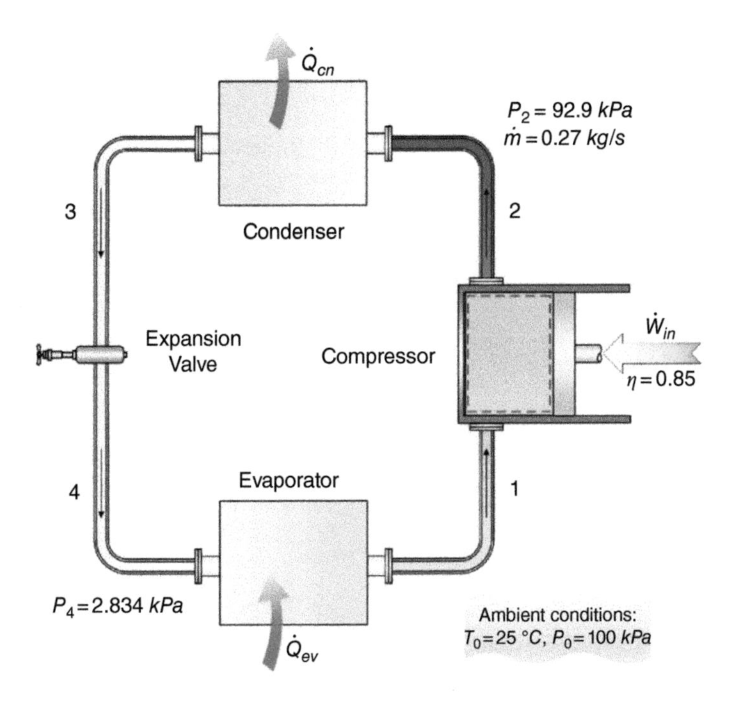

5 A commercial refrigerator with the working fluid Refrigerant 134*a* is shown in the figure. The refrigerant leaves the evaporator with a temperature and pressure of −33.5 °*C* and 60 *kPa*, respectively. It is then compressed to the pressure of 1.5 *MPa* and temperature of 64 °*C*. At the water-cooled condenser, the refrigerant is cooled down to 41 °*C*.

a Write the mass, energy, entropy, and exergy balance equations for each component.

b Determine the entropy generation and exergy destruction rate at each component and the entire system.

c Determine the quality of the refrigerant at the evaporator inlet.

d Determine the refrigeration load.

e Determine the COP_{en} and COP_{ex} of the refrigerator.

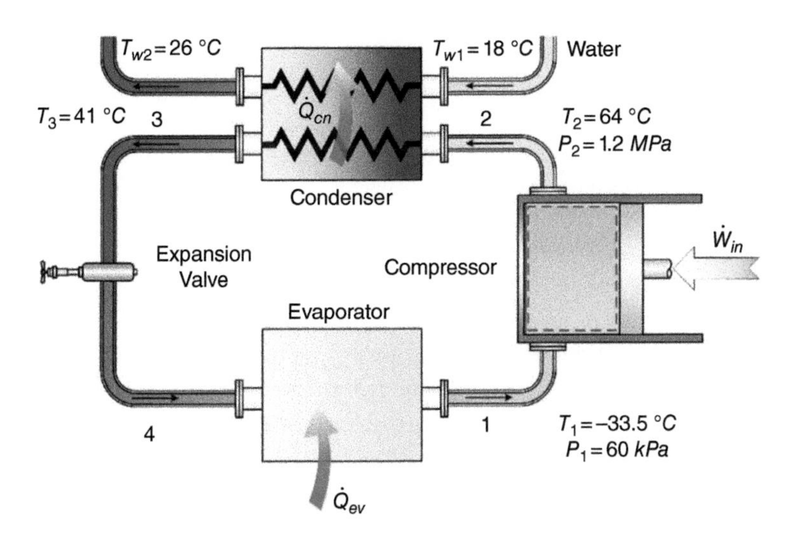

6 A vapor-compression refrigeration system with *R*22 as the refrigerant is shown in the figure. The evaporator and condenser temperatures are −4.8 °*C* and 46 °*C*, respectively. The *COP* of the refrigeration system as 4.67.

a Write the mass, energy, entropy, and exergy balance equations for each component in the vapor-compression refrigeration system and the entire cycle.

b Draw the *T*-*s* diagram for the vapor-compression refrigeration system.

c Determine the heat absorbed by the refrigerant, the work input to the compressor, and the heat rejected in the condenser per kilogram of the working fluid.

d Determine the COP_{ex} and the total exergy destruction rate for the entire refrigeration cycle.

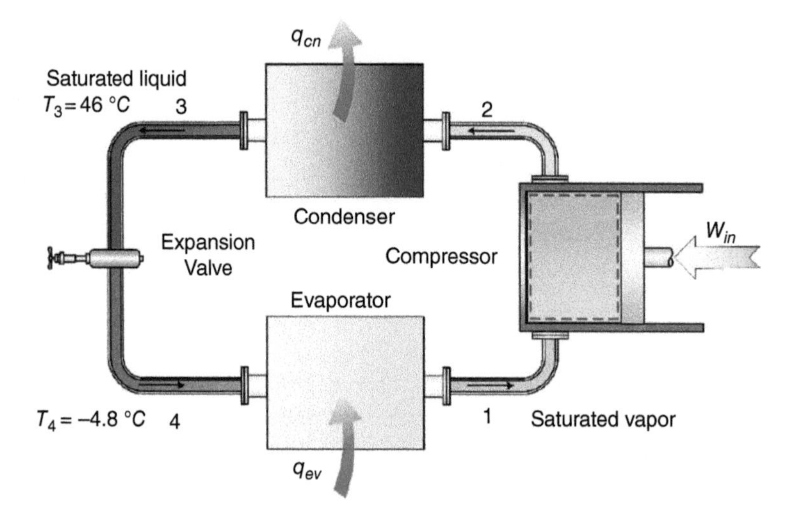

7 A warehouse uses a ammonia vapor-compression refrigeration system to keep dairy products at 4 °*C* while the surrounding temperature is 27 °*C*. The warehouse absorbs heat of 600 *kW* from the surroundings. The condenser operates at 1400 *kPa* and the evaporator operates at 250 *kPa*.

a Write mass, energy, entropy and exergy balance equations for each component.

b Write state point properties for each state.

c Calculate the quality of the evaporator inlet stream.

d Calculate the amount of power required and coefficient of performance for a Carnot refrigeration cycle.

e Calculate the amount of power required and coefficient of performance for an ideal ammonia vapor-compression refrigeration cycle.

8 A cooled space at a temperature of −8 °*C* needs to lose 40.0 *kW* by using an air-standard refrigeration cycle. The compressor receives air at a temperature of −24 °*C* and increases its pressure by a factor of seven, while the expander receives air at a temperature of 20 °*C*. The isentropic efficiencies of the compressor and turbine are 80 and 86%, respectively. Take the environment temperature and pressure to be 288 *K* and 100 *kPa*. Assume air to behave as an ideal gas with a specific heat ratio of 1.4 and constant pressure specific heat of 1.005 *kJ*/*kgK*.

a Write the mass, energy, entropy, and exergy balance equations for each component in the refrigeration cycle and the entire cycle.
b Determine the input power needed to run this cycle.
c Determine the COP_{en} and COP_{ex} of this refrigeration cycle.
d Determine the entropy generation rate of this cycle.
e Determine the exergy destruction rate of this cycle.
f Consider this problem again when the air is an actual fluid and use EES software to get the exact property values, then compare the results.

9 For an aircraft the inside cabin needs to be cooled at a rate of 105 kW using an air-standard refrigeration cycle. The cabin temperature is 20 °C. In the air-standard refrigeration cycle, the compressor receives air at a temperature of −5 °C and increases its pressure by a factor of four, while the expander receives air at a temperature of 40 °C. The isentropic efficiencies of the compressor and turbine are 68 and 78%, respectively. Take the environment temperature and pressure to be 303 K and 100 kPa. Assume air to behave as an ideal gas with a specific heat ratio of 1.4 and constant pressure specific heat of 1.005 kJ/kgK.
a Write the mass, energy, entropy, and exergy balance equations for each component in the refrigeration cycle and the entire cycle.
b Determine the mass flow rate running within this cycle.
c Determine the input power needed to run this cycle.
d Determine the COP_{en} and COP_{ex} of this refrigeration cycle.
e Determine the entropy generation rate of this cycle.
f Determine the exergy destruction rate of this cycle.

10 A cooled space at a temperature of −8 °C needs to lose 70 kW by using an air-standard refrigeration cycle. The compressor receives air at a temperature of −24 °C and increases its pressure by seven times. The expander input temperature is 20 °C. The isentropic efficiencies of the turbine and compressor are 86% and 80%. Assume the environment temperature and pressure to be 288 K and 100 kPa. Assume air to behave as an ideal gas with a specific heat ratio of 1.4 and constant pressure specific heat of 1.005 kJ/kgK.
a Write the mass, energy, entropy and exergy balance equations for each component of the refrigeration system.
b Determine the input power needed to run this cycle.
c Determine the energetic and exergetic $COPs$.
d Determine the exergy destruction rate of this cycle.

11 A heat pump using $R134a$ as the refrigerant of the cycle, which is the working fluid of the heat pump, takes up heat from an environment that is at a temperature of −10 °C to keep the temperature of a room at 22 °C. $R134a$ enters the compressor at 135 kPa at a flow rate of 380 L/min as a saturated vapor. The isentropic efficiency of the compressor is 80%. The refrigerant leaves the condenser at 46.3 °C as a saturated liquid.
a Write the mass, energy, entropy, and exergy balance equations of each component in the cycle.

b Find the rate of heating provided by the system.
c Find the *COP*.
d Find the exergy destruction in each component of the cycle.
e Draw the temperature variation of the refrigerant with its entropy throughout the cycle.

12 Consider an actual heat pump cycle with *R*134*a* as the working fluid. The working fluid has a temperature and pressure of 800 *kPa* and 55 °*C*, and undergoes a pressure loss of 50 *kPa* while passing through the condenser, as shown in the figure. Assume the mass flow rate of 0.5 *kg/s*.
a Write the mass, energy, entropy, and exergy balance equations for each component in the heat pump system.
b Determine the entropy generation and exergy destruction rate at each component and the entire system.
c Determine the COP_{en} and COP_{ex} of the heat pump cycle and determine the efficiency of the compressor.
d Draw *T-s* diagram of the heat pump cycle.
e Determine the rate of heat supplied to the heated room.

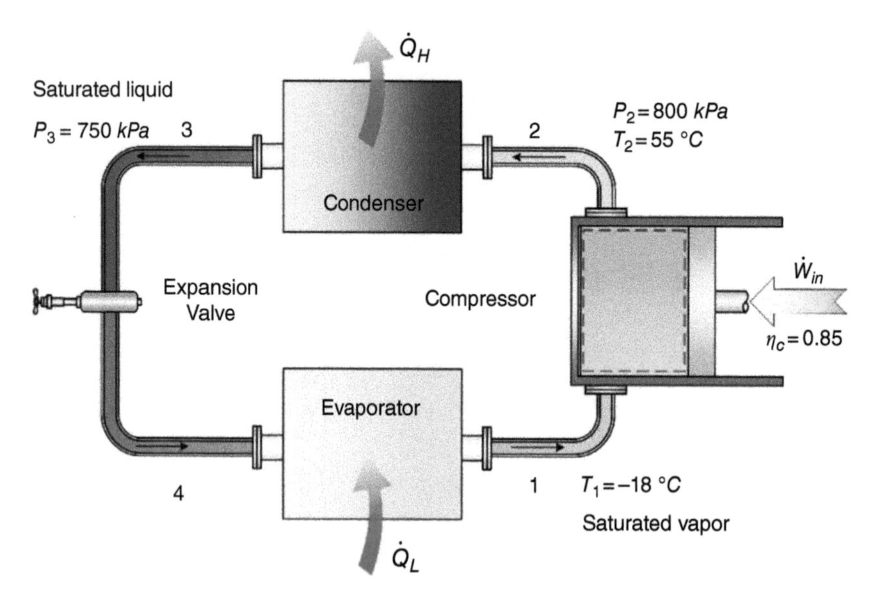

13 In the evaporator of geothermal heat pump cycle, the *R*134*a* working fluid with the temperature and quality of 20 °*C* and 0.23 turns into a fully saturated refrigerant, as shown in the figure. Through the isentropic compression, the saturated vapor is compressed to a pressure of 1.4 *MPa*. Assume the flow rate of the geothermal fluid to be 0.5 *kg/s* and ambient pressure to be 100 *kPa*.

a Write the mass, energy, entropy, and exergy balance equations for each component in the heat pump system.
b Find the entropy generation and exergy destruction rate at each unit and for the entire system.
c Determine the degree of subcooling done on the working fluid in the condenser.
d Calculate the mass flow rate of the refrigerant and the heating load.
e Determine the COP_{en} and COP_{ex} of the heat pump and the minimum power input.

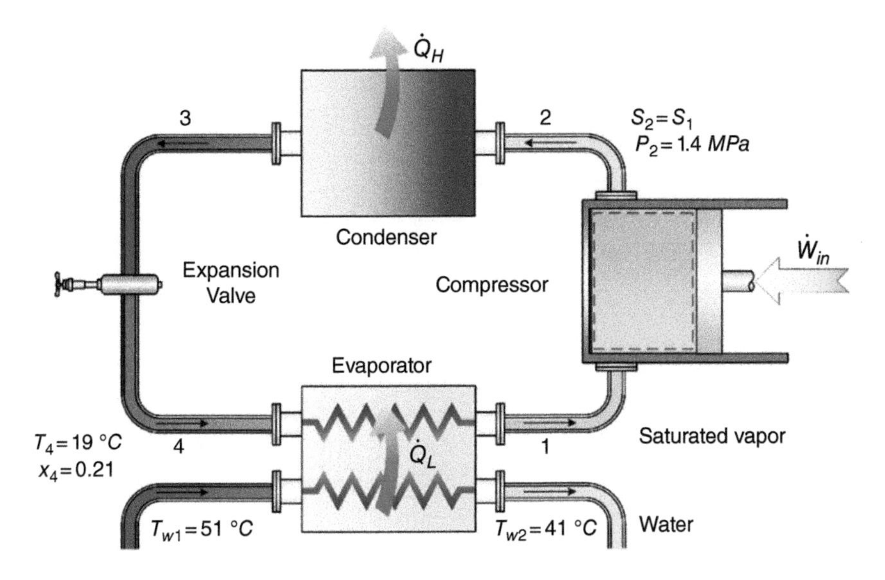

14 Consider a geothermal heat pump cycle with *R134a* working fluid operating the evaporator with the quality and temperature of 0.21 and 292 *K* turns into a fully saturated refrigerant. Assume the water flow rate of 0.6 *kg/s* with the input temperature of 324 *K* and output temperature of 314 *K* at 100 *kPa* pressure through the isentropic compression, the saturated vapor is compressed to a pressure of 1400 *kPa*.
a Write mass, energy, entropy and exergy balance equations for each component.
b Calculate all state point properties.
c Calculate the mass flow rate of the refrigerant and heating load.
d Find the entropy generation and exergy destruction rate of each component.
e Determine the energetic and exergetic *COPs* of the cycle.

15 A convention center building is losing heat and a heat pump, as shown in the figure, is used to maintain the temperature at 23 °*C* when the outside temperature is 18 °*C*. Refrigerant 134*a* leaves the evaporator with a flow rate of 0.15 *kg/s* and temperature of 12 °*C* in saturated vapor condition. The compressor works with 90% efficiency and pressurizes the refrigerant to 1200 *kPa*. Saturated liquid refrigerant leaves at the inlet pressure from the condenser and enters the expansion valve.

a Write the mass, energy, entropy, and exergy balance equations for each component.
b Find values for each state point.
c Determine the rejected heat at condenser and absorbed heat at evaporator.
d Calculate the power input to the compressor.
e Calculate the energetic *COP* and the exergetic *COP* for the heat pump (COP_{en}, COP_{ex}).

Assume that the changes in kinetic and potential energies are negligible, and that steady-state operating conditions exist. $T_{cn} = 40\,°C$ and $T_{ev} = 13\,°C$.

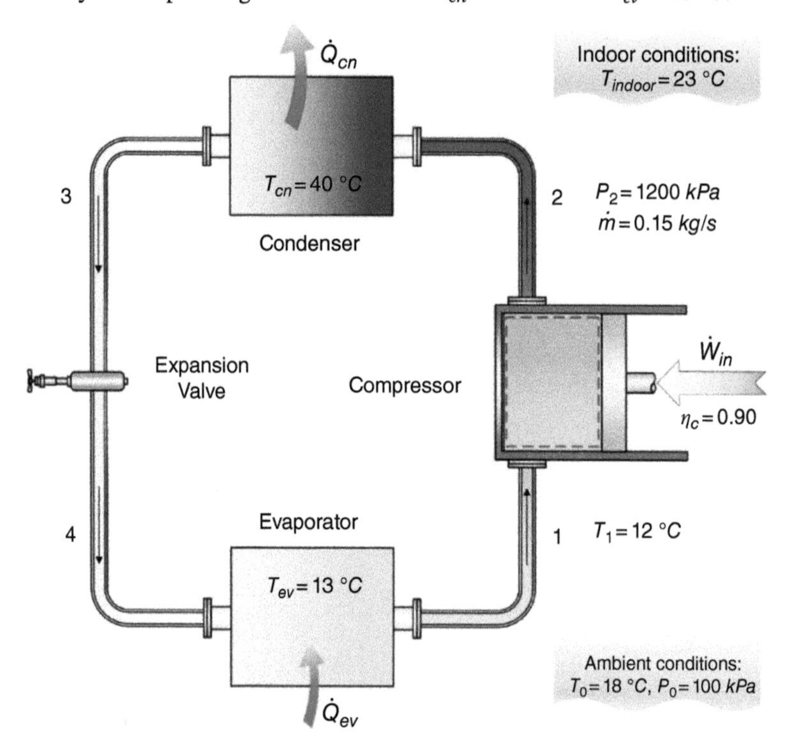

16 A heat pump system, as shown in the figure, maintains a greenhouse at 30 °*C* with refrigerant 152*a*. Geothermal water in saturated liquid condition enters the evaporator at 75 °*C* and leaves at 50 °*C* with a mass flow rate 0.0848 *kg/s* as a heat source. The refrigerant leaves the evaporator at 28 °*C* as saturated vapor with a flow rate of 0.017 *kg/s*. The compressor pressurizes the refrigerant to 1400 *kPa* with 85% efficiency. Saturated liquid refrigerant leaves at the inlet pressure from the condenser and enters the expansion valve for an isenthalpic process.

a Write the mass, energy, entropy, and exergy balance equations for each component.
b Find values for each state point.
c Determine the heat rejected at the condenser and heat absorbed at the evaporator.
d Calculate the power input to the compressor.
e Calculate the energetic *COP* and the exergetic *COP* for the heat pump (COP_{en}, COP_{ex}).

Assume that the changes in kinetic and potential energies are negligible and that steady-state operating conditions exist. $T_{cn} = 50\,°C$ and $T_o = 23\,°C$.

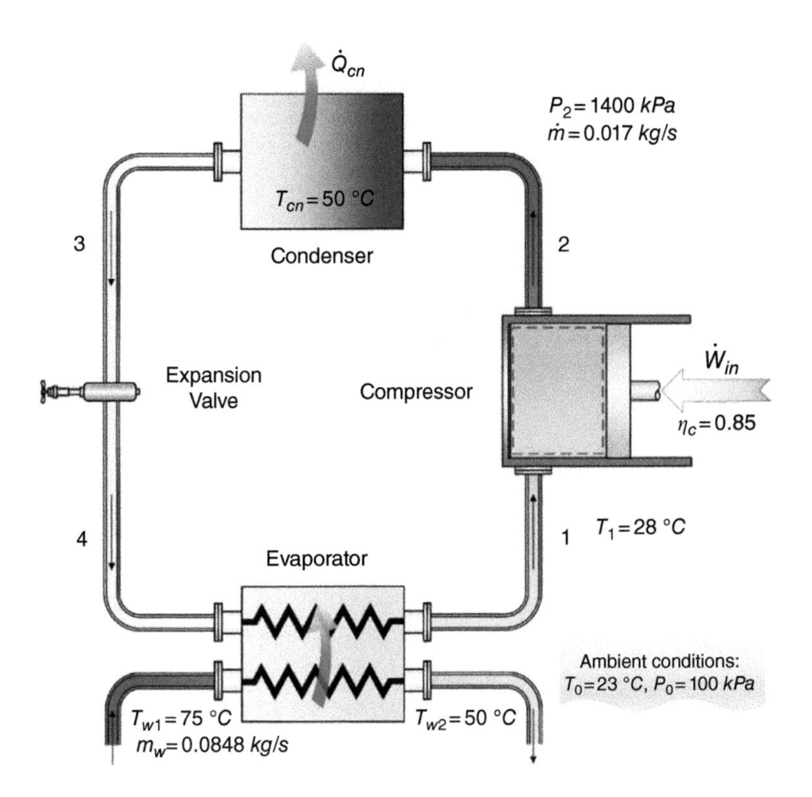

17 A heat pump is used to maintain the temperature of a house at 22 °*C* in winter when the outside temperature is 10 °*C*. Refrigerant *R*134*a* is used with a mass flow rate of 0.08 *kg/s*. The pressure of the refrigerant in the evaporator is measured to be 90 *kPa*. The isentropic efficiency of the compressor is 87% and the exit pressure of the compressor is 1400 *kPa*.

- **a** Write the mass, energy, entropy and exergy balance equations for each component.
- **b** Determine the thermophysical properties at each state point.
- **c** Find the rate of cooling heating provided to the house.
- **d** Determine the power input required by the compressor.
- **e** Calculate the energetic and exergetic coefficient of performance.

18 A heat pump, as shown in the figure, that works with Refrigerant 134*a* (*R*134*a*) is used with a reversing valve for both cooling and heating purposes. The compressor works with 86% efficiency and it pressurizes the refrigerant from 320 to 1700 *kPa*. The environment temperature assumed as 31 °*C* in summer and 7 °*C* in winter; the desired temperature is 21 °*C*. The flow rate of the refrigerant is 0.051 *kg/s*.

- **a** Write the mass, energy, entropy, and exergy balance equations for each component.
- **b** Find values for each state point.
- **c** Calculate the power input to the compressor and the heat transfer rates for the condenser and the evaporator.

d Calculate the energetic and exergetic *COP* for both seasons; compare and calculate the average *COP* energetic and exergetic (COP_{en}, COP_{ex}).

Assume that the changes in kinetic and potential energies are negligible and that steady-state operating conditions exist. $T_{cn} = 48\,°C$ and $T_{ev} = 4\,°C$.

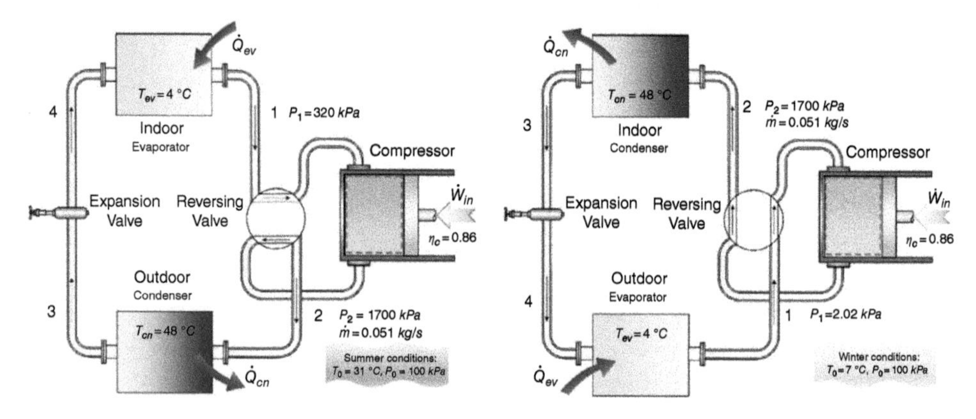

19 A heat pump, as shown in figure, that works with Refrigerant 113 (*R*113) is used with a reversing valve for both cooling and heating purposes. The compressor works with 88% efficiency and it pressurizes the refrigerant from 2.02 to 100 *kPa*. The environment temperature assumed as 32 °*C* in summer and 6 °*C* in winter; the desired temperature is 22 °*C*. The flow rate of the refrigerant is 0.15 *kg/s*.

a Write the mass, energy, entropy, and exergy balance equations for each component.

b Find values for each state point.

c Calculate the power input to the compressor and the heat transfer rates for the condenser and the evaporator.

d Calculate the energetic and exergetic *COP* for both seasons, compare and calculate the average *COP* energetic and exergetic (COP_{en}, COP_{ex}).

Assume that the changes in kinetic and potential energies are negligible and that steady-state operating conditions exist. $T_{cn} = 47\,°C$ and $T_{ev} = -12\,°C$.

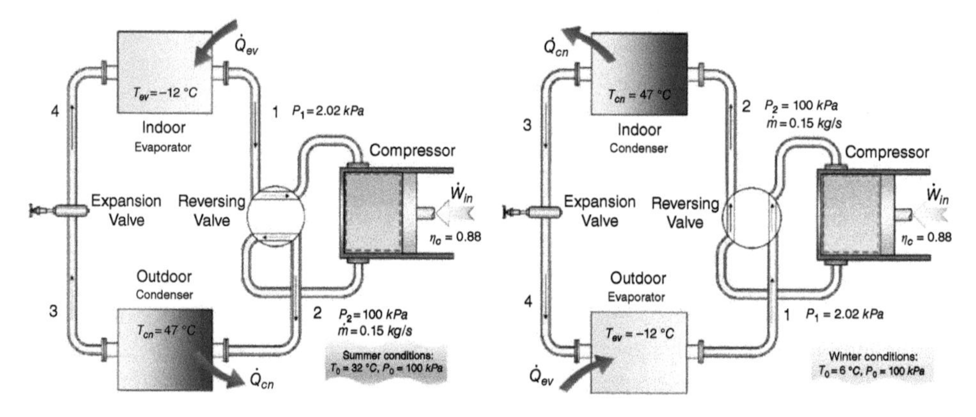

20 A two-stage cascade refrigeration system using *R*134 as refrigerant operates on an actual vapor-compression cycle within the pressure limits of 1 and 0.15 *MPa*. Heat is rejected from lower to upper cycle in a heat exchanger where both streams enter at about 0.4 *MPa*. The compression processes in both cycles take place with 85% efficiency and the mass flow rate of the refrigerant in the upper cycle is 0.08 *kg/s*.

a Write the mass, energy, entropy, and exergy balance equations.
b Determine the mass flow rate of refrigerant in the lower cycle.
c Calculate the rate of heat removal from refrigerant space.
d Calculate the energetic and exergetic *COPs* for this cascade refrigeration cycle.

21 A two-stage cascade refrigeration system operates on an ideal vapor-compression cycle using ammonia as refrigerant between the upper and lower pressure limits of 0.08 *MPa* and 1.2 *MPa*, respectively. Heat is transferred within the upper and lower cycles in a heat exchanger, where lower stream and upper streams enter at 0.6 and 0.5 *MPa*, respectively. The compression processes in both cycles take place with 80% efficiency and the mass flow rate of the refrigerant in the upper cycle is 0.1 *kg/s*,

a Write the mass, energy, entropy, and exergy balance equations.
b Determine the mass flow rate of refrigerant in the lower cycle.
c Calculate the rate of heat removal from refrigerant space.
d Calculate the energetic and exergetic *COPs* for this cascade refrigeration cycle.

22 Consider a two-stage cascaded refrigeration system with refrigerant *R*134*a*. The lowest pressure of the refrigerant in the cycle is 120 *kPa* and the highest pressure is 1400 *kPa*. Heat transfer takes place between the bottom and top cycles at a pressure of 500 *kPa*. The compressors of both cycles have an isentropic efficiency of 85%. The mass flow rate of the refrigerant in the upper cycle is 0.5 *kg/s*. The ambient temperature and pressure are 25 °*C* and 101 *kPa*. Also, the temperature of the refrigerated space is 20 °*C*.

a Write the mass, energy, entropy and exergy balance equations for all system components.
b Calculate the mass flow rate of refrigerant in the bottom cycle.
c Find the rate of cooling provided to the refrigerated space.
d Find the energetic and exergetic coefficient of performance of the cascaded refrigeration cycle.

23 The *T-s* diagram of a two-stage cascade refrigeration system with a flash chamber operating with the refrigerant 134*a* is shown in the figure. The working fluid minimum temperature and maximum operating pressure are −9.5 °*C* and 1.65 *MPa*, respectively, and the isentropic efficiency of the two compressors is 87%. The mass flow rate through the evaporator is 0.12 *kg/s*.

a Write the mass, energy, entropy, and exergy balance equations for each component in the two-stage cascade refrigeration cycle.
b Determine the entropy generation and exergy destruction rate at each component and the entire cascade refrigeration system.
c Determine the refrigerant mass flow rate at the high-pressure compressor.
d Calculate the COP_{en} and COP_{ex} of the cascade refrigeration system.

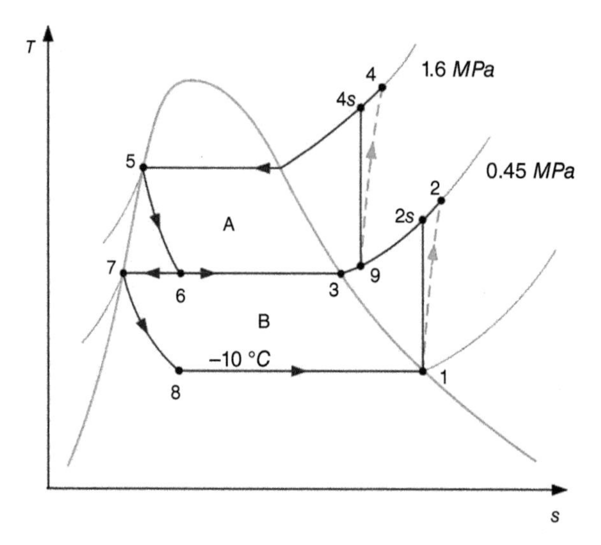

24 Consider a two-stage cascade refrigeration cycle operating with *R134a* as the refrigerant. The system operates between the pressure limits of 0.78 and 0.135 *MPa*. The rejected heat from the lower cycle is recovered in the upper cycle in an adiabatic counterflow heat exchanger at a pressure of 0.323 *MPa*. The mass flow rate of *R134a* through the upper cycle is 0.05 *kg/s*.

a Write the mass, energy, entropy, and exergy balance equations for each component in the two-stage cascade refrigeration system.

b Determine the entropy generation and exergy destruction rate at each component and the entire cascade refrigeration cycle.

c Calculate the COP_{en} and COP_{ex} of the cascade refrigeration system.

d Determine the mass flow rate of *R134a* in the lower cycle.

e Calculate the rate of removed heat from the cooled space and the power input to the compressor component.

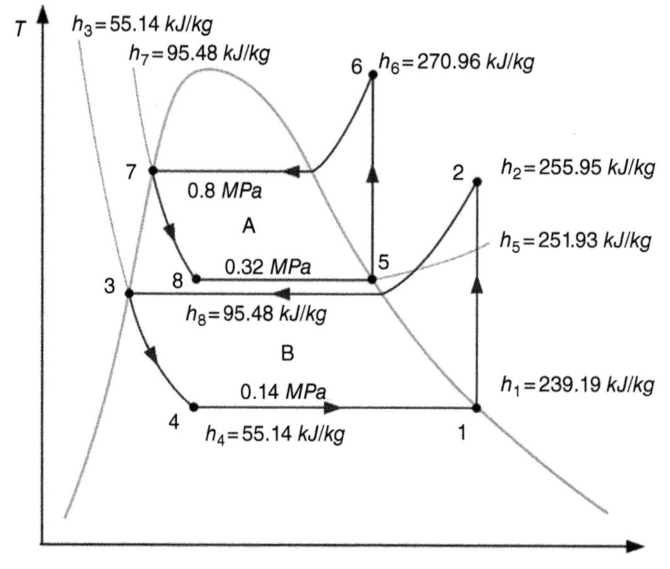

25 Consider a two-stage cascade refrigeration system as shown in the figure. In this system, *R410a* is used for the high-temperature cycle and *R23* for the low-temperature cycle. In the high-temperature stage, the saturated liquid leaves the condenser at a temperature of 41 $^{\circ}C$ and the saturated vapor leaves the evaporator at the temperature of -20 $^{\circ}C$. In the low-temperature side, the evaporator inlet and outlet temperatures are -79 and -9.5 $^{\circ}C$ with an enthalpy of 330 and 185 kJ/kg, respectively. The enthalpy of *R23* at the outlet compressor is 405 kJ/kg. Also, the outlet pressure of the *R23* compressor is 2200 kPa. The mass flow rate of top cycle is 0.004 kg/s, and the isentropic efficiency of the top cycle compressor is 85%.

a Write the mass, energy, entropy, and exergy balance equations for each component in the two-stage cascade refrigeration system.

b Determine the entropy generation and exergy destruction rate at each component and the entire cascade refrigeration cycle.

c Calculate the mass flow rate ratio of the bottom cycle.

d Calculate the COP_{en} and COP_{ex} of the cascade refrigeration system.

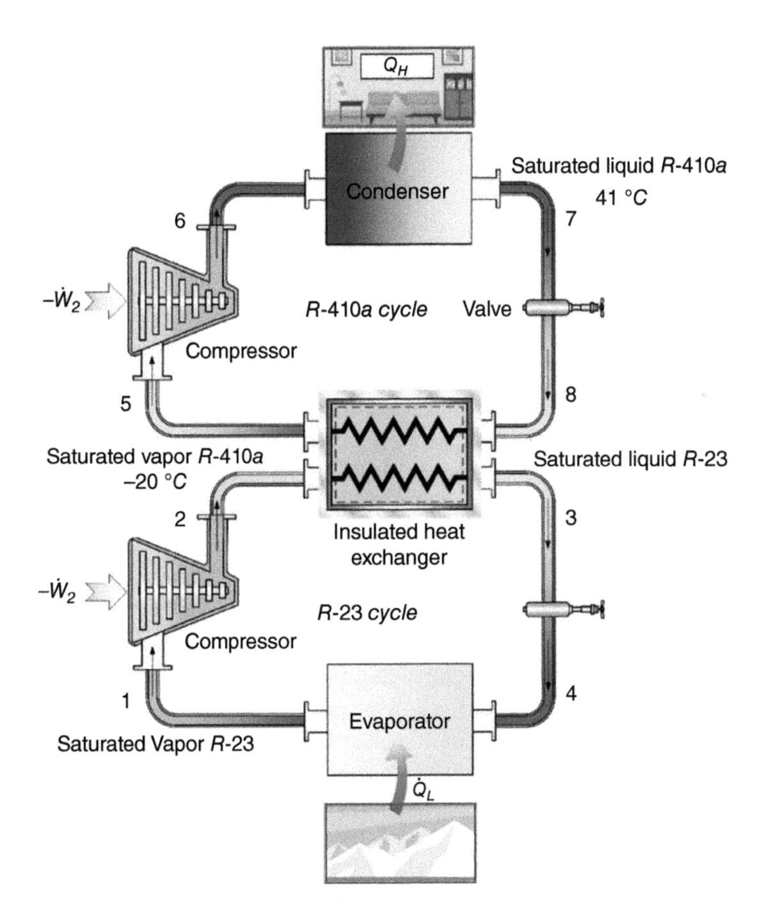

26 The generator of an ammonia absorption chiller receives heat of 252.6 kW from a 550 K heat source to keep a space cooled at 290 K using an evaporator that evaporates ammonia-water mixture at a pressure of 150 *kPa*. Both the absorber and the condenser release a total heat rate of 334 *kW*. The working fluid enters the evaporator with a quality of 0.165 and ammonia mass ratio of 0.999, and leaves with a quality of 0.783. Take the environment temperature and pressure to be 298 *K* and 100 *kPa*, respectively.

a Write the mass, energy, entropy, and exergy balance equations for each component in the ammonia absorption chiller and the entire cycle.
b Determine the heat transfer rate that takes place at the evaporator.
c Determine the COP_{en} and COP_{ex} of this refrigeration cycle.
d Determine the mass flow rate going through the evaporator.
e Determine the entropy generation rate of this evaporator.
f Determine the exergy destruction rate of this evaporator.

27 The generator of a water-LiBr absorption chiller receives heat of 25.12 *kW* from a 450 *K* heat source to keep a space cooled at 295 *K* using an evaporator that evaporates water-LiBr at a temperature of 280 *K*. Both the absorber and the condenser release a total heat rate of 36.45 *kW*. The working fluid enters the evaporator with a quality of 0.245 and water mass ratio of 0.999, and leaves with a quality of 0.98. Take the environment temperature and pressure to be 298 *K* and 100 *kPa*, respectively.

a Write the mass, energy, entropy, and exergy balance equations for each component in the absorption chiller and the entire cycle.
b Determine the heat transfer rate that takes place at the evaporator.
c Determine the COP_{en} and COP_{ex} of this refrigeration cycle.
d Determine the mass flow rate going through the evaporator.
e Determine the entropy generation rate of this evaporator.
f Determine the exergy destruction rate of this evaporator.

28 Consider an ammonia-based absorption refrigeration system (an air conditioning unit) powered by solar energy. The generator and evaporator operate at the temperatures of 49 and 10 °*C* respectively. The heat supplied by the solar collector in the generator is 7000 *kJ/kg* of produced ammonia vapor. The heat absorbed by the evaporator is known to 2000 *kJ/kg*, the heat rejected by the condenser is 7800 *kJ/kg*, and the heat rejected by the absorber is 1200 *kJ/kg*. The temperature of the condenser is 27 °*C*, and the temperature of the absorber is 7 °*C*.

a Write the mass, energy, entropy, and exergy balance equations for each component in the ammonia absorption refrigeration system.
b Determine the entropy generation and exergy destruction rate at each component and the entire ammonia absorption refrigeration system.
c Calculate the COP_{en} and COP_{ex} of the ammonia absorption refrigeration system.

29 Consider a small ammonia-based absorption refrigeration system with evaporator, condenser, and generator temperatures of −10, 50, and 150 °*C*, respectively. In the evaporator, 0.421 *kJ* is transferred to the ammonia-water solution for each kilojoule transferred from the high-temperature source to the ammonia-water solution in the generator. For comparison, assume that the reservoir temperature operates at 150 °*C* and the heat is transferred from this reservoir to the reversible engine that rejects heat to the environment at 23 °*C*. This work is used to power an ideal vapor-compression system operating with ammonia as the working fluid. Compare the amount of refrigeration load that can be yielded for each kilojoule from the high-temperature source with the 0.42 *kJ* that can be achieved in the ammonia absorption system.

30 An ammonia-water based absorption cooling cycle receives 500 *kW* of thermal energy from a heat source at 227 °*C*. The refrigerated space is maintained at 22 °*C* and the evaporator is operated at 200 *kPa*. The total rate of thermal energy released from the absorber and condenser is known to be 650 *kW*. The inlet of the evaporator is known to have a quality of 0.18 and an ammonia mass ratio of 0.999. The evaporator exit entails a quality of 0.73. The ambient temperature is at 30 °*C* and ambient pressure is known to be 100 *kPa*.

a Write the mass, energy, entropy and exergy balance equations for each system component.
b Find the rate of thermal energy absorbed by the evaporator.
c Determine the energetic and exergetic *COP* of the cycle.
d Find the entropy generation rate in the evaporator.
e Calculate the exergy destruction rate in the evaporator.

8

Fuel Combustion

8.1 Introduction

In almost every chapter the importance of energy has been emphasized as it plays a critical role in everything, ranging from industry to economy, education to politics, environment to sustainable development, and present generation to future generation. However, there is one more thing that is even more important than the way we produce energy: **thermodynamics**! This conceptual subject is extremely important to design, analyze, assess, evaluate and improve energy systems and applications. Learning this subject is a key to properly learn things and practice engineering right.

In energy production, heat requirement is a must for many systems ranging, for example, from power plants to vehicles where we have fuel combustion that is a chemical reaction. So, chemical energy production is generally recognized as a unique process where the goal is to generate thermal energy (so-called heat) through exothermic chemical reaction(s) taking place, such as fuel combustion. Such conversion processes play a significant role in the energy production sector of the world. Continuous efforts are being spent on developing higher capacity, lower cost, more efficient, and more environmentally friendly energy systems. These energy systems are best analyzed from the energy and exergy points of view. Energy analysis is not sufficient to assess comprehensively the performance of energy systems; exergy analysis is believed by many to be of greater importance for these systems. Exergy analysis compares processes to the best possible process, e.g. a process that can extract the maximum amount of useful energy from the same energy source.

Fossil fuels are the most widely used fuel by people and societies. In fact, these carbon-based fuels satisfy the great majority of the energy demands of most countries. Many energy conversion technologies convert carbon-based fuel to electricity or mechanical energy. However, the use of these carbon-based fuels is always associated with carbon-based emissions. The impacts of these emissions on the environment has raised concerns worldwide about the current reliance on fossil fuels and fostered the search for more sustainable sources of energy that can meet humanities large and continuously rising energy demands. Many organizations predict that global energy demands will increase by up to 50% of the current level by 2030. Finding and exploiting energy sources that can replace fossil fuels is challenging, especially over a short period, but such efforts are ongoing. In addition, people,

Thermodynamics: A Smart Approach, First Edition. Ibrahim Dincer.

industries, and societies are seeking more efficient and environmentally benign ways of using and extracting useful forms of energy from fossil fuels.

Many methods are employed today to increase the efficiency of energy production from fossil fuels and to reduce the emissions per *kW* or *kWh* of electricity generated. One of these methods is to continuously run power plants at a constant rate and at full capacity; another approach is the integration of power systems. Running power plants at full capacity is not practical due to the fluctuating nature of energy demands during night and day, summer and winter, etc. There is a need for storage as well.

Coming back to thermal energy (heat) requirement, the need is huge and is why many countries go around fossil fuels where combustion becomes the heart of the equation. A combustion process involves the oxidation of a fuel in order to release thermal energy. Fuels include not only fossil fuels but also many other materials that can be reacted exothermically, such as hydrogen and ammonia. A fuel is defined as any material that can be burned to release thermal energy. Fossil fuels are, nonetheless, the dominant fuel used currently globally, accounting for nearly 78% of the total energy used today. This chapter focuses its discussion on combustion reactions; however, the methodology discussed in this chapter is applicable to all chemical reactions. The main objectives of this chapter are the reader should be able to understand how to balance the chemical reactions in order to be able to understand the amounts of reactants and products, then how to apply the balance equations on chemical reactions, and, finally, to be able to assess the performance of these processes.

The aim of this chapter is to first introduce the basic information about fuels and their heating values (*HVs*), forms of chemical reactions, fuel combustion related parameters and concepts, and to discuss the use of the first law of thermodynamics (*FLT*) and the second law of thermodynamics (*SLT*) for combustion processes. It is also aimed to provide principles and methodologies about how to write balance equations and analyze the combustion processes and evaluate their performances.

8.2 Fuels

The basic definition of *fuel* is any material that is combustible and produces heat if combusted. There are many types of fuels used in daily life in various systems and applications, some of which are natural and some of which are derived or artificially (synthetically) manufactured.

Note that a common fuel property is the *HV*. The *HV* of a fuel is the amount of thermal energy released during its complete combustion with the condition that the combustion exhaust products (reaction products) are the same as those of the reactants. However, in industrial applications the water (H_2O) in the exhaust is usually rejected to the environment as superheated steam. This is mainly because water can corrode the exhaust stack if it is rejected as a saturated vapor–liquid mixture.

The reason we burn fuels is to get heat, which may be classified into two categories. These categories are lower heating value (*LHV*), which is the amount of heat obtained from the combustion of the fuel less the amount of heat released with the water/water vapor (the so called net *HV*), and higher heating value (*HHV*, which is the amount of heat obtained from combustion of the fuel including the amount of heat released with the water/water vapor (the so called gross *HV*). Both the *LHV* and *HHV* are illustrated in Figure 8.1 to clearly show the differences.

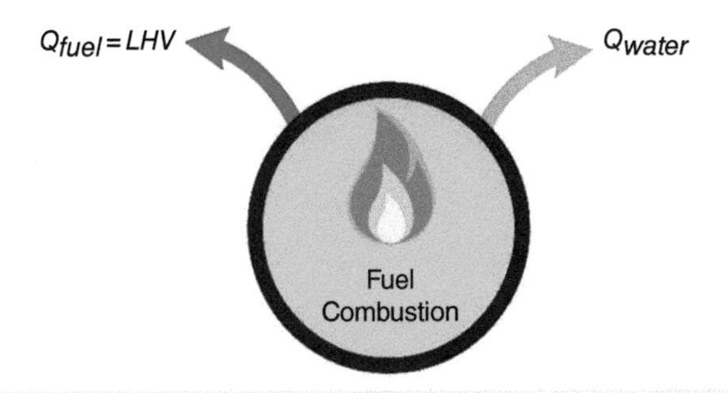

Figure 8.1 Illustration of *LHV* and *HHV*.

The relationship between the *LHV* and the *HHV* is:

$$HHV\left(\frac{kJ}{kg}\right) = LHV\left(\frac{kJ}{kg}\right) + x_{H_2O}h_{fg,H_2O} \tag{8.1}$$

Here, x_{H_2O} is the mass fraction of water produced by the combustion. The *HHV* of a fuel at the reference environment conditions can be calculated as:

$$HHV = \left|\sum n_p \overline{h}^o_{f,p} - \sum n_r \overline{h}^o_{f,r}\right| \tag{8.2}$$

where n_p is the number of moles of the product p and $\overline{h}^o_{f,p}$ is the molar enthalpy of formation of product p, while n_r is the number of moles of the reactant r and $\overline{h}^o_{f,r}$ is the molar enthalpy of formation of reactant r.

A common classification of fuels is made based on their occurrence and physical state. Based on the occurrence, they are categorized into two types, namely primary fuels (which are directly available in nature) and secondary fuels (which are derived from primary fuels). In this regard, a true classification with some examples is given in Figure 8.2 based on their physical states and occurrences. In this chapter, the remaining sections mainly discuss fossil fuels (which are the most familiar fuels) and their combustion.

In any fuel combustion process, air is needed to oxidize the fuel for combustion. Dry air is approximated as 21% oxygen and 79% nitrogen by mole numbers, which results in the following: each mole of oxygen entering a combustion chamber is accompanied by 0.79/0.21 = 3.76 *mol* of nitrogen. Both fuel and air are named as reactants and the compounds coming out as products (Figure 8.3. Products may vary depending on the fuel composition. Of course, hydrocarbon fuels under incomplete combustion will always produce water vapor, carbon dioxide, carbon monoxide, unreacted nitrogen, etc.

In combustion reactions, the air–fuel ratio (*AFR*) is an important parameter and defined as the ratio of the mass of air to the mass of fuel for a combustion process.

$$AFR = \frac{m_{air}}{m_{fuel}}$$

with $m = N \times M$

where m is mass (kg), N is number of moles, and M is the molecular weight ($kg/kmol$).

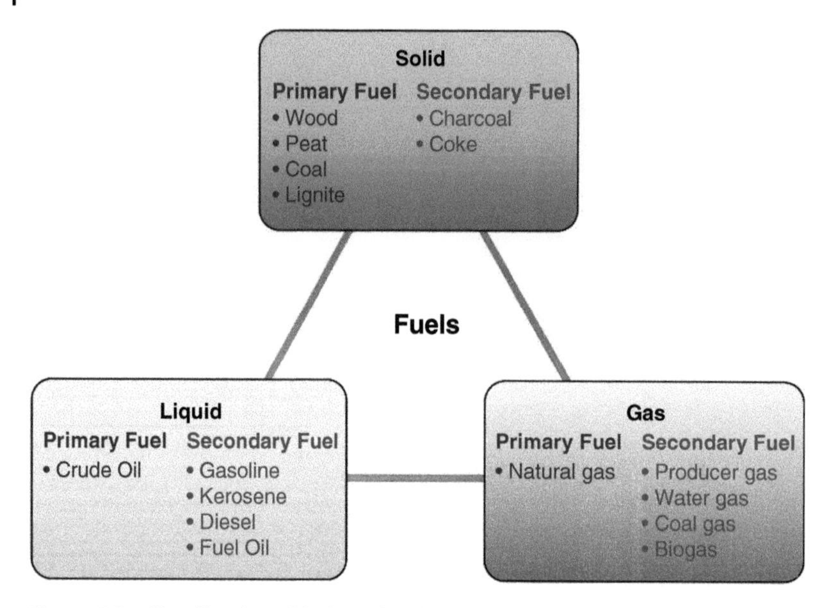

Figure 8.2 Classification of fuels under three physical states, namely solid, liquid, and gas.

8.3 Fossil Fuels

As stated in the introduction section of this chapter, fossil fuels have been around for a long time and played a critical role after the industrial revolution, with primarily coal, oil and natural gas; they continue to be the main energy source that most countries in the world depends on. A look back at the history of humankind and a prediction of the future based on the use of the fossil fuels are presented in Figure 8.4. As shown in the figure, the usage of

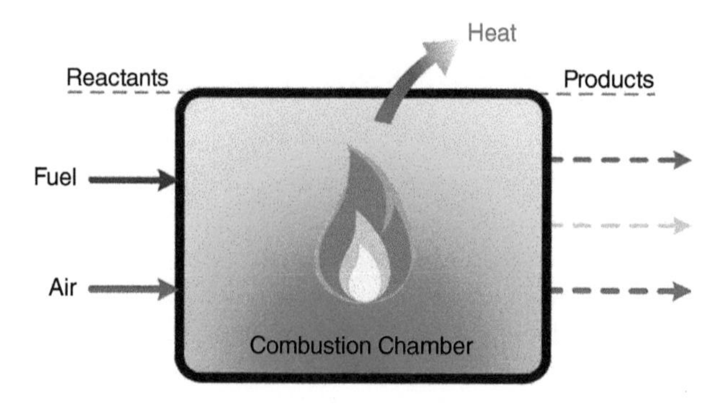

Figure 8.3 An illustration of a combustion reaction with reactants and products as well as heat.

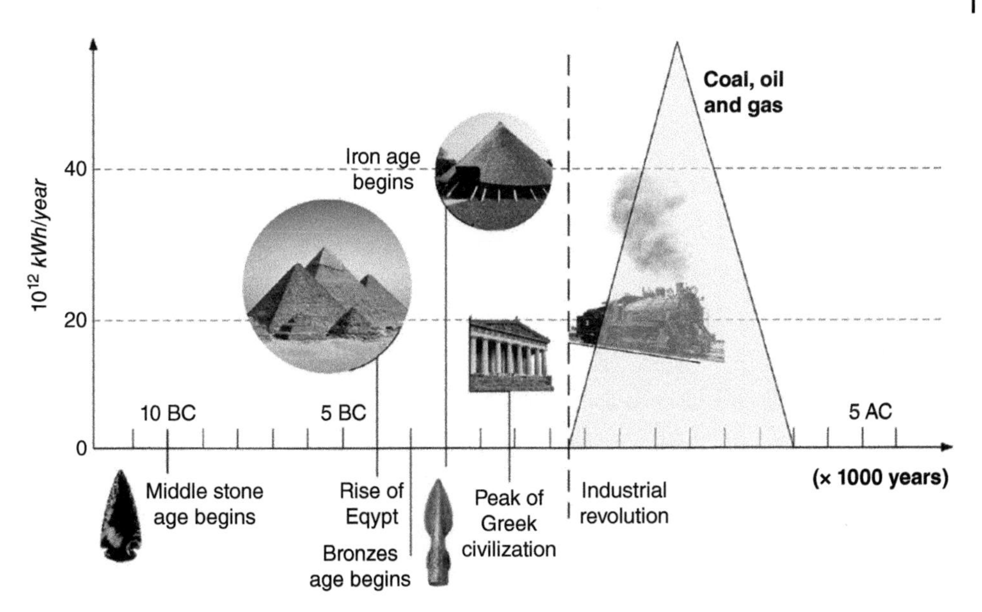

Figure 8.4 The utilization of the fossil fuels throughout the last 10 thousand years and the predicted over the next five thousand years in the future.

fossil fuels, mainly oil and natural gas, throughout the last 10 thousand years to the predicted next five thousand years in the future is given to show the fossil fuel era. Note that here the fossil fuel era is identified here in terms of the usage of oil and natural gas. We can see that the fossil fuel era will end when the resources of these fuels is depleted, and then it is projected that renewable energy resources era will come in. The historical background of fossil fuels and how they were formed is presented in brief next.

In brief, fossil fuels were formed many hundreds of million years ago and it is thought to have happened before the time of the dinosaurs. The age where the fossil of fuels were formed is referred to as the Carboniferous period, which is about 286–360 million years ago (Figure 8.4). The use of these fuels began with the industrial revolution (1760–1840). An example of how the fossil fuels were formed and became as we know them today is presented in Figure 8.5.

As shown in Figure 8.5, in the age before dinosaurs were believed to have walked the planet, giant plants died in swamps and were covered and buried under water and dirt for millions of years. These dead plants were subject to high heat from the earth's core and also to high pressures from the weight of the dirt and water covering them, which eventually turned them into coal.

In this section, there is a need to briefly discuss fossil fuels in the order of coal, oil, and natural gas and their heat contents (*HVs*), as well as some potential impacts. As mentioned earlier, the process of forming fossil fuels happened naturally millions of years ago; they also took millions of years to form and become the form that we use today. Figure 8.6 is an illustration of these three fossil fuels and general composition of what they contain. Since all

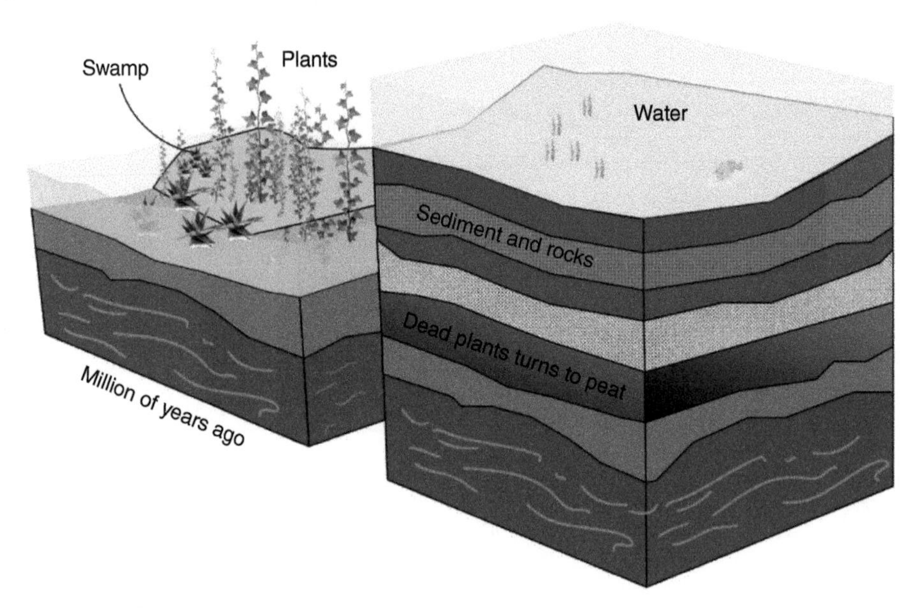

Figure 8.5 The process and the steps of how coal was formed summarized in mainly three steps.

three contain both carbon and hydrogen, they are so-called: hydrocarbon fuels. The next subsections will elaborate more on these fossil fuels.

8.3.1 Coal

Coal exists in a large variety of geologic forms and levels of quality and, more importantly, with specific *HVs*. Coal is characterized in a continuous sequence, beginning with peat, then lignite (brown coal), then bituminous (soft coal), and finally anthracite (hard coal) and graphite. The utilization of coal is primarily for steam and gas turbines for power generation and, furthermore, for coking in the iron and the steel industry.

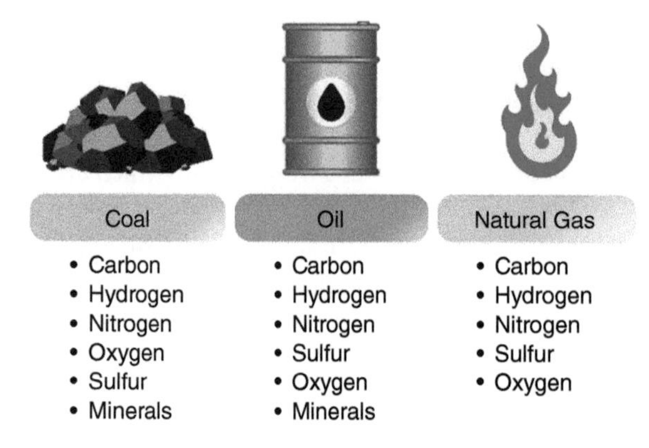

Figure 8.6 A comparative illustration of fossils fuels, namely coal, oil, and natural gas.

In addition to the categorization of coal presented previously, coal can be categorized based on the variation in the composition of its main elements. For coal, the central element is carbon. In addition, it contains varying amounts of other elements such as oxygen, hydrogen, nitrogen, sulfur, and chlorine as well as water and ash. The composition of coal varies with geographical location and coal type. Coals are herewith graded based on the weight percentages of their primary constitutes and the amount of heat that can be extracted from a given type, which also goes back to the coal composition. As per the story of the formation of coal, coals from different locations on earth end up with different compositions and thus are of different grades. To illustrate the effects of the location and the origin of the giant plants that made these coals, three types of coal from different regions of the world are compared based on their element composition; these are presented in Table 8.1. It is seen that even coals from the same

Table 8.1 Properties of three types of coal from three different parts of the world.

		Illinois #6[a]			Soma[a]			Elbistan[a]		
		Wet	Dry basis	Dry and ash free basis	Wet basis	Dry basis	Dry and Ash free basis	Wet basis	Dry basis	Dry and ash free basis
Proximate analysis	Moisture (*wt%*)	0.20	0.00	0.00	15.10	0.00	0.00	16.30	0.00	0.00
	Fixed carbon (*wt%*)	58.01	58.01	68.68	38.48	45.32	52.50	35.73	42.69	50.50
	Volatile matter (*wt%*)	26.46	26.46	31.32	34.82	41.01	47.50	35.03	41.85	49.50
	Ash (*wt%*)	15.53	15.53	0.00	11.60	13.66	0.00	12.93	15.45	0.00
Ultimate analysis(*wt%*)	*C*	73.90	74.05	87.66	52.48	61.81	71.61	48.76	58.25	68.92
	H	6.24	6.25	7.40	3.80	4.48	5.19	3.25	3.88	4.59
	N	0.71	0.71	0.84	1.32	1.55	1.80	1.27	1.52	1.80
	Cl	0.37	0.37	0.44	0.00	0.00	0.00	0.00	0.00	0.00
	S	1.77	1.77	2.10	2.71	3.19	3.70	3.67	4.39	5.19
	O	1.32	1.32	1.56	12.97	15.28	17.70	13.79	16.48	19.50
	Ash	15.53	15.53	0.00	11.60	13.66	0.00	12.93	14.45	0
Sulfur analysis	Pyritic (*wt%*)	0.59	0.59	0.70	0.08	0.09	0.10	0.04	0.05	0.06
	Sulfate (*wt%*)	0.59	0.59	0.70	0.00	0.00	0.00	0	0	0
	Organic (*wt%*)	0.59	0.59	0.70	2.63	3.10	3.60	3.63	4.34	5.13
Heating value	HHV (*MJ/kg*)	25.35	29.14	33.35	19.9	23.4	27.1	18.6	22.2	26.3
	LHV (*MJ/kg*)	24.12			18.98			17.8		

[a] Compiled from multiple sources.

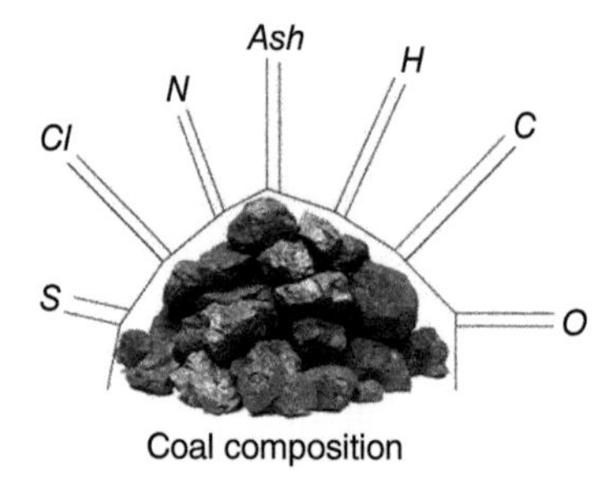

Figure 8.7 Illustration of coal composition.

country can have different compositions, depending on the geological formations. Figure 8.7 illustrates the coal with a common composition. Thus, the chemical elements that comprise coal are sulfur (*S*), chlorine (*Cl*), nitrogen (*N*), ash, hydrogen (*H*), carbon (*C*), and oxygen (*O*). Therefore, coal types are referred to as high-, medium-, or low-grade depending on their *HV*, which depends on their composition, as is discussed in more detail later.

8.3.2 Crude Oil

Another commonly used fossil fuel is crude oil (petroleum), which is made up of a mixture of hydrocarbons, mainly alkanes of formula C_nH_{2n+2}, and several impurities, with sulfur being the most important (sweet crude: oil with a low sulfur content and sour crude: oil with high sulfur content). Hydrocarbons are classified by their chain lengths. Those with longer chain lengths generally have lower boiling points, which allows for their separation by distillation in refineries. Most of the liquid hydrocarbon fuels used in vehicles, such as airplanes, automobiles, trucks, trains, and ships, are derived from crude oil, using chemical and distillation processes. Crude oil distillation, a critical step in the oil (petroleum) industry, is achieved in refineries. This distillation is one of the basic unit operations where the aim is to separate crude oil into different hydrocarbon components, or fractions (including fuels, chemicals, etc.), depending on their weights and boiling points. Heated crude oil is delivered into the distillation column after the furnace. Heavier fractions are recovered from the bottom part (such as, residual fuel oil and fuel oil) at the highest temperatures and lighter ones from the top (such as naphtha, gasoline, and butane) are recovered at the lowest temperatures. Figure 8.8 shows an illustration of crude oil distillation into various hydrocarbon components or fractions (or so-called petroleum products).

8.3.3 Natural Gas

Natural gas is the main hydrocarbon based gaseous fuel found in nature and may be extracted from gas wells directly or oil wells that are rich in natural gas. It can be used as a fuel for many systems and applications in various sectors, ranging from power generation to residential and from industrial to transportation. In such instances, it is

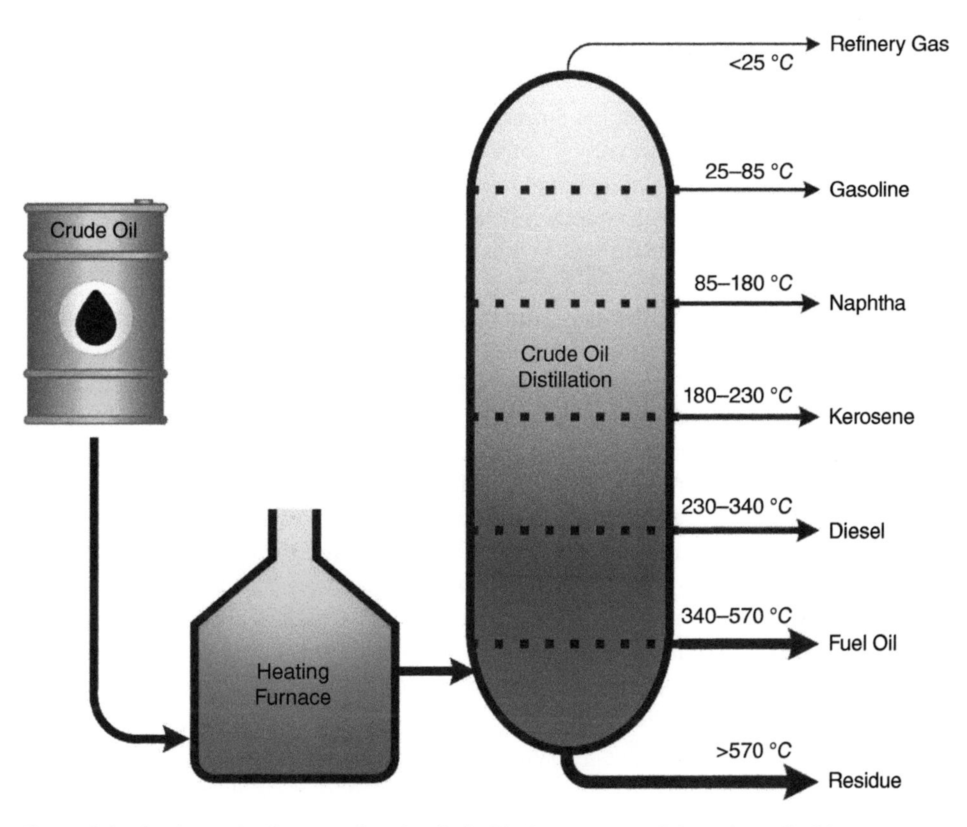

Figure 8.8 A schematic diagram of crude oil distillation to get useful products (fuels).

typically stored in tanks in one of two possible forms: compressed natural gas (*CNG*) or liquefied natural gas (*LNG*). The pressure of *CNG* varies between 150 and 250 *bar*, whereas for *LNG*, natural gas is in a liquid state at a temperature of $-162\,°C$ at ambient pressure. One should note that both liquid hydrocarbons and gaseous hydrocarbons are usually treated as if they have a single hydrocarbon composition. For example, although natural gas is primarily known as a gaseous hydrocarbon mainly composed of methane, it additionally contains ethane (C_2H_6), nitrogen (N_2), carbon dioxide (CO_2), propane (C_3H_8), and normal butane and isobutene (C_4H_{10}), having their shares ranging from 10 to 0.1%, respectively (Figure 8.9).

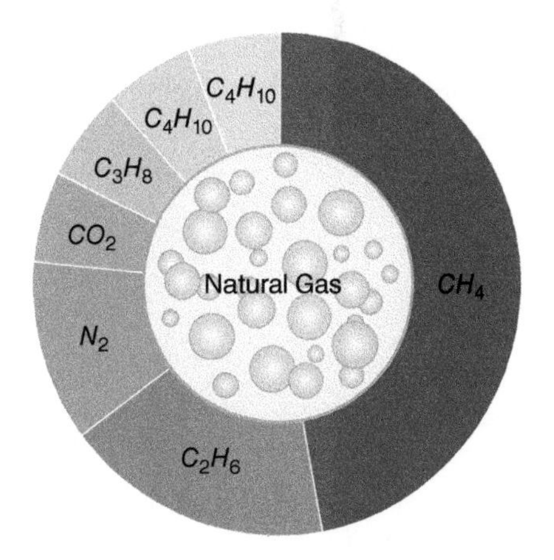

Figure 8.9 An approximate composition of natural gas.

8.4 Forms of Chemical Energy

The energy that is released or stored during a chemical reaction is essentially considered *chemical energy*. In any chemical reaction, there are reactants that are involved in the reaction and there are products coming out of the reaction (Figure 8.10). If the reaction needs heat to initiate, this is called an endothermic reaction, and if heat is generated during the reaction, this is called an exothermic reaction. One of the most common exothermic reaction is the fuel combustion reaction, which is the main focus of this chapter.

A chemical reaction may be defined as the characterization by rearrangement of the ionic or molecular structure of a substance or substances to form a new chemical structure that is different from that of all of the participating reactants. Remember that chemical reactions are unlike the change in the physical state of the material through the four material phases.

One should note that nuclear reactions are not categorized as chemical reactions, since they change the atomic structure of atoms, unlike chemical reactions which change the structure of the molecules and how the atoms are connected to each other. The most common form of chemical energy is in the form of bonds between the atoms that make up a molecule. Chemical bonds hold together the atoms that constitute a chemical compound. There are two types of chemical bond: covalent and ionic. Covalent bonds exist when the atoms share electrons to form a chemical compound in the form of a molecule. Ionic bonds are formed when one atom gives up its electron to another atom, which yields two electrically charged particles, so-called ions.

8.5 First Law of Thermodynamics Analysis

Further to the discussion about chemical reactions, thermodynamic laws have to be considered. The energy balance equations and analysis introduced in earlier chapters under the FLT are also applicable for energy systems where chemical reactions occur. However, the various systems and applications considered earlier did not undergo specific chemical reactions, they were rather physical and only physical quantities were involved in the systems based on the state point changes through the properties. This section presents the

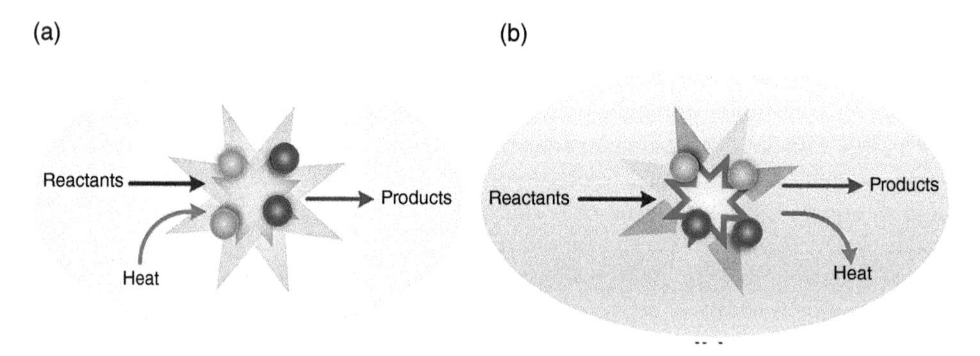

Figure 8.10 Illustrations of types of chemical reactions: (a) endothermic and (b) exothermic.

energy balance relations for energy systems that have chemical reactions and combustion is given as an example of this. Combustion was selected to be the example for chemical reactions since it is one of the most common and ancient processes that humans were able to use.

One can look at fossil fuels in essence as stores of chemical energy that were formed over thousands of years by the actions of thermal and mechanical energies. One of the most commonly used methods to extract energy from fossil fuels is combustion, which has been used for years to convert the stored chemical energy in fossil fuels to thermal energy.

One of the most common fossil fuels used is crude oil, which is a mixture of various types of liquid hydrocarbons. Most of the liquid hydrocarbons used in vehicles, airplanes, automobiles, trucks, trains, and ships, are essentially derived from crude oil, using chemical and distillation processes. The chemical and distillation processes are employed to separate the different hydrocarbons that makes up the crude oil. Note that crude oil composition depends on its source and is modified by the refinery operations to which it is subjected, including the distillation processes. Liquid hydrocarbons produced from the distillation of crude oil do not consist of a single type of hydrocarbon; rather, they are a mixture of different hydrocarbons. However, liquid hydrocarbons are usually treated as if they have a single hydrocarbon composition, based on their average make up. Similar to liquid hydrocarbons, gaseous hydrocarbons are usually treated as if they have a single hydrocarbon composition to simplify calculations with an acceptable accuracy. For example, natural gas is a gaseous hydrocarbon mainly composed of methane, and in most cases when natural gas is used in energy systems it is considered as methane and the results are within the acceptable range of accuracy. However, if we looked carefully, the actual composition is more complex, and also depends on the source from which the natural gas has been extracted. As seen from Figure 8.9, more than 90% of the content of natural gas is methane (CH_4), explaining why natural gas is often treated as methane to simplify analyses while providing acceptable accuracy. Later in the chapter we will calculate, using the mass and energy conservation principles, the percentage difference between the results of using a single composition of methane and using the exact composition of the natural gas. The approximation involved in treating natural gas as methane (CH_4) is discussed after Example 8.4.

The combustion process is the chemical reaction where the fuel is oxidized to convert part of the chemical energy stored in the fuel chemical bonds to thermal energy. For most of the combustion processes, air is used as the fuel oxidant since it is freely available in nature and comprises 21% oxygen, which is the fuel oxidant. In various cases, pure oxygen is used as the oxidant to achieve high combustion temperatures. For chemical reactions where air is used as an oxidant, a very important step in analyzing those chemical reactions is understanding the composition making up the air, which is (on a mole basis): 75.67% N_2, 20.35% O_2, 3.03% moisture, 0.91% Ar, 0.03% CO_2, and 0.01% H_2 at a temperature of 25 °C and a pressure of 101.325 kPa (1 atm). However, in the analysis of combustion processes, moisture-free (dry) air is often considered, and approximated as 79% N_2 and 21% O_2 on a mole basis, as mentioned in Section 8.2.

By using the approximation that 79% of air is N_2 and the remaining is O_2, means that for each mole of oxygen that enters a combustion chamber in air, 79/21 mol of nitrogen enter.

The large amount of nitrogen that enters the combustion chamber (3.76 times the amount of oxygen on a molar basis) does not contribute to the reactions, although small amounts of *NOx* are formed. However, the presence of nitrogen in air that is used in the combustion chamber reduces the temperature of the exhaust gases relative to the temperature that would exist if combustion occurred in pure oxygen. Having the nitrogen that enters the combustion chamber, usually at ambient pressure and temperature, absorb large amounts of thermal energy when heated to the exhaust gas temperature leads to a reduced potential for high-temperature exhausts. Unless otherwise stated, throughout this chapter nitrogen is assumed to be 100% inert, meaning that it does not contribute chemically in the combustion chemical reactions. In reality, however, small amounts of nitrogen react with oxygen at high temperatures forming *NOx*, which of concern environmentally. For example, if air is heated to 1000 °*C*, the resulting mole fraction of the reaction products, due to nitrogen reacting with oxygen, can be shown, based on the Gibbs free energy minimization approach, to be 4.63×10^{-4} (which is 0.0463%).

In any chemical reaction, the constituents on the left side of the chemical reaction equation, which are the feed to the reaction chamber, are called reactants. The products of the reaction, which are on the right side of the chemical reaction equation, are referred to as reaction products. The combustion of hydrocarbon fuels produces water (H_2O) and carbon dioxide (CO_2), as well as plus other products. Similar to what was stated for nitrogen, water is assumed to be nonreacting, although at very high temperatures water decomposes to some degree to H_2 and O_2. The water from a combustion reaction is usually released to the environment in gaseous form. If it is allowed to condense, it may result in corrosion of the exhaust piping and that is why it is made sure that the temperature of the exhaust gases leaving the combustion chamber is higher than the saturation temperature of water.

As an example, the combustion reaction for methane (CH_4) with pure oxygen can be written as:

$$CH_4(g) + 2O_2(g) \rightarrow 2H_2O(g) + CO_2(g) + Heat$$

where 1 *mol* of CH_4 requires 2 *mol* of O_2 to react and give 2 *mol* of H_2O and 1 *mol* of CO_2 under ideal (complete) combustion with full conversion.

It is important to note that reaction is balanced in terms of atoms, meaning that the number of atoms entering the reaction for each species is equal to those that leave the reaction. For example, in the combustion reaction of methane, we can see that one carbon atom enters the reaction in the mole of methane, and then one carbon atom leaves the reaction in a single mole of carbon dioxide. In a similar fashion, the numbers of all atoms have to balance with what is coming in and what is leaving the system, which is referred to as balancing the chemical reaction. The method of balancing will be introduced in detail in the coming examples in this chapter, and is often employed to determine the amount of oxygen needed for complete combustion. Note also that the presence of oxygen alone is not sufficient to cause combustion of a fuel. Fuel combustion requires, in addition to the availability of the correct amount of oxygen, an appropriate ignition temperature and an appropriate concentration of fuel in air. For example, the ignition temperature in atmospheric air of gasoline is 260 °*C* and that of hydrogen is 580 °*C*.

8.6 Combustion Reactions

This section focuses on chemical reactions and first starts with balancing chemical reactions through the thermodynamic analysis of energy systems that have chemical reactions. The example used here highlights the accuracy of assuming the natural gas as methane through the energy analysis of natural gas combustion systems versus considering the actual composition of natural gas.

Balancing the chemical reactions is an important part of the energy and exergy analysis of chemical reactors, since balanced chemical reactions makes sure that the masses of various species entering the system are equal to what leaves the reactor.

Example 8.1 This particular example is about establishing the combustion reaction and balancing it accordingly. The fuel considered here is methane, the simplest hydrocarbon fuel, which is made of a single carbon atom connected with a single bonds to four hydrogen atoms, e.g. CH_4. Methane is a flammable and is used mostly as fuel. Natural gas is often is approximated as methane since methane makes up more than 90% of natural gas. Write the chemical reaction equation that describes the combustion of methane with oxygen under ideal conditions and (b) balance this chemical reaction equation accordingly.

Solution

a) Write the chemical reaction equation that describe the combustion of methane in the presence of oxygen.

The oxygen combustion of methane under ideal (complete) combustion conditions is written in the following chemical reaction:

$$aCH_4(g) + bO_2(g) \rightarrow cCO_2(g) + dH_2O(l)$$

Note the above equation for the oxygen combustion of methane is an unbalanced reaction equation. The letters *a, b, c,* and *d* are constants that present the number of moles for each of the reactants and products.

b) Balance the chemical reaction equation written in part (a).

$$aCH_4(g) + bO_2(g) \rightarrow cCO_2(g) + dH_2O(l)$$

In order to determine the balanced chemical reaction equation for the combustion of methane by oxygen we need to find the values of the constants *a, b, c,* and *d*. Usually we aim to have the number of the moles of the main reactant, which is in this case is the methane, equal to one, although it is not necessary. This part of the question will be solved using two methods, where the first method highly depends on mathematical background to solve multiple equations and multiple unknowns. However, the first method can be difficult to use for large chemical reactions, which is why the second method is introduced in this example as well.

Method 1:

Balancing the chemical reaction means that the number of atoms of each species in the reactants side of the chemical reaction must equal those in the products side of the reaction, then we can write the relations between the constants *a, b, c,* and *d*. (Note that the number of equations will equal to the number of the different type of atom in the chemical reaction, for example for methane combustion in the presence of oxygen should have three equations, since we have carbon, hydrogen, and oxygen, i.e. three chemical species)

To balance the carbon atoms:

$$a = c$$

To balance the hydrogen atoms:

$$4a = 2d$$

To balance the oxygen atoms:

$$2b = 2c + d$$

Now we have three equations and four unknowns (a, b, c, and d), which means in order to solve for the unknowns and balance the methane combustion reaction equation we have to assume one of the unknowns. As mentioned earlier we usually prefer that the main chemical substance in the reaction, which in this case is methane, is equal to one:

$$a = 1$$

Then by solving the three equations with the three unknowns:

$$\left.\begin{aligned} c &= 1 \\ 4a &= 2d \\ 2b &= 2c + d \end{aligned}\right\} \rightarrow \begin{aligned} b &= 2 \\ c &= 1 \\ d &= 2 \end{aligned}$$

The balanced chemical reaction of the combustion of methane in the presence of oxygen only is then written as:

$$CH_4(g) + 2O_2(g) \rightarrow CO_2(g) + 2H_2O(l)$$

Method 2:

In order to determine the stoichiometric amount of oxygen required and the water product for complete combustion, the chemical reaction equation needs to be balanced. To balance the combustion equation for methane in the presence of stoichiometric oxygen (oxygen is the only reaction oxidant), we first balance any atom except for O and H, which are left to the end. The first step in balancing the chemical reaction of interest is to draw the first three rows of Table 8.2. There, the first compound column shows the number of atoms for each of

Table 8.2 Number of atoms of each element appearing in the methane combustion reaction.

Left side			Right side			
C	*H*	*O*	*C*	*H*	*O*	**Added/changed factor in front of each element**
1	4	2	1	2	3	
1	4	2	1	4	4	$2H_2O(l)$
1	4	4	1	4	4	$2O_2(g)$

the elements of all the chemical species in the reaction of interest, and the second compound column shows those on the right side of the chemical reaction. The next step is to compare the numbers of each element that are present on the right and the left sides of the chemical reaction. If the chemical reaction is not balanced, then one atom or more will have a different number on each side of the equation. The next step is to equate the numbers of the chemical elements/compounds in the chemical reaction by changing the coefficient of the chemical elements/compounds in the chemical reaction. The first change considered is multiplying the number of H_2O molecules by two (H_2O on the product side), in order to find the effect of that change on the number of the elements on each side of the reaction, as shown in the fourth row of Table 8.2. As can be seen in Table 8.2, hydrogen now has four atoms on the left side and four atoms on the right side of the chemical reaction. However, the oxygen atoms are still unbalanced. The next change, therefore, is to multiply the number of O_2 by two, and to find the result of that change on all elements in the reaction. As shown in the final row of Table 8.2, the number of the oxygen atoms is balanced on each side of the reaction as are the carbon and hydrogen atoms. After the element on each side of the equation (left and right sides) are balanced, then the final coefficients that appear in the first column from the right are transferred to the chemical equation. The balanced chemical equation is then used to develop an energy balance so that the system can be analyzed in detail.

From Table 8.2, the balanced chemical equation for methane combustion can therefore be written as:

$$CH_4(g) + 2O_2(g) \rightarrow CO_2(g) + 2H_2O(l)$$

From the balanced chemical reaction equation for the combustion of methane, it is seen that each mole of methane requires 2 *mol* of oxygen and produces 2 *mol* of water and 1 *mol* of CO_2.

Example 8.2 For simplicity, the composition of natural gas chemical is approximated as pure methane (CH_4), even though natural gas contains small amounts of other hydrocarbons including ethane, nitrogen, and others. Assuming complete combustion with a stoichiometric amount of oxygen, write the reaction equation of the combustion of the natural gas and balance it. See Figure 8.11 for further details.

Solution

The thermal energy released during natural gas combustion is calculated based on the mass and energy balance equations.

Two assumptions are made here: (i) natural gas enters the combustion reactor at ambient pressure and temperature and the resulting exhausts gases are at ambient pressure and temperature; and (ii) all gases are treated as ideal gases.

The properties and the chemical compositions of natural gas and methane are provided in Table 8.3, where the composition of natural gas is provided on a volume basis and the enthalpy of formation is provided on a mole basis. The properties of natural gas and methane (the main constituent of natural gas) are the basis of the next example. First, we

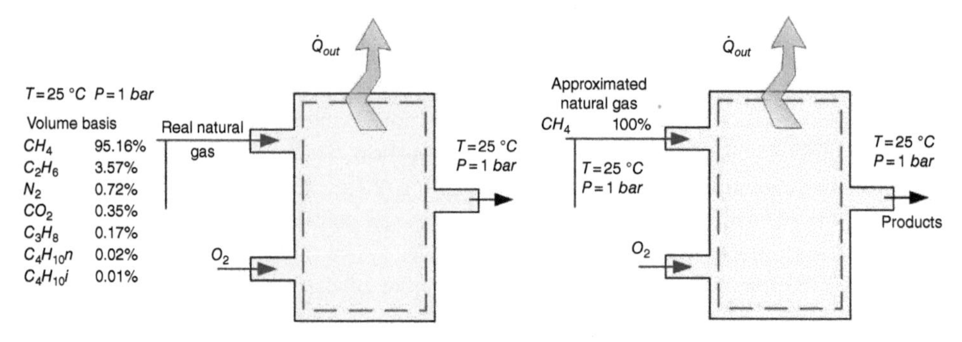

Figure 8.11 Schematic diagram for Example 8.2.

Table 8.3 Properties of actual natural gas and natural gas approximated as methane.

Constituent	Constituent chemical formula	Composition (*vol%*) Actual natural gas	Approximated natural gas	Enthalpy of formation (*kJ/mol*)
Methane	CH_4	95.16	100	−74.850
Ethane	C_2H_6	3.57	0	−84.680
Nitrogen	N_2	0.72	0	0
Carbon dioxide	CO_2	0.35	0	−393.520
Propane	C_3H_8	0.17	0	−103.850
Normal butane	C_4H_{10}	0.02	0	−125.790
Isobutane	C_4H_{10}	0.01	0	−134.99

determine for actual natural gas the stoichiometric amount of oxygen to achieve complete combustion with no excess oxygen remaining and the amount of water and CO_2 that results from the combustion reaction. This information is needed to solve the energy equation.

Since all gases are treated as ideal and since they enter the combustion chamber at ambient temperature and pressure, then based on Avogadro's law the molar composition of the actual natural gas is the same as the volume based composition.

The stoichiometric amount of oxygen required and the water produced are now determined for both cases.

a) Methane (CH_4):

Refer to Example 8.1 for the details of balancing the chemical reaction of methane. The balance equation can be written as:

$$CH_4(g) + 2O_2(g) \rightarrow CO_2(g) + 2H_2O(l)$$

b) Ethane (C_2H_6). The unbalanced chemical reaction for the combustion of ethane is:

$$C_2H_6(g) + O_2(g) \rightarrow CO_2(g) + H_2O(l)$$

To determine the stoichiometric amount of oxygen required and the water product for complete combustion, the chemical reaction equation needs to be balanced. To balance the combustion equation for ethane in the presence of a stoichiometric amount of oxygen (oxygen is the only reaction oxidant), first balance any atom except for O_2 and H_2, which are left to the end. We follow the same balancing procedure used in Example 8.1 (Table 8.2 to balance the ethane combustion equation, as shown in Table 8.4.

From Table 8.4 the balanced chemical equation for methane combustion is:

$$\mathbf{2}C_2H_6(g) + \mathbf{7}O_2(g) \rightarrow \mathbf{4}CO_2(g) + \mathbf{6}H_2O(l)$$

From the balanced chemical reaction equation of the combustion process of methane, every 1 *mol* of ethane requires 3.5 *mol* of oxygen and produces 3 *mol* of water and 2 *mol* of CO_2.

c) Propane (C_3H_8). The unbalanced chemical reaction for the combustion of propane is:

$$C_3H_8(g) + O_2(g) \rightarrow CO_2(g) + H_2O(l)$$

Following the same rules for balancing the above reaction as those used to balance the combustion reaction for propane. The balanced chemical reaction equation of the propane combustion is:

$$C_3H_8(g) + \mathbf{5}O_2(g) \rightarrow \mathbf{3}CO_2(g) + \mathbf{4}H_2O(l)$$

From the balanced chemical reaction equation of the combustion of propane with oxygen as the oxidation agent, each 1 *mol* of propane requires 5 *mol* of oxygen for a complete combustion with no remaining propane or oxygen. The combustion of 1 *mol* of propane produces 3 *mol* of CO_2 and 4 *mol* of water.

Table 8.4 Number of atoms of each element appearing in the ethane combustion reaction equation.

Left side			Right side			Added/changed factor in front of the element
C	*H*	*O*	*C*	*H*	*O*	
2	6	2	1	2	3	
2	6	2	2	2	5	$2CO_2$ (g)
4	12	2	2	2	5	$2C_2H_6$ (g)
4	12	2	2	6	7	$3H_2O$ (l)
4	12	2	4	6	11	$4CO_2$ (g)
4	12	2	4	12	14	$6H_2O$ (l)
4	12	14	4	12	14	$7O_2$ (g)

Table 8.5 Molar values of enthalpy of formation, Gibbs free energy, and absolute entropy at standard conditions (1 *atm* and 25 °C) for selected substances taken from the tables (such as Appendix G-1) and *EES*.

Substance (physical state)	$\overline{h}_f^o$ (kJ/kmol)	$\overline{g}^o$ (kJ/kmol)	$\overline{s}^o$ (kJ/kmolK)
$C(s)$	0	0	5.74
$CH_4(g)$	−74 850	−50 790	186.16
$C_2H_6(g)$	−84 680	−32 890	229.49
$C_3H_8(g)$	−103 850	−23 490	269.91
$C_6H_6(g)$	82 930	129 660	269.20
$CO(g)$	−110 530	−137 150	197.65
$CO_2(g)$	−393 520	−394 360	213.80
$H_2O(g)$	−241 820	−228 590	188.83
$H_2O(l)$	−285 830	−237 180	69.92
$H_2(g)$	0	0	130.68
$H(g)$	218 000	203 290	114.72
$O_2(g)$	0	0	205.04
$O(g)$	249 190	231 770	161.06

d) Normal butane and isobutane. The balanced chemical reaction for the combustion reaction of normal butane and isobutane is:

$$2C_4H_{10}(g) + 13O_2(g) \rightarrow 8CO_2(g) + 10H_2O(l)$$

The balanced reaction equation for the combustion of normal butane and isobutane shows that for the complete combustion of each mole of either normal butane or isobutane it is necessary to provide 6.5 *mol* of oxygen. The combustion of either normal butane or isobutane will produce 4 *mol* of CO_2 and 5 *mol* of water. Note that the balance equations of butane and isobutane are based on the combustion of 2 *mol* of butane or isobutane.

From the discussion and the definitions presented earlier regarding the enthalpy of the reactants and the products comes the $\overline{h}_f^o$, the enthalpy (or heat) of formation of the specified reactant. The value of the heat of formation of naturally occurring elements is zero. The enthalpy of formation of H_2 gas at 1 *atm* and 25 °*C* is 0 *kJ/kmol*. However, the energy of formation if *H* (monoatomic hydrogen) in the gaseous state at 1 *atm* and 25 °*C* is 218 000 *kJ/kmol*. For some common materials and fuels, molar values of the enthalpy of formation, the Gibbs free energy, and the absolute specific entropy at standard conditions (1 *atm* and 25 °*C*) are listed in Table 8.5.

8.7 Combustion in a Closed System

Combustion chambers are treated as reactors and can be analyzed either on the concept of closed systems. For closed combustion chamber analysis we usually use the concept of the HV since for this concept we are concerned with the amount of fuel burned only and how

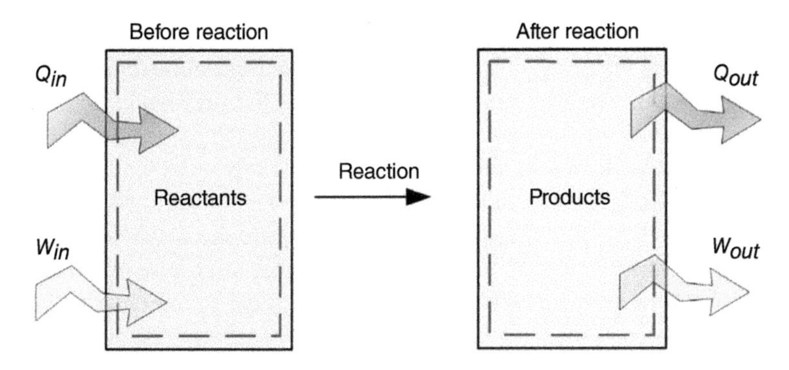

Figure 8.12 Schematic diagram of a closed energy system with chemical reactions.

much energy it will release; we still do not focus on the outputs of the reaction other than the heat energy. Next, an example is presented that describes the use of *LHV* and *HHV* when substituting gasoline engines with propane engines. Vehicles with propane engines can have lower environmental impacts.

After briefly discussing both *LHV* and *HHV*, one may wonder how to proceed to analyze the combustion process taking place in a closed system. In this regard, there is a need to write the combustion reaction with the fact that reactants (such as fuel and air) and products will not cross the boundary, as they will remain within the combustion chamber. A general energy balance equation for closed systems, as shown in Figure 8.12, can be written as:

$$Q_{in} + W_{in} + U_{react} = Q_{out} + W_{out} + U_{product} \tag{8.3}$$

Here, U denotes the internal energy of the reactants when it has the subscript react and that of the products when it has the subscript product. From previous chapters, the internal energy is related to enthalpy as follows:

$$\overline{u} = \overline{h} - P\overline{v}$$

Or we can present it in terms of formation properties as:

$$\overline{u}_f^o + \overline{u} - \overline{u}^o = \overline{h}_f^o + \overline{h} - \overline{h}^o - P\overline{v}$$

And by using the above expression of the internal energy we can present the energy balance equation for closed systems in terms of enthalpy rather than internal energy as:

$$\begin{aligned}&\underbrace{\dot{Q}_{in} + \dot{W}_{in} + \dot{n}_{r1}\left(\overline{h_f^o} + \overline{h} - \overline{h^o} - P\overline{v}\right)_{r1} + \dot{n}_{r2}\left(\overline{h_f^o} + \overline{h} - \overline{h^o} - P\overline{v}\right)_{r2}}_{\text{Energy flow rate of the reactants}} \\ &= \underbrace{\dot{Q}_{out} + \dot{W}_{out} + \dot{n}_{p1}\left(\overline{h_f^o} + \overline{h} - \overline{h^o} - P\overline{v}\right)_{p1} + \dot{n}_{p2}\left(\overline{h_f^o} + \overline{h} - \overline{h^o} - P\overline{v}\right)_{p2}}_{\text{Energy flow rate of the products}}\end{aligned} \tag{8.4}$$

Here, the term $P\overline{v}$ can be neglected for liquids and solids, while with the help of the ideal gas law it can be replaced with R_uT for ideal gases. In addition, the term $\overline{h} - P\overline{v}$ can also be replaced with the internal energy physical property $\overline{u}$.

Example 8.3 It is commonly known that propane has a lower environmental impact than some conventional vehicle fuels (e.g. diesel, gasoline), and some gasoline (benzene) fueled vehicles are herewith converted to propane fueled vehicles. One question in such a conversion is what would change if the vehicle is not reprogramed for propane. This example considers what is needed to obtain the same energy from propane as is provided by gasoline, i.e. the amount of propane required, and whether the same gasoline tank can still be used for propane. Consider a mid-size vehicle running on benzene that has a fuel consumption (volume basis) of $5.0\ L/100\ km$ on the highway. A mechanic is required to convert the car engine to run on propane while maintaining the same power output, as illustrated in Figure 8.13.

a) Calculate the volume of propane tank required to run the vehicle at the same performance, which means that the amount of propane must be able to produce thermal energy equal to that of the $5.0\ L$ tank of gasoline while maintaining an environmental pressure and temperature of the tanks for both fuels, to be done based on the *LHV* of both fuels

b) Determine if it is feasible to make the fuel switch as constrained by the problem, or if other tank pressures should be used

Solution

Based on the *LHV*, the volume of propane fuel required to produce the same thermal power provided by $5.0\ L$ of benzene is calculated. The following assumptions are made: (i) changing the fuel has no effect on the combustion process and (ii) the *LHV* can be considered in the analysis.

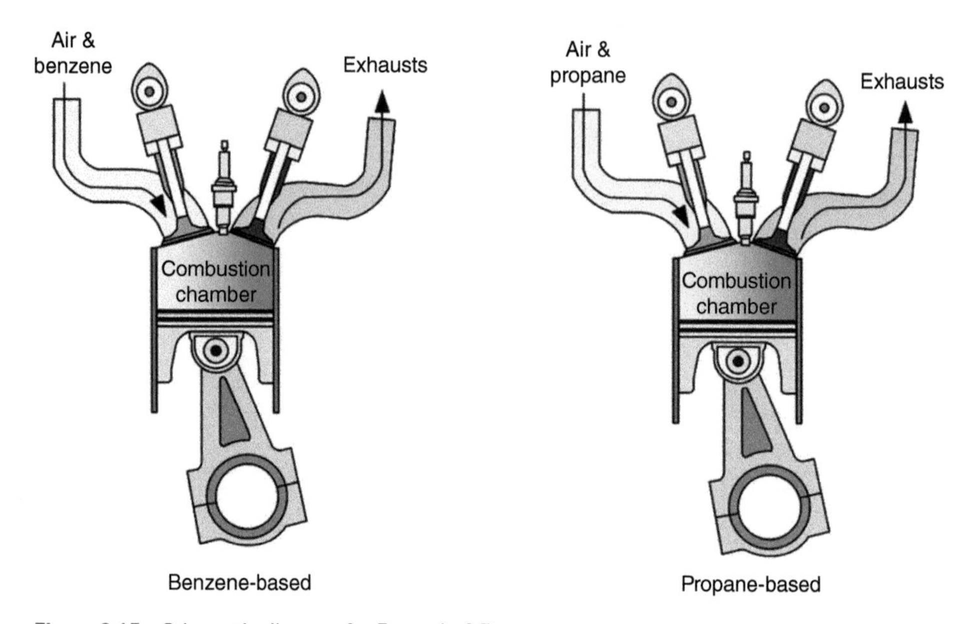

Figure 8.13 Schematic diagram for Example 8.3.

The LHV of benzene (C_6H_6) is 3169.6 kJ/mol and of propane (C_3H_8) it is 2043.4 kJ/mol. The molar density of C_6H_6 is 11.16 mol/L and of C_3H_8 it is 0.041 mol/L.

a) The thermal energy benzene produces when 5.0 L are completely burned is:

$$Q_{C_6H_6} = \overline{LHV}_{C_6H_6}\left(\frac{kJ}{mol}\right) \times \overline{\rho}_{C_6H_6}\left(\frac{mol}{L}\right) \times V_{C_6H_6}(L)$$

$$= 3,169.6\left(\frac{kJ}{mol}\right) \times 11.16\left(\frac{mol}{L}\right) \times 5.0\ (L)$$

$$= 176,864\ kJ$$

Since the same thermal energy is required to be released when using propane:

$$Q_{C_6H_6} = Q_{C_3H_8}$$

$$Q_{C_6H_6} = \overline{LHV}_{C_3H_8}\left(\frac{kJ}{mol}\right) \times \overline{\rho}_{C_3H_8}\left(\frac{mol}{L}\right) \times V_{C_3H_8}(L)$$

$$176,864 = 2043.4\left(\frac{kJ}{mol}\right) \times 0.041\left(\frac{mol}{L}\right) \times V_{C_3H_8}(L)$$

$$V_{C_3H_8} = \mathbf{2,111\ L}$$

Note that this is the volume of a propane tank that can produce the same thermal energy as a 5.0 L tank of benzene at environmental pressure and temperature.

b) The propane tank volume is extremely large, since the propane is in the gaseous phase at ambient temperature and pressure while gasoline at ambient temperature and pressure is in the liquid phase. So it is not feasible to use such tanks. Higher pressure tanks are suggested for propane so as to have a much higher density and permit a greatly reduced tank volume.

As shown in Example 8.3, switching a vehicle combustion fuel from gasoline to propane requires an unrealistic tank volume, which means that propane must be at an elevated pressure compared to the pressure of the gasoline tank of the vehicle. This permits the propane to be in the liquid state, increasing density sufficiently to allow the use of a regular size fuel tank.

Furthermore, natural gas was once burned (flared) in the environment in some situations without taking advantage of the thermal energy released during its combustion. Natural gas is currently used as a fuel for vehicles and residential heating applications. In analyzing the use of natural gas, engineers sometimes assess whether it is necessary to consider the actual composition of natural gas or whether the loss in accuracy associated with treating natural gas as only methane, which is the main constituent of natural gas, is acceptable.

8.8 Combustion in Open Systems

Chemical reactions also follow the laws of thermodynamics and the conservation of mass principle. In this section the mass balance and energy balance will be applied to systems that undergo chemical reactions.

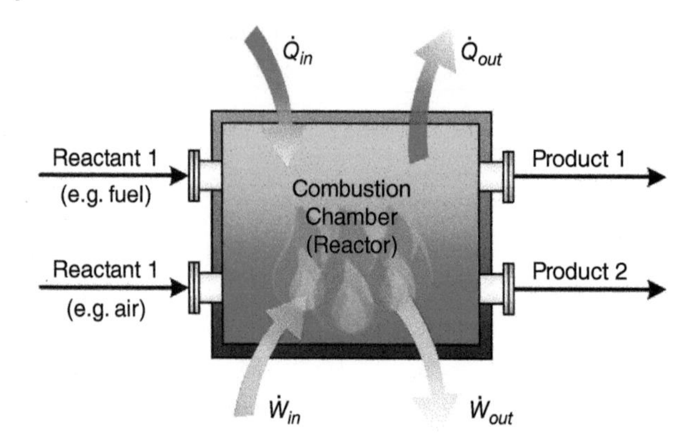

Figure 8.14 Schematic illustration of an open combustion chamber.

Here one can consider a combustion chamber (reactor), as shown in Figure 8.14 where there are in general two reactants (fuel and air) and two products coming out, with heat and work coming in and leaving out. A mass rate balance for the system in Figure 8.14 may be written as:

$$\dot{m}_{r1} + \dot{m}_{r2} = \dot{m}_p \tag{8.5}$$

Here, $\dot{m}$ denotes mass flow rate and the subscripts $r1$, $r2$, and p refer, respectively, to reactant 1, reactant 2, and products. An energy rate balance equation for the system in Figure 8.14 can be written in terms of moles as:

$$\begin{aligned} &\underbrace{\dot{Q}_{in} + \dot{W}_{in} + \dot{n}_{r1}\left(\overline{h_f^o} + \overline{h} - \overline{h^o}\right)_{r1} + \dot{n}_{r2}\left(\overline{h_f^o} + \overline{h} - \overline{h^o}\right)_{r2}}_{\text{Energy flow rate of the reactants}} \\ &= \underbrace{\dot{Q}_{out} + \dot{W}_{out} + \dot{n}_{p1}\left(\overline{h_f^o} + \overline{h} - \overline{h^o}\right)_{p1} + \dot{n}_{p2}\left(\overline{h_f^o} + \overline{h} - \overline{h^o}\right)_{p2}}_{\text{Energy flow rate of the products}} \end{aligned} \tag{8.6}$$

or in terms of mass as:

$$\begin{aligned} &\underbrace{\dot{Q}_{in} + \dot{W}_{in} + \dot{m}_{r1}\left(h_f^o + h - h^o\right)_{r1} + \dot{m}_{r2}\left(h_f^o + h - h^o\right)_{r2}}_{\substack{\text{Energy flow rate of the reactants} \\ \text{(energy entering the control volume)}}} \\ &= \underbrace{\dot{Q}_{out} + \dot{W}_{out} + \dot{m}_{p1}\left(h_f^o + h - h^o\right)_{p1} + \dot{m}_{p2}\left(h_f^o + h - h^o\right)_{p2}}_{\substack{\text{Energy flow rate of the products} \\ \text{(energy exiting the control volume)}}} \end{aligned} \tag{8.7}$$

Here, $\dot{Q}$ denotes heat transfer rate, $\dot{W}$ work rate, $\dot{n}$ mole flow rate, and h specific enthalpy, while the subscripts *in*, *out*, and *f* denote inflow, outflow, and formation, respectively. The superscript *o* denotes the property at standard conditions (usually typical ambient conditions) and the bar above the symbol indicates that it is on a molar basis. Note that, although the system in Figure 8.14 shows two inlets and two outlets, any number of inlets and outlets

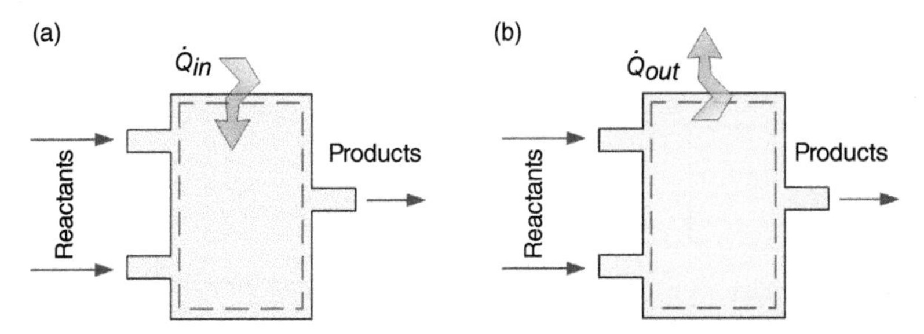

Figure 8.15 Illustration of steady-state steady-flow systems with (a) an endothermic chemical reaction and (b) an exothermic chemical reaction.

can be accommodated in the above energy equations. The properties presented in Eqs. (8.6) and (8.7) such as specific enthalpy are point functions, not path functions, which is why the initial and final properties are the only data required to analyze the energy interactions in the system.

Figure 8.15 shows an example of steady-flow reactors, where the categorization of chemical reactions in relation to their heat transfer interactions is also shown. Figure 8.15a shows a schematic of an endothermic reaction chamber, where the reaction absorbs thermal energy in order to complete the reaction. The energy balance of the endothermic reaction shown in Figure 8.15a can be written as:

$$\underbrace{\dot{n}_{r1}\left(\overline{h_f^o} + \overline{h} - \overline{h^o}\right)_{r1} + \dot{n}_{r2}\left(\overline{h_f^o} + \overline{h} - \overline{h^o}\right)_{r2}}_{\text{Energy flow rate of the reactants}} + \dot{Q}_{in} = \underbrace{\dot{n}_p\left(\overline{h_f^o} + \overline{h} - \overline{h^o}\right)_p}_{\text{Energy flow rate of the product}}$$

Figure 8.15b shows a schematic of an exothermic reaction, where the reaction releases thermal energy when in the reaction occurs. The energy balance of the exothermic reaction shown in Figure 8.15a can be written as:

$$\underbrace{\dot{n}_{r1}\left(\overline{h_f^o} + \overline{h} - \overline{h^o}\right)_{r1} + \dot{n}_{r2}\left(\overline{h_f^o} + \overline{h} - \overline{h^o}\right)_{r2}}_{\text{Energy flow rate of the reactants}} = \dot{Q}_{out} + \underbrace{\dot{n}_p\left(\overline{h_f^o} + \overline{h} - \overline{h^o}\right)_p}_{\text{Energy flow rate of the product}}$$

From this point on we will focus on combustion processes, including analyzing these processes in terms of mass and energy analyses.

Example 8.4 A steady-state combustion chamber, as shown in Figure 8.16, operates at a constant pressure of 1 *bar* and temperature of 25 °*C*. In the combustion chamber hydrogen (H_2) is used as a fuel and it is combusted using pure oxygen (O_2) as oxidant. Both the H_2 and O_2 are fed to the combustion chamber at the pressure and temperature of the combustion chamber. (a) Write the mass and energy balance equations of the combustion chamber, (b) calculate the mole flow rate of H_2O when 1 *mol/s* of H_2 and 0.5 *mol/s* of O_2 are fed to the reactor, and (c) find the heat rate released by the reactor.

Note that this simple example is hypothetical only, as it is unrealistic to maintain a hydrogen combustion reactor at a lower temperature than the ignition temperature of hydrogen

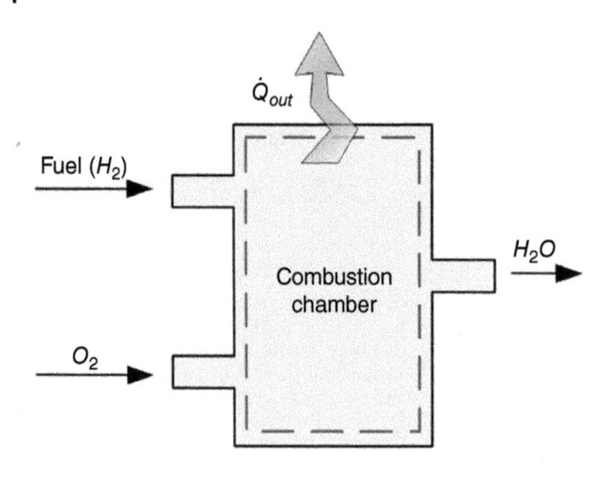

Figure 8.16 Schematic diagram for Example 8.4.

(580 °C) if combustion is to occur. One way to consider maintaining the temperature of the reactor at 25 °C is by continuously cooling it with water.

Solution

A specified amount of H_2 and O_2 react with each other in a constant pressure and temperature reaction chamber. The rate of heat released during the combustion reaction and the water production mole flow rate are to be determined. Some assumptions are made for analysis:

- The combustion of H_2 and O_2 occurs in a steady-state process.
- H_2O is produced as a liquid at 25 °C and 1 *bar*.
- The reactor material does not absorb any thermal energy and remains at constant temperature.
- Complete conversion of the reactants occurs in the reaction chamber (Figure 8.16) based on the stoichiometric chemical reaction.
- The reaction chamber has a constant volume.

Properties:
The molar mass ($\overline{m}$) of H_2 is 2.0 *g/mol*, of O_2 is 32.0 *g/mol*, and of H_2O is 18.0 *g/mol*. The enthalpy of formation of H_2 is 0 *kJ/mol*, of O_2 is 0 *kJ/mol*, and of H_2O in the liquid form is −285 830 *kJ/kmol*, as taken from the tables (such as Appendix G-1) and the Engineering Equation Solver (*EES*).

a) Write the mass and energy balance equations of the combustion chamber.

   Mass and energy balances for the reactor in the example are written based on Eqs. 8.5 and 8.6 as follows:

$$\text{MBE}: \dot{m}_{H_2} + \dot{m}_{O2} = \dot{m}_{H_2O}$$

$$\text{EBE}: \underbrace{\dot{n}_{r1}\left(\overline{h_f^o} + \overline{h} - \overline{h^o}\right)_{r1} + \dot{n}_{r2}\left(\overline{h_f^o} + \overline{h} - \overline{h^o}\right)_{r2}}_{\text{Energy flow rate of the reactants}} = \dot{Q}_{out} + \underbrace{\dot{n}_p\left(\overline{h_f^o} + \overline{h} - \overline{h^o}\right)_p}_{\text{Energy flow rate of the product}}$$

b) Calculate the mole flow rate of H_2O when 1 mol/s of H_2 and 0.5 mol/s of O_2 are fed to the reactor.

The mass balance equation is used to calculate the mole flow rate of the produced H_2O from the reactor:

$$\dot{m}_{H_2} + \dot{m}_{O2} = \dot{m}_{H_2O}$$

$$\dot{n}_{H_2}\overline{m}_{H_2} + \dot{n}_{O2}\overline{m}_{O_2} = \dot{n}_{H_2O}\overline{m}_{H_2O}$$

$$\left(1\frac{mol\,H_2}{s} \times 2.0\frac{g}{mol\,H_2}\right) + \left(0.5\,\frac{mol\,O_2}{s} \times 32\frac{g}{mol\,O_2}\right) = \dot{n}_{H_2O}\left(18\frac{g}{mol\,H_2O}\right)$$

$$\dot{n}_{H_2O} = \mathbf{1\ mol/s}$$

Thus, 1 mol/s of H_2O is produced when 1 mol/s of H_2 and 0.5 mol/s of O_2 are completely reacted.

c) Calculate the heat rate released by the reactor.

$$\underbrace{\dot{n}_{r1}\left(\overline{h}^o_f + \overline{h} - \overline{h}^o\right)_{r1} + \dot{n}_{r2}\left(\overline{h}^o_f + \overline{h} - \overline{h}^o\right)_{r2}}_{\text{Energy flow rate of the reactants}} = \dot{Q}_{out} + \underbrace{\dot{n}_p\left(\overline{h}^o_f + \overline{h} - \overline{h}^o\right)_p}_{\text{Energy flow rate of the product}}$$

By eliminating the null terms in the general energy equation and accounting for the fact that the reactants and the products are at ambient pressure and temperature, the energy equation reduces to:

$$\underbrace{\dot{n}_{H_2}\left(\overline{h}^o_f\right)_{H_2} + \dot{n}_{O_2}\left(\overline{h}^o_f\right)_{O_2}}_{\text{Energy flow rate of the reactants}} = \dot{Q}_{out} + \underbrace{\dot{n}_{H_2O}\left(\overline{h}^o_f\right)_{H_2O}}_{\text{Energy flow rate of the products}}$$

$$\left(1\,\frac{mol\,H_2}{s} \times 0\,\frac{kJ}{mol\,H_2}\right) + \left(0.5\,\frac{mol\,O_2}{s} \times 0\,\frac{kJ}{mol\,O_2}\right)$$
$$= \dot{Q}_{out} + \left(1\,\frac{mol\,H_2O}{s} \times (-285.830)\frac{kJ}{mol\,H_2O}\right)$$

$$\dot{Q}_{out} = \mathbf{285.830\ kW}$$

8.8.1 Incomplete Combustion

The combustion analysis presented in the previous examples is presented through the use of the energy balance equation to determine the amount of heat released by the combustion of fuels. Due to the common use of fuels and combustion analysis of fuels, a new fuel property has been defined, which is the *HV*. The next section discusses the *HV* of fuels and how they are used in the analysis of combustion processes. Before going to the next section though, a new concept – of theoretical air or what can also be referred to as stoichiometric air –is presented. All of the examples that have been considered in this chapter so far considered the combustion process or chemical reaction to be complete. A complete combustion, which is an example of a complete chemical reaction, is when all the carbon atoms in the fuel (for the case when the fuel is carbon based) are converted into CO_2, all the hydrogen atoms in the fuel are converted into H_2O, and the sulfur atoms if there is sulfur in the fuel are converted into SO_2. In general, a **complete combustion** is when all of the fuel is completely burned. In the other hand, an **incomplete combustion** is when the fuel is not completely burned or converted to products, which might be due to many different reasons, is mainly due to not having enough oxidants, which in many cases is the oxygen in the air, to oxidize the fuel. Other reasons for not having a complete combustion are not having enough contact time between the oxidant molecules and the fuel molecules, or that there is not appropriate mixing between the fuel and the oxidant.

Example 8.5 For simplicity, the composition of natural gas chemical is approximated as pure methane (CH_4), even though natural gas contains small amounts of other hydrocarbons, including ethane, nitrogen, and others. Assuming complete combustion with a stoichiometric amount of oxygen, determine the percentage difference between the thermal energy released when natural gas is approximated as methane and when the real composition of natural gas is considered. Parameters are given in Figure 8.17.

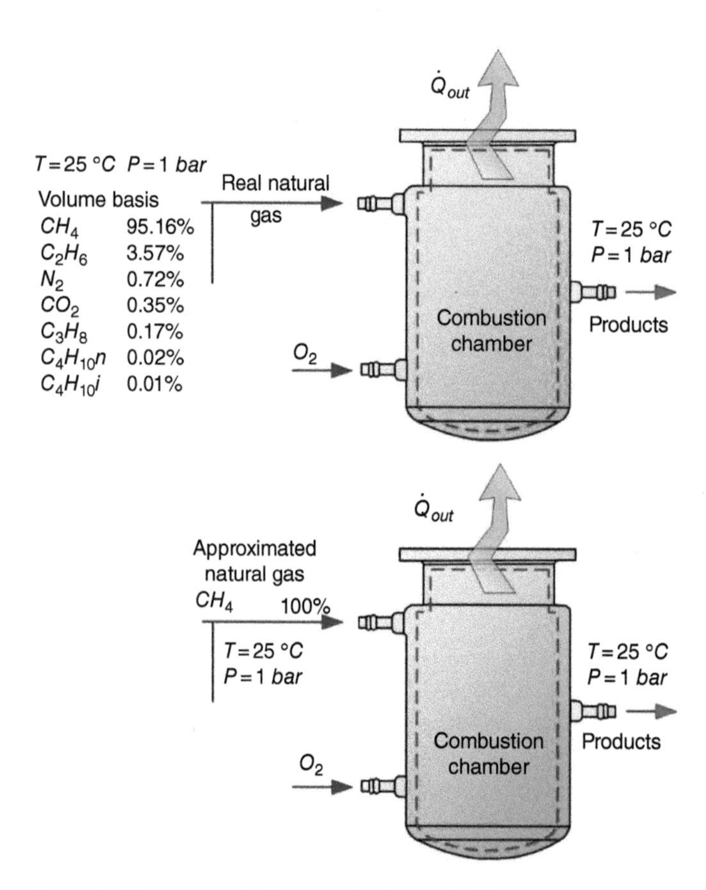

Figure 8.17 Schematic diagrams for Example 8.5.

Table 8.6 Required moles of oxygen for complete combustion of the selected 100 *mol* of real natural gas (note that 100 *mol* is selected for simplicity only).

Constituent	Constituent chemical formula	Composition (mole)	Moles of oxygen per each constituent
Methane	CH_4	95.16	190.32
Ethane	C_2H_6	3.57	12.495
Nitrogen	N_2	0.72	0
Carbon dioxide	CO_2	0.35	0
Propane	C_3H_8	0.17	0.85
Normal butane	C_4H_{10}	0.02	0.13
Isobutane	C_4H_{10}	0.01	0.065
Summation		**100**	**203.86**

Table 8.7 Required moles of oxygen a complete combustion of the selected 100 *mol* of the approximated natural gas (note that 100 *mol* is selected for simplicity only).

Constituent	Constituent chemical formula	Composition (mole)	Moles of oxygen per each constituent
Methane	CH_4	100	200

In Table 8.6, the analysis is based on 100 *mol* of natural gas to reduce the difficulty of calculations and to improve the clarity of the process since the natural gas chemical composition is presented in percentages indirectly.

Table 8.6 shows the specific amounts of oxygen needed for complete combustion of the real natural gas based on the results of balancing the combustion equations for all of the combustible components of the natural gas. Table 8.7 shows the specific amount of oxygen needed for a complete combustion of the approximated natural gas.

Tables 8.6 and 8.7 show that the combustion of the real natural gas will consume nearly 2% more moles of oxygen than the approximated natural gas. The resulting composition of the exhausts from the combustion reactions of 100 *mol* of the real natural gas and 100 *mol* of the approximated natural gas are shown in Tables 8.8 and 8.9. Table 8.8 presents part of the complicated analysis that accompanies the consideration of the original constituents of natural gas, where results of the combustion must consider each constituent of natural gas.

The second step, after finding out the exact amounts of oxygen needed for the complete combustion of both the real and the approximated natural gases, is applying the molar

Table 8.8 Resulting exhaust composition for complete combustion of the selected 100 *mol* of real natural gas (note that 100 *mol* are selected for simplicity only).

		Exhaust composition (mole)		
Fuel	**Fuel chemical symbol**	H_2O	CO_2	N_2
Methane	CH_4	190.32	95.16	0
Ethane	C_2H_6	10.71	7.14	0
Nitrogen	N_2	0	0	0.72
Carbon dioxide	CO_2	0	0.35	0
Propane	C_3H_8	0.68	0.51	0
Normal butane	C_4H_{10}	0.1	0.08	0
Isobutane	C_4H_{10}	0.05	0.04	0
Summation		**201.86**	**103.28**	**0.72**

Table 8.9 Resulting exhaust composition for complete combustion of the selected 100 *mol* of approximated natural gas (note that 100 *mol* are selected for simplicity only).

		Exhaust composition (mole)		
Constituent	**Constituent chemical formula**	H_2O	CO_2	N_2
Methane	CH_4	200	100	0

energy balance equations to find out the difference in the thermal energy released by each and how accurate the approximation is.

The energy balance equation for combustion of the real natural gas is (note the writing of the final balance equation will be presented in steps):

$$\underbrace{\dot{n}_{r1}\left(\overline{h_f^o}+\overline{h}-\overline{h^o}\right)_{r1}+\dot{n}_{r2}\left(\overline{h_f^o}+\overline{h}-\overline{h^o}\right)_{r2}}_{\text{Energy flow rate of the reactants}}=\dot{Q}_{out}+\underbrace{\dot{n}_p\left(\overline{h_f^o}+\overline{h}-\overline{h^o}\right)_p}_{\text{Energy flow rate of the products}}$$

The reactants are the constituents of the real natural gas and the products are those of the exhaust presented in Table 8.8. Since the reactants and the products are all at ambient pressure and temperature, then all of the $\overline{h}-\overline{h^o}$ terms cancel because in this instance $\overline{h}$ equals $\overline{h^o}$. After applying the approximations and expanding the reactants and the products, the energy balance becomes:

$$n_{CH_4}\overline{h_f^o}_{,CH_4}+n_{C_2H_6}\overline{h_f^o}_{,C_2H_6}+n_{N_2}\overline{h_f^o}_{,N_2}+n_{CO_2}\overline{h_f^o}_{,CO_2}+n_{C_3H_8}\overline{h_f^o}_{,C_3H_8}+n_{C_4H_{10}(n)}\overline{h_f^o}_{,C_4H_{10}(n)}$$
$$+n_{C_4H_{10}(iso)}\overline{h_f^o}_{,C_4H_{10}(iso)}+n_{O_2}\overline{h_f^o}_{,O_2}=Q_{out}+n_{CO_2}\overline{h_f^o}_{,CO_2}+n_{N_2}\overline{h_f^o}_{,N_2}+n_{H_2O}\overline{h_f^o}_{,H_2O}$$

$$(95.16 \times (-74.85)) + (3.57 \times (-84.68)) + (0.72 \times 0) + (0.35 \times (-393.520))$$
$$+ (0.17 \times (-103.85)) + (0.02 \times (-125.79)) + (0.01 \times (-134.99)) + (203.9 \times 0)$$
$$= Q_{out} + (103.28 \times (-393.520)) + (0.72 \times 0) + (201.86 \times (-285.83))$$

$$\boldsymbol{Q_{out,real}} = 90,756\ \boldsymbol{kJ}$$

For the approximate natural gas, the energy balance equation can be expressed as:

$$n_{CH_4}\overline{h_f^o}_{,CH_4} + n_{O_2}\overline{h_f^o}_{,O_2} = Q_{out} + n_{CO_2}\overline{h_f^o}_{,CO_2} + n_{H_2O}\overline{h_f^o}_{,H_2O}$$

$$(100 \times (-74.85)) + (200 \times 0) = Q_{out} + (100 \times (-393.52)) + (200 \times (-285.83))$$

$$\boldsymbol{Q_{out,app} = 89,033\ kJ}$$

The results of the combustion of 100 *mol* of each of the two natural gases shows that the approximated natural gas produces less thermal energy than the real natural gas. The percentage loss of accuracy (*LOA*) is calculated by:

$$\%LOA = \frac{Q_{out,real} - Q_{out,app}}{Q_{out,real}} \times 100$$

$$\%LOA = \frac{90,756 - 89,033}{90,756} \times 100 = \mathbf{1.89\%}$$

The analysis in this example shows that considering natural gas as a single hydrocarbon, specifically methane, provides results in term of thermal energy released during combustion that are 1.89% less than for real natural gas. This means that the approximation is acceptable in most engineering applications.

8.8.2 Adiabatic Flame Temperature

Adiabatic flame temperature or adiabatic combustion temperature is defined as the maximum temperature of products coming out of a combustion process where the system is adiabatic with no heat transfer in or out in the absence of work and kinetic and potential energies. This is, of course, considered for ideal situations. However, in practical combustion processes the temperature of products will be greatly lowered due to heat transfer, incomplete combustion, and dissociation of combustion gases. Figure 8.18 shows an illustration of an adiabatic combustion process with no heat transfer.

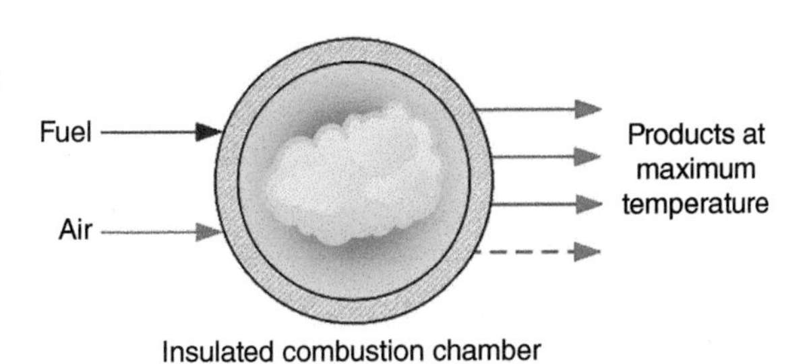

Figure 8.18 Adiabatic combustion chamber where the products come out at the maximum temperature.

If we consider the earlier balance equation, which was:

$$\underbrace{\dot{n}_{r1}\left(\overline{h_f^o}+\overline{h}-\overline{h^o}\right)_{r1}+\dot{n}_{r2}\left(\overline{h_f^o}+\overline{h}-\overline{h^o}\right)_{r2}}_{\text{Energy flow rate of the reactants}}=\dot{Q}_{out}+\underbrace{\dot{n}_p\left(\overline{h_f^o}+\overline{h}-\overline{h^o}\right)_p}_{\text{Energy flow rate of the products}}$$

$\dot{Q}_{out}=0$ due to adiabatic combustion, which will result in:

$$\underbrace{\dot{n}_{r1}\left(\overline{h_f^o}+\overline{h}-\overline{h^o}\right)_{r1}+\dot{n}_{r2}\left(\overline{h_f^o}+\overline{h}-\overline{h^o}\right)_{r2}}_{\text{Energy flow rate of the reactants}}=\underbrace{\dot{n}_p\left(\overline{h_f^o}+\overline{h}-\overline{h^o}\right)_p}_{\text{Energy flow rate of the products}}$$

It will finally result in:

$$\overline{H}_r=\overline{H}_p$$

Example 8.6 Consider an adiabatic combustion of gaseous hydrogen (H_2) at 25 °C and 100 kPa. Using enthalpy of formation data and assuming complete combustion with 100% theoretical air, calculate the adiabatic flame temperature.

Solution

One may first begin by writing the general combustion equation:

$$C_nH_m\,(g)+a_{th}\left(n+\frac{m}{4}\right)[O_2+3.76\,N_2]\rightarrow n\,CO_2+\frac{m}{2}\,H_2O+3.76\,a_{th}\left(n+\frac{m}{4}\right)N_2$$

where $a_{th}=1$, as the fuel is burnt with 100% theoretical air. The theoretical combustion equation of hydrogen can be written as:

$$H_2\,(g)+0.5\,[O_2+3.76\,N_2]\rightarrow H_2O+1.88\,N_2$$

In order to calculate the adiabatic flame temperature, we will proceed with $\dot{Q}_{out}=0$ due to adiabatic combustion, which will result in:

$$\sum_{\text{reactants}}\dot{n}_i\left[\overline{h}_f^0+\overline{h}_T-\overline{h}_{T_0}\right]_r=\sum_{\text{products}}\dot{n}_i\left[\overline{h}_f^0+\overline{h}_T-\overline{h}_{T_0}\right]_p$$

The heat of formation values $\overline{h}_f^0$ and the enthalpy values $\overline{h}_{T_0}$ at the reference condition are obtained:

Species	$\overline{h}_f^0$ [kJ/kmol]	$\overline{h}_{T_0=298\,K}$ [kJ/kmol]
H_2 (g)	0	8469
O_2	0	8682
N_2	0	8669
H_2O(g)	−241 820	9904

By substituting, one gets:

$$1\,[0]+0.5\,[0]+1.88\,[0]=\left[-241,820+\overline{h}_{H_2O}-9,904\right]+1.88\left[0+\overline{h}_{N_2}-8,669\right]$$

which yields to:

$$\bar{h}_{H_2O} + 1.88\,\bar{h}_{N_2} = \mathbf{260,393}$$

by trial and error method, one obtains:

at $T = 2000\,K$

$$1[82,593] + 1.88\,[64,810] = \mathbf{204,435.8}$$

at $T = 2600\,K$

$$1[114,273] + 1.88\,[86,650] = \mathbf{277,175}$$

by interpolation, the adiabatic temperature is obtained as:

$\boldsymbol{T_a} = \mathbf{2553.85}\ \boldsymbol{K}$ which is the maximum temperature that can be obtained.

Example 8.7 Consider an adiabatic combustion of gaseous ammonia (NH_3) at 25 °C and 100 kPa. Using enthalpy of formation data and assuming complete combustion with 100% theoretical air, calculate the adiabatic flame temperature.

Solution

One may first begin by writing the general combustion equation:

$$C_nH_m\,(g) + a_{th}\left(n + \frac{m}{4}\right)[O_2 + 3.76\,N_2] \rightarrow n\,CO_2 + \frac{m}{2}\,H_2O + 3.76\,a_{th}\left(n + \frac{m}{4}\right)N_2$$

where $a_{th} = 1$, as the fuel is burnt with 100% theoretical air. The theoretical combustion equation of ammonia can be written as:

$$NH_3\,(g) + 0.75\,[O_2 + 3.76\,N_2] \rightarrow 1.5\,H_2O + 3.82\,N_2$$

In order to calculate the adiabatic flame temperature, we will proceed with $\dot{Q}_{out} = 0$ due to adiabatic combustion which will result in

$$\sum_{reactants} \dot{n}_i\left[\overline{h}_f^0 + \bar{h}_T - \bar{h}_{T_0}\right]_r = \sum_{products} \dot{n}_i\left[\overline{h}_f^0 + \bar{h}_T - \bar{h}_{T_0}\right]_p$$

The heat of formation values $\overline{h}_f^0$ and the enthalpy values $\bar{h}_{T_0}$ at the reference condition are obtained:

Species	$\overline{h}_f^0$ [kJ/kmol]	$\bar{h}_{T_0=298\ K}$ [kJ/kmol]
NH_3 (g)	−46 190	26 335
O_2	0	8682
N_2	0	8669
H_2O(g)	−241 820	9904

By substituting, one obtains:

$$1[-46{,}190] + 0.75\,[0] + 2.82\,[0] = 1.5\left[-241{,}820 + \bar{h}_{H_2O} - 9{,}904\right] + 3.82\left[\bar{h}_{N_2} - 8{,}669\right]$$

which yields to:

$$1.5\,\overline{h}_{H_2O} + 3.82\,\overline{h}_{N_2} = \mathbf{364,511.58}$$

further, by trial and error:

at $T = 1800\,K$

$$1.5\,[72,513] + 3.82\,[57,651] = \mathbf{328,996.32}$$

at $T = 2000\,K$

$$1.5\,[82,593] + 3.82\,[64,810] = \mathbf{371,463.7}$$

after interpolation, the adiabatic flame temperature is calculated as:

$\boldsymbol{T_a} = \mathbf{1967.259}\ \boldsymbol{K}$ as to be the maximum temperature of the products.

Example 8.8 Determine the *HVs* (*HHV* and *LHV*) of gaseous octane C_8H_{18} at $25\,°C$ and $1\,atm$ using enthalpy of formation data. Assume that the H_2O in the product is in the liquid form.

Solution

First we begin by writing the general theoretical combustion equation:

$$C_nH_m\,(g) + \left(n + \frac{m}{4}\right)[O_2 + 3.76\,N_2] \rightarrow n\,CO_2 + \frac{m}{2}\,H_2O + 3.76\left(n + \frac{m}{4}\right)N_2$$

where the theoretical combustion equation of octane can be written as:

$$C_8H_{18}\,(g) + 12.5\,[O_2 + 3.76\,N_2] \rightarrow 8\,CO_2 + 9\,H_2O + 47\,N_2$$

a) To get the *HHV* of the fuel due to combustion:

$$HHV = H_p - H_r$$

$$HHV = \sum_{products} N_i\left[\overline{h}_f^0 + \overline{h}_T - \overline{h}_{T_0}\right]_p - \sum_{reactants} N_i\left[\overline{h}_f^0 + \overline{h}_T - \overline{h}_{T_0}\right]_r$$

The heat of formation values $\overline{h}_f^0$ and the enthalpy values $\overline{h}_{T_0}$ at the reference temperature at T_0 are obtained:

Species	$\overline{h}_f^0$ [kJ/kmol]	$\overline{h}_{T_0 = 298\ K}$ [kJ/kmol]
C_8H_{18} (g)	−208 450	−38.14
O_2	0	8682
N_2	0	8669
CO_2	−393 520	9364
$H_2O(l)$	−285 830	9904

By substituting, one gets:

$$HHV = |\,8\,[-393,520] + 9\,[-285,830] - [-208,450]|$$

$$HHV_{C_8H_{18}} = 5,512,180\ kJ/kmol$$

This can be written on the mass basis:

$$HHV_{C_8H_{18}} = \frac{5,512,180}{MW} = \frac{5,512,180}{8 \times 12 + 18} = \mathbf{48,352\ kJ/kg_{fuel}}$$

b) The LHV of the fuel is then obtained as:

$$LHV = HHV - \frac{(m\,h_{fg})_{H_2O@25^\circ C}}{m_{fuel}}$$

$$LHV_{C_8H_{18}} = 48,352 - \left[\frac{9 \times 18}{8 \times 12 + 18} \times 2442.3\right] = \mathbf{44,882\ kJ/kg_{fuel}}$$

The minus signs of the enthalpy of formation values reflect the exothermic reactions (which release heat).

8.9 SLT Analysis of Fuel Combustion Processes

The concept of entropy change of chemical reactions was introduced earlier with the addition of the change of enthalpy of the reaction to determine whether the chemical reaction wass spontaneous or not at a specified desired reaction temperature. All the chemical reactions discussed in this chapter so far have considered the mass and energy balance of these systems. In this section we look at the entropy and exergy balance equations of energy systems with chemical reactions.

The entropy balance for an energy system that has chemical reactions is the same as was introduced in earlier chapters and for this reason this section focuses on the property that presents in the second portion of thermodynamic science, exergy. Exergy is not conserved and, thus, it can be destroyed. Exergy destruction in a chemical reaction can be calculated through the entropy generation as follows:

$$Ex_d = T_o S_{gen} \tag{8.8}$$

Here, T_o denotes the ambient temperature. Similar to the energy balance relationship, the exergy balance relations introduced in the earlier chapters are reduced to their form since there are no chemical changes in the participating materials. For energy systems that have chemical reactions the exergy balance equation for open systems is presented next.

The general exergy rate balance equation for a steady-state process can be written as follows on a mole basis:

$$\underbrace{\dot{E}x_{\dot{Q}in} + \dot{W}_{in} + \sum \dot{n}_r\left(\overline{ex}_{ph} + \overline{ex}_{ch}\right)_r}_{\text{Exergy flow rate of the reactants}} = \underbrace{\dot{E}x_{\dot{Q}out} + \dot{W}_{out} + \sum \dot{n}_{p1}\left(\overline{ex}_{ph} + \overline{ex}_{ch}\right)_{p1} + \dot{E}x_d}_{\text{Exergy flow rate of the products}} \tag{8.9}$$

or on a mass basis:

$$\underbrace{\dot{E}x_{\dot{Q}in} + \dot{W}_{in} + \sum \dot{m}_r\left(ex_{ph} + ex_{ch}\right)_r +}_{\text{Exergy flow rate of the reactants}} = \underbrace{\dot{E}x_{\dot{Q}out} + \dot{W}_{out} + \sum \dot{m}_P\left(ex_{ph} + ex_{ch}\right)_p + \dot{E}x_d}_{\text{Exergy flow rate of the products}} \tag{8.10}$$

Here, ex_{ph} denotes specific physical exergy, which can be expressed as:

$$ex_{ph} = (h - h_o) - T_o(s - s_o) \tag{8.11}$$

where h and s are the specific enthalpy and entropy of the flow, h_O and s_O are the specific enthalpy and entropy of the flow at the reference environment conditions, and ex_{ch} is the specific chemical exergy. Note that in the earlier chapter the ex_{ch} cancels out as it is equal on both side of the reaction, since there were no chemical reactions and thus no change in the chemical structure of the materials contributing in the energy system. The standard chemical exergy values for selected elements are listed in Table 8.10, where the chemical exergy values of elements are given on a mole basis. The chemical exergy, as mentioned earlier, must be considered in energy conversion processes that have chemical reactions, such as combustion.

Table 8.10 Standard chemical exergy for selected substances obtained through the calculations.

Chemical element (phase)	$\overline{ex}_{ch}$ (kJ/mol)	Chemical element (phase)	$\overline{ex}_{ch}$ (kJ/mol)	Chemical element (phase)	$\overline{ex}_{ch}$ (kJ/mol)
Ag (s)	70.2	D_2 (g)	263.8	Ni (s)	232.7
Al (s)	888.4	F_2 (g)	466.3	O_2 (g)	3.97
Ar (s)	11.69	Fe (s_α)	376.4	P (s, red)	863.6
As (s)	494.6	H_2 (g)	236.1	Pb (s)	232.8
Au (s)	15.4	He (g)	30.37	Rb (s)	388.6
B (s)	628.8	Hg (l)	115.9	S (s, rhombic)	609.6
Ba (s)	747.4	I_2 (s)	174.7	Sb (s)	435.8
Bi (s)	274.5	K (s)	366.6	Se (s, black)	346.5
Br_2 (l)	101.2	Kr (g)	34.36	Si (s)	854.6
C (s, graphite)	410.26	Li (s)	393.0	Sn (s, white)	544.8
Ca (s)	712.4	Mg (s)	633.8	Sn (s)	730.2
Cd (s_α)	293.2	Mn (s_α)	482.3	Ti (s)	906.9
Cl_2 (g)	123.6	Mo (s)	730.3	U (s)	1190.7
Co (s_α)	265.0	N_2 (g)	0.72	V (s)	721.1
Cr (s)	404.4	Na (s)	336.6	W (s)	827.5
Cu (s)	134.2	Ne (g)	27.19	Xe (g)	40.33

The specific chemical exergy of a flow of a mixture of ideal gases can be expressed as:

$$ex_{ch} = \sum x_j ex^o_{ch} + RT_0 \sum x_j \ln(x_j) \tag{8.12}$$

Here, x_j is the mole fraction of constituent j in the flow and ex^0_{ch} is the standard chemical exergy of the constituent j in the flow. The exergy rate associated with a heat transfer rate can be expressed as:

$$\dot{E}x_{\dot{Q}} = \left(1 - \frac{T_o}{T_b}\right)\dot{Q} \tag{8.13}$$

Here, T_o denotes the reference environment temperature and T_b the temperature of the boundary of the control volume across which the heat transfer occurs. The quality of thermal energy in terms of exergy is determined based on the temperature at which this thermal energy (heat) is released.

8.10 Combustion Efficiency

To assess the efficiencies of a process, cycle or system from a thermodynamic point of view, it is insightful to use both energy and exergy efficiencies. These respective efficiencies can be expressed generally as follows:

$$\eta_{en,c} = \frac{Energy\ in\ product\ outputs}{Energy\ in\ inputs} \tag{8.14}$$

$$\eta_{ex,c} = \frac{Exergy\ in\ product\ outputs}{Exergy\ in\ inputs} \tag{8.15}$$

In the next example, a simple system is employed to demonstrate the calculations of the energy and exergy efficiencies for a system. The same principles apply for larger systems as subsequent case studies illustrate. The performance is considered of a steam generator plus superheater that uses propane as fuel and converts pressurized water to pressurized steam.

Example 8.9 A propane steam superheater, as shown in Figure 8.19, is considered in which complete combustion of propane with 900% excess air occurs. The steam superheater produces superheated steam $600\,°C$ and $5\,MPa$ from an original state of $320\,°C$ and $5\,MPa$. Calculate (a) the mass flow rate of propane per kg of steam generated to operate the propane steam superheater, (b) the energy efficiency, (c) the exergy efficiency, and (d) the exergy destruction rate for the superheating system considering the boiling of the pressurized water.

Solution

The complete combustion of propane with air as the oxidant produces the heat required to generate the superheated steam. It is required to determine the mass ratio of propane to steam.

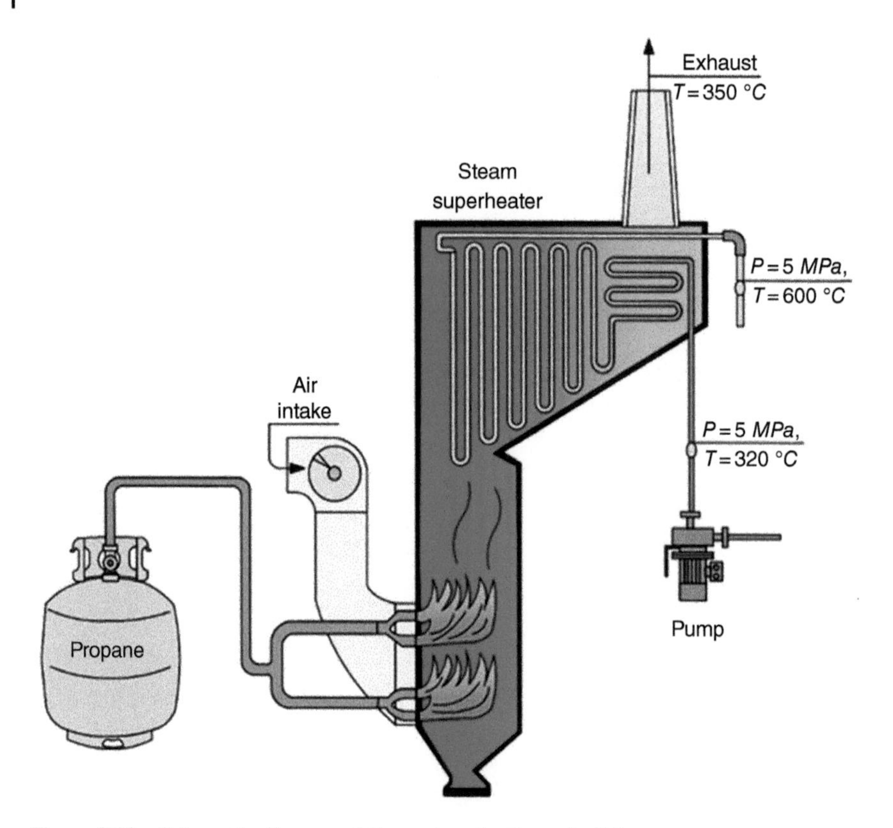

Figure 8.19 Schematic diagram of the system for Example 8.9.

Assumptions: **1** The combustion of propane with air occurs at steady state. **2** The H_2O product from the combustion process is in the vapor state (so the analysis will be based on *LHV*). **3** The boiler material does not absorb any thermal energy and remains at constant temperature. **4** A complete conversion of the reactants occurs in the reaction chamber based on the stoichiometric chemical reaction. **5** The reactor chamber has a constant volume. **5** No heat losses occur from the system. **6** The compressor has an isentropic efficiency of 72% and a mechanical efficiency of 100% (100% mechanical efficiency means that gas friction losses in the compressor plus losses associated with bearings, seals, gearbox, etc. are neglected).

Properties The molar mass ($\overline{m}$) for each of the relevant chemical compounds and elements follow:

$$\overline{m}_{C_3H_8} = 44\frac{g}{mol}, \overline{m}_{O_2} = 32\frac{g}{mol}, \overline{m}_{N_2} = 28\frac{g}{mol}, \overline{m}_{CO_2} = 44\frac{g}{mol}, \text{and } \overline{m}_{H_2O} = 18\frac{g}{mol}$$

Properties at 1 *bar* and 25 °*C*:

$$h_{o,O_2} = 0.000\ kJ/mol, \text{and } h_{o,H_2O} = 1.889\ kJ/mol$$

a) To determine the mass flow ratio of propane to water we first write the chemical balance equation of the reaction and then the energy balance equation for the system. Since air is

Table 8.11 Number of atoms of each element appearing in the propane combustion reaction.

Left side				Right side				
C	*H*	*O*	*N*	*C*	*H*	*O*	*N*	Added/Changed factor in front of the element
3	8	2	7.52	1	2	3	7.52	
3	8	2	7.52	3	2	7	7.52	$3CO_2(g)$
3	8	2	7.52	3	8	10	7.52	$4H_2O(g)$
3	8	10	37.6	3	8	10	7.52	$5(O_2(g) + 3.76N_2(g))$
3	8	10	37.6	3	8	10	37.6	$(5 \times 3.76)N_2(g)$

taken to be composed of 21% O_2 and 79% N_2 on a mole basis, the unbalanced chemical reaction for propane combustion in air is as follows:

$$C_3H_8(g) + (O_2(g) + 3.76N_2(g)) \rightarrow CO_2(g) + H_2O(g) + (3.76)N_2(g)$$

To balance this reaction, we balance any atom except for *O* and *H*, leaving them to the end. We write how much of each atom is present on each side of the equation, as show in Table 8.11.

From Table 8.11 the balanced chemical equation for propane combustion is:

$$C_3H_8(g) + \left(\frac{900 + 100}{100}\right)\mathbf{5}(O_2(g) + 3.76N_2(g))$$
$$\rightarrow \mathbf{3}CO_2(g) + \mathbf{4}H_2O(g) + \left(\frac{900 + 100}{100}\right)(\mathbf{5} \times 3.76)N_2(g) + \left(\frac{900}{100}\right)\mathbf{5}O_2(g)$$

From the balanced chemical equation, we can express all of the reactants and products in terms of the mass of the propane. Using the molar masses of the chemical substances in the propane combustion reaction, the masses in the above reaction equation are (for 1 *kmol* of propane):

$$m_{C_3H_8} = 44\,kg,\;\; m_{Air} = 686.4\,kg,\;\; m_{ex} = m_{C_3H_8} + m_{Air}$$
$$m_{Air} = 15.6 \times m_{C_3H_8},\;\; m_{ex} = m_{Air} + m_{C_3H_8} = 16.6 \times m_{C_3H_8}$$

The second step in solving part (a) involves presenting the schematic diagram for Example 8.9 in a block diagram format to show each step where a change occurs in the physical or chemical properties of the materials in the system. Then, we solve the combustion part of the propane steam superheater as shown in Figure 8.19 (streams names that start with S) first for 1 *kmol* of propane (C_3H_8). In this example, an important step in solving a real system is to translate the complicated diagram for the system to a block diagram, which consists of a group of blocks and lines. Each line represents a material stream and each block represents an energy interaction step during the operation of the system. Figure 8.20 presents an important step in solving any energy system, shows the block diagram and explains how mass and energy transfer through the system. There are three mass inlets to the system: propane, water, and compressed

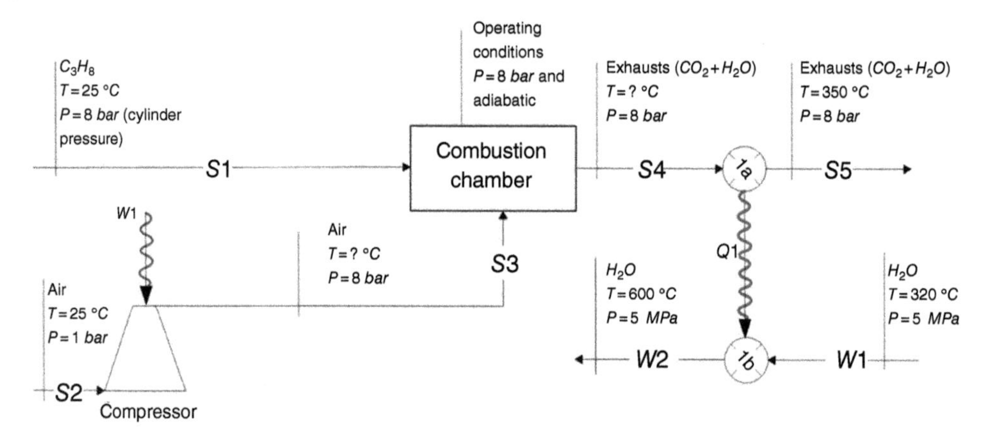

Figure 8.20 Block diagram presentation to show each step where a change in the physical or chemical properties occurs in the materials in the propane superheater system.

air. It is usually advantageous for clarity to differentiate between different main loops in the system by using different letters, such as *S* for propane and air (since they are mixed) and *W* for water streams. The following step describes the energy interactions between the different streams, which occur either through mixing (e.g. *S2* and *S3* mix in the combustion chamber) or through heat exchange. To avoid confusion, the energy streams are drawn in a different way than the mass streams; the method used here is a wavy line, with color coding (blue for work and red for thermal energy).

Water is heated to the combustion chamber reaction temperature (by Q_1 in Figure 8.20). The thermal energy to do this comes from the thermal energy released in the combustion chamber.

The third step in solving the problem is to list all given information (streams properties) in a table under the component properties and operating conditions, as listed in Table 8.12. By arranging the given data and those that can be found from property tables based on the given data, one can identify easily what is required and how to tackle the problem. A similar table is generated for the system devices (Table 8.13). A sample calculation is presented below of the properties of two streams in the system, to show the process of calculating the specific enthalpy ($h_f + h - h_o$) and the specific exergy:

$$h_{s1} = h_{f,C_3H_8} + h_{25\,°C,800\,kPa} - h_o$$

$$h_{s1} = -2,374 + 607.3 - 630.7$$

$$h_{s1} = -2,397\ kJ/kg$$

$$ex_{s1} = ex_{ph} + ex_{ch} = (h - h_o) - T_o(s - s_o) + ex_{ch}$$

$$ex_{s1} = (-2397 - (-2374)) - 298 \times (5.728 - 6.112) + 48,927$$

$$ex_{s1} = 91.43 + 48,927 = \mathbf{49,018\ kJ/kg}$$

We now write the mass and energy rate balance equations for the compressor in Figure 8.20. Since there is only physical change (and no chemical change) across the

Table 8.12 Known data for the example (derived from the problem statement and assumptions, and based on supplying the system with 1 kmol/s of propane).

State	Material	T (°C)	P (kPa)	$\dot{m}$ (kg/s)	$h_f + h - h_o$ (kJ/kg)	s (kJ/kgK)	ex (kJ/kg)
$S1$	C_3H_8	25	800	44.1	−2397	5.728	49,018
$S2$	Air[a]	25	100	1443	0	5.699	−1.19
$S3$,is	Air[a]	263[b]	800	1443	243.1[b]	5.699[b]	242[b]
$S3$	Air[a]	353[b]	800	1443	635.9[b]	5.862[b]	586[b]
$S4$	CO_2	1437[c]	800	132.0	−7257[c]	1.633[c]	1669[c]
	O_2	1437[c]	800	176.2	1510[c]	1.267[c]	1257[c]
	H_2O	1437[c]	800	72.06	−10 190[c]	0.358[c]	3439[c]
	N_2	1437[c]	800	1106	1635[c]	1.345[c]	1259[c]
	Total	1437[c]	800	1487	257[c]	1.549[c]	1309[c]
$S5$	Exhausts	350	800	1487	−1094	0.312	327
$W1$	H_2O	320	5000	3099[d]	−12 958	−3.071	1694
$W2$	H_2O	600	5000	3099[d]	−12 310	−2.174	2075

[a] Air = $O_2 + 3.76\,N_2$ (mole basis).
[b] Results of the compressor balance equations.
[c] Results of the combustion chamber balance equations.
[d] Results of the heat exchanger balance equations.

Table 8.13 Component properties and operating conditions.

Component	q (kJ/kg steam)	w (kJ/kg steam)	Operating parameters
Combustion chamber	0[a]	0[a]	Operating pressure = 8 *bar*
Compressor	0[b]	635[b]	Discharge pressure = 8 *bar*
Superheater	648[c]	0[c]	Operating pressure: Shell side = 8 *bar* Tube side = 50 *bar*

[a] Results of the combustion chamber balance equations.
[b] Results of the compressor balance equations.
[c] Results of the heat exchanger balance equations.

compressor, the energy balance equation need not include the enthalpy of formation, as it cancels on both sides of the equation. So, the mass rate balance is:

$$\dot{m}_{S3} = \dot{m}_{S2}$$

Since the isentropic efficiency of the compressor is given as 0.72, we can write:

$$s_{S3is} = s_{S2} \text{ (in an isentropic process)}$$

As s_{S3is} and P_{S3} are known, so $h_{S3,\ is}$ and $T_{S3,\ is}$ can be calculated:

$$\eta_{c,is} = \frac{\dot{W}_{is}}{\dot{W}_a} = \frac{\dot{m}_{S2}(h_{S3,is} - h_{S2})}{\dot{m}_{S2}(h_{S3} - h_{S2})} = \mathbf{0.72}$$

$$\frac{1{,}443 \times (541.4 - (298.4))}{1{,}443 \times (h_{S3} - (298.4))} = 0.72,\ h_{S3} = 635.9\ kJ/kg$$

$$\dot{W}_a = \dot{m}_{S2}(h_{S3} - h_{S2}) = \mathbf{917{,}603.7\ kW}$$

The energy rate balance equation for the combustion chamber is derived from the general energy rate balance equation, after which the appropriate terms are canceled, as follows:

$$\underbrace{\dot{Q}_{in} + \dot{W}_{in} + \dot{n}_{r1}\left(\overline{h_f^o} + \overline{h} - \overline{h^o}\right)_{r1} + \dot{n}_{r2}\left(\overline{h_f^o} + \overline{h} - \overline{h^o}\right)_{r2}}_{\text{Energy flow rate of the reactants}} = \underbrace{\dot{Q}_{out} + \dot{W}_{out} + \dot{n}_p\left(\overline{h_f^o} + \overline{h} - \overline{h^o}\right)_p}_{\text{Energy flow rate of the products}}$$

where

$\dot{Q}_{in} = 0$ (adiabatic combustion chamber)
$\dot{Q}_{out} = 0$ (adiabatic combustion chamber)
$\dot{W}_{in} = 0$ (no work is done on the combustion chamber)
$\dot{W}_{out} = 0$ (no work is done by the combustion chamber)

The mole flow rates of the chemical components of the exhaust gases leaving the combustion chamber are calculated based on the balanced chemical reaction equation, and are as follows:

$$\dot{n}_{S1} = 1\ kmol/s,\ \dot{n}_{S3,air} = 10.51\ kmol/s,$$
$$\dot{n}_{S4,H_2O} = 4.0\ kmol/s,\ \dot{n}_{S4,O_2} = 5.505\ kmol/s,$$
$$\dot{n}_{S4,N_2} = 39.495\ kmol/s,\ \dot{n}_{S4,CO_2} = 3.0\ kmol/s.$$

So,

$$\dot{n}_{S1}\left(\bar{h}_f^o + \bar{h} - \bar{h}^o\right)_{S1} + \dot{n}_{S3}\left(\bar{h}_f^o + \bar{h} - \bar{h}^o\right)_{S3} = \dot{n}_{S4,H_2O}\left(\bar{h}_f^o + \bar{h} - \bar{h}^o\right)_{S4,H_2O}$$
$$+ \dot{n}_{S4,O_2}\left(\bar{h}_f^o + \bar{h} - \bar{h}^o\right)_{S4,O_2} + \dot{n}_{S4,N_2}\left(\bar{h}_f^o + \bar{h} - \bar{h}^o\right)_{S4,N_2} + \dot{n}_{S4,CO_2}\left(\bar{h}_f^o + \bar{h} - \bar{h}^o\right)_{S4,CO_2}$$

Note that all products are at the same temperature and pressure, which means there is one equation and one unknown, the temperature of the products. Various approaches can be employed to solve the above equation. One approach is trial and error. Since the left side of the above equation is known, then by trying different temperatures until the temperature that makes right side of the equation similar to the left side will give the required temperature. Another approach involves expressing all the enthalpies as a function of temperature and c_p, and then solving a linear equation. A third approach is to use software that has embedded property tables, an example of which is *EES*. Table 8.14 lists the enthalpy of formation of the chemical species involved in the current example.

Table 8.14 Enthalpy of formation of the chemical species present in the combustion reaction of propane.

Chemical species	$\overline{h_f^o}$ (kJ/kmol)
$C_3H_8(g)$	−103 850
$CO_2(g)$	−393 520
$H_2O(g)$	−241 820
$O_2(g)$	0
$N_2(g)$	0

By solving the above equation, the properties for stream $S4$ are found. To calculate the overall exergy of stream $S4$, the following equation, which is based on the assumption that the flow consists of a mixture of ideal gases, is used:

$$ex_{ch,i} = \sum x_j ex_{ch}^o + RT_0 \sum x_j \, ln\left(x_j\right)$$

$$\begin{aligned} ex_{ch,S4} = &\left(\frac{\dot{n}_{CO_2}}{\dot{n}_{S4}} \times ex_{ch,CO_2}^o\right) + \left(\frac{\dot{n}_{O_2}}{\dot{n}_{S4}} \times ex_{ch,O_2}^o\right) + \left(\frac{\dot{n}_{H_2O}}{\dot{n}_{S4}} \times ex_{ch,H_2O}^o\right) \\ &+ \left(\frac{\dot{n}_{N_2}}{\dot{n}_{S4}} \times ex_{ch,N_2}^o\right) + RT_o\left(\left(\frac{\dot{n}_{CO_2}}{\dot{n}_{S4}} + ln\left(\frac{\dot{n}_{CO_2}}{\dot{n}_{S4}}\right)\right) + \left(\frac{\dot{n}_{O_2}}{\dot{n}_{S4}} + ln\left(\frac{\dot{n}_{O_2}}{\dot{n}_{S4}}\right)\right) \right. \\ &\left. + \left(\frac{\dot{n}_{H_2O}}{\dot{n}_{S4}} + ln\left(\frac{\dot{n}_{H_2O}}{\dot{n}_{S4}}\right)\right) + \left(\frac{\dot{n}_{N_2}}{\dot{n}_{S4}} + ln\left(\frac{\dot{n}_{N_2}}{\dot{n}_{S4}}\right)\right)\right) \end{aligned}$$

The hot exhausts (stream S4) are input to the steam reheater to reheat the steam. The reheater is an heat exchanger. It is modeled here based on energy and mass balances only without additional heat exchanger analysis, as follows.

$$\begin{aligned} \dot{m}_{S4}\left(h_f^o + h - h^o\right)_{S4} - \dot{m}_{S5}\left(h_f^o + h - h^o\right)_{S5} = &\ \dot{m}_{W2}\left(h_f^o + h - h^o\right)_{W2} \\ &- \dot{m}_{W1}\left(h_f^o + h - h^o\right)_{W1} \end{aligned}$$

$$\dot{m}_{S4}h_{S4} - \dot{m}_{S5}h_{S5} = \dot{m}_{W2}h_{W2} - \dot{m}_{W1}h_{W1}$$

$$(1487 \times 257) - (1487 \times (-1,094)) = (\dot{m}_{W2} \times (-12,310)) - (\dot{m}_{W1} \times (-12,958)),$$
$$\text{and } \dot{m}_{W2} = \dot{m}_{W1}$$

$$\dot{n}_{W2} = \dot{n}_{W1} = 1727 \ kmol/s$$

$$\dot{m}_{W1} = \dot{m}_{W1} = 3,099 \ kg/s$$

Thus, it is necessary to burn 44.1 *kg/s* of propane (C_3H_8) to superheat 3099 *kg/s* of steam from 320 to 600 °*C*. Or, **0.0142 kg/s** of propane is required to superheat 1 *kg/s* of H_2O.

b) The energy efficiency can be defined in general as:

$$\eta = \frac{Energy \ in \ product \ outputs}{Energy \ in \ inputs}$$

The superheater system has two energy inputs: propane (stream $S1$) and air compressor work. The output of the system is the superheated steam (from stream $S6$ to stream $S7$). Based on the energy inputs and outputs, the energy efficiency can be written as:

$$\eta_{en,c} = \frac{\dot{m}_{steam}(h_{W2} - h_{W1})}{\dot{m}_{C_3H_8} \times LHV_{C_3H_8} + \dot{W}_{compressor}} = \frac{(3,099 \times (-12,310 - (-12,958)))}{(44.1 \times 46,441) + (917,604)}$$

$$= \mathbf{0.6771}$$

Note that the hot exhaust at high pressure is not considered a product output because, based on the problem statement, it is rejected to the environment as a waste with no thermal energy recovery.

c) The exergy efficiency can be defined here as:

$$\eta_{ex,c} = \frac{Exergy\ in\ product\ outputs}{Exergy\ in\ inputs} = \frac{\dot{m}_{steam}(ex_{S7} - ex_{S6})}{\dot{m}_{C_3H_8} \times ex_{C_3H_8} + \dot{W}_{compressor}}$$

$$= \frac{(3,099 \times (2075 - (1694)))}{(44.1 \times 48930) + (917,604)} = \mathbf{0.384}$$

d) The exergy destruction rate of the system as a whole is calculated using the following expression:

$$\dot{E}x_c = \dot{E}x_d + \dot{E}x_{losses} = Exergy\ destroyed + Exergy\ lost\ as\ a\ result\ of\ waste\ emission$$

$$= Exergy\ in\ inputs - Exergy\ in\ product\ outputs$$

$$\dot{E}x_c = \dot{m}_{C_3H_8} \times ex_{C_3H_8} + \dot{W}_{compressor} - (\dot{m}_{steam}(ex_{W2} - ex_{W1}))$$

$$= (44.1 \times 48,930) + (917,604) - (3,099 \times (2,075 - (1,694))) = \mathbf{1.894,698\ kW}$$

$$ex_c = \frac{\dot{E}x_c}{\dot{m}_{steam}} = \frac{1,894,698}{3,099} = \mathbf{611.4}\ \frac{\mathbf{kJ}}{\mathbf{kg\ steam}}$$

Note that such a low exergy efficiency indicates that the system has large room for improvement. One way of improving the system is by using the hot exhausts that exit the steam reheater. For systems with chemical reactions, the enthalpy of formation and the chemical exergy of all streams must be taken into consideration. For more complicated systems, a more detailed exergy analysis of all system components (blocks in the system block diagram) may be required to provide additional insights.

Example 8.10 Consider a Rankine cycle based thermal power plant where a steam boiler is used. Feed water at a rate of 650 *ton/h* enters the boiler at a pressure of 12.5 *MPa* and a temperature of 200 °*C*. Steam leaves the boiler at 10 *MPa* and 500 °*C*. The net power developed by the power plant is 162 *MW*. Methane fuel is used at a rate of 44 000 *kg/h* and has a *HHV* of 42 900 *kJ/kg*. Determine (based on the *HHV*) (a) the combustion efficiency, (b) the thermal (energy) efficiency of the power plant, (c) the combustion exergy efficiency, and (d) repeat both (a) and (b) for a *LHV* = 39 600 *kJ/kg*.

Solution

Let us first start finding the enthalpy of the initial and final states at the boiler.

The following data are given:

For state 1:

$P_1 = 12{,}500\ kPa$ and $T_1 = 200\,°C$; the specific enthalpy and specific entropy are taken as $h_1 = 857\ kJ/kg$, $s_1 = 2.314\ kJ/kgK$.

For state 2:

$P_2 = 10{,}000\ kPa$ and $T_2 = 500\,°C$; the specific enthalpy and specific entropy are taken as $h_2 = 3374.6\ kJ/kg$, $s_2 = 6.597\ kJ/kgK$.

Additional data are:

$$\dot{m}^o_{st} = 650\ ton/h$$

$$\dot{W}^o_{net} = 162{,}000\ kW$$

$$\dot{m}^o_{\mathrm{fuel}} = 44{,}000\ kg/h$$

$$HHV = 42{,}900\ kJ/kg$$

a) The combustion efficiency of the boiler can be defined as:

$$\eta_{en,c} = \frac{\dot{Q}^o_{Steam}}{\dot{m}^o_{fuel} \times HHV}$$

$$\dot{Q}^o_{Steam} = \dot{m}^o_{st}\,(h_2 - h_1)$$

$$\dot{Q}^o_{Steam} = \frac{650 \times 10^3}{3600}\,(3374.6 - 857) = \mathbf{454{,}566.66\ kW}$$

$$\eta_{en,c} = \frac{454{,}566.66}{\frac{44{,}000}{3{,}600} \times 42{,}900} = \mathbf{86.7\%}$$

b) The thermal efficiency of the overall cycle is then calculated as:

$$\eta_{en,ov} = \frac{\dot{W}^o_{net}}{\dot{m}^o_{\mathrm{fuel}} \times HHV} = \frac{162{,}000}{\frac{44{,}000}{3{,}600} \times 42{,}900} = \mathbf{30.9\%}$$

c) It is now necessary to calculate the exergy values and exergy efficiency accordingly.

The exergy efficiency can be determined as follows:

$$\eta_{ex,c} = \frac{\Delta ex_{st}}{\Delta ex_c}$$

where st stands for steam.

Thus, the exergy gain of the steam is calculated as:

$$\Delta \dot{E}x_{steam} = \dot{m}_{st}[ex_2 - ex_1]$$

$$\Delta \dot{E}x_{steam} = \dot{m}_{st}[(h_2 - T_0 s_2) - (h_1 - T_0 s_1)]$$

$$\Delta\dot{E}x_{steam} = \frac{650 \times 10^3}{3600}[(3374.6 - 298 \times 6.597) - (857 - 298 \times 2.314)]$$

$$\boldsymbol{\Delta\dot{E}x_{steam} = 224,117.47\ kW}$$

Assume methane as a fuel and its product temperature at 500 K, we can then write the general chemical reaction:

$$C_nH_m\,(g) + a_{th}\left(n + \frac{m}{4}\right)[O_2 + 3.76\,N_2]$$
$$\rightarrow n\,CO_2 + \frac{m}{2}\,H_2O + \left(a_{th}\left(n + \frac{m}{4}\right) - n - \frac{m}{4}\right)O_2 + 3.76\,a_{th}\left(n + \frac{m}{4}\right)N_2$$

where $a_{th} = 1$, as the fuel is burnt with 100% theoretical air:

$$CH_4 + 2\,[O_2 + 3.76\,N_2] \rightarrow CO_2 + 2\,H_2O + 7.52\,N_2$$

In order to calculate the amount of heat released due to the combustion process:

$$\overline{q}_c = \overline{h}_p - \overline{h}_r$$

$$\overline{q}_c = \sum_p N_i\left[\overline{h}_f^0 + \overline{h}_T - \overline{h}_{T_0}\right]_p - \sum_r N_i\left[\overline{h}_f^0 + \overline{h}_T - \overline{h}_{T_0}\right]_r$$

The necessary heat of formation values $\overline{h}_f^0$ and the enthalpy values $\overline{h}_{T_0}$ at the reference condition can be obtained:

Species	$\overline{h}_f^0$ [kJ/kmol]	$\overline{h}_{T_0 = 298\ K}$ [kJ/kmol]	$\overline{h}_{T_0 = 500\ K}$ [kJ/kmol]
CH_4 (g)	−74 850	−21.03	8216
O_2	0	8682	14 770
N_2	0	8669	14 581
CO_2	−393 520	9364	17 678
H_2O(g)	−241 820	9904	16 828

It is now necessary to substitute these values and calculate the specific combustion heat as follows:

$$\overline{q}_c = 1\,[-393,520 + 17,678 - 9,364] + 2\,[-241,820 + 16,828 - 9,904]$$
$$+ 7.52\,[0 + 14,581 - 8,669] - [-74,850] = -735,689.76\ kJ/kmol$$

$$\overline{q}_c = \boldsymbol{-735.68\ MJ/kmol_{CH_4}}$$

Hence, the heat transfer per kg of fuel:

$$q_c = \frac{\overline{q}_c}{MW_{fuel}} = \frac{-735.68}{1 \times (12 + 4)} = \boldsymbol{-45.98\ MJ/kg_{CH_4}}$$

In order to find the specific exergy contents of the products and reactants, we need to obtain the specific entropy values of the products and reactants which are listed below (for the ith species in the combustion equation):

$$\overline{s}_i(T, P_i) = \overline{s}_i^0(T, P_0) - R_u\,ln\left(\frac{y_iP_m}{P_0}\right)\ kJ/kmolK$$

For reactant at 25 °C:

Reactants	N	Y_i	$\bar{s}_i^0$	$-R_u \ln\left(\frac{y_i P_m}{P_0}\right)$	$N_i \bar{s}_i$
CH_4	1	0.095	186.16	19.57	205.73
O_2	2	0.190	205.04	13.80	437.68
N_2	7.52	0.715	191.61	2.789	1461.88
$\sum_r N_i \bar{s}_i$					**2105.298**

For the products at $T = 773\ K$:

Reactants	N	Y_i	$\bar{s}_i^0$	$-R_u \ln\left(\frac{y_i P_m}{P_0}\right)$	$N_i \bar{s}_i$
CO_2	1	0.095	255.8	19.57	275.37
H_2O	2	0.190	222.5	13.80	472.6
N_2	7.52	0.715	220	2.789	1675.37
$\sum_p N_i \bar{s}_i$					**2423.34**

In order to find the exergy of combustion, first we find the exergy of products and reactions:

$$\overline{ex}_p = \bar{h}_p - T_0\, \bar{s}_p$$

Here, $\bar{s}_p$ is already calculated, while $\bar{h}_p$ can be determined as:

$$\bar{h}_p = \sum_p N_i \left[\bar{h}_f^0 + \bar{h}_T - \bar{h}_{T_0}\right]_p$$

$$\bar{h}_p = 1\,[-393,520 + 17,678 - 9,364] + 2\,[-241,820 + 16,828 - 9,904] + 7.52\,[0 + 14,581 - 8,669] = \mathbf{-810,539.76\ kJ/kmol}$$

Thus, the specific exergy of products becomes:

$$\overline{ex}_p = \bar{h}_p - T_0\, \bar{s}_p$$

$$\overline{ex}_p = -810,539.76 - (298)(2,423.34) = \mathbf{-1,532,695\ kJ/kmol}$$

Similarly, the exergy of the reactants:

$$\overline{ex}_r = \bar{h}_r - T_0\, \bar{s}_r$$

$$\bar{h}_r = \sum_r N_i \left[\bar{h}_f^0 + \bar{h}_T - \bar{h}_{T_0}\right]_r$$

$$\bar{h}_r = 1\,[-74,850] = \mathbf{-74,850\ kJ/kmol}$$

$$\overline{ex}_r = -74,850 - (298)(2105.29) = \mathbf{-702,226.42\ kJ/kmol}$$

Therefore, the exergy of combustion process is:

$$\Delta\overline{ex} = \overline{ex}_p - \overline{ex}_r$$

$$\Delta\overline{ex} = -1,532,695 + 702,226.42 = \mathbf{-830.468\ MJ/kmol}$$

The change of the exergy of combustion results in:

$$\Delta ex_c = \frac{\Delta \overline{ex}}{MW_{CH_4}} = \frac{-830,468}{16} = \mathbf{-51,904.25\ kJ/kg_{CH4}}$$

Thus, the exergy efficiency of the combustion process is defined as:

$$\eta_{ex,c} = \frac{\Delta \dot{E}x_{st}}{\Delta \dot{E}x_c} = \frac{\dot{m}_{st}[ex_2 - ex_1]}{\dot{m}_{fuel}[ex_p - ex_r]}$$

$$\eta_{ex,c} = \frac{224,117.47}{\frac{44,000}{3600} \times 51,904.25} = \mathbf{35.33\%}$$

d) For a $LHV = 39\,600\ kJ/kg$, the combustion efficiency becomes:

$$\eta_c = \frac{Q^o_{Steam}}{m^o_{fuel}.LHV} = \frac{454,566.66}{\frac{44,000}{3,600}\cdot 39,600} = \mathbf{93.92\%}$$

The thermal efficiency can be determined as:

$$\eta_{thermal} = \frac{W^o_{net}}{m^o_{fuel}.LHV} = \frac{162,000}{\frac{44,000}{3,600}\cdot 39,600} = \mathbf{33.47\%}$$

Example 8.11 Liquid octane, C_8H_{18}, enters an insulated combustion chamber of a gas turbine engine at 1 *atm* and 25 °*C*, as shown in Figure 8.21a, and it is combusted with air at the same conditions. Assume a complete combustion.

a) Determine the adiabatic flame temperature when fuel is burnt with 100% theoretical air.
b) Determine the specific entropy generation.
c) Determine the specific exergy destruction.
d) Assume this time nonadiabatic combustion as shown in Figure 8.21b. Determine the amount of specific heat released due to the combustion in *MJ*/*kg* of fuel if the product temperature is 530 *K*, specific entropy generation, and specific exergy destruction.

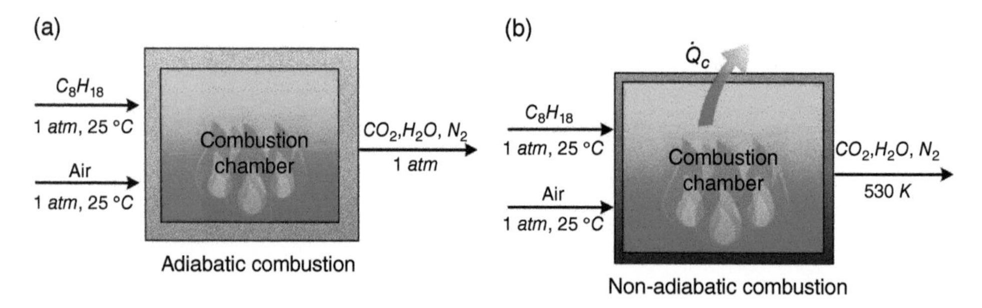

Figure 8.21 Schematic diagrams for (a) adiabatic and (b) nonadiabatic combustion for Example 8.11.

Solution

Let us first begin with writing the general combustion equation as follows:

$$C_nH_m\,(l) + a_{th}\left(n+\frac{m}{4}\right)[O_2 + 3.76N_2] \rightarrow n\,CO_2 + \frac{m}{2}H_2O$$
$$+\left(a_{th}\left(n+\frac{m}{4}\right) - n - \frac{m}{4}\right)O_2 + 3.76\,a_{th}\left(n+\frac{m}{4}\right)N_2$$

where $a_{th} = 1$. As the fuel is burnt with 100% theoretical air, the theoretical combustion equation of octane can be written as:

$$C_8H_{18}\,(l) + 12.5\,[O_2 + 3.76\,N_2] \rightarrow 8\,CO_2 + 9\,H_2O + 47\,N_2$$

a) In order to get the adiabatic flame temperature, the amount of heat released due to combustion to the surroundings $Q = 0$, thus:

$$H_r = H_p + Q \rightarrow H_r = H_p$$

In specific terms, $\bar{h}_r = \bar{h}_p$ which is formulated for the combustion as follows:

$$\sum_r N_i\left[\bar{h}_f^0 + \bar{h}_T - \bar{h}_{T_0}\right]_r = \sum_p N_i\left[\bar{h}_f^0 + \bar{h}_T - \bar{h}_{T_0}\right]_p$$

By obtaining the heat of formation values $\bar{h}_f^0$ and the enthalpy values $\bar{h}_{T_0}$ at the reference condition:

Species	$\bar{h}_f^0$ [kJ/kmol]	$\bar{h}_{T_0 = 298\ K}$ [kJ/kmol]
$C_8H_{18}\,(l)$	−249 950	0
O_2	0	8682
N_2	0	8669
CO_2	−393 520	9364
$H_2O(g)$	−241 820	9904

By substituting, we get:

$$1[-249{,}950] + 12.5\,[0] + 47\,[0] = 8\left[-393{,}520 + \bar{h}_{CO_2} + 9{,}364\right]$$
$$+\,9\left[-241{,}820 + \bar{h}_{H_2O} - 9{,}904\right] + 47\left[0 + \bar{h}_{N_2} - 8{,}669\right]$$

which yields to:

$$8\,\bar{h}_{CO_2} + 9\,\bar{h}_{H_2O} + 47\,\bar{h}_{N_2} = \mathbf{5,496,257}$$

by trial and error:
at $T = 2000\,K$

$$8\,[100,804] + 9\,[82,593] + 47\,[64,810] = \mathbf{4,595,839}$$

and
at $T = 2500\,K$

$$8\,[131,290] + 9\,[108,868] + 47\,[82,981] = 5,930,239$$

with interpolation, one finds the adiabatic temperature:

$$\boldsymbol{T_a = 2337.38\ K}$$

b) In order to find the entropy generation of the combustion system, the entropy generation equation should be used as follows:

$$\bar{s}_{gen} = \Delta\bar{s}_{system} + \Delta\bar{s}_{Surrondings}$$

or it can be written as:

$$\bar{s}_{gen} = \left(\bar{s}_p - \bar{s}_r\right) + \frac{\bar{q}_c}{T}$$

for an adiabatic combustion process where $\bar{q}_c = 0$, the entropy generation equation can be reduced as follows:

$$\bar{s}_{gen} = \left(\bar{s}_p - \bar{s}_r\right)$$

$$\bar{s}_{gen} = \sum_p N_i\bar{s}_i - \sum_r N_i\bar{s}_i$$

for the ith species in the combustion equation, the relation can be written as:

$$\bar{s}_i(T, P_i) = \bar{s}_i^0(T, P_0) - R_u\ ln\left(\frac{y_i P_m}{P_0}\right)\ kJ/kmolK$$

For reactant at 25 °C:

Reactants	N	Y_i	$\bar{s}_i^0$	$-R_u\ ln\left(\frac{y_i P_m}{P_0}\right)$	$N_i\ \bar{s}_i$
C_8H_{18}	1	0.0165	360.8	34.124	395.5
O_2	12.5	0.21	205.04	13	2725.2
N_2	47	0.778	191.61	2	9104
$\sum_r N_i S_i$					**12 223**

For the products at $T = 2337.38\ K$:

Products	N	Y_i	$\bar{s}_i^0$	$-R_u\ ln\left(\frac{y_i P_m}{P_0}\right)$	$N_i\ \bar{s}_i$
CO_2	8	0.125	318.8	17.3	2688.8
H_2O	9	0.141	273.1	16.3	2604.6
N_2	47	0.734	257.7	2.6	12 234
$\sum_p N_i\bar{s}_i$					**17 527.5**

Thus, the specific entropy generation can be found as:

$$\bar{s}_{gen} = \left(\bar{s}_p - \bar{s}_r\right) = (17,527.5 - 12,223) = \mathbf{5,304.5}\ \boldsymbol{kJ/kmolK}$$

c) In addition, the specific exergy destruction is found as:

$$\overline{ex}_d = T_0\,\bar{s}_{gen} = (298)(5,304.5) = \mathbf{1,580.74}\ \boldsymbol{MJ/kmol}$$

d) The above given solution is now considered for nonadiabatic combustion; the general and specific combustion reactions will remain same as above. However, the energy balance equation in specific form will change to the following version due to the nonadiabatic conditions this time based on Figure 8.21b:

$$H_r = H_p + Q_c$$

In specific terms, $\overline{h}_r = \overline{h}_p + \overline{q}_c$, which is formulated for the combustion as follows:

$$\overline{q} = \sum_r N_i\left[\overline{h}_f^0 + \overline{h}_T - \overline{h}_{T_0}\right]_r - \sum_p N_i\left[\overline{h}_f^0 + \overline{h}_T - \overline{h}_{T_0}\right]_p$$

By obtaining the heat of formation values $\overline{h}_f^0$ and the enthalpy values $\overline{h}_{T_0}$ at the reference condition:

Species	$\overline{h}_f^0$ [kJ/kmol]	$\overline{h}_{T_0=298\ K}$ [kJ/kmol]	$\overline{h}_{T_0=530\ K}$ [kJ/kmol]
C_8H_{18} (l)	−249 950	0	0
O_2	0	8682	15 708
N_2	0	8669	15 469
CO_2	−393 520	9364	19 029
H_2O(g)	−241 820	9904	17 889

By substituting all the table values, one obtains:

$$\overline{q}_c = -249{,}950 - 8[-393{,}520 + 19{,}029 - 9{,}364] - 9[-241{,}820 + 17{,}889 - 9{,}904]$$
$$-47[0 + 15{,}469 - 8{,}669] = 4{,}605{,}805\ kJ/kmol$$

$$\overline{\mathbf{q}}_c = \mathbf{4,605\ MJ/kmol_{C_8H_{18}}}$$

Hence, the heat transfer per kg of fuel consumed becomes:

$$\mathbf{q_c = \frac{\overline{q}_c}{MW_{fuel}} = \frac{4,605}{8 \times 12 + 18} = 40.39\ MJ/kg_{C_8H_{18}}}$$

In order to find the specific entropy generation of the nonadiabatic combustion we now end up with the following equation:

$$\overline{s}_{gen} = \sum_p N_i\overline{s}_i - \sum_r N_i\overline{s}_i + \frac{\overline{q}_c}{T}$$

for the ith species in the combustion equation, the relation can be written as:

$$\overline{s}_i(T, P_i) = \overline{s}_i^0(T, P_0) - R_u\ ln\left(\frac{y_iP_m}{P_0}\right)\ kJ/kmolK$$

For reactant at 25 °C

Reactants	N	Y_i	$\overline{s}_i^0$	$-R_u\ ln\left(\frac{y_iP_m}{P_0}\right)$	$N_i\ \overline{s}_i$
C_8H_{18}	1	0.0165	360.8	34.124	395.5
O_2	12.5	0.21	205.04	13	2725.2
N_2	47	0.778	191.61	2	9104
$\sum_r N_i\overline{s}_i$					**12 223**

For the products at T = 530 K:

Products	N	Y_i	$\bar{s}_i^0$	$-R_u\ ln\left(\frac{y_iP_m}{P_0}\right)$	$N_i\ \bar{s}_i$
CO_2	8	0.125	237.44	17.3	2037.92
H_2O	9	0.141	208.475	16.3	2022.975
N_2	47	0.734	208.4	2.6	9917
$\sum_p N_i\bar{s}_i$					**13 977.89**

Thus, the specific entropy generation is calculated as:

$$\bar{s}_{gen} = \sum_p N_i\bar{s}_i - \sum_r N_i\bar{s}_i + \frac{\overline{q}_c}{T} = (13,977.9 - 12,223) + \frac{4,605,805}{530}$$

$$= \mathbf{10,445.09\ kJ/kmolK}$$

The specific exergy destruction is then determined as:

$$\overline{ex}_d = T_0\,\bar{s}_{gen} = (298)(10,445.09) = \mathbf{3,112.6\ MJ/kmol}$$

Example 8.12 Consider gaseous ethane (C_2H_6) at 25 °C combusted with 400% theoretical air at 25°C during a combustion process. Assume a complete adiabatic combustion, as shown in Figure 8.22a with the fact that the produced water H_2O is in the vapor phase and a total pressure of 0.1 MPa.

a) Calculate the adiabatic flame temperature.
b) Calculate the specific entropy generation of the system.
c) Calculate the specific exergy of combustion products.
d) Calculate the reversible work of the combustion.
e) Assume nonadiabatic combustion this time, as shown in Figure 8.22b. Determine the amount of heat released due to combustion in MJ/kg of fuel if the product temperature is 700 K, along with the specific entropy generation and specific exergy destruction.

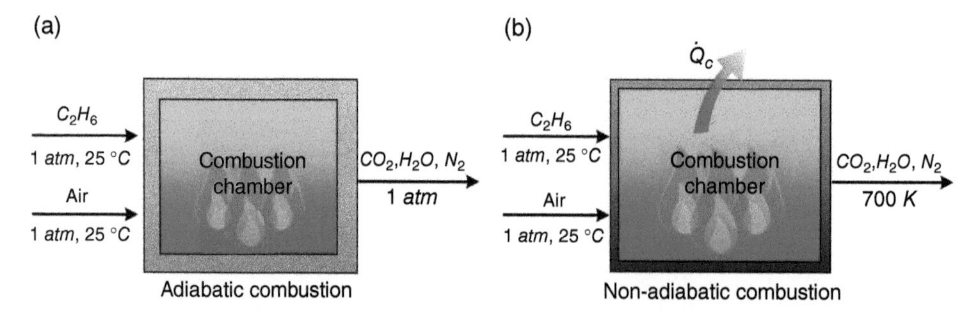

Figure 8.22 Schematic diagrams for (a) adiabatic and (b) nonadiabatic combustion for Example 8.12.

Solution

It is important to start by writing the general combustion equation:

$$C_nH_m\,(g) + a_{th}\left(n+\frac{m}{4}\right)[O_2+3.76N_2] \rightarrow n\,CO_2 + \frac{m}{2}H_2O$$
$$+\left(a_{th}\left(n+\frac{m}{4}\right)-n-\frac{m}{4}\right)O_2 + 3.76\,a_{th}\left(n+\frac{m}{4}\right)N_2$$

where $a_{th} = 4$. As the fuel is burnt with 400% theoretical air:

$$C_2H_6\,(g) + 14\,[O_2 + 3.76\,N_2] \rightarrow 2\,CO_2 + 3\,H_2O + 10.5\,O_2 + 52.64\,N_2$$

a) In order to calculate the adiabatic flame temperature, the amount of heat released due to combustion to the surroundings $\overline{q}_c = 0$

$$\overline{q}_c = \overline{h}_p - \overline{h}_r = 0$$

In this regard, it becomes $\overline{h}_r = \overline{h}_p$, which is further elaborated as:

$$\sum_r N_i\left[\overline{h}_f^0 + \overline{h}_T - \overline{h}_{T_0}\right]_r = \sum_p N_i\left[\overline{h}_f^0 + \overline{h}_T - \overline{h}_{T_0}\right]_p$$

By obtaining the heat of formation values $\overline{h}_f^0$ and the enthalpy values $\overline{h}_{T_0}$ at the reference condition:

Species	$\overline{h}_f^0$ [kJ/kmol]	$\overline{h}_{T_0=298\,K}$ [kJ/kmol]
C_2H_6 (g)	−84 680	0
O_2	0	8682
N_2	0	8669
CO_2	−393 520	9364
H_2O(g)	−241 820	9904

By substituting all values from the above table into the equation, one may get:

$$1[-84{,}680]+14[0]+52.64[0]=2\left[-393{,}520+\overline{h}_{CO_2}-9{,}364\right]$$
$$+3\left[-241{,}820+\overline{h}_{H_2O}-9{,}904\right]+10.5\left[0+\overline{h}_{O_2}-8{,}682\right]+52.64\left[0+\overline{h}_{N_2}-8{,}669\right]$$

which yields to:

$$2\,\overline{h}_{CO_2} + 3\,\overline{h}_{H_2O} + 10.5\,\overline{h}_{O_2} + 52.64\,\overline{h}_{N_2} = \mathbf{1,986,301}$$

by trial and error:
at $T = 1000\,K$

$$2\,[42,769] + 3\,[35,882] + 10.5\,[31,389] + 52.64\,[30,129] = \mathbf{2,118,759}$$

at $T = 800\,K$

$$2\,[32,179] + 3\,[27,896] + 10.5\,[24,523] + 52.64\,[23,714] = \mathbf{1,653,842}$$

After making the necessary interpolations, the adiabatic temperature is found to be:

$\boldsymbol{T_a = 943\ K}$

b) In order to find the specific entropy generation of the combustion system the entropy generation equation should be used as follows:

$$\bar{s}_{gen} = \Delta s_{system} + \Delta s_{Surrondings}$$

or it can be written as:

$$\bar{s}_{gen} = \left(\bar{s}_p - \bar{s}_r\right) + \frac{\bar{q}_c}{T}$$

for an adiabatic combustion process where $\bar{q}_c = 0$, the entropy generation equation can be reduced as follows:

$$\bar{s}_{gen} = \left(\bar{s}_p - \bar{s}_r\right)$$

$$s_{gen} = \sum_p N_i \bar{S}_i - \sum_r N_i \bar{S}_i$$

for the ith species in the combustion equation, the relation can be written as:

$$\bar{s}_i(T, P_i) = \bar{s}_i^0(T, P_0) - R_u\ ln\left(\frac{y_i P_m}{P_0}\right) kJ/kmolK$$

For reactant at 25 °C:

Reactants	N	Y_i	$\bar{s}_i^0$	$-R_u\ ln\left(\frac{Y_i P_m}{P_0}\right)$	$N_i\ \bar{s}_i$
C_2H_6	1	0.015	229.83	34.9	264.73
O_2	14	0.21	205.04	12.97	3052.2
N_2	52.64	0.778	191.61	2.087	10 196
$\sum_r N_i S_i$					**13 512.93**

For the products at $T = 943\ K$:

Products	N	Y_i	$\bar{s}_i^0$	$-R_u\ ln\left(\frac{Y_i P_m}{P_0}\right)$	$N_i\ \bar{s}_i$
CO_2	2	0.029	266	29.44	590.88
H_2O	3	0.044	230.3	26	768.9
O_2	10.5	0.156	241.6	15.44	2698.92
N_2	52.64	0.778	226.3	2.08	12 021.9
$\sum_p N_i\ \bar{s}_i$					**16 080.6**

Thus, the specific entropy generation becomes

$$\bar{s}_{gen} = \left(\bar{s}_p - \bar{s}_r\right) = (16{,}080.6 - 13{,}512.93) = \mathbf{2{,}568\ kJ/kmolK}$$

The specific exergy destruction is calculated as:

$$\overline{ex}_d = T_0\,\bar{s}_{gen} = (298)(2{,}568) = \mathbf{775{,}264\ kJ/kmol}$$

c) In order to find the exergy of combustion, first we find the exergy of products and reactions:

$$\overline{ex}_p = \overline{h}_p - T_0\,\overline{s}_p$$

$\overline{s}_p$ is already calculated above in (b), while $\overline{h}_p$ can be determined as:

$$\overline{h}_p = \sum_p N_i\left[\overline{h}_f^0 + \overline{h}_T - \overline{h}_{T_0}\right]_p$$

$$\overline{h}_p = 2[-393{,}520 + 39{,}535 - 9{,}364] + 3[-241{,}820 + 33{,}436 - 9{,}904]$$
$$+ 10.5[0 + 29{,}306 - 8{,}682] + 52.64[0 + 28{,}178 - 8{,}669] = \boldsymbol{-138{,}077\ kJ/kmol}$$

Thus, the exergy of products is calculated as:

$$\overline{ex}_p = \overline{h}_p - T_0\,\overline{s}_p$$

$$\overline{ex}_p = -138{,}077 - (298)(16{,}080.6) = \boldsymbol{-4{,}654\ MJ/kmol}$$

Similarly, the exergy of the reactants is calculated as:

$$\overline{ex}_r = \overline{h}_r - T_0\,\overline{s}_r$$

$$\overline{h}_r = \sum_r N_i\left[\overline{h}_f^0 + \overline{h}_T - \overline{h}_{T_0}\right]_r$$

$$\overline{h}_r = 1[-84{,}680] = \boldsymbol{-84{,}680\ kJ/kmol}$$

$$\overline{ex}_r = -84{,}680 - (298)(13{,}512.93) = \boldsymbol{-4{,}111\ MJ/kmol}$$

Therefore, the reversible work for this combustion process:

$$W_{rev} = \Delta\overline{ex} = \overline{ex}_p - \overline{ex}_R$$

$$W_{rev} = -4{,}654 + 4{,}111 = \boldsymbol{-543\ MJ/kmol}$$

d) Repeating the solution for nonadiabatic combustion process in Figure 8.22b.

The combustion equations will be same as above, except for the energy balance equation where we will have combustion heat to consider.

The amount of heat released during the combustion of the fuel is:

$$\overline{q}_c = \overline{h}_r - \overline{h}_p$$

$$\overline{q}_c = \sum_r N_i\left[\overline{h}_f^0 + \overline{h}_T - \overline{h}_{T_0}\right]_r - \sum_p N_i\left[\overline{h}_f^0 + \overline{h}_T - \overline{h}_{T_0}\right]_p$$

By obtaining the heat of formation values $\overline{h}_f^0$ and the enthalpy values $\overline{h}_{T_0}$ at the reference condition:

Species	$\overline{h}_f^0$ [kJ/kmol]	$\overline{h}_{T_0 = 298\ K}$ [kJ/kmol]	$\overline{h}_{T_0 = 700\ K}$ [kJ/kmol]
C_2H_6 (g)	−84 680	0	0
O_2	0	8682	21 184
N_2	0	8669	20 604
CO_2	−393 520	9364	27 125
H_2O(g)	−241 820	9904	24 088

By substituting the above listed values in the equation, one obtains

$$\bar{q}_c = -84,680 - 2[-393,520 + 27,125 - 9,364] - 3[-241,820 + 24,088 - 9,904]$$
$$- 10.5[0 + 21,184 - 8,682] - 52.64[0 + 20,604 - 8,669]$$
$$= 983,863\ kJ/kmol$$

$$\bar{q}_c = \mathbf{983.863\ MJ/kmol_{C_2H_6}}$$

Hence, the heat transfer per kg of fuel:

$$q_c = \frac{\bar{q}_c}{MW_{fuel}} = \frac{983.863}{2 \times 12 + 6} = \mathbf{32.79\ MJ/kg_{C_2H_6}}$$

In order to find the specific entropy generation of the combustion system through the entropy generation equation is found as follows:

$$\bar{s}_{gen} = \Delta\bar{s}_{system} + \Delta\bar{s}_{Surrondings}$$

or it can be written as:

$$\bar{s}_{gen} = (\bar{s}_p - \bar{s}_r) + \frac{\bar{q}_c}{T}$$

Thus, the entropy generation equation can be reduced as follows:

$$\bar{s}_{gen} = \sum_p N_i\bar{s}_i - \sum_r N_i\bar{s}_i + \frac{\bar{q}_c}{T}$$

for the ith species in the combustion equation, the relation can be written as:

$$\bar{s}_i(T, P_i) = \bar{s}_i^0(T, P_0) - R_u\ ln\left(\frac{y_iP_m}{P_0}\right) kJ/kmolK$$

For reactant at 25 °C:

Reactants	N	Y_i	$\bar{s}_i^0$	$-R_u\ ln\left(\frac{y_iP_m}{P_0}\right)$	$N_i\ \bar{s}_i$
C_2H_6	1	0.015	229.83	34.9	264.73
O_2	14	0.21	205.04	12.97	3052.2
N_2	52.64	0.778	191.61	2.087	10 196
$\sum_r N_i\bar{s}_i$					**13 512.93**

For the products at $T = 700\ K$:

Products	N	Y_i	$\bar{s}_i^0$	$-R_u\ ln\left(\frac{y_iP_m}{P_0}\right)$	$N_i\ \bar{s}_i$
CO_2	2	0.029	250.663	29.44	560.2
H_2O	3	0.044	218.6	26	733.8
O_2	10.5	0.156	231.358	15.44	2596.3
N_2	52.64	0.778	216.76	2.08	11 518.73
$\sum_r N_i\bar{s}_i$					**15 410.04**

Thus, the specific entropy generation is finally obtained as

$$\bar{s}_{gen} = \sum_{p} N_i \bar{s}_i - \sum_{r} N_i \bar{s}_i + \frac{\bar{q}_c}{T} = (15,410.04 - 13,512.93) + \frac{983,863}{700}$$

$$= \mathbf{3,302.63\ kJ/kmolK}$$

In closing, the specific exergy destruction is then determined as:

$$\overline{ex}_d = T_0\,\bar{s}_{gen} = (298)(3,302.63) = \mathbf{984,183\ kJ/kmol}$$

8.11 Concluding Remarks

This chapter has been concerned about the chemical reactions in very general form and in fuel combustion in specific with numerous examples. Fuel combustion is technically recognized as one of the most important applications of thermodynamics where thermal energy (heat) is needed for power generating cycles. A combustion process involves the oxidation of a fuel in order to release thermal energy. Fuels include not only fossil fuels but also many other materials that can be reacted exothermically, such as hydrogen and ammonia. Discussed have been some fundamental aspects of combustion, such as fuel, air, air fuel ratio, *LHV*, *HHV*, percentage theoretical air, adiabatic flame temperature, the *FLT* and *SLT* as applied to fuel combustion processes, closed and open systems, and combustion efficiencies to evaluate their performance energetically and exergetically.

Study Questions and Problems

Questions

1 Explain why fossil fuels have been the most widely used energy sources.

2 Give examples where each one of the fossil fuels is most commonly used.

3 Explain why coal has been a widely used fossil fuel and compare with other fuels.

4 Explain why crude oil has widely been used as a fossil fuel in transportation sector.

5 Explain why natural gas is proposed as a more acceptable fossil fuel.

6 Explain the drawbacks associated with fossil fuels.

7 Explain what fuel is and give examples.

8 Explain how fuels are classified and under what category.

9 Describe what the *LHV* and *HHV* are and illustrate these schematically.

10 Explain how crude oil distillation works and why it is used.

11 Draw a comparative environmental impact illustration of fossils fuels, namely coal, oil, and natural gas.

12 Describe the composition of natural gas.

13 Illustrate the composition of coal.

14 Describe chemical energy with an example.

15 Define endothermic and exothermic reactions and give examples.

16 Give an example for solid, liquid and gaseous fuels.

17 The balanced chemical equation for a combustion reaction provides which relation between reactants and products.
a Mole ratio
b Mass ratio
c Volume ratio
d None of the above

18 Explain what reactants and products are in a combustion reaction and give examples.

19 Compare endothermic and exothermic reactions and give examples from practical applications.

20 Is combustion an endothermic or exothermic process?

21 Apply the *FLT* and *SLT* to a fuel combustion process and write the balance equations accordingly.

22 Define the enthalpy of formation values for water and carbon dioxide and compare their values at 25 °*C*.

23 Define the enthalpy of formation values for H_2 and CH_4 and compare their values at 25 °*C*.

24 Describe air fuel ratio and give an example.

25 Describe percentage theoretical air and give an example.

26 What is adiabatic combustion? How does it take place?

27 Is adiabatic combustion chamber better than non-adiabatic one for practical applications? Explain Why.

28 Describe adiabatic flame temperature and explain where it leads to.

29 What is the difference between complete and incomplete combustion processes?

30 Describe both combustion energy and exergy efficiencies and solve a numerical example for a sample combustion process.

Problems

1 Methanol (CH_3OH) combustion takes place a novel thermal energy production system. Combustion occurs in the presence of air as the combustion oxidant and the supplied air is exactly the theoretical amount of air required.
   a Write the combustion reaction equation and balance it.
   b Write the mass and energy balance equations.
   c Calculate the amount of heat released per every 1 *mol* of methanol combusted.
   d Calculate the amount of carbon dioxide released in *kg* for every *kg* of methanol combusted.

2 Methane (CH_4) combustion takes place a novel thermal energy production system. Combustion occurs in the presence of air as the combustion oxidant and the supplied air is exactly the theoretical amount of air required.
   a Write the combustion reaction equation and balance it.
   b Write the mass and energy balance equations.
   c Calculate the amount of heat released per every 1 *mol* of methane combusted.
   d Calculate the amount of carbon dioxide released in *kg* for every *kg* of methane combusted.

3 Propane combustion takes place a novel thermal energy production system. Combustion occurs in the presence of air as the combustion oxidant and the supplied air is exactly the theoretical amount of air required.
   a Write the combustion reaction equation and balance it.
   b Write the mass and energy balance equations.
   c Calculate the amount of heat released per every 1 *mol* of propane combusted.
   d Calculate the amount of carbon dioxide released in *kg* for every *kg* of propane combusted.

4 Ammonia (NH_3) combustion takes place a novel thermal energy production system. Combustion occurs in the presence of air as the combustion oxidant and the supplied air is exactly the theoretical amount of air required.
   a Write the combustion reaction equation and balance it.
   b Write the mass and energy balance equations.
   c Calculate the amount of heat released per every 1 *mol* of methane combusted.

5 Ethanol (C_2H_5OH) combustion takes place a novel thermal energy production system. Combustion occurs in the presence of air as the combustion oxidant and the supplied air is exactly the theoretical amount of air required.
   a Write the combustion reaction equation and balance it.
   b Write the mass and energy balance equations.
   c Calculate the amount of heat released per every 1 *mol* of methane combusted.
   d Calculate the amount of carbon dioxide released in *kg* for every *kg* of methane combusted.

6 Dimethylether (CH_3OCH_3) combustion takes place a novel thermal energy production system. Combustion occurs in the presence of air as the combustion oxidant and the supplied air is exactly the theoretical amount of air required.
   a Write the combustion reaction equation and balance it.
   b Write the mass and energy balance equations.
   c Calculate the amount of heat released per every 1 *mol* of methane combusted.
   d Calculate the amount of carbon dioxide released in *kg* for every *kg* of methane combusted.

7 Isobutane (C_4H_{10}) combustion takes place a novel thermal energy production system. Combustion occurs in the presence of air as the combustion oxidant and the supplied air is exactly the theoretical amount of air required.
   a Write the combustion reaction equation and balance it.
   b Write the mass and energy balance equations.
   c Calculate the amount of heat released per every 1 *mol* of methane combusted.
   d Calculate the amount of carbon dioxide released in *kg* for every *kg* of methane combusted.

8 Repeat Example 8.1 for ethanol and methanol fuels and compare the results.

9 Repeat Example 8.2 for dimethylether (*DME*) fuel and compare the results.

10 Repeat Example 8.3 for the vehicles operating with hydrogen and ammonia and compare the results.

11 Repeat Example 8.4 for isobutane fuel and compare the results.

12 Repeat Example 8.5 for hydrogen and ammonia fuels and compare the results.

13 Repeat Example 8.6 for 150, 200, and 250% theoretical air and compares the results.

14 Repeat Example 8.7 for 150, 200, and 250% theoretical air and compare the results along with identified final adiabatic equations.

15 Identify three potential fuels and repeat Example 8.8. Compare the results.

16 Identify three potential fuels and repeat Example 8.9. Compare the results.

17 Identify three potential fuels and repeat Example 8.10. Compare the results.

18 Identify three potential fuels and repeat Example 8.11. Compare the results.

19 Identify three potential fuels and repeat Example 8.12. Compare the results.

20 Acetylene gas (C_2H_2) is burned completely with 20 percent excess air during a steady-flow combustion process. The fuel and air enter the combustion chamber at 25 °*C*, and the products leave at 1500 *K*. Determine the air–fuel ratio and the heat of combustion for this process.

21 Octane combustion takes place in a combustion chamber in the presence of air as the oxidant. If theoretical air is supplied and complete combustion takes place.

a Write the balanced chemical reaction depicting this combustion process.
b Write the mass and energy balance equations.
c Find the heat of combustion for every 1 *mol* of octane combusted.
d Find the mass of carbon dioxide released for every 1 *mol* of octane combusted.

22 A benzene-fueled car is to be converted to a diesel-fueled car at the same performance, which means that the amount of diesel must be able to produce the thermal energy equal to that of a 5 *L* tank of benzene. Considering environmental temperature and pressure for the overall process, determine the volume of the tank that would need to designed for operation with diesel fuel.

23 Consider a combustion chamber that burns 90% of the carbon and 100% of the hydrogen in liquid propane fuel. The remaining 10% of carbon in the fuel forms carbon monoxide. The input flow rate of fuel is 0.05 *kg/min* and 50% excess air is utilized that enters the combustion chamber at 7 °*C*. The temperature of the exhaust gases is measured to be 1500 *K*. (a) Find the air-fuel ratio and mass flow rate of air. (b) Find the rate of heat released during this combustion process.

24 A car engine is designed to operate with methyl alcohol fuel with 200% excess air. The input air is known to have a temperature of 25 °*C* and a pressure of 1 atm. Liquid fuel is mixed with this air before combustion. The temperature of the exhaust gases is measured to be 77 °*C*. (a) Determine the heat released during this process (b) Find the entropy generated during the process, (c) Calculated the exergy destroyed.

25 A type of natural gas is measured to have 65% CH_4, 25% N_2 and 10% O_2. Dry air is used as the oxidant and the combustion product gases are found to have the composition of 3.36% CO_2, 0.09% CO, 14.91% O_2 and 81.64% N_2. (a) Determine the air-fuel ratio, (b) Determine the percent theoretical air, (c) Find the volumetric flow rate of input air is the fuel input is 3.5 *kg/min*.

26 Carbon monoxide gas is allowed to burn in a combustion chamber. The inlet conditions of the gas are known to be 37 °*C* and 110 *kPa*. The volumetric flow rate of input gas is set at 0.4 m^3/min. Also, air inputs the combustion chamber at 25 °*C* and 110 *kPa* with a flow rate of 1.5 *kg/min*. The temperature of the exhaust combustion gases is measured to be 900 *K*. Assuming complete combustion, find the heat transfer rate from the combustion chamber.

27 Water at a rate of 650 *ton/h* enters a steam generator at a pressure of 12.5 *MPa* and a temperature of 200 °*C*. Steam leaves the boiler at 10 *MPa* and 500 °*C*. The net power developed by the power plant is 162 *MW*. Methanol fuel is used at a rate of 44 000 *kg/h* and has a *HHV* of 23000 *kJ/kg*. Determine (based on the *HHV*) (a) the combustion efficiency, (b) the thermal (energy) efficiency of the power plant, (c) the combustion exergy efficiency, and (d) repeat both (a) and (b) for a *LHV* = 19900 *kJ/kg*.

28 Feed water enters the boiler at a rate of 650 *ton/h* enters the boiler at a pressure of 12.5 *MPa* and a temperature of 200 °*C*. Steam leaves the boiler at 10 *MPa* and 500 °*C*. The net power developed by the power plant is 162 *MW*. Ethanol fuel is used at a rate of 44 000 *kg/h* and has a *HHV* of 29700 *kJ/kg*. Determine (based on the *HHV*) (a) the combustion efficiency, (b) the thermal (energy) efficiency of the power plant, (c) the combustion exergy efficiency, and (d) repeat both (a) and (b) for a *LHV* = 26700 *kJ/kg*.

29 Consider a Rankine cycle based thermal power plant where a steam boiler is used. Feed water at a rate of 650 *ton/h* enters the boiler at a pressure of 12.5 *MPa* and a temperature of 200 °*C*. Steam leaves the boiler at 10 *MPa* and 500 °*C*. The net power developed by the power plant is 162 *MW*. Propane fuel is used at a rate of 44 000 *kg/h* and has a *HHV* of 50400 *kJ/kg*. Determine (based on the *HHV*) (a) the combustion efficiency, (b) the thermal (energy) efficiency of the power plant, (c) the combustion exergy efficiency, and (d) repeat both (a) and (b) for a *LHV* = 46400 *kJ/kg*.

30 An automobile engine uses liquid octane as fuel. Air is used as the oxidant and 70% excess air is utilized during the combustion process. The exhaust gases exit the combustion chamber at 1500 *K* and air enters at 600 *K*. (a) Write the balanced chemical reaction equation. (b) Find the amount of heat released during this combustion process (c) Determine the entropy generated for every 1 *kmol* of fuel burnt and (d) Find the exergy destroyed for every 1 *kmol* of fuel burnt.

Nomenclature

A	area (m^2)
c	specific heat (kJ/kgK)
c_p	specific heat at constant pressure (kJ/kgK)
c_v	specific heat at constant volume (kJ/kgK)
C	cost ($)
e	specific energy (kJ/kg)
E	energy (kJ)
$\dot{E}$	energy rate (kW)
ex	specific exergy (flow or non-flow) (kJ/kg)
Ex	exergy (kJ)
$\dot{E}x$	exergy rate (kW)
ex_d	specific exergy destruction (kJ/kg)
Ex_d	exergy destruction (kJ)
$\dot{E}x_d$	exergy destruction rate (kW)
Ex_Q	exergy transfer associated with heat transfer (kJ)
$\dot{E}x_{\dot{Q}}$	exergy transfer rate associated with heat transfer rate (kW)
F	force (N)
g	gravitational acceleration (m/s^2)
g	Gibbs free energy (kJ/kg); natural gas consumption
h	specific enthalpy (kJ/kg); height (m)
$\overline{h}_f^0$	specific enthalpy of formation ($kJ/kmol$)
H	enthalpy (kJ)
HHV	higher heating value (kJ/kg)
HV	heating value (kJ/kg)
I	current (A)
k	specific heat ratio
KE	kinetic energy (kJ)
LHV	lower heating value (kJ/kg)
m	mass (kg)
M	molar mass ($kg/kmol$)
$\dot{m}$	mass flow rate (kg/s)
MEP	mean effective pressure
n	polytropic exponent
N	number of moles ($kmol$)
$\dot{n}$	mole flow rate ($kmol/s$)
P	pressure (kPa)
P_{cr}	critical pressure (kPa)
P_R	reduced pressure
PE	potential energy (kJ)
q	specific heat transfer (kJ/kg)
Q	heat transfer (kJ)

Thermodynamics: A Smart Approach, First Edition. Ibrahim Dincer.

$\dot{Q}$	heat transfer rate (kW)
Q_H	heat transfer rate with high temperature source (kJ)
Q_L	heat transfer rate with low temperature sink (kJ)
r	pressure ratio
r_{bw}	back work ratio
r_c	cut-off ratio
r_p	pressure ratio
r_v	volume ratio
R	gas constant ($kJ/kg.K$)
Ru	universal gas constant ($kJ/kmol.K$)
s	specific entropy ($kJ/kg.K$); distance (m)
S	entropy (kJ/K)
$\dot{S}$	entropy rate (kW/K)
SG	specific gravity
t	time (s)
T	temperature (K or $°C$)
T_{cr}	critical temperature (K)
T_H	high temperature (K)
T_L	low temperature (K)
T_R	reduced temperature
u	specific internal energy (kJ/kg)
U	total internal energy (kJ)
v	specific volume (m^3/kg)
v_r	relative specific volume
V	volume (m^3); velocity (m/s); *voltage* (V)
$\dot{V}$	volumetric flow rate (m^3/s)
w	specific work (kJ/kg)
W	total work (kJ)
$\dot{W}$	work rate or power (kW)
x	vapor quality; mass fraction (%); mole fraction (%)
y	molar fraction (%)
z	elevation (m)
Z	compressibility factor

Greek Letters

ϵ	regenerator effectiveness; ratio of efficiencies
η	energy efficiency
γ	Specific heat ratio
ρ	density
τ	exergetic temperature factor
Δ	difference

Subscripts

a	air; actual; adiabatic flame
abs	absolute; absorption, absorber
app	approximate
AR	absorption refrigeration cycle

avg	average
b	boundary
B	boiler
BC	Brayton cycle
Br	Brayton
bo	boiler
c	copper; compressor; compression
C	condenser
CC	combustion chamber
cf	cold fluid
cg	cogeneration
ch	charging; chemical
cn	condenser
comp	compressor
cons	consumption
d	destruction
df	diffuser
disch	discharging
e	electric
elec	electric
en	energy
EV	expansion valve
ev	evaporator
ex	exergy
exh	exhaust
exp	expansion
f	liquid phase
g	gas phase
ge	generator
gen	generation
GT	gas turbine
HE	heat exchanger
hf	hot fluid
HP	heat pump
HX	heat exchanger
i	iron
in	inlet condition
is	isentropic
ist	isothermal
l	loss
m	motor
MC	mixing chamber
n	number of moles (*kmol*)
o	oil
out	outlet
ov	overall
p	pump, process, product
ph	physical
PH	process heater
pol	polytropic process
r	reactant
RC	Rankine cycle
ref	reference
reg	regenerator
rev	reversible

s	source; steel; isentropic process
sat	saturated
sg	single generation
sh	shaft
st	steam turbine
surr	surrounding
sys	system
T	turbine
tot	total
tv	throttling valve
v	vapor
vac	vacuum
w	water
wcd	work-consuming devices
wpd	work-producing devices
0	ambient or reference condition

Superscripts

·	rate with respect to time
–	molar rate

Abbreviations

AFR	Air-fuel ratio
CC	combustion Chamber
CFCs	Chlorofluorocarbons
CNG	Compressed natural gas
COP	Coefficient of performance
EES	Engineering equation solver
EIF	Efficiency improvement factor
FLT	First law of thermodynamics
GHG	Greenhouse gas
GT	Gas turbine
HCs	Hydrocarbons
HX	Heat exchanger
LNG	Liquefied natural gas
LOA	Loss of accuracy
NIST	National Institute of Standards and Technology
REFROP	Reference fluid properties
SI	International System of Units
SLT	Second law of thermodynamics
SSSF	Steady-state Steady-flow
USCS	United States Customary System
USUF	Uniform-state Uniform-flow

Appendix 1: Thermodynamic Tables

TABLE A-1a	Properties of common liquids
TABLE A-1b	Properties of common solids
TABLE A-2	Molar mass, gas constant, specific heats and critical-point properties
TABLE A-3	Ideal-gas specific heats of various common gases at various temperatures
TABLE B-1a	Saturated Water - Temperature table
TABLE B-1b	Saturated Water - Pressure table
TABLE B-1c	Superheated Water
TABLE B-1d	Compressed Liquid Water
TABLE B-2a	Saturated Ammonia - Temperature table
TABLE B-2b	Saturated Ammonia - Pressure table
TABLE B-2c	Superheated Ammonia - Pressure table
TABLE B-3a	Saturated *R*-134*a* - Temperature table
TABLE B-3b	Saturated *R*-134*a* - Pressure table
TABLE B-3c	Superheated *R*-134*a* - Pressure table
TABLE C-1	Saturated ice-Water vapor
TABLE D-1	Compressibility factor at various reduced pressure and reduced temperature
TABLE E-1	Ideal gas properties of Air
TABLE E-2	Ideal gas properties of Carbon monoxide
TABLE E-3	Ideal gas properties of Carbon dioxide
TABLE E-4	Ideal gas properties of Hydrogen
TABLE E-5	Ideal gas properties of Water vapor
TABLE E-6	Ideal gas properties of Nitrogen
TABLE E-7	Ideal gas properties of Oxygen
TABLE E-8	Ideal gas properties of Oxygen
TABLE E-9	Ideal gas properties of Hydroxyl
TABLE F-1	Atmosphere properties at various altitude
TABLE G-1	Enthalpy of formation, Gibbs function of formation, and absolute entropy
TABLE H-1	Common fuels and hydrocarbons - Properties

Note: The above listed tables are prepared by utilizing the Engineering Equation Software (EES) available at Ontario Tech. University (formerly University of Ontario Institute of Technology) in Oshawa, Canada.

Thermodynamics: A Smart Approach, First Edition. Ibrahim Dincer.

Table A-1a Properties of common liquids

	Boiling data at 1 atm		Freezing data		Liquid properties		
Substance	Normal boiling point °C	Latent heat of vaporization, h_{fg} kJ/kg	Freezing point °C	Latent heat of fusion kJ/kg	Temperature °C	Density, ρ kg/m^3	Specific heat, C_p kJ/kg K
Ammonia	−33.3	1369.5	−77.7	332.4	−33.3	682.0	4.448
					−20.0	665.1	4.514
					0.0	638.6	4.617
					25.0	602.8	4.784
Argon	−185.8	169.3	−189.3	28.0	−185.8	1395.4	1.237
Benzene	80.1	395.6	5.5	0.1	80.1	813.4	1.905
n-Butane	−0.5	384.8	−138.3	80.2	−0.5	601.0	2.299
Carbon dioxide			−56.6	205.0	−56.6	1178.5	1.953
Ethanol	78.3	848.8	−114.1	109.0	78.3	736.4	3.124
Ethylene	−103.8	482.2	−169.2	119.4	−103.8	568.0	2.393
Helium	−268.9	20.7			−268.9	125.0	5.254
Hydrogen	−252.8	448.7	−259.2	59.5	−252.8	70.9	9.772
Isobutane	−11.7	365.9	−159.6	78.1	−11.7	593.9	2.236
Methane	−161.5	510.8	−182.5	58.6	−161.5	422.4	3.481
					−100.0	301.0	5.596
Methanol	65.0	1101.9	−97.5	99.2	65.0	747.8	2.831
Nitrogen	−195.8	199.2	−210.0	25.4	−195.8	806.1	2.041
					−160.0	600.6	3.039
n-Octane	125.5	302.4	−56.8	181.2	125.5	611.8	2.648
Oxygen	−183.0	213.2	−218.8	13.8	−183.0	1142.0	1.697
Propane	−42.1	425.8	−187.7	80.1	−42.1	581.5	2.248
					0.0	528.7	2.509
					50.0	448.5	3.099
R-134a	−26.1	217.0	−104.3		−50.0	1446.3	1.238
					−26.1	1376.8	1.280
					0.0	1294.8	1.341
					25.0	1206.7	1.425
Water	100.0	2256.7	0.0	333.6	0.0	958.4	4.217
					25.0	997.0	4.183
					50.0	988.0	4.182
					75.0	974.8	4.190
					100.0	958.4	4.217
Glycerine	289.9			198.5	20.0	1262.6	2.392
Oil					25.0	885.3	1.901

Table A-1b Properties of common solids

Substance	Temperature, *K*	Density, ρ *kg/m*^{3}	Specific heat, C_v *kJ/kg K*	Substance	Temperature, *K*	Density, ρ *kg/m*^{3}	Specific heat, C_v *kJ/kg K*
Metals				**Non Metals**			
Aluminum	200	2,717.0	0.802	Asphalt	297	2,115.0	0.920
	250	2,708.0	0.861	Brick	297	1,920.0	0.835
	300	2,699.0	0.901	Concrete stone mix	297	2,300.0	0.880
	350	2,689.0	0.926	Clay	297	1,460.0	0.880
	400	2,679.0	0.932	Glass Pyrex	297	2,225.0	0.835
	450	2,669.0	0.965	Glass Soda Lime	297	2,530.0	0.880
	500	2,659.0	0.998	Gypsum	297	1,680.0	1.085
Copper (in Celsius)	−175	9,036.7	0.250	Ice	200	926.0	1.569
	−100	9,010.7	0.340		220	923.8	1.712
	−50	8,990.4	0.369		240	921.3	1.859
	0	8,968.9	0.385		260	918.6	2.027
	25	8,957.8	0.389		273	916.7	2.160
	100	8,923.6	0.368	Plywood	297	545.0	1.215
	200	8,876.3	0.374	Rock, limestone	297	2,320.0	0.810
Bronze	297	8,802.1	0.384	Rubber, hard	297	1,190.0	1.680
Brass	297	8,528.3	0.378	Rubber, soft	297	1,100.0	2.010
Iron	297	7,872.9	0.441	Sand	297	1,515.0	0.800
Lead	297	11,335.9	0.129	Soill	297	2,050.0	1.840
Magnesium	297	1,737.4	1.021	Wood, hard	297	738.8	1.645
Nickel	297	8,900.9	0.442	Wood, soft	297	447.9	1.645
Silver	297	10,487.5	0.235				
Stainless Steel AISI304	297	7,997.4	0.475				
Tungsten	297	19,248.9	0.132				
Carbon Steel	297	7,849.0	2.082				
Zinc	297	7,137.3	0.389				
Zirconium	297	6,519.6	0.278				

Table A-2 Molar mass, gas constant, specific heats and critical-point properties

							Critical Properties		
Substance	**Formula**	**Molar mass, *M*** *kg/kmol*	***k***	**Gas constant, *R*** *kJ/kg K*	***Cp*** *kJ/kg K*	***Cv*** *kJ/kg K*	**Temperature,** *K*	**Pressure,** *kPa*	**Volume,** $m^3/kmol$
Air		28.967	1.400	0.287032	1.0047	0.7177	132.5	3,786	0.08446
Ammonia	NH_3	17.031	1.304	0.488196	2.0948	1.6066	405.4	11,333	0.07569
Argon	Ar	39.948	1.667	0.208132	0.5203	0.3122	150.7	4,863	0.07459
Benzene	C_6H_6	78.112	1.112	0.106443	1.0563	0.9499	562.0	4,894	0.25628
Bromine	Br_2	159.808	1.300	0.052028					
n-Butane	C_4H_{10}	58.124	1.092	0.143047	1.6950	1.5520	425.1	3,796	0.25510
CarbonDioxide	CO_2	44.010	1.289	0.188922	0.8435	0.6546	304.1	7,377	0.09412
CarbonMonoxide	CO	28.010	1.401	0.296839	1.0370	0.7402	132.8	3,494	0.09217
Carbon tetrachloride	CCl_4	153.823	1.111	0.054052					
Chlorine	Cl_2	70.906	1.324	0.117260					
Chloroform	$CHCl_3$	119.378	1.141	0.069648					
R-12	CCl_2F_2	120.910	1.253	0.068766	0.6180	0.5421	385.2	4,114	0.21663
Ethane	C_2H_6	30.070	1.185	0.276504	1.7643	1.4791	305.3	4,872	0.14584
Ethyl alcohol	C_2H_5OH	46.069	1.146	0.180479					
Ethylene	C_2H_4	28.043	1.241	0.296490	1.5430	1.2395	282.3	5,040	0.13099
Helium	He	4.003	1.667	2.077060	5.1930	3.1160	5.2	228	0.05748
n-Hexane	C_6H_{14}	86.172	1.062	0.096487	2.3545	1.4543	507.9	3,058	0.36820
Hydrogen	H_2	2.016	1.403	4.124241	14.3122	10.1859	33.1	1,296	0.06448
Krypton	Kr	83.800	1.667	0.099218	0.2492	0.1490	209.4	5,510	0.09230
Methane	CH_4	16.043	1.300	0.518262	2.2358	1.7128	190.6	4,599	0.09863
Methanol	CH_3OH	32.042	1.234	0.259487	2.5465	2.1183	513.4	8,104	0.11383

(Continued)

Table A-2 (Continued)

							Critical Properties		
Substance	**Formula**	**Molar mass, *M*** *kg/kmol*	***k***	**Gas constant, *R*** *kJ/kg K*	***Cp*** *kJ/kg K*	***Cv*** *kJ/kg K*	**Temperature,** *K*	**Pressure,** *kPa*	**Volume,** $m^3/kmol$
Methyl chloride	CH_3Cl	50.488	1.255	0.164682					
Neon	Ne	20.180	1.667	0.412015	1.0304	0.6181	44.5	2,680	0.04188
Nitrogen	N_2	28.013	1.401	0.296808	1.0414	0.7432	126.2	3,396	0.08941
NitrousOxide	N_2O	44.013	1.274	0.188909	0.8849	0.6916	309.5	7,245	0.09737
Oxygen	O_2	31.999	1.397	0.259835	0.9193	0.6581	154.6	5,043	0.07337
Propane	C_3H_8	44.097	1.128	0.188550	1.6987	1.4967	369.8	4,247	0.20181
Propylene	C_3H_6	42.080	1.355	0.197587	1.5567	1.3467	365.6	4,665	0.18837
SulfurDioxide	SO_2	64.063	1.264	0.129786	0.6560	0.5127	430.6	7,884	0.12203
R-134*a*	CF_3CH_2F	102.032	1.239	0.081489	0.8540	0.7634	374.2	4,059	0.20085
R-11	CCl_3F	137.380	1.144	0.060522	0.5896	0.5165	471.2	4,408	0.24809
Water	H_2O	18.016	1.328	0.461505	4.1831	4.1322	647.1	22,064	0.05595
Xenon	Xe	131.293	1.667	0.063328	0.1600	0.0954	289.7	5,840	0.11947

Table A-3 Ideal-gas specific heats of various common gases at various temperatures

Temperature, K	Gas	C_p kJ/kg K	C_v kJ/kg K	k	Gas	C_p kJ/kg K	C_v kJ/kg K	k	Gas	C_p kJ/kg K	C_v kJ/kg K	k
250		1.0030	0.7160	1.4008		0.9067	0.6469	1.4016		2.1243	1.6060	1.3227
300		1.0047	0.7177	1.3999		0.9143	0.6544	1.3972		2.2441	1.7258	1.3003
350		1.0080	0.7210	1.3981		0.9289	0.6690	1.3885		2.3957	1.8775	1.2760
400		1.0133	0.7262	1.3953		0.9453	0.6855	1.3790		2.5640	2.0458	1.2533
450		1.0204	0.7334	1.3913		0.9616	0.7018	1.3702		2.7407	2.2224	1.2332
500		1.0293	0.7423	1.3866		0.9772	0.7174	1.3621		2.9207	2.4025	1.2157
550		1.0396	0.7525	1.3815		0.9918	0.7319	1.3551		3.1012	2.5829	1.2007
600	Air	1.0508	0.7638	1.3758	Oxygen, O_2	1.0053	0.7454	1.3487	Methane, CH_4	3.2799	2.7617	1.1876
650		1.0626	0.7756	1.3700		1.0178	0.7580	1.3427		3.4557	2.9375	1.1764
700		1.0747	0.7876	1.3645		1.0295	0.7697	1.3375		3.6277	3.1094	1.1667
750		1.0867	0.7997	1.3589		1.0403	0.7805	1.3329		3.7951	3.2769	1.1581
800		1.0985	0.8114	1.3538		1.0505	0.7906	1.3287		3.9577	3.4395	1.1507
850		1.1098	0.8228	1.3488		1.0599	0.8001	1.3247		4.1152	3.5970	1.1441
900		1.1207	0.8337	1.3442		1.0688	0.8090	1.3211		4.2674	3.7491	1.1382
950		1.1311	0.8440	1.3402		1.0771	0.8173	1.3179		4.4141	3.8958	1.1330
1000		1.1408	0.8538	1.3361		1.0850	0.8252	1.3148		4.5554	4.0371	1.1284

(Continued)

Table A-3 (Continued)

Temperature, K	Gas	C_p kJ/kg K	C_v kJ/kg K	k	Gas	C_p kJ/kg K	C_v kJ/kg K	k	Gas	C_p kJ/kg K	C_v kJ/kg K	k
250		0.7888	0.5999	1.3149		0.5203	0.3122	1.6666		1.8595	1.3980	1.3301
300		0.8435	0.6546	1.2886		0.5203	0.3122	1.6666		1.8685	1.4070	1.3280
350		0.8947	0.7057	1.2678		0.5203	0.3122	1.6666		1.8813	1.4198	1.3250
400		0.9404	0.7515	1.2514		0.5203	0.3122	1.6666		1.8995	1.4380	1.3209
450		0.9808	0.7919	1.2385		0.5203	0.3122	1.6666		1.9226	1.4611	1.3159
500		1.0167	0.8278	1.2282		0.5203	0.3122	1.6666		1.9496	1.4881	1.3101
550		1.0487	0.8598	1.2197		0.5203	0.3122	1.6666		1.9797	1.5182	1.3040
600	Carbon dioxide, CO_2	1.0773	0.8884	1.2126	Argon, Ar	0.5203	0.3122	1.6666	Steam, H_2O	2.0121	1.5506	1.2976
650		1.1031	0.9142	1.2066		0.5203	0.3122	1.6666		2.0461	1.5846	1.2912
700		1.1265	0.9376	1.2015		0.5203	0.3122	1.6666		2.0812	1.6197	1.2849
750		1.1478	0.9589	1.1970		0.5203	0.3122	1.6666		2.1170	1.6555	1.2788
800		1.1673	0.9783	1.1932		0.5203	0.3122	1.6666		2.1533	1.6918	1.2728
850		1.1851	0.9962	1.1896		0.5203	0.3122	1.6666		2.1897	1.7282	1.2670
900		1.2016	1.0127	1.1865		0.5203	0.3122	1.6666		2.2261	1.7646	1.2615
950		1.2168	1.0278	1.1839		0.5203	0.3122	1.6666		2.2623	1.8008	1.2563
1000		1.2308	1.0419	1.1813		0.5203	0.3122	1.6666		2.2981	1.8366	1.2513

(Continued)

Table A-3 (Continued)

Temperature, K	Gas	C_p kJ/kg K	C_v kJ/kg K	k	Gas	C_p kJ/kg K	C_v kJ/kg K	k	Gas	C_p kJ/kg K	C_v kJ/kg K	k
250		1.0372	0.7403	1.4011		14.1315	10.0073	1.4121		1.0388	0.7420	1.4000
300		1.0370	0.7402	1.4010		14.3482	10.2240	1.4034		1.0377	0.7409	1.4006
350		1.0413	0.7445	1.3987		14.4169	10.2927	1.4007		1.0394	0.7426	1.3997
400		1.0486	0.7518	1.3948		14.4368	10.3127	1.3999		1.0437	0.7469	1.3974
450		1.0577	0.7609	1.3901		14.4456	10.3214	1.3996		1.0503	0.7535	1.3939
500		1.0681	0.7713	1.3848		14.4577	10.3335	1.3991		1.0585	0.7617	1.3897
550		1.0792	0.7823	1.3795		14.4785	10.3544	1.3983		1.0678	0.7710	1.3850
600		1.0907	0.7938	1.3740		14.5097	10.3855	1.3971		1.0779	0.7811	1.3800
650	Carbon monoxide, CO	1.1023	0.8055	1.3685	Hydrogen, H_2	14.5510	10.4268	1.3955	Nitrogen, N_2	1.0885	0.7917	1.3749
700		1.1140	0.8172	1.3632		14.6015	10.4773	1.3936		1.0994	0.8026	1.3698
750		1.1256	0.8288	1.3581		14.6603	10.5361	1.3914		1.1104	0.8136	1.3648
800		1.1370	0.8401	1.3534		14.7263	10.6021	1.3890		1.1214	0.8246	1.3599
850		1.1481	0.8512	1.3488		14.7984	10.6742	1.3864		1.1323	0.8355	1.3552
900		1.1589	0.8620	1.3444		14.8758	10.7516	1.3836		1.1430	0.8462	1.3507
950		1.1693	0.8725	1.3402		14.9576	10.8335	1.3807		1.1534	0.8566	1.3465
1000		1.1794	0.8826	1.3363		15.0432	10.9190	1.3777		1.1636	0.8668	1.3424

Table B-1a Saturated water - Temperature table

T °C	P_{sat} kPa	v_f m^3/kg	v_g m^3/kg	u_f kJ/kg	u_{fg} kJ/kg	u_g kJ/kg	h_f kJ/kg	h_{fg} kJ/kg	h_g kJ/kg	s_f kJ/kg K	s_{fg} kJ/kg K	s_g kJ/kg K
0.001	0.611	0.001000	205.9870	−0.038	2374.55	2374.52	−0.04	2500.56	2500.52	−0.00014	9.15449	9.15435
5	0.873	0.001000	147.0239	21.020	2360.40	2381.42	21.02	2488.69	2509.71	0.07626	8.94729	9.02355
10	1.228	0.001000	106.3229	41.986	2346.32	2388.31	41.99	2476.90	2518.89	0.15096	8.74766	8.89862
15	1.706	0.001001	77.8971	62.915	2332.27	2395.18	62.92	2465.12	2528.04	0.22423	8.55500	8.77923
20	2.339	0.001002	57.7777	83.833	2318.20	2402.04	83.84	2453.33	2537.17	0.29620	8.36887	8.66507
25	3.169	0.001003	43.3566	104.750	2304.12	2408.87	104.75	2441.52	2546.27	0.36695	8.18889	8.55584
30	4.246	0.001004	32.8955	125.666	2290.01	2415.68	125.67	2429.67	2555.34	0.43653	8.01473	8.45126
35	5.627	0.001006	25.2204	146.582	2275.87	2422.46	146.59	2417.78	2564.37	0.50496	7.84612	8.35108
40	7.381	0.001008	19.5283	167.496	2261.71	2429.21	167.50	2405.86	2573.36	0.57228	7.68276	8.25504
45	9.590	0.001010	15.2634	188.406	2247.51	2435.93	188.42	2393.89	2582.30	0.63853	7.52439	8.16292
50	12.344	0.001012	12.036664	209.32	2233.28	2442.60	209.33	2381.86	2591.19	0.70374	7.37076	8.07450
55	15.752	0.001015	9.572609	230.22	2219.00	2449.24	230.24	2369.78	2600.02	0.76794	7.22165	7.98959
60	19.932	0.001017	7.674324	251.13	2204.67	2455.82	251.15	2357.63	2608.79	0.83118	7.07680	7.90798
65	25.022	0.001020	6.199571	272.05	2190.28	2462.36	272.08	2345.41	2617.48	0.89350	6.93600	7.82950
70	31.176	0.001023	5.044649	292.98	2175.82	2468.83	293.01	2333.09	2626.10	0.95494	6.79904	7.75398
75	38.563	0.001026	4.133267	313.92	2161.28	2475.24	313.96	2320.67	2634.63	1.01552	6.66573	7.68125
80	47.373	0.001029	3.408812	334.88	2146.66	2481.58	334.93	2308.14	2643.07	1.07529	6.53588	7.61117
85	57.815	0.001032	2.828927	355.86	2131.94	2487.85	355.92	2295.49	2651.41	1.13429	6.40929	7.54358
90	70.117	0.001036	2.361668	376.86	2117.11	2494.04	376.93	2282.70	2659.63	1.19253	6.28583	7.47836
95	84.529	0.001040	1.982764	397.89	2102.16	2500.14	397.98	2269.76	2667.74	1.25005	6.16532	7.41537
100	101.322	0.001043	1.6736350	418.96	2087.09	2506.15	419.06	2256.66	2675.73	1.30688	6.04760	7.35448
105	120.788	0.001047	1.4199640	440.06	2071.88	2512.06	440.18	2243.39	2683.57	1.36305	5.93254	7.29559
110	143.241	0.001052	1.2106400	461.19	2056.52	2517.86	461.34	2229.93	2691.27	1.41856	5.82001	7.23857
115	169.019	0.001056	1.0369880	482.36	2041.01	2523.54	482.54	2216.28	2698.82	1.47346	5.70986	7.18332
120	198.483	0.001060	0.8921930	503.57	2025.33	2529.11	503.78	2202.42	2706.19	1.52776	5.60197	7.12973

(Continued)

Table B-1a (Continued)

T °C	P_{sat} kPa	v_f m^3/kg	v_g m^3/kg	u_f kJ/kg	u_{fg} kJ/kg	u_g kJ/kg	h_f kJ/kg	h_{fg} kJ/kg	h_g kJ/kg	s_f kJ/kg K	s_{fg} kJ/kg K	s_g kJ/kg K
125	232.014	0.001065	0.7708660	524.82	2009.48	2534.54	525.07	2188.33	2713.40	1.58148	5.49624	7.07772
130	270.020	0.001070	0.6687260	546.12	1993.43	2539.84	546.41	2174.00	2720.41	1.63464	5.39254	7.02718
135	312.930	0.001075	0.5823520	567.46	1977.20	2544.99	567.80	2159.43	2727.23	1.68726	5.29077	6.97803
140	361.195	0.001080	0.5089930	588.85	1960.75	2549.99	589.24	2144.59	2733.84	1.73936	5.19083	6.93019
150	475.717	0.001090	0.3928590	631.80	1927.18	2559.49	632.32	2114.06	2746.38	1.84207	4.99602	6.83809
160	617.663	0.001102	0.3070920	674.97	1892.62	2568.27	675.65	2082.30	2757.95	1.94293	4.80734	6.75027
170	791.471	0.001114	0.2428280	718.40	1856.98	2576.26	719.28	2049.17	2768.45	2.04207	4.62410	6.66617
180	1001.927	0.001127	0.1940260	762.12	1820.14	2583.38	763.25	2014.54	2777.78	2.13965	4.44563	6.58528
190	1254.165	0.001141	0.1565040	806.17	1781.96	2589.56	807.60	1978.25	2785.84	2.23583	4.27129	6.50712
200	1553.650	0.001156	0.1273200	850.59	1742.33	2594.71	852.38	1940.14	2792.52	2.33076	4.10048	6.43124
210	1906.173	0.001173	0.1043760	895.43	1701.08	2598.75	897.66	1900.04	2797.71	2.42460	3.93262	6.35722
220	2317.846	0.001190	0.0861570	940.75	1658.06	2601.57	943.51	1857.76	2801.27	2.51752	3.76713	6.28465
230	2795.097	0.001209	0.0715520	986.62	1613.07	2603.07	990.00	1813.07	2803.07	2.60971	3.60343	6.21314
240	3344.673	0.001229	0.0597420	1033.13	1565.90	2603.14	1037.24	1765.72	2802.96	2.70134	3.44095	6.14229
250	3973.649	0.001251	0.0501110	1080.35	1516.31	2601.63	1085.32	1715.43	2800.75	2.79264	3.27904	6.07168
260	4689.450	0.001276	0.0421940	1128.40	1463.99	2598.37	1134.38	1661.85	2796.23	2.88382	3.11704	6.00086
270	5499.875	0.001303	0.0356360	1177.41	1408.58	2593.15	1184.57	1604.58	2789.15	2.97514	2.95420	5.92934
280	6413.154	0.001332	0.0301640	1227.53	1349.64	2585.72	1236.08	1543.09	2779.17	3.06691	2.78964	5.85655
290	7438.015	0.001366	0.0255630	1278.98	1286.60	2575.74	1289.14	1476.74	2765.88	3.15949	2.62229	5.78178
300	8583.784	0.001404	0.0216670	1332.00	1218.71	2562.77	1344.05	1404.70	2748.75	3.25335	2.45084	5.70419
310	9860.538	0.001447	0.0183400	1386.96	1144.96	2546.19	1401.23	1325.81	2727.04	3.34911	2.27352	5.62263
320	11279.317	0.001498	0.0154760	1444.35	1063.92	2525.17	1461.25	1238.48	2699.73	3.44760	2.08796	5.53556
330	12852.451	0.001560	0.0129850	1504.93	973.44	2498.42	1524.98	1140.32	2665.30	3.55011	1.89061	5.44072
340	14594.085	0.001637	0.0107880	1569.93	870.06	2463.89	1593.83	1027.51	2621.34	3.65873	1.67579	5.33452
350	16521.128	0.001740	0.0088120	1641.69	747.44	2417.88	1670.44	893.03	2563.47	3.77740	1.43309	5.21049
360	18655.301	0.001894	0.0069620	1725.64	591.19	2352.15	1760.97	721.06	2482.03	3.91532	1.13884	5.05416
370	21029.867	0.002207	0.0049930	1843.33	345.43	2235.17	1889.74	450.42	2340.16	4.10943	0.70034	4.80977

Table B-1b Saturated Water - Pressure table

P kPa	T_{sat} °C	v_f m^3/kg	v_g m^3/kg	u_f kJ/kg	u_{fg} kJ/kg	u_g kJ/kg	h_f kJ/kg	h_{fg} kJ/kg	h_g kJ/kg	s_f kJ/kg K	s_{fg} kJ/kg K	s_g kJ/kg K
1	6.97	0.0010001	129.194313	29.28664710	2354.847112	2384.133760	29.28764720	2484.040426	2513.328073	0.10586910	8.867785	8.973654
2	17.50	0.0010013	66.997228	73.36404220	2325.244916	2398.608958	73.36604480	2459.237370	2532.603415	0.26033740	8.461238	8.721575
3	24.08	0.0010028	45.661373	100.91538310	2306.704662	2407.620045	100.91839130	2443.685774	2544.604165	0.35407240	8.221436	8.575508
4	28.97	0.0010041	34.797892	121.34137430	2292.933221	2414.274595	121.34539070	2432.120774	2553.466165	0.42223450	8.050284	8.472519
5	32.88	0.0010053	28.191058	137.71881630	2281.873056	2419.591872	137.72384290	2422.823321	2560.547164	0.47609530	7.916918	8.393014
6	36.17	0.0010065	23.738462	151.46287190	2272.577581	2424.040453	151.46891060	2415.002316	2566.471226	0.52076660	7.807537	8.328303
7	39.01	0.0010075	20.528996	163.34777320	2264.528655	2427.876428	163.35482580	2408.224576	2571.579402	0.55901430	7.714755	8.273770
8	41.52	0.0010085	18.102953	173.84542850	2257.410296	2431.255725	173.85349640	2402.225855	2576.079351	0.59250910	7.634154	8.226663
9	43.77	0.0010094	16.202963	183.26570210	2251.014935	2434.280637	183.27478680	2396.832521	2580.107308	0.62233970	7.562876	8.185216
10	45.82	0.0010103	14.673562	191.82348050	2245.198570	2437.022051	191.83358330	2391.924091	2583.757674	0.64925580	7.498966	8.148222
15	53.98	0.0010141	10.022514	225.96968410	2221.923084	2447.892768	225.98489490	2372.245578	2598.230473	0.75496100	7.251626	8.006587
20	60.07	0.0010172	7.649874	251.44098100	2204.478078	2455.919059	251.46132420	2357.455210	2608.916535	0.83210740	7.074697	7.906804
25	64.98	0.0010199	6.204796	271.96650040	2190.362116	2462.328616	271.99199660	2345.456511	2617.448507	0.89325530	6.936554	7.829810
30	69.11	0.0010222	5.229756	289.26638480	2178.419377	2467.685762	289.29705170	2335.281393	2624.578445	0.94410870	6.823050	7.767159
40	75.88	0.0010264	3.994005	317.59396520	2158.764632	2476.358598	317.63501980	2318.483765	2636.118785	1.02606720	6.642702	7.668769
50	81.34	0.0010299	3.240851	340.49236480	2142.777218	2483.269583	340.54386030	2304.768263	2645.312124	1.09116590	6.501667	7.592833
75	91.78	0.0010372	2.217509	384.35728660	2111.868813	2496.226099	384.43507700	2278.104164	2662.539241	1.21312000	6.242533	7.455653
100	99.63	0.0010431	1.694318	417.40835970	2088.303192	2505.711552	417.51267240	2257.630631	2675.143303	1.30272830	6.056161	7.358890
125	105.99	0.0010482	1.375156	444.24991440	2068.968681	2513.218595	444.38093990	2240.732116	2685.113056	1.37412540	5.909987	7.284113
150	111.38	0.0010527	1.159528	467.01912520	2052.418699	2519.437825	467.17702980	2226.189990	2693.367020	1.43375880	5.789404	7.223163

(Continued)

Table B-1b (Continued)

P kPa	T_{sat} °C	v_f m^3/kg	v_g m^3/kg	u_f kJ/kg	u_{fg} kJ/kg	u_g kJ/kg	h_f kJ/kg	h_{fg} kJ/kg	h_g kJ/kg	s_f kJ/kg K	s_{fg} kJ/kg K	s_g kJ/kg K
175	116.07	0.0010568	1.003781	486.89252930	2037.854320	2524.746849	487.07746230	2213.331114	2700.408576	1.48513020	5.686583	7.171713
200	120.24	0.0010605	0.885855	504.58958040	2024.783764	2529.373345	504.80167820	2201.742701	2706.544379	1.53035810	5.596836	7.127194
250	127.44	0.0010672	0.718791	535.22129590	2001.928381	2537.149677	535.48809380	2181.359209	2716.847302	1.60752320	5.445324	7.052847
300	133.56	0.0010732	0.605864	561.28879800	1982.230533	2543.519331	561.61074210	2163.667895	2725.278637	1.67210790	5.319992	6.992100
350	138.89	0.0010786	0.524266	584.10364900	1964.791654	2548.895303	584.48114200	2147.907190	2732.388332	1.72784400	5.212853	6.940697
400	143.64	0.0010835	0.462456	604.47162810	1949.058093	2553.529721	604.90504040	2133.607022	2738.512062	1.77699750	5.119110	6.896108
450	147.94	0.001088	0.413960	622.925515	1934.662805	2557.588319	623.415191	2120.455250	2743.870441	1.821051	5.035656	6.856707
500	151.87	0.001093	0.374861	639.837097	1921.349029	2561.186126	640.383362	2108.233195	2748.616557	1.861032	4.960357	6.821389
600	158.86	0.001101	0.315626	670.047738	1897.263570	2567.311308	670.708087	2085.978973	2756.687059	1.931544	4.828513	6.760057
700	164.98	0.001108	0.272810	696.577731	1875.778522	2572.356253	697.353288	2065.970230	2763.323518	1.992534	4.715394	6.707928
800	170.44	0.001115	0.240370	720.334592	1856.261502	2576.596094	721.226400	2047.665402	2768.891802	2.046433	4.616080	6.662514
900	175.39	0.001121	0.214912	741.919365	1838.292176	2580.211541	742.928406	2030.703723	2773.632128	2.094835	4.527380	6.622215
1000	179.92	0.001127	0.194383	761.752071	1821.575826	2583.327897	762.879273	2014.831733	2777.711006	2.138842	4.447101	6.585943
1100	184.10	0.001133	0.177466	780.138554	1805.897181	2586.035735	781.384808	1999.863944	2781.248752	2.179251	4.373673	6.552924
1200	188.00	0.001138	0.163277	797.308906	1791.093809	2588.402715	798.675065	1985.660151	2784.335216	2.216657	4.305929	6.522586
1300	191.64	0.001144	0.151199	813.440869	1777.039883	2590.480752	814.927759	1972.111559	2787.039317	2.251516	4.242981	6.494498
1400	195.08	0.001149	0.140789	828.674798	1763.635774	2592.310572	830.283217	1959.131888	2789.415106	2.284186	4.184136	6.468322
1500	198.33	0.001154	0.131721	843.123590	1750.801133	2593.924723	844.854318	1946.651451	2791.505769	2.314953	4.128840	6.443793
1600	201.41	0.001159	0.123748	856.879505	1738.470134	2595.349639	858.733300	1934.613069	2793.346369	2.344050	4.076648	6.420698
1750	205.76	0.001166	0.113441	876.378	1720.800816	2597.18	878.418	1917.282434	2795.700	2.38497	4.003392	6.38837
2000	212.42	0.00117667	0.09958839	906.3324809	1693.209731	2599.542212	908.6858275	1890.033163	2798.71899	2.4471335	3.892428	6.3395617
2250	218.45	0.00118722	0.08871646	933.6999643	1667.514319	2601.214283	936.3712032	1864.455105	2800.826308	2.5031935	3.792610	6.2958034

(Continued)

Table B-1b (Continued)

P kPa	T_{sat} °C	v_f m³/kg	v_g m³/kg	u_f kJ/kg	u_{fg} kJ/kg	u_g kJ/kg	h_f kJ/kg	h_{fg} kJ/kg	h_g kJ/kg	s_f kJ/kg K	s_{fg} kJ/kg K	s_g kJ/kg K
2500	223.99	0.00119733	0.07994872	958.9794824	1643.352971	2602.332453	961.9728058	1840.231443	2802.204249	2.554377	3.701642	6.256019
2750	229.11	0.00120709	0.07272296	982.5350331	1620.460449	2602.995482	985.8545286	1817.129096	2802.983624	2.6015676	3.617876	6.2194431
3000	233.89	0.00121656	0.06666161	1004.640879	1598.635778	2603.276657	1008.290553	1794.970946	2803.261499	2.6454257	3.540085	6.1855107
3500	242.60	0.00123481	0.05705364	1045.307799	1557.596556	2602.904354	1049.629623	1752.962480	2802.592102	2.7250638	3.398891	6.1239543
4000	250.39	0.00125236	0.04977088	1082.211963	1519.320618	2601.532581	1087.221416	1713.394693	2800.616109	2.7962088	3.272700	6.0689087
4500	257.47	0.00126943	0.04405303	1116.172902	1483.191969	2599.36487	1121.88532	1675.718167	2797.603487	2.8607783	3.158016	6.0187942
5000	263.98	0.00128614	0.03943963	1147.766039	1448.772454	2596.538493	1154.196753	1639.539899	2793.736652	2.9201039	3.052423	5.972527
6000	275.62	0.00131897	0.0324416	1205.430881	1383.835028	2589.265909	1213.34468	1570.570820	2783.9155	3.026642	2.861980	5.8886219
7000	285.86	0.00135151	0.02737208	1257.523917	1322.678458	2580.202375	1266.984505	1504.822406	2771.806911	3.1210716	2.691923	5.812995
8000	295.04	0.00138428	0.02351968	1305.497115	1264.108245	2569.60536	1316.571383	1441.191398	2757.762781	3.2066251	2.536451	5.743076
9000	303.38	0.0014177	0.02048484	1350.335902	1207.268965	2557.604867	1363.095214	1378.873248	2741.968462	3.28546	2.391681	5.6771413
10000	311.03	0.00145216	0.01802546	1392.753933	1151.496070	2544.250002	1407.275566	1317.229053	2724.504619	3.3591126	2.254832	5.6139444
12500	327.85	0.00154571	0.01349383	1491.563473	1013.182824	2504.746297	1510.884804	1162.534382	2673.419186	3.5276089	1.934344	5.4619528
15000	342.19	0.0016571	0.01033913	1584.968848	869.997541	2454.966388	1609.825359	1000.228001	2610.05336	3.6836891	1.625483	5.3091716
17500	354.72	0.00180334	0.00793173	1679.172724	711.042109	2390.214833	1710.731255	818.288825	2529.02008	3.8390347	1.303287	5.1423221
20000	365.80	0.00203596	0.00587382	1786.027506	510.051733	2296.079239	1826.746773	586.808787	2413.555559	4.0145774	0.918396	4.9329731
22000	373.75	0.00274435	0.00357619	1960.153885	124.993913	2085.147798	2020.529627	143.294392	2163.824019	4.3084526	0.221522	4.5299741

Table B-1c Superheated Water

T °C	P kPa	v m³/kg	u kJ/kg	h kJ/kg	s kJ/kg K	P kPa	v m³/kg	u kJ/kg	h kJ/kg	s kJ/kg K	P kPa	v m³/kg	u kJ/kg	h kJ/kg	s kJ/kg K
50.000		14.8692	2443.0606	2591.7530	8.1731										
100.000		17.1965	2514.9893	2686.9541	8.4471										
150.000		19.5131	2587.3633	2782.4940	8.6873		0.9597	2576.6762	2768.6121	7.2793					
200.000		21.8256	2660.7932	2879.0489	8.9030		1.0803	2653.9492	2870.0103	7.5059		0.4249	2642.5031	2854.9390	7.0585
250.000		24.1362	2735.5379	2976.8995	9.0995		1.1988	2730.7526	2970.5096	7.7078		0.4743	2722.9529	2960.1123	7.2699
300.000		26.4457	2811.7323	3076.1897	9.2808		1.3162	2808.1831	3071.4177	7.8920		0.5225	2802.4747	3063.7496	7.4591
350.000		28.7547	2889.4584	3177.0056	9.4494		1.4329	2886.7120	3173.3001	8.0624		0.5701	2882.3260	3167.3848	7.6324
400.000		31.0633	2968.7742	3279.4074	9.6075		1.5493	2966.5795	3276.4443	8.2216		0.6173	2963.0877	3271.7310	7.7935
450.000		33.3717	3049.7265	3383.4433	9.7565		1.6655	3047.9270	3381.0185	8.3714		0.6642	3045.0703	3377.1694	7.9446
500.000		35.6799	3132.3560	3489.1546	9.8979		1.7814	3130.8493	3487.1330	8.5133		0.7109	3128.4602	3483.9280	8.0873
550.000	10	37.9879	3216.6980	3596.5774	10.0325		1.8973	3215.4140	3594.8658	8.6483		0.7576	3213.3796	3592.1547	8.2230
600.000		40.2959	3302.7816	3705.7408	10.1612	200	2.0130	3301.6709	3704.2731	8.7773		0.8041	3299.9120	3701.9497	8.3524
650.000		42.6038	3390.6280	3816.6663	10.2847		2.1287	3389.6548	3815.3941	8.9011	500	0.8505	3388.1142	3813.3811	8.4765
700.000		44.9117	3480.2491	3929.3660	10.4036		2.2443	3479.3868	3928.2532	9.0201		0.8969	3478.0224	3926.4931	8.5958
750.000		47.2195	3571.6469	4043.8421	10.5183		2.3599	3570.8756	4042.8612	9.1350		0.9433	3569.6554	4041.3102	8.7109
800.000		49.5273	3664.8136	4160.0867	10.6292		2.4755	3664.1178	4159.2163	9.2460		0.9896	3663.0175	4157.8406	8.8221
850.000		51.8351	3759.7319	4278.0827	10.7367		2.5910	3759.0997	4277.3061	9.3535		1.0360	3758.1001	4276.0790	8.9298
900.000		54.1428	3856.3763	4397.8046	10.8410		2.7066	3855.7982	4397.1084	9.4579		1.0822	3854.8842	4396.0085	9.0342
950.000		56.4506	3954.7141	4519.2198	10.9423		2.8221	3954.1824	4518.5931	9.5593		1.1285	3953.3420	4517.6034	9.1357
1000.000		58.7583	4054.7070	4642.2898	11.0409		2.9375	4054.2155	4641.7239	9.6580		1.1748	4053.4388	4640.8302	9.2345
1050.000		61.0660	4156.3120	4766.9719	11.1370		3.0530	4155.8556	4766.4593	9.7541		1.2210	4155.1346	4765.6502	9.3306
1100.000		63.3737	4259.4828	4893.2196	11.2307		3.1685	4259.0573	4892.7543	9.8477		1.2673	4258.3852	4892.0199	9.4244

(Continued)

Table B-1c (Continued)

T °C	P kPa	v m^3/kg	u kJ/kg	h kJ/kg	s kJ/kg K	P kPa	v m^3/kg	u kJ/kg	h kJ/kg	s kJ/kg K	P kPa	v m^3/kg	u kJ/kg	h kJ/kg	s kJ/kg K
1150.000		65.6814	4364.1703	5020.9840	11.3220		3.2839	4363.7723	5020.5608	9.9392		1.3135	4363.1436	5019.8931	9.5158
1200.000		67.9890	4470.3240	5150.2145	11.4113		3.3994	4469.9504	5149.8291	10.0284		1.3597	4469.3604	5149.2210	9.6051
1250.000		70.2967	4577.8920	5280.8592	11.4985		3.5148	4577.5402	5280.5077	10.1157		1.4059	4576.9849	5279.9534	9.6924
1300.000		72.6044	4686.8215	5412.8654	11.5838		3.6303	4686.4895	5412.5447	10.2010		1.4521	4685.9654	5412.0391	9.7777
100.000		3.4188	2511.1976	2682.1355	7.6941										
150.000		3.8895	2585.1853	2779.6609	7.9394		0.6339	2570.6867	2760.8519	7.0779					
200.000		4.3560	2659.3772	2877.1771	8.1572		0.7163	2650.2238	2865.0989	7.3108		0.3520	2638.5014	2849.6824	6.9658
250.000		4.8205	2734.5401	2975.5665	8.3548		0.7963	2728.1866	2967.0865	7.5157		0.3938	2720.2838	2956.5593	7.1806
300.000		5.2840	2810.9893	3075.1904	8.5367		0.8753	2806.2946	3068.8800	7.7015		0.4344	2800.5427	3061.1563	7.3716
350.000		5.7469	2888.8822	3176.2281	8.7057		0.9536	2885.2568	3171.3372	7.8729		0.4742	2880.8502	3165.3952	7.5459
400.000		6.2094	2968.3132	3278.7850	8.8640		1.0315	2965.4192	3274.8779	8.0327		0.5137	2961.9166	3270.1504	7.7076
450.000		6.6717	3049.3483	3382.9336	9.0132		1.1092	3046.9769	3379.7383	8.1830		0.5529	3044.1138	3375.8808	7.8591
500.000		7.1338	3132.0392	3488.7296	9.1547		1.1867	3130.0543	3486.0665	8.3252		0.5920	3127.6611	3482.8562	8.0021
550.000		7.5958	3216.4280	3596.2174	9.2894		1.2641	3214.7368	3593.9633	8.4604		0.6309	3212.6996	3591.2486	8.1380
600.000	50	8.0577	3302.5480	3705.4321	9.4182	300	1.3414	3301.0853	3703.4995	8.5895	600	0.6697	3299.3243	3701.1735	8.2676
650.000		8.5195	3390.4233	3816.3987	9.5417		1.4186	3389.1418	3814.7237	8.7134		0.7085	3387.5996	3812.7090	8.3918
700.000		8.9813	3480.0677	3929.1319	9.6606		1.4958	3478.9324	3927.6669	8.8325		0.7472	3477.5668	3925.9056	8.5112
750.000		9.4430	3571.4846	4043.6357	9.7754		1.5729	3570.4692	4042.3445	8.9474		0.7859	3569.2481	4040.7926	8.6264
800.000		9.9047	3664.6672	4159.9035	9.8863		1.6500	3663.7513	4158.7580	9.0585		0.8246	3662.6502	4157.3816	8.7376
850.000		10.3664	3759.5989	4277.9193	9.9938		1.7271	3758.7667	4276.8972	9.1661		0.8632	3757.7665	4275.6696	8.8453
900.000		10.8281	3856.2546	4397.6581	10.0981		1.8042	3855.4937	4396.7418	9.2705		0.9018	3854.5792	4395.6418	8.9498
950.000		11.2897	3954.6022	4519.0878	10.1995		1.8812	3953.9024	4518.2632	9.3719		0.9404	3953.0616	4517.2734	9.0514
1000.000		11.7513	4054.6036	4642.1707	10.2981		1.9582	4053.9567	4641.4260	9.4706		0.9789	4053.1797	4640.5324	9.1501
1050.000		12.2130	4156.2160	4766.8640	10.3941		2.0352	4155.6153	4766.1896	9.5667		1.0175	4154.8941	4765.3805	9.2463

(Continued)

Table B-1c (Continued)

T °C	P kPa	v m³/kg	u kJ/kg	h kJ/kg	s kJ/kg K	P kPa	v m³/kg	u kJ/kg	h kJ/kg	s kJ/kg K	P kPa	v m³/kg	u kJ/kg	h kJ/kg	s kJ/kg K
1100.000		12.6746	4259.3932	4893.1216	10.4878		2.1123	4258.8333	4892.5094	9.6605		1.0560	4258.1611	4891.7752	9.3401
1150.000		13.1362	4364.0865	5020.8949	10.5792		2.1893	4363.5627	5020.3382	9.7519		1.0946	4362.9339	5019.6706	9.4315
1200.000		13.5978	4470.2454	5150.1334	10.6684		2.2662	4469.7538	5149.6263	9.8412		1.1331	4469.1637	5149.0185	9.5209
1250.000		14.0593	4577.8179	5280.7852	10.7557		2.3432	4577.3551	5280.3229	9.9284		1.1716	4576.7997	5279.7688	9.6082
1300.000		14.5209	4686.7516	5412.7978	10.8409		2.4202	4686.3148	5412.3761	10.0137		1.2101	4685.7907	5411.8707	9.6935
100.000		1.6961	2506.2822	2675.8940	7.3609										
150.000		1.9364	2582.4098	2776.0537	7.6129		0.4708	2564.4247	2752.7526	6.9300					
200.000		2.1723	2657.5888	2874.8143	7.8335		0.5342	2646.4094	2860.0765	7.1699		0.2607	2630.1971	2838.7930	6.8151
250.000		2.4061	2733.2857	2973.8911	8.0325		0.5951	2725.5869	2963.6210	7.3779		0.2931	2714.8376	2949.3172	7.0373
300.000		2.6388	2810.0574	3073.9373	8.2152		0.6548	2804.3919	3066.3241	7.5654		0.3241	2796.6341	3055.9126	7.2319
350.000		2.8709	2888.1605	3175.2543	8.3846		0.7139	2883.7948	3169.3655	7.7378		0.3544	2877.8777	3161.3889	7.4084
400.000		3.1027	2967.7362	3278.0059	8.5432		0.7726	2964.2553	3273.3069	7.8982		0.3843	2959.5633	3266.9748	7.5713
450.000		3.3342	3048.8751	3382.2959	8.6926		0.8311	3046.0247	3378.4552	8.0489		0.4139	3042.1944	3373.2952	7.7237
500.000	100	3.5655	3131.6429	3488.1978	8.8342	400	0.8894	3129.2579	3484.9982	8.1913		0.4433	3126.0588	3480.7071	7.8673
550.000		3.7968	3216.0902	3595.7672	8.9690		0.9475	3214.0587	3593.0596	8.3268	800	0.4726	3211.3367	3589.4329	8.0036
600.000		4.0279	3302.2558	3705.0459	9.0979		1.0056	3300.4990	3702.7250	8.4561		0.5018	3298.1468	3699.6188	8.1335
650.000		4.2590	3390.1672	3816.0640	9.2215		1.0636	3388.6282	3814.0527	8.5801		0.5310	3386.5689	3811.3629	8.2579
700.000		4.4900	3479.8408	3928.8391	9.3405		1.1215	3478.4776	3927.0802	8.6993		0.5601	3476.6543	3924.7293	8.3775
750.000		4.7210	3571.2817	4043.3776	9.4553		1.1794	3570.0625	4041.8275	8.8143		0.5892	3568.4324	4039.7566	8.4928
800.000		4.9519	3664.4841	4159.6745	9.5662		1.2373	3663.3845	4158.2994	8.9254		0.6182	3661.9150	4156.4631	8.6041
850.000		5.1828	3759.4325	4277.7149	9.6737		1.2951	3758.4335	4276.4881	9.0331		0.6472	3757.0988	4274.8507	8.7120
900.000		5.4137	3856.1025	4397.4748	9.7781		1.3530	3855.1890	4396.3752	9.1375		0.6762	3853.9689	4394.9081	8.8165

(Continued)

Table B-1c (Continued)

T °C	P kPa	v m^3/kg	u kJ/kg	h kJ/kg	s kJ/kg K	P kPa	v m^3/kg	u kJ/kg	h kJ/kg	s kJ/kg K	P kPa	v m^3/kg	u kJ/kg	h kJ/kg	s kJ/kg K
950.000		5.6446	3954.4623	4518.9229	9.8794		1.4108	3953.6222	4517.9333	9.2389		0.7051	3952.5005	4516.6134	8.9181
1000.000		5.8755	4054.4742	4642.0217	9.9781		1.4686	4053.6978	4641.1281	9.3377		0.7341	4052.6613	4639.9367	9.0169
1050.000		6.1063	4156.0959	4766.7291	10.0741		1.5264	4155.3750	4765.9198	9.4338		0.7630	4154.4129	4764.8413	9.1132
1100.000		6.3372	4259.2813	4892.9992	10.1678		1.5841	4258.6093	4892.2646	9.5275		0.7920	4257.7127	4891.2860	9.2070
1150.000		6.5680	4363.9818	5020.7835	10.2592		1.6419	4363.3532	5020.1156	9.6190		0.8209	4362.5146	5019.2259	9.2985
1200.000		6.7988	4470.1471	5150.0319	10.3485		1.6997	4469.5571	5149.4236	9.7083		0.8498	4468.7702	5148.6136	9.3878
1250.000		7.0297	4577.7254	5280.6927	10.4357		1.7574	4577.1700	5280.1381	9.7955		0.8787	4576.4295	5279.3999	9.4751
1300.000		7.2605	4686.6642	5412.7134	10.5210		1.8152	4686.1401	5412.2075	9.8808		0.9076	4685.4413	5411.5342	9.5605
200.000		0.2059	2621.4639	2827.3680	6.6932										
250.000		0.2326	2709.2419	2941.8868	6.9234		0.1418	2691.4850	2918.3750	6.6718		0.0870	2661.6880	2879.1335	6.4069
300.000		0.2579	2792.6644	3050.5909	7.1219		0.1586	2780.3714	3034.1367	6.8833		0.0989	2760.7594	3007.9643	6.6424
350.000		0.2825	2874.8769	3157.3458	7.3005		0.1746	2865.6995	3144.9902	7.0688		0.1097	2851.4157	3125.7895	6.8395
400.000		0.3066	2957.1952	3263.7798	7.4648		0.1900	2950.0008	3254.0769	7.2372		0.1201	2938.9472	3239.1813	7.0146
450.000		0.3304	3040.2664	3370.6983	7.6180		0.2053	3034.4303	3362.8396	7.3930		0.1301	3025.5272	3350.8571	7.1746
500.000		0.3541	3124.4511	3478.5509	7.7622		0.2203	3119.5947	3472.0398	7.5390		0.1400	3112.2163	3462.1525	7.3235
550.000		0.3776	3209.9700	3587.6124	7.8989		0.2352	3205.8473	3582.1226	7.6770		0.1497	3199.5994	3573.8083	7.4634
600.000	1000	0.4011	3296.9666	3698.0607	8.0292	1600	0.2500	3293.4096	3693.3670	7.8082	2500	0.1593	3288.0280	3686.2718	7.5960
650.000		0.4245	3385.5361	3810.0144	8.1538		0.2647	3382.4254	3805.9553	7.9335		0.1688	3377.7250	3799.8283	7.7225
700.000		0.4478	3475.7403	3923.5513	8.2736		0.2794	3472.9886	3920.0076	8.0539		0.1783	3468.8348	3914.6650	7.8436
750.000		0.4711	3567.6156	4038.7194	8.3890		0.2940	3565.1575	4035.6010	8.1697		0.1878	3561.4500	4030.9044	7.9601
800.000		0.4944	3661.1787	4155.5437	8.5005		0.3086	3658.9642	4152.7810	8.2815		0.1972	3655.6265	4148.6242	8.0724
850.000		0.5176	3756.4303	4274.0312	8.6084		0.3232	3754.4203	4271.5697	8.3897		0.2066	3751.3929	4267.8693	8.1810
900.000		0.5408	3853.3580	4394.1740	8.7130		0.3378	3851.5217	4391.9703	8.4946		0.2160	3848.7578	4388.6601	8.2862
950.000		0.5640	3951.9390	4515.9533	8.8147		0.3523	3950.2518	4513.9722	8.5964		0.2253	3947.7137	4510.9987	8.3884

(Continued)

Table B-1c (Continued)

T °C	P kPa	v m^3/kg	u kJ/kg	h kJ/kg	s kJ/kg K	P kPa	v m^3/kg	u kJ/kg	h kJ/kg	s kJ/kg K	P kPa	v m^3/kg	u kJ/kg	h kJ/kg	s kJ/kg K
1000.000		0.5872	4052.1425	4639.3410	8.9135		0.3669	4050.5843	4637.5539	8.6954		0.2347	4048.2416	4634.8736	8.4876
1050.000		0.6104	4153.9315	4764.3022	9.0098		0.3814	4152.4859	4762.6855	8.7918		0.2440	4150.3136	4760.2624	8.5842
1100.000		0.6335	4257.2642	4890.7969	9.1037		0.3959	4255.9176	4889.3309	8.8858		0.2533	4253.8950	4887.1351	8.6783
1150.000		0.6567	4362.0951	5018.7814	9.1952		0.4104	4360.8362	5017.4495	8.9774		0.2626	4358.9461	5015.4557	8.7701
1200.000		0.6798	4468.3767	5148.2091	9.2846		0.4249	4467.1958	5146.9972	9.0669		0.2719	4465.4237	5145.1841	8.8597
1250.000		0.7030	4576.0592	5279.0313	9.3719		0.4394	4574.9483	5277.9273	9.1543		0.2812	4573.2817	5276.2768	8.9472
1300.000		0.7261	4685.0920	5411.1981	9.4573		0.4538	4684.0440	5410.1917	9.2397		0.2905	4682.4725	5408.6880	9.0327
200.000		0.1692	2612.2640	2815.3600	6.5890										
250.000		0.1923	2703.4897	2934.2594	6.8281		0.1249	2685.2151	2910.0964	6.6052					
300.000		0.2138	2788.6319	3045.1891	7.0307		0.1402	2776.1396	3028.4813	6.8214		0.0811	2749.1895	2992.5685	6.5375
350.000		0.2345	2871.8473	3153.2654	7.2115		0.1546	2862.5803	3140.7942	7.0093		0.0905	2843.1947	3114.7556	6.7420
400.000		0.2548	2954.8122	3260.5653	7.3771		0.1685	2947.5720	3250.8026	7.1792		0.0994	2932.6654	3230.7231	6.9210
450.000		0.2748	3038.3298	3368.0901	7.5312		0.1821	3032.4674	3360.1971	7.3359		0.1079	3020.5020	3344.0977	7.0835
500.000		0.2946	3122.8378	3476.3877	7.6760		0.1955	3117.9648	3469.8551	7.4825		0.1162	3108.0681	3456.5968	7.2339
550.000	1200	0.3143	3208.5996	3585.7872	7.8131	1800	0.2088	3204.4655	3580.2832	7.6209	3000	0.1244	3196.0953	3569.1483	7.3750
600.000		0.3339	3295.7837	3696.4994	7.9437		0.2220	3292.2184	3691.7959	7.7524		0.1324	3285.0146	3682.3020	7.5084
650.000		0.3535	3384.5013	3808.6637	8.0686		0.2351	3381.3844	3804.5977	7.8781		0.1404	3375.0960	3796.4049	7.6355
700.000		0.3730	3474.8246	3922.3717	8.1885		0.2482	3472.0682	3918.8231	7.9986		0.1484	3466.5137	3911.6831	7.7571
750.000		0.3924	3566.7974	4037.6810	8.3041		0.2612	3564.3357	4034.5592	8.1145		0.1563	3559.3799	4028.2857	7.8739
800.000		0.4118	3660.4415	4154.6236	8.4156		0.2742	3658.2241	4151.8586	8.2264		0.1642	3653.7642	4146.3084	7.9865
850.000		0.4312	3755.7610	4273.2112	8.5236		0.2872	3753.7488	4270.7482	8.3347		0.1720	3749.7049	4265.8093	8.0954
900.000		0.4506	3852.7464	4393.4397	8.6284		0.3002	3850.9085	4391.2352	8.4397		0.1799	3847.2174	4386.8187	8.2008

(Continued)

Table B-1c (Continued)

T °C	P kPa	v m^3/kg	u kJ/kg	h kJ/kg	s kJ/kg K	P kPa	v m^3/kg	u kJ/kg	h kJ/kg	s kJ/kg K	P kPa	v m^3/kg	u kJ/kg	h kJ/kg	s kJ/kg K
950.000		0.4699	3951.3770	4515.2931	8.7301		0.3131	3949.6885	4513.3117	8.5416		0.1877	3946.3000	4509.3458	8.3031
1000.000		0.4893	4051.6235	4638.7453	8.8290		0.3261	4050.0643	4636.9583	8.6406		0.1955	4046.9374	4633.3846	8.4024
1050.000		0.5086	4153.4499	4763.7632	8.9253		0.3390	4152.0036	4762.1468	8.7371		0.2033	4149.1048	4758.9172	8.4992
1100.000		0.5279	4256.8155	4890.3081	9.0192		0.3519	4255.4684	4888.8426	8.8311		0.2110	4252.7701	4885.9167	8.5934
1150.000		0.5472	4361.6756	5018.3372	9.1108		0.3648	4360.4164	5017.0060	8.9227		0.2188	4357.8953	5014.3502	8.6852
1200.000		0.5665	4467.9831	5147.8048	9.2002		0.3777	4466.8021	5146.5938	9.0122		0.2266	4464.4388	5144.1793	8.7749
1250.000		0.5858	4575.6889	5278.6629	9.2875		0.3905	4574.5779	5277.5600	9.0997		0.2343	4572.3558	5275.3625	8.8625
1300.000		0.6051	4684.7426	5410.8623	9.3729		0.4034	4683.6947	5409.8570	9.1851		0.2421	4681.5996	5407.8555	8.9480
200.000	1400	0.1430	2602.5525	2802.7132	6.4966	2000					4000				
250.000		0.1635	2697.5735	2926.4255	6.7454		0.1114	2678.7539	2901.5774	6.5438					
300.000		0.1823	2784.5348	3039.7052	6.9523		0.1254	2771.8375	3022.7363	6.7651		0.0588	2724.3910	2959.6789	6.3598
350.000		0.2003	2868.7883	3149.1471	7.1354		0.1386	2859.4301	3136.5583	6.9556		0.0664	2826.0911	3091.8407	6.5811
400.000		0.2178	2952.4141	3257.3311	7.3024		0.1512	2945.1276	3247.5080	7.1269		0.0734	2919.7850	3213.3977	6.7688
450.000		0.2351	3036.3844	3365.4706	7.4573		0.1635	3030.4957	3357.5430	7.2845		0.0800	3010.2780	3330.3547	6.9364
500.000		0.2521	3121.2190	3474.2173	7.6027		0.1757	3116.3293	3467.6633	7.4318		0.0864	3099.6655	3445.3501	7.0902
550.000		0.2691	3207.2253	3583.9572	7.7403		0.1877	3203.0799	3578.4391	7.5706		0.0927	3189.0164	3559.7406	7.2335
600.000		0.2860	3294.5980	3694.9348	7.8712		0.1996	3291.0245	3690.2216	7.7024		0.0988	3278.9378	3674.3032	7.3687
650.000		0.3027	3383.4644	3807.3107	7.9963		0.2114	3380.3414	3803.2378	7.8283		0.1049	3369.8012	3789.5169	7.4970
700.000		0.3195	3473.9074	3921.1905	8.1164		0.2232	3471.1463	3917.6371	7.9490		0.1110	3461.8436	3905.6906	7.6195
750.000		0.3362	3565.9781	4036.6416	8.2321		0.2350	3563.5127	4033.5164	8.0651		0.1170	3555.2183	4023.0282	7.7371
800.000		0.3529	3659.7033	4153.7027	8.3438		0.2467	3657.4831	4150.9354	8.1771		0.1229	3650.0229	4141.6633	7.8503
850.000		0.3695	3755.0910	4272.3907	8.4519		0.2584	3753.0766	4269.9262	8.2855		0.1288	3746.3159	4261.6809	7.9596
900.000		0.3861	3852.1344	4392.7052	8.5567		0.2701	3850.2947	4390.4998	8.3905		0.1348	3844.1270	4383.1311	8.0654
950.000		0.4027	3950.8146	4514.6327	8.6585		0.2818	3949.1248	4512.6510	8.4925		0.1406	3943.4651	4506.0381	8.1680

(Continued)

Table B-1c (Continued)

T °C	P kPa	v m^3/kg	u kJ/kg	h kJ/kg	s kJ/kg K	P kPa	v m^3/kg	u kJ/kg	h kJ/kg	s kJ/kg K	P kPa	v m^3/kg	u kJ/kg	h kJ/kg	s kJ/kg K
1000.000		0.4193	4051.1041	4638.1496	8.7574		0.2934	4049.5439	4636.3626	8.5916		0.1465	4044.3234	4630.4070	8.2676
1050.000		0.4359	4152.9680	4763.2243	8.8538		0.3050	4151.5210	4761.6083	8.6881		0.1524	4146.6833	4756.2287	8.3646
1100.000		0.4525	4256.3666	4889.8194	8.9477		0.3167	4255.0190	4888.3545	8.7821		0.1582	4250.5175	4883.4834	8.4590
1150.000		0.4690	4361.2560	5017.8932	9.0393		0.3283	4359.9964	5016.5628	8.8738		0.1641	4355.7920	5012.1433	8.5510
1200.000		0.4856	4467.5895	5147.4008	9.1288		0.3399	4466.4083	5146.1907	8.9633		0.1699	4462.4682	5142.1747	8.6408
1250.000		0.5021	4575.3186	5278.2950	9.2161		0.3515	4574.2076	5277.1929	9.0508		0.1758	4570.5040	5273.5396	8.7285
1300.000		0.5187	4684.3933	5410.5268	9.3016		0.3631	4683.3455	5409.5226	9.1363		0.1816	4679.8545	5406.1964	8.8142
300.000		0.0453	2696.9886	2923.5049	6.2067							0.0013	1288.0401	1327.9905	3.1743
350.000		0.0519	2808.0241	3067.6973	6.4482		0.0161	2623.2994	2824.8115	5.7097		0.0016	1561.4887	1608.0533	3.6421
400.000		0.0578	2906.4605	3195.5014	6.6456		0.0200	2788.6988	3038.7667	6.0409		0.0028	2066.8901	2150.6762	4.4723
450.000		0.0633	2999.8159	3316.3058	6.8187		0.0230	2912.7983	3200.1677	6.2724		0.0067	2620.1684	2822.2577	5.4435
500.000		0.0686	3091.1196	3433.9218	6.9760		0.0256	3022.2251	3342.2470	6.4625		0.0087	2823.1729	3083.4566	5.7936
550.000		0.0737	3181.8431	3550.2170	7.1218		0.0280	3125.0437	3475.1639	6.6291		0.0102	2972.0426	3276.7815	6.0362
600.000		0.0787	3272.7949	3666.2267	7.2586		0.0303	3224.6996	3603.3073	6.7802		0.0114	3100.1243	3443.0556	6.2324
650.000	5000	0.0836	3364.4580	3782.5752	7.3882	12500	0.0325	3322.9460	3728.9389	6.9202	30000	0.0126	3218.0245	3595.5085	6.4022
700.000		0.0885	3457.1370	3899.6608	7.5117		0.0346	3420.7802	3853.3624	7.0514		0.0137	3330.4089	3740.0561	6.5547
750.000		0.0933	3551.0288	4017.7449	7.6300		0.0367	3518.8132	3977.3881	7.1757		0.0147	3439.7093	3879.8313	6.6948
800.000		0.0981	3646.2602	4137.0007	7.7438		0.0387	3617.4372	4101.5449	7.2942		0.0156	3547.3372	4016.6817	6.8254
850.000		0.1029	3742.9103	4257.5414	7.8536		0.0407	3716.9105	4226.1891	7.4077		0.0166	3654.1732	4151.7795	6.9484
900.000		0.1077	3841.0238	4379.4373	7.9598		0.0427	3817.4042	4351.5635	7.5169		0.0175	3760.7947	4285.9068	7.0653
950.000		0.1124	3940.6207	4502.7278	8.0627		0.0447	3919.0300	4477.8344	7.6223		0.0184	3867.5924	4419.6040	7.1769
1000.000		0.1171	4041.7024	4627.4297	8.1626		0.0467	4021.8579	4605.1135	7.7243		0.0193	3974.8359	4553.2532	7.2840

(Continued)

Table B-1c (Continued)

T °C	P kPa	v m³/kg	u kJ/kg	h kJ/kg	s kJ/kg K	P kPa	v m³/kg	u kJ/kg	h kJ/kg	s kJ/kg K	P kPa	v m³/kg	u kJ/kg	h kJ/kg	s kJ/kg K
1050.000		0.1219	4144.2568	4753.5428	8.2598		0.0486	4125.9280	4733.4734	7.8232		0.0201	4082.7128	4687.1282	7.3871
1100.000		0.1266	4248.2615	4881.0543	8.3543		0.0505	4231.2584	4862.9583	7.9192		0.0210	4191.3527	4821.4263	7.4867
1150.000		0.1313	4353.6867	5009.9420	8.4465		0.0525	4337.8509	4993.5907	8.0127		0.0218	4300.8438	4956.2894	7.5832
1200.000		0.1359	4460.4968	5140.1767	8.5365		0.0544	4445.6955	5125.3772	8.1037		0.0227	4411.2436	5091.8182	7.6768
1250.000		0.1406	4568.6521	5271.7239	8.6243		0.0563	4554.7728	5258.3123	8.1924		0.0235	4522.5869	5228.0828	7.7678
1300.000		0.1453	4678.1100	5404.5452	8.7101		0.0582	4665.0568	5392.3813	8.2790		0.0243	4634.8914	5365.1295	7.8563
300.000		0.0267	2611.8687	2812.2244	5.8605		0.0014	1316.6917	1337.3567	3.2261		0.0013	1246.4151	1322.6047	3.0970
350.000		0.0324	2757.8184	3000.9578	6.1771		0.0115	2519.2669	2691.3033	5.4404		0.0014	1482.3235	1566.7199	3.5051
400.000		0.0369	2871.0051	3148.0278	6.4043		0.0157	2739.9283	2974.7091	5.8799		0.0016	1745.0612	1843.0282	3.9312
450.000		0.0410	2972.5616	3279.7852	6.5932		0.0185	2879.8981	3156.6488	6.1410		0.0021	2054.9860	2180.0203	4.4134
500.000		0.0447	3069.1134	3404.5460	6.7600		0.0208	2997.2596	3309.2590	6.3452		0.0030	2394.5883	2571.8707	4.9373
550.000		0.0483	3163.4981	3525.9051	6.9122		0.0229	3104.9345	3448.7568	6.5201		0.0040	2663.7380	2901.0287	5.3505
600.000		0.0518	3257.1539	3645.7037	7.0535		0.0249	3207.9109	3581.4756	6.6767		0.0048	2862.6833	3152.2619	5.6471
650.000		0.0552	3350.8952	3764.9954	7.1863		0.0268	3308.5901	3710.5035	6.8204		0.0056	3025.0105	3359.9462	5.8786
700.000	7500	0.0586	3445.2181	3884.4303	7.3123	15000	0.0286	3408.2908	3837.5598	6.9544	60000	0.0063	3167.9164	3543.4586	6.0723
750.000		0.0619	3540.4394	4004.4295	7.4325		0.0304	3507.8030	3963.6893	7.0808		0.0069	3299.6514	3712.4775	6.2417
800.000		0.0651	3636.7650	4125.2727	7.5479		0.0321	3607.6274	4089.5654	7.2009		0.0075	3424.5740	3872.2553	6.3942
850.000		0.0684	3734.3287	4247.1473	7.6589		0.0338	3708.0930	4215.6390	7.3158		0.0080	3545.2051	4025.8975	6.5341
900.000		0.0716	3833.2145	4370.1773	7.7660		0.0355	3809.4190	4342.2187	7.4260		0.0085	3663.1101	4175.3714	6.6643
950.000		0.0748	3933.4711	4494.4420	7.8697		0.0372	3911.7513	4469.5175	7.5323		0.0090	3779.3146	4321.9909	6.7867
1000.000		0.0780	4035.1217	4619.9881	7.9703		0.0388	4015.1851	4597.6819	7.6350		0.0095	3894.5182	4466.6697	6.9027
1050.000		0.0812	4138.1706	4746.8387	8.0681		0.0405	4119.7796	4726.8111	7.7345		0.0100	4009.2122	4610.0622	7.0132
1100.000		0.0843	4242.6086	4874.9992	8.1631		0.0421	4225.5672	4856.9693	7.8310		0.0105	4123.7483	4752.6476	7.1189
1150.000		0.0875	4348.4159	5004.4618	8.2557		0.0437	4332.5614	4988.1951	7.9249		0.0109	4238.3820	4894.7822	7.2206

(Continued)

Table B-1c (Continued)

T °C	P kPa	v m^3/kg	u kJ/kg	h kJ/kg	s kJ/kg K	P kPa	v m^3/kg	u kJ/kg	h kJ/kg	s kJ/kg K	P kPa	v m^3/kg	u kJ/kg	h kJ/kg	s kJ/kg K
1200.000		0.0906	4455.5651	5135.2087	8.3460		0.0453	4440.7610	5120.5079	8.0163		0.0114	4353.2996	5036.7340	7.3186
1250.000		0.0938	4564.0231	5267.2148	8.4342		0.0469	4550.1539	5253.9124	8.1053		0.0118	4468.6369	5178.7052	7.4134
1300.000		0.0969	4673.7526	5400.4492	8.5202		0.0485	4660.7199	5388.4022	8.1922		0.0123	4584.4920	5320.8485	7.5052
350.000	10000	0.0224	2698.0533	2922.2109	5.9425	20000	0.0017	1612.0750	1645.3653	3.7277					
400.000		0.0264	2832.0130	3096.0923	6.2114		0.0099	2617.9464	2816.8628	5.5521					
450.000		0.0298	2943.6151	3241.1363	6.4193		0.0127	2806.7673	3060.7949	5.9026					
500.000		0.0328	3046.1618	3373.9980	6.5971		0.0148	2944.0529	3239.4265	6.1417					
550.000		0.0356	3144.5649	3500.8838	6.7561		0.0165	3062.9538	3393.9335	6.3355					
600.000		0.0384	3241.1181	3624.7236	6.9022		0.0182	3173.2718	3536.6534	6.5039					
650.000		0.0410	3337.0527	3747.1097	7.0385		0.0197	3279.1884	3672.9214	6.6556					
700.000		0.0436	3433.0941	3868.9917	7.1671		0.0211	3382.8402	3805.5026	6.7955					
750.000		0.0461	3529.6963	3990.9727	7.2893		0.0225	3485.4489	3936.0018	6.9263					
800.000		0.0486	3627.1534	4113.4507	7.4062		0.0239	3587.7676	4065.4240	7.0498					
850.000		0.0511	3725.6588	4236.6942	7.5184		0.0252	3690.2838	4194.4315	7.1673					
900.000		0.0536	3825.3386	4360.8846	7.6266		0.0265	3793.3230	4323.4757	7.2797					
950.000		0.0560	3926.2720	4486.1434	7.7311		0.0278	3897.1054	4452.8703	7.3877					
1000.000		0.0584	4028.5051	4612.5491	7.8324		0.0291	4001.7800	4582.8357	7.4919					
1050.000		0.0608	4132.0596	4740.1492	7.9307		0.0303	4107.4459	4713.5262	7.5925					
1100.000		0.0632	4236.9398	4868.9678	8.0263		0.0315	4214.1663	4845.0488	7.6901					
1150.000		0.0656	4343.1365	4999.0122	8.1193		0.0328	4321.9785	4977.4750	7.7848					
1200.000		0.0680	4450.6308	5130.2766	8.2100		0.0340	4430.8999	5110.8508	7.8769					
1250.000		0.0703	4559.3963	5262.7454	8.2984		0.0352	4540.9336	5245.2018	7.9666					
1300.000		0.0727	4669.4013	5396.3959	8.3847		0.0364	4652.0716	5380.5389	8.0541					

Table B-1d Compressed Liquid Water

T °C	P kPa	v m^3/kg	u kJ/kg	h kJ/kg	s kJ/kg K	T °C	P kPa	v m^3/kg	u kJ/kg	h kJ/kg	s kJ/kg K	T °C	P kPa	v m^3/kg	u kJ/kg	h kJ/kg	s kJ/kg K
0		0.00099769	0.060	5.05	0.00019	0		0.00099037	0.277	20.08	0.00066	0		0.00098114	0.382	39.63	0.00003
20		0.00099954	83.527	88.52	0.29514	20		0.00099291	82.618	102.48	0.29176	20		0.00098452	81.422	120.80	0.28682
40		0.00100568	166.893	171.92	0.57034	40		0.00099924	165.140	185.13	0.56449	40		0.00099109	162.928	202.57	0.55664
60		0.00101489	250.261	255.34	0.82855	60		0.00100836	247.732	267.90	0.82071	60		0.00100011	244.569	284.57	0.81048
80		0.00102669	333.737	338.87	1.07205	80		0.00101986	330.422	350.82	1.06242	80		0.00101128	326.297	366.75	1.05002
100		0.00104093	417.543	422.75	1.30307	100		0.00103361	413.397	434.07	1.29172	100		0.00102447	408.258	449.24	1.27721
120		0.00105759	501.872	507.16	1.52343	120		0.00104958	496.815	517.81	1.51031	120		0.00103964	490.578	532.16	1.49369
140		0.00107679	586.876	592.26	1.73455	140		0.00106784	580.794	602.15	1.71956	140		0.00105684	573.339	615.61	1.70072
160		0.00109878	672.727	678.22	1.93773	160		0.00108859	665.458	687.23	1.92065	160		0.00107619	656.619	699.67	1.89939
180	5000	0.00112395	759.662	765.28	2.13421	180	20000	0.00111211	750.982	773.22	2.11473	180	40000	0.00109792	740.533	784.45	2.09074
200		0.00115293	848.027	853.79	2.32533	200		0.00113887	837.618	860.40	2.30296	200		0.00112233	825.255	870.15	2.27579
220		0.00118666	938.321	944.25	2.51257	220		0.00116953	925.725	949.12	2.48660	220		0.00114986	911.025	957.02	2.45561
240		0.00122659	1031.264	1037.40	2.69770	240		0.00120505	1015.788	1039.89	2.66702	240		0.00118112	998.161	1045.41	2.63129
260		0.00127513	1127.951	1134.33	2.88297	260		0.00124689	1108.483	1133.42	2.84581	260		0.00121698	1087.068	1135.75	2.80398
280		0.04223020	2644.780	2855.93	6.08667	280		0.00129737	1204.785	1230.73	3.02497	280		0.00125864	1178.266	1228.61	2.97496
300		0.04530326	2696.989	2923.50	6.20672	300		0.00136050	1306.236	1333.45	3.20734	300		0.00130791	1272.444	1324.76	3.14569
320		0.04809044	2743.805	2984.26	6.31093	320		0.00144424	1415.629	1444.51	3.39777	320		0.00136759	1370.564	1425.27	3.31804
340		0.05068834	2787.218	3040.66	6.40447	340		0.00156855	1539.361	1570.73	3.60696	340		0.00144238	1474.076	1531.77	3.49460
360		0.05315262	2828.367	3094.13	6.49029	360		0.00182480	1703.226	1739.72	3.87782	360		0.00154088	1585.381	1647.02	3.67951
380		0.05551817	2867.959	3145.55	6.57026	380		0.00825572	2493.444	2658.56	5.31309	380		0.00168127	1708.969	1776.22	3.88034

(Continued)

Table B-1d (Continued)

T °C	P kPa	v m^3/kg	u kJ/kg	h kJ/kg	s kJ/kg K	T °C	P kPa	v m^3/kg	u kJ/kg	h kJ/kg	s kJ/kg K	T °C	P kPa	v m^3/kg	u kJ/kg	h kJ/kg	s kJ/kg K
0		0.00099521	0.147	10.10	0.00044	0		0.00098800	0.322	25.02	0.00062	0		0.00097674	0.364	49.20	−0.00076
20		0.00099730	83.223	93.20	0.29404	20		0.00099077	82.317	107.09	0.29057	20		0.00098049	80.829	129.85	0.28418
40		0.00100350	166.299	176.33	0.56839	40		0.00099716	164.575	189.50	0.56253	40		0.00098718	161.868	211.23	0.55270
60		0.00101267	249.402	259.53	0.82592	60		0.00100625	246.920	272.08	0.81813	60		0.00099617	243.068	292.88	0.80544
80		0.00102437	332.609	342.85	1.06881	80		0.00101767	329.361	354.80	1.05928	80		0.00100719	324.347	374.71	1.04397
100		0.00103844	416.131	426.52	1.29924	100		0.00103127	412.072	437.85	1.28803	100		0.00102012	405.837	456.84	1.27020
120		0.00105486	500.147	510.70	1.51899	120		0.00104702	495.204	521.38	1.50607	120		0.00103494	487.651	539.40	1.48571
140		0.00107374	584.797	595.53	1.72947	140		0.00106500	578.864	605.49	1.71474	140		0.00105167	569.856	622.44	1.69173
160		0.00109528	670.236	681.19	1.93192	160		0.00108537	663.163	690.30	1.91519	160		0.00107043	652.514	706.04	1.88932
180	10000	0.00111988	756.678	767.88	2.12755	180	25000	0.00110841	748.259	775.97	2.10854	180	50000	0.00109138	735.715	790.28	2.07946
200		0.00114806	844.433	855.91	2.31766	200		0.00113453	834.380	862.74	2.29591	200		0.00111481	819.604	875.34	2.26313
220		0.00118068	933.945	945.75	2.50361	220		0.00116432	921.850	950.96	2.47851	220		0.00114106	904.384	961.44	2.44134
240		0.00121898	1025.842	1038.03	2.68702	240		0.00119863	1011.102	1041.07	2.65761	240		0.00117065	990.318	1048.85	2.61509
260		0.00126498	1121.043	1133.69	2.86987	260		0.00123874	1102.722	1133.69	2.83466	260		0.00120424	1077.733	1137.95	2.78540
280		0.00132217	1221.009	1234.23	3.05496	280		0.00128657	1197.529	1229.69	3.01141	280		0.00124279	1167.032	1229.17	2.95336
300		0.00139746	1328.402	1342.38	3.24697	300		0.00134532	1296.749	1330.38	3.19020	300		0.00128759	1258.716	1323.10	3.12015
320		0.01924846	2588.158	2780.64	5.70933	320		0.00142082	1402.420	1437.94	3.37462	320		0.00134058	1353.434	1420.46	3.28711
340		0.02146396	2665.510	2880.15	5.87445	340		0.00152562	1518.541	1556.68	3.57144	340		0.00140464	1452.060	1522.29	3.45594
360		0.02330012	2728.012	2961.01	6.00428	360		0.00169651	1655.653	1698.07	3.79820	360		0.00148446	1555.847	1630.07	3.62888
380		0.02492289	2782.485	3031.71	6.11425	380		0.00222212	1880.842	1936.40	4.16785	380		0.00158804	1666.691	1746.09	3.80925

(Continued)

Table B-1d (Continued)

T °C	P kPa	v m^3/kg	u kJ/kg	h kJ/kg	s kJ/kg K	T °C	P kPa	v m^3/kg	u kJ/kg	h kJ/kg	s kJ/kg K	T °C	P kPa	v m^3/kg	u kJ/kg	h kJ/kg	s kJ/kg K
0		0.00099277	0.219	15.11	0.00060	0		0.00098568	0.354	29.92	0.00051						
20		0.00099509	82.920	97.85	0.29292	20		0.00098865	82.018	111.68	0.28935						
40		0.00100136	165.715	180.74	0.56644	40		0.00099511	164.017	193.87	0.56057						
60		0.00101050	248.559	263.72	0.82331	60		0.00100417	246.123	276.25	0.81557						
80		0.00102210	331.505	346.84	1.06560	80		0.00101550	328.320	358.79	1.05616						
100		0.00103601	414.749	430.29	1.29546	100		0.00102896	410.775	441.64	1.28438						
120		0.00105219	498.462	514.24	1.51462	120		0.00104451	493.629	524.96	1.50189						
140		0.00107075	582.771	598.83	1.72447	140		0.00106222	576.980	608.85	1.70999						
160		0.00109189	667.815	684.19	1.92623	160		0.00108224	660.927	693.39	1.90983						
180		0.00111593	753.786	770.53	2.12107	180		0.00110481	745.612	778.76	2.10248						
200	15000	0.00114338	840.967	858.12	2.31021	200	30000	0.00113033	831.244	865.15	2.28904						
220		0.00117498	929.752	947.38	2.49496	220		0.00115931	918.115	952.89	2.47066						
240		0.00121182	1020.694	1038.87	2.67681	240		0.00119253	1006.613	1042.39	2.64853						
260		0.00125561	1114.574	1133.41	2.85752	260		0.00123107	1097.252	1134.18	2.82401						
280		0.00130916	1212.576	1232.21	3.03942	280		0.00127659	1190.730	1229.03	2.99862						
300		0.00137766	1316.692	1337.36	3.22611	300		0.00133168	1288.040	1327.99	3.17435						
320		0.00147254	1430.951	1453.04	3.42443	320		0.00140079	1390.727	1432.75	3.35398						
340		0.00163070	1567.149	1591.61	3.65403	340		0.00149249	1501.506	1546.28	3.54217						
360		0.01257125	2579.667	2768.24	5.56295	360		0.00162692	1626.127	1674.93	3.74856						
380		0.01427510	2669.538	2883.66	5.74258	380		0.00187278	1781.660	1837.84	4.00166						

Table B-2a Saturated Ammonia - Temperature table

T °C	P_{sat} kPa	v_f m^3/kg	v_g m^3/kg	u_f kJ/kg	u_{fg} kJ/kg	u_g kJ/kg	h_f kJ/kg	h_{fg} kJ/kg	h_g kJ/kg	s_f kJ/kg K	s_{fg} kJ/kg K	s_g kJ/kg K
−50	40.82	0.0014243	2.6289114	−24.786	1308.676	1283.890	−24.728	1415.927	1391.199	0.0944982	6.3453887	6.4398869
−45	54.47	0.0014364	2.0079972	−2.926	1293.166	1290.241	−2.847	1402.455	1399.607	0.1913799	6.1472867	6.3386666
−40	71.66	0.0014490	1.5539219	19.066	1277.351	1296.418	19.170	1388.607	1407.777	0.2867334	5.9560447	6.2427781
−35	93.07	0.0014619	1.2171795	41.185	1261.223	1302.408	41.321	1374.368	1415.689	0.3806041	5.7711698	6.1517739
−30	119.40	0.0014753	0.9641702	63.427	1244.771	1308.198	63.603	1359.721	1423.324	0.4730361	5.5922075	6.0652436
−25	151.46	0.0014891	0.7717400	85.788	1227.988	1313.776	86.013	1344.650	1430.663	0.5640737	5.4187366	5.9828103
−20	190.09	0.0015035	0.6237039	108.265	1210.862	1319.127	108.551	1329.134	1437.686	0.6537624	5.2503646	5.9041270
−15	236.21	0.0015183	0.5085958	130.858	1193.382	1324.240	131.217	1313.157	1444.373	0.7421494	5.0867246	5.8288740
−10	290.79	0.0015336	0.4181890	153.565	1175.536	1329.101	154.011	1296.695	1450.706	0.8292842	4.9274711	5.7567553
−5	354.88	0.0015495	0.3465108	176.388	1157.309	1333.697	176.938	1279.728	1456.665	0.9152184	4.7722783	5.6874967
0	429.55	0.0015660	0.2891757	199.329	1138.685	1338.014	200.002	1262.229	1462.231	1.0000066	4.6208361	5.6208427
5	515.97	0.0015831	0.2429295	222.392	1119.648	1342.040	223.209	1244.174	1467.383	1.0837058	4.4728483	5.5565541
10	615.30	0.0016009	0.2053331	245.583	1100.176	1345.759	246.568	1225.533	1472.101	1.1663760	4.3280298	5.4944058
15	728.80	0.0016195	0.1745410	268.909	1080.247	1349.156	270.089	1206.273	1476.362	1.2480802	4.1861038	5.4341840
20	857.75	0.0016388	0.1491441	292.379	1059.836	1352.215	293.785	1186.360	1480.144	1.3288845	4.0468001	5.3756846
25	1003.49	0.0016590	0.1280573	316.004	1038.914	1354.918	317.669	1165.752	1483.421	1.4088587	3.9098514	5.3187101
30	1167.37	0.0016802	0.1104384	339.798	1017.445	1357.243	341.759	1144.406	1486.166	1.4880767	3.7749910	5.2630677
35	1350.82	0.0017024	0.0956286	363.775	995.393	1359.169	366.075	1122.271	1488.346	1.5666168	3.6419497	5.2085665
40	1555.31	0.0017258	0.0831088	387.955	972.713	1360.668	390.639	1099.288	1489.928	1.6445626	3.5104527	5.1550153
45	1782.34	0.0017505	0.0724668	412.358	949.351	1361.710	415.478	1075.392	1490.870	1.7220042	3.3802151	5.1022193
50	2033.46	0.0017766	0.0633736	437.009	925.250	1362.259	440.622	1050.505	1491.127	1.7990389	3.2509382	5.0499771
55	2310.29	0.0018044	0.0555647	461.938	900.336	1362.273	466.106	1024.537	1490.644	1.8757737	3.1223029	4.9980766

(Continued)

Table B-2a (Continued)

T °C	P_{sat} kPa	v_f m^3/kg	v_g m^3/kg	u_f kJ/kg	u_{fg} kJ/kg	u_g kJ/kg	h_f kJ/kg	h_{fg} kJ/kg	h_g kJ/kg	s_f kJ/kg K	s_{fg} kJ/kg K	s_g kJ/kg K
60	2614.48	0.0018340	0.0488260	487.177	874.526	1361.703	491.972	997.386	1489.358	1.9523263	2.9939643	4.9462906
65	2947.77	0.0018658	0.0429832	512.766	847.722	1360.488	518.266	968.927	1487.193	2.0288289	2.8655424	4.8943713
70	3311.96	0.0019000	0.0378938	538.754	819.803	1358.557	545.047	939.013	1484.060	2.1054313	2.7366113	4.8420426
75	3708.90	0.0019371	0.0334404	565.196	790.625	1355.821	572.380	907.467	1479.848	2.1823063	2.6066844	4.7889907
80	4140.56	0.0019776	0.0295256	592.160	760.011	1352.171	600.348	874.075	1474.424	2.2596564	2.4751946	4.7348510
85	4608.99	0.0020221	0.0260683	619.731	727.741	1347.472	629.050	838.570	1467.621	2.3377243	2.3414652	4.6791895
90	5116.34	0.0020715	0.0230004	648.013	693.537	1341.550	658.611	800.616	1459.227	2.4168085	2.2046675	4.6214760
95	5664.89	0.0021269	0.0202639	677.142	657.037	1334.178	689.190	759.782	1448.972	2.4972871	2.0637559	4.5610430
100	6257.07	0.0021899	0.0178093	707.298	617.759	1325.057	721.001	715.490	1436.491	2.5796604	1.9173602	4.4970206
105	6895.44	0.0022629	0.0155929	738.737	575.030	1313.767	754.340	666.946	1421.287	2.6646266	1.7635980	4.4282246
110	7582.77	0.0023494	0.0135750	771.841	527.858	1299.700	789.656	612.980	1402.636	2.7532365	1.5997149	4.3529514
115	8322.03	0.0024556	0.0117171	807.252	474.653	1281.905	827.688	551.727	1379.415	2.8472435	1.4213086	4.2685521
120	9116.47	0.0025937	0.0099756	846.215	412.509	1258.724	869.860	479.807	1349.666	2.9500558	1.2203423	4.1703981
125	9969.69	0.0027950	0.0082861	891.846	334.799	1226.645	919.712	389.543	1309.255	3.0702332	0.9783905	4.0486237
130	10885.95	0.0032208	0.0064792	959.100	215.314	1174.415	994.161	250.786	1244.947	3.2491349	0.6221604	3.8712953

Table B-2b Saturated ammonia - Pressure table

P kPa	T_{sat} °C	v_f m^3/kg	v_g m^3/kg	u_f kJ/kg	u_{fg} kJ/kg	u_g kJ/kg	h_f kJ/kg	h_{fg} kJ/kg	h_g kJ/kg	s_f kJ/kg K	s_{fg} kJ/kg K	s_g kJ/kg K
30	−55.07	0.00141245	3.50542563	−46.820	1324.104	1277.284	−46.778	1429.225	1382.447	−0.0053859	6.553896	6.548510
40	−50.34	0.00142351	2.67915885	−26.276	1309.726	1283.450	−26.219	1416.835	1390.616	0.0878144	6.359203	6.447017
50	−46.51	0.00143274	2.17504218	−9.539	1297.880	1288.341	−9.467	1406.560	1397.093	0.1622970	6.206342	6.368639
60	−43.26	0.00144074	1.83443230	4.692	1287.713	1292.405	4.779	1397.692	1402.471	0.2246445	6.080162	6.304807
70	−40.44	0.00144785	1.58839689	17.136	1278.748	1295.884	17.237	1389.835	1407.072	0.2784464	5.972524	6.250971
80	−37.92	0.00145428	1.40208025	28.235	1270.694	1298.928	28.351	1382.744	1411.095	0.3258847	5.878541	6.204426
90	−35.66	0.00146019	1.25593221	38.280	1263.354	1301.634	38.412	1376.256	1414.668	0.3683889	5.795044	6.163433
100	−33.58	0.00146566	1.13812447	47.477	1256.593	1304.070	47.624	1370.258	1417.883	0.4069487	5.719860	6.126809
125	−29.05	0.00147787	0.92372509	67.645	1241.625	1309.269	67.829	1356.906	1424.735	0.4903496	5.559009	6.049359
150	−25.21	0.00148855	0.77877129	84.855	1228.693	1313.548	85.078	1345.286	1430.364	0.5603113	5.425853	5.986165
175	−21.85	0.00149811	0.67401208	99.951	1217.227	1317.178	100.213	1334.917	1435.130	0.6207945	5.311974	5.932769
200	−18.85	0.00150682	0.59464726	113.453	1206.872	1320.325	113.754	1325.500	1439.255	0.6742110	5.212310	5.886521
225	−16.14	0.00151485	0.53237122	125.705	1197.393	1323.098	126.046	1316.835	1442.881	0.7221426	5.123575	5.845717
250	−13.66	0.00152234	0.48215658	136.949	1188.623	1325.571	137.329	1308.781	1446.111	0.7656856	5.043512	5.809198
275	−11.36	0.00152937	0.44077910	147.358	1180.442	1327.800	147.779	1301.236	1449.015	0.8056323	4.970504	5.776137
300	−9.23	0.00153601	0.40607414	157.067	1172.758	1329.825	157.528	1294.120	1451.647	0.8425748	4.903351	5.745926
350	−5.35	0.00154836	0.35107359	174.767	1158.613	1333.380	175.309	1280.947	1456.256	0.9091674	4.783148	5.692316
400	−1.89	0.00155969	0.30939682	190.640	1145.775	1336.415	191.264	1268.910	1460.174	0.9680769	4.677679	5.645756
450	1.25	0.00157021	0.27668851	205.080	1133.968	1339.048	205.786	1257.771	1463.558	1.0210196	4.583547	5.604567
500	4.13	0.00158008	0.25031165	218.361	1122.998	1341.359	219.151	1247.364	1466.515	1.0691853	4.498420	5.567605
550	6.79	0.00158941	0.22857459	230.684	1112.724	1343.408	231.558	1237.566	1469.124	1.1134345	4.420623	5.534057
600	9.27	0.00159829	0.21034135	242.201	1103.037	1345.237	243.160	1228.282	1471.442	1.1544107	4.348912	5.503322

(Continued)

Table B-2b (Continued)

P kPa	T_{sat} °C	v_f m^3/kg	v_g m^3/kg	u_f kJ/kg	u_{fg} kJ/kg	u_g kJ/kg	h_f kJ/kg	h_{fg} kJ/kg	h_g kJ/kg	s_f kJ/kg K	s_{fg} kJ/kg K	s_g kJ/kg K
650	11.60	0.00160677	0.19482032	253.028	1093.853	1346.881	254.072	1219.442	1473.514	1.1926081	4.282338	5.474946
700	13.79	0.00161491	0.18144264	263.257	1085.108	1348.365	264.388	1210.987	1475.375	1.2284148	4.220161	5.448576
750	15.87	0.00162276	0.16978881	272.964	1076.747	1349.711	274.181	1202.871	1477.053	1.2621415	4.161790	5.423932
800	17.84	0.00163034	0.15954261	282.209	1068.726	1350.935	283.514	1195.055	1478.569	1.2940409	4.106747	5.400788
850	19.72	0.00163769	0.15046102	291.043	1061.008	1352.051	292.435	1187.508	1479.943	1.3243216	4.054640	5.378962
900	21.51	0.00164483	0.14235419	299.508	1053.563	1353.071	300.988	1180.202	1481.190	1.3531576	4.005143	5.358300
950	23.23	0.00165178	0.13507160	307.639	1046.366	1354.005	309.209	1173.114	1482.323	1.3806959	3.957980	5.338676
1000	24.89	0.00165856	0.12849234	315.469	1039.392	1354.861	317.128	1166.226	1483.353	1.4070616	3.912920	5.319981
1100	28.01	0.00167166	0.11706839	330.324	1026.042	1356.366	332.163	1152.978	1485.141	1.4566913	3.828334	5.285026
1200	30.93	0.00168425	0.10748551	344.248	1013.385	1357.633	346.269	1140.347	1486.615	1.5027485	3.750091	5.252840
1300	33.67	0.00169639	0.09932775	357.374	1001.322	1358.697	359.580	1128.243	1487.823	1.5457729	3.677200	5.222973
1400	36.25	0.00170816	0.09229620	369.811	989.774	1359.585	372.202	1116.598	1488.800	1.5861886	3.608886	5.195074
1500	38.70	0.00171960	0.08617040	381.643	978.677	1360.320	384.223	1105.353	1489.576	1.6243351	3.544532	5.168868
1600	41.03	0.00173076	0.08078416	392.943	967.977	1360.920	395.712	1094.462	1490.174	1.6604889	3.483640	5.144128
1700	43.25	0.00174167	0.07600977	403.767	957.631	1361.398	406.728	1083.887	1490.615	1.6948781	3.425796	5.120674
1800	45.37	0.00175237	0.07174744	414.166	947.602	1361.767	417.320	1073.593	1490.913	1.7276934	3.370660	5.098353
1900	47.40	0.00176287	0.06791807	424.181	937.857	1362.038	427.530	1063.552	1491.082	1.7590958	3.317943	5.077039
2000	49.36	0.00177321	0.06445810	433.847	928.370	1362.218	437.394	1053.740	1491.134	1.7892226	3.267401	5.056624
2250	53.95	0.00179844	0.05710771	456.683	905.633	1362.317	460.730	1030.079	1490.809	1.8596951	3.149252	5.008947
2500	58.17	0.00182298	0.05117702	477.912	884.071	1361.983	482.469	1007.456	1489.925	1.9243551	3.040868	4.965223
2750	62.09	0.00184704	0.04628636	497.816	863.462	1361.278	502.895	985.670	1488.566	1.9842658	2.940379	4.924645

(Continued)

Table B-2b (Continued)

P kPa	T_{sat} °C	v_f m³/kg	v_g m³/kg	u_f kJ/kg	u_{fg} kJ/kg	u_g kJ/kg	h_f kJ/kg	h_{fg} kJ/kg	h_g kJ/kg	s_f kJ/kg K	s_{fg} kJ/kg K	s_g kJ/kg K
3000	65.74	0.00187075	0.04218075	516.610	843.639	1360.248	522.222	964.569	1486.790	2.0402249	2.846388	4.886613
3250	69.18	0.00189424	0.03868249	534.460	824.466	1358.927	540.616	944.028	1484.645	2.0928474	2.757823	4.850670
3500	72.42	0.00191762	0.03566388	551.500	805.838	1357.339	558.212	923.950	1482.162	2.1426187	2.673841	4.816459
3750	75.49	0.00194097	0.03303076	567.837	787.665	1355.503	575.116	904.252	1479.368	2.1899293	2.593763	4.783693
4000	78.42	0.00196437	0.03071211	583.561	769.873	1353.433	591.418	884.863	1476.282	2.2350990	2.517037	4.752136
4250	81.20	0.00198791	0.02865339	598.745	752.396	1351.141	607.194	865.724	1472.918	2.2783935	2.443201	4.721595
4500	83.87	0.00201165	0.02681194	613.453	735.179	1348.632	622.506	846.780	1469.286	2.3200371	2.371867	4.691904
4750	86.43	0.00203566	0.02515395	627.740	718.172	1345.912	637.410	827.983	1465.393	2.3602212	2.302700	4.662921
5000	88.89	0.00206000	0.02365226	641.654	701.330	1342.983	651.954	809.291	1461.245	2.3991110	2.235412	4.634523
5500	93.54	0.00210995	0.02103330	668.522	667.979	1336.501	680.127	772.057	1452.184	2.4735708	2.105481	4.579052
6000	97.88	0.00216206	0.01882062	694.347	634.823	1329.169	707.319	734.774	1442.093	2.5443946	1.980331	4.524725
6500	101.95	0.00221697	0.01692022	719.370	601.577	1320.947	733.780	697.148	1430.928	2.6123885	1.858491	4.470880
7000	105.78	0.00227544	0.01526406	743.810	567.952	1311.762	759.738	658.873	1418.610	2.6782580	1.738635	4.416893
7500	109.42	0.00233844	0.01380137	767.877	533.632	1301.510	785.416	619.604	1405.020	2.7426678	1.619469	4.362137
8000	112.87	0.00240729	0.01249296	791.799	498.238	1290.038	811.058	578.924	1389.982	2.8063046	1.499613	4.305917
8500	116.15	0.00248391	0.01130743	815.845	461.282	1277.127	836.958	536.282	1373.240	2.8699613	1.377436	4.247398
9000	119.29	0.00257128	0.01021816	840.377	422.070	1262.447	863.518	490.892	1354.411	2.9346777	1.250793	4.185470
9500	122.29	0.00267456	0.00920048	865.956	379.520	1245.477	891.365	441.517	1332.881	3.0020331	1.116475	4.118508
10000	125.17	0.00280396	0.00822779	893.614	331.691	1225.305	921.653	385.930	1307.583	3.0748959	0.968902	4.043798
10500	127.94	0.00298533	0.00726300	925.797	274.294	1200.091	957.143	319.209	1276.353	3.1600611	0.795925	3.955986
11000	130.60	0.00333784	0.00623006	973.063	191.960	1165.023	1009.780	223.774	1233.554	3.2869196	0.554340	3.841259

Table B-2c Superheated Ammonia - Pressure table

T °C	P kPa	v m^3/kg	u kJ/kg	h kJ/kg	s kJ/kg K	T °C	P kPa	v m^3/kg	u kJ/kg	h kJ/kg	s kJ/kg K	T °C	P kPa	v m^3/kg	u kJ/kg	h kJ/kg	s kJ/kg K
−40.00		3.7605073	1301.703	1414.519	6.69073	−20.00		0.7977980	1322.938	1442.608	6.03504						
−35.00		3.8443914	1309.723	1425.054	6.73544	−10.00		0.8336949	1340.586	1465.640	6.12428						
−30.00		3.9279963	1317.718	1435.558	6.77909	0.00		0.8689451	1357.882	1488.224	6.20851						
−10.00		4.2603234	1349.612	1477.422	6.94455	10.00		0.9037110	1374.959	1510.515	6.28866						
−5.00		4.3430112	1357.591	1487.882	6.98393	20.00		0.9381065	1391.910	1532.626	6.36541						
0.00		4.4255798	1365.581	1498.349	7.02260	30.00		0.9722136	1408.804	1554.636	6.43924	30.00		0.1320378	1367.006	1499.044	5.37219
10.00		4.5904140	1381.608	1519.321	7.09801	40.00		1.0060924	1425.694	1576.608	6.51055	40.00		0.1386577	1389.523	1528.180	5.46676
20.00		4.7549176	1397.716	1540.364	7.17104	50.00		1.0397877	1442.619	1598.587	6.57963	50.00		0.1449633	1410.864	1555.828	5.55368
30.00		4.9191586	1413.925	1561.499	7.24194	60.00		1.0733339	1459.609	1620.609	6.64675	60.00		0.1510366	1431.405	1582.442	5.63480
40.00		5.0831884	1430.249	1582.744	7.31089	70.00		1.1067571	1476.689	1642.703	6.71209	70.00		0.1569325	1451.395	1608.328	5.71136
50.00	30.00	5.2470464	1446.701	1604.113	7.37806	80.00	150.00	1.1400781	1493.879	1664.891	6.77583	80.00	1000.00	0.1626892	1471.008	1633.697	5.78423
60.00		5.4107631	1463.294	1625.617	7.44359	90.00		1.1733131	1511.195	1687.192	6.83810	90.00		0.1683343	1490.369	1658.703	5.85406
70.00		5.5743625	1480.036	1647.267	7.50762	100.00		1.2064754	1528.651	1709.622	6.89902	100.00		0.1738885	1509.570	1683.459	5.92131
80.00		5.7378636	1496.936	1669.072	7.57025	110.00		1.2395754	1546.257	1732.193	6.95872	110.00		0.1793674	1528.683	1708.051	5.98635
90.00		5.9012816	1514.001	1691.039	7.63159	120.00		1.2726220	1564.023	1754.917	7.01726	120.00		0.1847830	1547.761	1732.544	6.04945
100.00		6.0646288	1531.237	1713.176	7.69173	130.00		1.3056222	1581.958	1777.802	7.07474	130.00		0.1901451	1566.847	1756.993	6.11086
110.00		6.2279154	1548.652	1735.489	7.75073	140.00		1.3385820	1600.069	1800.857	7.13123	140.00		0.1954613	1585.976	1781.437	6.17076
120.00		6.3911497	1566.249	1757.984	7.80869	150.00		1.3715064	1618.363	1824.089	7.18679	150.00		0.2007379	1605.174	1805.912	6.22929
130.00		6.5543385	1584.034	1780.664	7.86566	160.00		1.4043997	1636.844	1847.504	7.24148	160.00		0.2059801	1624.464	1830.444	6.28659
140.00		6.7174876	1602.012	1803.536	7.92170	170.00		1.4372653	1655.518	1871.108	7.29536	170.00		0.2111920	1643.865	1855.057	6.34277
150.00		6.8806019	1620.185	1826.603	7.97686	180.00		1.4701064	1674.389	1894.905	7.34846	180.00		0.2163773	1663.394	1879.771	6.39791

(*Continued*)

Table B-2c (Continued)

T °C	P kPa	v m³/kg	u kJ/kg	h kJ/kg	s kJ/kg K	T °C	P kPa	v m³/kg	u kJ/kg	h kJ/kg	s kJ/kg K	T °C	P kPa	v m³/kg	u kJ/kg	h kJ/kg	s kJ/kg K
−40.00		2.2424742	1299.197	1411.320	6.43053												
−35.00		2.2938694	1307.453	1422.147	6.47647	−10.00		0.6192913	1336.627	1460.485	5.96859						
−30.00		2.3449733	1315.656	1432.904	6.52118	0.00		0.6464906	1354.540	1483.838	6.05570						
−10.00		2.5472202	1348.151	1475.512	6.68959	10.00		0.6731756	1372.095	1506.730	6.13801						
−5.00		2.5973760	1356.240	1486.109	6.72948	20.00		0.6994717	1389.425	1529.319	6.21642						
0.00		2.6474097	1364.328	1496.699	6.76861	30.00		0.7254676	1406.624	1551.717	6.29155						
10.00		2.7471675	1380.521	1517.879	6.84477	40.00		0.7512274	1423.762	1574.008	6.36389						
20.00		2.8465884	1396.764	1539.093	6.91839	50.00		0.7767986	1440.892	1596.252	6.43382	50.00		0.0647292	1364.092	1493.550	5.06411
30.00		2.9457427	1413.082	1560.369	6.98976	60.00		0.8022170	1458.054	1618.498	6.50161	60.00		0.0687546	1391.730	1529.240	5.17291
40.00		3.0446831	1429.498	1581.732	7.05909	70.00		0.8275101	1475.280	1640.782	6.56752	70.00		0.0724639	1416.985	1561.913	5.26955
50.00	50.00	3.1434498	1446.027	1603.199	7.12657	80.00	200.00	0.8526993	1492.594	1663.134	6.63172	80.00	2000.00	0.0759525	1440.668	1592.572	5.35764
60.00		3.2420739	1462.684	1624.788	7.19236	90.00		0.8778013	1510.017	1685.578	6.69439	90.00		0.0792785	1463.272	1621.829	5.43934
70.00		3.3405798	1479.482	1646.511	7.25661	100.00		0.9028296	1527.566	1708.131	6.75566	100.00		0.0824801	1485.120	1650.080	5.51608
80.00		3.4389868	1496.429	1668.378	7.31942	110.00		0.9277950	1545.253	1730.812	6.81564	110.00		0.0855837	1506.435	1677.602	5.58887
90.00		3.5373102	1513.535	1690.401	7.38091	120.00		0.9527065	1563.091	1753.632	6.87443	120.00		0.0886082	1527.374	1704.590	5.65841
100.00		3.6355625	1530.808	1712.586	7.44118	130.00		0.9775712	1581.089	1776.604	6.93213	130.00		0.0915678	1548.054	1731.190	5.72522
110.00		3.7337540	1548.254	1734.942	7.50030	140.00		1.0023952	1599.257	1799.736	6.98881	140.00		0.0944732	1568.563	1757.509	5.78971
120.00		3.8318929	1565.879	1757.474	7.55835	150.00		1.0271836	1617.601	1823.038	7.04454	150.00		0.0973327	1588.968	1783.634	5.85219
130.00		3.9299862	1583.689	1780.189	7.61540	160.00		1.0519407	1636.127	1846.516	7.09937	160.00		0.1001531	1609.324	1809.630	5.91291
140.00		4.0280396	1601.689	1803.091	7.67152	170.00		1.0766700	1654.842	1870.176	7.15338	170.00		0.1029396	1629.671	1835.551	5.97207
150.00		4.1260582	1619.882	1826.185	7.72675	180.00		1.1013747	1673.751	1894.025	7.20660	180.00		0.1056967	1650.046	1861.439	6.02984

(*Continued*)

Table B-2c (Continued)

T °C	P kPa	v m³/kg	u kJ/kg	h kJ/kg	s kJ/kg K	T °C	P kPa	v m³/kg	u kJ/kg	h kJ/kg	s kJ/kg K	T °C	P kPa	v m³/kg	u kJ/kg	h kJ/kg	s kJ/kg K
−40.00		1.5916964	1296.634	1408.052	6.25518												
−35.00		1.6292048	1305.138	1419.182	6.30241												
−30.00		1.6664087	1313.555	1430.204	6.34822												
−10.00		1.8129748	1346.674	1473.582	6.51968	10.00		0.2575461	1353.586	1482.359	5.62415						
−5.00		1.8491979	1354.875	1484.319	6.56010	20.00		0.2694613	1373.604	1508.334	5.71432						
0.00		1.8852958	1363.063	1495.034	6.59969	30.00		0.2809870	1392.898	1533.392	5.79838						
10.00		1.9571753	1379.425	1516.428	6.67661	40.00		0.2922209	1411.707	1557.817	5.87765						
20.00		2.0287116	1395.805	1537.815	6.75084	50.00		0.3032301	1430.192	1581.807	5.95307						
30.00		2.0999771	1412.235	1559.234	6.82269	60.00		0.3140626	1448.471	1605.503	6.02528						
40.00		2.1710259	1428.743	1580.715	6.89240	70.00		0.3247534	1466.630	1629.007	6.09480						
50.00	70.00	2.2418991	1445.350	1602.283	6.96020	80.00	500.00	0.3353290	1484.734	1652.398	6.16199		5000.00				
60.00		2.3126285	1462.073	1623.957	7.02625	90.00		0.3458093	1502.831	1675.736	6.22715	90.00		0.0239554	1348.212	1467.989	4.65312
70.00		2.3832387	1478.926	1645.753	7.09071	100.00		0.3562102	1520.961	1699.066	6.29053	100.00		0.0263618	1389.297	1521.106	4.79748
80.00		2.4537493	1495.921	1667.684	7.15371	110.00		0.3665440	1539.154	1722.426	6.35231	110.00		0.0284029	1423.738	1565.753	4.91559
90.00		2.5241759	1513.069	1689.761	7.21536	120.00		0.3768206	1557.435	1745.845	6.41265	120.00		0.0302275	1454.405	1605.542	5.01813
100.00		2.5945311	1530.378	1711.995	7.27575	130.00		0.3870481	1575.825	1769.349	6.47168	130.00		0.0319069	1482.663	1642.197	5.11021
110.00		2.6648252	1547.856	1734.393	7.33499	140.00		0.3972331	1594.339	1792.956	6.52952	140.00		0.0334812	1509.283	1676.689	5.19473
120.00		2.7350665	1565.509	1756.964	7.39314	150.00		0.4073810	1612.993	1816.684	6.58627	150.00		0.0349758	1534.743	1709.621	5.27350
130.00		2.8052621	1583.344	1779.712	7.45027	160.00		0.4174965	1631.798	1840.546	6.64201	160.00		0.0364075	1559.363	1741.400	5.34773
140.00		2.8754177	1601.365	1802.645	7.50646	170.00		0.4275834	1650.763	1864.555	6.69680	170.00		0.0377886	1583.365	1772.308	5.41828
150.00		2.9455383	1619.579	1825.766	7.56176	180.00		0.4376449	1669.899	1888.721	6.75073	180.00		0.0391278	1606.913	1802.552	5.48577

(Continued)

Table B-2c (Continued)

T °C	P kPa	v m³/kg	u kJ/kg	h kJ/kg	s kJ/kg K	T °C	P kPa	v m³/kg	u kJ/kg	h kJ/kg	s kJ/kg K	T °C	P kPa	v m³/kg	u kJ/kg	h kJ/kg	s kJ/kg K
−30.00		1.1573119	1310.332	1426.063	6.16071												
−10.00		1.2622120	1344.425	1470.646	6.33695												
−5.00		1.2879987	1352.801	1481.601	6.37819	20.00		0.1734195	1358.999	1489.064	5.46520						
0.00		1.3136554	1361.143	1492.509	6.41849	30.00		0.1818754	1380.493	1516.899	5.55858						
10.00		1.3646420	1377.766	1514.231	6.49659	40.00		0.1899757	1400.980	1543.462	5.64480						
20.00		1.4152755	1394.356	1535.883	6.57175	50.00		0.1978125	1420.785	1569.144	5.72553						
30.00		1.4656317	1410.957	1557.520	6.64432	60.00		0.2054477	1440.123	1594.209	5.80193						
40.00		1.5157670	1427.606	1579.183	6.71463	70.00		0.2129250	1459.150	1618.843	5.87479						
50.00		1.5657239	1444.330	1600.903	6.78290	80.00		0.2202760	1477.974	1643.181	5.94470						
60.00	100.00	1.6155349	1461.152	1622.706	6.84935	90.00	750.00	0.2275241	1496.679	1667.323	6.01211		10000.00				
70.00		1.6652254	1478.090	1644.612	6.91414	100.00		0.2346874	1515.328	1691.343	6.07736						
80.00		1.7148153	1495.158	1666.639	6.97741	110.00		0.2417796	1533.968	1715.303	6.14072						
90.00		1.7643204	1512.368	1688.800	7.03929	120.00		0.2488117	1552.638	1739.247	6.20242						
100.00		1.8137537	1529.731	1711.107	7.09988	130.00		0.2557926	1571.369	1763.213	6.26261	130.00		0.0099346	1290.624	1389.970	4.24955
110.00		1.8631254	1547.257	1733.570	7.15929	140.00		0.2627294	1590.184	1787.231	6.32146	140.00		0.0119351	1361.932	1481.283	4.47352
120.00		1.9124441	1564.953	1756.197	7.21758	150.00		0.2696279	1609.106	1811.327	6.37909	150.00		0.0133813	1411.530	1545.343	4.62680
130.00		1.9617168	1582.825	1778.997	7.27485	160.00		0.2764930	1628.150	1835.519	6.43560	160.00		0.0145886	1452.331	1598.217	4.75033
140.00		2.0109494	1600.880	1801.975	7.33115	170.00		0.2833286	1647.330	1859.827	6.49107	170.00		0.0156558	1488.253	1644.811	4.85670
150.00		2.0601469	1619.123	1825.138	7.38655	180.00		0.2901383	1666.660	1884.263	6.54560	180.00		0.0166289	1521.081	1687.370	4.95169

Table B-3a Saturated R-134a - Temperature table

T °C	P_{sat} kPa	v_f m^3/kg	v_g m^3/kg	u_f kJ/kg	u_{fg} kJ/kg	u_g kJ/kg	h_f kJ/kg	h_{fg} kJ/kg	h_g kJ/kg	s_f kJ/kg K	s_{fg} kJ/kg K	s_g kJ/kg K
−40	51.245	0.0007054	0.3608115	−0.036	207.402	207.366	0.000	225.855	225.855	0.0000000	0.9686563	0.9686563
−36	62.949	0.0007112	0.2975052	4.992	204.668	209.660	5.037	223.351	228.388	0.0213848	0.9417628	0.9631476
−32	76.706	0.0007172	0.2471107	10.048	201.909	211.958	10.103	220.809	230.912	0.0425290	0.9156015	0.9581305
−28	92.760	0.0007234	0.2066643	15.133	199.122	214.255	15.200	218.225	233.425	0.0634443	0.8901207	0.9535650
−24	111.374	0.0007297	0.1739497	20.248	196.302	216.550	20.329	215.595	235.924	0.0841423	0.8652715	0.9494138
−20	132.820	0.0007362	0.1472942	25.394	193.448	218.842	25.491	212.914	238.405	0.1046345	0.8410077	0.9456422
−18	144.694	0.0007396	0.1358282	27.979	192.006	219.985	28.086	211.553	239.638	0.1148067	0.8290818	0.9438885
−16	157.385	0.0007430	0.1254244	30.571	190.555	221.127	30.688	210.178	240.866	0.1249316	0.8172858	0.9422174
−14	170.930	0.0007464	0.1159693	33.173	189.093	222.266	33.300	208.788	242.089	0.1350105	0.8056149	0.9406254
−12	185.368	0.0007499	0.1073626	35.783	187.621	223.403	35.922	207.383	243.305	0.1450448	0.7940638	0.9391086
−10	200.739	0.0007535	0.0995163	38.401	186.136	224.537	38.552	205.962	244.514	0.1550357	0.7826278	0.9376635
−8	217.083	0.0007571	0.0923524	41.028	184.641	225.669	41.193	204.524	245.717	0.1649846	0.7713019	0.9362865
−6	234.439	0.0007608	0.0858020	43.664	183.133	226.797	43.843	203.070	246.912	0.1748929	0.7600814	0.9349743
−4	252.851	0.0007646	0.0798041	46.310	181.612	227.922	46.503	201.597	248.100	0.1847619	0.7489614	0.9337233
−2	272.359	0.0007684	0.0743043	48.965	180.078	229.042	49.174	200.106	249.280	0.1945930	0.7379375	0.9325305
0	293.007	0.0007723	0.0692545	51.629	178.530	230.159	51.856	198.596	250.451	0.2043875	0.7270050	0.9313925
2	314.839	0.0007763	0.0646118	54.304	176.968	231.271	54.548	197.065	251.613	0.2141467	0.7161594	0.9303061
4	337.897	0.0007804	0.0603378	56.988	175.390	232.378	57.252	195.515	252.766	0.2238721	0.7053961	0.9292682
6	362.228	0.0007845	0.0563982	59.682	173.798	233.480	59.967	193.943	253.909	0.2335650	0.6947107	0.9282757

(*Continued*)

Table B-3a (Continued)

T °C	P_{sat} kPa	v_f m^3/kg	v_g m^3/kg	u_f kJ/kg	u_{fg} kJ/kg	u_g kJ/kg	h_f kJ/kg	h_{fg} kJ/kg	h_g kJ/kg	s_f kJ/kg K	s_{fg} kJ/kg K	s_g kJ/kg K
8	387.877	0.0007887	0.0527624	62.388	172.189	234.577	62.694	192.348	255.042	0.2432268	0.6840988	0.9273256
10	414.889	0.0007930	0.0494030	65.104	170.563	235.667	65.433	190.731	256.164	0.2528590	0.6735557	0.9264147
12	443.312	0.0007975	0.0462952	67.831	168.920	236.751	68.184	189.090	257.274	0.2624630	0.6630771	0.9255401
14	473.194	0.0008020	0.0434169	70.569	167.259	237.828	70.949	187.424	258.373	0.2720403	0.6526585	0.9246988
16	504.581	0.0008066	0.0407482	73.319	165.579	238.899	73.726	185.733	259.459	0.2815924	0.6422952	0.9238876
18	537.523	0.0008113	0.0382711	76.082	163.879	239.961	76.518	184.015	260.533	0.2911208	0.6319827	0.9231035
20	572.069	0.0008161	0.0359693	78.856	162.159	241.015	79.323	182.269	261.592	0.3006271	0.6217165	0.9223436
22	608.270	0.0008210	0.0338281	81.644	160.417	242.061	82.143	180.494	262.637	0.3101129	0.6114917	0.9216046
24	646.176	0.0008261	0.0318343	84.444	158.653	243.097	84.978	178.690	263.668	0.3195799	0.6013036	0.9208835
26	685.838	0.0008313	0.0299758	87.258	156.865	244.123	87.828	176.854	264.682	0.3290299	0.5911471	0.9201770
28	727.309	0.0008366	0.0282417	90.086	155.053	245.139	90.695	174.985	265.680	0.3384645	0.5810174	0.9194819
30	770.642	0.0008421	0.0266220	92.929	153.216	246.144	93.578	173.083	266.660	0.3478857	0.5709091	0.9187948
32	815.890	0.0008478	0.0251079	95.786	151.351	247.137	96.478	171.145	267.622	0.3572954	0.5608170	0.9181124
34	863.108	0.0008536	0.0236909	98.659	149.458	248.118	99.396	169.170	268.565	0.3666956	0.5507354	0.9174310
36	912.351	0.0008595	0.0223636	101.548	147.536	249.084	102.332	167.156	269.488	0.3760885	0.5406585	0.9167470
38	963.676	0.0008657	0.0211193	104.454	145.583	250.037	105.288	165.101	270.389	0.3854762	0.5305804	0.9160566
40	1017.140	0.0008720	0.0199515	107.377	143.597	250.974	108.264	163.003	271.268	0.3948611	0.5204946	0.9153557
42	1072.800	0.0008786	0.0188547	110.319	141.576	251.895	111.261	160.861	272.122	0.4042457	0.5103943	0.9146400
44	1130.716	0.0008854	0.0178235	113.279	139.520	252.798	114.280	158.672	272.952	0.4136326	0.5002726	0.9139052
46	1190.948	0.0008924	0.0168531	116.259	137.424	253.683	117.321	156.433	273.754	0.4230247	0.4901218	0.9131465
48	1253.558	0.0008996	0.0159392	119.259	135.288	254.547	120.387	154.141	274.528	0.4324249	0.4799340	0.9123589

(Continued)

Table B-3a (Continued)

T °C	P_{sat} kPa	v_f m^3/kg	v_g m^3/kg	u_f kJ/kg	u_{fg} kJ/kg	u_g kJ/kg	h_f kJ/kg	h_{fg} kJ/kg	h_g kJ/kg	s_f kJ/kg K	s_{fg} kJ/kg K	s_g kJ/kg K
50	1318.608	0.0009072	0.0150776	122.282	133.109	255.390	123.478	151.794	275.272	0.4418365	0.4697004	0.9115369
52	1386.162	0.0009150	0.0142646	125.327	130.883	256.210	126.595	149.388	275.983	0.4512631	0.4594117	0.9106748
56	1529.047	0.0009317	0.0127710	131.490	126.282	257.771	132.914	144.384	277.299	0.4701764	0.4386278	0.9088042
60	1682.756	0.0009498	0.0114336	137.760	121.455	259.216	139.359	139.097	278.455	0.4892000	0.4174890	0.9066890
64	1847.862	0.0009697	0.0102313	144.152	116.368	260.520	145.944	133.482	279.426	0.5083767	0.3958817	0.9042584
68	2024.977	0.0009917	0.0091458	150.684	110.974	261.658	152.692	127.486	280.178	0.5277604	0.3736633	0.9014237
72	2214.756	0.0010165	0.0081609	157.378	105.214	262.592	159.629	121.037	280.667	0.5474216	0.3506480	0.8980696
76	2417.916	0.0010446	0.0072623	164.268	99.004	263.272	166.794	114.038	280.832	0.5674560	0.3265844	0.8940404
80	2635.247	0.0010772	0.0064364	171.400	92.227	263.627	174.239	106.350	280.589	0.5880007	0.3011149	0.8891156
84	2867.650	0.0011160	0.0056701	178.845	84.702	263.547	182.046	97.761	279.807	0.6092652	0.2736965	0.8829617
88	3116.182	0.0011641	0.0049488	186.723	76.130	262.853	190.351	87.923	278.274	0.6315996	0.2434267	0.8750263
92	3382.159	0.0012274	0.0042527	195.262	65.938	261.200	199.413	76.170	275.584	0.6556769	0.2085771	0.8642540
96	3667.385	0.0013215	0.0035433	205.021	52.721	257.741	209.867	60.869	270.736	0.6831580	0.1648711	0.8480291
100	3975.056	0.0015269	0.0026304	218.716	29.195	247.911	224.786	33.581	258.367	0.7221680	0.0899854	0.8121534

Table B-3b Saturated R-134a - Pressure table

P kPa	T_{sat} °C	v_f m^3/kg	v_g m^3/kg	u_f kJ/kg	u_{fg} kJ/kg	u_g kJ/kg	h_f kJ/kg	h_{fg} kJ/kg	h_g kJ/kg	s_f kJ/kg K	s_{fg} kJ/kg K	s_g kJ/kg K
60	−36.950	0.00070980	0.31120525	3.798	205.318	209.116	3.841	223.948	227.789	0.0163412	0.948065	0.964407
70	−33.870	0.00071437	0.26928533	7.680	203.203	210.883	7.730	222.003	229.733	0.0326675	0.927751	0.960419
80	−31.130	0.00071852	0.23753460	11.154	201.304	212.458	11.211	220.250	231.461	0.0471063	0.909992	0.957098
90	−28.650	0.00072235	0.21262527	14.310	199.574	213.884	14.375	218.646	233.020	0.0600798	0.894194	0.954274
100	−26.370	0.00072591	0.19254359	17.208	197.980	215.188	17.280	217.162	234.442	0.0718802	0.879950	0.951830
110	−24.280	0.00072925	0.17599835	19.894	196.498	216.392	19.974	215.778	235.752	0.0827191	0.866969	0.949688
120	−22.320	0.00073241	0.16212286	22.401	195.111	217.511	22.488	214.477	236.966	0.0927543	0.855035	0.947789
130	−20.500	0.00073541	0.15031338	24.755	193.803	218.558	24.850	213.249	238.099	0.1021071	0.843983	0.946090
140	−18.770	0.00073828	0.14013620	26.976	192.566	219.542	27.080	212.082	239.161	0.1108727	0.833685	0.944558
150	−17.150	0.00074102	0.13127156	29.083	191.389	220.472	29.194	210.969	240.162	0.1191273	0.824039	0.943166
160	−15.600	0.00074365	0.12347850	31.086	190.266	221.353	31.205	209.904	241.109	0.1269328	0.814963	0.941896
170	−14.130	0.00074620	0.11657188	32.999	189.191	222.190	33.126	208.881	242.007	0.1343403	0.806389	0.940729
180	−12.730	0.00074865	0.11040710	34.830	188.159	222.989	34.965	207.897	242.862	0.1413926	0.798260	0.939653
190	−11.380	0.00075103	0.10486952	36.588	187.165	223.753	36.731	206.947	243.678	0.1481256	0.790531	0.938656
200	−10.090	0.00075334	0.09986707	38.278	186.206	224.485	38.429	206.029	244.458	0.1545699	0.783160	0.937730
220	−7.660	0.00075777	0.09118182	41.482	184.382	225.864	41.649	204.275	245.924	0.1666947	0.769361	0.936056
240	−5.380	0.00076199	0.08389704	44.479	182.665	227.144	44.662	202.618	247.279	0.1779389	0.756643	0.934582
260	−3.250	0.00076602	0.07769624	47.299	181.041	228.340	47.499	201.043	248.541	0.1884348	0.744837	0.933272
280	−1.250	0.00076989	0.07235196	49.967	179.496	229.463	50.183	199.539	249.722	0.1982862	0.733809	0.932096
300	0.650	0.00077362	0.06769653	52.501	178.021	230.522	52.733	198.098	250.831	0.2075763	0.723456	0.931032
320	2.460	0.00077723	0.06360354	54.916	176.608	231.525	55.165	196.713	251.878	0.2163729	0.713692	0.930065
340	4.180	0.00078073	0.05997585	57.226	175.250	232.476	57.492	195.376	252.868	0.2247318	0.704447	0.929179

(Continued)

Table B-3b (Continued)

P kPa	T_{sat} °C	v_f m^3/kg	v_g m^3/kg	u_f kJ/kg	u_{fg} kJ/kg	u_g kJ/kg	h_f kJ/kg	h_{fg} kJ/kg	h_g kJ/kg	s_f kJ/kg K	s_{fg} kJ/kg K	s_g kJ/kg K
360	5.820	0.00078414	0.05673759	59.441	173.941	233.382	59.723	194.084	253.807	0.2326997	0.695663	0.928363
380	7.400	0.00078745	0.05382861	61.571	172.676	234.247	61.870	192.832	254.701	0.2403161	0.687292	0.927608
400	8.910	0.00079069	0.05120059	63.623	171.451	235.074	63.939	191.615	255.554	0.2476146	0.679292	0.926906
450	12.460	0.00079847	0.04561904	68.455	168.543	236.998	68.814	188.712	257.526	0.2646505	0.660695	0.925345
500	15.710	0.00080590	0.04111759	72.926	165.821	238.746	73.328	185.976	259.305	0.2802284	0.643773	0.924002
550	18.730	0.00081303	0.03740761	77.097	163.251	240.348	77.545	183.378	260.923	0.2946086	0.628214	0.922822
600	21.550	0.00081993	0.03429529	81.017	160.810	241.827	81.509	180.895	262.404	0.3079860	0.613783	0.921769
650	24.200	0.00082662	0.03164557	84.720	158.478	243.198	85.258	178.510	263.768	0.3205106	0.600303	0.920813
700	26.690	0.00083315	0.02936136	88.236	156.240	244.477	88.819	176.210	265.030	0.3323005	0.587635	0.919935
750	29.060	0.00083953	0.02737110	91.588	154.084	245.672	92.218	173.983	266.201	0.3434505	0.575667	0.919118
800	31.310	0.00084580	0.02562082	94.795	152.000	246.795	95.471	171.820	267.291	0.3540379	0.564311	0.918348
850	33.450	0.00085196	0.02406907	97.872	149.979	247.851	98.596	169.713	268.309	0.3641264	0.553491	0.917617
900	35.510	0.00085804	0.02268343	100.833	148.014	248.847	101.605	167.657	269.262	0.3737696	0.543147	0.916916
950	37.480	0.00086405	0.02143823	103.690	146.099	249.788	104.510	165.644	270.155	0.3830125	0.533226	0.916239
1000	39.370	0.00087001	0.02031285	106.451	144.229	250.679	107.321	163.671	270.992	0.3918938	0.523685	0.915579
1200	46.290	0.00089343	0.01671538	116.698	137.113	253.811	117.771	156.099	273.870	0.4244057	0.488627	0.913033
1400	52.400	0.00091662	0.01410734	125.939	130.431	256.371	127.223	148.899	276.121	0.4531533	0.457343	0.910497
1600	57.880	0.00093998	0.01212348	134.427	124.040	258.467	135.931	141.934	277.864	0.4791128	0.428730	0.907843
1800	62.870	0.00096386	0.01055855	142.332	117.835	260.167	144.067	135.105	279.172	0.5029375	0.402045	0.904982
2000	67.450	0.00098858	0.00928808	149.781	111.733	261.513	151.758	128.332	280.090	0.5250925	0.376747	0.901840
2500	77.540	0.00105657	0.00693590	166.987	96.466	263.453	169.628	111.164	280.792	0.5753089	0.316953	0.892262
3000	86.160	0.00114061	0.00527543	183.042	80.222	263.264	186.464	92.627	279.091	0.6211811	0.257760	0.878941

Table B-3c Superheated R-134a - Pressure table

T °C	P kPa	v m^3/kg	u kJ/kg	h kJ/kg	s kJ/kg K	P kPa	v m^3/kg	u kJ/kg	h kJ/kg	s kJ/kg K	P kPa	v m^3/kg	u kJ/kg	h kJ/kg	s kJ/kg K
−20		0.3360780	220.598	240.763	1.01745										
−10		0.3504845	227.551	248.580	1.04773										
0		0.3647589	234.657	256.543	1.07743		0.0728187	230.443	250.832	0.93617					
10		0.3789314	241.924	264.659	1.10661		0.0764596	238.274	259.683	0.96799					
20		0.3930229	249.354	272.935	1.13533		0.0799656	246.128	268.519	0.99866					
30		0.4070485	256.950	281.373	1.16363		0.0833774	254.062	277.407	1.02848					
40		0.4210197	264.712	289.973	1.19154		0.0867180	262.103	286.384	1.05761		0.0233746	253.134	274.171	0.93271
50		0.4349454	272.640	298.736	1.21909		0.0900025	270.266	295.467	1.08616		0.0248095	262.440	284.769	0.96602
60		0.4488328	280.733	307.663	1.24629		0.0932413	278.561	304.668	1.11420		0.0261457	271.599	295.130	0.99760
70		0.4626876	288.991	316.752	1.27317		0.0964423	286.993	313.997	1.14179		0.0274133	280.720	305.392	1.02795
80	60	0.4765145	297.413	326.004	1.29975	280	0.0996115	295.568	323.459	1.16897	900	0.0286298	289.862	315.629	1.05736
90		0.4903172	305.998	335.417	1.32603		0.1027539	304.286	333.058	1.19577		0.0298063	299.060	325.886	1.08600
100		0.5040991	314.744	344.990	1.35203		0.1058732	313.151	342.796	1.22222		0.0309506	308.335	336.191	1.11399
110		0.5178628	323.651	354.723	1.37777		0.1089727	322.163	352.676	1.24835		0.0320684	317.703	346.565	1.14142
120		0.5316106	332.718	364.614	1.40326		0.1120549	331.324	362.699	1.27417		0.0331640	327.175	357.023	1.16837
130		0.5453444	341.942	374.663	1.42849		0.1151221	340.633	372.867	1.29971		0.0342409	336.758	367.575	1.19487
140		0.5590658	351.324	384.868	1.45350		0.1181761	350.090	383.179	1.32498		0.0353017	346.459	378.231	1.22098
150		0.5727763	360.861	395.228	1.47827		0.1212185	359.696	393.637	1.34999		0.0363486	356.283	388.996	1.24673
160		0.5864769	370.553	405.742	1.50283		0.1242506	369.450	404.241	1.37475		0.0373835	366.232	399.877	1.27214
170		0.6001689	380.399	416.409	1.52718		0.1272735	379.353	414.989	1.39929		0.0384078	376.310	410.877	1.29725
180		0.6138530	390.397	427.228	1.55132		0.1302883	389.402	425.883	1.42359		0.0394228	386.519	421.999	1.32207

(*Continued*)

Table B-3c (Continued)

T °C	P kPa	v m^3/kg	u kJ/kg	h kJ/kg	s kJ/kg K	P kPa	v m^3/kg	u kJ/kg	h kJ/kg	s kJ/kg K	P kPa	v m^3/kg	u kJ/kg	h kJ/kg	s kJ/kg K
−20		0.1984080	219.662	239.503	0.97207										
−10		0.2074310	226.748	247.491	1.00302										
0		0.2163007	233.954	255.584	1.03321										
10		0.2250584	241.299	263.805	1.06276		0.0515058	235.974	256.577	0.93053					
20		0.2337294	248.794	272.167	1.09178		0.0542132	244.176	265.862	0.96275					
30		0.2423311	256.444	280.677	1.12033		0.0567960	252.355	275.074	0.99365					
40		0.2508762	264.252	289.340	1.14844		0.0592924	260.584	284.301	1.02360		0.0204063	251.305	271.711	0.91788
50		0.2593742	272.219	298.157	1.17616		0.0617237	268.899	293.589	1.05279		0.0217960	260.942	282.738	0.95254
60		0.2678328	280.347	307.130	1.20350		0.0641039	277.320	302.962	1.08136		0.0230676	270.315	293.383	0.98499
70		0.2762580	288.635	316.261	1.23051		0.0664426	285.860	312.437	1.10938		0.0242613	279.592	303.854	1.01595
80	100	0.2846546	297.083	325.549	1.25718	400	0.0687471	294.526	322.025	1.13692	1000	0.0253985	288.855	314.254	1.04583
90		0.2930267	305.691	334.994	1.28356		0.0710227	303.324	331.733	1.16403		0.0264924	298.150	324.643	1.07484
100		0.3013775	314.459	344.596	1.30964		0.0732740	312.259	341.568	1.19075		0.0275521	307.506	335.058	1.10313
110		0.3097099	323.384	354.355	1.33545		0.0755044	321.332	351.534	1.21710		0.0285838	316.942	345.526	1.13081
120		0.3180261	332.467	364.270	1.36099		0.0777167	330.547	361.634	1.24312		0.0295923	326.472	356.064	1.15796
130		0.3263282	341.706	374.339	1.38628		0.0799134	339.905	371.870	1.26883		0.0305812	336.105	366.687	1.18465
140		0.3346177	351.101	384.563	1.41133		0.0820964	349.406	382.244	1.29425		0.0315536	345.851	377.404	1.21090
150		0.3428961	360.651	394.941	1.43615		0.0842675	359.051	392.758	1.31939		0.0325116	355.713	388.224	1.23678
160		0.3511647	370.354	405.471	1.46075		0.0864279	368.840	403.412	1.34428		0.0334573	365.697	399.154	1.26231
170		0.3594244	380.210	416.152	1.48512		0.0885789	378.775	414.206	1.36891		0.0343922	375.806	410.198	1.28752
180		0.3676763	390.217	426.985	1.50930		0.0907215	388.854	425.142	1.39332		0.0353175	386.042	421.359	1.31242

(Continued)

Table B-3c (Continued)

T °C	P kPa	v m^3/kg	u kJ/kg	h kJ/kg	s kJ/kg K	P kPa	v m^3/kg	u kJ/kg	h kJ/kg	s kJ/kg K	P kPa	v m^3/kg	u kJ/kg	h kJ/kg	s kJ/kg K
−10		0.1460550	225.908	246.356	0.97236										
0		0.1526309	233.228	254.597	1.00310										
10		0.1590809	240.660	262.931	1.03307										
20		0.1654371	248.224	271.385	1.06240		0.0421152	242.402	263.460	0.93828					
30		0.1717199	255.930	279.971	1.09120		0.0443380	250.841	273.010	0.97032					
40		0.1779435	263.786	288.698	1.11953		0.0464558	259.257	282.485	1.00107					
50		0.1841184	271.794	297.571	1.14742		0.0484990	267.715	291.965	1.03087					
60		0.1902527	279.957	306.593	1.17491		0.0504853	276.253	301.495	1.05991		0.0150052	264.465	285.472	0.93889
70		0.1963527	288.276	315.765	1.20204		0.0524265	284.889	311.102	1.08832		0.0160596	274.620	297.103	0.97330
80	140	0.2024234	296.751	325.091	1.22882	500	0.0543310	293.637	320.802	1.11619	1400	0.0170225	284.506	308.337	1.00557
90		0.2084691	305.383	334.569	1.25529		0.0562049	302.506	330.608	1.14357		0.0179230	294.276	319.368	1.03637
100		0.2144932	314.171	344.200	1.28145		0.0580532	311.501	340.528	1.17051		0.0187776	304.011	330.299	1.06607
110		0.2204986	323.115	353.985	1.30733		0.0598796	320.629	350.569	1.19706		0.0195967	313.758	341.193	1.09488
120		0.2264876	332.215	363.923	1.33293		0.0616873	329.891	360.735	1.22326		0.0203876	323.549	352.092	1.12296
130		0.2324622	341.470	374.014	1.35828		0.0634788	339.290	371.030	1.24911		0.0211554	333.406	363.024	1.15041
140		0.2384241	350.878	384.258	1.38337		0.0652562	348.829	381.457	1.27466		0.0219039	343.344	374.010	1.17733
150		0.2443748	360.440	394.653	1.40823		0.0670212	358.508	392.018	1.29992		0.0226363	353.375	385.066	1.20377
160		0.2503156	370.155	405.199	1.43287		0.0687754	368.328	402.715	1.32490		0.0233547	363.507	396.204	1.22979
170		0.2562475	380.020	415.895	1.45728		0.0705199	378.289	413.549	1.34963		0.0240613	373.747	407.433	1.25542
180		0.2621714	390.037	426.741	1.48148		0.0722559	388.393	424.521	1.37411		0.0247574	384.101	418.762	1.28070

(Continued)

Table B-3c (Continued)

T °C	P kPa	v m^3/kg	u kJ/kg	h kJ/kg	s kJ/kg K	P kPa	v m^3/kg	u kJ/kg	h kJ/kg	s kJ/kg K	P kPa	v m^3/kg	u kJ/kg	h kJ/kg	s kJ/kg K
−10		0.0999138	224.555	244.538	0.93803										
0		0.1048103	232.088	253.050	0.96978										
10		0.1095512	239.669	261.579	1.00045										
20		0.1141842	247.346	270.183	1.03031										
30		0.1187363	255.144	278.891	1.05952		0.0299658	247.476	268.452	0.93129					
40		0.1232247	263.075	287.720	1.08817		0.0316960	256.388	278.575	0.96414					
50		0.1276616	271.147	296.680	1.11633		0.0333218	265.201	288.526	0.99542					
60		0.1320558	279.365	305.776	1.14406		0.0348748	274.011	298.423	1.02559		0.0123725	260.893	280.689	0.91635
70		0.1364142	287.731	315.014	1.17138		0.0363732	282.866	308.327	1.05488		0.0134302	271.761	293.249	0.95350
80		0.1407425	296.248	324.397	1.19833		0.0378287	291.796	318.276	1.08346		0.0143619	282.089	305.069	0.98746
90	200	0.1450449	304.916	333.925	1.22493	700	0.0392495	300.819	328.294	1.11143	1600	0.0152154	292.170	316.515	1.01942
100		0.1493252	313.737	343.602	1.25122		0.0406417	309.947	338.396	1.13887		0.0160144	302.140	327.763	1.04998
110		0.1535862	322.710	353.427	1.27720		0.0420100	319.189	348.596	1.16584		0.0167729	312.073	338.910	1.07946
120		0.1578305	331.835	363.401	1.30290		0.0433579	328.552	358.902	1.19240		0.0174997	322.016	350.016	1.10807
130		0.1620602	341.113	373.525	1.32832		0.0446885	338.040	369.322	1.21857		0.0182013	332.000	361.122	1.13597
140		0.1662769	350.542	383.797	1.35349		0.0460040	347.657	379.860	1.24439		0.0188820	342.046	372.257	1.16325
150		0.1704823	360.122	394.219	1.37842		0.0473064	357.406	390.520	1.26988		0.0195454	352.169	383.442	1.19000
160		0.1746776	369.854	404.789	1.40311		0.0485974	367.289	401.307	1.29508		0.0201941	362.382	394.692	1.21628
170		0.1788639	379.735	415.508	1.42757		0.0498783	377.307	412.222	1.31999		0.0208303	372.693	406.022	1.24213
180		0.1830421	389.766	426.374	1.45182		0.0511503	387.462	423.267	1.34464		0.0214555	383.110	417.439	1.26761

Table C-1 Saturated ice-Water vapor

T °C	P_{sat} kPa	v_f m^3/kg	v_g m^3/kg	u_f kJ/kg	u_{fg} kJ/kg	u_g kJ/kg	h_f kJ/kg	h_{fg} kJ/kg	h_g kJ/kg	s_f kJ/kg K	s_{fg} kJ/kg K	s_g kJ/kg K
0.001	0.6117	0.001091	205.99	−333.43	2707.96	2374.53	−333.43	2833.97	2500.54	−1.2204	10.3745	9.154
0	0.6112	0.001091	206.17	−333.43	2707.95	2374.52	−333.43	2833.95	2500.52	−1.2204	10.3749	9.154
−2	0.5177	0.001091	241.62	−337.63	2709.39	2371.76	−337.63	2834.47	2496.84	−1.2358	10.4533	9.218
−4	0.4375	0.001090	283.84	−341.80	2710.79	2368.99	−341.80	2834.96	2493.17	−1.2513	10.5329	9.282
−6	0.3687	0.001090	334.27	−345.94	2712.17	2366.23	−345.94	2835.42	2489.48	−1.2667	10.6134	9.347
−8	0.3100	0.001090	394.66	−350.04	2713.51	2363.46	−350.04	2835.84	2485.80	−1.2821	10.6951	9.413
−10	0.2599	0.001089	467.17	−354.12	2714.81	2360.69	−354.12	2836.23	2482.11	−1.2976	10.7779	9.480
−12	0.2173	0.001089	554.47	−358.17	2716.09	2357.92	−358.17	2836.59	2478.42	−1.3130	10.8618	9.549
−14	0.1812	0.001088	659.88	−362.19	2717.34	2355.15	−362.19	2836.92	2474.73	−1.3285	10.9469	9.618
−16	0.1507	0.001088	787.51	−366.17	2718.55	2352.38	−366.17	2837.21	2471.04	−1.3439	11.0332	9.689
−18	0.1249	0.001088	942.51	−370.13	2719.73	2349.60	−370.13	2837.47	2467.34	−1.3594	11.1207	9.761
−20	0.1033	0.001087	1131.30	−374.06	2720.88	2346.83	−374.06	2837.70	2463.65	−1.3748	11.2095	9.835
−22	0.0851	0.001087	1361.96	−377.95	2722.00	2344.05	−377.95	2837.90	2459.95	−1.3903	11.2996	9.909
−24	0.0699	0.001087	1644.66	−381.82	2723.09	2341.27	−381.82	2838.07	2456.25	−1.4057	11.3910	9.985
−26	0.0573	0.001087	1992.24	−385.66	2724.15	2338.49	−385.66	2838.21	2452.55	−1.4212	11.4837	10.063
−28	0.0467	0.001086	2421.02	−389.47	2725.18	2335.72	−389.47	2838.32	2448.85	−1.4367	11.5778	10.141
−30	0.0380	0.001086	2951.73	−393.25	2726.18	2332.94	−393.25	2838.39	2445.15	−1.4521	11.6734	10.221
−32	0.0308	0.001086	3610.87	−397.00	2727.15	2330.16	−397.00	2838.44	2441.45	−1.4676	11.7704	10.303
−34	0.0249	0.001085	4432.43	−400.72	2728.09	2327.38	−400.72	2838.46	2437.74	−1.4831	11.8690	10.386
−36	0.0200	0.001085	5460.13	−404.41	2729.00	2324.60	−404.41	2838.45	2434.04	−1.4986	11.9690	10.470
−38	0.0161	0.001085	6750.46	−408.07	2729.88	2321.81	−408.07	2838.40	2430.34	−1.5141	12.0706	10.557
−40	0.0129	0.001084	8376.69	−411.70	2730.73	2319.03	−411.70	2838.33	2426.63	−1.5296	12.1739	10.644

Table D-1 Compressibility factor at various reduced pressure and reduced temperature

Pr	Tr	Air Z	CO_2 Z	CO Z	H_2 Z	N_2 Z	O_2 Z	Pr	Tr	Air Z	CO_2 Z	CO Z	H_2 Z	N_2 Z	O_2 Z
0.01	1	0.99622	0.99656	0.99656	0.99675	0.99668	0.99667	4	1.0	0.64571	0.54524	0.57985	0.60205	0.57962	0.57590
	1.1	0.99717	0.9975	0.99745	0.99748	0.99751	0.9975		1.1	0.65020	0.56928	0.59155	0.59168	0.59103	0.58628
	1.2	0.99784	0.99815	0.99808	0.99801	0.9981	0.99809		1.2	0.67299	0.62027	0.62442	0.59819	0.62416	0.61796
	1.3	0.99833	0.99861	0.99853	0.99841	0.99853	0.99852		1.3	0.71273	0.69250	0.67875	0.62328	0.67769	0.67002
	1.5	0.99898	0.99919	0.99911	0.99896	0.9991	0.99909		1.5	0.81366	0.83088	0.80183	0.71512	0.79956	0.79076
	2	0.99971	0.99982	0.99975	0.9996	0.99975	0.99973		2.0	0.97702	0.99497	0.97190	0.89941	0.96916	0.96151
0.5	1	0.77477	0.80163	0.80223	0.81201	0.80773	0.80682	4.5	1.0	0.71389	0.60269	0.64079	0.66459	0.64058	0.63639
	1.1	0.84383	0.8654	0.8625	0.86133	0.8647	0.86358		1.1	0.71202	0.62052	0.64648	0.64783	0.64572	0.64058
	1.2	0.88605	0.904	0.8999	0.89397	0.9006	0.89944		1.2	0.72638	0.66054	0.66998	0.64645	0.66938	0.66292
	1.3	0.9145	0.92964	0.92512	0.91702	0.92511	0.92396		1.3	0.75539	0.71910	0.71126	0.66088	0.71021	0.70241
	1.5	0.94968	0.9607	0.95616	0.94687	0.95572	0.95467		1.5	0.83676	0.84193	0.81532	0.72870	0.81279	0.80356
	2	0.98663	0.99209	0.98845	0.98035	0.98815	0.98739		2.0	0.98847	1.00406	0.97936	0.89983	0.97658	0.96818
1	1	0.21951	0.25038	0.2447	0.31895	0.25404	0.27316	5	1.0	0.78111	0.65937	0.70088	0.72617	0.70071	0.69604
	1.1	0.63888	0.70232	0.69453	0.68058	0.69571	0.69279		1.1	0.77331	0.67172	0.70111	0.70343	0.70016	0.69460
	1.2	0.75863	0.80089	0.79167	0.77164	0.79089	0.78807		1.2	0.78054	0.70322	0.71716	0.69534	0.71618	0.70943
	1.3	0.82607	0.85875	0.8491	0.82705	0.8477	0.84499		1.3	0.80097	0.75104	0.74842	0.70148	0.74722	0.73921
	1.5	0.90189	0.92402	0.91465	0.89281	0.91322	0.91081		1.5	0.86542	0.85850	0.83509	0.74980	0.83228	0.82273
	2	0.97579	0.98618	0.97885	0.96166	0.97824	0.97653		2.0	1.00259	1.01511	0.98914	0.90337	0.98632	0.97725

(Continued)

Table D-1 (Continued)

Pr	Tr	Air Z	CO_2 Z	CO Z	H_2 Z	N_2 Z	O_2 Z	Pr	Tr	Air Z	CO_2 Z	CO Z	H_2 Z	N_2 Z	O_2 Z
1.5	1	0.28714	0.24544	0.26121	0.27381	0.26134	0.25987	5.5	1.0	0.84745	0.71534	0.76018	0.78689	0.76007	0.75490
	1.1	0.40825	0.49957	0.46726	0.40916	0.47168	0.4664		1.1	0.83401	0.72265	0.75529	0.75845	0.75421	0.74821
	1.2	0.62592	0.69443	0.67533	0.62836	0.6732	0.66845		1.2	0.83492	0.74714	0.76503	0.74437	0.76371	0.75661
	1.3	0.74075	0.79038	0.77442	0.732	0.77094	0.76655		1.3	0.84811	0.78619	0.78824	0.74361	0.78678	0.77850
	1.5	0.85917	0.89105	0.87702	0.84004	0.87409	0.87022		1.5	0.89792	0.87939	0.85941	0.77585	0.85640	0.84656
	2	0.96781	0.98227	0.9715	0.9446	0.97047	0.96771		2.0	1.01910	1.02795	1.00110	0.90985	0.99818	0.98851
2	1	0.36115	0.30639	0.32608	0.34076	0.32608	0.32413	6	1.0	0.91299	0.77067	0.81876	0.84680	0.81873	0.81306
	1.1	0.41204	0.40239	0.39762	0.37745	0.39709	0.39319		1.1	0.89412	0.77324	0.80899	0.81287	0.80782	0.80135
	1.2	0.54244	0.60298	0.57151	0.50391	0.5726	0.56642		1.2	0.88924	0.79167	0.81314	0.79327	0.81151	0.80403
	1.3	0.6741	0.73108	0.70598	0.64164	0.70322	0.69733		1.3	0.89608	0.82334	0.82960	0.78650	0.82781	0.81923
	1.5	0.82494	0.86341	0.84459	0.79155	0.84052	0.83526		1.5	0.93303	0.90358	0.88702	0.80513	0.88385	0.87373
	2	0.96296	0.98042	0.96658	0.92974	0.96506	0.96122		2.0	1.03765	1.04237	1.01503	0.91900	1.01196	1.00173
2.5	1	0.43426	0.36749	0.39105	0.4078	0.39094	0.38855	6.5	1.0	0.97779	0.82540	0.87667	0.90598	0.87675	0.87056
	1.1	0.46548	0.42515	0.43217	0.42403	0.43247	0.42868		1.1	0.95364	0.82345	0.86220	0.86671	0.86097	0.85402
	1.2	0.54023	0.55824	0.53588	0.48219	0.53537	0.52903		1.2	0.94335	0.83646	0.86123	0.84192	0.85934	0.85145
	1.3	0.64414	0.68975	0.65881	0.58419	0.6582	0.65126		1.3	0.94447	0.86174	0.87183	0.82974	0.86970	0.86078
	1.5	0.80258	0.8428	0.81894	0.75142	0.81496	0.80849		1.5	0.96993	0.93024	0.91700	0.83655	0.91370	0.90326
	2	0.96144	0.98071	0.96417	0.91753	0.96217	0.95727		2.0	1.05795	1.05819	1.03073	0.93054	1.02743	1.01670

(Continued)

Table D-1 (Continued)

Pr	Tr	Air Z	CO_2 Z	CO Z	H_2 Z	N_2 Z	O_2 Z
3	1	0.50601	0.4277	0.45504	0.47374	0.45487	0.45204
	1.1	0.52599	0.46921	0.48261	0.47878	0.48271	0.47865
	1.2	0.57469	0.5597	0.54898	0.50996	0.54865	0.54264
	1.3	0.6493	0.67183	0.64434	0.5736	0.64357	0.63607
	1.5	0.79396	0.83044	0.80242	0.7242	0.79939	0.79194
	2	0.96332	0.98323	0.96427	0.90829	0.96188	0.95598
3.5	1	0.57645	0.48694	0.51796	0.53847	0.51774	0.51448
	1.1	0.58803	0.51845	0.53664	0.53517	0.53641	0.53202
	1.2	0.62151	0.58481	0.5825	0.55172	0.58241	0.57638
	1.3	0.67559	0.67487	0.6544	0.59181	0.65329	0.64567
	1.5	0.79846	0.82666	0.79685	0.71248	0.79452	0.7863
	2	0.96856	0.988	0.96686	0.90222	0.96423	0.95741

Pr	Tr	Air Z	CO_2 Z	CO Z	H_2 Z	N_2 Z	O_2 Z
7	1.0	1.04191	0.87959	0.93397	0.96447	0.93418	0.92745
	1.1	1.01260	0.87326	0.91491	0.91999	0.91366	0.90620
	1.2	0.99717	0.88132	0.90917	0.89025	0.90707	0.89873
	1.3	0.99302	0.90093	0.91453	0.87309	0.91207	0.90278
	1.5	1.00803	0.95873	0.94871	0.86937	0.94525	0.93448
	2.0	1.07974	1.07522	1.04794	0.94416	1.04438	1.03318

Table E-1 Ideal gas properties of Air

T K	*u* kJ/kg	*h* kJ/kg	*s* kJ/kg.K	*T* K	*u* kJ/kg	*h* kJ/kg	*s* kJ/kg.K	*T* K	*u* kJ/kg	*h* kJ/kg	*s* kJ/kg.K	*T* K	*u* kJ/kg	*h* kJ/kg	*s* kJ/kg.K
200	142.708	200.114	1.2949	780	576.435	800.318	2.6895	1360	1077.261	1467.620	3.3261	1940	1620.420	2177.256	3.7599
210	149.861	210.137	1.3438	790	584.514	811.267	2.7034	1370	1086.352	1479.582	3.3349	1950	1630.016	2189.722	3.7663
220	157.016	220.162	1.3904	800	592.617	822.240	2.7172	1380	1095.455	1491.555	3.3436	1960	1639.618	2202.194	3.7727
230	164.171	230.188	1.4350	810	600.743	833.236	2.7309	1390	1104.569	1503.540	3.3522	1970	1649.226	2214.673	3.7791
240	171.328	240.215	1.4777	820	608.892	844.255	2.7444	1400	1113.695	1515.536	3.3608	1980	1658.841	2227.158	3.7854
250	178.486	250.243	1.5186	830	617.063	855.297	2.7578	1410	1122.833	1527.544	3.3694	1990	1668.462	2239.649	3.7917
260	185.647	260.274	1.5580	840	625.258	866.362	2.7710	1420	1131.981	1539.563	3.3779	2000	1678.089	2252.147	3.7979
270	192.810	270.308	1.5958	850	633.474	877.449	2.7842	1430	1141.141	1551.593	3.3863	2010	1688.071	2264.999	3.8042
280	199.977	280.345	1.6323	860	641.713	888.558	2.7972	1440	1150.312	1563.634	3.3947	2020	1697.701	2277.500	3.8104
290	207.147	290.385	1.6676	870	649.975	899.690	2.8100	1450	1159.493	1575.685	3.4030	2030	1707.337	2290.006	3.8165
300	214.321	300.430	1.7016	880	658.258	910.843	2.8228	1460	1168.686	1587.748	3.4113	2040	1716.979	2302.518	3.8227
310	221.501	310.480	1.7346	890	666.562	922.018	2.8354	1470	1177.888	1599.821	3.4195	2050	1726.627	2315.037	3.8288
320	228.686	320.535	1.7666	900	674.888	933.214	2.8479	1480	1187.101	1611.904	3.4277	2060	1736.281	2327.560	3.8349
330	235.877	330.596	1.7976	910	683.236	944.432	2.8603	1490	1196.325	1623.998	3.4359	2070	1745.940	2340.090	3.8410
340	243.075	340.665	1.8276	920	691.604	955.671	2.8726	1500	1205.559	1636.102	3.4440	2080	1755.605	2352.625	3.8470
350	250.281	350.741	1.8568	930	699.993	966.930	2.8847	1510	1214.802	1648.216	3.4520	2090	1765.275	2365.166	3.8530
360	257.495	360.826	1.8852	940	708.403	978.210	2.8968	1520	1224.056	1660.340	3.4600	2100	1774.951	2377.712	3.8590
370	264.719	370.919	1.9129	950	716.833	989.511	2.9088	1530	1233.320	1672.474	3.4680	2110	1784.632	2390.263	3.8650
380	271.952	381.024	1.9398	960	725.284	1000.831	2.9206	1540	1242.593	1684.618	3.4759	2120	1794.319	2402.821	3.8709
390	279.197	391.138	1.9660	970	733.754	1012.172	2.9324	1550	1251.876	1696.771	3.4837	2130	1804.011	2415.383	3.8768
400	286.453	401.265	1.9916	980	742.244	1023.532	2.9440	1560	1261.168	1708.933	3.4915	2140	1813.709	2427.951	3.8827
410	293.722	411.404	2.0166	990	750.753	1034.911	2.9556	1570	1270.470	1721.105	3.4993	2150	1823.412	2440.524	3.8886

(Continued)

Table E-1 (Continued)

T K	u kJ/kg	h kJ/kg	s kJ/kg.K	T K	u kJ/kg	h kJ/kg	s kJ/kg.K	T K	u kJ/kg	h kJ/kg	s kJ/kg.K	T K	u kJ/kg	h kJ/kg	s kJ/kg.K
420	301.004	421.556	2.0410	1000	759.281	1046.310	2.9670	1580	1279.781	1733.287	3.5070	2160	1833.120	2453.102	3.8944
430	308.300	431.722	2.0649	1010	767.829	1057.728	2.9784	1590	1289.101	1745.477	3.5147	2170	1842.833	2465.686	3.9002
440	315.610	441.903	2.0883	1020	776.395	1069.164	2.9896	1600	1298.431	1757.677	3.5224	2180	1852.551	2478.274	3.9060
450	322.936	452.099	2.1112	1030	784.979	1080.619	3.0008	1610	1307.769	1769.886	3.5300	2190	1862.275	2490.868	3.9118
460	330.278	462.312	2.1336	1040	793.582	1092.092	3.0119	1620	1317.116	1782.103	3.5375	2200	1872.003	2503.467	3.9175
470	337.637	472.541	2.1555	1050	802.203	1103.584	3.0229	1630	1326.472	1794.330	3.5450	2210	1881.737	2516.071	3.9232
480	345.014	482.788	2.1771	1060	810.842	1115.093	3.0338	1640	1335.837	1806.565	3.5525	2220	1891.475	2528.680	3.9289
490	352.408	493.052	2.1982	1070	819.499	1126.620	3.0446	1650	1345.210	1818.808	3.5600	2230	1901.219	2541.294	3.9346
500	359.821	503.336	2.2189	1080	828.172	1138.164	3.0553	1660	1354.592	1831.061	3.5674	2240	1910.967	2553.912	3.9402
510	367.254	513.639	2.2393	1090	836.863	1149.725	3.0660	1670	1363.983	1843.321	3.5747	2250	1920.721	2566.536	3.9459
520	374.706	523.961	2.2593	1100	845.571	1161.303	3.0765	1680	1373.381	1855.590	3.5820	2300	1969.559	2629.726	3.9736
530	382.179	534.304	2.2790	1110	854.296	1172.898	3.0870	1690	1382.788	1867.867	3.5893	2350	2018.513	2693.031	4.0009
540	389.672	544.668	2.2984	1120	863.037	1184.510	3.0974	1700	1392.203	1880.153	3.5966	2400	2067.580	2756.449	4.0276
550	397.187	555.053	2.3174	1130	871.795	1196.137	3.1077	1710	1401.627	1892.446	3.6038	2450	2116.754	2819.975	4.0538
560	404.724	565.460	2.3362	1140	880.568	1207.781	3.1180	1720	1411.058	1904.748	3.6109	2500	2166.032	2883.605	4.0795
570	412.282	575.888	2.3546	1150	889.358	1219.441	3.1282	1730	1420.497	1917.057	3.6181	2550	2215.412	2947.336	4.1047
580	419.863	586.339	2.3728	1160	898.163	1231.116	3.1383	1740	1429.944	1929.374	3.6252	2600	2264.889	3011.164	4.1295
590	427.466	596.813	2.3907	1170	906.983	1242.807	3.1483	1750	1439.399	1941.699	3.6322	2650	2314.461	3075.088	4.1539
600	435.092	607.310	2.4083	1180	915.819	1254.513	3.1582	1760	1448.861	1954.032	3.6393	2700	2364.125	3139.103	4.1778
610	442.742	617.829	2.4257	1190	924.670	1266.234	3.1681	1770	1458.331	1966.372	3.6462	2750	2413.879	3203.209	4.2013
620	450.414	628.372	2.4428	1200	933.536	1277.971	3.1779	1780	1467.809	1978.720	3.6532	2800	2463.721	3267.402	4.2244
630	458.111	638.939	2.4597	1210	942.416	1289.721	3.1877	1790	1477.294	1991.076	3.6601	2850	2513.648	3331.680	4.2472
640	465.831	649.530	2.4764	1220	951.311	1301.487	3.1974	1800	1486.786	2003.438	3.6670	2900	2563.658	3396.042	4.2696
650	473.575	660.144	2.4929	1230	960.220	1313.266	3.2070	1810	1496.286	2015.808	3.6738	2950	2613.750	3460.486	4.2916

(Continued)

Table E-1 (Continued)

T K	u kJ/kg	h kJ/kg	s kJ/kg.K	T K	u kJ/kg	h kJ/kg	s kJ/kg.K	T K	u kJ/kg	h kJ/kg	s kJ/kg.K	T K	u kJ/kg	h kJ/kg	s kJ/kg.K
660	481.343	670.782	2.5091	1240	969.144	1325.060	3.2165	1820	1505.793	2028.186	3.6807	3000	2663.922	3525.009	4.3133
670	489.135	681.444	2.5252	1250	978.081	1336.867	3.2260	1830	1515.307	2040.570	3.6874	3050	2714.173	3589.611	4.3347
680	496.951	692.131	2.5410	1260	987.032	1348.689	3.2354	1840	1524.829	2052.962	3.6942	3100	2764.501	3654.290	4.3557
690	504.791	702.841	2.5566	1270	995.997	1360.523	3.2447	1850	1534.357	2065.361	3.7009	3150	2814.904	3719.046	4.3764
700	512.656	713.576	2.5721	1280	1004.975	1372.372	3.2540	1860	1543.892	2077.766	3.7076	3200	2865.383	3783.875	4.3968
710	520.544	724.335	2.5874	1290	1013.966	1384.233	3.2633	1870	1553.435	2090.179	3.7142	3250	2915.935	3848.779	4.4170
720	528.457	735.118	2.6024	1300	1022.970	1396.108	3.2724	1880	1562.984	2102.598	3.7209	3300	2966.560	3913.756	4.4368
730	536.393	745.925	2.6174	1310	1031.987	1407.995	3.2815	1890	1572.540	2115.025	3.7275	3350	3017.257	3978.804	4.4564
740	544.354	756.756	2.6321	1320	1041.017	1419.895	3.2906	1900	1582.103	2127.458	3.7340	3400	3068.025	4043.923	4.4757
750	552.339	767.610	2.6467	1330	1050.060	1431.808	3.2995	1910	1591.672	2139.898	3.7405	3450	3118.863	4109.113	4.4947
760	560.347	778.489	2.6611	1340	1059.114	1443.733	3.3085	1920	1601.248	2152.344	3.7470	3500	3169.772	4174.373	4.5135
770	568.379	789.392	2.6753	1350	1068.182	1455.671	3.3173	1930	1610.831	2164.797	3.7535				

Table E-2 Ideal gas properties of Carbon monoxide

T K	u kJ/kg	h kJ/kg	s kJ/kg.K	T K	u kJ/kg	h kJ/kg	s kJ/kg.K	T K	u kJ/kg	h kJ/kg	s kJ/kg.K	T K	u kJ/kg	h kJ/kg	s kJ/kg.K
200	4147.259	5809.542	185.9235	780	16742.328	23226.929	226.3935	1350	30997.816	42221.590	244.5990	1940	47095.852	63225.087	257.4891
210	4356.281	6101.707	187.3490	790	16976.702	23544.446	226.7980	1360	31261.757	42568.674	244.8551	1950	47375.651	63588.029	257.6757
220	4564.771	6393.341	188.7057	800	17211.709	23862.596	227.1982	1370	31526.085	42916.146	245.1097	1960	47655.613	63951.134	257.8614
230	4772.830	6684.543	190.0001	810	17447.347	24181.377	227.5942	1380	31790.796	43264.000	245.3627	1970	47935.734	64314.398	258.0463
240	4980.549	6975.405	191.2380	820	17683.612	24500.785	227.9861	1390	32055.885	43612.233	245.6141	1980	48216.011	64677.819	258.2303
250	5188.013	7266.013	192.4244	830	17920.500	24820.818	228.3740	1400	32321.348	43960.839	245.8640	1990	48496.442	65041.394	258.4135
260	5395.303	7556.447	193.5635	840	18158.010	25141.471	228.7580	1410	32587.179	44309.814	246.1124	2000	48777.025	65405.120	258.5958
270	5602.491	7846.778	194.6592	850	18396.137	25462.741	229.1382	1420	32853.376	44659.153	246.3593	2010	49057.755	65768.993	258.7773
280	5809.644	8137.074	195.7149	860	18634.879	25784.626	229.5147	1430	33119.932	45008.853	246.6047	2020	49338.631	66133.013	258.9579
290	6016.823	8427.397	196.7337	870	18874.230	26107.121	229.8875	1440	33386.845	45358.909	246.8486	2030	49619.649	66497.174	259.1378
300	6224.085	8717.802	197.7182	880	19114.188	26430.223	230.2568	1450	33654.109	45709.317	247.0911	2040	49900.808	66861.476	259.3168
310	6431.483	9008.343	198.6709	890	19354.750	26753.927	230.6226	1460	33921.720	46060.071	247.3322	2050	50182.103	67225.915	259.4950
320	6639.062	9299.066	199.5939	900	19595.910	27078.231	230.9849	1470	34189.674	46411.169	247.5718	2060	50463.534	67590.489	259.6724
330	6846.867	9590.015	200.4892	910	19837.666	27403.130	231.3439	1480	34457.966	46762.604	247.8101	2070	50745.096	67955.195	259.8490
340	7054.939	9881.230	201.3586	920	20080.013	27728.621	231.6997	1490	34726.593	47114.374	248.0470	2080	51026.788	68320.031	260.0248
350	7263.313	10172.747	202.2036	930	20322.947	28054.698	232.0522	1500	34995.549	47466.474	248.2825	2090	51308.608	68684.993	260.1999
360	7472.024	10464.601	203.0258	940	20566.464	28381.359	232.4016	1510	35264.831	47818.899	248.5167	2100	51590.551	69050.080	260.3741
370	7681.101	10756.822	203.8264	950	20810.561	28708.599	232.7479	1520	35534.434	48171.646	248.7495	2110	51872.617	69415.289	260.5476
380	7890.575	11049.439	204.6068	960	21055.232	29036.414	233.0911	1530	35804.355	48524.710	248.9810	2120	52154.802	69780.618	260.7204
390	8100.469	11342.477	205.3680	970	21300.474	29364.799	233.4314	1540	36074.589	48878.087	249.2112	2130	52437.105	70146.064	260.8923

(Continued)

Table E-2 (Continued)

T K	u kJ/kg	h kJ/kg	s kJ/kg.K	T K	u kJ/kg	h kJ/kg	s kJ/kg.K	T K	u kJ/kg	h kJ/kg	s kJ/kg.K	T K	u kJ/kg	h kJ/kg	s kJ/kg.K
400	8310.809	11635.960	206.1110	980	21546.282	29693.750	233.7688	1550	36345.131	49231.773	249.4402	2140	52719.522	70511.625	261.0636
410	8521.616	11929.911	206.8368	990	21792.652	30023.264	234.1033	1560	36615.978	49585.764	249.6678	2150	53002.053	70877.299	261.2340
420	8732.910	12224.348	207.5464	1000	22039.580	30353.335	234.4351	1570	36887.126	49940.055	249.8942	2160	53284.693	71243.082	261.4038
430	8944.710	12519.291	208.2404	1010	22287.061	30683.959	234.7641	1580	37158.570	50294.643	250.1193	2170	53567.441	71608.974	261.5728
440	9157.031	12814.755	208.9196	1020	22535.090	31015.132	235.0903	1590	37430.307	50649.523	250.3432	2180	53850.296	71974.972	261.7411
450	9369.889	13110.757	209.5848	1030	22783.664	31346.849	235.4140	1600	37702.333	51004.692	250.5659	2190	54133.254	72341.073	261.9086
460	9583.299	13407.310	210.2366	1040	23032.778	31679.107	235.7350	1610	37974.642	51360.145	250.7874	2200	54416.313	72707.276	262.0754
470	9797.272	13704.427	210.8756	1050	23282.427	32011.899	236.0534	1620	38247.232	51715.878	251.0076	2210	54699.472	73073.579	262.2416
480	10011.820	14002.118	211.5023	1060	23532.606	32345.222	236.3694	1630	38520.099	52071.888	251.2267	2220	54982.728	73439.978	262.4070
490	10226.954	14300.396	212.1174	1070	23783.312	32679.071	236.6829	1640	38793.238	52428.171	251.4446	2230	55266.080	73806.473	262.5717
500	10442.684	14599.269	212.7212	1080	24034.539	33013.442	236.9939	1650	39066.646	52784.722	251.6614	2240	55549.525	74173.061	262.7357
510	10659.017	14898.745	213.3142	1090	24286.284	33348.329	237.3026	1660	39340.319	53141.538	251.8770	2250	55833.061	74539.741	262.8991
520	10875.961	15198.833	213.8969	1100	24538.540	33683.729	237.6089	1670	39614.252	53498.615	252.0914	2300	57252.044	76374.441	263.7056
530	11093.524	15499.539	214.4697	1110	24791.304	34019.637	237.9129	1680	39888.443	53855.949	252.3048	2350	58673.034	78211.148	264.4956
540	11311.712	15800.870	215.0330	1120	25044.571	34356.047	238.2146	1690	40162.887	54213.537	252.5170	2400	60095.824	80049.655	265.2697
550	11530.529	16102.831	215.5870	1130	25298.337	34692.956	238.5141	1700	40437.581	54571.374	252.7281	2450	61520.230	81889.778	266.0285
560	11749.981	16405.426	216.1323	1140	25552.596	35030.359	238.8113	1710	40712.521	54929.457	252.9381	2500	62946.089	83731.354	266.7726
570	11970.072	16708.661	216.6690	1150	25807.345	35368.251	239.1064	1720	40987.702	55287.782	253.1471	2550	64373.260	85574.242	267.5025
580	12190.806	17012.538	217.1975	1160	26062.577	35706.627	239.3994	1730	41263.123	55646.346	253.3549	2600	65801.628	87418.327	268.2187
590	12412.185	17317.061	217.7180	1170	26318.290	36045.483	239.6903	1740	41538.778	56005.144	253.5617	2650	67231.097	89263.513	268.9216
600	12634.213	17622.232	218.2309	1180	26574.477	36384.813	239.9791	1750	41814.664	56364.174	253.7675	2700	68661.595	91109.728	269.6118
610	12856.892	17928.054	218.7364	1190	26831.135	36724.615	240.2658	1760	42090.778	56723.431	253.9722	2750	70093.074	92956.924	270.2897

(Continued)

Table E-2 (Continued)

T K	u kJ/kg	h kJ/kg	s kJ/kg.K	T K	u kJ/kg	h kJ/kg	s kJ/kg.K	T K	u kJ/kg	h kJ/kg	s kJ/kg.K	T K	u kJ/kg	h kJ/kg	s kJ/kg.K
620	13080.222	18234.528	219.2348	1200	27088.258	37064.881	240.5506	1770	42367.116	57082.913	254.1759	2800	71525.507	94805.074	270.9557
630	13304.207	18541.656	219.7262	1210	27345.843	37405.609	240.8333	1780	42643.675	57442.615	254.3785	2850	72958.891	96654.175	271.6103
640	13528.846	18849.438	220.2109	1220	27603.883	37746.793	241.1141	1790	42920.450	57802.534	254.5801	2900	74393.246	98504.247	272.2538
650	13754.140	19157.876	220.6891	1230	27862.376	38088.429	241.3930	1800	43197.440	58162.667	254.7808	2950	75828.616	100355.334	272.8867
660	13980.090	19466.969	221.1610	1240	28121.315	38430.511	241.6700	1810	43474.639	58523.009	254.9804	3000	77265.066	102207.501	273.5093
670	14206.695	19776.717	221.6268	1250	28380.696	38773.036	241.9451	1820	43752.045	58883.559	255.1791	3050	78702.686	104060.838	274.1220
680	14433.955	20087.121	222.0867	1260	28640.515	39115.999	242.2184	1830	44029.655	59244.312	255.3767	3100	80141.589	105915.458	274.7251
690	14661.869	20398.179	222.5408	1270	28900.767	39459.394	242.4899	1840	44307.465	59605.265	255.5734	3150	81581.910	107771.496	275.3190
700	14890.437	20709.890	222.9893	1280	29161.448	39803.218	242.7595	1850	44585.471	59966.415	255.7692	3200	83023.808	109629.111	275.9041
710	15119.658	21022.254	223.4324	1290	29422.552	40147.465	243.0274	1860	44863.671	60327.758	255.9640	3250	84467.466	111488.486	276.4807
720	15349.529	21335.269	223.8701	1300	29684.075	40492.132	243.2936	1870	45142.061	60689.292	256.1578	3300	85913.091	113349.828	277.0491
730	15580.050	21648.934	224.3028	1310	29946.013	40837.213	243.5580	1880	45420.638	61051.012	256.3507	3350	87360.910	115213.364	277.6095
740	15811.219	21963.246	224.7304	1320	30208.360	41182.704	243.8207	1890	45699.399	61412.917	256.5427	3400	88811.179	117079.350	278.1624
750	16043.034	22278.204	225.1532	1330	30471.113	41528.600	244.0818	1900	45978.341	61775.002	256.7338	3450	90264.171	118948.059	278.7080
760	16275.491	22593.805	225.5712	1340	30734.266	41874.897	244.3412	1910	46257.460	62137.264	256.9240	3500	91720.188	120819.793	279.2467
770	16508.591	22910.047	225.9846	1350	30997.816	42221.590	244.5990	1920	46536.754	62499.702	257.1132				

Table E-3 Ideal gas properties of Carbon Dioxide

T K	*u* kJ/kg	*h* kJ/kg	*s* kJ/kg.K	*T* K	*u* kJ/kg	*h* kJ/kg	*s* kJ/kg.K	*T* K	*u* kJ/kg	*h* kJ/kg	*s* kJ/kg.K	*T* K	*u* kJ/kg	*h* kJ/kg	*s* kJ/kg.K
200	5947.709	4284.426	199.8653	780	31161.806	24676.206	256.1150	1360	62906.522	51598.604	286.3776	1940	97177.420	81047.185	307.3365
210	6275.235	4528.809	201.4632	790	31673.023	25104.280	256.7663	1370	63481.865	52090.804	286.7991	1950	97780.490	81567.112	307.6465
220	6606.577	4777.007	203.0045	800	32185.910	25534.023	257.4114	1380	64057.906	52583.702	287.2180	1960	98383.860	82087.339	307.9552
230	6942.151	5029.438	204.4961	810	32700.436	25965.406	258.0506	1390	64634.635	53077.287	287.6344	1970	98987.528	82607.863	308.2624
240	7282.237	5286.380	205.9435	820	33216.575	26398.401	258.6839	1400	65212.043	53571.552	288.0483	1980	99591.488	83128.680	308.5682
250	7627.014	5548.014	207.3508	830	33734.299	26832.982	259.3114	1410	65790.119	54066.485	288.4598	1990	100195.736	83649.784	308.8726
260	7976.586	5814.443	208.7218	840	34253.583	27269.122	259.9333	1420	66368.856	54562.078	288.8688	2000	100800.267	84171.172	309.1756
270	8331.002	6085.715	210.0594	850	34774.400	27706.796	260.5497	1430	66948.243	55058.321	289.2754	2010	101405.078	84692.839	309.4773
280	8690.268	6361.838	211.3659	860	35296.725	28145.977	261.1606	1440	67528.271	55555.206	289.6796	2020	102010.163	85214.781	309.7775
290	9054.358	6642.785	212.6435	870	35820.533	28586.642	261.7662	1450	68108.932	56052.724	290.0814	2030	102615.520	85736.995	310.0765
300	9423.225	6928.508	213.8939	880	36345.800	29028.766	262.3665	1460	68690.216	56550.865	290.4809	2040	103221.143	86259.474	310.3741
310	9796.805	7218.945	215.1188	890	36872.504	29472.326	262.9616	1470	69272.115	57049.620	290.8781	2050	103827.029	86782.217	310.6704
320	10175.020	7514.016	216.3196	900	37400.620	29917.299	263.5517	1480	69854.620	57548.982	291.2731	2060	104433.174	87305.219	310.9653
330	10557.785	7813.638	217.4974	910	37930.127	30363.662	264.1368	1490	70437.723	58048.942	291.6657	2070	105039.574	87828.475	311.2590
340	10945.009	8117.718	218.6533	920	38461.002	30811.394	264.7170	1500	71021.415	58549.490	292.0562	2080	105646.225	88351.982	311.5513
350	11336.597	8426.163	219.7884	930	38993.224	31260.473	265.2924	1510	71605.689	59050.620	292.4444	2090	106253.123	88875.737	311.8424
360	11732.452	8738.875	220.9036	940	39526.772	31710.877	265.8630	1520	72190.535	59552.323	292.8304	2100	106860.264	89399.735	312.1322
370	12132.476	9055.756	221.9996	950	40061.625	32162.587	266.4290	1530	72775.946	60054.590	293.2143	2110	107467.645	89923.973	312.4208
380	12536.572	9376.708	223.0772	960	40597.763	32615.582	266.9904	1540	73361.913	60557.415	293.5960	2120	108075.263	90448.447	312.7081
390	12944.643	9701.636	224.1371	970	41135.167	33069.842	267.5473	1550	73948.430	61060.788	293.9757	2130	108683.113	90973.154	312.9941
400	13356.594	10030.443	225.1801	980	41673.816	33525.348	268.0998	1560	74535.488	61564.703	294.3532	2140	109291.192	91498.090	313.2789
410	13772.329	10363.035	226.2066	990	42213.693	33982.081	268.6479	1570	75123.079	62069.151	294.7286	2150	109899.497	92023.251	313.5625

(*Continued*)

Table E-3 (Continued)

T K	u kJ/kg	h kJ/kg	s kJ/kg.K	T K	u kJ/kg	h kJ/kg	s kJ/kg.K	T K	u kJ/kg	h kJ/kg	s kJ/kg.K	T K	u kJ/kg	h kJ/kg	s kJ/kg.K
420	14191.759	10699.321	227.2173	1000	42754.778	34440.023	269.1917	1580	75711.197	62574.125	295.1021	2160	110508.024	92548.634	313.8449
430	14614.793	11039.212	228.2127	1010	43297.053	34899.154	269.7313	1590	76299.833	63079.617	295.4734	2170	111116.770	93074.237	314.1261
440	15041.344	11382.620	229.1933	1020	43840.500	35359.458	270.2667	1600	76888.980	63585.621	295.8428	2180	111725.731	93600.055	314.4061
450	15471.328	11729.460	230.1596	1030	44385.102	35820.917	270.7980	1610	77478.631	64092.129	296.2102	2190	112334.904	94126.085	314.6849
460	15904.662	12079.651	231.1120	1040	44930.842	36283.513	271.3253	1620	78068.779	64599.134	296.5756	2200	112944.287	94652.324	314.9625
470	16341.267	12433.112	232.0510	1050	45477.702	36747.230	271.8486	1630	78659.417	65106.628	296.9391	2210	113553.876	95178.769	315.2389
480	16781.065	12789.767	232.9769	1060	46025.666	37212.051	272.3680	1640	79250.538	65614.605	297.3006	2220	114163.667	95705.417	315.5142
490	17223.982	13149.540	233.8902	1070	46574.718	37677.960	272.8836	1650	79842.134	66123.058	297.6603	2230	114773.659	96232.265	315.7884
500	17669.944	13512.359	234.7911	1080	47124.842	38144.940	273.3953	1660	80434.199	66631.980	298.0180	2240	115383.847	96759.310	316.0614
510	18118.883	13878.155	235.6801	1090	47676.023	38612.977	273.9033	1670	81026.727	67141.364	298.3739	2250	115994.229	97286.549	316.3333
520	18570.730	14246.858	236.5575	1100	48228.244	39082.055	274.4076	1680	81619.711	67651.204	298.7279	2300	119048.949	99925.552	317.6761
530	19025.420	14618.405	237.4236	1110	48781.490	39552.158	274.9083	1690	82213.143	68161.494	299.0801	2350	122108.106	102568.992	318.9919
540	19482.889	14992.730	238.2787	1120	49335.748	40023.272	275.4054	1700	82807.019	68672.226	299.4305	2400	125171.390	105216.559	320.2817
550	19943.076	15369.774	239.1231	1130	49891.001	40495.382	275.8990	1710	83401.330	69183.394	299.7790	2450	128238.522	107867.974	321.5466
560	20405.921	15749.476	239.9571	1140	50447.237	40968.474	276.3890	1720	83996.072	69694.992	300.1258	2500	131309.252	110522.987	322.7873
570	20871.368	16131.779	240.7809	1150	51004.440	41442.534	276.8757	1730	84591.237	70207.014	300.4708	2550	134383.360	113181.378	324.0048
580	21339.360	16516.628	241.5948	1160	51562.597	41917.547	277.3589	1740	85186.821	70719.454	300.8141	2600	137460.654	115842.955	325.1999
590	21809.844	16903.968	242.3991	1170	52121.694	42393.501	277.8389	1750	85782.815	71232.305	301.1557	2650	140540.971	118507.555	326.3734
600	22282.768	17293.749	243.1939	1180	52681.717	42870.381	278.3155	1760	86379.216	71745.562	301.4955	2700	143624.174	121175.041	327.5260
610	22758.080	17685.918	243.9796	1190	53242.655	43348.175	278.7888	1770	86976.016	72259.219	301.8336	2750	146710.156	123845.306	328.6585
620	23235.733	18080.428	244.7562	1200	53804.493	43826.870	279.2590	1780	87573.210	72773.270	302.1701	2800	149798.834	126518.267	329.7716
630	23715.679	18477.230	245.5242	1210	54367.218	44306.452	279.7260	1790	88170.792	73287.709	302.5048	2850	152890.154	129193.870	330.8659
640	24197.872	18876.280	246.2835	1220	54930.819	44786.909	280.1899	1800	88768.757	73802.530	302.8380	2900	155984.086	131872.085	331.9421

(Continued)

Table E-3 (Continued)

T K	*u* kJ/kg	*h* kJ/kg	*s* kJ/kg.K	*T* K	*u* kJ/kg	*h* kJ/kg	*s* kJ/kg.K	*T* K	*u* kJ/kg	*h* kJ/kg	*s* kJ/kg.K	*T* K	*u* kJ/kg	*h* kJ/kg	*s* kJ/kg.K
650	24682.268	19277.532	247.0345	1230	55495.283	45268.229	280.6507	1810	89367.099	74317.728	303.1695	2950	159080.629	134552.911	333.0008
660	25168.823	19680.944	247.7774	1240	56060.597	45750.400	281.1084	1820	89965.812	74833.298	303.4993	3000	162179.804	137236.369	334.0425
670	25657.496	20086.473	248.5122	1250	56626.750	46233.410	281.5631	1830	90564.891	75349.234	303.8276	3050	165281.661	139922.509	335.0680
680	26148.245	20494.079	249.2393	1260	57193.729	46717.246	282.0149	1840	91164.330	75865.530	304.1543	3100	168386.273	142611.404	336.0776
690	26641.032	20903.722	249.9587	1270	57761.524	47201.897	282.4638	1850	91764.125	76382.181	304.4794	3150	171493.737	145303.151	337.0720
700	27135.818	21315.365	250.6706	1280	58330.122	47687.352	282.9097	1860	92364.270	76899.182	304.8029	3200	174604.179	147997.876	338.0517
710	27632.565	21728.969	251.3752	1290	58899.513	48173.600	283.3528	1870	92964.759	77416.529	305.1249	3250	177717.746	150695.726	339.0172
720	28131.238	22144.498	252.0727	1300	59469.685	48660.628	283.7931	1880	93565.589	77934.214	305.4453	3300	180834.610	153396.873	339.9689
730	28631.801	22561.918	252.7631	1310	60040.628	49148.427	284.2306	1890	94166.752	78452.235	305.7642	3350	183954.968	156101.514	340.9074
740	29134.221	22981.194	253.4467	1320	60612.329	49636.986	284.6654	1900	94768.246	78970.585	306.0816	3400	187079.042	158809.871	341.8330
750	29638.463	23402.293	254.1236	1330	61184.780	50126.293	285.0974	1910	95370.064	79489.260	306.3976	3450	190207.076	161522.188	342.7463
760	30144.496	23825.182	254.7938	1340	61757.969	50616.339	285.5268	1920	95972.203	80008.255	306.7120	3500	193339.340	164238.735	343.6477
770	30652.287	24249.830	255.4576	1350	62331.887	51107.113	285.9535	1930	96574.656	80527.565	307.0250				

Table E-4 Ideal gas properties of Hydrogen

T K	u kJ/kg	h kJ/kg	s kJ/kg.K	T K	u kJ/kg	h kJ/kg	s kJ/kg.K	T K	u kJ/kg	h kJ/kg	s kJ/kg.K	T K	u kJ/kg	h kJ/kg	s kJ/kg.K
200	4024.716	5686.856	119.2706	780	16070.095	22554.551	158.6364	1360	28999.575	40306.349	175.5988	1940	43222.069	59351.160	187.2384
210	4215.356	5960.639	120.6064	790	16283.420	22851.020	159.0140	1370	29233.586	40623.503	175.8311	1950	43478.608	59690.843	187.4131
220	4409.490	6237.916	121.8962	800	16497.020	23147.763	159.3873	1380	29467.993	40941.054	176.0621	1960	43735.516	60030.894	187.5870
230	4606.419	6517.989	123.1411	810	16710.898	23444.785	159.7563	1390	29702.796	41259.001	176.2916	1970	43992.793	60371.314	187.7603
240	4805.592	6800.305	124.3426	820	16925.061	23742.091	160.1211	1400	29937.996	41577.344	176.5198	1980	44250.436	60712.101	187.9328
250	5006.565	7084.422	125.5024	830	17139.512	24039.686	160.4818	1410	30173.593	41896.084	176.7467	1990	44508.446	61053.254	188.1047
260	5208.986	7369.986	126.6224	840	17354.257	24337.574	160.8386	1420	30409.586	42215.220	176.9722	2000	44766.820	61394.772	188.2759
270	5412.570	7656.714	127.7045	850	17569.299	24635.760	161.1914	1430	30645.976	42534.754	177.1965	2010	45025.558	61736.653	188.4464
280	5617.087	7944.374	128.7507	860	17784.644	24934.248	161.5406	1440	30882.763	42854.684	177.4194	2020	45284.660	62078.898	188.6162
290	5822.351	8232.781	129.7627	870	18000.295	25233.042	161.8860	1450	31119.947	43175.012	177.6411	2030	45544.123	62421.505	188.7854
300	6028.211	8521.784	130.7425	880	18216.255	25532.146	162.2278	1460	31357.528	43495.736	177.8615	2040	45803.948	62764.473	188.9539
310	6234.546	8811.263	131.6917	890	18432.530	25831.565	162.5662	1470	31595.506	43816.857	178.0807	2050	46064.132	63107.801	189.1218
320	6441.258	9101.119	132.6119	900	18649.123	26131.301	162.9011	1480	31833.881	44138.376	178.2987	2060	46324.675	63451.487	189.2891
330	6648.267	9391.271	133.5048	910	18866.037	26431.358	163.2326	1490	32072.652	44460.291	178.5155	2070	46585.576	63795.532	189.4557
340	6855.509	9681.656	134.3716	920	19083.277	26731.741	163.5609	1500	32311.821	44782.602	178.7311	2080	46846.834	64139.932	189.6217
350	7062.934	9972.224	135.2139	930	19300.845	27032.452	163.8860	1510	32551.386	45105.311	178.9455	2090	47108.447	64484.689	189.7870
360	7270.500	10262.934	136.0329	940	19518.744	27333.495	164.2080	1520	32791.347	45428.415	179.1588	2100	47370.415	64829.800	189.9517
370	7478.175	10553.753	136.8297	950	19736.978	27634.873	164.5269	1530	33031.704	45751.916	179.3709	2110	47632.736	65175.265	190.1159
380	7685.936	10844.657	137.6055	960	19955.551	27936.589	164.8428	1540	33272.458	46075.813	179.5819	2120	47895.409	65521.082	190.2794
390	7893.764	11135.628	138.3613	970	20174.464	28238.645	165.1558	1550	33513.607	46400.106	179.7918	2130	48158.434	65867.249	190.4423
400	8101.644	11426.652	139.0981	980	20393.720	28541.045	165.4660	1560	33755.152	46724.794	180.0006	2140	48421.808	66213.768	190.6046

(Continued)

Table E-4 (Continued)

T K	u kJ/kg	h kJ/kg	s kJ/kg.K	T K	u kJ/kg	h kJ/kg	s kJ/kg.K	T K	u kJ/kg	h kJ/kg	s kJ/kg.K	T K	u kJ/kg	h kJ/kg	s kJ/kg.K
410	8309.568	11717.719	139.8168	990	20613.324	28843.792	165.7734	1570	33997.091	47049.877	180.2083	2150	48685.532	66560.634	190.7663
420	8517.529	12008.823	140.5183	1000	20833.276	29146.887	166.0780	1580	34239.426	47375.355	180.4150	2160	48949.603	66907.849	190.9274
430	8725.523	12299.961	141.2034	1010	21053.580	29450.335	166.3799	1590	34482.155	47701.227	180.6206	2170	49214.021	67255.410	191.0879
440	8933.551	12591.132	141.8728	1020	21274.238	29754.136	166.6792	1600	34725.279	48027.494	180.8251	2180	49478.784	67603.317	191.2479
450	9141.612	12882.337	142.5272	1030	21495.252	30058.294	166.9760	1610	34968.796	48354.155	181.0287	2190	49743.892	67951.568	191.4073
460	9349.710	13173.578	143.1673	1040	21716.625	30362.810	167.2702	1620	35212.706	48681.208	181.2312	2200	50009.343	68300.162	191.5661
470	9557.848	13464.860	143.7937	1050	21938.359	30667.687	167.5619	1630	35457.009	49008.655	181.4327	2210	50275.135	68649.098	191.7243
480	9766.032	13756.187	144.4071	1060	22160.455	30972.927	167.8513	1640	35701.705	49336.494	181.6332	2220	50541.268	68998.375	191.8820
490	9974.266	14047.564	145.0079	1070	22382.917	31278.532	168.1382	1650	35946.793	49664.725	181.8327	2230	50807.741	69347.991	192.0392
500	10182.558	14339.000	145.5966	1080	22605.745	31584.504	168.4229	1660	36192.272	49993.348	182.0313	2240	51074.552	69697.945	192.1957
510	10390.915	14630.500	146.1739	1090	22828.943	31890.845	168.7052	1670	36438.142	50322.362	182.2289	2250	51341.700	70048.237	192.3518
520	10599.344	14922.072	146.7401	1100	23052.510	32197.556	168.9853	1680	36684.403	50651.766	182.4256	2300	52682.454	71804.707	193.1239
530	10807.853	15213.724	147.2956	1110	23276.450	32504.639	169.2632	1690	36931.054	50981.560	182.6213	2350	54031.447	73569.418	193.8829
540	11016.450	15505.465	147.8409	1120	23500.764	32812.097	169.5390	1700	37178.094	51311.743	182.8161	2400	55388.521	75342.209	194.6294
550	11225.143	15797.301	148.3764	1130	23725.453	33119.929	169.8126	1710	37425.522	51642.315	183.0100	2450	56753.511	77122.916	195.3637
560	11433.940	16089.242	148.9025	1140	23950.519	33428.139	170.0841	1720	37673.339	51973.275	183.2029	2500	58126.251	78911.372	196.0863
570	11642.851	16381.297	149.4194	1150	24175.964	33736.726	170.3537	1730	37921.543	52304.623	183.3950	2550	59506.568	80707.407	196.7976
580	11851.883	16673.472	149.9275	1160	24401.788	34045.694	170.6212	1740	38170.134	52636.358	183.5862	2600	60894.290	82510.846	197.4980
590	12061.045	16965.778	150.4272	1170	24627.992	34355.042	170.8867	1750	38419.112	52968.478	183.7766	2650	62289.239	84321.512	198.1878
600	12270.346	17258.221	150.9187	1180	24854.580	34664.772	171.1503	1760	38668.474	53300.984	183.9660	2700	63691.236	86139.225	198.8674
610	12479.792	17550.811	151.4024	1190	25081.550	34974.886	171.4120	1770	38918.222	53633.876	184.1546	2750	65100.097	87963.804	199.5369
620	12689.393	17843.556	151.8784	1200	25308.905	35285.384	171.6718	1780	39168.354	53967.151	184.3424	2800	66515.639	89795.062	200.1969

(Continued)

Table E-4 (Continued)

T K	u kJ/kg	h kJ/kg	s kJ/kg.K	T K	u kJ/kg	h kJ/kg	s kJ/kg.K	T K	u kJ/kg	h kJ/kg	s kJ/kg.K	T K	u kJ/kg	h kJ/kg	s kJ/kg.K
630	12899.157	18136.463	152.3470	1210	25536.645	35596.268	171.9298	1790	39418.869	54300.809	184.5293	2850	67937.672	91632.813	200.8474
640	13109.091	18429.540	152.8086	1220	25764.772	35907.539	172.1860	1800	39669.767	54634.851	184.7154	2900	69366.009	93476.866	201.4888
650	13319.203	18722.796	153.2633	1230	25993.287	36219.197	172.4404	1810	39921.047	54969.274	184.9007	2950	70800.456	95327.030	202.1214
660	13529.501	19016.237	153.7113	1240	26222.190	36531.243	172.6931	1820	40172.708	55304.078	185.0851	3000	72240.820	97183.111	202.7453
670	13739.992	19309.871	154.1528	1250	26451.482	36843.678	172.9441	1830	40424.749	55639.263	185.2688	3050	73686.904	99044.913	203.3608
680	13950.683	19603.706	154.5881	1260	26681.164	37156.504	173.1933	1840	40677.170	55974.827	185.4517	3100	75138.511	100912.237	203.9680
690	14161.582	19897.749	155.0174	1270	26911.236	37469.720	173.4409	1850	40929.969	56310.770	185.6338	3150	76595.442	102784.884	204.5673
700	14372.696	20192.005	155.4408	1280	27141.700	37783.327	173.6869	1860	41183.147	56647.091	185.8151	3200	78057.494	104662.654	205.1587
710	14584.030	20486.483	155.8585	1290	27372.557	38097.327	173.9312	1870	41436.702	56983.789	185.9956	3250	79524.465	106545.342	205.7425
720	14795.593	20781.189	156.2707	1300	27603.806	38411.719	174.1740	1880	41690.632	57320.863	186.1754	3300	80996.150	108432.743	206.3188
730	15007.389	21076.129	156.6775	1310	27835.448	38726.505	174.4152	1890	41944.939	57658.313	186.3544	3350	82472.343	110324.653	206.8878
740	15219.426	21371.309	157.0791	1320	28067.484	39041.684	174.6549	1900	42199.619	57996.137	186.5327	3400	83952.835	112220.863	207.4497
750	15431.709	21666.736	157.4757	1330	28299.914	39357.258	174.8931	1910	42454.673	58334.334	186.7102	3450	85437.419	114121.164	208.0045
760	15644.244	21962.414	157.8673	1340	28532.739	39673.226	175.1298	1920	42710.100	58672.905	186.8870	3500	86925.884	116025.345	208.5525
770	15857.038	22258.351	158.2542	1350	28765.959	39989.590	175.3650	1930	42965.899	59011.847	187.0631				

Table E-5 Ideal gas properties of Water vapor

T K	u kJ/kg	h kJ/kg	s kJ/kg.K	T K	u kJ/kg	h kJ/kg	s kJ/kg.K	T K	u kJ/kg	h kJ/kg	s kJ/kg.K	T K	u kJ/kg	h kJ/kg	s kJ/kg.K
200	4961.362	6624.645	175.346	780	20638.043	27123.643	222.690	1360	40314.599	51622.517	245.986	1940	63710.395	79840.630	263.210
210	5210.850	6957.276	176.969	790	20940.874	27509.618	223.181	1370	40689.455	52080.515	246.322	1950	64140.000	80353.378	263.473
220	5461.115	7290.684	178.520	800	21245.015	27896.902	223.669	1380	41065.418	52539.622	246.656	1960	64570.381	80866.902	263.736
230	5711.966	7624.679	180.005	810	21550.466	28285.496	224.151	1390	41442.485	52999.832	246.988	1970	65001.532	81381.197	263.998
240	5963.276	7959.132	181.428	820	21857.229	28675.403	224.630	1400	41820.648	53461.139	247.319	1980	65433.448	81896.256	264.259
250	6214.964	8293.964	182.795	830	22165.304	29066.621	225.104	1410	42199.904	53923.538	247.648	1990	65866.123	82412.075	264.518
260	6466.984	8629.127	184.109	840	22474.692	29459.152	225.574	1420	42580.245	54387.022	247.975	2000	66299.552	82928.647	264.777
270	6719.315	8964.602	185.376	850	22785.392	29852.996	226.040	1430	42961.665	54851.587	248.301	2010	66733.731	83445.969	265.035
280	6971.955	9300.385	186.597	860	23097.405	30248.152	226.502	1440	43344.161	55317.225	248.626	2020	67168.652	83964.034	265.292
290	7224.915	9636.489	187.776	870	23410.729	30644.620	226.961	1450	43727.725	55783.933	248.949	2030	67604.311	84482.836	265.549
300	7478.216	9972.933	188.917	880	23725.366	31042.400	227.415	1460	44112.351	56251.703	249.270	2040	68040.703	85002.372	265.804
310	7731.888	10309.748	190.021	890	24041.312	31441.490	227.866	1470	44498.035	56720.530	249.590	2050	68477.822	85522.634	266.058
320	7985.963	10646.967	191.092	900	24358.568	31841.889	228.314	1480	44884.770	57190.408	249.909	2060	68915.664	86043.619	266.312
330	8240.478	10984.625	192.131	910	24677.131	32243.596	228.757	1490	45272.550	57661.332	250.226	2070	69354.221	86565.320	266.565
340	8495.471	11322.761	193.140	920	24997.001	32646.609	229.198	1500	45661.370	58133.295	250.542	2080	69793.491	87087.733	266.816
350	8750.982	11661.416	194.122	930	25318.176	33050.927	229.635	1510	46051.224	58606.292	250.856	2090	70233.466	87610.852	267.067
360	9007.052	12000.630	195.078	940	25640.652	33456.547	230.069	1520	46442.105	59080.317	251.169	2100	70674.143	88134.672	267.317
370	9263.720	12340.441	196.009	950	25964.429	33863.467	230.499	1530	46834.009	59555.365	251.480	2110	71115.515	88659.188	267.566
380	9521.026	12680.890	196.916	960	26289.505	34271.686	230.927	1540	47226.930	60031.428	251.791	2120	71557.578	89184.394	267.815
390	9779.008	13022.016	197.803	970	26615.875	34681.200	231.351	1550	47620.861	60508.503	252.099	2130	72000.327	89710.286	268.062
400	10037.703	13363.854	198.668	980	26943.538	35092.007	231.773	1560	48015.796	60986.582	252.407	2140	72443.755	90236.858	268.309
410	10297.148	13706.442	199.514	990	27272.492	35504.103	232.191	1570	48411.731	61465.659	252.713	2150	72887.859	90764.105	268.555

(Continued)

Table E-5 (Continued)

T K	u kJ/kg	h kJ/kg	s kJ/kg.K	T K	u kJ/kg	h kJ/kg	s kJ/kg.K	T K	u kJ/kg	h kJ/kg	s kJ/kg.K	T K	u kJ/kg	h kJ/kg	s kJ/kg.K
420	10557.377	14049.815	200.341	1000	27602.732	35917.487	232.606	1580	48808.658	61945.730	253.018	2160	73332.634	91292.023	268.800
430	10818.424	14394.006	201.151	1010	27934.256	36332.155	233.019	1590	49206.572	62426.788	253.321	2170	73778.073	91820.606	269.044
440	11080.322	14739.046	201.945	1020	28267.061	36748.103	233.429	1600	49605.468	62908.827	253.623	2180	74224.172	92349.848	269.287
450	11343.099	15084.967	202.722	1030	28601.144	37165.329	233.836	1610	50005.340	63391.842	253.924	2190	74670.926	92879.746	269.530
460	11606.787	15431.798	203.484	1040	28936.500	37583.828	234.240	1620	50406.181	63875.826	254.224	2200	75118.330	93410.293	269.771
470	11871.412	15779.567	204.232	1050	29273.126	38003.598	234.642	1630	50807.985	64360.774	254.522	2210	75566.380	93941.486	270.012
480	12137.001	16128.299	204.966	1060	29611.018	38424.634	235.041	1640	51210.748	64846.681	254.820	2220	76015.069	94473.319	270.252
490	12403.578	16478.020	205.687	1070	29950.173	38846.932	235.438	1650	51614.463	65333.539	255.116	2230	76464.393	95005.786	270.492
500	12671.168	16828.753	206.396	1080	30290.587	39270.489	235.832	1660	52019.124	65821.343	255.410	2240	76914.348	95538.884	270.730
510	12939.794	17180.522	207.093	1090	30632.255	39695.301	236.223	1670	52424.725	66310.088	255.704	2250	77364.928	96072.608	270.968
520	13209.475	17533.347	207.778	1100	30975.174	40121.363	236.612	1680	52831.262	66799.768	255.996	2300	79627.038	98750.435	272.145
530	13480.232	17887.247	208.452	1110	31319.339	40548.671	236.999	1690	53238.727	67290.376	256.287	2350	81904.061	101443.175	273.303
540	13752.085	18242.243	209.115	1120	31664.746	40977.222	237.383	1700	53647.115	67781.908	256.577	2400	84195.414	104150.245	274.443
550	14025.050	18598.352	209.769	1130	32011.390	41407.009	237.765	1710	54056.420	68274.356	256.866	2450	86500.527	106871.075	275.565
560	14299.144	18955.590	210.412	1140	32359.267	41838.030	238.145	1720	54466.637	68767.716	257.154	2500	88818.848	109605.113	276.670
570	14574.384	19313.973	211.047	1150	32708.373	42270.279	238.523	1730	54877.759	69261.982	257.440	2550	91149.841	112351.823	277.758
580	14850.783	19673.515	211.672	1160	33058.702	42703.752	238.898	1740	55289.781	69757.148	257.726	2600	93492.983	115110.682	278.829
590	15128.356	20034.232	212.289	1170	33410.250	43138.443	239.271	1750	55702.697	70253.207	258.010	2650	95847.773	117881.189	279.885
600	15407.116	20396.135	212.897	1180	33763.013	43574.349	239.642	1760	56116.502	70750.155	258.293	2700	98213.723	120662.856	280.924
610	15687.074	20759.236	213.497	1190	34116.985	44011.465	240.011	1770	56531.189	71247.986	258.575	2750	100590.363	123455.213	281.949
620	15968.242	21123.547	214.090	1200	34472.162	44449.785	240.378	1780	56946.753	71746.694	258.856	2800	102977.241	126257.808	282.959
630	16250.630	21489.079	214.674	1210	34828.538	44889.304	240.742	1790	57363.189	72246.272	259.136	2850	105373.922	129070.206	283.955
640	16534.248	21855.841	215.252	1220	35186.109	45330.018	241.105	1800	57780.489	72746.716	259.415	2900	107779.989	131891.990	284.936

(Continued)

Table E-5 (Continued)

T K	u kJ/kg	h kJ/kg	s kJ/kg.K	T K	u kJ/kg	h kJ/kg	s kJ/kg.K	T K	u kJ/kg	h kJ/kg	s kJ/kg.K	T K	u kJ/kg	h kJ/kg	s kJ/kg.K
650	16819.106	22223.842	215.823	1230	35544.869	45771.922	241.466	1810	58198.649	73248.020	259.693	2950	110195.042	134722.760	285.904
660	17105.211	22593.090	216.386	1240	35904.813	46215.010	241.825	1820	58617.664	73750.178	259.969	3000	112618.699	137562.134	286.858
670	17392.571	22963.594	216.943	1250	36265.937	46659.277	242.182	1830	59037.526	74253.184	260.245	3050	115050.597	140409.749	287.800
680	17681.194	23335.360	217.494	1260	36628.234	47104.718	242.536	1840	59458.232	74757.032	260.520	3100	117490.390	143265.259	288.728
690	17971.086	23708.396	218.039	1270	36991.700	47551.327	242.890	1850	59879.774	75261.718	260.793	3150	119937.752	146128.338	289.645
700	18262.253	24082.706	218.577	1280	37356.330	47999.100	243.241	1860	60302.148	75767.235	261.066	3200	122392.374	148998.677	290.549
710	18554.700	24458.297	219.110	1290	37722.118	48448.031	243.590	1870	60725.347	76273.578	261.337	3250	124853.965	151875.985	291.441
720	18848.433	24835.173	219.637	1300	38089.058	48898.115	243.938	1880	61149.367	76780.741	261.608	3300	127322.254	154759.991	292.322
730	19143.456	25213.339	220.159	1310	38457.145	49349.346	244.283	1890	61574.201	77288.719	261.877	3350	129796.988	157650.442	293.191
740	19439.772	25592.799	220.675	1320	38826.375	49801.718	244.627	1900	61999.844	77797.505	262.146	3400	132277.933	160547.104	294.049
750	19737.386	25973.556	221.186	1330	39196.740	50255.227	244.970	1910	62426.291	78307.095	262.413	3450	134764.872	163449.760	294.897
760	20036.301	26355.614	221.692	1340	39568.236	50709.867	245.310	1920	62853.535	78817.483	262.680	3500	137257.610	166358.215	295.734
770	20336.519	26738.976	222.193	1350	39940.858	51165.632	245.649	1930	63281.572	79328.663	262.945				

Table E-6 Ideal gas properties of Nitrogen

T K	u kJ/kg	h kJ/kg	s kJ/kg.K	T K	u kJ/kg	h kJ/kg	s kJ/kg.K	T K	u kJ/kg	h kJ/kg	s kJ/kg.K	T K	u kJ/kg	h kJ/kg	s kJ/kg.K
200	4147.593	5809.876	179.8701	780	16615.620	23100.220	220.1300	1360	30887.506	42194.424	238.3461	1940	46518.506	62647.741	250.8515
210	4355.782	6101.208	181.2925	790	16845.697	23413.441	220.5290	1370	31147.908	42537.969	238.5977	1950	46795.196	63007.574	251.0365
220	4563.966	6392.536	182.6482	800	17076.389	23727.276	220.9238	1380	31408.708	42881.912	238.8476	1960	47072.060	63367.582	251.2207
230	4772.083	6683.796	183.9428	810	17307.695	24041.725	221.3145	1390	31669.902	43226.249	239.0961	1970	47349.095	63727.760	251.4041
240	4980.098	6974.955	185.1817	820	17539.613	24356.787	221.7011	1400	31931.485	43570.976	239.3431	1980	47626.299	64088.107	251.5866
250	5187.999	7265.999	186.3693	830	17772.142	24672.459	222.0837	1410	32193.454	43916.088	239.5885	1990	47903.667	64448.618	251.7683
260	5395.789	7556.933	187.5098	840	18005.280	24988.740	222.4625	1420	32455.803	44261.581	239.8326	2000	48181.198	64809.293	251.9492
270	5603.482	7847.769	188.6068	850	18239.025	25305.629	222.8375	1430	32718.528	44607.449	240.0751	2010	48458.888	65170.126	252.1293
280	5811.099	8138.529	189.6637	860	18473.375	25623.122	223.2088	1440	32981.625	44953.690	240.3163	2020	48736.734	65531.116	252.3085
290	6018.666	8429.240	190.6833	870	18708.328	25941.219	223.5766	1450	33245.089	45300.297	240.5560	2030	49014.735	65892.260	252.4870
300	6226.214	8719.931	191.6683	880	18943.881	26259.916	223.9407	1460	33508.916	45647.268	240.7943	2040	49292.887	66253.555	252.6646
310	6433.774	9010.634	192.6211	890	19180.033	26579.210	224.3015	1470	33773.102	45994.597	241.0313	2050	49571.187	66614.999	252.8414
320	6641.378	9301.382	193.5438	900	19416.779	26899.100	224.6589	1480	34037.642	46342.280	241.2668	2060	49849.632	66976.588	253.0175
330	6849.060	9592.207	194.4384	910	19654.118	27219.582	225.0129	1490	34302.532	46690.313	241.5011	2070	50128.221	67338.320	253.1928
340	7056.852	9883.142	195.3067	920	19892.046	27540.654	225.3637	1500	34567.767	47038.692	241.7340	2080	50406.951	67700.193	253.3673
350	7264.786	10174.220	196.1503	930	20130.561	27862.312	225.7114	1510	34833.343	47387.411	241.9656	2090	50685.818	68062.204	253.5410
360	7472.894	10465.471	196.9707	940	20369.659	28184.554	226.0560	1520	35099.256	47736.468	242.1959	2100	50964.821	68424.350	253.7140
370	7681.205	10756.926	197.7692	950	20609.338	28507.376	226.3975	1530	35365.502	48085.857	242.4249	2110	51243.957	68786.629	253.8863
380	7889.749	11048.613	198.5471	960	20849.593	28830.775	226.7360	1540	35632.076	48435.575	242.6526	2120	51523.223	69149.039	254.0577
390	8098.553	11340.560	199.3055	970	21090.422	29154.747	227.0716	1550	35898.974	48785.616	242.8790	2130	51802.617	69511.577	254.2285
400	8307.643	11632.794	200.0454	980	21331.821	29479.289	227.4044	1560	36166.192	49135.978	243.1042	2140	52082.138	69874.240	254.3985

(Continued)

Table E-6 (Continued)

T K	u kJ/kg	h kJ/kg	s kJ/kg.K	T K	u kJ/kg	h kJ/kg	s kJ/kg.K	T K	u kJ/kg	h kJ/kg	s kJ/kg.K	T K	u kJ/kg	h kJ/kg	s kJ/kg.K
410	8517.045	11925.339	200.7679	990	21573.786	29804.398	227.7343	1570	36433.726	49486.655	243.3282	2150	52361.781	70237.027	254.5677
420	8726.781	12218.219	201.4738	1000	21816.314	30130.069	228.0615	1580	36701.572	49837.644	243.5510	2160	52641.546	70599.935	254.7363
430	8936.875	12511.457	202.1640	1010	22059.401	30456.300	228.3860	1590	36969.724	50188.940	243.7725	2170	52921.429	70962.962	254.9041
440	9147.348	12805.073	202.8393	1020	22303.044	30783.086	228.7078	1600	37238.181	50540.540	243.9929	2180	53201.429	71326.105	255.0712
450	9358.219	13099.087	203.5002	1030	22547.238	31110.423	229.0270	1610	37506.936	50892.438	244.2121	2190	53481.544	71689.363	255.2376
460	9569.507	13393.518	204.1476	1040	22791.979	31438.308	229.3437	1620	37775.986	51244.632	244.4301	2200	53761.771	72052.734	255.4033
470	9781.229	13688.384	204.7820	1050	23037.265	31766.737	229.6578	1630	38045.328	51597.117	244.6469	2210	54042.108	72416.214	255.5683
480	9993.402	13983.701	205.4040	1060	23283.090	32095.705	229.9695	1640	38314.956	51949.889	244.8626	2220	54322.553	72779.803	255.7326
490	10206.042	14279.483	206.0142	1070	23529.450	32425.209	230.2787	1650	38584.868	52302.944	245.0771	2230	54603.104	73143.498	255.8962
500	10419.161	14575.746	206.6130	1080	23776.342	32755.244	230.5856	1660	38855.058	52656.278	245.2906	2240	54883.760	73507.296	256.0591
510	10632.774	14872.503	207.2010	1090	24023.761	33085.807	230.8901	1670	39125.524	53009.887	245.5029	2250	55164.518	73871.198	256.2214
520	10846.894	15169.766	207.7785	1100	24271.703	33416.892	231.1923	1680	39396.261	53363.767	245.7141	2300	56569.773	75692.170	257.0226
530	11061.531	15467.546	208.3460	1110	24520.165	33748.497	231.4922	1690	39667.266	53717.915	245.9243	2350	57977.315	77515.429	257.8077
540	11276.696	15765.855	208.9039	1120	24769.141	34080.617	231.7899	1700	39938.534	54072.327	246.1333	2400	59386.942	79340.773	258.5772
550	11492.400	16064.702	209.4526	1130	25018.627	34413.246	232.0854	1710	40210.061	54426.998	246.3413	2450	60798.481	81168.029	259.3316
560	11708.650	16364.096	209.9923	1140	25268.620	34746.382	232.3787	1720	40481.845	54781.924	246.5482	2500	62211.784	82997.049	260.0715
570	11925.456	16664.045	210.5235	1150	25519.114	35080.020	232.6700	1730	40753.880	55137.103	246.7541	2550	63626.726	84827.708	260.7974
580	12142.826	16964.558	211.0465	1160	25770.105	35414.155	232.9591	1740	41026.164	55492.531	246.9589	2600	65043.210	86659.909	261.5098
590	12360.766	17265.642	211.5614	1170	26021.590	35748.782	233.2461	1750	41298.693	55848.203	247.1627	2650	66461.163	88493.579	262.2091
600	12579.283	17567.302	212.0687	1180	26273.563	36083.899	233.5312	1760	41571.462	56204.116	247.3655	2700	67880.540	90328.673	262.8959
610	12798.383	17869.545	212.5685	1190	26526.020	36419.499	233.8142	1770	41844.469	56560.266	247.5673	2750	69301.319	92165.169	263.5705
620	13018.070	18172.376	213.0612	1200	26778.956	36755.579	234.0952	1780	42117.709	56916.650	247.7680	2800	70723.507	94003.074	264.2334

(Continued)

Table E-6 (Continued)

T K	*u* kJ/kg	*h* kJ/kg	*s* kJ/kg.K	*T* K	*u* kJ/kg	*h* kJ/kg	*s* kJ/kg.K	*T* K	*u* kJ/kg	*h* kJ/kg	*s* kJ/kg.K	*T* K	*u* kJ/kg	*h* kJ/kg	*s* kJ/kg.K
630	13238.351	18475.800	213.5469	1210	27032.368	37092.135	234.3744	1790	42391.180	57273.264	247.9678	2850	72147.136	95842.420	264.8849
640	13459.229	18779.821	214.0259	1220	27286.251	37429.161	234.6516	1800	42664.877	57630.104	248.1666	2900	73572.264	97683.265	265.5255
650	13680.707	19084.443	214.4984	1230	27540.600	37766.653	234.9269	1810	42938.797	57987.168	248.3645	2950	74998.976	99525.694	266.1556
660	13902.790	19389.669	214.9646	1240	27795.411	38104.607	235.2004	1820	43212.937	58344.451	248.5613	3000	76427.385	101369.820	266.7754
670	14125.480	19695.503	215.4247	1250	28050.678	38443.018	235.4720	1830	43487.293	58701.950	248.7572	3050	77857.627	103215.779	267.3855
680	14348.780	20001.946	215.8789	1260	28306.399	38781.882	235.7418	1840	43761.861	59059.661	248.9522	3100	79289.868	105063.737	267.9861
690	14572.692	20309.001	216.3273	1270	28562.567	39121.194	236.0099	1850	44036.638	59417.582	249.1462	3150	80724.301	106913.887	268.5775
700	14797.217	20616.670	216.7702	1280	28819.180	39460.950	236.2762	1860	44311.622	59775.709	249.3393	3200	82161.142	108766.445	269.1601
710	15022.358	20924.954	217.2076	1290	29076.231	39801.145	236.5407	1870	44586.808	60134.038	249.5315	3250	83600.638	110621.658	269.7343
720	15248.114	21233.854	217.6398	1300	29333.717	40141.774	236.8036	1880	44862.193	60492.567	249.7227	3300	85043.062	112479.799	270.3003
730	15474.488	21543.371	218.0668	1310	29591.633	40482.834	237.0648	1890	45137.774	60851.291	249.9130	3350	86488.713	114341.167	270.8584
740	15701.480	21853.506	218.4889	1320	29849.975	40824.319	237.3243	1900	45413.547	61210.208	250.1025	3400	87937.917	116206.088	271.4089
750	15929.089	22164.259	218.9061	1330	30108.738	41166.225	237.5822	1910	45689.511	61569.315	250.2911	3450	89391.028	118074.916	271.9522
760	16157.315	22475.629	219.3186	1340	30367.917	41508.547	237.8384	1920	45965.660	61928.608	250.4787	3500	90848.427	119948.032	272.4885
770	16386.159	22787.616	219.7265	1350	30627.508	41851.282	238.0931	1930	46241.993	62288.084	250.6655				

Table E-7 Ideal gas properties of Monatomic Oxygen

T K	u kJ/kg	h kJ/kg	s kJ/kg.K	T K	u kJ/kg	h kJ/kg	s kJ/kg.K	T K	u kJ/kg	h kJ/kg	s kJ/kg.K	T K	u kJ/kg	h kJ/kg	s kJ/kg.K
200	3002.738	4665.338	152.1439	780	10617.360	17102.080	181.5733	1360	17920.121	29226.961	193.1988	1940	25184.432	41313.392	200.6012
210	3146.359	4892.099	153.2503	790	10744.116	17311.976	181.8407	1370	18045.548	29435.528	193.3516	1950	25309.545	41521.645	200.7082
220	3288.901	5117.781	154.3001	800	10870.825	17521.825	182.1047	1380	18170.968	29644.088	193.5033	1960	25434.656	41729.896	200.8147
230	3430.433	5342.453	155.2989	810	10997.486	17731.626	182.3653	1390	18296.379	29852.639	193.6539	1970	25559.766	41938.146	200.9207
240	3571.023	5566.183	156.2511	820	11124.103	17941.383	182.6227	1400	18421.783	30061.183	193.8034	1980	25684.873	42146.393	201.0262
250	3710.736	5789.036	157.1608	830	11250.675	18151.095	182.8769	1410	18547.180	30269.720	193.9518	1990	25809.980	42354.640	201.1311
260	3849.631	6011.071	158.0317	840	11377.206	18360.766	183.1280	1420	18672.569	30478.249	194.0992	2000	25935.085	42562.885	201.2355
270	3987.766	6232.346	158.8668	850	11503.696	18570.396	183.3761	1430	18797.950	30686.770	194.2455	2010	26060.189	42771.129	201.3393
280	4125.191	6452.911	159.6689	860	11630.147	18779.987	183.6212	1440	18923.325	30895.285	194.3908	2020	26185.292	42979.372	201.4427
290	4261.956	6672.816	160.4406	870	11756.560	18989.540	183.8635	1450	19048.692	31103.792	194.5351	2030	26310.395	43187.615	201.5455
300	4398.105	6892.105	161.1840	880	11882.936	19199.056	184.1029	1460	19174.052	31312.292	194.6784	2040	26435.497	43395.857	201.6478
310	4533.680	7110.820	161.9012	890	12009.277	19408.537	184.3396	1470	19299.405	31520.785	194.8207	2050	26560.599	43604.099	201.7497
320	4668.719	7328.999	162.5939	900	12135.583	19617.983	184.5737	1480	19424.751	31729.271	194.9621	2060	26685.700	43812.340	201.8510
330	4803.258	7546.678	163.2637	910	12261.856	19827.396	184.8051	1490	19550.091	31937.751	195.1025	2070	26810.802	44020.582	201.9518
340	4937.329	7763.889	163.9122	920	12388.097	20036.777	185.0339	1500	19675.423	32146.223	195.2419	2080	26935.905	44228.825	202.0522
350	5070.963	7980.663	164.5405	930	12514.307	20246.127	185.2602	1510	19800.749	32354.689	195.3804	2090	27061.008	44437.068	202.1521
360	5204.187	8197.027	165.1501	940	12640.486	20455.446	185.4841	1520	19926.068	32563.148	195.5180	2100	27186.111	44645.311	202.2515
370	5337.028	8413.008	165.7418	950	12766.635	20664.735	185.7056	1530	20051.380	32771.600	195.6547	2110	27311.216	44853.556	202.3504
380	5469.509	8628.629	166.3169	960	12892.756	20873.996	185.9247	1540	20176.686	32980.046	195.7905	2120	27436.322	45061.802	202.4489
390	5601.652	8843.912	166.8761	970	13018.849	21083.229	186.1415	1550	20301.985	33188.485	195.9254	2130	27561.429	45270.049	202.5469
400	5733.479	9058.879	167.4203	980	13144.915	21292.435	186.3561	1560	20427.277	33396.917	196.0595	2140	27686.538	45478.298	202.6444
410	5865.007	9273.547	167.9504	990	13270.954	21501.614	186.5685	1570	20552.564	33605.344	196.1927	2150	27811.649	45686.549	202.7415

(Continued)

Table E-7 (Continued)

T K	u kJ/kg	h kJ/kg	s kJ/kg.K	T K	u kJ/kg	h kJ/kg	s kJ/kg.K	T K	u kJ/kg	h kJ/kg	s kJ/kg.K	T K	u kJ/kg	h kJ/kg	s kJ/kg.K
420	5996.255	9487.935	168.4670	1000	13396.967	21710.767	186.7787	1580	20677.844	33813.764	196.3250	2160	27936.762	45894.802	202.8381
430	6127.239	9702.059	168.9709	1010	13522.950	21919.890	186.9867	1590	20803.117	34022.177	196.4565	2170	28061.877	46103.057	202.9343
440	6257.974	9915.934	169.4626	1020	13648.900	22128.980	187.1927	1600	20928.385	34230.585	196.5871	2180	28186.995	46311.315	203.0301
450	6388.474	10129.574	169.9427	1030	13774.819	22338.039	187.3967	1610	21053.646	34438.986	196.7170	2190	28312.116	46519.576	203.1254
460	6518.753	10342.993	170.4117	1040	13900.710	22547.070	187.5987	1620	21178.902	34647.382	196.8460	2200	28437.239	46727.839	203.2203
470	6648.821	10556.201	170.8703	1050	14026.573	22756.073	187.7987	1630	21304.151	34855.771	196.9743	2210	28562.366	46936.106	203.3147
480	6778.692	10769.212	171.3187	1060	14152.410	22965.050	187.9968	1640	21429.394	35064.154	197.1017	2220	28687.497	47144.377	203.4088
490	6908.375	10982.035	171.7576	1070	14278.223	23174.003	188.1930	1650	21554.632	35272.532	197.2284	2230	28812.631	47352.651	203.5024
500	7037.880	11194.680	172.1872	1080	14404.013	23382.933	188.3873	1660	21679.864	35480.904	197.3543	2240	28937.769	47560.929	203.5955
510	7167.217	11407.157	172.6079	1090	14529.781	23591.841	188.5799	1670	21805.090	35689.270	197.4794	2250	29062.911	47769.211	203.6883
520	7296.393	11619.473	173.0202	1100	14655.529	23800.729	188.7706	1680	21930.310	35897.630	197.6038	2300	29688.695	48810.695	204.1461
530	7425.417	11831.637	173.4243	1110	14781.257	24009.597	188.9596	1690	22055.525	36105.985	197.7275	2350	30314.622	49852.322	204.5942
540	7554.297	12043.657	173.8206	1120	14906.967	24218.447	189.1470	1700	22180.735	36314.335	197.8504	2400	30940.724	50894.124	205.0328
550	7683.039	12255.539	174.2094	1130	15032.659	24427.279	189.3326	1710	22305.939	36522.679	197.9726	2450	31567.033	51936.133	205.4625
560	7811.651	12467.291	174.5910	1140	15158.335	24636.095	189.5166	1720	22431.138	36731.018	198.0941	2500	32193.584	52978.384	205.8837
570	7940.137	12678.917	174.9655	1150	15283.995	24844.895	189.6989	1730	22556.332	36939.352	198.2149	2550	32820.410	54020.910	206.2966
580	8068.504	12890.424	175.3334	1160	15409.640	25053.680	189.8797	1740	22681.521	37147.681	198.3349	2600	33447.545	55063.745	206.7016
590	8196.758	13101.818	175.6948	1170	15535.271	25262.451	190.0589	1750	22806.705	37356.005	198.4543	2650	34075.025	56106.925	207.0990
600	8324.904	13313.104	176.0499	1180	15660.888	25471.208	190.2366	1760	22931.884	37564.324	198.5730	2700	34702.883	57150.483	207.4891
610	8452.945	13524.285	176.3989	1190	15786.491	25679.951	190.4127	1770	23057.059	37772.639	198.6910	2750	35331.155	58194.455	207.8722
620	8580.888	13735.368	176.7422	1200	15912.082	25888.682	190.5874	1780	23182.229	37980.949	198.8084	2800	35959.876	59238.876	208.2486
630	8708.736	13946.356	177.0798	1210	16037.661	26097.401	190.7606	1790	23307.394	38189.254	198.9251	2850	36589.081	60283.781	208.6185
640	8836.493	14157.253	177.4119	1220	16163.228	26306.108	190.9324	1800	23432.556	38397.556	199.0412	2900	37218.803	61329.203	208.9821
650	8964.163	14368.063	177.7387	1230	16288.784	26514.804	191.1027	1810	23557.712	38605.852	199.1566	2950	37849.078	62375.178	209.3397

(Continued)

Table E-7 (Continued)

T K	u kJ/kg	h kJ/kg	s kJ/kg.K	T K	u kJ/kg	h kJ/kg	s kJ/kg.K	T K	u kJ/kg	h kJ/kg	s kJ/kg.K	T K	u kJ/kg	h kJ/kg	s kJ/kg.K
660	9091.749	14578.789	178.0605	1240	16414.329	26723.489	191.2717	1820	23682.865	38814.145	199.2713	3000	38479.939	63421.739	209.6915
670	9219.256	14789.436	178.3772	1250	16539.863	26932.163	191.4393	1830	23808.014	39022.434	199.3854	3050	39111.420	64468.920	210.0377
680	9346.685	15000.005	178.6892	1260	16665.387	27140.827	191.6056	1840	23933.159	39230.719	199.4990	3100	39743.553	65516.753	210.3785
690	9474.041	15210.501	178.9965	1270	16790.901	27349.481	191.7705	1850	24058.301	39439.001	199.6118	3150	40376.373	66565.273	210.7140
700	9601.326	15420.926	179.2993	1280	16916.406	27558.126	191.9342	1860	24183.439	39647.279	199.7241	3200	41009.909	67614.509	211.0445
710	9728.543	15631.283	179.5976	1290	17041.901	27766.761	192.0965	1870	24308.573	39855.553	199.8358	3250	41644.195	68664.495	211.3701
720	9855.694	15841.574	179.8918	1300	17167.387	27975.387	192.2577	1880	24433.704	40063.824	199.9469	3300	42279.261	69715.261	211.6909
730	9982.783	16051.803	180.1817	1310	17292.864	28184.004	192.4175	1890	24558.832	40272.092	200.0574	3350	42915.136	70766.836	212.0072
740	10109.811	16261.971	180.4677	1320	17418.332	28392.612	192.5762	1900	24683.958	40480.358	200.1673	3400	43551.850	71819.250	212.3190
750	10236.780	16472.080	180.7497	1330	17543.792	28601.212	192.7336	1910	24809.080	40688.620	200.2766	3450	44189.433	72872.533	212.6265
760	10363.693	16682.133	181.0279	1340	17669.243	28809.803	192.8898	1920	24934.200	40896.880	200.3853	3500	44827.910	73926.710	212.9299
770	10490.553	16892.133	181.3024	1350	17794.686	29018.386	193.0449	1930	25059.317	41105.137	200.4935				

Table E-8 Ideal gas properties of Oxygen

T K	u kJ/kg	h kJ/kg	s kJ/kg.K	T K	u kJ/kg	h kJ/kg	s kJ/kg.K	T K	u kJ/kg	h kJ/kg	s kJ/kg.K	T K	u kJ/kg	h kJ/kg	s kJ/kg.K
200	4155.027	5817.310	193.3956	780	17382.560	23867.161	235.0115	1360	32847.562	44154.479	254.3883	1940	49504.810	65634.044	267.5261
210	4365.746	6111.173	194.8294	790	17634.605	24202.349	235.4385	1370	33126.427	44516.487	254.6535	1950	49799.516	66011.894	267.7204
220	4574.775	6403.345	196.1886	800	17887.284	24538.171	235.8609	1380	33405.629	44878.833	254.9170	1960	50094.446	66389.967	267.9138
230	4782.703	6694.416	197.4824	810	18140.589	24874.620	236.2788	1390	33685.166	45241.513	255.1789	1970	50389.598	66768.263	268.1063
240	4989.995	6984.852	198.7185	820	18394.512	25211.686	236.6924	1400	33965.034	45604.525	255.4391	1980	50684.973	67146.781	268.2980
250	5197.018	7275.018	199.9030	830	18649.045	25549.362	237.1017	1410	34245.230	45967.865	255.6977	1990	50980.568	67525.520	268.4888
260	5404.064	7565.207	201.0412	840	18904.179	25887.640	237.5069	1420	34525.753	46331.530	255.9547	2000	51276.383	67904.478	268.6787
270	5611.364	7855.651	202.1373	850	19159.908	26226.512	237.9079	1430	34806.598	46695.519	256.2101	2010	51572.415	68283.654	268.8678
280	5819.103	8146.533	203.1952	860	19416.222	26565.969	238.3049	1440	35087.763	47059.828	256.4640	2020	51868.666	68663.048	269.0561
290	6027.428	8438.001	204.2180	870	19673.115	26906.006	238.6980	1450	35369.247	47424.455	256.7164	2030	52165.132	69042.658	269.2436
300	6236.455	8730.172	205.2085	880	19930.580	27246.614	239.0873	1460	35651.045	47789.396	256.9672	2040	52461.814	69422.483	269.4302
310	6446.277	9023.137	206.1691	890	20188.609	27587.787	239.4728	1470	35933.155	48154.650	257.2165	2050	52758.710	69802.522	269.6161
320	6656.965	9316.969	207.1020	900	20447.195	27929.516	239.8546	1480	36215.575	48520.214	257.4643	2060	53055.820	70182.775	269.8011
330	6868.577	9611.724	208.0090	910	20706.332	28271.797	240.2329	1490	36498.303	48886.085	257.7107	2070	53353.142	70563.240	269.9853
340	7081.156	9907.447	208.8918	920	20966.013	28614.621	240.6075	1500	36781.335	49252.260	257.9557	2080	53650.675	70943.917	270.1688
350	7294.735	10204.169	209.7519	930	21226.230	28957.982	240.9787	1510	37064.670	49618.738	258.1992	2090	53948.419	71324.804	270.3515
360	7509.338	10501.915	210.5907	940	21486.979	29301.873	241.3465	1520	37348.305	49985.516	258.4413	2100	54246.372	71705.901	270.5334
370	7724.981	10800.702	211.4093	950	21748.251	29646.289	241.7110	1530	37632.237	50352.592	258.6820	2110	54544.534	72087.206	270.7145
380	7941.677	11100.541	212.2089	960	22010.042	29991.223	242.0722	1540	37916.464	50719.963	258.9213	2120	54842.904	72468.719	270.8949
390	8159.431	11401.439	212.9905	970	22272.345	30336.669	242.4302	1550	38200.985	51087.627	259.1593	2130	55141.480	72850.440	271.0745

(Continued)

Table E-8 (Continued)

T K	u kJ/kg	h kJ/kg	s kJ/kg.K	T K	u kJ/kg	h kJ/kg	s kJ/kg.K	T K	u kJ/kg	h kJ/kg	s kJ/kg.K	T K	u kJ/kg	h kJ/kg	s kJ/kg.K
400	8378.246	11703.397	213.7550	980	22535.153	30682.621	242.7850	1560	38485.796	51455.581	259.3959	2140	55440.263	73232.366	271.2534
410	8598.121	12006.415	214.5032	990	22798.462	31029.074	243.1367	1570	38770.895	51823.824	259.6312	2150	55739.252	73614.498	271.4316
420	8819.052	12310.489	215.2360	1000	23062.265	31376.020	243.4854	1580	39056.281	52192.353	259.8652	2160	56038.445	73996.834	271.6090
430	9041.033	12615.615	215.9539	1010	23326.557	31723.455	243.8311	1590	39341.951	52561.166	260.0979	2170	56337.841	74379.374	271.7857
440	9264.059	12921.783	216.6578	1020	23591.332	32071.374	244.1739	1600	39627.903	52930.262	260.3293	2180	56637.441	74762.117	271.9617
450	9488.118	13228.986	217.3482	1030	23856.585	32419.770	244.5138	1610	39914.135	53299.637	260.5594	2190	56937.243	75145.063	272.1369
460	9713.202	13537.214	218.0256	1040	24122.310	32768.639	244.8509	1620	40200.644	53669.290	260.7883	2200	57237.247	75528.210	272.3115
470	9939.300	13846.454	218.6907	1050	24388.503	33117.975	245.1852	1630	40487.429	54039.219	261.0160	2210	57537.452	75911.558	272.4853
480	10166.399	14156.697	219.3438	1060	24655.158	33467.773	245.5167	1640	40774.489	54409.421	261.2424	2220	57837.856	76295.106	272.6585
490	10394.488	14467.930	219.9856	1070	24922.269	33818.028	245.8456	1650	41061.820	54779.896	261.4676	2230	58138.461	76678.854	272.8310
500	10623.555	14780.140	220.6163	1080	25189.833	34168.735	246.1719	1660	41349.420	55150.640	261.6916	2240	58439.264	77062.801	273.0028
510	10853.585	15093.313	221.2365	1090	25457.844	34519.889	246.4955	1670	41637.289	55521.652	261.9144	2250	58740.266	77446.946	273.1739
520	11084.567	15407.439	221.8464	1100	25726.297	34871.486	246.8166	1680	41925.424	55892.930	262.1361	2300	60248.227	79370.624	274.0195
530	11316.487	15722.502	222.4466	1110	25995.189	35223.521	247.1352	1690	42213.823	56264.473	262.3566	2350	61761.061	81299.175	274.8490
540	11549.331	16038.490	223.0372	1120	26264.513	35575.989	247.4513	1700	42502.485	56636.278	262.5760	2400	63278.709	83232.540	275.6631
550	11783.087	16355.389	223.6187	1130	26534.266	35928.885	247.7650	1710	42791.407	57008.343	262.7942	2450	64801.120	85170.668	276.4623
560	12017.741	16673.186	224.1913	1140	26804.443	36282.206	248.0763	1720	43080.588	57380.668	263.0113	2500	66328.256	87113.521	277.2473
570	12253.280	16991.869	224.7554	1150	27075.040	36635.946	248.3852	1730	43370.026	57753.249	263.2273	2550	67860.087	89061.069	278.0187
580	12489.691	17311.423	225.3111	1160	27346.053	36990.102	248.6919	1740	43659.719	58126.086	263.4422	2600	69396.592	91013.291	278.7768
590	12726.961	17631.836	225.8589	1170	27617.477	37344.670	248.9962	1750	43949.666	58499.176	263.6560	2650	70937.761	92970.177	279.5223
600	12965.077	17953.096	226.3988	1180	27889.308	37699.644	249.2983	1760	44239.866	58872.519	263.8687	2700	72483.592	94931.725	280.2556
610	13204.027	18275.190	226.9312	1190	28161.542	38055.022	249.5982	1770	44530.315	59246.112	264.0804	2750	74034.092	96897.942	280.9772

(Continued)

Table E-8 (Continued)

T K	u kJ/kg	h kJ/kg	s kJ/kg.K	T K	u kJ/kg	h kJ/kg	s kJ/kg.K	T K	u kJ/kg	h kJ/kg	s kJ/kg.K	T K	u kJ/kg	h kJ/kg	s kJ/kg.K
620	13443.799	18598.105	227.4563	1200	28434.175	38410.798	249.8959	1780	44821.014	59619.954	264.2910	2800	75589.278	98868.845	281.6875
630	13684.380	18921.829	227.9743	1210	28707.204	38766.970	250.1915	1790	45111.959	59994.043	264.5006	2850	77149.175	100844.459	282.3868
640	13925.758	19246.350	228.4853	1220	28980.624	39123.533	250.4850	1800	45403.150	60368.377	264.7091	2900	78713.816	102824.817	283.0756
650	14167.921	19571.657	228.9897	1230	29254.431	39480.484	250.7764	1810	45694.586	60742.956	264.9166	2950	80283.243	104809.961	283.7543
660	14410.858	19897.738	229.4875	1240	29528.622	39837.818	251.0657	1820	45986.264	61117.778	265.1231	3000	81857.506	106799.941	284.4232
670	14654.558	20224.581	229.9790	1250	29803.193	40195.533	251.3530	1830	46278.183	61492.840	265.3286	3050	83436.665	108794.817	285.0827
680	14899.010	20552.176	230.4644	1260	30078.140	40553.624	251.6384	1840	46570.342	61868.143	265.5332	3100	85020.787	110794.656	285.7331
690	15144.201	20880.511	230.9437	1270	30353.460	40912.087	251.9217	1850	46862.739	62243.683	265.7367	3150	86609.946	112799.532	286.3747
700	15390.123	21209.576	231.4172	1280	30629.150	41270.920	252.2032	1860	47155.373	62619.460	265.9393	3200	88204.226	114809.529	287.0077
710	15636.764	21539.361	231.8850	1290	30905.206	41630.119	252.4827	1870	47448.242	62995.473	266.1409	3250	89803.718	116824.738	287.6326
720	15884.114	21869.854	232.3472	1300	31181.624	41989.681	252.7604	1880	47741.346	63371.720	266.3416	3300	91408.522	118845.259	288.2496
730	16132.164	22201.047	232.8040	1310	31458.401	42349.601	253.0362	1890	48034.682	63748.200	266.5413	3350	93018.745	120871.199	288.8589
740	16380.902	22532.929	233.2556	1320	31735.534	42709.878	253.3102	1900	48328.250	64124.911	266.7401	3400	94634.502	122902.673	289.4608
750	16630.321	22865.491	233.7020	1330	32013.020	43070.507	253.5823	1910	48622.048	64501.852	266.9380	3450	96255.915	124939.803	290.0556
760	16880.409	23198.723	234.1433	1340	32290.855	43431.486	253.8527	1920	48916.075	64879.022	267.1349	3500	97883.114	126982.719	290.6435
770	17131.159	23532.616	234.5798	1350	32569.037	43792.811	254.1214	1930	49210.329	65256.420	267.3310				

Table E-9 Ideal gas properties of Hydroxyl

T K	u kJ/kg	h kJ/kg	s kJ/kg.K	T K	u kJ/kg	h kJ/kg	s kJ/kg.K	T K	u kJ/kg	h kJ/kg	s kJ/kg.K	T K	u kJ/kg	h kJ/kg	s kJ/kg.K
200	4565.808	6229.408	171.6789	780	16970.430	23456.150	212.2089	1360	30142.081	41449.921	229.3914	1940	44768.462	60898.422	241.2785
210	4787.327	6534.067	173.1653	790	17185.943	23754.803	212.5894	1370	30382.625	41773.605	229.6285	1950	45031.255	61244.355	241.4564
220	5007.948	6837.828	174.5784	800	17401.757	24053.757	212.9654	1380	30623.622	42097.742	229.8643	1960	45294.361	61590.601	241.6335
230	5227.759	7140.779	175.9251	810	17617.880	24353.020	213.3372	1390	30865.072	42422.332	230.0986	1970	45557.777	61937.157	241.8098
240	5446.837	7442.997	177.2114	820	17834.323	24652.603	213.7048	1400	31106.972	42747.372	230.3316	1980	45821.502	62284.022	241.9855
250	5665.247	7744.547	178.4424	830	18051.095	24952.515	214.0683	1410	31349.320	43072.860	230.5633	1990	46085.532	62631.192	242.1604
260	5883.049	8045.489	179.6227	840	18268.206	25252.766	214.4279	1420	31592.113	43398.793	230.7936	2000	46349.867	62978.667	242.3345
270	6100.294	8345.874	180.7563	850	18485.663	25553.363	214.7836	1430	31835.349	43725.169	231.0227	2010	46614.503	63326.443	242.5080
280	6317.030	8645.750	181.8469	860	18703.477	25854.317	215.1356	1440	32079.025	44051.985	231.2504	2020	46879.440	63674.520	242.6807
290	6533.297	8945.157	182.8976	870	18921.654	26155.634	215.4840	1450	32323.140	44379.240	231.4769	2030	47144.674	64022.894	242.8528
300	6749.132	9244.132	183.9112	880	19140.202	26457.322	215.8287	1460	32567.691	44706.931	231.7021	2040	47410.205	64371.565	243.0241
310	6964.570	9542.710	184.8902	890	19359.130	26759.390	216.1701	1470	32812.675	45035.055	231.9261	2050	47676.029	64720.529	243.1948
320	7179.640	9840.920	185.8370	900	19578.443	27061.843	216.5080	1480	33058.090	45363.610	232.1488	2060	47942.145	65069.785	243.3647
330	7394.370	10138.790	186.7536	910	19798.148	27364.688	216.8426	1490	33303.934	45692.594	232.3704	2070	48208.552	65419.332	243.5340
340	7608.786	10436.346	187.6419	920	20018.252	27667.932	217.1741	1500	33550.203	46022.003	232.5907	2080	48475.247	65769.167	243.7026
350	7822.910	10733.610	188.5036	930	20238.760	27971.580	217.5023	1510	33796.895	46351.835	232.8099	2090	48742.228	66119.288	243.8705
360	8036.766	11030.606	189.3402	940	20459.676	28275.636	217.8275	1520	34044.008	46682.088	233.0279	2100	49009.493	66469.693	244.0378
370	8250.372	11327.352	190.1533	950	20681.006	28580.106	218.1497	1530	34291.539	47012.759	233.2447	2110	49277.041	66820.381	244.2044
380	8463.748	11623.868	190.9441	960	20902.754	28884.994	218.4690	1540	34539.486	47343.846	233.4604	2120	49544.870	67171.350	244.3703
390	8676.911	11920.171	191.7137	970	21124.923	29190.303	218.7854	1550	34787.845	47675.345	233.6749	2130	49812.977	67522.597	244.5356
400	8889.880	12216.280	192.4634	980	21347.516	29496.036	219.0989	1560	35036.615	48007.255	233.8884	2140	50081.361	67874.121	244.7002
410	9102.668	12512.208	193.1941	990	21570.535	29802.195	219.4098	1570	35285.792	48339.572	234.1007	2150	50350.021	68225.921	244.8643

(Continued)

Table E-9 (Continued)

T K	u kJ/kg	h kJ/kg	s kJ/kg.K	T K	u kJ/kg	h kJ/kg	s kJ/kg.K	T K	u kJ/kg	h kJ/kg	s kJ/kg.K	T K	u kJ/kg	h kJ/kg	s kJ/kg.K
420	9315.293	12807.973	193.9068	1000	21793.982	30108.782	219.7179	1580	35535.374	48672.294	234.3120	2160	50618.954	68577.994	245.0276
430	9527.769	13103.589	194.6024	1010	22017.851	30415.791	220.0234	1590	35785.359	49005.419	234.5222	2170	50888.159	68930.339	245.1904
440	9740.110	13399.070	195.2817	1020	22242.137	30723.217	220.3263	1600	36035.743	49338.943	234.7313	2180	51157.634	69282.954	245.3525
450	9952.330	13694.430	195.9455	1030	22466.851	31031.071	220.6266	1610	36286.525	49672.865	234.9393	2190	51427.376	69635.836	245.5140
460	10164.443	13989.683	196.5944	1040	22691.996	31339.356	220.9245	1620	36537.701	50007.181	235.1463	2200	51697.386	69988.986	245.6749
470	10376.462	14284.842	197.2292	1050	22917.581	31648.081	221.2199	1630	36789.269	50341.889	235.3523	2210	51967.660	70342.400	245.8352
480	10588.400	14579.920	197.8505	1060	23143.610	31957.250	221.5130	1640	37041.227	50676.987	235.5573	2220	52238.198	70696.078	245.9948
490	10800.271	14874.931	198.4587	1070	23370.089	32266.869	221.8037	1650	37293.571	51012.471	235.7612	2230	52508.997	71050.017	246.1539
500	11012.086	15169.886	199.0546	1080	23597.021	32576.941	222.0921	1660	37546.299	51348.339	235.9641	2240	52780.056	71404.216	246.3124
510	11223.858	15464.798	199.6386	1090	23824.411	32887.471	222.3783	1670	37799.409	51684.589	236.1661	2250	53051.373	71758.673	246.4703
520	11435.601	15759.681	200.2112	1100	24052.263	33198.463	222.6623	1680	38052.898	52021.218	236.3671	2300	54411.776	73534.776	247.2510
530	11647.326	16054.546	200.7729	1110	24280.578	33509.918	222.9442	1690	38306.763	52358.223	236.5671	2350	55778.402	75317.102	248.0176
540	11859.046	16349.406	201.3241	1120	24509.360	33821.840	223.2240	1700	38561.002	52695.602	236.7661	2400	57151.067	77105.467	248.7706
550	12070.772	16644.272	201.8651	1130	24738.612	34134.232	223.5016	1710	38815.612	53033.352	236.9642	2450	58529.595	78899.695	249.5106
560	12282.519	16939.159	202.3965	1140	24968.334	34447.094	223.7773	1720	39070.591	53371.471	237.1614	2500	59913.821	80699.621	250.2378
570	12494.297	17234.077	202.9185	1150	25198.530	34760.430	224.0509	1730	39325.936	53709.956	237.3576	2550	61303.589	82505.089	250.9529
580	12706.118	17529.038	203.4314	1160	25429.199	35074.239	224.3226	1740	39581.645	54048.805	237.5529	2600	62698.750	84315.950	251.6561
590	12917.997	17824.057	203.9358	1170	25660.343	35388.523	224.5924	1750	39837.715	54388.015	237.7473	2650	64099.165	86132.065	252.3480
600	13129.943	18119.143	204.4317	1180	25891.963	35703.283	224.8603	1760	40094.144	54727.584	237.9408	2700	65504.700	87953.300	253.0289
610	13341.970	18414.310	204.9196	1190	26124.059	36018.519	225.1263	1770	40350.929	55067.509	238.1334	2750	66915.232	89779.532	253.6991
620	13554.090	18709.570	205.3997	1200	26356.631	36334.231	225.3905	1780	40608.068	55407.788	238.3251	2800	68330.641	91610.641	254.3589
630	13766.316	19004.936	205.8723	1210	26589.680	36650.420	225.6529	1790	40865.558	55748.418	238.5159	2850	69750.816	93446.516	255.0088
640	13978.658	19300.418	206.3376	1220	26823.205	36967.085	225.9135	1800	41123.397	56089.397	238.7059	2900	71175.652	95287.052	255.6490
650	14191.130	19596.030	206.7960	1230	27057.206	37284.226	226.1724	1810	41381.582	56430.722	238.8950	2950	72605.048	97132.148	256.2798

(Continued)

Table E-9 (Continued)

T K	u kJ/kg	h kJ/kg	s kJ/kg.K	T K	u kJ/kg	h kJ/kg	s kJ/kg.K	T K	u kJ/kg	h kJ/kg	s kJ/kg.K	T K	u kJ/kg	h kJ/kg	s kJ/kg.K
660	14403.743	19891.783	207.2475	1240	27291.683	37601.843	226.4296	1820	41640.112	56772.392	239.0832	3000	74038.910	98981.710	256.9015
670	14616.509	20187.689	207.6925	1250	27526.634	37919.934	226.6851	1830	41898.984	57114.404	239.2706	3050	75477.150	100835.650	257.5144
680	14829.441	20483.761	208.1311	1260	27762.059	38238.499	226.9390	1840	42158.194	57456.754	239.4572	3100	76919.683	102693.883	258.1187
690	15042.551	20780.011	208.5636	1270	27997.957	38557.537	227.1912	1850	42417.742	57799.442	239.6429	3150	78366.430	104556.330	258.7147
700	15255.849	21076.449	208.9901	1280	28234.326	38877.046	227.4418	1860	42677.625	58142.465	239.8278	3200	79817.316	106422.916	259.3026
710	15469.349	21373.089	209.4109	1290	28471.166	39197.026	227.6908	1870	42937.840	58485.820	240.0119	3250	81272.268	108293.568	259.8827
720	15683.061	21669.941	209.8261	1300	28708.474	39517.474	227.9382	1880	43198.385	58829.505	240.1952	3300	82731.220	110168.220	260.4551
730	15896.997	21967.017	210.2359	1310	28946.250	39838.390	228.1841	1890	43459.258	59173.518	240.3777	3350	84194.107	112046.807	261.0201
740	16111.168	22264.328	210.6404	1320	29184.492	40159.772	228.4285	1900	43720.456	59517.856	240.5594	3400	85660.869	113929.269	261.5779
750	16325.586	22561.886	211.0398	1330	29423.198	40481.618	228.6714	1910	43981.978	59862.518	240.7404	3450	87131.448	115815.548	262.1286
760	16540.262	22859.702	211.4343	1340	29662.366	40803.926	228.9129	1920	44243.821	60207.501	240.9205	3500	88605.788	117705.588	262.6726
770	16755.206	23157.786	211.8239	1350	29901.994	41126.694	229.1528	1930	44505.983	60552.803	241.0999				

Table F-1 Atmosphere properties at various altitude

Altitude *m*	Temperature °*C*	Pressure *kPa*	Gravity *m/s²*	Density, ρ *kg/m³*	Viscosity, μ *kg/m.s*	thermal conductivity *W/m.K*
0	15.0	101.3250	9.8067	1.2253	0.00001796	0.025499
200	13.7	98.9448	9.8060	1.2019	0.00001790	0.025400
400	12.4	96.6101	9.8054	1.1789	0.00001783	0.025301
600	11.1	94.3202	9.8048	1.1562	0.00001777	0.025202
800	9.8	92.0745	9.8042	1.1339	0.00001770	0.025103
1000	8.5	89.8723	9.8036	1.1119	0.00001764	0.025004
1200	7.2	87.7129	9.8030	1.0902	0.00001758	0.024905
1400	5.9	85.5957	9.8024	1.0689	0.00001751	0.024805
1600	4.6	83.5201	9.8017	1.0478	0.00001745	0.024706
1800	3.3	81.4854	9.8011	1.0271	0.00001738	0.024606
2000	2.0	79.4911	9.8005	1.0067	0.00001732	0.024506
2200	0.7	77.5365	9.7999	0.9866	0.00001725	0.024406
2400	−0.6	75.6209	9.7993	0.9668	0.00001718	0.024306
2600	−1.9	73.7439	9.7987	0.9474	0.00001712	0.024205
2800	−3.2	71.9048	9.7981	0.9282	0.00001705	0.024105
3000	−4.5	70.1030	9.7975	0.9093	0.00001699	0.024004
3200	−5.8	68.3380	9.7969	0.8907	0.00001692	0.023903
3400	−7.1	66.6091	9.7962	0.8724	0.00001686	0.023802
3600	−8.4	64.9158	9.7956	0.8544	0.00001679	0.023701
3800	−9.7	63.2575	9.7950	0.8367	0.00001672	0.023600
4000	−11.0	61.6337	9.7944	0.8193	0.00001666	0.023499
4200	−12.3	60.0438	9.7938	0.8021	0.00001659	0.023397
4400	−13.6	58.4873	9.7932	0.7852	0.00001652	0.023296
4600	−14.9	56.9636	9.7926	0.7686	0.00001646	0.023194
4800	−16.2	55.4722	9.7920	0.7523	0.00001639	0.023092
5000	−17.5	54.0126	9.7913	0.7362	0.00001632	0.022990
5200	−18.8	52.5843	9.7907	0.7204	0.00001626	0.022887
5400	−20.1	51.1867	9.7901	0.7049	0.00001619	0.022785
5600	−21.4	49.8194	9.7895	0.6896	0.00001612	0.022682
5800	−22.7	48.4817	9.7889	0.6746	0.00001605	0.022580
6000	−24.0	47.1733	9.7883	0.6598	0.00001599	0.022477
6200	−25.3	45.8937	9.7877	0.6452	0.00001592	0.022374
6400	−26.6	44.6423	9.7871	0.631	0.00001585	0.022270
6600	−27.9	43.4186	9.7865	0.6169	0.00001578	0.022167
6800	−29.2	42.2223	9.7858	0.6031	0.00001571	0.022063
7000	−30.5	41.0528	9.7852	0.5895	0.00001565	0.021960
8000	−37.0	35.5918	9.7822	0.5252	0.00001530	0.021439
9000	−43.5	30.7346	9.7791	0.4663	0.00001495	0.020913
10000	−50.0	26.4287	9.7761	0.4127	0.00001460	0.020384

Table G-1 Enthalpy of formation, Gibbs function of formation, and absolute entropy

Substance	Formula	h_f kJ/kmol	g_f kJ/kmol	s kJ/kmol K
Acetylene	C_2H_2	228186.00	210679.99	200.80
Ammonia	NH_3	−45937.00	−16484.32	192.76
Benzene	C_6H_6	82875.00	129663.98	269.14
Butane	C_4H_{10}	−147379.00	−38299.38	309.75
Carbon	C	0.00	0.00	5.74
Carbon Dioxide	CO_2	−393486.00	−394348.41	213.67
Carbon Monoxide	CO	−110528.00	−137134.04	197.54
Ethane	C_2H_6	−83846.00	−31970.16	229.10
Hexane	C_6H_{14}	−108135.00	58652.32	388.72
Hydrogen	H_2	0.00	0.00	130.56
Hydrogen	H	217985.00	203255.75	114.71
Hydroxyl	OH	37276.00	32528.86	183.73
Methane	CH_4	−74595.00	−50570.24	186.25
Nitrogen	N_2	0.00	0.00	191.49
Nitrogen	N	472651.00	455502.14	153.29
Octane	C_8H_{18}	−260302.00	−35667.22	467.21
Oxygen	O_2	0.00	0.00	205.03
Oxygen	O	249160.00	231716.57	161.05
Propane	C_3H_8	−104674.00	−24423.79	270.19
Water	H_2O	−285813.00	−272591.04	188.71

Table H-1 Common fuels and hydrocarbons - Properties

Substance	Formula	Molar mass kg/kmol	Enthalpy of vaporization kJ/kg	Cp kJ/kg K	Higher heating value kJ/kg	Lower heating value kJ/kg
Acetylene	C_2H_2	26.04		1.695	49960	48271
Benzene	C_6H_6	78.11	433.6	1.049	42263	40573
Butane	C_4H_{10}	58.12	360.8	1.686	49132	45348
Carbon Monoxide	CO	28.01	119.6	1.037	10102	10102
Diesel nr.2		233		1.934	45579	42601
Ethane	C_2H_6	30.07		1.76	51896	47508
Hexane	C_6H_{14}	86.17		1.655	48316	44742
Methane	CH_4	16.04	510.8	2.239	55516	50023
Octane	C_8H_{18}	114.2		1.644	47896	44430
Propane	C_3H_8	44.1	335.3	1.657	50321	46330

Appendix 2: *T-s* Diagrams

Figure A1: *T-s* diagram for water
Figure A2: *T-s* diagram for ammonia
Figure A3: *T-s* diagram for *R*-134*a*
Figure A4: *T-s* diagram for air

Note: The above listed graphs are prepared by utilizing the Engineering Equation Software (EES) available at Ontario Tech. University (formerly University of Ontario Institute of Technology) in Oshawa, Canada.

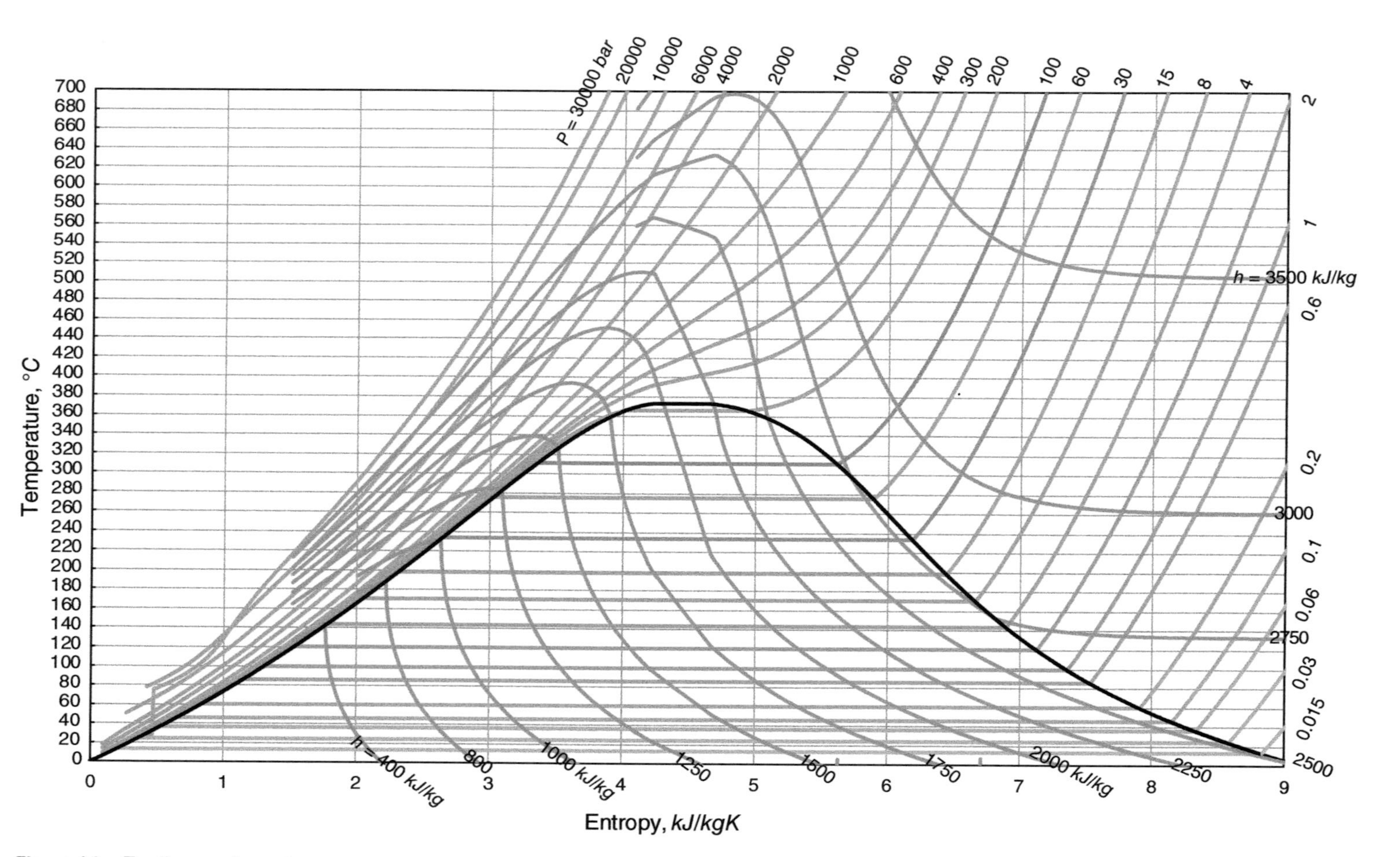

Figure A1 *T-s* diagram for water.

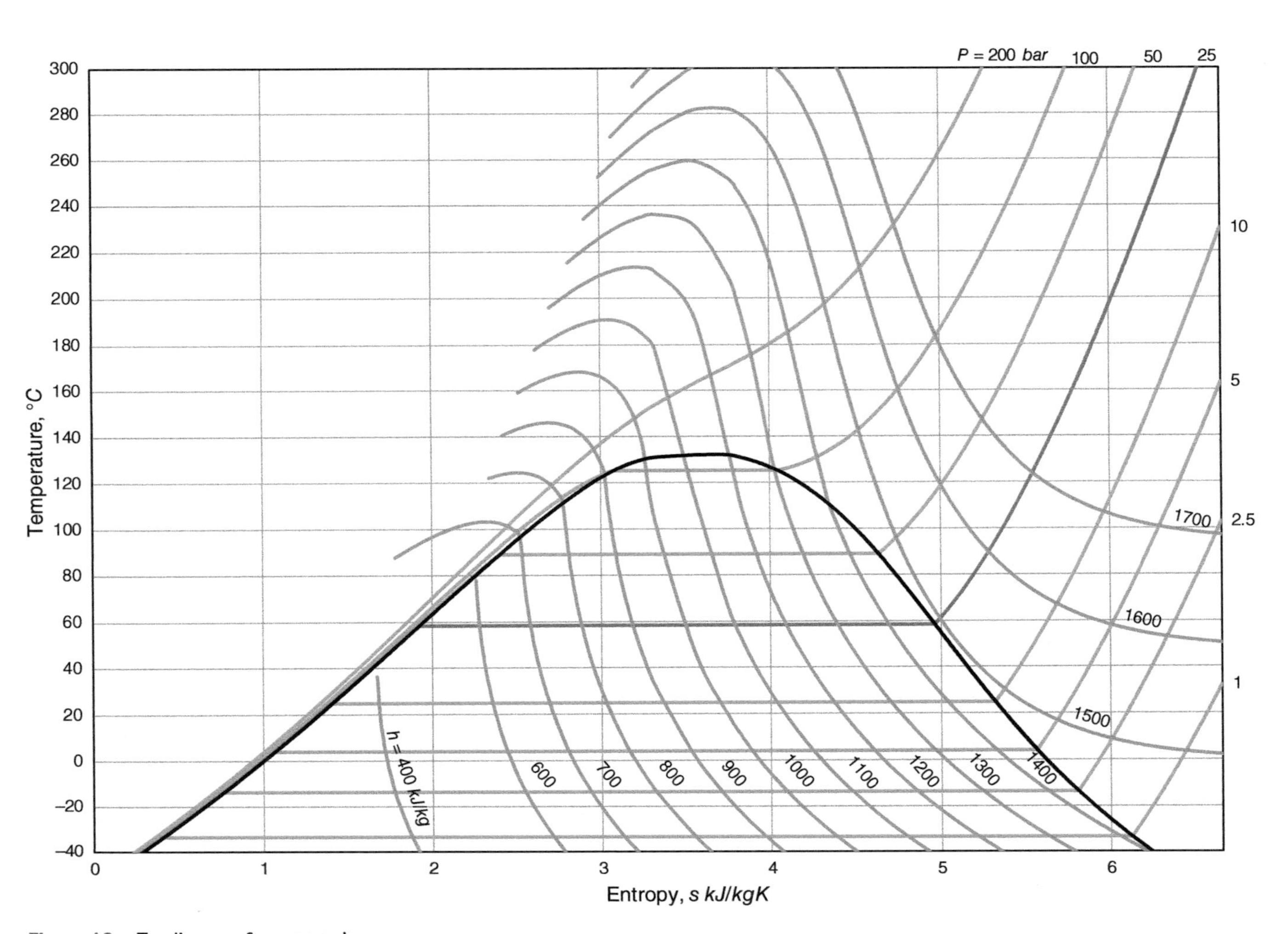

Figure A2 *T-s* diagram for ammonia.

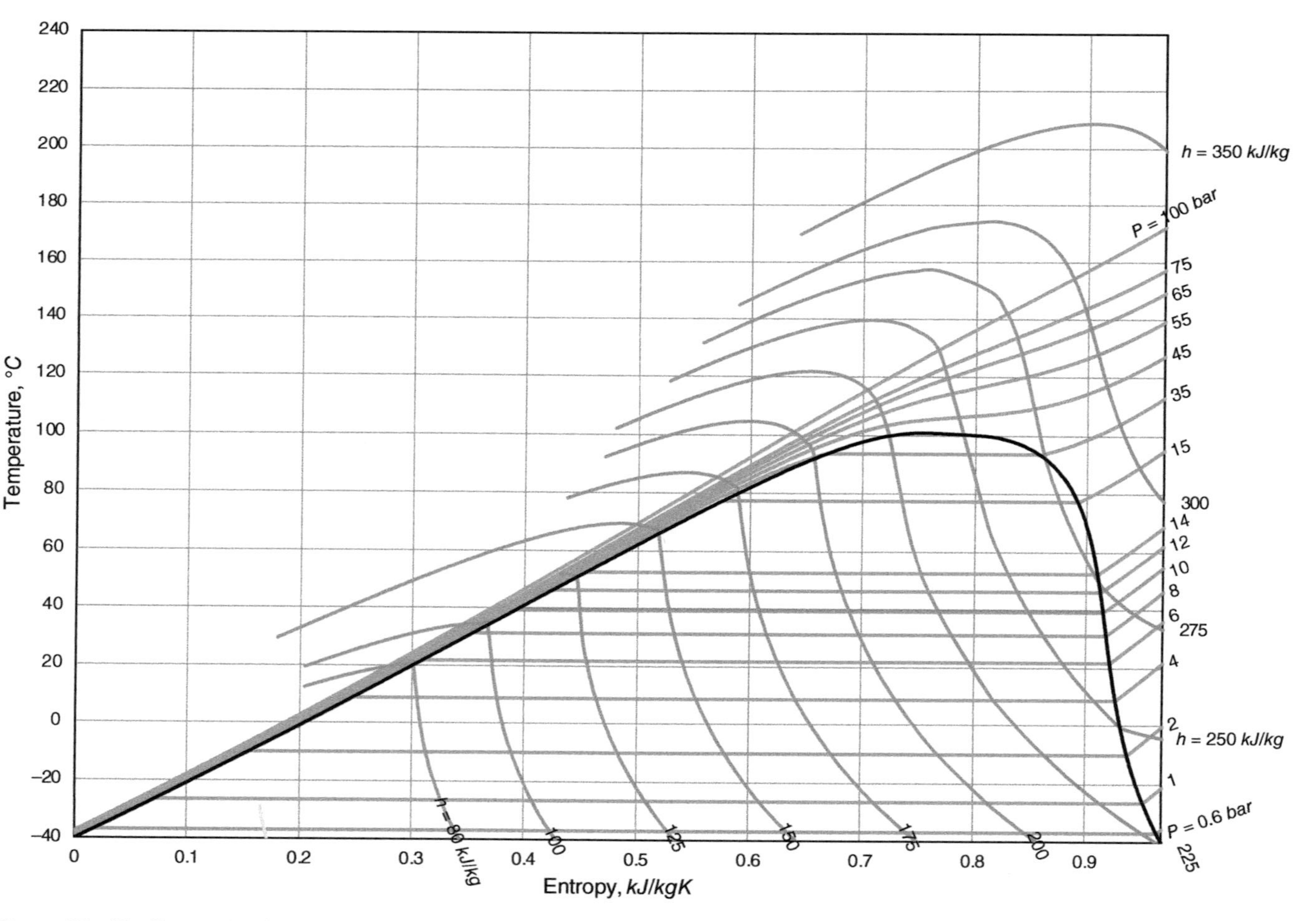

Figure A3 *T-s* diagram for *R-134a*.

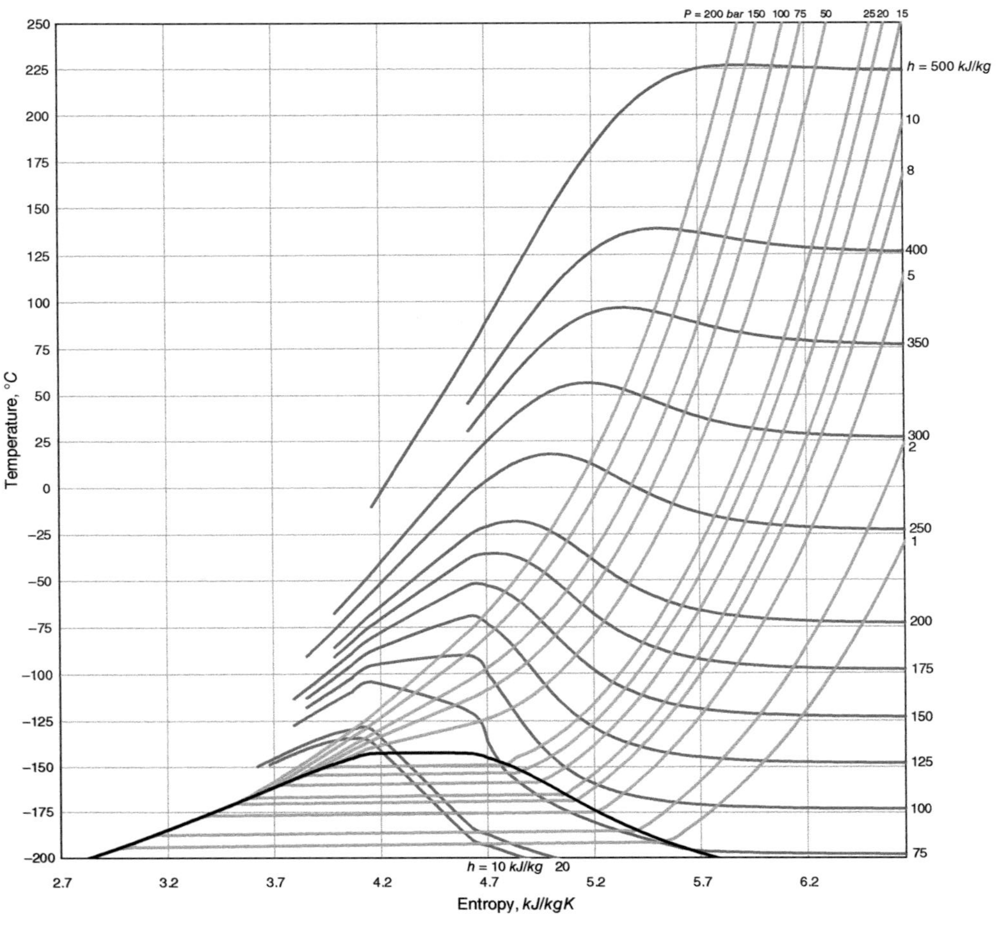

Figure A4 *T-s* diagram for air.

Index

a

absorption refrigeration cycles 482–491
adiabatic compression and expansion 126, 127
adiabatic flame temperature 537–541
adiabatic process 68
air–fuel ratio (AFR) 511
air refrigeration 453
air standard Brayton cycle 337, 338, 340, 341
air-standard refrigeration systems 453–458
ammonia 91
anthracite 514
autoignition temperature 309
azeotropic refrigerants 439

b

bituminous coal 514
boiling point 20, 516
boundary (thermodynamic system) 13
boundary movement work 36
  adiabatic compression and expansion 126, 127
  consumed by piston–cylinder device 123–124, 127–131
  isobaric compression and expansion 125
  isothermal compression and expansion 125
  polytropic compression and expansion 126
Brayton cycle 291, 343–351
  air standard 337, 338, 340, 341
  balance equations for 341, 342
  closed-cycle configuration 338, 339
  energy efficiency 340
  open-cycle real gas turbine engine 336, 337
  pressure ratio 340
  processes 337
  property diagrams 338, 339
  regenerative 352–362
  temperature-entropy diagram 340–341
  working conditions 337, 338
brown coal 514

c

carbon-based fuels 509
carbon/hydrogen ratio *vs.* fuel types 1, 2
Carnot concept
  principle 170
  temperature ratio 170–172
Carnot cycle 168–169
  heat engine cycle 35, 36
  for power generation 296–297
  refrigeration cycle 297
Carnot heat engine 300
cascade refrigeration systems 458–465
Celsius scales 20, 22
CFCs *see* chlorofluorocarbons (CFCs)
chemical energy 10, 64, 77, 195, 509, 518, 519
chemical equilibrium 32
chemical reactions 518
chlorofluorocarbons (CFCs) 438
Clausius inequality 166–167
Clausius statement 165–167

closed combustion chamber 526–529
closed systems 13, 14, 119, 120, 230–235
  classification of 121
  energy balance equation for 131–133
  with fixed mass 32–33
  with moving boundary 33, 239–248
  nonflow exergy with specific heat capacity 235–239
coal 514–516
coefficient of performance (COP) 210
  absorption refrigeration system 474
  heat pump 41, 466–468
  refrigeration cycle 41, 442
cogeneration Rankine cycle 387–403
combined Brayton–Rankine cycles 403–417
combined cycle 336, 404
combustion chambers 526
combustion efficiency 543–550
combustion reactions 521–526
complete combustion 534
compressed liquid
  phase changes 86
  property tables 97–99
compressed natural gas (CNG) 517
compressibility factor 110–112
compressible substances 16
compression ratio, for diesel cycle 314
compressor, isentropic efficiency 188–190
conduction 68, 69
conservation of energy principle *see* first law of thermodynamics (FLT)
constant entropy 183
constant pressure specific heat 133
constant volume specific heat 133
convection 68, 69
cooling with pressure drop 89–90
critical point 91
  vs. triple point 94–95
critical pressure, of pure substances 91, 93, 94
critical temperature, of pure substances 93, 94
crude oil 516
cyclic integration 166

d

derived dimensions 6
Diesel cycle 316–326
  compression ratio 314
  cut-off ratio 315
  diesel oil engine 315, 316
  energy efficiency of 315
  processes 314–315
dimensional homogeneity 7
dimensions 6, 7
dual cycle 326–336

e

efficiency, defined 40
efficiency improvement factor (EIF) 354
electrical work 122
electricity, exergy of 198
energy 61
  for closed and open systems 66
  and environment 64–65
  vs. exergy 194–198
  forms of 62
    flow energy 73–75
    heat transfer 67–72
    work transfer 71–73
  importance of 1
  quality 197
  quality diagram, forms of heat and work 12
  sectors driven by 61, 62
  spectrum of 3–4
energy balance equation (EBE) 78–80, 299
  for closed systems 131–133
energy conservation concept 75–77
energy efficiency
  of coal fueled power plant 44–45
  of compressor 41–42
  defined 40, 41
  of heat engine 299–303
  of motor–pump system 45–47
  of propane fueled water heater 42–44
energy losses 80–83
energy sources, development of 39
energy transfer, modes of 121, 122
energy transformation/conversion 291

Engineering Equation Solver
(EES) 47–49, 132
English System of Units 21
enthalpy 6, 32, 41, 98, 133, 134, 189, 195
as flow energy 73–74
entropy 4, 5, 172–175
balance equations 175–182
generation 173, 174
graphic illustration 5
magnitude of 175
equations of state 108
equilibrium states 31–32
exergetic temperature factor 198
exergy 3–5, 193–194
analysis 40
destruction 101
efficiency 210–213, 279–283
vs. energy 194–198
reference environment 199
exergy balance equation
for closed systems 199–204
for open systems 205–210
exergy forms
chemical 552
electricity 198
flow exergy 198
thermal exergy 198
work 198
extensive properties 15, 16, 172

f

Fahrenheit scales 20, 21
first law of thermodynamics (FLT)
2–5, 9–11, 35–37, 163, 295, 442, 518–520
energy balance equations 78–80
energy conservation concept 75–77
energy losses 80–83
fixed boundary systems 121
flow exergy 198 *see also* exergy
FLT *see* first law of thermodynamics (FLT)
fossil fuels 509
coal 514–516
crude oil 516
natural gas 516–517
prediction 513
fuel combustion
chemical energy 518
closed system 526–529
combustion efficiency 543–550
combustion reactions 521–526
first law of thermodynamics (FLT)
analysis 518–520
fossil fuels (*see* fossil fuels)
fuels 510–512
open system 529–541
second law of thermodynamics
(SLT) 541–543
fuels
classification of 511, 512
definition of 510
fossil (*see* fossil fuels)
heating value of 510–511
fundamental dimensions 6

g

gas cycles 296
gaseous hydrocarbon 517
gas refrigeration 453
gas turbine cycle *see* air standard Brayton cycle
gas turbine engines 195
governing laws 8
greenhouse gas (GHG) emissions 65
guiding policies 8

h

hard coal 514
heat 509 *see also* thermal energy
heated crude oil 516
heat engines 171, 295
classification of 297, 298
energy and exergy interactions of 298, 299
performance assessment 299–300
heating values (HVs) 510–511
heat loss 37–38
heat pump (HP) 436, 471
coefficient of performance 466–468
vs. heat engine 465
house temperature 469
refrigeration operation mode 468
with reversing valve unit 479
T-s diagram 464

heat transfer 12, 36
 adiabatic process 68
 case illustration 70
 conduction 68, 69
 convection 68, 69
 heat transfer rate 68
 radiation 68, 69
 system boundary considerations 69–70
 temperature difference 67
higher heating value (HHV) 511
HVs *see* heating values (HVs)
hydrocarbons (HCs) 438–439

i
ideal gas
 constant calculation 108–109
 isentropic efficiencies for (*see* isentropic efficiency, for ideal gases)
 specific heat capacity for 134–139
ideal gas equation
 compressibility factor 110–112
 ideal gas constant calculation 108–109
 water vapor as ideal gas 109–110
ideality, concept of 168
ignition temperature, of fuel 309
incomplete combustion 533–537
incompressible substances 16
intensive properties 15–17, 32, 95, 98, 172, 199
internal energy 15, 63, 64, 78, 79, 134, 527
International System of Units 21
irreversible process 167–168
isenthalpic process 32
isentropic efficiency
 for ideal gases
  compressor 188–190
  diffusers 191–193
  nozzle 190–191
  pump 190
  turbine 187–188
 of nozzle 191
isentropic process 182–185
isobaric compression and expansion 125
isobaric process 32, 187, 314, 337
isochoric process 32, 187, 304
isothermal compression and expansion 125
isothermal process 32, 35, 125, 187, 189

k
Kelvin–Plank statement 165
Kelvin scales 21, 22
kinetic energy (KE) 9, 63, 64, 75, 76, 119

l
latent energy 64
latent heat of fusion 89
latent heat of vaporization 89
lignite 514
liquefied natural gas (LNG) 517
lower heating value (LHV) 511

m
macroscopic thermodynamics 62
manometer 27–30
mass balance equation (MBE) 134
mean effective pressure (MEP) 305
mechanical equilibrium 31
microscopic thermodynamics 62
mixtures *see* azeotropic refrigerants
moving boundary systems 33, 121, 123, 131, 230, 239, 300

n
natural gas 516–517
nonazeotropic mixtures 439
nozzle, isentropic efficiency of 190–191
nuclear energy 64, 195, 227

o
open combustion chamber 530
 adiabatic flame temperature 537–541
 incomplete combustion 533–537
 steady-state steady-flow systems 531
open systems 13, 14, 119, 120
 with mass flows and energy transfers 33
 steady-state steady-flow process 140, 248–271
 thermal energy storage systems 155–157
 types of 139, 140
 uniform-state uniform-flow process 140–155, 271–279

Otto cycle
  energy and exergy efficiency of 304, 305
  example 305–313
  mean effective pressure 305
  piston–cylinder operation 304
  schematic illustration 303
  spark ignition engine operation 304

p

Pascal's law/principle 25–27
performance assessment 40–47
petroleum products 516
phase changes, of water 85
  compressed liquid 86
  saturated liquid
    defined 86–87
    and vapor mixture 87
  saturated vapor 87–88
  saturation pressure 88–89
  saturation temperature 86, 88–89
  superheated vapor 88
phase, defined 84
phase equilibrium 31
piston–cylinder mechanism 203
polytropic compression and expansion 126
polytropic process 189
potential energy (PE) 10, 63, 64, 75, 76, 78, 79, 119
power cycles 295–296 *see also specific cycles*
power generation 295
  Carnot concept for 296–297
pressure
  defined 22
  manometer 27–30
  Pascal's law/principle 25–27
  pressure gage example 23–24
pressure-specific volume (P-v) diagram 94–95
pressure-specific volume-temperature (P-v-T) surface diagrams 95–96
primary dimensions 6
process
  defined 32, 33
  isobaric process 32
  isochoric process 32
  isothermal process 32
  SSSF process 33, 34
  USUF process 34, 35
propane, saturation temperature of 88
property
  defined 15
  density 17–19
  extensive and intensive 15, 16
  pressure 22–30
  specific volume 15–17
  temperature 19–22
property diagrams 34, 35
  pressure-specific volume diagram 94–95
  pressure-specific volume-temperature surface diagrams 95–96
  temperature-specific volume diagram 90–94
property tables
  compressed liquid 97–99
  saturated liquid and vapor mixtures 99–103
  for saturated water 96, 98
  superheated vapor 103–108
pump, isentropic efficiency 190
pure substance
  definition 83
  examples of 83–84
  phase changes 85–90
  phase, defined 84
  property diagrams 90–96
  property tables 96–108

q

quasi-equilibrium process 32
quasi-static process 32

r

radiation 68, 69
Rankine cycle 295, 366–374
  balance equations for 364, 365
  cogeneration 387–403
  components 363
  energy and exergy efficiencies 365
  operating principle of 365
  processes of 363
  reheat 374–387
  schematic illustration of 363

Rankine cycle (*cont'd*)
  steam 363, 365
Rankine scales 21
reference-environment modeling 199
refrigerants 436–439
refrigeration
  absorption refrigeration cycles 482–491
  air-standard refrigeration systems 453–458
  cascade refrigeration systems 458–465
  component of 445
  cooling and heating processes 436, 438
  heat pumps (*see* heat pump (HP))
  household fridge 441
  refrigerants 436–439
  vapor-compression refrigeration cycle 440
regenerative Brayton cycle 352–362
reheat Rankine cycle 374–387
reverse Brayton cycle 453
reverse cycle refrigeration systems 466
reversible adiabatic compression 168, 296
reversible adiabatic expansion 168, 296
reversible heat engine 300
reversible isothermal compression 168, 296
reversible isothermal expansion 168, 296
reversible process 167–168

S
saturated liquid
  defined 86–87
  and vapor mixtures 87, 99–103
saturated vapor 87–88
saturated water, property tables for 96, 98
saturation pressure 88–89
saturation temperature 86, 88–89
secondary dimensions 6
second law efficiency 168
second law of thermodynamics (SLT) 2, 3, 5, 11–13, 38–40, 295, 436, 442, 541–543
  Clausius statement 165–167
  entropy (*see* entropy)
  exergy (*see* exergy)
  Kelvin–Plank statement 165
sensible heat 89
seven-step approach
  first law of thermodynamics 35–38
  performance assessment 40–47
  process 32–34
  property 14–31
  schematic illustration 14, 15
  second law of thermodynamics 38–40
  state 31–32
  thermodynamic cycle 34–35
shaft work 36–37, 122
SI system 21
SI units 7
SLT *see* second law of thermodynamics (SLT)
smart energy solutions 4
soft coal 514
spark-ignition internal combustion engine 195
specific gravity 17–19
specific heat capacity
  definition 133
  for ideal gases 134–139
specific internal energy 64
specific volume 15, 101, 102, 190
  of liquid water *vs.* air 16, 17
  for water 96
  of water vapor 109
SSSF process *see* steady-state steady-flow (SSSF) process
state
  defined 31
  equilibrium states 31–32
steady-state steady-flow (SSSF) process 33–34, 140
steam Rankine cycle 363, 365
stoichiometric air 533
subcooled liquid *see* compressed liquid
superheated vapor 88
  property tables for 103–108
  water vapor 110
surroundings (thermodynamic system) 13, 33
sustainable development 1, 3

t

temperature
  defined 19
  scales 20–22
temperature-specific volume (T-v) diagram 90–94
  of water heating and cooling 87
thermal energy 12, 68, 195, 298, 482, 509, 510, 519, 531, 543
  storage systems 155–157
thermal equilibrium 31
thermal exergy 198
thermal refrigeration cycles *see* absorption refrigeration cycles
thermodynamic cycle 34–35
thermodynamics 61, 63, 435–436
  defined 2–3, 228
  energy 4
  entropy 4, 5
  exergy 4
thermodynamic systems 13 *see also* closed systems; open systems
thermometer 19
total energy 64, 76, 78, 79, 510
Trinkler cycle *see* dual cycle
triple point
  vs. critical point 94–95
  temperature 20
turbine, isentropic efficiency 187–188

u

uniform-state uniform-flow (USUF) process 34, 35
  in air compressor 148–149
  in connected turbines 153–154
  defined 140
  in diffuser and nozzle 141–142
  in heat exchanger 145–148
  in insulated rigid tank 155
  in mixing chamber 151–153
  in pipes 140–141
  in pump–motor system 149–150
  refrigerant R-134a 150–151
  in steam turbine 143–145
United State Customary System (USCS) 21
units 6, 7
USUF process *see* uniform-state uniform-flow (USUF) process

v

vacuum cooling 89–90
vacuum freezing 89–90
vacuum pressure 24
vapor-compression heat pump cycle 475
vapor-compression refrigeration cycle 440
vapor cycles 296
vapor quality 101–102

w

water
  compressed liquid property table 97, 98
  critical point for 91
  phase changes of (*see* phase changes, of water)
  pressure-specific volume diagram for 94, 95
  saturation temperature 88
  temperature-specific volume diagram of 96, 97
water vapor
  as ideal gas 109–110
  superheated 110
work 36
  boundary movement work 123–131
  electrical work 122
  mechanical work 11
  shaft work 122
  transfer 71–73
work-consuming devices 300
work-producing devices 300

z

zeotropic blends 439
zeotropic mixtures 439
zeroth law of thermodynamics (ZLT) 8–9